D1749994

AN **ULLMANN'S**
**ENCYCLO**PEDIA

# INDUSTRIAL ORGANIC CHEMICALS

Starting Materials
and Intermediates

VOLUME **4**

**WILEY-VCH**

Weinheim · New York · Chichester · Brisbane · Singapore · Toronto

# INDUSTRIAL ORGANIC CHEMICALS

**AN ULLMANN'S ENCYCLOPEDIA**

## VOLUME 1
**Acetaldehyde** to **Aniline**

## VOLUME 2
**Anthracene** to **Cellulose Ethers**

## VOLUME 3
**Chlorinated Hydrocarbons** to **Dicarboxylic Acids, Aliphatic**

## VOLUME 4
**Dimethyl Ether** to **Fatty Acids**

## VOLUME 5
**Fatty Alcohols** to **Melamine and Guanamines**

## VOLUME 6
**Mercaptoacetic Acid and Derivatives** to **Phosphorus Compounds, Organic**

## VOLUME 7
**Phthalic Acid and Derivatives** to **Sulfones and Sulfoxides**

## VOLUME 8
**Sulfonic Acids, Aliphatic** to **Xylidines**

**Index**

**AN ULLMANN'S ENCYCLOPEDIA**

# INDUSTRIAL ORGANIC CHEMICALS

## Starting Materials and Intermediates

### VOLUME 4
**Dimethyl Ether** to **Fatty Acids**

**WILEY-VCH**

Weinheim · New York · Chichester · Brisbane · Singapore · Toronto

This book was carefully produced. Nevertheless, authors and publisher do not warrant the information contained therein to be free of errors. Readers are advised to keep in mind that statements, data, illustrations, procedural details or other items may inadvertently be inaccurate.

Library of Congress Card No.: Applied for.
British Library Cataloguing-in-Publication Data: A catalogue record for this book is available from the British Library.

Die Deutsche Bibliothek – CIP-Einheitsaufnahme
**Industrial organic chemicals** : starting materials and intermediates ;
an Ullmann's encyclopedia. – Weinheim ; New York ;
Chichester ; Brisbane ; Singapore ; Toronto : Wiley-VCH
    ISBN 3-527-29645-X
Vol. 4. Dimethyl Ether to Fatty Acids. – 1. Aufl. – 1999.

© WILEY-VCH Verlag GmbH, D-69469 Weinheim (Federal Republic of Germany), 1999
Printed on acid-free and chlorine-free paper.
All rights reserved (including those of translation in other languages). No part of this book may be reproduced in any form – by photoprinting, microfilm, or any other means – nor transmitted or translated into machine language without written permission from the publishers. Registered names, trademarks, etc. used in this book, even when not specifically marked as such, are not to be considered unprotected by law.

Composition and Printing: Rombach GmbH, Druck- und Verlagshaus, D-79115 Freiburg
Bookbinding: Wilhelm Osswald & Co., D-67433 Neustadt (Weinstraße)
Cover design: mmad, Michel Meyer, D-69469 Weinheim
Printed in the Federal Republic of Germany

# Contents

## 1 Dimethyl Ether

1. Introduction .................. 1931
2. Properties .................... 1931
3. Production ................... 1933
4. Uses .......................... 1935
5. Toxicology .................... 1935
6. References .................... 1935

## 2 Dioxane

1. Physical Properties ............ 1937
2. Chemical Properties ........... 1938
3. Production .................... 1940
4. Environmental Protection ...... 1941
5. Quality Specifications ......... 1941
6. Storage and Transportation .... 1942
7. Legal Aspects ................. 1942
8. Uses .......................... 1942
9. Economic Aspects ............. 1943
10. Toxicology and Occupational Health .... 1943
11. References .................... 1945

## 3 Dithiocarbamic Acid and Derivatives

1. Introduction .................. 1947
2. Physical and Chemical Properties .. 1948
3. Manufacture of Dithiocarbamate Salts  1949
4. Transformation Products ....... 1953
5. Uses .......................... 1967
6. Toxicology and Occupational Health .... 1975
7. References .................... 1975

## 4 Epoxides

1. Introduction .................. 1987
2. Reactions of Epoxides ......... 1988
3. Production of Epoxides ........ 1992
4. Industrially Important Epoxides ... 1999
5. Analysis ...................... 2005
6. Economic Aspects ............. 2005
7. Toxicology .................... 2006
8. References .................... 2007

## 5 Esters, Organic

1. Introduction .................. 2011
2. Physical Properties ............ 2013
3. Chemical Properties ........... 2015
4. Natural Sources ............... 2021
5. Production .................... 2022
6. Environmental Protection ...... 2034
7. Quality Specifications and Analysis . 2034
8. Storage and Transportation .... 2035
9. Uses and Economic Aspects .... 2036
10. Toxicology and Occupational Health .... 2042
11. References .................... 2044

V

## 6 Ethanol

1. Introduction .................. 2048
2. Physical Properties ............. 2049
3. Chemical Properties ........... 2050
4. Synthesis .................... 2052
5. Fermentation ................. 2062
6. Recovery and Purification ....... 2112
7. Comparison of Process Economics for Synthetic and Fermentation Ethanol 2124
8. Analysis ..................... 2130
9. Uses ........................ 2131
10. Economic Aspects .............. 2133
11. Toxicology ................... 2137
12. References ................... 2138

## 7 Ethanolamines and Propanolamines

1. Introduction .................. 2149
2. Ethanolamines ................ 2150
3. N-Alkylated Ethanolamines ...... 2158
4. Isopropanolamines ............. 2163
5. N-Alkylated Propanolamines and 3-Alkoxypropylamines .......... 2170
6. Storage and Transportation ...... 2176
7. Environmental Protection ....... 2176
8. Toxicology and Occupational Health 2177
9. References ................... 2180

## 8 Ethers, Aliphatic

1. Introduction .................. 2185
2. Properties .................... 2186
3. Synthesis .................... 2188
4. Individual Aliphatic Ethers ....... 2189
5. Toxicology ................... 2201
6. References ................... 2203

## 9 Ethylbenzene

1. Introduction .................. 2207
2. Physical Properties ............. 2208
3. Chemical Properties ........... 2209
4. Production ................... 2209
5. Environmental Protection ....... 2215
6. Quality Specifications .......... 2215
7. Handling, Storage, and Transportation ...................... 2216
8. Uses ........................ 2217
9. Economic Aspects .............. 2217
10. Toxicology and Occupational Health 2218
11. References ................... 2219

## 10 Ethylene

1. Introduction .................. 2221
2. Physical Properties ............. 2222
3. Chemical Properties ........... 2223
4. Raw Materials ................ 2224
5. Production ................... 2226
6. Environmental Protection ....... 2285
7. Quality Specifications .......... 2286
8. Chemical Analysis ............. 2286
9. Storage and Transportation ...... 2288
10. Uses and Economic Aspects ...... 2289
11. Toxicology and Occupational Health 2290
12. References ................... 2291

## 11 Ethylenediaminetetraacetic Acid and Related Chelating Agents

1. Introduction . . . . . . . . . . . . . . . . . 2295
2. Physical Properties . . . . . . . . . . . . . 2296
3. Chemical Properties . . . . . . . . . . . 2296
4. Production . . . . . . . . . . . . . . . . . . 2299
5. Chemical Analysis . . . . . . . . . . . . . 2300
6. Storage and Transportation . . . . . . 2300
7. Uses . . . . . . . . . . . . . . . . . . . . . . . 2300
8. Trade Names . . . . . . . . . . . . . . . . 2302
9. Economic Aspects . . . . . . . . . . . . . 2302
10. Toxicology and Occupational Health 2303
11. References . . . . . . . . . . . . . . . . . . 2303

## 12 Ethylene Oxide

1. Introduction . . . . . . . . . . . . . . . . . 2329
2. Physical Properties . . . . . . . . . . . . . 2330
3. Chemical Properties . . . . . . . . . . . 2332
4. Production . . . . . . . . . . . . . . . . . . 2335
5. Environmental Protection and Ecology . . . . . . . . . . . . . . . . . . . . 2343
6. Quality Specifications . . . . . . . . . . 2344
7. Analysis . . . . . . . . . . . . . . . . . . . . 2344
8. Handling, Storage, and Transportation . . . . . . . . . . . . . . . . . . . . . . . 2345
9. Uses . . . . . . . . . . . . . . . . . . . . . . . 2349
10. Economic Aspects . . . . . . . . . . . . . 2350
11. Toxicology and Occupational Health 2350
12. References . . . . . . . . . . . . . . . . . . 2354

## 13 Ethylene Glycol

1. Introduction . . . . . . . . . . . . . . . . . 2305
2. Physical Properties . . . . . . . . . . . . . 2306
3. Chemical Properties . . . . . . . . . . . 2308
4. Production . . . . . . . . . . . . . . . . . . 2311
5. Environmental Protection and Ecology . . . . . . . . . . . . . . . . . . . . . 2316
6. Quality Specifications and Analysis . 2316
7. Storage and Transportation . . . . . . 2316
8. Derivatives . . . . . . . . . . . . . . . . . . 2317
9. Uses . . . . . . . . . . . . . . . . . . . . . . . 2321
10. Economic Aspects . . . . . . . . . . . . . 2322
11. Toxicology and Occupational Health 2323
12. References . . . . . . . . . . . . . . . . . . 2326

## 14 2-Ethylhexanol

1. Introduction . . . . . . . . . . . . . . . . . 2361
2. Properties . . . . . . . . . . . . . . . . . . . 2362
3. Industrial Production . . . . . . . . . . 2362
4. Uses . . . . . . . . . . . . . . . . . . . . . . . 2366
5. Economic Aspects . . . . . . . . . . . . . 2367
6. Toxicology . . . . . . . . . . . . . . . . . . 2367
7. References . . . . . . . . . . . . . . . . . . 2368

## 15 Fats and Fatty Oils

1. Introduction . . . . . . . . . . . . . . . . . 2372
2. Composition . . . . . . . . . . . . . . . . 2374
3. Physical Properties . . . . . . . . . . . . 2385
4. Chemical Properties . . . . . . . . . . . 2395
5. Manufacture and Processing . . . . . 2400
6. Refining . . . . . . . . . . . . . . . . . . . . 2412
7. Fractionation . . . . . . . . . . . . . . . . . 2423
8. Hydrogenation . . . . . . . . . . . . . . . . 2425
9. Interesterification . . . . . . . . . . . . . 2428
10. Environmental Aspects . . . . . . . . . 2431
11. Standards and Quality Control . . . . 2431
12. Storage and Transportation . . . . . 2440
13. Individual Vegetable Oils and Fats . 2441
14. Individual Animal Fats . . . . . . . . . 2467
15. Economic Aspects . . . . . . . . . . . . . 2473
16. Toxicology and Occupational Health 2473
17. References . . . . . . . . . . . . . . . . . . 2477

## 16 Fatty Acids

1. Introduction . . . . . . . . . . . . . . . . . 2481
2. Properties . . . . . . . . . . . . . . . . . . . 2482
3. Production of Natural Fatty Acids . . 2496
4. Production of Synthetic Fatty Acids. 2521
5. Analysis . . . . . . . . . . . . . . . . . . . . 2523
6. Storage and Transportation . . . . . . 2524
7. Environmental Protection . . . . . . . 2526
8. Uses . . . . . . . . . . . . . . . . . . . . . . . 2527
9. Toxicology and Occupational Health 2529
10. References . . . . . . . . . . . . . . . . . . 2529

# Dimethyl Ether

HERMANN HÖVER, Union Rheinische Braunkohlen Kraftstoff AG, Wesseling, Federal Republic of Germany

1. Introduction ............. 1931
2. Properties ............... 1931
3. Production .............. 1933
4. Uses .................... 1935
5. Toxicology .............. 1935
6. References .............. 1935

# 1. Introduction

Dimethyl ether [115-10-6], $C_2H_6O$, is the simplest aliphatic ether.

$CH_3—O—CH_3$

For a long time it was obtained on an industrial scale as a byproduct in the high-pressure production of methanol. Recently the high-pressure methanol plants have been almost completely replaced by low-pressure plants; this has led to the erection of special plants for the synthesis of dimethyl ether. Western Europe produces approximately 25 000 t of dimethyl ether per year. Dimethyl ether is industrially important as the starting material in the production of the methylating agent dimethyl sulfate and is used increasingly as an aerosol propellant.

# 2. Properties

Dimethyl ether (DME, methoxymethane), $M_r$ 46.07, is a colorless, almost odorless gas at room temperature and atmospheric pressure and has the following physical properties:

| | |
|---|---|
| *mp* | − 141 °C [1] |
| *bp* at 0.1 MPa | − 24.8 °C |
| Critical pressure | 5.37 MPa (53.7 bar) [1] |
| Critical temperature | 126.9 °C |
| Critical density | 0.271 g/cm$^3$ |
| Heat of combustion (gas) | 31.75 MJ/kg |

| | |
|---|---|
| Heat of formation | −183 kJ/mol |
| Specific heat capacity (at −24 °C) | 2.26 kJ kg$^{-1}$ K$^{-1}$ |
| Heat of vaporization (at −20 °C) | 410.2 kJ/kg |
| Autoignition temperature | 235 °C [2] |
| Explosive limits | 3.0–17 vol% in air |
| Flash point | −41 °C |
| Relative density (gaseous, air = 1) | 1.59 |
| Density (at 0.51 MPa) | ca. 670 kg/m$^3$ |

Density of liquid dimethyl ether [1]:

| $t$, °C | 10 | 20 | 30 | 40 | 50 | 60 | 70 | 80 |
|---|---|---|---|---|---|---|---|---|
| $\varrho$, kg/m$^3$ | 682 | 666 | 649 | 631 | 612 | 593 | 573 | 552 |

Vapor pressure [1]:

| $t$, °C | −40 | −20 | 0 | 20 | 50 |
|---|---|---|---|---|---|
| $p$, MPa | 0.04 | 0.14 | 0.24 | 0.51 | 1.14 |

Dimethyl ether is miscible with most polar and nonpolar organic solvents. It is also partly miscible with water (76 g in 1 L of water at 18 °C). In addition, numerous polar and nonpolar substances readily dissolve in dimethyl ether.

**Chemical Properties.** Unlike most other aliphatic ethers, dimethyl ether is not susceptible to autoxidation. This is of considerable importance for industrial applications. Numerous studies have confirmed that dimethyl ether is stable to atmospheric oxygen and does not form peroxides [3].

Dimethyl ether is converted to dimethyl sulfate [77-78-1] by sulfur trioxide (→ Dialkyl Sulfates and Alkylsulfuric Acids).

$$SO_3 + CH_3-O-CH_3 \longrightarrow (CH_3)_2SO_4$$

Addition of boron trifluoride to dimethyl ether results in the formation of BF$_3$· CH$_3$OCH$_3$ [353-42-4], mp −14 °C, bp 127 °C, a distillable liquid which fumes when exposed to moist air. This addition compound is easier to handle than gaseous boron trifluoride.

In the presence of CoI$_2$, dimethyl ether reacts with carbon monoxide and water to form acetic acid [64-19-7] (→ Acetic Acid) [4].

$$CH_3-O-CH_3 + H_2O + 2\,CO \longrightarrow 2\,CH_3COOH$$

The reaction of dimethyl ether with hydrogen sulfide in the presence of a catalyst, e.g., tungsten sulfide (WS$_2$), gives dimethyl sulfide [75-18-3] [5].

$$CH_3-O-CH_3 + H_2S \longrightarrow CH_3-S-CH_3 + H_2O$$

The reaction of dimethyl ether to give unsaturated and saturated hydrocarbons proceeds in the presence of zeolitic catalysts [6].

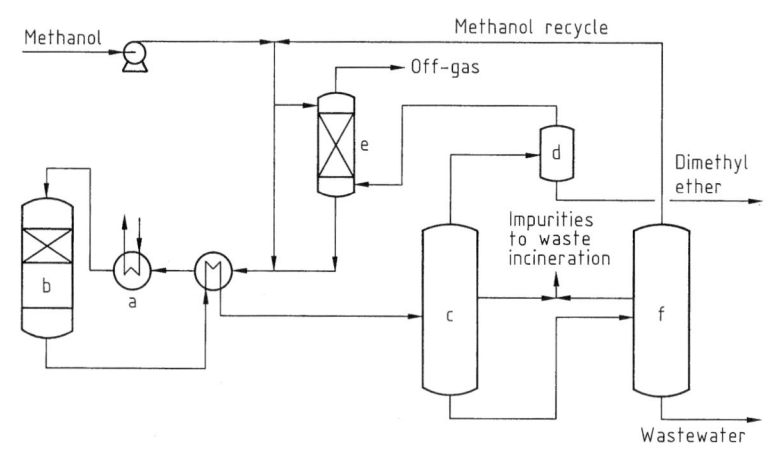

**Figure 1.** Dimethyl ether production by dehydration of methanol
a) Vaporizer; b) Reactor; c) Dimethyl ether column; d) Condenser; e) Scrubber; f) Methanol column

$$CH_3-O-CH_3 \longrightarrow H_2C=CH_2 + H_2O$$

The catalytic conversion of dimethyl ether to formaldehyde has also been observed [7].

$$CH_3-O-CH_3 + O_2 \longrightarrow 2\,CH_2O + H_2O$$

# 3. Production

Until about 1975 dimethyl ether was obtained as a byproduct in the high-pressure production of methanol [8]. In this process up to 3–5 wt% dimethyl ether is formed. Dimethyl ether can be recovered in pure form by distillation of crude methanol. The development of the low-pressure methanol synthesis, particularly by Lurgi [9] and ICI [10], has resulted in the almost complete replacement of all high-pressure plants by 1980. The low-pressure processes, which require less severe conditions, produce only very small amounts of dimethyl ether (→ Methanol). As a result, special catalytic processes have been developed for the production of dimethyl ether.

The preparation of dimethyl ether from methanol in the presence of acidic catalysts on a laboratory scale has been known for many years [11]. Numerous methods have been discussed in the patent literature. For instance, aliphatic ethers can be prepared by heating alcohols in the presence of zinc chloride [12]. Other suitable catalysts are iron chloride, copper sulfate, copper chloride, manganese chloride, aluminum chloride, aluminum sulfate, chromium sulfate, alums, thorium compounds, aluminum oxide, titanium oxide, barium oxide, silica gel, and aluminum phospate [13]. Aluminum oxide and aluminum silicate, with or without doping, are the most important catalysts for industrial use [14]. Figure 1 illustrates the process developed by Union Rheinische

Braunkohlen Kraftstoff AG for the production of high-purity dimethyl ether (purity > 99.9 %).

The catalytic dehydration of pure, gaseous methanol is carried out in a fixed-bed reactor. The product is cooled in two stages and subsequently distilled to yield pure dimethyl ether. Small amounts of dimethyl ether are recovered from the off-gas in a scrubber and recycled to the reactor. Unreacted methanol is separated from water in a second column and also recycled.

Very pure dimethyl ether which is suitable as an aerosol propellant is obtained by special rectification processes.

Direct synthesis of dimethyl ether from synthesis gas ($CO + H_2$) has also been described [15].

**Safety and Environmental Aspects.** Dimethyl ether is a flammable gas. Water, foam, carbon dioxide, and dry chemical powders can be employed to control dimethyl ether fires. Fire extinguishers suitable for class C (Europe) or class B (United States) can be used. Endangered vessels must be cooled.

Although dimethyl ether is soluble in water and biologically easily degradable, large amounts of this compound should not enter the wastewater system because evaporation can cause the formation of explosive mixtures over the surface of the water.

**Quality Specifications and Analysis.** Dimethyl ether is available in two commercial qualities:

- Technical grade contains up to 0.05 % methanol and contaminations with a strong odor.
- High-purity dimethyl ether, practically free of sulfur-containing and other substances with an unpleasant odor, can be used in the aerosol industry; the methanol content should not exceed 10 mg/kg for this quality.

The purity of dimethyl ether is determined by GC analysis. Oil and ash content are measured by special methods of evaporation and combustion.

**Storage and Transportation.** Dimethyl ether is usually stored as a liquid under pressure. It is transported in railway pressure tanks, tank trucks, and other containers. Tank trucks used to transport dimethyl ether have capacities reaching up to 18.2 t; railway pressure tanks have capacities of ca. 10, 30, and 35 t. Individual containers (tank containers) with a capacity of 14 t are shipped even to Australia. Dimethyl ether is also available in small quantities of 45 – 520 kg.

Transportation by rail and road is subject to the following regulations: GGVE and GGVS (Federal Republic of Germany), class 2, no. 3b; RID and ADR (international), class 2, no. 3b; IMDG Code (Amendment 20-82), class 2.1, p. 2052.

Suppliers are Union Rheinische Braunkohlen Kraftstoff AG in the Federal Republic of Germany, Aerofako B.V. in the Netherlands, and Du Pont in the United States.

# 4. Uses

For a long time, the only industrial use for dimethyl ether was its conversion to dimethyl sulfate by treatment with sulfur trioxide. Dimethyl sulfate is employed as a methylating agent. Of the 20 000 t of dimethyl ether produced in Western Europe in 1986, about 9000 t was used in the production of dimethyl sulfate.

The use of dimethyl ether as a comparatively safe aerosol propellant is an increasingly important commercial application which was developed in the 1980s. Dimethyl ether is miscible with water and numerous substances dissolve in it; it is nontoxic and has a relatively low flammability. In Western Europe, about 10 000–11 000 t/a are used as aerosol propellant.

The reaction of dimethyl ether with carbon monoxide and water can be used in the large-scale production of acetic acid in place of the methanol–carbon monoxide reaction.

Future industrial uses of dimethyl ether could include the production of olefins, especially ethene, propene, and butenes, in the presence of zeolitic catalysts. The production of saturated hydrocarbons can be carried out by an analogous process. These processes, starting with methanol, can involve two steps, dimethyl ether being produced in the first step; alternatively, a one-step process may be used, with dimethyl ether being formed as an intermediate.4.

# 5. Toxicology

Pure dimethyl ether is nontoxic. Inhalation experiments conducted on rats using concentrations of up to 20 000 ppm (2 vol% in air) over a period of 8 months did not lead to any deaths [3]. Contact with dimethyl ether is not irritating to the skin.

# 6. References

[1]  K. Nabert, G. Schön: *Sicherheitstechnische Kennzahlen brennbarer Gase und Dämpfe*, 2nd ed., Deutscher Eich-Verlag GmbH, Braunschweig, 1970.
[2]  Union Rheinische Braunkohlen Kraftstoff AG, unpublished results.
[3]  L. J. M. Bohnenn, *Manuf. Chem. Aerosol News* 1978, August, 39.
[4]  *Hydrocarbon Process.* **50** (1971) no. 11, 115.
[5]  Union Rheinische Braunkohlen Kraftstoff AG, Wesseling, DE 1 016 261, 1956 (F. Hübenett, N. Schnack).
[6]  F. X. Cormerais, G. Perot, M. Guisnet, *Zeolites* **1** (1981) October, 141.W. Lee, J. Mazink, V. W. Weckman Jr., S. Yurchak, *Biomass Wastes* **4** (1980) 721. C. D. Chang, C. T.-W. Chu, *J. Catal.* **74** (1982)203. J. Haggin, *Chem. Eng. News* **63** (1985) March 25, 39.

[7] Shell Oil Company, US 4 439 624, 1982 (R. M. Lewis, R. V. Ryan, L. H. Slaugh); US 4 442 307, 1982 (R. M. Lewis, R. C. Ryan, L. H. Slaugh).
[8] *Winnacker-Küchler*, 4th ed., vol. **5,** Carl Hanser Verlag, München–Wien 1981, p. 512.
[9] E. Supp, *Chem. Technol.* **3** (1973) 430.
[10] P. L. Rogerson, *Chem. Eng. (N.Y.)* **80** (1973) no. 19, 112.
[11] *Houben-Weyl*, vol. **VI/3,** part 3, Georg Thieme Verlag, Stuttgart 1965, pp. 17, 18.
[12] Dr. Alexander Wacker, Gesellschaft für elektrochemische Industrie GmbH, DE 680 328, 1934 (P. Halbig, O. Moldenhauer).
[13] R. L. Brown, W. W. Odells, US 1 873 537, 1927.N. V. de Bataafsche Petroleum Maatschappij, FR 701 335, 1930;GB 332 756, 1929; GB 350 010, 1931; GB 403 402, 1932.
[14] Mobil Oil Corp., DE 2 818 831, 1978 (F. G. Dwyer, A. B. Schwartz); DE-OS 3 201 155, 1982 (W. K. Bell, C. Chang). Du Pont, EP-A 99 676, 1983 (D. L. Brake). Mitsubishi Chemical Industries, EP-A 124 078, 1984 (N. Murai, K. Nakamichi, M. Otake, T. Ushikubo).
[15] Snamprogetti, SpA., DE 2 362 944, 1973 (G. Giorgio); DE 2 757 788, 1977 (G. Manara, B. Notari, V. Fattore); DE 3 220 547, 1982 (G. Manara).

# Dioxane

KENNETH S. SURPRENANT, Dow Chemical Co., Midland, Michigan 48640, United States

| | | | | | |
|---|---|---|---|---|---|
| 1. | Physical Properties | 1937 | 7. | Legal Aspects | 1942 |
| 2. | Chemical Properties | 1938 | 8. | Uses | 1942 |
| 3. | Production | 1940 | 9. | Economic Aspects | 1943 |
| 4. | Environmental Protection | 1941 | 10. | Toxicology and Occupational Health | 1943 |
| 5. | Quality Specifications | 1941 | | | |
| 6. | Storage and Transportation | 1942 | 11. | References | 1945 |

## 1. Physical Properties

The most important *physical and flammable properties* of dioxane are as follows:

| | |
|---|---|
| bp | 101.3 °C |
| fp | 11.8 °C |
| $d_{20}^{20}$ | 1.0356 |
| Latent heat of evaporation | 413 kJ/kg |
| Heat of fusion | 141 kJ/kg |
| Specific heat (20 °C) | 1.76 kJ kg$^{-1}$ K$^{-1}$ |
| $n_D^{20}$ | 1.4224 |
| Vapor pressure (20 °C) | 4.13 kPa (41.3 mbar) |
| Dielectric constant (20 °C) | 2.23 |
| Surface tension (20 °C) | 36.9 mN/m |
| $t_{crit}$ | 312 °C |
| $p_{crit}$ | 5.14 MPa |
| Flash point (closed cup) | 11 °C |
| Flammable limits (in air) | |
|    Lower limit | 2 vol% |
|    Upper limit | 22 vol% |
| Heat of combustion | 27 600 kJ/kg |
| Autoignition temperature | 180 °C |

1,4-Dioxane [*123-91-1*], $C_4H_8O_2$, $M_r$ 88.11, is an extraordinary solvent, capable of solubilizing most organic compounds, water in all proportions, and many inorganic compounds.

**Table 1.** Dioxane azeotropic data

| Component | Component bp, °C | Dioxane, wt% | Azeotropic bp, °C |
|---|---|---|---|
| Formic acid | 100.7 | 57 | 113.35 |
| Nitromethane | 100.8 | 43.5 | 100.55 |
| Acetic acid | 117.9 | 20.5–23 | 119.4 |
| Ethanol | 78.5 | 9.3 | 78.1 |
| 3-Iodo-1-propene | 103 | 44 | 98.5 |
| 1-Propanol | 97.4 | 45 | 95.3 |
| tert-Amyl alcohol | 102.5* | 20 | 100.65 |
| Cyclohexane | 80.7 | 24.6 | 79.5 |
| Heptane | 98.4 | 44 | 91.85 |
| Water | 100 | 82 | 87.8 |

*At 8 kPa.

Dioxane is a cyclic diether forming a six-membered ring with the following structure:

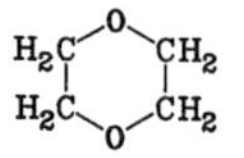

$$\begin{array}{c}\text{H}_2\text{C}\text{—O—}\text{CH}_2\\ |\qquad\qquad|\\ \text{H}_2\text{C—O—}\text{CH}_2\end{array}$$

*Synonyms* for dioxane are *p*-dioxane, diethylene oxide, diethylene dioxide, diethylene ether, 1,4-dioxacyclohexane, dioxyethylene ether, and dioxan.

At ambient temperature, dioxane is a clear liquid with an ether-like odor. Its relative stability is similar to other aliphatic ethers, and it forms peroxides on exposure to air as they do. In spite of its exceptional solvent properties, its largest use is in stabilizing 1,1,1-trichloroethane and keep it from reacting with aluminum (see Chap. 8).

Reportedly, 1,4-dioxane was prepared by A. V. LOURENÇO in 1863 by reacting ethylene glycol and 1,2-dibromoethane [1]. However, commercial production did not occur until 1929.

Infinite *solubility* in dioxane can be expected from lower molecular mass aliphatic and aromatic liquid hydrocarbons, ethers, alcohols, ketones, and chlorinated hydrocarbons. Dioxane has solubility for animal and vegetable oils, paraffin oils, synthetic and natural resins, and some inorganic compounds and elements, such as iron chlorides, mercuric chloride, hydrochloric, sulfuric, and phosphoric acids, bromine, chlorine, and iodine.

Dioxane forms *azeotropic mixtures* with many compounds. Some common binary systems are given in Table 1.

# 2. Chemical Properties

As is characteristic of ethers, dioxane forms *peroxides* by air oxidation. These peroxides are often highly explosive and unstable. The peroxide of dioxane is 2-hydroperoxide 1,4-dioxane [*4722-59-2*] (**1**) [2], which decomposes to dioxanone [*3041-16-5*] (**2**) and water, and subsequently is hydrolyzed to form 2-(2-hydroxyethoxy)acetic acid [*13382-47-3*] (**3**):

$$\text{1} \xrightarrow[-H_2O]{\text{Decomp.}} \text{2} \xrightarrow{H_2O} HOCH_2CH_2OCH_2COOH \quad \text{3}$$

The reaction rates favor formation of **3** and limit the peroxide concentration. However, dioxane should be treated with the same precautions as other ethers to prevent peroxide formation. Inert gas padding (e.g., nitrogen atmosphere) may be used. Peroxide-containing ethers should not be evaporated or distilled to dryness. Peroxides can be destroyed by reducing agents, such as iron (II) or tin (II) chlorides or sodium bisulfite. Activated carbon, activated basic alumina, and the hydroxyl form of anion-exchange resins also remove peroxides [3].

Dioxane solubilizes and complexes with $Cl_2$, $Br_2$, $I_2$, and the halogen acids. These *complexes* can moderate halogenation or halogen acid reactions. For example, dioxane dibromide complex can brominate phenols, aromatic aldehydes or amines, and ketones. Conversely, chlorination can produce chlorinated dioxanes containing from one to eight chlorine atoms, with isomeric yields controlled by temperature, reaction conditions, and catalysts employed. Dioxanes containing fewer chlorine atoms can be used to prepare dioxane addition compounds, e.g., addition of ethoxy or ethyl groups using $NaOCH_2CH_3$ or $CH_3CH_2MgBr$, respectively.

2,3-Dichlorodioxane is particularly reactive. It condenses with aliphatic and aromatic acids, alcohols, and glycols to give diesters and ethers [4]–[6]. With ethylene glycol, the dicyclic isomers of 1,4,5,8-tetraoxadecahydronaphthalene are formed [7]. 2-Chloro-1,4-dioxene, made by dehydrohalogenation of 2,3-dichlorodioxane, can be polymerized or hydrolyzed to give dioxanone [8].

Although the *dioxane ring* is quite stable, it can be cleaved by strong acids at high temperature and pressure. Under such conditions, dioxane forms ethylene glycol diacetate and diethylene glycol diacetate when reacted with acetic acid anhydride [9]. In addition, benzoyl chloride and acetyl chloride catalyzed by $TiCl_4$ react with dioxane to form the 2-chloroethyl esters of benzoic and acetic acids [10], respectively.

Benzoyl peroxide catalyzes the reaction between α-olefins and dioxane to form *polymeric condensation products* [11]. The reaction requires a temperature of 80–140 °C and a reaction time of 5–10 h. The polymers are useful as lubricants. Similarly, tetrafluoroethylene reacted with dioxane at 110 °C with peroxide catalysis under pressure for 8 h forms analogous polymers of the following general structure [12]:

$$\text{dioxane}-(C_2F_4)_xH$$

Numerous alternative catalysts are reported to be effective, including organic and inorganic peroxides, bases (e.g., borax, disodium phosphate, and hydrazine), zinc chloride, and phosphoric acid. Chlorofluoroethylenes containing at least two fluorine atoms and only one hydrogen atom yield similar products. The polymers formed can contain values of $x$ from 1 to 25. A second polymer chain at another site on the dioxane ring can be formed by repeating the reaction with the first polymer product. The lower molecular mass products are volatile liquids of potential use as solvents. Higher molecular mass polymers are waxy solids and are regarded as possible lubricants.

Many *stable complexes* are reported in addition to those of the halogens mentioned previously. Some of the agents that give dioxane complexes are the following:

| | | |
|---|---|---|
| $AlBr_3$ | LiBr | HCl |
| $AlCl_3$ | LiCl | $H_2SO_4$ |
| $AuCl_3$ | LiI | $H_3PO_4$ |
| $BBr_3$ | $MnCl_2$ | $SO_2$ |
| $BCl_3$ | $NiCl_2$ | $SO_3$ |
| $BF_3$ | $PF_5$ | |
| $CuCl_2$ | $PtCl_4$ | 2,4,6-trinitrophenol |
| $CoCl_2$ | $SbCl_5$ | |
| $FeCl_3$ | $ZnCl_2$ | |
| $HgCl_2$ | $ZnF_2$ | |

These complexes can be employed as catalysts, as moderating agents in carrying out reactions, such as bromination or sulfonation, or as anhydrous systems for acid reactions.

# 3. Production

1,4-Dioxane is manufactured commercially by *dehydration and ring closure of diethylene glycol.* Concentrated sulfuric acid (ca. 5%) is used as a catalyst, although phosphoric acid, *p*-toluenesulfonic acid, and strongly acidic ion-exchange resins are recognized alternatives. Operating conditions vary; temperatures range from 130 to 200 °C and pressures range from a slight pressure to a partial vacuum (25–110 kPa). A favorable temperature is reported to be 160 °C. The process is continuous, with dioxane vaporized from the reaction vessel. The vapors are passed through an acid trap and two distillation columns to remove water and to purify the product. Yields of ca. 90% are possible. 2-Methyl-1,3-dioxolane, acetaldehyde, crotonaldehyde, and polyglycol are unwanted byproducts.

Dioxane can also be prepared by dehydrohalogenation of 2-chloro-2'-hydroxydiethyl ether, by reacting ethylene glycol with 1,2-dibromoethane, and by dimerizing ethylene oxide either over $NaHSO_4$ [1], $SiF_4$, or $BF_3$ [13], or at elevated temperature with an acidic cation-exchange resin [14].

# 4. Environmental Protection

Dioxane, emitted into the atmosphere, is decomposed rapidly by hydroxyl radical oxidation, as are other hydrocarbons. The half-life is estimated to be 3.9 d [15]. The intermediates are probably short-lived hydrocarbon radicals; the ultimate products are carbon dioxide and water.

Dioxane is highly resistant to biotransformation, although it may possibly be oxidized by acclimated sewage microorganisms. Carbon adsorption purification of water containing dioxane has limited practicality because of the high (ca. 100-fold) ratio of carbon to adsorbed dioxane required for complete removal. Aeration of water containing dioxane is inefficient because of dioxane's extreme solubility and low vapor pressure. Dioxane can be decomposed by $H_2O_2$ by using $FeSO_4$ to catalyze the reaction at 25 °C. Oxidation of dioxane by chlorination is effective at 75 °C, but not at 25 °C [16].

# 5. Quality Specifications

The American Chemical Society (ACS) has published specifications for 1,4-dioxane [17]. A summary of the ACS requirements is as follows:

| | |
|---|---|
| Color (APHA), max. | 20 |
| Peroxide (as $H_2O_2$), max. | 0.005 % |
| $fp$, min. | 11.0 °C |
| Nonvolatile residue, max. | 0.005 % |
| Neutrality | to pass test |
| | A maximum of 1.0 mL of 0.01 M $H_3PO_4$ to neutralize a 50-mL sample. |
| Carbonyl (as HCHO), max. | 0.01 % |
| Water, max. | 0.05 % |

Industrial grades are sold under producer specifications that are more or less demanding than for ACS grade. These specifications often include boiling and relative density ranges and a maximum acidity. Industrial specifications are accompanied by analytical methods or referenced methods.

Higher purity dioxane may be required for research or special industrial processes. Techniques have been described in the literature to remove traces of peroxides, acids, stabilizers, and aldehydes [3].

Samples should be handled to prevent air contact, and nonvolatile residue tests should not be performed on material that contains peroxides.

# 6. Storage and Transportation

The most important consideration in shipping and storing dioxane is to prevent peroxide formation. This is achieved by shipping and storing with a nitrogen pad. Failure to maintain an inert atmosphere causes the material to exceed acid, peroxide, and possibly other specifications.

Mild steel drums and tanks are adequate for shipping and storage. To limit the fire hazard, all containers should be grounded to prevent static electrical charges from accumulating during transfer operations. All areas of storage must be equipped with explosion-proof equipment and be well ventilated. The material must be maintained above its freezing point in cases where temperatures may be < 12 °C by use of insulation or heat.

The United Nations' recommendations include the shipping name "Dioxane," number UN 1165, and labeling using the warning as a flammable liquid hazard class 3. The International Maritime identifies dioxane in class 3.2.

# 7. Legal Aspects

The U.S. National Toxicology Program (NTP) and the International Agency for Research on Cancer (IARC) have listed dioxane as a possible carcinogen on the basis of extreme high-dose lifetime animal studies discussed in Chapter 10. These listings may require special labeling or restrictions; national and local regulations should be consulted [18].

# 8. Uses

The largest single use of 1,4-dioxane is the *stabilization of 1,1,1-trichloroethane* against chemical attack by aluminum [19], [20]. The aluminum oxide film prevents the metal from reacting with 1,1,1-trichloroethane until the film is disturbed. When the bare metal contacts 1,1,1-trichloroethane, chlorine is extracted from the solvent to form $AlCl_3$, which causes dehydrohalogenation of 1,1,1-trichloroethane to produce HCl and vinylidene chloride ($CH_2 = CCl_2$). Dioxane is the foremost inhibitor of these reactions. An insoluble dioxane–aluminum chloride complex is formed, which deactivates the catalyst and seals any openings in the $Al_2O_3$ film.

Dioxane is used as a *solvent* in the formulation of inks, coatings, and adhesives, and as solvent for extracting animal and vegetable oils. As a *chemical intermediate*, dioxane reaction products are useful as insecticides, herbicides, plasticizers, and monomers. The oxonium complexes of dioxane with salts, mineral acids, halogens, and sulfur trioxide are used as *catalysts* and as *reagents* for anhydrous acid reactions, brominations, and sulfonations. In the laboratory, dioxane is useful as a *cryoscopic solvent* for molecular mass determinations and as a *stable reaction medium* for diverse reactions. Historically, dioxane is reported to be valuable in *polymer manufacture* and as a *solvent* for natural and synthetic resins, including cellulose derivatives, polyvinyl acetal resins, acrylonitrile – methyl vinyl pyridine copolymers, and others.

# 9. Economic Aspects

Recognized *manufacturers* of dioxane are

BASF Aktiengesellschaft, Ludwigshafen, Federal Republic of Germany
CPS Chemical Co., Old Bridge, New Jersey, United States
Dow Chemical Co., Terneuzen, The Netherlands
Grant Chemical Co., Baton Rouge, Louisiana, United States
Osaka Organic Chemical Industries Co., Japan
Sanraku-Ocean Co., Japan
Toho Chemical Industry Co., Japan
Ugine-Kuhlmann, Frankfurt, Federal Republic of Germany
Union Carbide, South Charleston, West Virginia, United States

No information is available on producers in the Eastern-bloc countries. *Capacity* of known producers was estimated to be 11 000 – 14 000 t in 1985.

# 10. Toxicology and Occupational Health

The *acute toxicity* of dioxane is relatively low. The oral $LD_{50}$ (rats) is 5170 mg/kg; by skin absorption, the $LD_{50}$ (rabbits) is 7600 mg/kg. The $LC_{50}$ (rats) with a 4-h exposure is 14 260 ppm (51 880 mg/m$^3$) [21]. *Acceptable occupational standards* are as follows:

American Conference of Governmental Industrial Hygienists (ACGIH), TLV–TWA
   25 ppm (90 mg/m$^3$) [22] "skin" notation
U.S. Occupational Safety and Health Administration (OSHA)
   100 ppm (360 mg/m$^3$)
Deutsche Forschungsgemeinschaft (MAK)
   50 ppm (180 mg/m$^3$) [23] IIIB notation: suspected of carcinogenic potential
Sweden (1975)   25 ppm (90 mg/m$^3$)
USSR (1980)   3 ppm (10 mg/m$^3$)

Processes that employ dioxane should be designed to protect workers against skin and eye exposure, as well as excessive vapor inhalation. The odor warning properties of dioxane are not adequate to afford protection. Exposure to high concentrations may cause respiratory irritation and damage to the kidney and liver.

Early, chronic *animal testing* (rodents) at extremely high doses (7–18 000 mg/kg in water) resulted in nasal and liver tumors and in liver and kidney damage. A follow-up study with exposures at 10 000, 1000, and 100 mg/kg in water confirmed the earlier results at high dose, but the lower dose testing showed definite indications of a threshold [24]. Animals exposed to 1000 mg/kg in water suffered varying degrees of kidney and liver injury but no treatment-related tumors. The animals dosed with 100 mg/kg (in water) experienced no adverse health effects.

A number of tests have been employed to determine whether or not dioxane caused these nasal and liver *tumors* by direct interaction with deoxyribonucleic acid (DNA) or by some other mechanism. Included were mutagenicity tests on microorganisms to observe if genetic changes occurred, e.g., the Ames bacterial mutagenicity test and a test to detect DNA repair in liver cells [25], [26]. Further tests using relatively high doses of $^{14}$C-radiolabeled dioxane were conducted; the liver DNA was isolated and measured for trace quantities of radioactivity from DNA–dioxane reaction products [26]. Strong positive responses in these tests would have provided evidence of possible carcinogenicity. However, no mutagenicity or DNA reaction products were discovered [25], [26]. Therefore, dioxane may cause tumors in rodents via an alternate mechanism, possibly by recurrent tissue damage [26].

Pharmacokinetic and metabolic studies using $^{14}$C-radioactive dioxane demonstrated that low doses of dioxane are eliminated in the urine as 2-hydroxyethoxyacetic acid (HEAA) [27]–[30]. Conversion of dioxane to HEAA is nearly complete, as indicated by the observed 118:1 ratio of HEAA to dioxane; the reaction is rapid, with a dioxane half-life of ca. 1 h. However, the process has a saturation point, and when this is exceeded by a high dose, elimination requires much more time.

The threshold effect is further supported by a 2-year inhalation study conducted on rats at a concentration of 111 ppm [31]. Unlike the high dose drinking water studies, no adverse health effects were observed, including no increase in tumor incidence.

*Human exposures,* estimated to have been at 470 ppm, have resulted in deaths. Three epidemiologic studies have been made of plant workers exposed to concentrations of up to 51 ppm. Each of these studies found no evidence of any adverse health effects.

These data were reviewed by the ACGIH with the following conclusion: ". . . In view of the findings of tumors in both liver and lung at or near the 10 000 ppm dietary level, and the lack of such findings at inhalation exposure levels slightly above 100 ppm for two years, dioxane has been classed by the TLV Committee as an animal tumorigen of such low potential as to be of no practical significance as an occupational carcinogen at or around the former TLV of 50 ppm." The carcinogenicity of dioxane at high-dose exposures appears to be the result of repeated tissue injury. Therefore, controlling exposures to levels below those that would cause organ injury should prevent any significant risk of cancer.

# 11. References

[1] M. Sainsbury in S. Coffey (ed.): *Rodd's Chemistry of Carbon Compounds*, 2nd ed., vol. **4**, Elsevier Scientific Publ., Amsterdam 1978, part H, p. 382.
[2] W. Stumpf: *Chemie und Anwendungen des 1,4-Dioxans*, Verlag Chemie, Weinheim 1956, p. 33.
[3] J. F. Coetzee, T. H. Chang, *J. Pure Appl. Chem.* **57** (1985) no. 4, 633–638.
[4] Dow Chemical, US 2 195 386, 1940 (E. C. Britton, H. R. Slagh).
[5] Dow Chemical, US 2 843 602, 1958 (N. B. Lorette).
[6] Dow Chemical, US 2 887 491, 1959 (A. E. Gurgiolo).
[7] J. Boeseken et al., *Recl. Trav. Chim. Pays-Bas* **50** (1931) 909.
[8] Olin Mathieson, US 2 756 240, 1956 (M. J. Astle, W. C. Gergel).
[9] M. MacLeod, *J. Chem. Soc.* **50** (1928) part 2, 3092.
[10] L. M. Smorgonski, Y. L. Goldfarb, *J. Gen. Chem. USSR* **8** (1938) 1516. *Chem. Abstr.* **33** (1939) 4593.
[11] Socony Mobil Oil, US 2 743 281, 1956 (C. F. Feasley).
[12] Du Pont, US 2 433 844, 1948 (W. E. Hanford).
[13] H. Schecker, W. Kochler, B. Sander, US 3 825 568, 1974.
[14] Mathieson, CA 540 278, 1957 (M. J. Astle, J. A. Zaslowsky).
[15] L. T. Cupitt, EPA report no. 600/S3-80-084, 1980.
[16] G. M. Klecka, S. J. Gonsior, *J. Hazard. Mater.* **13** (1968) 161–168.
[17] W. E. Schmidt: *Reagent Chemicals*, 5th ed., Amer. Chem. Soc., Washington, D.C., 1974, pp. 237–239.
[18] CFR 29, U.S. Hazard Communication Standard, 1910.1200, Washington, D.C., 1983.
[19] W. L. Archer, *Ind. Eng. Chem. Prod. Res. Dev.* **18** (1979) no. 2, 131–135.
[20] Dow Chemical, US 2 811 252, 1957 (H. J. Batchel).
[21] V. K. Rowe, M. A. Wolf in G. D. Clayton, F. E. Clayton (eds.): *Patty's Industrial Hygiene and Toxicology*, 3rd ed., vol. **2 C**, Wiley-Interscience, New York 1981, pp. 3947–3956.
[22] ACGIH (ed.): *Threshold Limit Values (TLV) and Biological Exposure Indices*, ACGIH, Cincinnati, Ohio, 1986–1987.
[23] DFG (ed.): *Maximum Concentrations at the Workplace (MAK)*, VCH Verlagsgesellschaft, Weinheim 1986.
[24] R. J. Kociba, S. B. McCollister, C. Park, T. R. Torkelson, *Toxicol. Appl. Pharmacol.* **30** (1974) 275–286.
[25] S. Haworth, T. Lawlor, K. Mortelmans, W. Speck et al., *Environ. Mutagenesis* **5** (1983) Suppl. 1, 3–142.
[26] W. T. Stott, J. F. Quast, P. G. Watanabe, *Toxicol. Appl. Pharmacol.* **60** (1981) 287–300.
[27] J. D. Young, W. H. Braun, L. W. Rampy, M. B. Chenoweth et al., *J. Toxicol. Environ. Health* **3** (1977) 507–520.
[28] J. D. Young, W. H. Braun, P. J. Gehring, *J. Toxicol. Environ. Health* **4** (1978) 709–726.
[29] W. H. Braun, J. D. Young, *Toxicol. Appl. Pharmacol.* **39** (1977) 33–38.
[30] F. K. Dietz, W. T. Stott, J. C. Ramsey, *Drug Metab. Rev.* **13** (1982) no. 6, 963–981.
[31] T. R. Torkelson, B. K. J. Leong, R. J. Kociba, W. A. Richter et al., *Toxicol. Appl. Pharmacol.* **30** (1974) 287–298.

# Dithiocarbamic Acid and Derivatives

RÜDIGER SCHUBART, Bayer AG, Leverkusen, Federal Republic of Germany

| | | | | |
|---|---|---|---|---|
| 1. | Introduction | 1947 | 4.6.1. | Dithiolanylium Salts . . . . . . . . . 1965 |
| 2. | Physical and Chemical Properties . . . . . . . . . . . . . . . 1948 | | 4.6.2. | Alkylthioformimidic Chlorides . . 1965 |
| | | | 4.6.3. | Isothiocyanates, Isocyanates, and Thioureas . . . . . . . . . . . . . . . . 1966 |
| 3. | Manufacture of Dithiocarbamate Salts . . . . . . 1949 | | 4.7. | Photolysis of Dithiocarbamates . . . . . . . . . . . . . . . . . . 1967 |
| 4. | Transformation Products . . . . 1953 | | 5. | Uses . . . . . . . . . . . . . . . . . . . . 1967 |
| 4.1. | Thiuram Sulfides . . . . . . . . . . 1953 | | 5.1. | Vulcanization Accelerators . . . 1967 |
| 4.1.1. | Thiuram Disulfides . . . . . . . . . . 1953 | | | |
| 4.1.2. | Thiuram Monosulfides . . . . . . . 1955 | | 5.2. | Pesticides . . . . . . . . . . . . . . . 1971 |
| 4.1.3. | Thiuram Trisulfides and Tetrasulfides . . . . . . . . . . . . . . 1955 | | 5.3. | Medical Applications . . . . . . . 1971 |
| | | | 5.4. | Radioprotective Agents . . . . . 1971 |
| 4.2. | Thiocarbamoylsulfenamides . . 1956 | | 5.5. | Imaging Technology . . . . . . . . 1971 |
| 4.3. | Thiocarbamoyl Chlorides . . . . 1957 | | 5.6. | Other Uses . . . . . . . . . . . . . . 1974 |
| 4.4. | Dithiocarbamic Acid Esters . . 1957 | | 6. | Toxicology and Occupational Health . . . . . . . . . . . . . . . . . . 1975 |
| 4.5. | Heterocyclic Compounds . . . . 1961 | | | |
| 4.6. | Other Derivatives . . . . . . . . . 1965 | | 7. | References . . . . . . . . . . . . . . . 1975 |

## 1. Introduction

Dithiocarbamic acids are monoamides of dithiocarbonic acid. As a rule, one or both of the hydrogen atoms on the nitrogen atom are replaced by alkyl, aralkyl, cycloalkyl, or aryl radicals; the amide nitrogen atom may also belong to a ring:

Acyl- and thioacyldithiocarbamic acids have also been described:

1947

$$\underset{\substack{|\\R}}{R-\overset{O}{\overset{\|}{C}}-N-\overset{S}{\overset{\|}{C}}-SH} \qquad \underset{\substack{|\\R}}{R-\overset{S}{\overset{\|}{C}}-N-\overset{S}{\overset{\|}{C}}-SH}$$

Industrially important derivatives of dithiocarbamic acids are ammonium and metal salts, esters, thiuram sulfides, thiocarbamoylsulfenamides, and a number of other products. They are used mainly as vulcanization accelerators in the rubber industry and as fungicides in agriculture.

## 2. Physical and Chemical Properties

Free dithiocarbamic acid is unstable in the solid state (*mp* 35.7 °C) and in solution. Thermal decomposition results in carbon disulfide, hydrogen sulfide, and ammonium thiocyanate [11].

$$2\ H_2N-C\overset{S}{\underset{SH}{\diagup}} \xrightarrow{\text{Heat}} CS_2 + H_2N-\overset{S}{\overset{\|}{C}}-S^-\ NH_4^+ \xrightarrow{\text{Heat}} H_2S + NH_4SCN$$

Free N-substituted dithiocarbamic acids are also unstable. However, diphenyldithiocarbamic acid [12] is stable, as are several acyldithiocarbamic acids, such as formyl- [13], 2-ethoxybenzoyl- [14], and furoyldithiocarbamic acid [14], as well as 2-pyrrolidone-1-dithiocarbamic acid [15] (Table 1); these acids can be produced from the alkali salts by acidification with HCl gas in solvents such as ether. N-Thiobenzoyldithiocarbamic acid cannot be obtained by this method [16].

Salts of dithiocarbamic acids are more stable than the free acids [17]. The physical properties of some alkali-metal, heavy-metal, and ammonium dithiocarbamates are listed in Tables 2, 3, 4. Salts of unsubstituted and N-monosubstituted dithiocarbamic acids decompose on heating and are sensitive to oxidation (see Section 4.6.3). Derivatives of the tautomeric imide form of N-monosubstituted dithiocarbamic acids are also known [14], [18]–[22].

$$\underset{H}{\overset{R}{\diagdown}}N-C\overset{S}{\underset{SH}{\diagup}} \rightleftharpoons R-N=C\overset{SH}{\underset{SH}{\diagup}}$$

Salts and derivatives of N,N-disubstituted dithiocarbamic acids are very stable. These salts (especially the heavy-metal salts) are strongly colored; they are often volatile and soluble in organic solvents. For example, the copper, nickel, and cobalt salts of N,N-diisobutyldithiocarbamic acid can be distilled at low pressure. These compounds are apparently internal complexes.

**Table 1.** Physical properties of dithiocarbamic acids

$$\begin{array}{c} R^1 \\ \diagdown \\ R^2 \end{array} N-\overset{\overset{\displaystyle S}{\|}}{C}-SH$$

| $R^1$ | $R^2$ | CAS registry number | Physical form | Molecular formula | $M_r$ | mp, °C |
|---|---|---|---|---|---|---|
| H– | 2-(ethoxycarbonyl)phenyl (o-C(=O)-O-C₂H₅ phenyl) | [79388-79-7] | yellow crystals | $C_{10}H_{11}NO_2S_2$ | 241.3 | 122–123 |
| H– | 2-furoyl | [79388-80-0] | orange crystals | $C_6H_5NO_2S_2$ | 187.2 | 77–79 |
| $C_6H_5–$ | $C_6H_5–$ | [7283-79-6] | | $C_{13}H_{11}NS_2$ | 245.36 | 147 |
| H– | HC(=O)– | [100647-03-8] | bright yellow | $C_2H_3NOS_2$ | 121.18 | 66–67 |
| | 2-oxopyrrolidin-1-yl | [71408-15-6] | yellow crystals | $C_5H_7NOS_2$ | 161.25 | 101–102 |

Reactions of dithiocarbamate salts to form esters, thiuram sulfides, and other derivatives are discussed in detail in Chapter 4. All dithiocarbamic acids and their derivatives are cleaved by mineral acids to carbon disulfide and ammonium salts of the mineral acid. This reaction can be used to determine the structure of unknown derivatives [23].

# 3. Manufacture of Dithiocarbamate Salts

The parent compound in this series, ammonium dithiocarbamate [513-74-6], is prepared most easily from ammonia and carbon disulfide in alcohol or an ester solvent [24]. N-Alkyldithiocarbamic acids can be obtained as ammonium salts from two moles of a primary amine and one mole of carbon disulfide:

$$2\ R-NH_2 + CS_2 \longrightarrow \underset{H}{\overset{R}{\diagdown}} N-\overset{\overset{\displaystyle S}{\|}}{C}-S^-\ RNH_3^+$$

If the reaction is carried out in the presence of an alkali or alkaline-earth hydroxide, the corresponding metal salts are formed. To prepare N-aryldithiocarbamates a strong base is required; otherwise N,N-diarylthioureas are formed. Heavy-metal dithiocarbamates are obtained from ammonium or alkali dithiocarbamates and the corresponding heavy-metal salts in aqueous solution, or directly by reaction of an aromatic amine and carbon disulfide in the presence of finely divided metal oxide [25]. The deeply colored

# Dithiocarbamic Acid and Derivatives

**Table 2.** Physical properties of ammonium salts of dithiocarbamic acids

$$\begin{matrix} R^1 \\ R^2 \end{matrix} N-\overset{\overset{S}{\|}}{C}-S^- \quad \overset{+}{N}R^3R^4R^5R^6$$

| $R^1$ | $R^2$ | $R^3$ | $R^4$ | $R^5$ | $R^6$ | CAS registry number | Physical form | Molecular formula | $M_r$ | $d$ | mp, °C | Ref. |
|---|---|---|---|---|---|---|---|---|---|---|---|---|
| $CH_3-$ | $CH_3-$ | $CH_3-$ | $CH_3-$ | H– | H– | [598-64-1] | colorless crystals | $C_5H_{14}N_2S_2$ | 166.3 | | 115–120 | a |
| $C_2H_5-$ | $C_2H_5-$ | $C_2H_5-$ | $C_2H_5-$ | H– | H– | [1518-58-7] | colorless crystals | $C_9H_{22}N_2S_2$ | 222.4 | | 88 | a |
| $C_2H_5-$ | $C_6H_{11}-$ | $C_2H_5-$ | $C_6H_{11}-$ | H– | H– | [13167-44-7] | yellowish powder | $C_{17}H_{34}N_2S_2$ | 360.6 | 1.08 | 92 | b |
| \multicolumn{2}{N–} | \multicolumn{2}{N–} | H– | H– | [98-77-1] | yellowish powder | $C_{11}H_{22}N_2S_2$ | 246.45 | 1.15 | 179 | c |
| H– | HCO– | H– | H– | H– | $n\text{-}C_4H_9-$ | [100647-07-2] | yellow | $C_2H_6N_2OS_2$ | 138.22 | 1.48 | 114 (decomp.) | a |
| H– | HCO– | $n\text{-}C_4H_9-$ | $n\text{-}C_4H_9-$ | $n\text{-}C_4H_9-$ | $n\text{-}C_4H_9-$ | [100647-09-4] | | $C_{18}H_{38}N_2OS_2$ | 362.63 | 1.06 | 79–81 | a |
| $CH_3-$ | HCO– | H– | H– | H– | $n\text{-}C_4H_9-$ | [102564-66-9] | | $C_3H_8N_2OS_2$ | 152.24 | | 87–88 | b |
| H– | HCS– | $n\text{-}C_4H_9-$ | $n\text{-}C_4H_9-$ | $n\text{-}C_4H_9-$ | $n\text{-}C_4H_9-$ | [97960-84-4] | brilliant red | $C_{18}H_{38}N_2S_3$ | 378.69 | 1.06 | 90.5 | c |
| $CH_3-$ | HCS– | $n\text{-}C_4H_9-$ | $n\text{-}C_4H_9-$ | $n\text{-}C_4H_9-$ | $n\text{-}C_4H_9-$ | [101901-15-9] | blackish red | $C_{19}H_{40}N_2S_3$ | 392.87 | 1.09 | 89–91 (decomp.) | d |
| H– | HCS– | H– | H– | H– | H– | | orange | $C_2H_6N_2S_3$ | 122.22 | | 73–74 (decomp.) | c |
| $C_6H_5-$ | $C_6H_5-$ | H– | H– | H– | HO– | [97960-89-9] | yellow | $C_{13}H_{14}N_2OS_2$ | 278.39 | | 65 (decomp.) | c |

**Table 3.** Metal salts of ethylenebis(dithiocarbamic acids)

$$R\begin{matrix} -NH-\overset{\overset{S}{\|}}{C}-S \\ \phantom{-}\\ -NH-\overset{\overset{}{\|}}{C}-S \\ \phantom{-}\overset{}{\underset{S}{\|}} \end{matrix} M$$

| R | M | CAS registry number | Physical form | mp, °C |
|---|---|---|---|---|
| H– | 2 Na | [142-59-6] | white crystals | 82 |
| H– | Zn | [12122-67-7] | light-colored powder, polymer | decomp. on heating |
| H– | Mn | [12427-38-2] | yellow powder, polymer | decomp. on heating |
| $CH_3-$ | Zn | [12071-83-9] | pale yellow powder, polymer | decomp. above 160 °C |

**Table 4.** Physical properties of metal salts of dithiocarbamic acids

$$\left[ \begin{array}{c} R^1 \\ R^2 \end{array} \!\!\!\!> \!\! N - \overset{\overset{\displaystyle S}{\|}}{C} - S - \right]_x \!\! M$$

| $R^1$ | $R^2$ | M | x | CAS registry number | Form | Molecular formula | $M_r$ | d | mp, °C |
|---|---|---|---|---|---|---|---|---|---|
| $C_2H_5-$ | $C_2H_5-$ | Na | 1 | [148-18-5] | yellow crystals | $C_5H_{10}NNaS_2$ | 171.27 | | 97–98 |
| $CH_3-$ | $H-$ | Na | 1 | [137-42-8] | colorless crystals | $C_2H_4NNaS_2$ | 129.19 | | decomp. |
| $CH_3-$ | $CH_3-$ | Zn | 2 | [137-30-4] | white powder | $C_6H_{12}N_2S_4Zn$ | 305.82 | | 248–250 |
| $CH_3-$ | $CH_3-$ | Fe | 3 | [14484-64-1] | black powder | $C_9H_{18}FeN_3S_6$ | 416.61 | 1.65 | 180 (decomp.) |
| $C_2H_5-$ | $C_2H_5-$ | Zn | 2 | [14324-55-1] | white powder | $C_{10}H_{20}N_2S_4Zn$ | 361.93 | 1.47 | 178–180 |
| $n-C_3H_7-$ | $n-C_3H_7-$ | Zn | 2 | [15694-56-1] | white powder | $C_{14}H_{28}N_2S_4Zn$ | 418.03 | 1.35 | 107–108 |
| $n-C_4H_9-$ | $n-C_4H_9-$ | Zn | 2 | [136-23-2] | white powder | $C_{18}H_{36}N_2S_4Zn$ | 474.15 | 1.24 | 105–108 |
| $n-C_5H_{11}-$ | $n-C_5H_{11}-$ | Zn | 2 | [15337-18-5] | yellowish powder | $C_{22}H_{44}N_2S_4Zn$ | 530.25 | 1.14 | 65.5 |
| ⬡N– | | Zn | 2 | [13878-54-1] | white powder | $C_{12}H_{20}N_2S_4Zn$ | 385.95 | 1.55 | 222–225 |
| $C_6H_5CH_2-$ | $C_6H_5CH_2-$ | Zn | 2 | [14726-36-4] | white powder | $C_{30}H_{28}N_2S_4Zn$ | 610.21 | 1.41 | 165–175 |
| $C_6H_5-$ | $C_2H_5-$ | Zn | 2 | [14634-93-6] | light-colored | $C_{18}H_{20}N_2S_4Zn$ | 458.02 | 1.46 | 205 |
| $CH_3-$ | $CH_3-$ | Pb | 2 | [19010-66-3] | pale gray powder | $C_6H_{12}N_2PbS_4$ | 447.65 | 2.38 | 320–323 |
| $C_2H_5-$ | $C_2H_5-$ | Pb | 2 | [17549-30-3] | pale gray powder | $C_{10}H_{20}N_2PbS_4$ | 503.76 | 1.87 | 206–207 |
| $n-C_3H_7-$ | $n-C_3H_7-$ | Pb | 2 | [70995-63-0] | grayish brown powder | $C_{14}H_{28}N_2PbS_4$ | 559.87 | 1.65 | 97–98 |
| $n-C_4H_9-$ | $n-C_4H_9-$ | Pb | 2 | [27803-77-6] | light-colored powder | $C_{18}H_{36}N_2PbS_4$ | 615.98 | 1.42 | 72–73 |
| $n-C_5H_{11}-$ | $n-C_5H_{11}-$ | Pb | 2 | [36501-84-5] | pale yellow liquid | $C_{22}H_{44}N_2PbS_4$ | 672.78 | 1.40 | – |
| ⬡N– | | Pb | 2 | [41556-46-1] | pale gray powder | $C_{12}H_{20}N_2PbS_4$ | 527.78 | 2.79 | 249–250 |
| ⬡N– | | Cd | 2 | [14949-59-8] | yellow powder | $C_{12}H_{20}CdN_2S_4$ | 432.98 | 1.82 | 240–245 |
| $n-C_4H_9-$ | $n-C_4H_9-$ | Ni | 2 | [13927-77-0] | green powder | $C_{18}H_{36}N_2NiS_4$ | 467.48 | 1.29 | ~88 |
| $CH_3-$ | $CH_3-$ | Cu | 2 | [137-29-1], [23726-35-4] | pale brown powder | $C_6H_{12}CuN_2S_4$ | 303.98 | 1.7 | 360 |
| $C_2H_5-$ | $C_2H_5-$ | Cu | 2 | [13681-87-3] | dark brown powder | $C_{10}H_{20}CuN_2S_4$ | 360.09 | 1.49 | 196–197 |
| $n-C_3H_7-$ | $n-C_3H_7-$ | Cu | 2 | [14354-08-6], [2102-05-8] | black powder | $C_{14}H_{28}CuN_2S_4$ | 416.20 | 1.24 | 100–101 |
| $n-C_4H_9-$ | $n-C_4H_9-$ | Cu | 2 | [52691-95-9], [13927-71-4] | black powder | $C_{18}H_{36}CuN_2S_4$ | 472.31 | 1.26 | 77–78 |
| $n-C_5H_{11}-$ | $n-C_5H_{11}-$ | Cu | 2 | [36190-66-6] | black powder | $C_{22}H_{44}CuN_2S_4$ | 528.41 | 1.15 | 50 |
| $CH_3-$ | $CH_3-$ | Se | 4 | [21559-13-7], [144-34-3] | yellow powder | $C_{12}H_{24}N_4S_8Se$ | 559.84 | 1.57 | 146 |
| $C_2H_5-$ | $C_2H_5-$ | Se | 4 | [105618-13-1], [21559-14-8] | dark yellow powder | $C_{20}H_{40}N_4S_8Se$ | 672.06 | 1.32 | 62–86 |
| $n-C_3H_7-$ | $n-C_3H_7-$ | Se | 4 | | orange powder | $C_{28}H_{56}N_4S_8Se$ | 784.28 | 1.23 | 45 |
| $n-C_4H_9-$ | $n-C_4H_9-$ | Se | 4 | | dark red liquid | $C_{36}H_{72}N_4S_8Se$ | 896.49 | 1.11 | – |
| $n-C_5H_{11}-$ | $n-C_5H_{11}-$ | Se | 4 | | dark red liquid | $C_{44}H_{88}N_4S_8Se$ | 1008.71 | 1.03 | – |
| $C_2H_5-$ | $C_2H_5-$ | Bi | 3 | [20673-31-8], [96126-13-5] | yellow powder | $C_{15}H_{30}BiN_3S_6$ | 653.83 | 2.06 | 230 (decomp.) |
| $H-$ | $HCO-$ | K | 1 | [97288-11-4] | yellow | $C_2H_2KNOS_2$ | 159.27 | 1.83 | 186 (decomp.) |
| $CH_3-$ | $HCO-$ | Na | 1 | [102118-28-5] | yellow | $C_3H_4NNaOS_2$ | 157.2 | | 161–162 |
| $CH_3-$ | $HCS-$ | K | 1 | [102137-68-9] | pale red | $C_3H_4KNS_3$ | 189.36 | 1.57 | 94–95 |
| $H-$ | $HCS-$ | K | 1 | [97960-86-6] | orange–yellow | $C_2H_2KNS_3$ | 175.34 | | 142 |

salts of formyldithiocarbamic acid are produced from potassium formyldithiocarbamate [18].

Amine and ammonium salts of N,N-disubstituted dithiocarbamic acids are produced from carbon disulfide by reaction with two equivalents of amine or one each of amine and ammonia. An example is the piperidinium salt of pentamethylenedithiocarbamic acid, which is used as a vulcanization accelerator.

$$2 \; C_5H_{10}NH + CS_2 \longrightarrow C_5H_{10}N-C(=S)-S^- \; H_2N^+C_5H_{10}$$

Guanidine and arylguanidine react analogously [26]; hydroxylamine derivatives have also been described [27].

Alkali and heavy-metal salts of N,N-disubstituted dithiocarbamic acids are produced like the monosubstituted salts [28], [29].

$$2 \; (C_2H_5)(C_6H_5)N-H + 2\;CS_2 + 2\;NH_4OH \xrightarrow{H_2O}$$

$$2 \; (C_2H_5)(C_6H_5)N-C(=S)-S^-\;NH_4^+ \xrightarrow[-2\;NH_4Cl]{ZnCl_2} [(C_2H_5)(C_6H_5)N-C(=S)-S-]_2 Zn$$

Diamines react with two moles of carbon disulfide to yield bis(dithiocarbamates). Zinc, iron, and copper ethylenebis(dithiocarbamates) are used as fungicides and nematocides [30]. With one mole of carbon disulfide, diamines form inner salts or mixtures of inner salts and amine salts of bis(dithiocarbamic acids). Such mixtures are produced by piperazines [31].

$$3 \; HN(C_4H_8N-CH_3) + 3\;CS_2 \longrightarrow$$

$$H_2N^+(C_4H_8)N-C(=S)S^-\;\;+\;\;{}^-S-C(=S)-N(C_4H_8)N-C(=S)S^-\;\;\cdot\;\;H_2N^+(C_4H_8)NH_2^+$$
(CH_3 groups on the ring nitrogens)

Hydrazine reacts with one mole of carbon disulfide in the presence of alkali hydroxide to form alkali salts of aminodithiocarbamic acid (dithiocarbazic acid) [32]:

$$H_2N-NH_2 + NaOH + CS_2 \xrightarrow{H_2O} H_2N-NH-C(=S)SNa$$

With two moles of carbon disulfide, salts derived from the tautomeric imide form are obtained [33]:

$$H_2N-NH_2 + 3\ NaOH + 2\ CS_2 \xrightarrow{H_2O}$$

$$Na_3\left[ -S-\overset{S}{\underset{\|}{C}}-NH-N=C\overset{S-}{\underset{S-}{\diagup}} \right] \cdot 7\ H_2O$$

Metal salts, particularly zinc dithiocarbamates, are used as vulcanization accelerators, oil additives, bactericides, and fungicides.

# 4. Transformation Products

## 4.1. Thiuram Sulfides

### 4.1.1. Thiuram Disulfides

The oxidation of salts of dithiocarbamic acids usually gives thiuram disulfides (bis(aminothiocarbonyl)disulfanes) [34]–[36]. Table 5 lists physical properties of some thiuram sulfides. Only the tetraalkyl derivatives are stable. Oxidizing agents include iodine, bromine, hydrogen peroxide, and potassium peroxodisulfate [37]; ammonium peroxodisulfate [38]; sodium tetrathionate [39]; and chlorine or bromine in the presence of buffers [40]. The colorless compound tetramethylthiuram disulfide, for example, is obtained from sodium dimethylthiocarbamate and hydrogen peroxide [41]:

$$2\ \overset{H_3C}{\underset{H_3C}{\diagdown}}N-C\overset{S}{\underset{SNa}{\diagup}} \xrightarrow[H_2SO_4]{H_2O_2} \overset{H_3C}{\underset{H_3C}{\diagdown}}N-C\overset{S}{\underset{S-S}{\diagup}}\overset{S}{\diagdown}C-N\overset{CH_3}{\underset{CH_3}{\diagup}}$$

Other methods use hypochlorites [42], sodium nitrite [43], oxygen in the presence of catalysts such as copper or manganese salts [44] (see also Section 4.6.3), or electrolytic oxidation [45]. Tetraalkylthiuram disulfides are powerful vulcanization accelerators [46].

N-Acyl derivatives are also known. For example, oxidation of N-methyl-N-formyl-dithiocarbamate with iodine in ether yields the corresponding thiuram [47]:

$$\overset{H_3C}{\underset{\underset{\|}{\overset{H}{C}}}{\diagdown}}N-C\overset{S}{\underset{SK}{\diagup}} \xrightarrow[0\ ^\circ C]{I_2} \overset{H_3C}{\underset{\underset{\|}{\overset{H}{C}}}{\diagdown}}N-C\overset{S}{\underset{S-S}{\diagup}}\overset{S}{\diagdown}C-N\overset{CH_3}{\underset{\underset{\|}{\overset{C}{O}}}{\diagup}}$$

N-Methyl-N-thioformyldithiocarbamate gives a cyclic, pale yellow thiuram (*mp* 108.5 °C) [48]:

**Table 5.** Physical properties of thiuram sulfides

$$R^1_{R^2}\!\!>\!\!N\text{-}\overset{S}{\overset{\|}{C}}\text{-}(S)_x\text{-}\overset{S}{\overset{\|}{C}}\text{-}N\!\!<\!\!^{R^1}_{R^2}$$

| $R^1$ | $R^2$ | CAS registry number | Physical form | Molecular formula | $M_r$ | d | mp, °C |
|---|---|---|---|---|---|---|---|
| *Monosulfides (x = 1)* | | | | | | | |
| $CH_3-$ | $CH_3-$ | [97-74-5] | yellow powder | $C_6H_{12}N_2S_3$ | 208.38 | 1.39 | 107 |
| $C_2H_5-$ | $C_2H_5-$ | [95-05-6] | brown liquid | $C_{10}H_{20}N_2S_3$ | 264.48 | 1.13 | 28.5–29.8 |
| $n\text{-}C_3H_7-$ | $n\text{-}C_3H_7-$ | [2556-41-4] | brown liquid | $C_{14}H_{28}N_2S_3$ | 320.59 | 1.04 | |
| $n\text{-}C_4H_9-$ | $n\text{-}C_4H_9-$ | [93-73-2] | dark red liquid | $C_{18}H_{36}N_2S_3$ | 376.70 | 0.99 | |
| $n\text{-}C_5H_{11}-$ | $n\text{-}C_5H_{11}-$ | [6405-88-5] | dark red liquid | $C_{22}H_{44}N_2S_3$ | 432.81 | 0.96 | |
| $H-$ | $C_2H_4-$ | [3082-38-0] | monomer | $C_4H_6N_2S_3$ | 178.31 | | 125–126 |
| $H-$ | $C_2H_4-$ | [27321-95-5] | polymer | $(C_4H_6N_2S_3)_n$ | $(178.31)_n$ | | 145–147 |
| *Disulfides (x = 2)* | | | | | | | |
| $CH_3-$ | $CH_3-$ | [137-26-8] | white powder | $C_6H_{12}N_2S_4$ | 240.44 | 1.29 | 146–148 |
| $C_2H_5-$ | $C_2H_5-$ | [97-77-8] | almost colorless powder | $C_{10}H_{20}N_2S_4$ | 296.55 | 1.17 | 65–70 |
| $n\text{-}C_3H_7-$ | $n\text{-}C_3H_7-$ | [2556-42-5] | pale yellowish powder | $C_{14}H_{28}N_2S_4$ | 352.66 | 1.09 | 49.5–50.5 |
| $n\text{-}C_4H_9-$ | $n\text{-}C_4H_9-$ | [1634-02-2] | dark red liquid | $C_{18}H_{36}N_2S_4$ | 408.77 | 1.03 | |
| $n\text{-}C_5H_{11}-$ | $n\text{-}C_5H_{11}-$ | [5721-31-3] | dark red liquid | $C_{22}H_{44}N_2S_4$ | 464.87 | 0.99 | |
| $CH_3-$ | $C_6H_5-$ | [10591-84-1] | white powder | $C_{16}H_{16}N_2S_4$ | 364.58 | 1.33 | 206 |
| $CH_3-$ | $HCO-$ | [102127-57-1] | pale yellow | $C_6H_8N_2O_2S_4$ | 268.40 | 1.088 | 69–70 |
| $H-$ | 2-ethoxybenzoyl | [79388-81-1] | yellow crystals | $C_{20}H_{20}N_2O_4S_4$ | 480.6 | | 137–138 |
| *Tetrasulfides (x = 4)* | | | | | | | |
| $CH_3-$ | $CH_3-$ | [97-91-6] | pale yellowish powder | $C_6H_{12}N_2S_6$ | 304.57 | 1.52 | ~90 |
| $CH_3-$ | $C_6H_5-$ | [103597-02-0] | pale yellowish powder | $C_{16}H_{16}N_2S_6$ | 428.72 | | 117 |
| $C_2H_5-$ | $C_6H_5-$ | | pale yellowish powder | $C_{18}H_{20}N_2S_6$ | 456.77 | | 142 |
| piperidino | | [120-54-7] | white powder | $C_{12}H_{20}N_2S_6$ | 384.70 | 1.41 | 117 |
| morpholino | | [103597-04-2] | white powder | $C_{10}H_{16}N_2O_2S_6$ | 388.65 | | 111.5–114.5 |

$$2\ \underset{CH_3}{\underset{|}{H\text{-}N}}\text{-}\overset{S}{\overset{\|}{C}}\text{-}SK \xrightarrow[-15\,°C]{I_2,\ CH_2Cl_2} \text{1,2,4-dithiazole derivative}$$

Yield 32%

*N*-Thioformyldithiocarbamate, however, yields 1,2,4-dithiazole-3-thione (*mp* 73–75 °C) [47]:

$$\underset{H}{\underset{|}{H\text{-}\overset{S}{\overset{\|}{C}}\text{-}N}}\text{-}\overset{S}{\overset{\|}{C}}\text{-}SK \xrightarrow{I_2} \text{1,2,4-dithiazole-3-thione}$$

Yield 41%

Oxidation of *N*-(2-ethoxybenzoyl)dithiocarbamate gives the corresponding disulfide (*mp* 138 °C) which, when heated in the presence of pyridine, yields 2-ethoxybenzoyl disulfide (*mp* 137–139 °C) [14].

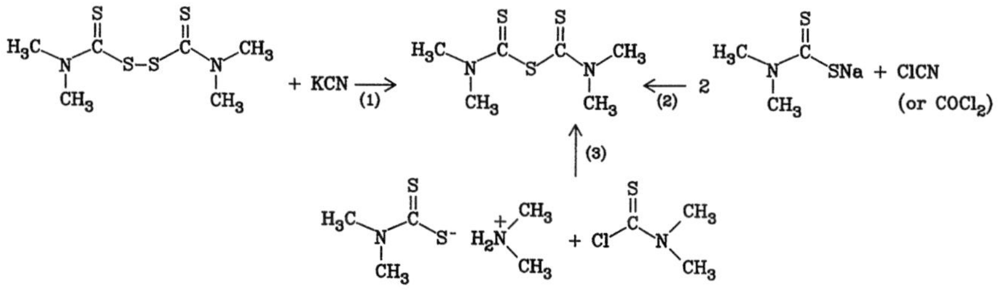

**Figure 1.** Syntheses of tetramethylthiuram monosulfide

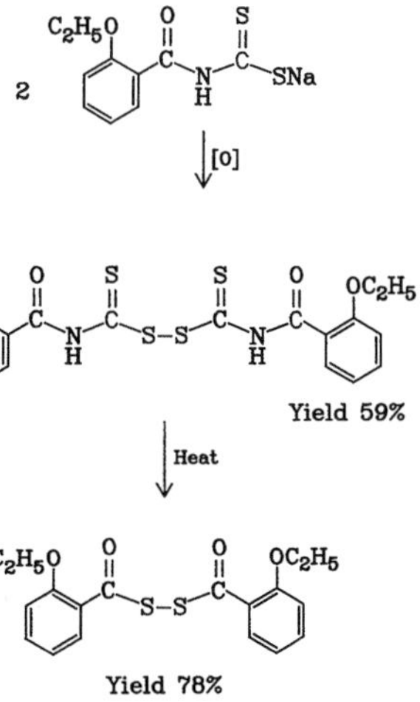

## 4.1.2. Thiuram Monosulfides

Thiuram disulfides may lose one sulfur atom to form thiuram monosulfides (bis(aminothiocarbonyl)sulfanes), which are also used as vulcanization accelerators [46].

Tetramethylthiuram disulfide gives the yellow monosulfide on heating with triphenylphosphine [49] or potassium cyanide [50] (Fig. 1, Reaction 1). The monosulfide can also be obtained directly from sodium dimethyldithiocarbamate by reaction with phosgene [51], [52] or cyanogen chloride [51], [53] (Reaction 2). Another method uses the ammonium dithiocarbamate and thiocarbamoyl chloride [54] (Reaction 3).

## 4.1.3. Thiuram Trisulfides and Tetrasulfides

Sulfur dichloride reacts with alkali dithiocarbamates to form thiuram trisulfides (bis(aminothiocarbonyl)trisulfanes). Disulfur dichloride gives the corresponding tetrasulfides, e.g., the tetrasulfide of pentamethylenedithiocarbamic acid, an important vulcanization accelerator [55]:

$$2 \; \text{(piperidine)}N-C(=S)SNa + S_2Cl_2 \xrightarrow{Na_2CO_3}$$

$$\text{(piperidine)}N-C(=S)-S-S-S-S-C(=S)-N\text{(piperidine)}$$

## 4.2. Thiocarbamoylsulfenamides

Thiocarbamoylsulfenamides are obtained from N,N-disubstituted dithiocarbamates by mild oxidation with iodine in the presence of primary or secondary amines [56] or by electrolytic oxidation [57]:

$$(H_3C)_2N-C(=S)-S^- + H_2N(CH_3)_2 \xrightarrow{I_2} (H_3C)_2N-C(=S)-S-N(CH_3)_2$$

$$\xrightarrow{\text{Heat, }-S} (H_3C)_2N-C(=S)-N(CH_3)_2$$

These compounds are used as vulcanization accelerators, fungicides [58], and bactericides [59]. Thermolysis removes one atom of sulfur to give thioureas. The morpholine derivative is manufactured by oxidation with sodium hypochlorite [60]–[69].

$$\text{(morpholine)}N-C(=S)-S-N\text{(morpholine)}$$

Substituted thiocarbamoylsulfenamides are obtained by reaction with isocyanates [70]:

$$\text{Ph}-NCO + H_2N-S-C(=S)-N(CH_3)_2 \xrightarrow{\text{Heat}}$$

$$\text{Ph}-N(H)-C(=O)-N(H)-S-C(=S)-N(CH_3)_2$$

Polymeric thiocarbamoylsulfenamides have also been reported [71].

## 4.3. Thiocarbamoyl Chlorides

Thiuram disulfides react with chlorine in organic solvents to give disubstituted thiocarbamoyl chlorides, which are useful intermediates [72], [73].

$$(H_3C)_2N-C(=S)-S-S-C(=S)-N(CH_3)_2 + 3\,Cl_2 \longrightarrow 2\,(H_3C)_2N-C(=S)-Cl + 2\,SCl_2$$

The methyl groups can be replaced by ethyl, propyl, butyl, or phenyl groups; the nitrogen can also be part of a piperidine ring [74]. Alternatively, thiocarbamoyl chlorides can be obtained from amines and thiophosgene [75] or by chlorination of thioformamides [76]. Further chlorination leads to iminium salts [77]:

$$(H_3C)_2N-C(=S)-Cl \xrightarrow[-SCl_2]{Cl_2} (H_3C)_2N^+=CCl_2 \; Cl^-$$

## 4.4. Dithiocarbamic Acid Esters

Dithiocarbamic acid esters (dithiourethanes) are used in agriculture as nematocides, fungicides, and herbicides. Physical properties of some dithiocarbamates are given in Table 6. The most important methods of preparation are (1) alkylation of alkali dithiocarbamates with halogen compounds or other alkylating agents and (2) insertion reactions of carbon disulfide with tertiary amines.

Suitable halogen compounds include alkyl chlorides, benzyl chloride [78], phenacyl chloride [79], acyloxymethyl chloride [80], and allyl chlorides [81]. For example, the reaction of sodium N,N-dimethyldithiocarbamate with 1,3-dichloro-2-butene gives dimethyldithiocarbamic acid 3-chloro-2-butenyl ester, which is used as a herbicide:

$$(H_3C)_2N-C(=S)-SNa + Cl-CH_2-CH=C(Cl)(CH_3) \xrightarrow{Heat} (H_3C)_2N-C(=S)-S-CH_2-CH=C(Cl)(CH_3)$$

Yield 87%

Other alkylating agents are (1) halogenated aldehydes, ketones, and chlorohydrin [78]; (2) dihalogen compounds such as benzal chloride [82] and 1,3-dichloropropane [83]; (3) trihalogen compounds such as cyanuric chloride [84] and chloroform [85]; and (4) polychlorinated compounds [86]. The reaction of an alkali dithiocarbamate with chloroform, for example, yields an amino diester and carbon disulfide:

**Table 6.** Physical properties of dithiocarbamic acid esters

$$\left[\begin{array}{c}R^1\\ \diagdown\\ R^2\diagup\end{array} N-\overset{\overset{S}{\|}}{C}-S-\right]_x R^3$$

| R¹ | R² | R³ | x | CAS registry number | Physical form | Molecular formula | $M_r$ | d | mp, °C |
|---|---|---|---|---|---|---|---|---|---|
| CH₃– | CH₃– | O₂N–C₆H₃(NO₂)– | 1 | [89-37-2] | yellow powder | C₉H₉N₃O₄S₂ | 287.33 | 1.75 | 140–145 |
| piperidyl–N– | | piperidyl–N–CH₂– | 1 | [10254-56-5] | brown crystals | C₁₂H₂₂N₂S₂ | 258.46 | 1.11 | 59 |
| CH₃– | CH₃– | C₆H₅–CH< | 2 | [49773-60-6] | white powder | C₁₃H₁₈N₂S₄ | 330.57 | | 180–182 |
| CH₃– | CH₃– | benzothiazolyl | 1 | [3432-25-5] | yellowish powder | C₁₀H₁₀N₂S₃ | 254.4 | | 106–112 |
| C₂H₅– | C₂H₅– | benzothiazolyl | 1 | [95-30-7] | yellowish powder | C₁₂H₁₄N₂S₃ | 282.46 | 1.27 | 70–73 |
| C₂H₅– | C₂H₅– | benzothiazolyl | 1 | | yellowish powder | C₁₆H₁₄N₂S₃ | 330.50 | | 105–108 |
| piperidyl–N– | | benzothiazolyl | 1 | [86443-84-7] | yellowish powder | C₁₃H₁₄N₂S₃ | 294.47 | | 133–137 |
| CH₃– | CH₃–CO– | C₂H₅– | 1 | [102127-59-3] | pale yellow | C₅H₉NOS₂ | 163.26 | 1.179 | 73.5 |
| H– | HCO– | C₂H₅– | 1 | [102933-65-3] | white–yellow | C₄H₇NOS₂ | 149.24 | 1.292 | 77–81 |
| H– | HC(=S)– | C₂H₅– | 1 | [102118-26-3] | orange | C₄H₇NS₃ | 165.30 | 1.33 | 67.5 |
| H– | H₂N– | CH₃– | 1 | [5397-03-5] | yellowish | C₂H₆N₂S₂ | 122.22 | | |

* Boiling point at 0.01 mm Hg (1.3 Pa).

$$3\ \underset{H_3C}{\overset{H_3C}{\diagdown}}N-C\underset{SNa}{\overset{S}{\diagup\!\!\!\diagdown}} + CHCl_3 \longrightarrow \underset{H_3C}{\overset{H_3C}{\diagdown}}N-\overset{\overset{S}{\|}}{C}-S-\overset{H}{\underset{N(CH_3)_2}{C}}-S-\overset{\overset{S}{\|}}{C}-N\underset{CH_3}{\overset{CH_3}{\diagup}}$$

High yields are also obtained from alkali acyldithiocarbamates and alkyl halides [48], [87], [88]:

$$\underset{H_3C(O)}{\overset{H}{\diagdown}}N-C\underset{SK}{\overset{S}{\diagup\!\!\!\diagdown}} + C_2H_5I \xrightarrow[24h,\ 20\,°C]{C_2H_5OH/(C_2H_5)_2O} \underset{H_3C(O)}{\overset{H}{\diagdown}}N-C\underset{S-C_2H_5}{\overset{S}{\diagup\!\!\!\diagdown}}$$

Yield 77%

$$\underset{H_3C(S)}{\overset{H_3C}{\diagdown}}N-C\underset{SK}{\overset{S}{\diagup\!\!\!\diagdown}} + CH_3I \xrightarrow{Benzene} \underset{H_3C(S)}{\overset{H_3C}{\diagdown}}N-C\underset{S-CH_3}{\overset{S}{\diagup\!\!\!\diagdown}}$$

Arylation is possible with strongly activated halobenzenes such as 2,4-dinitrochlorobenzene [82] or with diazonium salts [89]. Phenyl esters are also produced by reaction of phenyldithioformic acid chloride with amines [90]. *o*-Phenylene diesters are obtained from *o*-nitrophenyl monoesters by substitution of the nitro group [91].

Sulfenyl halides [92] and thiosulfates [93] react similarly with alkali dithiocarbamates to form thiocarbamoylalkyldisulfanes:

$$(H_3C)_2N-C(=S)-SNa + R-S-SO_3Na \xrightarrow{CH_2O} (H_3C)_2N-C(=S)-S-S-R + Na_2SO_3$$

Dithiocarbamic acid esters are also obtained by reaction of thiocarbamoyl chlorides with thiophosphates [94] or by addition of thiols to isothiocyanates [95], [96].

Vinyl esters can be prepared from phenolates [97]:

$$(C_2H_5)_2N-C(=S)-SNa + CH_3-SO_2-CH=CH-O-C_6H_5 \xrightarrow[3h, 85°C]{DMSO}$$

$$CH_3-SO_2-CH=CH-S-C(=S)-N(C_2H_5)_2 + C_6H_5ONa$$

Alkylation of ammonium dithiocarbamates with epoxides is also possible [98]:

$$(H_3C)_2N-C(=S)-S^- \ \overset{+}{H}N(CH_3)_3 + CH_3-CH-C(CH_3)_2 \ (epoxide)$$

$$\longrightarrow (H_3C)_2N-C(=S)-S-CH(CH_3)-C(OH)(CH_3)_2 + N(CH_3)_3$$

Styrene oxide and sodium diethyldithiocarbamate react similarly [99]. At low temperature, epichlorohydrin reacts like an epoxide to form the dithiocarbamic acid ester; when heated in the presence of a tertiary amine, this ester rearranges to the thiocarbamate [100]:

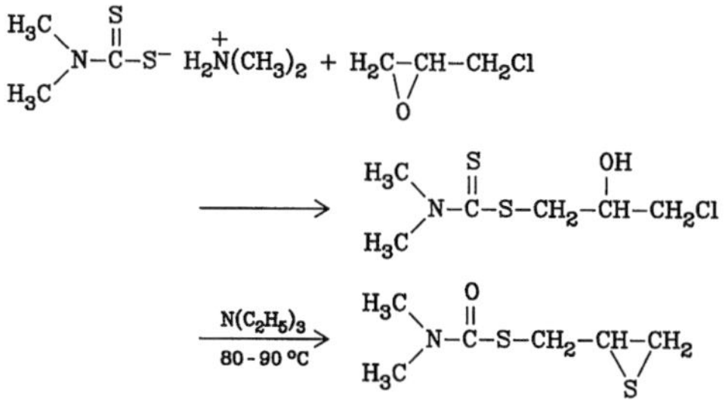

$$\underset{H_3C}{\overset{H_3C}{>}}N-\overset{\overset{S}{\|}}{C}-S^-\ H_2\overset{+}{N}(CH_3)_2 + H_2C-CH-CH_2Cl$$
$$\overset{\diagdown\!\diagup}{O}$$

$$\longrightarrow \underset{H_3C}{\overset{H_3C}{>}}N-\overset{\overset{S}{\|}}{C}-S-CH_2-\overset{\overset{OH}{|}}{CH}-CH_2Cl$$

$$\xrightarrow[80-90\,°C]{N(C_2H_5)_3} \underset{H_3C}{\overset{H_3C}{>}}N-\overset{\overset{O}{\|}}{C}-S-CH_2-CH-CH_2$$
$$\overset{\diagdown\!\diagup}{S}$$

Whereas primary or secondary amines react with carbon disulfide to form mono- or disubstituted salts of dithiocarbamic acids (see Chap. 3), some tertiary amines such as aziridines form cyclic esters. For example, N-phenylthiazolidine-2-thione is obtained from N-phenylaziridine [101]:

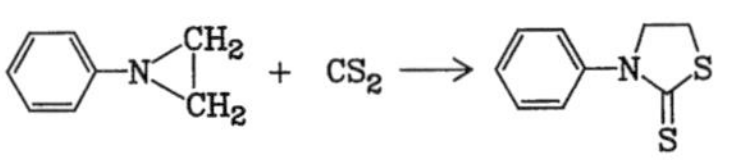

β-Amino ketones [102] and benzylamines [103] also react with insertion of carbon disulfide; for example, 2,6-di-*tert*-butyl-4-dialkylaminomethylphenols form esters that are used as stabilizers and intermediates [104]:

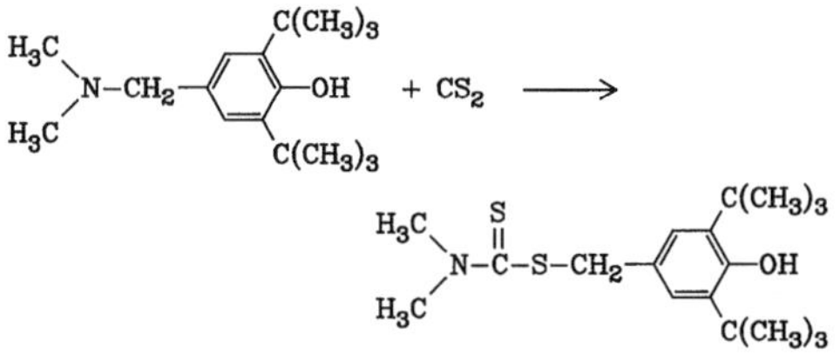

Silylamines also give insertion products, but no insertion is observed with silylaminophosphine [105]:

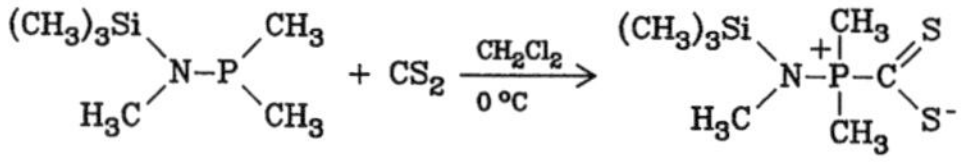

Methylenediamines, such as bis(piperidino)-methane, and carbon disulfide form aminomethyl dithiocarbamates [106]:

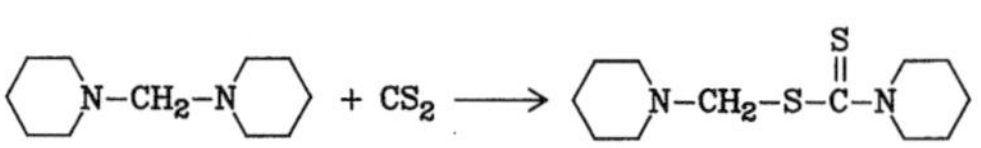

Piperidinomethyl carbamates can also be produced from methylolpiperidine and salts of dithiocarbamic acids [107]. Methylene bis(dithiocarbamates) are obtained from formaldehyde and dithiocarbamic acids in acidic solution [108]:

$$2 \;\text{(piperidine)}N-C(=S)SH + CH_2O \longrightarrow \text{(piperidine)}N-C(=S)-S-CH_2-S-C(=S)-N\text{(piperidine)}$$

Bis(dithiocarbamates) are also formed by cleavage of ethylene trithiocarbonate with amines [4], [109]. Benzothiazolyl dithiocarbamates, which are used as accelerators for rubber, can be obtained in high yield from benzothiazole disulfide and thiuram disulfide in the presence of sodium cyanide [110]:

$$\text{(benzothiazolyl)-S-S-(benzothiazolyl)} + (H_3C)_2N-C(=S)-S-S-C(=S)-N(CH_3)_2 \xrightarrow[-\text{NaSCN}]{\text{NaCN}} 2\;\text{(benzothiazolyl)-S-C(=S)-N(CH_3)_2}$$

Dithiocarbamates are also prepared by addition of dithiocarbamic acids to unsaturated compounds such as vinylsulfonates [111], vinyl ether [112], acrylonitrile [113], vinylpyridine [114], mesityl oxide [115], acetylenedicarboxylate [116], acetylene ketones [117], and even acetylene [118]:

$$R_2NH + CS_2 + HC\equiv CH \xrightarrow[1.4-1.9\;\text{MPa}]{130\;°C} R_2N-C(=S)-S-CH=CH_2$$

Yield 60%

Unsaturated dithiocarbamates have been used in the stereoselective synthesis of polyenes [119]. The chlorination of monosubstituted dithiocarbamates leads to isocyanide dichlorides and removal of the sulfur [120], [121]:

$$R-NH-C(=S)S-R \xrightarrow[-SCl_2]{Cl_2} R-N=C(Cl)S-Cl \xrightarrow[-SCl_2]{Cl_2} R-N=CCl_2$$

## 4.5. Heterocyclic Compounds

**Five-membered Heterocycles.** Dithiocarbamic acid and its salts react with monochloroacetic acid to yield first the dithiocarbamic acid ester and then rhodanine [122], [123]:

$$H_2N-C(=S)-S^- \;NH_4^+ + ClCH_2-C(=O)-OH \longrightarrow$$

$$H_2N-C(=S)-S-CH_2-C(=O)-OH \longrightarrow \text{rhodanine}$$

Rhodanine reacts with aldehydes [122], [124]; the products are stimulants [125].

Thiorhodanine (thiazolidinedithione) [126], N-aminorhodanine [127], and other substituted rhodanines [128], [129] have also been prepared. Certain rhodanines act as cross-linking agents [130].

Ammonium dithiocarbamate and α-halogenated ketones or aldehydes give 2-mercaptothiazoles; N-alkyldithiocarbamates form N-alkylthiazoline-2-thiones [129], [131]:

$$H_2N-\overset{S}{\underset{\|}{C}}-S^- \; NH_4^+ + ClCH_2-\overset{O}{\underset{\|}{C}}-CH_3 \longrightarrow \underset{H_3C}{\text{thiazole}}-SH$$

$$\underset{H_3C}{\overset{H}{N}}-\overset{S}{\underset{SNa}{C}} + ClCH_2-CHO \longrightarrow \underset{CH_3}{\text{thiazole}}=S$$

The reaction of dithiocarbamic acid methyl ester and chloroacetaldehyde leads to 2-methylthiothiazole, which is easily converted to thiazole with lithium in liquid ammonia [132]. Further thiazoline-2-thione syntheses have been described [133], [134].

Thiazolidine-2-thione is prepared by treating methyl thioxanthate with bromoethylamine [135]; substituted thiazolidine-2-thiones are obtained by heating aminoethanols with carbon disulfide [136]:

$$2 \; CH_3-\underset{CH_2-CH_2-OH}{\overset{H}{N}} + CS_2 \longrightarrow$$

$$CH_3-\underset{CH_2-CH_2-OH}{\overset{\overset{S}{\underset{\|}{C}}-S^-}{N}} \cdot H_2\overset{+}{N}\underset{CH_2-CH_2-OH}{\overset{CH_3}{\diagdown}} \xrightarrow[-COS, -H_2S]{+ CS_2}$$

$$2 \; \underset{CH_3}{\text{thiazolidine}}=S$$

N-Methylthiazolidine-2-thione [137] is a good accelerator for polychloroprene.

Bis(carboxyaminoethyl)disulfane and carbon disulfide yield carboxythiazolidine-2-thione [138]:

$$\begin{array}{c} S-CH_2-\overset{NH_2}{\underset{|}{CH}}-COOH \\ | \\ S-CH_2-\underset{NH_2}{\overset{|}{CH}}-COOH \end{array} + CS_2 \longrightarrow 2 \; \underset{HOOC \;\; H}{\text{thiazolidine}}=S$$

The reaction between acylaminodithiocarboxylic acid esters and aziridine leads to 2-acyliminothiazolidines [139]:

$$\underset{\underset{O}{\overset{\|}{C}}}{R}\overset{H}{\underset{}{\diagdown}}N-\overset{S}{\underset{S-CH_3}{C}} + H_2C\underset{\underset{H}{N}}{-}CH_2 \longrightarrow R-\overset{O}{\underset{\|}{C}}-N=\underset{S}{\overset{H}{\underset{N}{\diagup}}}$$

Thiazolotriazinium salts are prepared from dithiocarbazic acid methyl ester [140].

Aroylcarbimidodithioates easily cyclize with hydrazine to triazoles [14]:

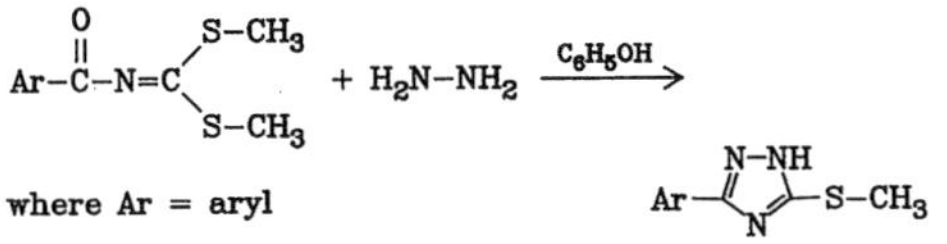

where Ar = aryl

The cyclization of N-methylquinoxalinium iodide with dithiocarbamates gives tetrahydrothiazoloquinoxalines [141]:

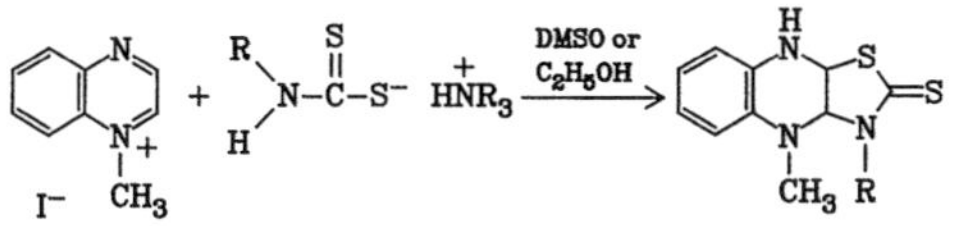

**Six-membered Heterocycles.** The reaction of N-methyldithiocarbamic acid with propiolic acid after several steps leads to 3-methyl-3,4-dihydro-4-oxo-2-thioxo-2H-1,3-thiazine [142]:

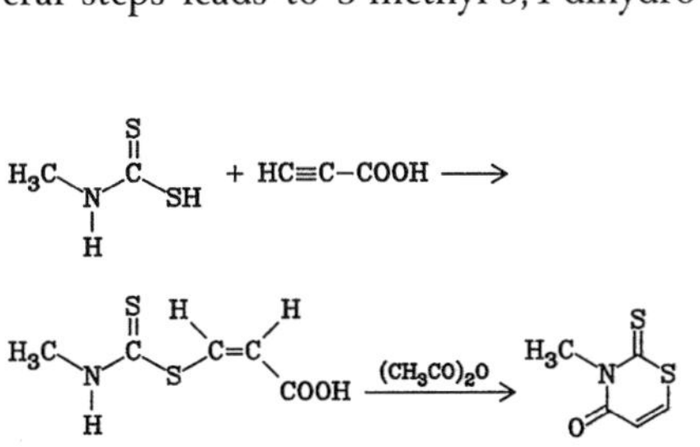

Heating this compound with primary amines gives 2-thiouracyls. 3H-1,3-Thiazine-2,6-dithiones have also been reported [143]. Saturated six-membered rings are synthesized by reaction of dithiocarbamic acids with acrylonitrile [144] or propiolactone [145]. Acrylamides and carbon disulfide give 2-thioxoperhydro-1,3-thiazin-4-ones [146] and other derivatives [147]:

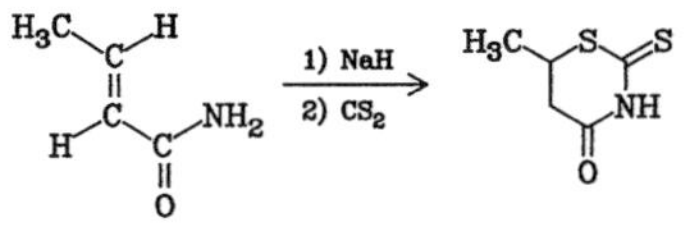

Oxidation removes the thione sulfur [148]:

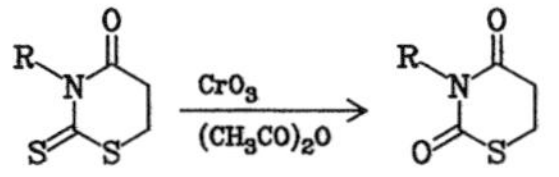

3-Amino-2-thioxotetrahydro-1,3-thiazin-4-ones are obtained from hydrazine or its derivatives [149]:

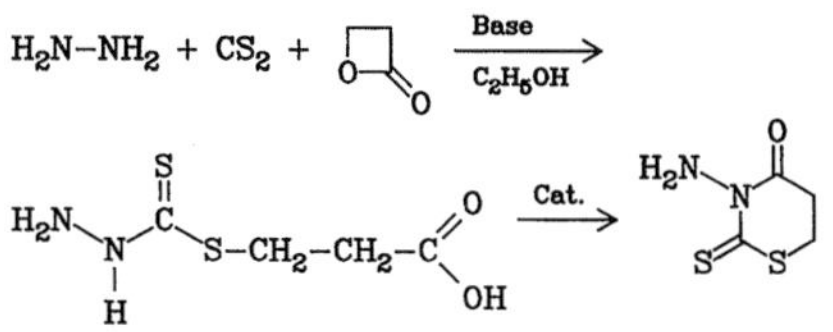

1,3-Thiazines can also be prepared from thioureas [150]:

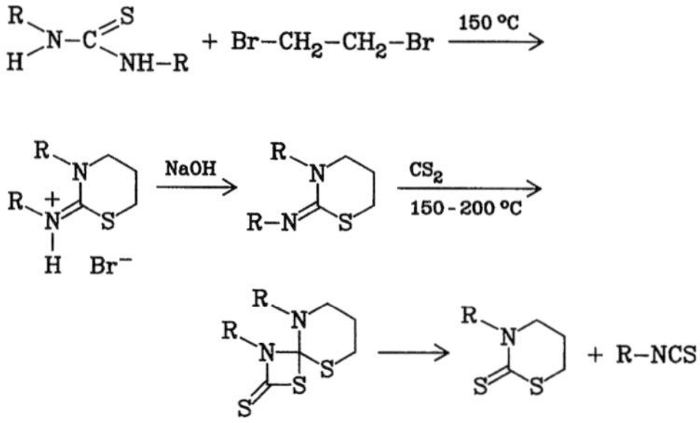

Reactions at the nitrogen atom of unsubstituted 1,3-thiazine-2-thione have been reported [151]. Mercaptothiazines are produced from ammonium dithiocarbamate and $\alpha,\beta$-unsaturated ketones; the reaction of mesityl oxide with dithiocarbamic acid gives 2-mercapto-4,4,6-trimethyl-4H-1,3-thiazine [152]:

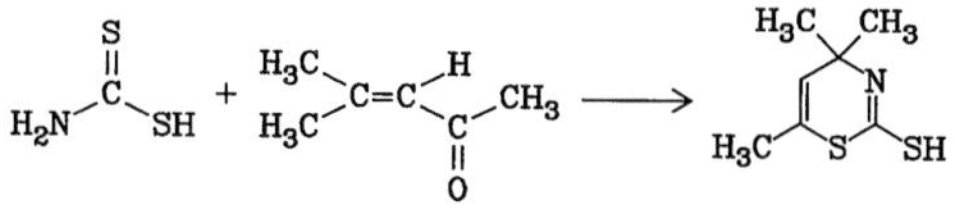

Thiadiazinethiones are produced in high yield by condensation of N-monosubstituted ammonium dithiocarbamates with aldehydes. For example, the reaction of methylammonium N-methyldithiocarbamate with formaldehyde in water gives 3,5-dimethyltetrahydro-2H-1,3,5-thiadiazine-2-thione, an accelerator for polychloroprene [176], in 90% yield [153]:

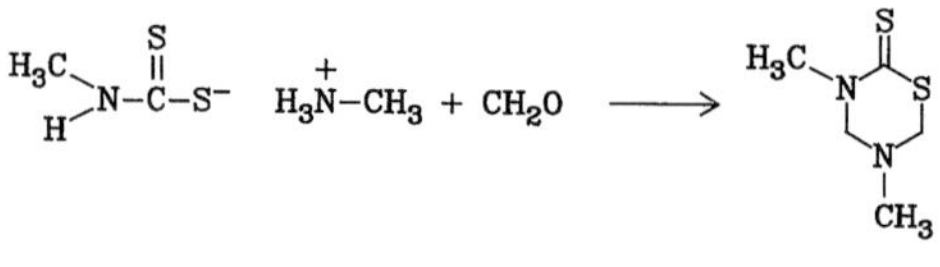

The corresponding N-benzyl derivative, known as sulbentine, is an antimycotic agent [154]. The reaction of methylamine and aminoacetic acid with carbon disulfide and formaldehyde yields 3-methyl-5-carboxymethyltetrahydro-2H-1,3,5-thiadiazine-2-thione, a nematocide [155].

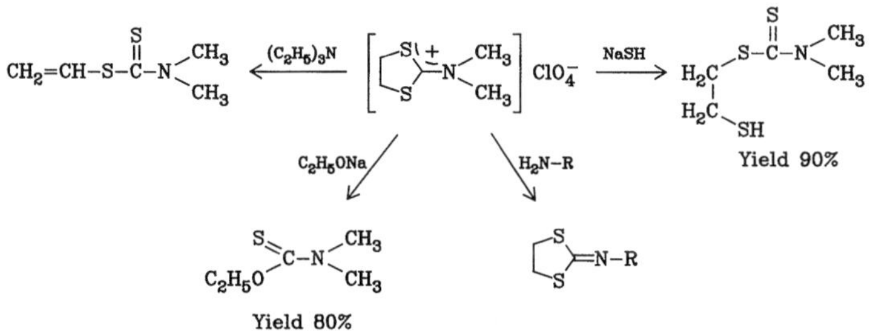

**Figure 2.** Reactions of 2-dialkylamino-1,3-dithiolanylium salts with nucleophiles

## 4.6. Other Derivatives

### 4.6.1. Dithiolanylium Salts

2-Dialkylamino-1,3-dithiolanylium salts are prepared by reacting sodium N,N-dimethyldithiocarbamate with 1,2-dichloroethane in tetrahydrofuran in the presence of sodium perchlorate [156]:

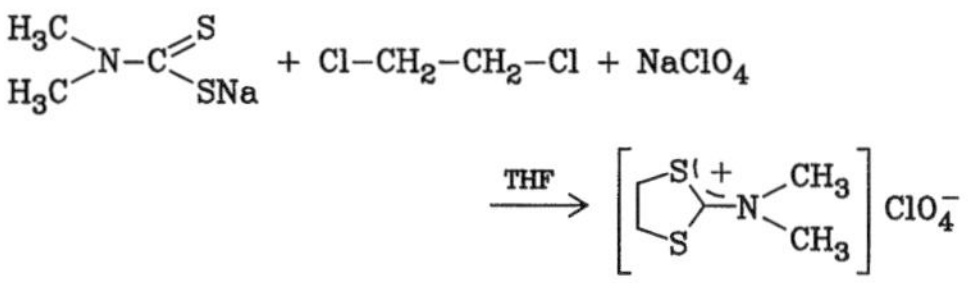

These compounds react with many nucleophiles [156], [157] (see Fig. 2). Unsaturated dithiolium salts can be used to prepare tetrathiafulvenes [158] and 1,2-dimercaptoalkenes [159].

### 4.6.2. Alkylthioformimidic Chlorides

Alkylthioformimidic chlorides are obtained from N,N-dialkyldithiocarbamic acid esters and phosgene [160]:

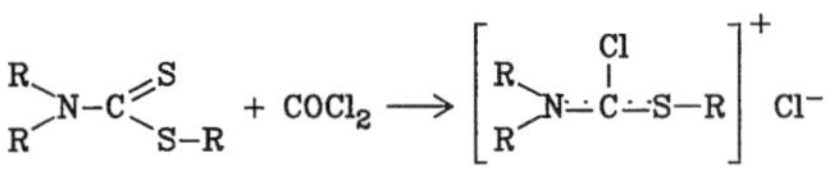

These salts are very reactive; for example, methylthioformimidic chloride and pyrrole give a 2-pyrrolyl derivative of methylthiocarboximidic chloride [161].

## 4.6.3. Isothiocyanates, Isocyanates, and Thioureas

*Isothiocyanates* (mustard oils) can be produced from monosubstituted salts or esters of dithiocarbamic acids by oxidation or heat (see also Section 4.1.1) [162], [163]:

$$\underset{H}{\overset{R}{>}}N-C\overset{S}{\underset{SNa}{<}} \xrightarrow{[O]} R-N=C=S$$

*N*-Acyl and *N*-thioacyl derivatives react differently (see Section 4.1.1).

*Isocyanates* are produced from monosubstituted dithiocarbamates by reaction with phosgene or chloroformates (→ Isocyanates, Organic) [164]:

$$\underset{H}{\overset{R}{>}}N-C\overset{S}{\underset{SNa}{<}} + Cl-\overset{O}{\underset{}{C}}-O-CH_2-\text{Ph} \longrightarrow$$

$$\underset{H}{\overset{R}{>}}N-\overset{S}{\underset{S}{C}}-\overset{O}{\underset{}{C}}-O-CH_2-\text{Ph} \longrightarrow$$

$$R-NCO + CS_2 + HO-CH_2-\text{Ph}$$

*Thioureas* can be obtained from isothiocyanates and primary or secondary amines, by pyrolysis of thiuram disulfides [165], [166], or from dithiocarbamate salts [167]:

$$H\overset{+}{N}\text{-Ph-}NH-C\overset{S}{\underset{S^-}{<}} + H_2N-CN \xrightarrow[CH_3OH]{(C_2H_5)_3N} H\overset{+}{N}\text{-Ph-}NH-C\overset{S^-}{\underset{N-CN}{<}}$$

The synthesis of ethylenethiourea, an accelerator for polychloroprene, also proceeds via a dithiocarbamate salt:

$$H_2N-CH_2-CH_2-NH_2 + CS_2 \longrightarrow H_3\overset{+}{N}-CH_2-CH_2-NH-C\overset{S}{\underset{S^-}{<}}$$

$$\xrightarrow{\text{Heat}} \underset{S}{\overset{}{HN}\diagup\diagdown NH}$$

*N*-Substituted ethylenethiourea derivatives have also been described [168].

## 4.7. Photolysis of Dithiocarbamates

Irradiation of acyl dithiocarbamates gives acyl and dithiocarbamic acid radicals [169]:

$$\begin{matrix} H_3C \\ H_3C \end{matrix} N-\overset{S}{\underset{\|}{C}}-S-\overset{O}{\underset{\|}{C}}-R \xrightarrow{h\nu} \begin{matrix} H_3C \\ H_3C \end{matrix} N-\overset{S}{\underset{\|}{C}}-S\cdot\cdot\overset{O}{\underset{\|}{C}}-R$$

These radicals can initiate polymerizations of olefins. The living radical photopolymerization of styrene with the aid of benzyl N,N-dialkyldithiocarbamates has also been described [170]:

Ph-CH$_2$-S-C(=S)-N(CH$_3$)$_2$

(C$_2$H$_5$)$_2$N-C(=S)-S-CH$_2$-C$_6$H$_4$-CH$_2$-S-C(=S)-N(C$_2$H$_5$)$_2$

Other dithiocarbamates have been proposed as photoinitiators for methyl acrylate [171].

# 5. Uses

## 5.1. Vulcanization Accelerators

Thiuram sulfides and salts, especially zinc salts, of N,N-dialkyldithiocarbamic acids are vulcanization accelerators for natural and synthetic rubber [46], [172], [173]; some examples of commercial products are given in Tables 7 and 8. Water-soluble derivatives are used in low-temperature processing of latex. Thiuram disulfides and polysulfides give heat-resistant rubber products. Sodium and potassium dimethyldithiocarbamates are modifiers in emulsion polymerization [174]. 4-[(Morpholinothiocarbonyl)thio]morpholine is a vulcanization accelerator which gives good resistance to reversion and low compression set [175]. Thiadiazines are also utilized as accelerators [176].

Dithiocarbamic acids are intermediates in the production of thioureas, particularly ethylene-, N,N'-diphenyl-, and tetramethylthiourea, which are used as accelerators for polychloroprene. Dust production has been reduced by formulation as polymeric granules. Under certain circumstances, thioureas can be injurious to health (see Chap. 6) [177]; 3-methylthiazolidine-2-thione is a polychloroprene accelerator that does not have this disadvantage [137]. Other nonhazardous vulcanization accelerators are currently being studied [178].

**Table 7.** Metal dithiocarbamates used as vulcanization accelerators

$$\left[ \begin{array}{c} R^1 \\ R^2 \end{array} \!\!\! N\text{-}\!\!\!\overset{\overset{\displaystyle S}{\|}}{C}\text{-}S\text{-} \right]_x \!\! M$$

| $R^1$ | $R^2$ | M | x | CAS registry number | Trade name | Manufacturer |
|---|---|---|---|---|---|---|
| $CH_3-$ | $CH_3-$ | Na | 1 | [128-04-1] | ACCEL SDD<br>KUMAC SDD | Kawaguchi<br>KUMHO (Korea) |
| $CH_3-$ | $CH_3-$ | Zn | 2 | [137-30-4] | Vulkacit L<br>Vulcafor ZDMC<br>Methazate<br>ACCEL PZ | Bayer<br>Vulnax<br>Uniroyal<br>Kawaguchi |
| $CH_3-$ | $CH_3-$ | Pb | 2 | [19010-66-3] | Methyl Ledate | R. T. Vanderbilt |
| $CH_3-$ | $CH_3-$ | Cu | 2 | [137-29-1] | Ekaland CDMC<br>Methyl Cumate | Atochem<br>R. T. Vanderbilt |
| $CH_3-$ | $CH_3-$ | Fe | 3 | [67055-36-1] | Royalac 133 | Uniroyal |
| $CH_3-$ | $CH_3-$ | Bi | 3 | [21260-46-8] | Robac BJDD<br>Bismate | Robinson Brothers<br>R. T. Vanderbilt |
| $CH_3-$ | $CH_3-$ | Se | 4 | [19632-73-6] | Methyl Selenac | R. T. Vanderbilt |
| $C_2H_5-$ | $C_2H_5-$ | Na | 1 | [148-18-5] | Oricel ESL<br>Eveite L<br>Vulcafor SDC | Oriental<br>ACNA (Italy)<br>Vulnax |
| $C_2H_5-$ | $C_2H_5-$ | Zn | 2 | [14324-55-1] | ACCEL EZ<br>Eveite Z<br>Robac ZDC<br>Vulkacit LDA<br>Ethasan<br>Ethazate<br>Ethyl Zimate<br>Perkacit ZDEC | Kawaguchi<br>ACNA<br>Robinson Brothers<br>Bayer<br>Monsanto<br>Uniroyal<br>R. T. Vanderbilt<br>Akzo |
| $C_2H_5-$ | $C_2H_5-$ | Se | 4 | [5456-28-0] | Ethyl Selenac | R. T. Vanderbilt |
| $C_2H_5-$ | $C_2H_5-$ | Te | 4 | [20941-65-5] | Ethyl Tellurac<br>Ekaland TE DEC<br>Ethyl Telluram | R. T. Vanderbilt<br>Atochem<br>Prochimie |
| $C_2H_5-$ | $C_2H_5-$ | Cd | 2 | [14239-68-0] | Ethyl Cadmate | R. T. Vanderbilt |
| $C_4H_9-$ | $C_4H_9-$ | Zn | 2 | [136-23-2] | Butasan<br>Vulcafor ZDBC<br>Butazate<br>Butyl Ziram<br>Butyl Zimate<br>Vulkacit LDB | Monsanto<br>Vulnax<br>Uniroyal<br>Prochimie<br>R. T. Vanderbilt<br>Bayer |
| $C_4H_9-$ | $C_4H_9-$ | Na | 1 | [136-30-1] | Oricel TP | Oriental |
| $C_4H_9-$ | $C_4H_9-$ | Ni | 2 | [13927-77-0] | NDBC<br>Antage NBC<br>Ekaland NBC<br>Antigene NBC<br>NJBUD | Prochimie<br>Kawaguchi<br>Atochem<br>Sumitomo<br>Akron |
| $C_4H_9-$ | $C_4H_9-$ | Cu | 2 | [52691-95-9] | Ekaland CDBC | Atochem |
| $C_5H_{11}-$ | $C_5H_{11}-$ | Pb | 2 | [36501-84-5] | Amyl Ledate | R. T. Vanderbilt |
| $C_5H_{11}-$ | $C_5H_{11}-$ | Zn | 2 | [15337-18-5] | Amyl Zimate | R. T. Vanderbilt |
| piperidino-N- | | Cd | 2 | [14949-59-8] | Robac CPD | Robinson Brothers |
| piperidino-N- | | Pb | 2 | [81195-36-0] | Robac LPD | Robinson Brothers |

**Table 7.** continued.

| $R^1$ | $R^2$ | M | x | CAS registry number | Trade name | Manufacturer |
|---|---|---|---|---|---|---|
| piperidinyl (N-) | | Zn | 2 | [13878-54-1] | Vulkacit ZP<br>Nocceler ZP | Bayer<br>Ouchi |
| $C_2H_5-$ | $C_6H_5-$ | Zn | 2 | [14634-93-6] | Hermat FEDK<br>Vulkacit P extra N<br>Eveite P<br>ACCEL PX | Chemapol<br>Bayer<br>ACNA<br>Kawaguchi |
| $C_6H_5CH_2-$ | $C_6H_5CH_2-$ | Zn | 2 | [14726-36-4] | Robac ZBED<br>Arazate<br>Perkacit ZBEC | Robinson Brothers<br>Uniroyal<br>Akzo |

**Table 8.** Other dithiocarbamate derivatives used as vulcanization accelerators

| Compound | CAS registry number | Trade name | Manufacturer |
|---|---|---|---|
| $(CH_3)_2N-C(=S)-S^-$  $H_3N^+-C_6H_{11}$ (cyclohexyl) | [78600-26-7] | AZ-100 | Monsanto |
| $(C_2H_5)_2N-C(=S)-S^-$  $H_2N^+(C_2H_5)_2$ | [1518-58-7] | Vulcafor DDCN | Vulnax |
| $(C_4H_9)_2N-C(=S)-SH \cdot$ $C_6H_5-NH-C(=NH)-NH-C_6H_5$ | [61792-03-8] | Ultex | C. P. Hall |
| $(C_4H_9)_2N-C(=S)-S^-$  $HN^+(CH_3)_2$-cyclohexyl | [149-82-6] | RZ-100 | Monsanto |
| piperidinyl-C(=S)-S$^-$  $H_2N^+$-piperidinyl | [98-77-1] | Nocceler PPD<br>Vulkacit P<br>Robac PPD | Ouchi (Japan)<br>Bayer<br>Robinson Brothers |
| benzothiazolyl-S-C(=S)-N$(C_2H_5)_2$ | [95-30-7] | Ethylac | Pennwalt |
| $(H_3C)_2N-C(=S)-S-C(=S)-N(CH_3)_2$ | [97-74-5] | Eveite MST<br>Vulkacit Thiuram MS<br>Vulkafor TMTM<br>Mono Thiurad<br>Monex<br>Unads<br>ACCEL TS<br>Sanceller TS<br>Soxinol TS<br>Nocceler TS | ACNA<br>Bayer<br>Vulnax<br>Monsanto<br>Uniroyal<br>R. T. Vanderbilt<br>Kawaguchi<br>Sanshin<br>Sumitomo<br>Ouchi (Japan) |
| $(C_4H_9)_2N-C(=S)-S-C(=S)-N(C_4H_9)_2$ | [93-73-2] | Pentex | Uniroyal |

**Table 8.** continued

| Compound | CAS registry number | Trade name | Manufacturer |
|---|---|---|---|
| piperidine-N-C(=S)-S-C(=S)-N-piperidine | [725-32-6] | Robac PTM | Robinson Brothers |
| (H$_3$C)$_2$N-C(=S)-S-S-C(=S)-N(CH$_3$)$_2$ | [137-26-8] | ACCEL TMT<br>Eveite 4 MT<br>Perkacit TMTD<br>Vulkacit Thiuram<br>Vulcafor TMTD<br>Thiurad<br>Tuex<br>Robac TMT<br>Accelerator TMTD<br>Methyl-Tuads<br>Super Accélerator 501<br>Sanceller TT | Kawaguchi<br>ACNA<br>Akzo<br>Bayer<br>Vulnax<br>Monsanto<br>Premier (Taiwan)<br>Robinson Brothers<br>Bann (Brazil)<br>R. T. Vanderbilt<br>Rhône-Poulenc<br>Sanshin |
| (C$_2$H$_5$)$_2$N-C(=S)-S-S-C(=S)-N(C$_2$H$_5$)$_2$ | [97-77-8] | Eveite T<br>Thiuram E<br>Ethylthiurad<br>Ethyl Tuex<br>Ethyl Tuads<br>Etiurac | ACNA<br>Du Pont<br>Monsanto<br>Uniroyal<br>R. T. Vanderbilt<br>Pennwalt |
| (C$_4$H$_9$)$_2$N-C(=S)-S-S-C(=S)-N(C$_4$H$_9$)$_2$ | [1634-02-2] | Butyl Tuads<br>Oricel TBT<br>Robac TBTU | R. T. Vanderbilt<br>Oriental<br>Robinson Brothers |
| H$_3$C(Ph)N-C(=S)-S-S-C(=S)-N(Ph)CH$_3$ | [10591-84-1] | Vulkacit J | Bayer |
| C$_2$H$_5$(Ph)N-C(=S)-S-S-C(=S)-N(Ph)C$_2$H$_5$ | [41365-24-6] | PETD | Bozzetto |
| piperidine-N-C(=S)-S-S-S-S-C(=S)-N-piperidine | [120-54-7] | Tetrone A<br>DPTT<br>Ekaland TSPM | Du Pont<br>Akron<br>Atochem |
| N-methyl thiazolidine-2-thione | [1908-87-8] | Vulkacit CRV | Bayer |
| morpholine-N-C(=S)-S-N-morpholine | [13752-51-7] | Cure-rite 18 | B. F. Goodrich |

## 5.2. Pesticides

Salts and derivatives of various dithiocarbamic acids are used as fungicides, nematocides, or bactericides (Table 9). These compounds liberate mustard oils or dialkyldithiocarbamic acids [179], [180]. Dithiocarbamate pesticides have been the subject of intensive metabolic [181], [182] and environmental [183]–[186] investigations. Analytical questions are discussed in [187].

## 5.3. Medical Applications

Tetraethylthiuram disulfide, known as disulfiram, is a withdrawal agent used in the treatment of alcoholism (Table 10). Nontoxic penicillins containing a dithiocarbamic acid structure are effective against penicillin-resistant organisms [188]. Dithiocarbamic acid derivatives are also used as antimycotics [154], [189] and nickel-poisoning antidotes [190] (see Table 10).

A broad spectrum of activity is shown by 2,6-bis-(thiocarbamoylthiomethyl)pyridine [191]. Dithiocarbamic acids affect the biosynthesis of catecholamines [192] and hepatosine [193], cause immune responses when used with levamisole [194], and reduce cholesterol production in rat liver [195]. Tetraethylthiuram monosulfide is used in conjunction with salicylic acid in the treatment of leishmaniasis [196]. Fungal infections of the skin can be treated with zinc ethylenebis(dithiocarbamate) [197]. Dithiocarbamates are anticaries agents [198]. 3-Aminotetrahydro-1,3-thiazine-2,4-diones [199], produced from cyclic dithiocarbamic acids, are skin-protecting agents. Certain tetrahydro-$\beta$-carbolines containing a dithiocarbamic acid structure are effective in liver therapy [200].

## 5.4. Radioprotective Agents

Dithiocarbamic acid esters and other derivatives provide protection against $\gamma$-radiation [201]–[206]. The metabolism of some of these agents in mice has been studied [207]. Polymers, e.g., poly(vinyl chloride), that contain dithiocarbamic acid substituents are more radiation resistant than unsubstituted polymers [208].

## 5.5. Imaging Technology

Dithiocarbamic acids and their derivatives have many applications in photographic and recording materials [209]–[211].

Heat-sensitive copying materials contain, for example, tetramethylthiuram disulfide and zinc dithiocarbamates [212]. Light-sensitive photographic materials for high-contrast negatives and di-

**Table 9.** Dithiocarbamic acid derivatives used in agriculture

| Compound | CAS registry number | Application | Trade name | Manufacturer |
|---|---|---|---|---|
| $CH_3-NH-C(=S)-SNa \cdot 2\,H_2O$ | [137-42-8] | soil fungicide, herbicide | Vapam | Stauffer Chemical |
| $(H_3C)_2N-C(=S)-SNa \cdot 2\,H_2O$ | [72140-17-1] | herbicide | Na-DMDT, Diram | Stauffer Chemical |
| $[(H_3C)_2N-C(=S)-S-]_2 Zn$ | [137-30-4] | protective fungicide | Zerlate, Milban / Fuklasin / Pomarsol-Z / Zinc carbamate | Du Pont / Schering / Bayer / Bayer |
| $[(H_3C)_2N-C(=S)-S-]_3 Fe$ | [14484-64-1] | fungicide | Fermate | Du Pont |
| $[(H_3C)_2N-C(=S)-S-]_2 Ni$ | [15521-65-0] | rice bactericide | Sankel | Sankyo |
| $(H_3C)_2N-C(=S)-S-S-C(=S)-N(CH_3)_2$ | [137-26-8] | fungicide | Arasan, Tersan / Nomersan / Pomarsol | Du Pont / Plant Protection / Bayer |
| $(H_3C)_2N-C(=O)-S-S-C(=S)-N(CH_3)_2$ | [1115-06-6] | fungicide | Niagara 9130 | R. T. Vanderbilt |
| $NaS-C(=S)-NH-CH_2CH_2-NH-C(=S)-SNa$ | [142-59-6] | soil fungicide | Dithane D-14 / Parzate, DSE / Amobam | Rohm and Haas / Du Pont / Roberts |
| $[-S-C(=S)-NH-CH_2CH_2-NH-C(=S)-S-]\,Zn$ | [12122-67-7] | leaf fungicide, soil fungicide | Dithane Z-78 / Parzate, Zineb / Lonacol | Rohm and Haas / Du Pont / Bayer |
| $[-S-C(=S)-NH-CH_2CH_2-NH-C(=S)-S-]\,Mn$ | [12427-38-2] | leaf fungicide, soil fungicide, seed disinfectant | Dithane M-22 / Manzate, Maneb / Dithane M-22 / Maneb Wettable Powder | Du Pont / Du Pont / Rohm and Haas / Bayer |
| $[-S-C(=S)-NH-CH_2CH_2-NH-C(=S)-S-]\,(Mn + Zn)\;20\%\;2.5\%$ | [8018-01-7] | leaf fungicide, fungicide | Dithane M-45 (mancozeb) | Rohm and Haas |
| $[-Zn-S-C(=S)-NH-CH_2-CH(CH_3)-NH-C(=S)-S-]_n$ | [12071-83-9] | fungicide | Antracol | Bayer |

**Table 9.** continued

| Compound | CAS registry number | Application | Trade name | Manufacturer |
|---|---|---|---|---|
| [-C(=S)-NH-CH$_2$-CH$_2$-NH-C(=S)-S-]$_n$ | [58855-93-9] | fungicide | Thioneb | Du Pont |
| [-S-C(=S)-NH-CH$_2$-CH$_2$-NH-C(=S)-S-]$_n$ | [9006-42-2] | fungicide | Polyram | BASF |
| (C$_2$H$_5$)$_2$N-C(=S)-S-CH$_2$-CCl=CH$_2$ | [95-06-7] | soil herbicide | Vegadex | Monsanto Chemical |
| H$_3$C-N(CH$_3$)-CH$_2$-S-C(=S)- (dimethyl thiadiazinethione) | [533-74-4] | nematocide fungicide | Mylone, Dazomet N-521 Basamid Drozine | Union Carbide Stauffer Chemical BASF |
| H$_3$C-thiadiazinethione-CH$_2$-CO$_2$Na | [3655-88-7] | nematocide soil fungicide | Terracur | Bayer |
| bis(dimethyl-thiadiazinethione)-CH$_2$-CH$_2$ | [3773-49-7] | fungicide | Du Pont Fungicide 328 Milneb Banlate | Du Pont |

**Table 10.** Medical applications of dithiocarbamic acid derivatives

| Compound | CAS registry number | Application | Trade name | Manufacturer |
|---|---|---|---|---|
| (C$_2$H$_5$)$_2$N-C(=S)-S-S-C(=S)-N(C$_2$H$_5$)$_2$ | [97-77-8] | alcoholism withdrawal agent | Antabuse (disulfiram) | Tosse |
| dibenzyl-thiadiazinethione | [350-12-9] | antimycotic | Fungiplex (sulbentine) | Hermal |
| (C$_2$H$_5$)$_2$N-C(=S)-SNa | [148-18-5] | nickel-poisoning antidote | Dithiocarb | |
| 4-phenyl-3-(4-methylbenzylideneamino)thiazoline-2-thione | [15387-18-5] | antimycotic | Polyodin (fezatione) | Takeda |

rect-positive photographic materials employ certain dithiocarbamates [213], [214]. Dithiocarbamates are used in color photography [215] and in electrophotographic materials [216]. Photopolymerizable printing compositions contain thiuram sulfides [217], and latent image developers contain zinc dithiocarbamates [218]. Developer compositions for electrolyte photography may contain small amounts of dimethyldithiocarbamates [219]. Dithiocarbamates are likewise used in the bleaching of photographic materials [220] and in photothermographic materials [221]. Thermodeveloping photographic materials contain metal derivatives of dithiocarbamic acid [222], [223].

Dialkyldithiocarbamates [224] or rhodanines [225] are employed in photosensitive resin compositions, and certain rhodanines are used in sensitizing dyes [226]. Photosensitive materials for high resolution may contain cyclic dithiocarbamic acid esters [227]. Solutions containing dithiocarbamic acid or its salts are used in the production of photographic relief images [228], [229]. Dialkyldithiocarbamic acid derivatives are used in the bleaching of silver halide-containing photographic materials [230]. Small amounts of dithiocarbamic acid derivatives increase the brilliance of photographic silver images [231]. Ammonium dithiocarbamate is a heat and moisture stabilizer for photographic emulsions [232]. Combinations of sulfinic acids and ammonium salts of dithiocarbamic acid, as well as thiuram disulfides, increase the sensitivity of photographic emulsions [233]–[238]. Certain dithiocarbamic acid derivatives increase the sensitivity of diazole compounds in diazotype printing [239]. Dialkyldithiocarbamic acid derivatives stabilize silver halide emulsions [240]. Gel formation in photopolymerization is prevented by dithiocarbamates [241], which produce positive prints with small amounts of silver halides [242]. Color fading in silver halide images can be prevented by treatment with a dithiocarbamate salt solution [243]. Color images are produced in direct-positive processes with the aid of dithiocarbamic acid derivatives [244]. Silver ions and a solution containing sodium *N*-methyldithiocarbamate [245] participate in the photocuring of high molecular mass substances containing amino or amido groups. Dithiocarbamates are used in the recovery of silver from photographic processing solutions [246].

## 5.6. Other Uses

*Curing Agents.* Heavy-metal salts of dibutyldithiocarbamic acid in epoxy resins improve adhesion properties and moisture resistance, and inhibit rust formation [247]. Treatment of halogen-containing elastomers with a dimercaptothiadiazole – dithiocarbamate mixture improves flow behavior and cure rates [248].

*Separation of Heavy Metals.* Dithiocarbamates precipitate heavy metals such as cadmium, nickel, and zinc from aqueous solution [249], [250]. Polydithiocarbamate resins remove heavy-metal ions from wastewater [251]. Dithiocarbamic acid derivatives are employed as flotation agents for copper, zinc, nickel, lead, and iron ores [252].

*Complexation.* Dithiocarbamic acid derivatives stabilize transition metals in high states of oxidation [253]. The resulting metal complexes have been the subject of electrochemical [254], crystallographic [255], thermochemical [256], and structural [257] studies.

*Stabilizers for Polymers and Oils.* Zinc dithiocarbamates improve the thermal stability of polymers, especially polyurethane elastomers [258]; combinations with sterically hindered phenols show synergistic stabilizing effects [259], [260]. Zinc dialkyldithiocarbamates are antioxidants [261]. Nickel dithiocarbamate inhibits the photodegradation of polyethylene [259], [262]. Dithiocarbamates improve the ozone resistance of printing plates [263] and the stability of photosensitive resins [264]. Protective effects are found in the radio-frequency range [265]. Certain metal derivatives serve as oil [266] and lubricant additives [267]. Aqueous synthetic resin dispersions containing tetramethylthiuram disulfide as a fungicide have improved storage stability [268]. Organopolysiloxane-based sealants used in the construction industry are protected from fungus attack by tetraalkylthiuram disulfide [269]. Biocides consisting of copper salts of alkylenebis(dithiocarbamates) and triphenyltin hydroxide considerably extend the life of underwater paints [270].

*Analysis.* Dithiocarbamic acids are useful in inorganic [271], [272] and other analyses [273], especially to determine trace amounts of heavy metals in organic and biological materials [186], [274]–[276]; they can also be used as chelating agents in HPLC [277].

# 6. Toxicology and Occupational Health

The toxicity of dithiocarbamic acid derivatives [278]–[280] and oxidation products of the dithiocarbamates [281] has been investigated extensively. The biological activity of some dithiocarbamic acid derivatives is treated in [282]. The embryotoxicity of dithiocarbamates is being studied, with particular attention to the derivatives used in the rubber industry [283] and in pesticides [284]. The teratogenicity potential of such compounds [285] and their metabolites [286] has also been investigated. Mutagenicity studies of dithiocarbamic acid pesticides [287]–[289] and of products used in the rubber industry [290]–[292] have been reported. Cytological and genetic effects of ethylenebis(dithiocarbamates) used as fungicides have been investigated with the aid of *Allium cepa* [293]. Genotoxic effects have been observed for both synthetic elastomer additives (especially thiuram sulfides and dithiocarbamates) and pesticides [294].

*Formation of Nitrosamines.* During the vulcanization of rubber, dithiocarbamic acid derivatives used as accelerators form trace amounts of amines [295]. These amines react with nitrogen oxides from the air or with nitrosating ingredients of rubber compounds to form highly toxic nitrosamines [296]. The carcinogenic potential of these nitrosamines is being studied.

The TLV-TWAs and MAK values are 5 mg/m$^3$ for thiram (tetramethylthiuram disulfide) and 2 mg/m$^3$ for disulfiram (tetraethylthiuram disulfide).

# 7. References

General References

[1]  Houben-Weyl, **9**, 824; E 4, 458.
[2]  G. D. Thorn, R. A. Ludwig: *The Dithiocarbamates and Related Compounds*, Elsevier, Amsterdam-New York 1962.
[3]  G. Scheuerer, Fortschr. Chem. Forsch. **9** (1967) 254.
[4]  W. Walter, K.-D. Bode, Angew. Chem. **79** (1967) 285; Angew. Chem. Int. Ed. Engl. **6** (1967) 281.
[5]  L. A. Summers, Rev. Pure Appl. Chem. **18** (1968) 1.
[6]  G. Gattow, W. Behrendt: "Carbon Sulfides and Their Inorganic and Complex Chemistry" in A. Senning (ed.): *Topics in Sulfur Chemistry*, vol. **2**, Thieme-Verlag, Stuttgart 1977.

[7] E. E. Reid: *Organic Chemistry of Bivalent Sulfur,* vol. **4,** Chemical Publishing Co., New York 1962, p. 131 ff.

[8] F. Duus: "Thionocarbamic and Dithiocarbamic Acids and their Derivatives," in D. Barton, W. D. Ollis (eds.): *Comprehensive Organic Chemistry,* vol. 3, Pergamon Press, Oxford 1979, p. 469.

[9] R. Zahradnik, *Chem. Tech. (Leipzig)* **10** (1958) 546.

[10] D. Coucouvanis, *Prog. Inorg. Chem.* **11** (1970) 233; **26** (1979) 301.

Specific References

[11] G. Gattow, V. Hahnkamm, *Z. Anorg. Allg. Chem.* **364** (1969) 161. G. Gattow, V. Hahnkamm, *Angew. Chem.* **78** (1966) 334; *Angew. Chem. Int. Ed. Engl.* **5** (1966) 316. J.-L. Fourquet, *Bull. Soc. Chim. Fr.* 1969, no. 9, 3001. D. De Filippo, P. Deplano, F. Devillanova, E. F. Trogu, G. Verani, *J. Org. Chem.* **38** (1973) 560. S. J. Joris, K. J. Aspila, Ch. L. Chakrabarti, *J. Phys. Chem.* **74** (1970) 860.

[12] D. Graig, A. E. Juve, W. L. Davidson, W. L. Semon, D. C. Hay, *J. Polym. Sci.* **8** (1952) 321.

[13] R. Gerner, G. Gattow, *Z. Anorg. Allg. Chem.* **524** (1985) 111.

[14] M. Sato, N. Fukada, M. Kurauchi, *Synthesis* 1981, 554.

[15] T. Takeshima, M. Ikeda, M. Yokoyama, N. Fukada, *J. Chem. Soc. Perkin Trans. I* 1979, 692.

[16] R. Gerner, G. Gattow, *Z. Anorg. Allg. Chem.* **525** (1985) 112.

[17] V. Hahnkamm, G. Kiel, G. Gattow, *Z. Anorg. Allg. Chem.* **368** (1969) 127. F. Christiani, F. A. Devillanova, G. Verani, *Int. Symp. Interact. Mol. Ions [Proc.] 3rd,* vol. **1,** 103; *Chem. Abstr.* **88** (1978) 104 189.

[18] R. Gerner, G. Gattow, *Z. Anorg. Allg. Chem.* **522** (1985) 145.

[19] R. Gerner, G. Gattow, *Z. Anorg. Allg. Chem.* **526** (1985) 122.

[20] R. Gerner, G. Gattow, *Z. Anorg. Allg. Chem.* **524** (1985) 122.

[21] R. Gerner, G. Gattow, *Z. Anorg. Allg. Chem.* **528** (1985) 157.

[22] C. W. Voigt, G. Gattow, *Z. Anorg. Allg. Chem.* **437** (1977) 226.

[23] T. Callan, M. Strafford, *J. Soc. Chem. Ind. Trans.* **43** (1924) 7.

[24] B. F. Goodrich, US 2 117 619, 1937; US 2 123 370, 1937; US 2 123 373, 1937. Amer. Cyanamid, US 2 235 747, 1939.

[25] Sharples Chem., US 2 492 314, 1945.

[26] G. Gattow, W. Eul, *Z. Anorg. Allg. Chem.* **483** (1981) 103.

[27] C. W. Voigt, G. Gattow, *Z. Anorg. Allg. Chem.* **437** (1977) 233.

[28] Sharples Chem., US 2 443 160, 1944.

[29] R. T. Vanderbilt Co., US 2 347 128, 1943. J. Stary et al., *Talanta* **15** (1968) 505. N. K. Wilson, L. Fishbein, *J. Agric. Food Chem.* **20** (1972) 847. O. Foss, *Acta Chem. Scand.* **5** (1951) 115.

[30] Rohm & Haas Co., US 2 317 765, 1941.

[31] H. Nishimura, T. Kinugasa, *Chem. Pharm. Bull.* **17** (1969) 94. J. Dunderdale, T. J. Watkins, *Chem. Ind. (London)* 1956, 174.

[32] G. Gattow, S. Lotz, *Z. Anorg. Allg. Chem.* **531** (1985) 101.

[33] G. Gattow, S. Lotz, *Z. Anorg. Allg. Chem.* **531** (1985) 97.

[34] G. D. Thorn, R. A. Ludwig: *The Dithiocarbamates and Related Compounds,* Elsevier, Amsterdam 1962.

[35] G. Gattow, W. Behrendt, "Carbon Sulfides and Their Inorganic and Complex Chemistry," in A. Senning (ed.), *Topics in Sulfur Chemistry,* Thieme-Verlag, Stuttgart 1977.

[36] Du Pont, US 3 276 950, 1965 (A. W. Engelhard); *Chem. Abstr.* **66** (1967) 1871.

[37]  J. v. Braun, W. Kaiser, *Ber. Dtsch. Chem. Ges.* **56** (1923) 550. Naugatuck Chem. Co., US 1 782 111, 1925.

[38]  Wingfoot Corp., US 2 014 353, 1931.

[39]  Silesia, DE 444 014, 1925.

[40]  Monsanto Chem. Co., US 2 375 083, 1943. J. v. Braun, *Ber. Dtsch. Chem. Ges.* **35** (1902) 819.

[41]  *Fiat* **1018** (1947) 52, 58. Goodyear, DE-OS 2 349 313, 1973 (J. J. Tazuma, B. A. Bergorni). UCB, DE-OS 2 527 898, 1975 (J. M. G. Lietard, G. Matthijis).

[42]  Roessler & Hasslacher Chem. Co., US 1 796 977, 1928. Monsanto Chem. Co., GB 555 874, 1942.

[43]  Naugatuck Chem. Co., US 1 782 111, 1925 (H. S. Adams, L. Meuser). SU 56 086, 1939. Sharples Chem. Inc., US 2 325 194, 1941.

[44]  Akzo, DE 3 105 587, 1981 (L. Eisenhut, H. G. Zengel, M. Bergfeld); BE 892 143, 1981; *Chem. Abstr.* **97** (1982) 144 394. Akzo, DE 3 105 622, 1981 (L. Eisenhuth, H. G. Zengel, M. Bergfeld); BE 892 144, 1981; *Chem. Abstr.* **97** (1982) 144 395. C. J. Swan, D. L. Trimm, *J. Appl. Chem.* **18** (1968) 340.

[45]  S. Torii, H. Tanaka, K. Mishima, *Bull. Chem. Soc. Jpn.* **51** (1978) 1575; *Chem. Abstr.* **89** (1978) 67 437.

[46]  *Bayer Manual for the Rubber Industry,* Bayer, Leverkusen, Rubber Division, Technical Service Section, 1971.

[47]  R. Gerner, G. Gattow, *Z. Anorg. Allg. Chem.* **527** (1985) 125.

[48]  R. Gerner, G. Gattow, *Z. Anorg. Allg. Chem.* **528** (1985) 168.

[49]  A. Schönberg, *Ber. Dtsch. Chem. Ges.* **68** (1935) 163.

[50]  J. v. Braun, F. Stechele, *Ber. Dtsch. Chem. Ges.* **36** (1903) 2280. Naugatuck Chem. Co., US 1 682 920, 1926.

[51]  I.G. Farben, DE 519 445, 1930.

[52]  Du Pont, US 2 048 043, 1931.

[53]  Du Pont, US 1 788 632, 1928.

[54]  A. Cambron, *Can. J. Res.* **2** (1930) 341.

[55]  Roessler & Hasslacher Chem. Co., US 1 681 717, 1926. Du Pont US 2 414 014, 1943.

[56]  E. L. Carr, G. E. P. Smith, G. Alliger, *J. Org. Chem.* **14** (1949) 921. R. A. Donia, J. A. Shotton, L. O. Beutz, G. E. P. Smith jr., *J. Org. Chem.* **14** (1949) 946. G. E. P. Smith, G. Alliger, E. L. Carr, K. C. Young, *J. Org. Chem.* **14** (1949) 935. Monsanto Chem. Co., US 3 732 222, 1970; *Chem. Abstr.* **79** (1973) 54 669.

[57]  S. Torii, H. Tanaka, M. Ukita, *Jpn. Kokai Tokkyo Koho* **79** (1979) 115, 323; *Chem. Abstr.* **92** (1980) 84 927.

[58]  Denki Kagaku Kogyo K., Kanesha Co. Ltd., JP 75/7134, 1975; *Chem. Abstr.* **83** (1975) 142 999.

[59]  Vanderbilt R. T. Co. Inc., DE-OS 2 446 555, 1975 (K. S. Karsten); *Chem. Abstr.* **82** (1975) 173 000.

[60]  R. N. Datta, M. M. Das, D. K. Basu, A. K. Chaudhuri, *Rubber Chem. Technol.* **57** (1984) 879.

[61]  W. Hofmann, *Plast. Rubber Process Appl.* **5** (1985) 209.

[62]  B. Adhikari et al., *Rubber Chem. Technol.* **56** (1983) 327.

[63]  B. F. Goodrich Co., DE-OS 2 325 027, 1973 (R. D. Taylor); Chem. Abstr. **80** (1974) 109 604.

[64]  B. F. Goodrich Co., DE-OS 2 324 934, 1973 (R. D. Taylor); Chem. Abstr. **81** (1974) 26 970.

[65]  B. F. Goodrich Co., DE-OS 2 324 933, 1973 (R. D. Taylor); Chem. Abstr. **80** (1974) 146 841.

[66]  Amer. Cyanamid, DE-OS 2 827 933, 1979 (Franz, Curtis Allen); Chem. Abstr. **90** (1979) 122 845.

[67]  B. F. Goodrich Co., US 3 985 743, 1972, 1976 (R. D. Taylor); *Chem. Abstr.* **86** (1977) 30 837.

[68] Th. Kempermann, *Kautsch. Gummi Kunstst.* **20** (1967) 126. R. D. Taylor, "Thiocarbamyl-Sulfenamide als Vulkanisationsbeschleuniger," *Rubber Chem. Technol.* **47** (1974) 906.

[69] B. F. Boodrich Co., DE-OS 2 324 981, 1973 (R. N. Taylor); *Chem. Abstr.* **81** (1974) 26 969.

[70] Uniroyal Inc., EP 154 437, 1984 (J. K. Stieber); *Chem. Abstr.* **104** (1986) 111 175. G. E. P. Smith, G. Alliger, E. L. Corr, K. C. Young, *J. Org. Chem.* **14** (1949) 935.

[71] A. K. Khamrai, B. Adhikari, M. M. Maiti, S. Maiti, *Angew. Makromol. Chem.* **143** (1986) 39.

[72] Sharples Chem. Inc., US 2 466 276, 1946; FR 960 006, 1948.

[73] O. C. Billeter, H. Rivier, *Ber. Dtsch. Chem. Ges.* **37** (1904) 4319. G. M. Dyson, H. J. George, *J. Chem. Soc.* **125** (1924) 1702.

[74] M. S. Newmann, F. W. Hetzel, *J. Org. Chem.* **34** (1969) 3604. H. G. Viehe, Z. Janousek, *Angew. Chem.* **83** (1971) 614; *Angew. Chem. Int. Ed. Engl.* **10** (1971) 573. R. H. Goshorn, W. W. Levis jr., E. Jaul, E. J. Ritter, T. L. Cairns, H. E. Cupery: *Organic Syntheses*, Coll. vol. **IV,** Wiley, New York 1963, pp. 307, 310.

[75] *Houben-Weyl*, **E 4**, 416.

[76] S. Scheithauer, R. Mayer: "Thio- and Dithiocarboxylic Acids and Their Derivatives," in A. Senning (ed.): Topics in Sulfur Chemistry, vol. **4,** Thieme, Stuttgart 1979, p. 293. *Houben Weyl*, **E 4**, 416. W. Walter, R. F. Becker, *Justus Liebigs Ann. Chem.* **755** (1972) 145.

[77] N. N. Yarovenko, A. S. Vasil'eva, *Zh. Obshch. Khim.* **29** (1959) 3786; *J. Gen. Chem. USSR (Engl. Transl.)* **29** (1959) 3747. L. M. Yagupol'skii, M. I. Dronkina, *Zh. Obshch. Khim.* **36** (1966) 1309; *J. Gen. Chem. USSR (Engl. Transl.)* **36** (1966) 1323. Bayer, DE-OS 3 044 216, 1982 (B. Baasner, G. M. Petruck, H. Hagemann, E. Klauke); *Chem. Abstr.* **97** (1982) 181 932.

[78] G. Nochmias, *Ann. Chim.* **12** (1952) 584.

[79] A. Bhai, A. Das, S. Medheker, K. S. Boparai, *J. Indian Chem. Soc.* **58** (1981) 295.

[80] A. M. Kuliev, M. A. Kulieva, T. N. Kulibekova, A. K. Ibazade, *Neftekhimiya* **25** (1985) 679.

[81] M. W. Harman, J. J. D'Amico, *J. Am. Chem. Soc.* **75** (1953) 4081.

[82] Naugatuck Chem. Co., US 1 726 647, 1928.

[83] J. v. Braun, *Ber. Dtsch. Chem. Ges.* **42** (1909) 4568.

[84] I.G. Farben, DE 575 372, 1931; Monsanto Chem. Co., US 2 695 901, 1953.

[85] Bayer, DE-AS 1 245 358, 1964.

[86] A. I. Griogorieva, I. F. Titova, V. N. Konygin, SU 1 065 396, 1982; *Chem. Abstr.* **101** (1984) 38 146.

[87] R. Gerner, G. Gattow, *Z. Anorg. Allg. Chem.* **524** (1985) 117.

[88] R. Gerner, G. Gattow, *Z. Anorg. Allg. Chem.* **527** (1985) 130.

[89] A. M. Clifford, J. G. Lichty, *J. Am. Chem. Soc.* **54** (1932) 1163.

[90] A. Rieche, G. Hilgetag, D. Martin, I. Kreyzi, *Arch. Pharm. (Weinheim Ger.)* **296** (1963) 310. Y. Iwakura, A. Nabeya, F. Nishigushi, K. H. Okkawa, *J. Org. Chem.* **31** (1966) 3352. J. v. Braun, *Ber. Dtsch. Chem. Ges.* **35** (1902) 3377. A. Kaji, *Bull. Chem. Soc. Jpn.* **34** (1961) 1147.

[91] Bayer, DE 1 768 874, 1968.

[92] Phillips Petroleum Co., US 2 690 440, 1951.

[93] A. A. Watson, *J. Chem. Soc.* **1964,** 2100. Phillips Petroleum Co., US 2 862 850, 1954.

[94] M. G. Zimin, M. M. Afanasev, A. V. Mironov, A. N. Pudovik, *Zh. Obshch. Khim.* **51** (1981) 470.

[95] M. Roshdestwenski, *Zh. Russ. Fiz. Khim. Ova. Chast. Khim.* **41** (1909) 1438; *Chem. Zentralbl.* 1910, 910.

[96] Olin Mathieson Chem. Corp., US 2 940 978, 1958.

[97] T. I. Bychkova, M. A. Vasil'eva, L. B. Krivdin, A. V. Kalabina, *Zh. Org. Khim.* **20** (1984) no. 10, 2114.

[98]   Amer. Cyanamid, US 3 407 222, 1965; *Chem. Abstr.* **70** (1969) 37 247. T. Nakai, M. Okawara, *Bull. Chem. Soc. Jpn.* **41** (1968) 707.
[99]   U. Stanior, M. Weißler, *Arch. Pharm. (Weinheim Ger.)* **317** (1984) 1042.
[100]  Gulf Research and Development Co., US 3 676 479, 1969; *Chem. Abstr.* **77** (1972) 100 872.
[101]  A. P. Sineokov, F. N. Gladsheva, V. S. Etlis, *Chem. Heterocycl. Comp. (Engl. Transl.)* 1970, 562.H. Stamm, *Pharm. Zentralhalle* **107** (1968) 440. T. A. Foglia, L. M. Gregory, G. Maerker, S. F. Osman, *J. Org. Chem.* **36** (1971) 1068.
[102]  N. Kreutzkamp, H. Y. Oei, H. Peschel, *Arch. Pharm. (Weinheim Ger.)* **304** (1971) 649.
[103]  A. O. Fitton et al., *J. Chem. Soc. C* **1969**, 230; **1968**, 996; **1971**, 1245. A. O. Fitton, J. Hill, M. Clutob, A. Thompon, *J. Chem. Soc. Perkin Trans. I* **1972**, 2658. N. Kreutzkamp, H. Y. Oei, H. Peschel, *Arch. Pharm. (Weinheim Ger.* ) **304** (1971) 648. ICI, FR 1 559 120, 1968; *Chem. Abstr.* **71** (1969) 113 737. Uniroyal Inc., DE-OS 1 815 221, 1968; *Chem. Abstr.* **72** (1970) 3218.
[104]  Uniroyal F. X. O'Shea, US 3 330 804, 1963; NL 64/8883, 1963; *Chem. Abstr.* **63** (1965) 17 974 E.
[105]  D. M. Morton, R. H. Neilson, *Phosphorus Sulfur* **25** (1985) 315.
[106]  Rubber Service Laboratories Co., US 1 586 121, 1925.
[107]  R. A. Donia, J. A. Shotton, L. O. Beutz, G. E. P. Smith jr., *J. Org. Chem.* **14** (1949) 952.
[108]  Wingfoot Corp., US 2 238 331, 1936.
[109]  T. P. Johnston, C. R. Stringfellow jr., A. Gallagher, *J. Org. Chem.* **27** (1962) 4068. R. Dalaby et al., *C. R. Hebd. Séances Acad. Sci.* **232** (1951) 1676. Monsanto Chem. Co., US 3 372 150, 1966; *Chem. Abstr.* **67** (1967) 3088. US 3 361 752, 1964; BE 6 704 434, 1964; *Chem. Abstr.* **65** (1966) 15 383 E.
[110]  Wingfoot Corp., GB 675 450, 1950; US 2 597 988, 1950. Monsanto Chem. Co., US 3 726 866, 1970; *Chem. Abstr.* **79** (1973) 20 064.
[111]  N. K. Blizuyuk et al., SU 175 054, 1966.
[112]  Bayer, DE-AS 1 178 417, 1964; DE-AS 1 178 418, 1964. G. Buchmann, O. Wolniak, *Pharmazie* **21** (1966) 650. E. G. Novikov, J. N. Tugarinova, *Chem. Heterocycl. Comp.* 1968, 207.
[113]  R. Delaby, R. Daumiens, R. Seyden-Penne, *C. R. Hebd. Séances Acad. Sci.* **238** (1954) 121.
[114]  E. Profft, R. Schmuck, *Arch. Pharm. (Weinheim Ger.)* **296** (1963) 209.
[115]  J. L. Garraway, *J. Chem. Soc.* 1964, 4004, 4008.
[116]  C. S. Angadiyavar, M. N. Gudi, M. V. George, *Indian J. Chem. Soc.* **10** (1972) 888.
[117]  V. N. Elokhina, A. S. Nakhmanovich, A. E. Aleksandrova, B. J. Vishnevskii, J. D. Kalikhman, *Khim. Farm. Zh.* **20** (1986) 1061.
[118]  J. C. Sauer, *J. Org. Chem.* **24** (1959) 1592.
[119]  T. Hayashi, T. Oishi, *Chem. Lett.* 1985, 413.
[120]  *Houben-Weyl,* **E 11,** 82; E 4, 551.
[121]  T. Olijusma, I. B. F. N. Engberts, *Synth. Commun.* 3(1973) 1.H. C. Hansen, A. Senning, *Org. Prep. Proced. Int.* **17** (1985) 275. *Houben Weyl,* **E 4,** 524.
[122]  W. J. Stephen, A. Townshend, *J. Chem. Soc.* 1965, 5127.
[123]  E. Cherbuliez, J. Marszalek, J. Rabinowitz, *Helv. Chim. Acta.* **47** (1964) 1666.
[124]  L. Musial, J. Staniec, *Rocz. Chem.* **44** (1970) 1801.
[125]  L. Y. Ladnaya, E. M. Protsenko, *Pharm. Chem. J. (Engl. Transl.)* 1968, 257.
[126]  S. O. Abdallah, H. H. Hammouda, *J. Heterocycl. Chem.* **22** (1985) 497.
[127]  E. K. Mikitenko, N. N. Romanov, *Khim. Geterotsikl. Soedin.* 1981, 141.
[128]  J. Kinugawa, H. Nagase, *Yakugaku Zasshi* **86** (1966) 95, 101.

[129] A. Miolati, *Gazz. Chim. Ital.* **23** (1894) 578. H. Erlenmeyer, M. Simon, *Helv. Chim. Acta* **25** (1942) 362. M. O. Kolosowa, V. I. Stavroskaya, *J. Gen. Chem. USSR (Engl. Transl.)* **33** (1963) 2706.

[130] Agency of Ind. Sci. Tech., JP 60 096 604-A, 1983; *Chem. Abstr.* **103** (1985) 179 168.

[131] C. M. Roussel, R. Gallo, M. Chanon, J. Metzger, *Bull. Chim. Fr.* 1971, 1902. J. J. D'Amico, T. W. Bartman, *J. Org. Chem.* **25** (1960) 1336.

[132] L. Brandsma, R. L. P. De Jong, H. D. Verkruijsse, *Synthesis* 1985, 948.

[133] T. Chiba, H. Sato, T. Kato, *Heterocycles* **21** (1984) 613.

[134] D. H. R. Barton, D. Bridon, S. Z. J. Zard, *J. Chem. Soc., Chem. Commun.* 1985, 1066.

[135] G. J. Pustoshkin, F. Yu. Rachinski, *J. Org. Chem. USSR (Engl. Transl.)* 1966, 1254.

[136] Amer. Cyanamid, US 3 215 704, 1963.

[137] Amer. Cyanamid, US 3 215 703, 1962 (F. A. V. Sullivan, A. C. Lindaw); *Chem. Abstr.* **64** (1966) 20 910. Bayer, DE 2 701 215, 1977 (R. Schubart, U. Eholzer); *Chem. Abstr.* **89** (1978) 179 990. Bayer, EP 17 039, 1980 (R. Schubart, U. Eholzer); DE 2 911 662, 1979; *Chem. Abstr.* **94** (1981) 65 669. Bayer, EP 16 420, 1980 (D. Hüllstrung, J. Trimbach); DE 2 911 661, 1979; *Chem. Abstr.* **94** (1981) 84107.

[138] J. Kopecký, J. Smejkal, *Bull. Soc. Chim. Belg.* **93** (1984) 231.

[139] Luther-Universität Halle, DD 200 958-A, 1984; *Chem. Abstr.* **104** (1986) 68 845.

[140] Yu. P. Kovtun, N. N. Romanov, *Khim. Geterotsikl. Soedin.* 1985, 498.

[141] V. N. Charushin, V. G. Baklykov, O. N. Chupakhin, G. M. Petrova, E. O. Sidorov, *Khim. Geterotsikl, Soedin.* 1984, 680.

[142] R. N. Warrener, E. N. Cain, *Chem. Ind. (London)* **48** (1964) 1989. E. N. Cain, R. N. Warrener, *Aust. J. Chem.* **23** (1970) 51.

[143] W. Schroth, A. Hildebrandt, U. Becker, S. Freitag, M. Akram, R. Spitzner, *Z. Chem.* **25** (1985) 20.

[144] N. M. Turkevic, B. S. Zimenkovskii, *Khim. Geterotsikl. Soedin.* 1967, 845.

[145] T. L. Gresham, J. E. Jansen, E. W. Shawer, *J. Amer. Chem. Soc.* **70** (1948) 1001. E. Cherbuliez, A. Buchs, J. Marszalek, J. Rabinowitz, *Helv. Chim. Acta* **48** (1965) 1414.

[146] T. Takeshima, N. Fukada, E. Ohki, M. Muroaka, *J. Chem. Research* 1979, 212.

[147] W. Hanefeld, G. Glaeske, *Liebigs Ann. Chem.* 1981, 1388.

[148] Wella AG, DE 3 304 871-A, 1983; *Chem. Abstr.* **101** (1984) 211 158.

[149] H. Hanefeld, G. Glaeske, H. J. Staude, *Arch. Pharm. (Weinheim Ger.)* **315** (1982) 103.

[150] W. Hanefeld, E. Bercin, *Arch. Pharm. (Weinheim Ger.)* **314** (1981) 413.

[151] W. Hanefeld, E. Bercin, *Arch. Pharm. (Weinheim Ger.)* **318** (1985) 848. Wella AG, DE 3 403 147, 1984; *Chem. Abstr.* **104** (1986) 34 095.

[152] B. F. Goodrich Co., US 2 440 095, 1944 (J. E. Jansen). P. L. Ovechkin, L. A. Ignatova, B. V. Unkovskii, *Chem. Heterocycl. Comp. (Engl. Transl.)* 1971, 882.

[153] K. Bodendorf, *J. Prakt. Chem.* **126** (1930) 233. P. Kristian, J. Bernat, *Tetrahedron Lett.* **1968**, 679. Union Carbide Corp., US 2 838 389, 1952. Henkel, DE-AS 1 284 043, 1967; *Chem. Abstr.* **70** (1969) 87 861. A. Rieche, D. Martin, W. Schade, *Arch. Pharm. (Weinheim Ger.)* **296** (1963) 770. Geigy, FR 6521 M 1967; *Chem. Abstr.* **74** (1971) 88 072.

[154] A. Rieche, G. Hilgetag, A. Martini, O. Nejedly, J. Schlegel, *Arch. Pharm. (Weinheim Ger.)* **293** (1960) 957.

[155] Bayer, DE-AS 1 120 801, 1957.

[156] T. Nakai, Y. Ueno, M. Okawara, *Tetrahedron Lett.* 1967, 3831, 3835.

[157] T. Nakai, M. Okawara, *Bull. Chem. Soc. Jpn.* **43** (1970) 1864; **43** (1970) 3882, 3528. K. Hartke, E. Schmidt, M. Castillo, J. Bartulin, *Chem. Ber.* **99** (1966) 3268.

[158] C. Polycarpe, E. Torreilles, L. Giral, A. Babeau, N. H. Tinh, H. Gasparoux, *J. Heterocycl. Chem.* **21** (1984) 1741.

[159] D. J. Rowe, C. D. Garner, J. A. Joule, *J. Chem. Soc. Perkin Trans. I,* 1985, 1907.

[160] BASF, BE 660 941, 1965; *Chem. Abstr.* **64** (1966) 3364 H. H. Eilingsfeld, L. Möbius, *Chem. Ber.* **98** (1965) 1293.

[161] R. L. N. Harris, *Tetrahedron Lett.* 1970, 5217.

[162] *Houben Weyl,* **E 4,** 853. Story Chem. Corp., US 3 923 852, 1974; *Chem. Abstr.* **84** (1974) 58 620.

[163] G. Btotny, *Liebigs Ann. Chem.* 1982, no. 10, 1927–1932.

[164] A. Danopoulos, M. Avouri, S. Paraskewas, *Synthesis* 1985, 682.

[165] J. v. Braun, K. Weissbach, *Ber. Dtsch. Chem. Ges.* **63** (1930) 2846.

[166] Berezniki Polymer, SU 1 154 275-A, 1983; *Chem. Abstr.* **103** (1985) 70 954.

[167] E. T. Hansen, J. J. Petersen, *Synth. Commun.* **14** (1984) 1275.

[168] Farma Khim, Sofia, BE 868 745, 1978 (A. Georgiew, Kh. Dasholov, V. Mikhailov, K. Konstantinova); *Chem. Abstr.* **90** (1979) 168 327.

[169] Research Institute for Medicine and Chemistry, CH 467 225, 1963.

[170] T. Otsu, A. Kuryama, *Polym. J.* **17** (1985) 97. T. Otsu et al., *Polym. Bull.* **7** (1982) no. 1, 45. T. Otsu et al., *Macromolecules* **19** (1986) 287.

[171] Wako Pure Chemical Industries, Ltd., JP 55/3344, 1980; *Chem. Abstr.* **94** (1981) 104 125.

[172] *Ullmann,* 4th ed., **13,** 637. C. F. Martin, T. O. Martin Jr., US 4 339 506, 1981; *Chem. Abstr.* **97** (1982) 183 807. B. Banerjee, "Thiuram-Vulkanisation von Naturkautschuk in Gegenwart von Aminen," *Kautsch. Gummi. Kunstst.* **37** (1984) 21. Osaka Soda KK, JP 59 142 239, 1983; *Chem. Abstr.* **102** (1985) 114 936. Japan Synthetic Rubber, JP 60 047 040, 1983; *Chem. Abstr.* **103** (1985) 7627. Du Pont, JP 56 122 812, 1980; *Chem. Abstr.* **96** (1982) 70 243. P. K. Bandyopadhyay, S. Banerjee, *Kautsch. Gummi, Kunstst.* **32** (1979) 961. Bando Chem. Ind., JP 59 124 942, 1982; *Chem. Abstr.* **101** (1984) 231 782. Japan Synthetic Rubber, JP 60 047 040, 1983; *Chem. Abstr.* **103** (1985) 7627. Asahi Chemical, JP 59 197 449, 1983; *Chem. Abstr.* **102** (1985) 133 696. Osaka Soda KK, JP 59 142 239, 1983; *Chem. Abstr.* **102** (1985) 114 936. Uniroyal Inc., EP 93 500, 1983 (A. E. Crepeau); *Chem. Abstr.* **100** (1984) 35 959. W. R. Grace & Co., US 4 348 515, 1980, 1982 (C. R. Morgan); *Chem. Abstr.* **97** (1982) 199 101.

[173] J. Kelm, D. Gross, "Untersuchung über Zersetzungsprodukte von Vulkanisationsbeschleunigern," *Rubber Chem. Technol.* **58** (1985) 37.

[174] Japan Synthetic Rubber Co. Ltd., GB 2 092 604, 1982 (Y. Nakanishi); NL 82/341, 1981; *Chem. Abstr.* **97** (1982) 183 428.

[175] M. M. Das, D. K. Basu, A. K. Chaudhuri, *Kautsch. Gummi, Kunstst.* **36** (1983) 569. R. D. Taylor, *Rubber Chem. Technol.* **47** (1974) 900. W. Hofmann, *Kautsch. Gummi, Kunstst.* **36** (1983) 1044. N. G. Chiew Sum, *J. Rubber Res. Inst. Malays.* **29** (1981) 146.

[176] Bayer, DE-OS 2 227 338, 1972 (R. Schubart, U. Eholzer, E. Roos, Th. Kempermann); *Chem. Abstr.* **80** (1974) 109 596.

[177] D. M. Smith, *Br. J. Ind. Med.* **41** (1984) 362. J. A. Ruddick, W. H. Newsome, L. Nash, *Teratology* **13** (1976) 263. Du Pont, *Material Safety Data Sheet* (NA-22 F). Bayer, *Sicherheitsdaten-Blatt zu Vulkacit NPV/C.* Du Pont, NA-22 Handling Precautions and Toxicity. M. Bruze, S. Fregert, *Contact Dermatitis* **9** (1983) 208.

[178] G. Eisenbrand, R. Preussmann, B. Spiegelhalder, DE 3 243 141, 1982; *Chem. Abstr.* **101** (1984) 132 259. Sumitomo, EP 184 179, 1985 (A. Okamoto, T. Yamaguchi, H. Okamura, E. Okino).

[179] R. Wegler: *Chemie der Pflanzenschutz- und Schädlingsbekämpfungsmittel*, Springer Verlag, Berlin, Heidelberg, New York, vol. 2, p. 4 (1970); vol. 4, p. 120 (1977); vol. **6,** p. 390 (1981).

[180] K.-H. Büchel: *Pflanzenschutz- und Schädlingsbekämpfungsmittel*, Thieme, Stuttgart 1977. N. M. Golyshin, *Zashch. Rast. (Moscow)* **8** (1979) 31; *Chem. Abstr.* **91** (1979) 205 502. K. Yoneyama, *Nippon Noyaku Gakkaishi* **6** (1981) 452; *Chem. Abstr.* **96** (1982) 157 159. O. A. Korotkova, *Khim. Sel'sk Khoz.* **12** (1974) 869; *Chem. Abstr.* **82** (1975) 81 240. T. Ragemath, *Pesticides* **12** (1978) XIX; *Chem. Abstr.* **89** (1978) 210 231.

[181] I. Nitsche, J. Siemrova, K. Ballschmiter, F. Selenka, *Environ. Qual. Saf. Suppl.* **3** (1975) (Pesticides) 292.

[182] R. Engst, W. Schnaak: *Pesticides*, Thieme Verlag, Stuttgart 1975, p. 62. W. R. Lyman, R. J. Lacoste:
*Pesticides*, Thieme Verlag, Stuttgart 1975, p. 67.A. Kaars Sijpesteijn, J. W. Vonk:
*Pesticides*, Thieme Verlag, Stuttgart 1975, p. 57.

[183] L. Fishbein, *J. Toxicol. Environ. Health* **1** (1975) 713. N. Shindo, *Shokubutsu Bocki*, **30** (1976) 107; *Chem. Abstr.* **85** (1976) 187 557.

[184] R. Engst, W. Schnaak, *Environ. Qual. Saf. Suppl.* **3** (1975) (Pesticides) 62. M. Sh. Vekshtein, M. A. Klisenko, *Probl. Anal. Khim.* **2** (1972) 21; *Chem. Abstr.* **79** (1973) 14 277. H. M. Dekhuijzen, J. W. Vonk, A. Kaars Sijpesteijn, *Pestic. Terminal Residues Invited Pap. Int. Symp.* 1971, 233. *Pure Appl. Chem.* **49** (1977) 675; *Chem. Abstr.* **87** (1977) 195 207.

[185] Food and Agriculture Organization, *WHO Pestic. Residues Ser.* **4** (1975) 545; *Chem. Abstr.* **84** (1976) 103 877.

[186] A. R. C. Hill, J. W. Edmunds, *Anal. Proc. (London)* **19** (1982) 433.

[187] G. Pfeifer, *Mag. Kem. Lapja* **8** (1983) 333.

[188] ICI, GB 916 204, 1960; GB 916 205, 1960.

[189] E. Mutschler: *Arzneimittelwirkungen*, Wissenschaftliche Verlagsgesellschaft, Stuttgart 1981, p. 624.

[190] S. Ebel: *Synthetische Arzneimittel*, Verlag Chemie, Weinheim 1979, p. 571.

[191] Banyu Pharmaceutical Co., Ltd., NL 7 207 077, 1971; DE 2 225 482, 1971; *Chem. Abstr.* **78** (1973) 58 249.

[192] M. V. Korablev, N. M. Kurbat, *Farmakol., Toksikol. (Moscow)* 40 (1977) 230; Chem. Abstr. **86** (1977) 182 718.

[193] Newport Pharmaceuticals International, Inc., EP 58 857, 1982 (G. E. Renoux, M. J. Renoux); *Chem. Abstr.* **97** (1982) 222 930.

[194] G. Renoux, *Trends Pharmacol. Sci.* **2** (1981) 248.

[195] Kumiai Chemical Industry Co., Ltd., JP 57/46 916, 1982; *Chem. Abstr.* **97** (1982) 28 602.

[196] M. Gh. Alwafaie, GB 2 038 626, 1980; *Chem. Abstr.* **94** (1981) 90 344.

[197] M. Morioka, JP 54/12 674, 1974; *Chem. Abstr.* **92** (1980) 135 423.

[198] Colgate-Palmolive Co., US 4 007 281, 1966, 1977 (R. A. Baumann); *Chem. Abstr.* **86** (1977) 177 317. US 3 956 479, 1966, 1976; *Chem. Abstr.* **85** (1976) 83 229.

[199] Wella AG, DE 3 309 400, 1983; *Chem. Abstr.* **102** (1985) 62 252.

[200] Tanabe Seiyaku Co. Ltd., EP 165 017, 1985 (I. Iijiima, Y. Seiga, T. Miyagishima, Y. Matsuka, M. Matsumato).

[201] W. O. Foye, D. H. Kay, P. R. Amin, *J. Pharm. Sci.* **57** (1963) 1793.C. van der Meer, D. W. Bekkum, *Int. J. Radiat. Biol.* **1** (1959) 5. J. R. Piper et al., *J. Med. Chem.* **9** (1966) 911.

[202] M. J. Allalunis-Turner, J. D. Chapman, *Int. J. Radiat. Oncol. Biol. Phys.* **10** (1984) 1569.

[203] J. H. Barnes, M. Fatone, G. F. Esslemont, L. Andrien, E, Bargy, *Eur. J. Med. Chem. Chim. Ther.* **10** (1975) 619.

[204]   W. O. Foye, D. H. Kay, *J. Pharm. Sci.* **57** (1968) 345.
[205]   J. R. Piper, C. R. Springfield jr., Th. P. Johnston, *J. Med. Chem.* **9** (1966) 911.
[206]   J. Barnes, G. Esslemont, P. Holt, *Makromol. Chem.* **176** (1975) 275.
[207]   W. O. Foye, R. S. F. Chu, K. A. Shah, W. H. Parsons, *J. Pharm. Sci.* **60** (1971) 1839.
[208]   T. Nakagawa, Y. Fujiwara, *J. Appl. Polym. Sci.* **20** (1976) 753.
[209]   Canon KK, JP 60 195 731-A, 1984.
[210]   TDK Corp., JP 60 201 988-A, 1984; *Chem. Abstr.* **85** (1976) 126 945.
[211]   Asahi Chemical Industry Co., Ltd., JP 59/135 187, 1984; *Chem. Abstr.* **102** (1985) 176 585. Mitsubishi Paper Mills, Ltd., JP 59/41 296, 1984; *Chem. Abstr.* **100** (1984) 219 119. Honshu Paper Co., Ltd., JP 56/148 587, 1981; *Chem. Abstr.* **97** (1982) 14 828. Mitsubishi Electric. Corp., JP 55/39340, 1980 (K. Enmanji, K. Takahashi, T. Kitagawa); *Chem. Abstr.* **93** (1980) 85 201. Canon KK, JP 51/114 124, 1976 (T. Ohta, S. Togano); *Chem. Abstr.* **89** (1978) 51 405.
[212]   H. W. Reinhard, O. Riessner, H. J. Smolenski, W. Walther, DD 123 282, 1976; *Chem. Abstr.* **87** (1977) 60 796. Orchard Paper Co., BE 631 000, 1962.
[213]   H. Mifune, S. Takada, Y. Akimura, S. Hirano, DE-OS 2 941 428, 1980; *Chem. Abstr.* **94** (1981) 9957. S. Hirano, Y. Takagi, DE-OS 3 203 661, 1982; *Chem. Abstr.* **99** (1983) 46 004.
[214]   Fuji Photo Film Co., Ltd., JP 55/151 635, 1980; *Chem. Abstr.* **94** (1981) 200 769.
[215]   Fuji Photo Film Co., Ltd., JP 58/95 347, 1983; *Chem. Abstr.* **101** (1984) 201 327.
[216]   Mita Industrial Co., Ltd., EP 98 089, 1984 (K. Nakatani, N. Miyakawa, T. Higashiguchi, T. Edo); *Chem. Abstr.* **100** (1984) 129 871.
[217]   Fuji Photo Film KK, DE 3 534 527, 1984; *Chem. Abstr.* **105** (1986) 105 790.
[218]   Canon KK, JP 60 196 775, 1984; *Chem. Abstr.* **104** (1985) 99 511.
[219]   Y. Ohba, JP 50/63 938, 1975; *Chem. Abstr.* **83** (1975) 170 880.
[220]   Fuji Photo Film Co., Ltd., JP 55/26 506, 1980 (M. Yamada, I. Ito, K. Watase); *Chem. Abstr.* **93** (1980) 85 147. Konishiroku Photo Industry Co., Ltd., JP 52/20 832, 1977 (K. Sakamoto, I. Fushiki, S. Koposhi, E. Sakamoto); *Chem. Abstr.* **87** (1977) 175 640.
[221]   Canon KK, JP 50/99 720, 1975 (T. Minegishi); *Chem. Abstr.* **84** (1976) 114 193.
[222]   Canon KK, DE-OS 2 537 935, 1976 (N. Hasegawa, H. Kobayashi, J. Endo, K. Kinjo); *Chem. Abstr.* **85** (1976) 184 825.
[223]   Canon KK, US 4 245 033, 1974 (T. Eida, I. Endo); *Chem. Abstr.* **94** (1981) 183 422.
[224]   Japan Synthetic Rubber Co., Ltd., JP N 54/31 727, 1977 (Y. Hosaka, Y. Haruta, M. Kurokawa, K. Harada); *Chem. Abstr.* **91** (1979) 166 399. Pittsburgh Plate Glass Co., US 2 673 151, 1954 (H. L. Gerhart).
[225]   Agency Ind. Sci. Tech., JP 60 076 740, 1983; *Chem. Abstr.* **103** (1985) 96 392. Agency Ind. Sci. Tech., JP 60 076 735, 1983; *Chem. Abstr.* **103** (1985) 79 496.
[226]   Konishiroku Photo Industry Co., Ltd., DE 2 734 335, 1977 (M. Fujiwhara, S. Matsuo, T. Masukawa, Y. Kaneko, M. Kawasaki); *Chem. Abstr.* **88** (1978) 161 442.
[227]   Fuji Photo Film KK, JP 55 090 943, 1979; *Chem. Abstr.* **94** (1981) 74 729.
[228]   Fuji Photo Film Co., Ltd., JP 50/40 645, 1973; *Chem. Abstr.* **83** (1975) 211 256.
[229]   Horizons, US 4 018 604, 1976, 1977 (P. L. Bachmann); *Chem. Abstr.* **86** (1977) 198 032. Asahi Chemical Industry Co., Ltd., JP 51/58 327, 1976 (T. Shiga, I. Kikuchi, M. Yoshino); *Chem. Abstr.* **85** (1976) 200 565.
[230]   Fuji Photo Film KK, JP N 55 026 506, 1978; *Chem. Abstr.* **93** (1980) 85 147. Konishiroku, JP N 5 220 832, 1975; *Chem. Abstr.* **87** (1977) 175 640.
[231]   Agfa, US 2 124 159, 1934, 1938 (E. Weyde).
[232]   Kodak, US 2 453 346, 1945, 1948 (H. D. Russel).
[233]   Eastman Kodak Comp., US 3 144 336 1961, 1964 (A. H. Herz).

[234] Du Pont, US 3 505 069, 1966, 1970 (J. H. Bigelow); *Chem. Abstr.* **73** (1970) 40 458.
[235] Konishiroku Photo, JP 22 943/69, 1966.
[236] Konishiroku Photo, JP 22 944/69, 1966.
[237] Du Pont, US 3 597 210, 1968, 1971 (J. H. Bigelow); *Chem. Abstr.* **75** (1971) 103 662.
[238] Fuji Photo KK, DE 3 534 527, 1984; *Chem. Abstr.* **105** (1986) 105 790.
[239] Canon, DE-OS 2 117 044, 1971 (E. Inoue, T. Yamase); Chem. Abstr. 76 (1972) 8 953.
[240] Agfa-Gevaert, DE-OS 2 043 205, 1970 (W. Himmelmann, A. v. König, F. Moll, W. Saleck); *Chem. Abstr.* **77** (1972) 54 873. Agfa-Gevaert, DE-OS 2 044 622, 1970 (W. Himmelmann, A. v. König, F. Moll, D. Rücker, W. Saleck); *Chem. Abstr.* **77** (1972) 95 349.
[241] Continental Can., DE-OS 2 505 448, 1975 (G. Pasternack, T. P. Johndal); *Chem. Abstr.* **84** (1976) 46 258.
[242] Du Pont, DE-OS 2 651 941, 1976 (R. K. Blake); *Chem. Abstr.* **87** (1977) 109 399.
[243] Oriental Photo, JP 7 729 611, 1972.
[244] Canon, US 3 718 467, 1971, 1973 (E. Inoue, T. Yamase); *Chem. Abstr.* **78** (1973) 167 050.
[245] Fuji, JP 5 040 645, 1973; *Chem. Abstr.* **83** (1975) 211 256.
[246] Oriental Photo Industrial Co., Ltd., JP 51/111 756, 1976 (T. Koura); *Chem. Abstr.* **87** (1977) 58 112.
[247] Toyota Cent. Res. & Dev., JP 60/110 721-A, 1983; *Chem. Abstr.* **104** (1986) 20 389.
[248] Federal Mogul Corp., US 4 524 185, 1983 (R. F. Hinderer); *Chem. Abstr.* **103** (1985) 124 799.
[249] Unie van Kunstmestfab. BV, EP 116 988, 1983; *Chem. Abstr.* **101** (1984) 173 936.
[250] Y. Kato, O. Katsuki, *Kogyo Yosui* **290** (1982) 37; *Chem. Abstr.* **98** (1983) 95 142.
[251] Olin. Corp., US 4 518 760, 1984 (M. R. Smith, H. B. Cochran); *Chem. Abstr.* **103** (1985) 161 294.
[252] Phillips Petroleum Co., US 4 554 108, 1983; *Chem. Abstr.* **104** (1986) 190 296. Phillips Petroleum Co., US 4554-068-A, 1983; *Chem. Abstr.* **104** (1986) 190 296. Bresson CR, US 4514-293-A, 1984; *Chem. Abstr.* **103** (1985) 40 425.
[253] J. J. Steggerda, J. A. Cras, J. Willemse, *Recl. Trav. Chim. Pays Bas* **100** (1981) 41.
[254] A. M. Bond, R. L. Martin, *Coord. Chem. Rev.* **54** (1984) 23. V. F. Toropova, G. K. Budnikov, N. A. Ulakhovich, *Elektrodyne Protsessy Adsorbtsiya* **2** (1978) 3; *Chem. Abstr.* **91** (1979) 183 779. R. L. Martin, *Coord. Chem. – Invited Lect. Int. Conf. 20th* 1979, 255; *Chem. Abstr.* **94** (1981) 75 848.
[255] J. Garaj, *Proc. Conf. Coord. Chem. 9th* 1983, 83; *Chem. Abstr.* **99** (1983) 61 800.
[256] J. O.Hill, R. J. Magee, *Rev. Inorg. Chem.* **3** (1981) 141.
[257] J. Garaj, V. Kettmann, *Int. Semin. Cryst. Chem. Coord. Organomet. Compd. [Proc.]*, *3rd* 1977, 166; *Chem. Abstr.* **93** (1980) 36 047.
[258] Mitsui-Nisso Urethane KK, *Chem. Abstr.* **96** (1982) 201 081. Mitsui-Nisso Urethane KK, JP 56 135 545, 1980; *Chem. Abstr.* **96** (1982) 53 594. Mitsui-Nisso Urethane KK, JP 57 159 846, 1981; *Chem. Abstr.* **98** (1983) 127 531.
[259] ICI, FR 1 559 120, 1968; *Chem. Abstr.* **71** (1969) 113 737. A. O. Fitton, A. Rigby, R. J. Hurlock, *J. Chem. Soc.* C1969, 230 Uniroyal Inc., DE-OS 1 815 221, 1968; *Chem. Abstr.* **72** (1970) 3218.
[260] E. G. Kolawole, *J. Appl. Polym. Sci.* **27** (1982) 3437.
[261] A. Al-Malaika et al., *J. Appl. Polym. Sci.* **30** (1985) 789.
[262] G. Scott, D. Gilead, DE-OS 2 839 867, 1979; *Chem. Abstr.* **90** (1979) 205 265.
[263] Du Pont, US 4 452 879, 1983 (M. G. Fickes, P. F. Warfield); *Chem. Abstr.* **101** (1984) 63 698.
[264] Japan Synthetic Rubber Co., Ltd., JP 54/31 727, 1977 (Y. Hosaka, Y. Haruta, M. Kurokawa, K. Harada); *Chem. Abstr.* **91** (1979) 166 399.

[265]  New England Nuclear Corp., US 4 411 881, 1982 (N. R. Tzodikov); *Chem. Abstr.* **100** (1984) 39 659.

[266]  A. T. Polishuk, H. H. Farmer, *NLGI Spokesman* **43** (1979) 200; *Chem. Abstr.* **92** (1980) 44 111. Idemitsu Kosan KK, JP 60 084 394, 1983; *Chem. Abstr.* **103** (1985) 180 736. Nitto Chem. Ind. KK, JP 61 028 694, 1984; *Chem. Abstr.* **105** (1986) 100 209. Lengd. Petro Oil WKS, SU 1 065 396, 1982; *Chem. Abstr.* **101** (1984) 38 146. General Sekiyu KK, JP 59/199 796-A, 1983; *Chem. Abstr.* **102** (1985) 169 528. Pennwalt Corp., EP 128 262-A, 1983; *Chem. Abstr.* **102** (1985) 116 430.

[267]  T. Sakurai et al., *Bull. Jpn. Pet. Inst.* **13** (1971) 243.

[268]  VEB Chemiekombinat Bitterfeld, DD 214 855, 1983; *Chem. Abstr.* **103** (85) 106 382.

[269]  Rhône-Poulenc Specialités Chimiques, EP 141 685, 1984 (G. Fabre, P. Perrin); FR 2 550 541, 1983; *Chem. Abstr.* **103** (1985) 55 277.

[270]  Denki Kagaku Kogyo KK, JP 59/053 560, 1982; *Chem. Abstr.* **101** (1984) 112 538.

[271]  K. Gleu, R. Schwab, *Angew. Chem.* **62** (1950) 320. E. Eckert, *Z. Anal. Chem.* **155** (1957) 23. H. Bode, *Z. Anal. Chem.* **143** (1954) 182; **144** (1955) 90, 166. H. Bode, *Z. Anorg. Allg. Chem.* **289** (1957) 207.

[272]  B. Ch. Verma, S. Kumar, *Rev. Anal. Chem.* **4** (1978) 61.

[273]  R. J. Magee, *Rev. Anal. Chem.* **1** (1973) 335. R. J. Magee, J. O. Hill, *Rev. Anal. Chem.* **8** (1985) 5.

[274]  W. D. McFarlane, *Biochem. J.* **26** (1932) 1022. B. Eisler, K. G. Rosdahl, R. Theorell, *Biochem. Z.* **285** (1936) 76. M. Picotti, G. Baldari, *Mikrochim. Acta* **30** (1947) 77.

[275]  R. M. Barnes, *Biol. Trace Chem. Res.* **61** (1984) 93.

[276]  R. Neeb, *Pure Appl. Chem.* **54** (1982) 847.

[277]  Ch. Fu, B. Zuo, *Fen Hsi Hua Hsueh* **9** (1981) 737; *Chem. Abstr.* **97** (1982) 84 123.

[278]  M. V. Korablev, *Farmakol. Toksikol. (Moscow)* **32** (1969) 356; *Chem. Abstr.* **71** (1969) 68 928.

[279]  A. Rolandi, E. De Marinis, M. De Caterina, *Mutat. Res.* **135** (1984) 193.

[280]  O. P. Chepinoga, *Tr. S'ezda Gig. Ukr. SSR. 8th 1970* 1971, 209; *Chem. Abstr.* **77** (1972) 70 953.

[281]  I. Schuphan, Y. Segall, J. D. Rosen, J. E. Casida, *ACS Symp. Ser.* **158** (Sulfur Pestic. Action Metab.), 65 (1981); *Chem. Abstr.* **95** (1981) 126 730.

[282]  A. P. Shitskova, R. A. Ryanzanova; *Vsestoronnii Anal. Okruzh. Prir. Sredy, 3rd* **1978,** 96; *Chem. Abstr.* **93** (1980) 39 117.

[283]  A. Korhonen, K. Hemminki, H. Vainio, *Teratog., Carcinog., Mutagen.* **3** (1983) 163. A. Korhonen, K. Hemminki, H. Vainio, *Scand. J. Work Environ. Health* **9** (1983) 115. A. Korhonen, K. Hemminki, H. Vainio, *Scand. J. Work Environ. Health* **8** (1982) 63. L. Ivanova-Chemishanska, T. Petrova-Vergieva, E. Mirkova, *Eksp. Med. Morfol.* **14** (1975) 29; *Chem. Abstr.* **83** (1975) 158 773. F. B. Kuzan, K. V. Prahlad, *Poult. Sci.* **54** (1975) 1054; *Chem. Abstr.* **84** (1976) 70 028.

[284]  M. Dirimanov, A. Mateeva, *Nauchni Tr. Vissh Selskostop. Inst. Vasil Koralov, Plovdiv* **24** (1979) 125; *Chem. Abstr.* **94** (1981) 1035. H. V. Ghate, L. Mulherkar, *Indian J. Exp. Biol.* **18** (1980) 1040; *Chem. Abstr.* **93** (1980) 232 143. H. V. Ghate, *Toxicol. Lett.* **19** (1983) 253.

[285]  L. Vicari, G. De Dominicis, M. Vito, C. Placida, E. De Marinis, *Boll. Soc. Ital. Biol. Sper.* **61** (1985) 271. H. V. Ghate, J. K. Pal, *Curr. Sci.* **53** (1984) 662. R. D. Short, J. L. Ninor, T. M. Unger, *Report EPA-600/1-80-017,* Order no. PB80-181 175, 28 pp., Chem. Abstr. 93 (1980) 180 594. E. Arias, T. Zavanella, *Bull. Environ. Contam. Toxicol.* **22** (1979) 297. N. Chernoff, R. J. Kavlock, E. H. Rogers, D. B. Carver, S. Murray, *J. Toxicol. Environ. Health* **5** (1979) 821. K. S. Larsson, Cl. Arnander E. Cekanova, M. Kjellberg, *Teratology* **14** (1976) 171. M. A. Pilinskaya, A. I. Kurinnyi, T. I. Kondratenko, *Mol. Mekh. Genet. Protsessov* **295** (1976); *Chem. Abstr.* **87** (1977) 34 142. T. Petrova-Vergieva, *Khig. Zdraveopaz.* **19** (1976) 435; *Chem. Abstr.* **87** (1977) 16 806.

[286]  D. J. Clegg, K. Khera, *Pestic. Environ. Contin. Controversy Pap. Inter Am. Conf. Toxicol. Occup. Med., 8th* 1973, 276; *Chem. Abstr.* **84** (1976) 145 460.

[287]  A. Rannug, U. Rannug, *Chem. Biol. Interact.* **49** (1984) 329. V. A. Kiryushiu, *Gig. Sanit.* 1975, no. 9, 43; *Chem. Abstr.* **83** (1975) 202 555. Y. Shirasu, M. Moriya, H. Tezuka, S. Teramoto, T. Ohta, T. Inoue, *Environ. Sci. Res.* **31** (1984) 617. M. Moriya, T. Ohta, K. Watanabe, T. Miyazawa, K. Kato, Y. Shirasu, *Mutat. Res.* **116** (1983) 185. L. Fishbein, *Environ. Mutagens Carcinog. Proc. Int. Conf. 3rd* 1981, 371; *Chem. Abstr.* **98** (1983) 12 428. Y. Shirasu, M. Moriya, K. Kato, A. Furuhashi, T. Kada, *Mutat. Res.* **40** (1976) 19. M. E. Gerzoni, L. Del Cupolo, I. Ponti, *Riv. Sci. Tecnol. Alimenti Nutr. Um.* **6** (1976) 161. H. C. Sikka, P. Florczyk, *J. Agric. Food Chem.* **26** (1978) 146. F. De Lorenzo, N. Staiano, L. Silengo, R. Cortese, *Cancer Res.* **38** (1978) 13. V. A. Kiryushin, *Gig. Aspekty Okruzh. Zdorov'ya Naseleniya* **196**; *Chem. Abstr.* **89** (1978) 85537. M. I. Kulik, *Genet. Posledstviya Zagryaz. Okruzh. Sredy* **177**; *Chem. Abstr.* **88** (1978) 131 630. M. Moriya, T. Ohta, K. Watanabe, Y. Watanabe, F. Sugiyama, T. Miyazawa, Y. Shirasu, *Cancer Lett. (Shannon Irel.)* **7** (1979) 325. V. Vasudev, N. B. Krishnamurthy, *Mutat. Res.* **77** (1980) 189; *Chem. Abstr.* **92** (1980) 122 984. R. A. Ryazanova, T. V. Gafurova, *Gig. Sanit.* 1980, no. 1, 80; *Chem. Abstr.* **92** (1980) 105 449. M. Tsoneva, V. Georgieva, Chan Chi Lien, *Genet. Sel.* **12** (1979) 118; *Chem. Abstr.* **93** (1980) 144 099. J. D. Rosen, I. Schuphan, Y. Segall, J. E. Casida, *J. Agric. Food Chem.* **28** (1980) 880. Chii-hing Jeang, Gwo-Chen Li, *K'o Hsueh Fa Chan Yueh K'an* **8** (1980) 551; *Chem. Abstr.* **93** (1980) 180 338. Y. Kato, M. Tanaka, K. Umezawa, S. Takayama, *Toxicol. Lett.* **7** (1980) 125. F. Decloitre, G. Hamon, *Mutat. Res.* **79** (1980) 185. K. Hemminki, K. Falck, H. Vainio, *Arch. Toxicol.* **46** (1980) 277. P. Principe, E. Dogliotti, M. Bignami, R. Crebelli, E. Falcone, M. Fabrini, G. Conti, P. Comba, *J. Sci. Food Agric.* **32** (1981) 826. M. Zdzienicka, M. Zielenska, M. Trojanowska, T. Szymczyk, *Bromatol. Chem. Toksykol.* **15** (1982) 83; *Chem. Abstr.* **98** (1983) 29 291.

[288]  C. J. van Leeuwen, J. L. Maas Diepeveen, G. Niebeek, W. H. A. Vergouw, P. S. Griffioen, M. W. Luijken, *Aquat. Toxicol.* **7** (1985) 145. L. Fishbein, *J. Toxicol. Environ. Health* **1** (1976) 713.

[289]  C. J. van Leeuwen, F. Moberts, G. Niebeek, *Aquatic Toxicology* **7** (1985) 165. *Carbamates, Thiocarbamates and Carbazides,* IARC Monographs on the Evaluation of the Carcinogenic Risk of Chemicals to Man, vol. **12,** Lyon 1976.

[290]  A. Hedenstedt, U. Rannug, C. Ramel, C. A. Wachtmeister, *Mutat. Res.* **68** (1979) 313. X. You, Y. Zhou, Y. Hu, *Huan Ching K'o Hsueh* **3** (1982) 39; *Chem. Abstr.* **98** (1983) 84 705. M. Donner, K. Husgafvel-Pursiainen, D. Jenssen, A. Rannug, *Scand. J. Work Environ. Health* **9** (1983) (Suppl. 2), 27.

[291]  A. Hedenstedt, *SGF Publ.* **57** (1981) 39; *Chem. Abstr.* **97** (1982) 105 398.

[292]  M. Donner, *SGF Publ.* **57** (1981) 47; *Chem. Abstr.* **97** (1982) 86 819.

[293]  S. K. Mann, *Environ. Exp. Bot.* **17** (1977) 7.

[294]  A. Rannug, U. Rannug, C. Ramel, *Prog. Clin. Biol. Res.* **141** (1984) 407. S. De Flora, P. Zanacchi, A. Camoirano, C. Bennicelli, G. S. Badolati, *Mutat. Res.* 133 (1984) 161. M. D. Waters et al., *Basic Life Sci.* **21** (1982) 275.

[295]  J. Kelm et al., *Kautsch. Gummi Kunstst.* **36** (1983) 274.

[296]  H. Druckrey, R. Preussmann, S. Ivankovic, D. Schmähl, *Z. Krebsforsch.* **69** (1967) 103. B. L. Pool, B. Bertram, M. Wiessler, *Carcinogenesis (London)* **3** (1982) 563.

# Epoxides

*Individual keywords:* → *Ethylene Oxide;* → *Propylene Oxide.*

GUENTER SIENEL, Peroxid-Chemie GmbH, Höllriegelskreuth, Federal Republic of Germany (Chaps. 2, 3, 4.2 – 7)

ROBERT RIETH, Peroxid-Chemie GmbH, Höllriegelskreuth, Federal Republic of Germany (Chaps. 2, 3, 4.2 – 7)

KENNETH T. ROWBOTTOM, Laporte Industries Ltd., Widnes, Cheshire WA8 0JU, United Kingdom (Section 4.1)

| | | | | |
|---|---|---|---|---|
| 1. | Introduction | 1987 | 3.4. | Epoxidation with Halohydrins 1997 |
| 2. | Reactions of Epoxides | 1988 | 3.5. | Epoxidation with Oxygen .... 1998 |
| 2.1. | Reactions with Compounds Containing Ionizable Hydrogen | 1988 | 3.6. | Epoxidation by Other Methods 1998 |
| | | | 4. | Industrially Important Epoxides . . . . . . . . . . . . . . . . 1999 |
| 2.2. | Reactions with Nitrogen or Phosphorus Compounds. . . . . | 1991 | 4.1. | Epichlorohydrin . . . . . . . . . . 1999 |
| 2.3. | Rearrangements. . . . . . . . . . . | 1991 | 4.2. | Other Important Monoepoxides . . . . . . . . . . . . 2001 |
| 2.4. | Other Reactions. . . . . . . . . . . | 1992 | 4.3. | Functional Epoxides. . . . . . . . 2003 |
| 3. | Production of Epoxides . . . . . | 1992 | 4.4. | Diepoxides . . . . . . . . . . . . . . 2004 |
| 3.1. | Epoxidation with Percarboxylic Acids | 1992 | 4.5. | Polyepoxides . . . . . . . . . . . . . 2005 |
| | | | 5. | Analysis . . . . . . . . . . . . . . . . 2005 |
| 3.2. | Epoxidation with Hydrogen Peroxide | 1995 | 6. | Economic Aspects . . . . . . . . . 2005 |
| | | | 7. | Toxicology. . . . . . . . . . . . . . . 2006 |
| 3.3. | Epoxidation with Hydroperoxides . . . . . . . . . . | 1996 | 8. | References. . . . . . . . . . . . . . . 2007 |

# 1. Introduction

Epoxides, also known as oxiranes, are characterized by the following group:

$$H_2C\overset{O}{-\!\!\!\triangle\!\!\!-}CH-$$

The epoxy group is a highly reactive moiety (see Chap. 2), which makes epoxides an important group of industrial organic intermediates. The most significant members of this group, ethylene oxide and propylene oxide, are treated in separate articles. Epichlorohydrin (Section 4.1) dominates among the raw materials for epoxy resins.

## 2. Reactions of Epoxides

Polarity and ring strain make the oxirane ring highly reactive. Thus, epoxides participate in numerous reactions, which makes these compounds useful building blocks in organic synthesis (see Table 1) [1]–[6]. Often epoxides formed in an initial step react further to provide industrially important products, such as surfactants or detergents (tensides), antistatic- or corrosion-protection agents, additives to laundry deter-gents, lubricating oils, textiles, and cosmetics [1].

## 2.1. Reactions with Compounds Containing Ionizable Hydrogen

Reactions of epoxides with oxygen, sulfur, or carbon anions, usually in the presence of either acid or alkaline catalysts, affords β-hydroxy compounds (see Table 1). Unsymmetrically substituted epoxides may yield two isomers the ratio of which is controlled by pH [7].

The *base-catalyzed reaction* follows an $S_N2$ substitution mechanism; attack of the nucleophile $X^-$ occurs predominantly at the sterically less hindered and more electron-deficient carbon atom. This substitution leads to Walden inversion at this carbon atom. The major product obtained from **1** is isomer **2**:

$$\underset{\mathbf{1}}{\text{R-CH-CH}_2 \atop \backslash\!\!\diagup \atop O} \xrightarrow[S_N 2]{X^-} \text{R-CH-CH}_2 \atop |\quad\ | \atop O^-\ X} \xrightarrow[-OH^-]{H_2O} \underset{\mathbf{2}}{\text{R-CH-CH}_2 \atop |\quad\ | \atop OH\ X}$$

The *acid-catalyzed reaction* proceeds via an intermediate oxonium ion (**3**), which may yield either the same isomer (**2**) obtained by the base-catalyzed reaction via the $S_N2$ substitution mechanism or isomer **4** via the more stable secondary carbocation intermediate by the $S_N1$ substitution mechanism with retention of configuration.

**Table 1.** Reactions of epoxides

| Reaction number | Reagent* | Product | Name |
|---|---|---|---|
| 1) | $H_2O$ | -C(OH)-C(OH)- | vicinal diol |
| 2) | ROH | -C(OH)-C(OR)- | β-hydroxyether |
| 3) | RCOOH | -C(OH)-C(O-C(=O)-R)- | β-hydroxyester |
| 4) | HX | -C(OH)-C(X)- | halohydrin |
| 5) | $H_2S$ | -C(OH)-C(SH)- | β-hydroxythiol |
| 6) | $SO_3^{2-}$ | -C(OH)-C(OSO$_2^-$)- | β-hydroxysulfonate |
| 7) | $H_2N-C(=S)-NH_2$ or RSCN | -C-C- with S bridge | episulfide |
| 8) | HCN | -C(OH)-C(CN)- | β-hydroxynitrile |
| 9) | $R-CH_2-CO_2Et$ | lactone ring with R | lactone |
| 10) | $NH_3$ | -C(OH)-C(NH$_2$)- | monoalkanolamine |
| 11) | $NR_3$/HX | -C(OH)-C(N$^+$R$_3$X$^-$)- | betaine |
| 12) | $PR_3$ | C=C | olefin |
| 13) | $RP(OR')(OH)$ (with P=O) | cyclic phosphonate | phosphonic acid diester |
| 14) |  | -C-C(=O) | aldehyde or ketone |

Table 1. continued

| Reaction number | Reagent* | Product | Name |
|---|---|---|---|
| 15) | (Heat) | C=C-CH₂OH | allyl alcohol |
| 16) | $H_2$/Cat. | -C-C- with HO, H | alcohol |
| 17) | Oxidizing agent | -C(OH)-COOH | α-hydroxycarboxylic acid |
| 18) | $CH_3SCH_3$ | -C(OH)-C(=O)- | α-hydroxyketone |
| 19) | Ph-NCO | oxazolidone ring | N-phenyloxazolidone |
| 20) | $CO_2$ | cyclic carbonate | carbonate |
| 21) | R-C(=O)-R' | 1,3-dioxolane ring | 1,3-dioxolane |
| 22) | $n(-\overset{\diagup O \diagdown}{C-C}-)$ | -C-C-O-(-C-C-O-)$_{n-2}$-C-C-O- | polymers |

* Reagent reacts with epoxide $-\overset{\diagup O \diagdown}{C-C}-$

$$R-CH-CH_2 \text{ (epoxide, 1)} \xrightarrow{H^+} R-\overset{+}{CH}-CH_2-OH \text{ (3)}$$

Scheme: epoxide 1 → via $S_N2$ with HX → R-CH(OH)-CH₂-X⁺H → (−H⁺) → R-CH(OH)-CH₂-X (2); or via H⁺ to protonated epoxide → $S_N1$ → R-CH⁺-CH₂-OH (3) → HX, −H⁺ → R-CHX-CH₂-OH (4).

Thus, acid-catalyzed addition generally yields an isomeric mixture.

Hydrolysis of epoxides in the presence of acid or alkali leads to the corresponding vicinal diols (Table 1, no. 1). Monofunctional alcohols afford β-hydroxyethers (Table 1, no. 2), which may be further reacted to yield polyoxyethylated compounds. β-Hydroxyesters are formed with monocarboxylic acids (Table 1, no. 3). In the presence of basic aluminum oxide, these ring-opening reactions frequently proceed under mild conditions with a good yield [3], [8]. Halohydrins are obtained by reaction with hydrogen halides (Table 1, no. 4). This equilibrium reaction usually proceeds rapidly and smoothly. Base-catalyzed reaction with hydrogen sulfide yields β-hydroxythiols (Table 1, no. 5), which readily react further to bis(β-hydroxyalkyl) sulfides. Sulfites yield β-hydroxysulfonates (Table 1, no. 6). Episulfides (Table 1, no. 7) can be obtained readily with thioureas [9] or thiocyanates [10]. The reactions with hydrocyanic acid to provide β-hydroxynitriles (Table 1, no. 8) and with malonic acid derivatives or acetoacetic ester to form isomeric five-membered lactones (Table 1, no. 9) [11] are examples of the reaction of epoxides with carbanion species.

## 2.2. Reactions with Nitrogen or Phosphorus Compounds

Ammonia reacts with epoxides (Table 1, no. 10) to form mono-, di-, or trialkanolamines, depending on the molar ratio. Tertiary amines ($R_3N$) and tertiary phosphines ($R_3P$) are sufficiently nucleophilic to open the oxirane ring with initial formation of a betaine. Betaines of tertiary amines can be captured under acidic conditions (Table 1, no. 11), whereas the reaction of tertiary phosphines generates phosphine oxide and leads to olefins (Table 1, no. 12). Reaction with phosphonic acid or its esters leads to the formation of cyclic phosphonic acid diesters (Table 1, no. 13) [12].

## 2.3. Rearrangements

Epoxides are capable of undergoing intramolecular rearrangement to form carbonyl compounds (Table 1, no. 14) or allyl alcohols (Table 1, no. 15). The rearrangement resulting in carbonyl compounds is generally initiated by Lewis acids (e.g., $ZnCl_2$, $SnCl_2$, $AlCl_3$, $FeCl_3$, and $TiCl_4$) [13] or Brønsted acids (e.g., $H_2SO_4$, HCOOH, 4-toluenesulfonic acid, HF, HCl, and HI) [14], [15]. Epoxides on cyclic systems may lead to ring enlargement or contraction. Epoxides with a sufficiently acidic hydrogen atom at the oxirane ring may rearrange to carbonyl compounds in the presence of strong bases (e.g., lithium diethylamide, LiBr, LiI, and $LiClO_4$) or transition-metal complexes (e.g., $Mo(CO)_6$ [16] and $Co_2(CO)_8$ [17]). Rearrangement in the presence of strong nonnucleophilic bases (e.g., n-butyl lithium and lithium dialkylamide) may also lead to allyl alcohols [18].

## 2.4. Other Reactions

Reduction of epoxides to alcohols can be accomplished by using a variety of reducing agents. In addition to catalytic hydrogenation (Table 1, no. 16), which is frequently accompanied by deoxygenation of the epoxide, hydrogenolysis with complex metal hydrides (e.g., $LiAlH_4$) or borohydrides is also possible.

Catalytic oxidation with oxygen yields α-hydroxycarboxylic acids (Table 1, no. 17), which may be followed by C—C bond cleavage. Oxidation with dimethyl sulfoxide can be used to prepare α-hydroxyketones (Table 1, no. 18) [19].

If epoxides are reacted with a molecule containing polar double or triple bonds, either saturated or unsaturated five-membered heterocycles are obtained. Cycloaddition with phenylisocyanate affords N-phenyloxazolidones (Table 1, no. 19) [20]; cycloaddition with $CO_2$ yields alkylene carbonates (Table 1, no. 20), and cycloaddition with carbonyl compounds gives 1,3-dioxolanes (Table 1, no. 21) [21].

Because of the high activity of the oxirane ring, epoxides are readily polymerized (Table 1, no. 22). Polymerization of short-chain diepox-ides is industrially important for the production of polymers cross-linked from an epoxide and a hardener. Examples of hardeners include di- and polyamines, polyphenols, di- and polycarboxylic anhydrides, and various di- and tricarboxylic acids [22].

# 3. Production of Epoxides

Epoxides are produced by the addition of oxygen to alkenes. The electrophilic agent may be either molecular or chemically bound oxygen such as percarboxylic acids, hydrogen peroxide, or hydroperoxides.

## 3.1. Epoxidation with Percarboxylic Acids [23]

Of all the available production methods for converting alkenes to epoxides, the most widely used method is the Prilezhaev (Prileschajew) reaction [24].

The *mechanism* proposed by P. D. BARTLETT is generally accepted and involves a bicyclic transition state [25].

Epoxidation is a second-order reaction and is very exothermic (ca. 250 kJ/mol); great

**Table 2.** Relative rates of epoxidation of selected olefins with peracetic acid at 25.8°C

| Olefin | CAS registry number | Formula | Relative rate of epoxidation |
|---|---|---|---|
| Ethylene | [74-85-1] | $CH_2=CH_2$ | 1 |
| Propene | [115-07-1] | $CH_3CH=CH_2$ | |
| Hex-1-ene | [592-41-6] | $CH_3(CH_2)_3CH=CH_2$ | 25 |
| Oct-1-ene | [111-66-0] | $CH_3(CH_2)_5CH=CH_2$ | |
| But-2-ene | [590-18-1] | $CH_3CH=CHCH_3$ | |
| Hex-2-ene | [7688-21-3] | $CH_3CH=CH(CH_2)_2CH_3$ | 500–600 |
| Methyl oleate | [2462-84-2] | $CH_3(CH_2)_7CH=CH(CH_2)_7COOCH_3$ | |
| 2-Methylbut-2-ene | [513-35-9] | $(CH_3)_2C=CHCH_3$ | 6000 |
| Vinyl acetate | [108-05-4] | $CH_3COOCH=CH_2$ | 0.2 |
| Allyl acetate | [591-87-7] | $CH_3COOCH_2CH=CH_2$ | 1–2 |
| Cyclohexene | [110-83-8] | ⬡ | 675 |
| Styrene | [100-42-5] | $C_6H_5CH=CH_2$ | 60 |

care must be taken at all times to ensure safe operation. Safety is especially critical when the speed of epoxidation is fast. Electron-donating groups, e.g., alkyl groups at the double bond carbon atoms, greatly enhance the reaction rate (see Table 2). Electron-withdrawing groups have the opposite effect and may sometimes stop the reaction entirely. 2-Butene reacts considerably faster than propene, whereas allyl chloride reacts slower [26]. Cyclic olefins react faster than dialkyl-substituted open-chain olefins, and substitution by a phenyl group exerts only a mild accelerating effect.

The reactivity of the percarboxylic acid is also subject to electronic effects. Electron-withdrawing groups increase the reaction rate of epoxidation by enhancing its electrophilicity.

Conversely, electron-donating groups reduce the reaction rate. 3-Chloroperbenzoic acid is a more rapid epoxidizing agent than perbenzoic acid, and trifluoroperacetic acid is similarly more reactive than peracetic acid.

Solvents also have a marked effect on the reaction rate. Hydrophilic solvents retard the rate by interfering with intramolecular hydrogen bonding of the percarboxylic acid. Chlorinated and aromatic solvents are more suitable for faster reaction rates.

Epoxidation is stereospecific; cis olefins give cis epoxides and trans olefins give trans epoxides. An excellent review of the stereochemistry of percarboxylic acid epoxidation is available [27].

The selection of a percarboxylic acid for epoxidation is not entirely dependent on the olefin and percarboxylic acid structure. The stability of the epoxide ring in the system that is chosen also influences product yield. Electronic effects of substituents attached to the epoxide ring markedly affect stability.

Carboxylic acids and water attack epoxides to produce mono esters and vicinal diols (Table 1, nos. 3 and 1). Strong acids have a marked catalytic effect on ring cleavage. Loss of epoxide in the presence of acid is generally inevitable, although careful choice of the percarboxylic acid and reaction conditions can minimize loss. A solvent is often employed to moderate the reaction and to facilitate subsequent isolation of the epoxide in high yield. Some evidence also suggests that use of a solvent reduces the extent of side reactions. Several excellent reviews on epoxidation with percarboxylic acids are available [28], [29].

The percarboxylic acid that is employed may be prepared in situ or made in advance (preformed). In both procedures, the percarboxylic acid or hydrogen peroxide is always added slowly to the olefin. The reverse procedure can be hazardous and is not recommended.

**In Situ Percarboxylic Acids.** In this simplest method of epoxidation, the olefinic compound is dissolved in the carboxylic acid and hydrogen peroxide is added. A low level of carboxylic acid (0.25 mol per mole of olefin) is required. Because of water from the aqueous hydrogen peroxide and from the reaction, the process occurs in two phases. Epoxidation with in situ performic acid or in situ peracetic acid (a strong acid catalyst is desirable) are established commercial processes. Major applications are in epoxidation of long-chain olefins [28], vegetable oils (e.g., soya bean oil), polybutadiene, natural and synthetic rubbers, and polyesters.

**Preformed Percarboxylic Acid.** Both equilibrium and solvent-extracted percarboxylic acids are used. Equilibrium peracetic acid contains 1 wt % sulfuric acid, which is neutralized shortly before use in epoxidation. The resulting solution, which is generally called buffered equilibrium peracetic acid, has been used to epoxidize stilbene [30]:

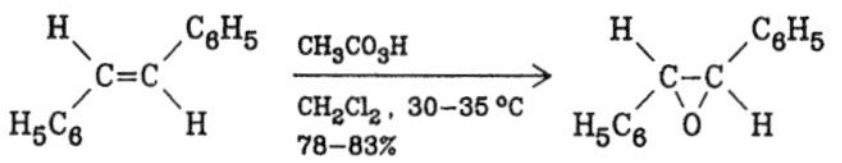

Greater selectivity is obtained by neutralization of acetic acid present in equilibrium peracetic acid, using stoichiometric quantities of sodium carbonate. The best method is to combine buffered peracetic acid with sodium carbonate. This technique has been applied to bicyclic terpenes, e.g., α -pinene [31], [32]:

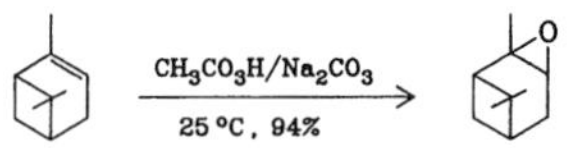

**Other Percarboxylic Acid Systems.** Distilled aqueous peracetic acid can be used in much the same way as equilibrium peracetic acid. Anhydrous peracetic and perpropionic acids (extracted percarboxylic acids) in aprotic solvents, such as ethyl acetate [33] or dichloropropane [34], have been developed. A wide range of other aliphatic and aromatic percarboxylic acids have been used in epoxidation reactions [27], [28].

The commercial availability of 3-chloroperbenzoic acid and the ecologically beneficial magnesium monoperoxyphthalate hexahydrate (INTEROX MMPP) has resulted in wide application [35], [36].

Cholesterol → (MMPP/PTC, CHCl$_3$/H$_2$O, 50 °C, 88%) → epoxide product

Unsaturated esters that are difficult to epoxidize have given good yields with trifluoroperacetic acid [37]. The use of this very strong epoxidizing agent is of academic interest.

Water-soluble olefins can be epoxidized readily in aqueous solution in the presence of tungstic or molybdic acids. The reaction is industrially important for the preparation of glycidol from allyl alcohol [38], [39], and *cis*-epoxysuccinic acid from maleic acid [40].

Benzene perselenic acid, $C_6H_5SeO_3H$, is produced in situ in two-phase systems from catalytic quantities (typically 1%) of benzeneselenic acid and 30% hydrogen peroxide [41], [42].

## 3.2. Epoxidation with Hydrogen Peroxide

**Alkaline Hydrogen Peroxide** [23]. Olefins with electron-withdrawing substituents may be epoxidized by alkaline hydrogen peroxide. The active epoxidizing agent is the hydroperoxy anion, and the generally accepted mechanism for the epoxidation of an $\alpha,\beta$-unsaturated carbonyl compound is as follows:

$$H_2O_2 + OH^- \longrightarrow H_2O + HOO^-$$

[mechanism showing addition of HOO⁻ to C=C–C=O, forming intermediate, then loss of OH⁻ to give epoxide]

These reactions are stereoselective, in contrast to epoxidation with percarboxylic acids, which is stereospecific. Only one isomer is obtained when both isomers of 3-methyl-2-penten-4-one are epoxidized by alkaline hydrogen peroxide [43].

**Akaline Nitrile Hydrogen Peroxide** [23]. In the absence of electron-withdrawing groups adjacent to the double bond, hydrogen peroxide must be activated for epoxidation to take place, Conversion to a percarboxylic acid is one method of achieving this. A similar method of activation is the reaction of nitriles in alkaline media to produce peroxyimidic acids (Payne system) [44].

$$R-C\equiv N + H_2O_2 \xrightarrow{HO^-} R-\underset{\underset{H}{\overset{\overset{NH}{\|}}{C}}}{}-O-OH \xrightarrow{>C=C<} $$
$$>\underset{\underset{O}{\diagdown /}}{C-C}< + R-\overset{\overset{O}{\|}}{C}-NH_2$$

Acetonitrile and benzonitrile are generally used. The reaction is particularly applicable to acid-labile epoxides, e.g., glycidaldehyde diethylacetal [45]. The method has been adapted for large-scale production of hexafluoropropene oxide [46].

**New Catalytic Systems.** Although present methods of epoxidation are versatile, highly efficient systems and simpler processes, particularly for industrial production, are desirable. Many catalytic systems have been developed recently. Various catalytic systems and their reaction mechanisms have been compared with other oxidizing agents in a review article [47].

The catalytic tungstate/phosphate/tetraalkylammonium salt system (QX) is an example of a highly selective system that gives excellent yields [48].

$$>C=C< + H_2O_2 \xrightarrow[H_2O/ClCH_2CH_2Cl]{H^+/WO_4^{2-}/PO_4^{3-}/QX} >\underset{\underset{O}{\diagdown /}}{C-C}< + H_2O$$

1-Dodecene can be epoxidized in 1 h at 70 °C in 87% yield, using this system. Electron-rich double bonds can be converted with 85% $H_2O_2$ and $SeO_2$ in an alcohol–alkali system with equally short reaction time and high selectivity [49].

For a catalyst to be effective in epoxidation reactions, it must be stable, highly reactive, and easily recyclable [50], [51]. Because many catalysts lose activity in the presence of water, water must be removed continuously during the process.

## 3.3. Epoxidation with Hydroperoxides

Catalytic epoxidation of olefins with hydroperoxides gained popularity only after the introduction of asymmetric epoxidation [52]. This method is restricted to allyl alcohols. In addition to reliability and ease of handling, it has the advantage of effective asymmetric induction and predictability of product. For instance, the reaction proceeds as follows depending on the stereochemistry of the diethyl tartrate employed [53]:

$$\underset{R^2\;\overset{\uparrow}{b}\;R^3}{\overset{R^1\;\overset{\downarrow a}{\;}\;CH_2OH}{C=C}} \xrightarrow[CH_2Cl_2,\,-20\,°C]{(CH_3)_3COOH,\;Ti[OCH(CH_3)_2]_4} \underset{R^2\diagup\overset{}{O}\diagdown R^3}{\overset{R^1\diagdown\;\;\diagup CH_2OH}{}} $$

70 - 90%

In the presence of (−)-diethyl D-tartrate (unnatural isomer), the active oxygen attacks from "above" (a) but attacks from "below" (b) in the presence of (+)-diethyl L-tartrate (natural isomer). The optical selectivity of the reaction is $\geq 90\%$.

***tert*-Butyl hydroperoxide** [*75-91-2*] (TBHP) has been employed in catalytic epoxidation of linear and cyclic olefins [54]. High yields are obtained only if the reagents used in the reaction, especially the TBHP reagent, are absolutely dry [54], [55]. Epoxidation with TBHP has been used industrially to produce propylene oxide [56], [57].

Some of the advantages of TBHP as an oxidizing agent in the epoxidation of olefins include the following [47]:

1) high thermal stability
2) safer handling than $H_2O_2$ or percarboxylic acids
3) noncorrosive
4) selective oxidizing agent
5) good solubility in nonpolar solvents
6) neutral pH
7) the byproduct *tert*-butanol can be easily separated by distillation.

## 3.4. Epoxidation with Halohydrins

Hypohalous acids and their salts are suitable agents for the epoxidation of olefins with electron-deficient double bonds. Oxiranes are formed according to the following mechanism [58]:

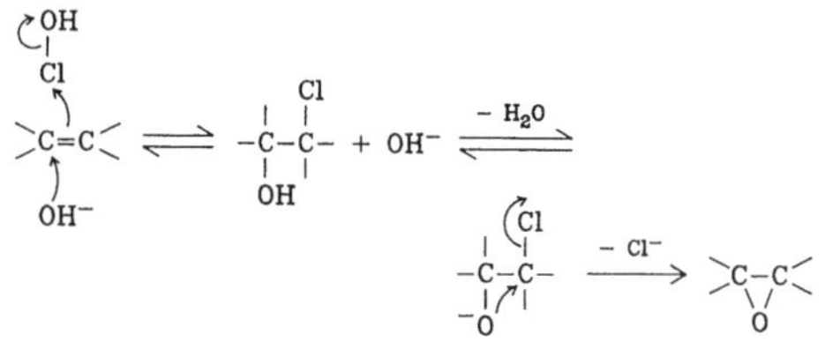

Dehydrohalogenation in the presence of base of the trans halohydrin formed by adding hypohalous acid to the olefin gives an epoxide with an inversion of configuration. Because addition of hypohalous acid already involves an inversion, the olefin undergoes at net double Walden inversion [24].

Aqueous chlorine is the preferred reactant for industrial application. Water-insoluble olefins are either emulsified in water or dissolved in a solvent. Alkali or alkaline-earth hydroxides are used as the base for the dehydrohalogenation reaction. Alkyl substituents on the olefin increase the rate of oxirane formation.

One difference between epoxide production by percarboxylic acids and synthesis via halohydrin is that dichlorides and halogen ethers are byproducts in the latter case. This restricts use of halohydrins to low molecular mass monoxides.

## 3.5. Epoxidation with Oxygen

In contrast to epoxidation processes described so far, selectivity and yield of epoxidation with oxygen are low. The reaction mechanism is presumed to be a radical-chain reaction. The actual oxidizing agent is an intermediate hydroperoxide.

The direct oxidation of olefins with oxygen can be carried out in the presence or absence of catalyst. Compounds from groups 4B–6B of the periodic table (Mo, V, W, Cr, and Ti) are used as catalysts and are highly selective, but have little activity. On the other hand, compounds from groups 1B, 7B, and 8B (Co, Ni, Mn, Cu, Ir, Rh, Pt, and Ru) are more active, but show less selectivity [58].

The following process is important for the industrial production of ethylene oxide (→ Ethylene Oxide):

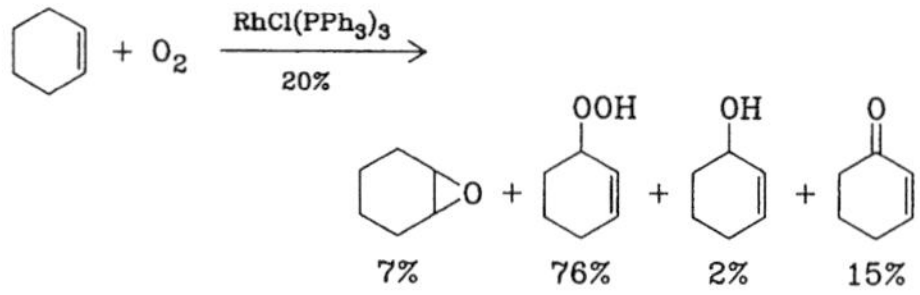

## 3.6. Epoxidation by Other Methods

**Darzens Reaction.** Esters of epoxycarboxylic acid can be prepared from esters of α-halocarboxylic acids and carbonyl compounds in the presence of alkaline condensing agents. The carbanion formed by reaction of the α-halocarboxylic acid ester with alkali reacts with the carbonyl moiety and then stabilizes by forming the glycidol ester [24]:

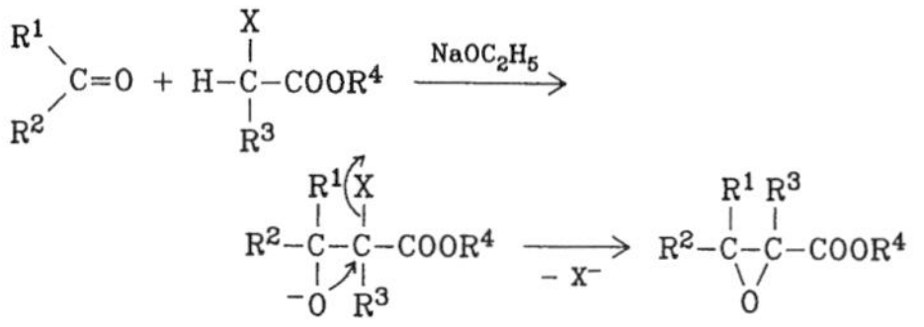

Esters of α-chloro- or α-bromo-α-phenylacetic acid and ketones in tautomeric keto–enol equilibrium do not undergo the Darzens reaction [24].

**Reaction with Epichlorohydrin.** Halohydrins are formed if compounds with active hydrogen atoms are reacted with epichlorohydrin [*106-89-8*] (see Section 4.1). From these halohydrins, subsequent treatment with alkali removes hydrogen halide, and a new epoxide ring is formed [24]:

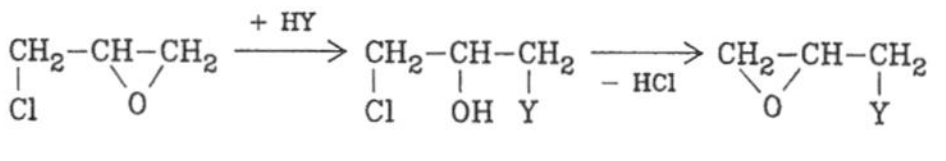

This method can be used, for example, to produce 2,3-epoxyalkyl ethers in a yield of 40–80% [24].

**Reaction with Glycidol.** Glycidol [556-52-5] (5) is an important intermediate for the production of functional epoxides (see Section 4.3) [24]. For example, reaction of phosgene with glycidol yields 2,3-epoxypropyloxy chloroformate [24]:

$$Cl-C(=O)Cl + HO-CH_2-CH-CH_2 \underset{5 \quad \diagdown O \diagup}{\xrightarrow{-HCl}} Cl-C(=O)-O-CH_2-CH-CH_2 \atop \diagdown O \diagup$$

Reaction of glycidol (5) with an isocyanate affords the commercially important glycidyl urethanes (2,3-epoxypropyl urethanes) [59].

$$R-N=C=O + 5 \longrightarrow R-NH-C(=O)-O-CH_2-CH-CH_2 \atop \diagdown O \diagup$$

# 4. Industrially Important Epoxides

## 4.1. Epichlorohydrin

Epichlorohydrin [106-89-8], $C_3H_5ClO$, $M_r$ 92.53, 1-chloro-2,3-epoxypropane, chloromethyl-oxirane, 1,2-epoxy-3-chloropropane, is the most important material for the production of epoxy resins. It is also used for the industrial production of glycerol (→ Glycerol).

$$CH_2-CH-CH_2Cl \atop \diagdown O \diagup$$

**Physical Properties.** Epichlorohydrin, bp 116.56 °C, bp 30–32 °C at 1.35 kPa, mp −48 °C, $d_4^{20}$ 1.18066 g/cm³, $n_D^{20}$ 1.4382, viscosity 1.03 mPa · s at 25 °C, flash point 40.6 °C (Tagliabue Open Cup), autoignition temperature 415.6 °C, heat of vaporization (calculated) 37.9 kJ/mol, heat of combustion 18.943 kJ/mol, specific conductivity $34 \times 10^{-9}$ $\Omega^{-1}$ cm$^{-1}$, vapor density 3.29 (air = 1, at boiling point of epichlorohydrin), vapor pressure 1.333 kPa at 16.6 °C, is a colorless mobile liquid with a characteristic chloroform-like, irritating odor. The lower flammable limit of epichlorohydrin in air is 3.8 vol%; the upper limit is 21.0.

Epichlorohydrin is soluble in low molecular mass alcohols, esters, ethers, ketones, and aromatic hydrocarbons; it is sparingly soluble in water (6.6 wt% at 20 °C). Epichlorohydrin forms binary azeotropes with the following solvents (percent solvent; remaining percentage is epichlorohydrin):

| | |
|---|---|
| water (25%) | bp 88 °C |
| 1-propanol (68%) | bp 96 °C |
| isobutanol (54%) | bp 105 °C |
| isoamyl alcohol (19%) | bp 115 °C |

**Chemical Properties.** As a pure compound, epichlorohydrin is stable. The epoxide and chlorine groups of epichlorohydrin are both potential reactive sites. In reactions with compounds containing active hydrogens, such as alcohols, acids, phenols, amines, and thiols, reaction occurs at the more reactive epoxide group:

$$RH + CH_2\text{-}CH\text{-}CH_2 \longrightarrow RCH_2CH\text{-}CH_2$$
$$\quad\quad\quad \underset{O}{\smile}\ \ Cl \quad\quad\quad\quad\quad\quad OH\ \ Cl$$

where $R = R'O,\ R'CO_2,\ R'_2N$ or $R'S$ and $R' =$ alkyl or aryl

The monochlorohydrin can be converted to a glycidyl compound by reaction with sodium hydroxide or to the glycerol derivative by acid hydrolysis:

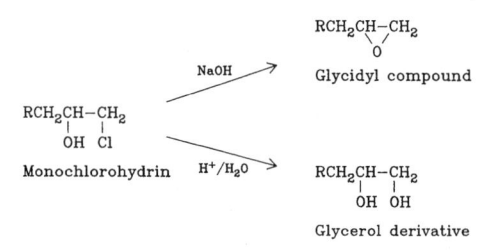

$$RCH_2CH\text{-}CH_2 \xrightarrow{NaOH} RCH_2CH\text{-}CH_2\ (\text{Glycidyl compound})$$
$$RCH_2CH\text{-}CH_2\ (OH, Cl)\ \text{Monochlorohydrin} \xrightarrow{H^+/H_2O} RCH_2CH\text{-}CH_2\ (OH, OH)\ \text{Glycerol derivative}$$

**Uses.** Epichlorohydrin is used mainly for the production of glycerol, unmodified epoxy resins (epichlorohydrin–bisphenol A resin), and elastomers. Several other products made from epichlorohydrin are glycidyl ethers, epichlorohydrin–polyamide resins, and alkyl glyceryl ether sulfonate salts.

**Production** [60]. Epichlorohydrin is made by chlorohydrination of allyl chloride (→ Allyl Compounds), which is obtained by high-temperature chlorination of propylene (Eq. 1). Byproducts of chlorination are *cis*- and *trans*-1,3-dichloropropene and 1,2-dichloropropane. Glycerol dichlorohydrins are made from allyl chloride (Eq. 2), with 1,2,3-trichloropropane being obtained as a byproduct. Finally, epichlorohydrin is produced from the glycerol–dichlorohydrin mixture by treatment with base (Eq. 3). The reactions are as follows:

$$CH_2=CHCH_3 + Cl_2 \xrightarrow{500\ °C} CH_2=CHCH_2Cl + HCl \quad\quad (1)$$

$$2\ CH_2=CHCH_2Cl + 2\ HOCl \longrightarrow \begin{array}{c} HOCH_2CHClCH_2Cl \\ + \\ ClCH_2CHOHCH_2Cl \end{array} \quad (2)$$

$$\xrightarrow{Base} CH_2\text{-}CH\text{-}CH_2Cl \quad\quad (3)$$
$$\quad\quad\quad \underset{O}{\smile}$$

In 1985 Showa Denko KK (Japan) introduced an epichlorohydrin manufacturing process starting from allyl alcohol. Current major producers include Shell Chemical and Dow Chemical.

**Toxicology** [61], [62]. Epichlorohydrin is intensely irritating and moderately toxic by oral, percutaneous, and subcutaneous routes as well as by inhalation of the vapor. The ACGIH has established a TLV-TWA for epichlorohydrin of 2 ppm, 10 mg/m$^2$, and a TLV-STEL of 5 ppm; the compound has a "skin" notation [63]. Direct contact of the liquid with skin or eyes causes severe burns and permanent injury. Inhalation of the vapor produces at least temporary sterility and causes lasting liver, lung, and kidney damage. A moderate degree of carcinogenic activity has been demonstrated in rats when exposed to epichlorohydrin. Persons working with epichlorohydrin require continuous medical supervision.

## 4.2. Other Important Monoepoxides

**1,2-Epoxydodecane** [*2855-19-8*], 1-dodecene oxide, decyloxirane, $C_{12}H_{24}O$, $M_r$ 184.32, is a colorless liquid with a sweet, "soapy" odor; color (Hazen scale) $\leq$ 10; bp 250 °C (101.3 kPa), 125 °C (1.6 kPa), 80 °C (0.17 kPa); mp −5 °C; flash point (Seta flash) 104 °C; $n_D^{20}$ 1.435; $\varrho_4^{20}$ 0.840 g/cm$^3$; $\eta^{20}$ 3.0 mPa · s; insoluble in water and soluble in most organic solvents; spontaneous exothermic reaction with strong acids and bases.

$$CH_2-CH-(CH_2)_9CH_3$$
$$\diagdown O \diagup$$

Typical of long-chain α-epoxides, 1,2-epoxydodecane can enter into numerous reactions (see Chap. 2) [1]. It is produced industrially by established peracid processes either in the presence or absence of solvents [64], [65]. 1,2-Epoxydodecane is stable when stored in cool, dry conditions.

Long-chain α-epoxides are not used as such in most instances but rather as well-defined derivatives. Examples of direct application are as stabilizers for halogen hydrocarbons, as reactive diluents (e.g., for epoxy resins), as resin modifiers, or as a coating material. The various industrial applications of 1,2-epoxydodecane have been reviewed [1].

**Isoamylene oxide** [*5076-19-7*], 2-methyl-2-butene oxide, 2,3-epoxy-2-methylbutane, $C_5H_{10}O$, $M_r$ 86.14, is a clear, colorless liquid with a stringent, ether-like odor; color (Hazen scale) 5; bp 73 °C (101.3 kPa); mp −83 °C; flash point (Seta flash) < 15 °C; $n_D^{20}$ 1.3860; $\varrho_4^{20}$ 0.805 g/cm$^3$ ; $\eta^{20}$ 0.9 mPa·s; 8% soluble in water (unstable) and soluble in polar organic solvents [66].

$$CH_3-\underset{\underset{O}{|}}{\overset{\overset{CH_3}{|}}{C}}-CH-CH_3$$

The major reaction of isoamylene oxide of industrial importance is ring-opening with acid. Isoamylene oxide isomerizes to methyl isopropyl ketone when heated to 250 °C over $Al_2O_3$.

Isoamylene oxide can be prepared by various methods, e.g., via the chlorohydrin route, with percarboxylic acid, or with hydrogen peroxide in a catalyzed reaction.

Isoamylene oxide is used commercially for stabilizing the chlorinated hydrocarbons 1,1,1-trichloroethane and trichloroethylene [67], [68]. These hydrocarbons are frequently used in metal cleaning (including light metal alloys) and must, therefore, be stabilized against decomposition and formation of acidic substances during storage and application. The basic stabilizer system consists of either an organic amine or a nitro compound and a low-molecular mass epoxide, to which additional components may be added.

**α-Pinene oxide** [1686-14-2], 2,3-epoxypinane, 2,7,7-trimethyl-3-oxatricyclo[4.1.1.0$^{2.4}$]-octane, $C_{10}H_{16}O$, $M_r$ 152.24, is a colorless liquid with a characteristic camphor–menthol odor; color (Hazen scale) 5–10; $bp$ 185 °C (101.3 kPa), 82 °C (20 kPa); flash point (Seta flash) 62 °C; $n_D^{20}$ 1.4690; $\varrho_4^{20}$ 1.008 g/cm$^3$ ; $\eta^{25}$ 4.01 mPa · s; $[\alpha]_D^{20}$ (+)-α-pinene oxide + 51 °, $[\alpha]_D^{20}$ (–)-α-pinene oxide –98 °; insoluble in water and soluble in most organic solvents; spontaneous exothermic reaction with concentrated acids.

**6**

α-Pinene oxide is prepared industrially by using percarboxylic acid [31], [32]. Processes based on air oxidation are also known.

An industrially important reaction of optically active α-pinene oxide (**6**) is rearrangement to campholene aldehyde (**7**) (e.g., in the presence of zinc bromide). Subsequent aldol condensation with lower aliphatic aldehydes or ketones (e.g., propionaldehyde, butyraldehyde, or acetone) provides unsaturated aldehydes (e.g., **8**) or ketones that can be reduced to the corresponding saturated alcohols (e.g., **9**) [69]. These are used in the fragrance industry as a sandalwood scent [70].

Hydrolysis of racemic α-pinene oxide and simultaneous rearrangement yields therapeutically active substances (e.g., **10**), which are used in compounding of pharmaceutical preparations [71].

[structure: cyclohexene with OH groups]

**10**

**Styrene oxide** [*96-09-3*], phenylethylene oxide, phenyloxirane, $C_8H_8O$, $M_r$ 120.15, is a colorless liquid with an aromatic odor; *bp* 194 °C (101.3 kPa), 91 °C (3.33 kPa); *mp* −36.7 °C; $\varrho_4^{25}$ 1.0469 g/cm$^3$ ; $n_D^{20}$ 1.5350; vapor pressure < 1 kPa (20 °C);flash point 79 °C (open pan); soluble up to 0.3% in water and miscible in all proportions with toluene, benzene, ether, chlorohydrocarbons, and ethyl acetate.

[structure: phenyl-CH-CH₂ with epoxide O]

Similar to glycidol, styrene oxide is one of the most reactive epoxides [72]. In acid media, the rate of ring opening of the oxirane ring of styrene oxide is higher than for glycidyl ethers.

Styrene oxide is used industrially as a reactive diluent for epoxy resins. It is prepared either via the chlorohydrin route or catalytically with hydrogen peroxide.

## 4.3. Functional Epoxides

**Glycidol** [*556-52-5*], (**5,** p. 1999), glycidyl alcohol, epihydrine alcohol, glycide, glycerin anhydride, 2,3-epoxypropanol, hydroxymethyloxirane, $C_3H_6O_2$, $M_r$ 74.081, is a colorless and odorless liquid; *bp* 162 °C (101.3 kPa), 61 °C (0.13 kPa); *mp* −53 °C; $n_D^{20}$ 1.431; $\varrho_4^{20}$ 1.115 g/ cm$^3$; $\eta^{20}$ 4.0 mPa · s; $\eta^{60}$ 1.70 mPa · s; $[\alpha]_D^{21}$ L-(+)-glycidol +15°; flash point (DIN 51 758) 71 °C; ignition temperature (DIN 51 794) 415 °C; $c_p$ (40 °C) 2.177 kJ kg$^{-1}$ K$^{-1}$; miscible in all proportions with water, alcohols, ketones, esters, ethers, aromatics, etc. and almost insoluble in aliphatic hydrocarbons.

Because it is a bifunctional compound, glycidol tends to condense with itself to form glycerol glycidol ether, which can then react further with itself or with another glycidol molecule. The half-life for the polymerization of glycidol is 7.0 min (at 200 °C) and 4.0 min (at 250 °C). The polymerization energy at 200 °C has been reported to be 1170 kJ/kg. Glycidol is somewhat unstable during storage even in its purest form. The oxirane oxygen content decreases at 25 °C by 2% and at 0 °C by 0.5% per month. When heated, glycidol forms acetol and hydraacrylaldehyde. The latter yields acrolein by elimination of water.

Solutions (up to 70%) in toluene, ketones, esters, ethers, chlorohydrocarbons, and certain alcohols, such as 2-propanol and *tert*-butanol, are relatively stable up to 80 °C. For information on the numerous reactions and the preparation of glycidol, see [59].

**Epoxidized soybean oil** [*8013-07-8*], (ESBO) is a mixture of glycerol esters of epoxidized linoleic, linolenic, and oleic acid. It is a colorless to pale yellow oily liquid; mp 0 °C, $\varrho_4^{20}$ ca. 0.99 g/cm$^3$, $n_D^{20}$ ca. 1.473, $\eta^{20}$ < 600 mPa · s; oxirane oxygen 6.5 %; iodine number < 6 (g of iodine/100 g); saponification number ca. 180 (mg of KOH/g); acid number < 0.4 (mg of KOH/g); iodine color number < 2; flash point < 300 °C (open pan); ignition temperature 430 °C (DIN 51 794); insoluble in water and miscible with toluene, benzene, chlorohydrocarbons, and ethyl acetate.

Epoxidized soybean oil is prepared commercially by the in situ peroxycarboxylic acid method either with or without a solvent. Flow charts for the continuous solventless processes used by Du Pont and Henkel may serve as representative examples of all current production methods [73], [74].

The ESBO market is second to that of epoxy resins; ESBO is used mainly as a plasticizer and stabilizer for poly(vinyl chloride) (PVC). Use as a costabilizer predominates in Europe.

## 4.4. Diepoxides

3,4-Epoxycyclohexylmethyl-3',4'-epoxycyclohexane carboxylate [*2386-87-0*], (**13**), 3,4-epoxytetrahydrobenzyl-3',4'-epoxytetrahydrobenzoate, 3,4-epoxycyclohexanecarboxylic acid-3',4'-epoxycyclohexylmethyl ester, 7-oxabicyclo[4.1.0]heptane-3-carboxylic acid-7'-oxabicyclo[4.1.0]hept-3'-ylmethyl ester, $C_{14}H_{20}O_4$, $M_r$ 252.3, is a clear, colorless, viscous liquid, almost odorless; color (Hazen scale) 5; bp 350 °C (10.13 kPa); glass transition temperature −16 to −18 °C; flash point (Seta flash) 197 °C; $n_D^{20}$ 1.4977; $\varrho_4^{20}$ 1.1680 g/cm$^3$ ; $\eta^{20}$ 255 mPa · s, $\eta^{40}$ 100 mPa · s, $\eta^{80}$ 25 mPa · s; insoluble in water and miscible with polar organic solvents.

*Production.* If 1,2,5,6-tetrahydrobenzaldehyde (**11**) (obtained via the Diels–Alder reaction [75]) is used as the starting material, 3-cyclohexenylmethyl-3'-cyclohexene carboxylate (**12**) is formed in the presence of a catalyst (e.g., aluminum *sec*-butoxide) according to the Tishchenko reaction.

Compound **12** can be epoxidized to the product (**13**) by using the well-established peracid process [76].

This material is a low-viscosity bifunctional epoxy resin of great industrial importance because of its cross-linking capabilities, in most instances with cycloaliphatic dicarboxylic anhydrides. This diepoxide is used as a casting resin that can be hardened either thermally or by UV radiation. It includes among its distinguishing features good dielectric characteristics, excellent stability to environmental conditions, and shape retention when heated. It has also been used as a stabilizer in acid-releasing systems.

## 4.5. Polyepoxides

Polybutadiene oxide, epoxidized polyoils, $M_r$ ca. 900–5000, are clear to pale yellow liquids, odorless to slightly sweet odor; degree of epoxidation 10–50%; $\eta^{20}$ 1300–17 000 mPa · s; flash point (Seta flash) 110 °C; insoluble in water, low solubility in toluene, benzene, and ethyl acetate, and miscible with tetrahydrofuran and chlorohydrocarbons.

$$R\text{---}(CH_2\text{---}CH\text{---}CH\text{---}CH_2CH_2CHCH_2CH=CHCH_2)_n\text{---}R'$$
$$\text{\hspace{2em}}\underset{O}{\diagdown\diagup}\text{\hspace{4em}}|$$
$$\text{\hspace{6em}}CH=CH_2$$

Polybutadiene oxides are prepared by in situ percarboxylic acid processes and are used to manufacture polymers that are used as architectural sealants and in the electronics industry.

## 5. Analysis

Titrimetric methods are generally used for the quantitative determination of epoxide groups. The method is based on the quick ring-opening reaction of the epoxide groups to form halohydrins when the epoxide is reacted with hydrohalides [77], [78].

When the direct titration method with hydrogen bromide in glacial acetic acid is used, maintenance of a constant HBr titer is difficult [79]. A variant of this method involves generating hydrogen bromide from a quaternary ammonium salt (e.g., tetraethylammonium bromide) during the titration by the action of perchloric acid in glacial acetic acid [80]. This very reliable method can also be used for functional epoxides. Peroxides do not interfere. Other methods include titration with hydrogen chloride in various solvents [81].

## 6. Economic Aspects

Reliable market data are available only for certain epoxides, such as epichlorohydrin and ESBO. In 1983, United States production of ESBO was 40 000 t, the same as that in Europe [82]. The bulk prices in the United States ranged between $ 1.52 and 1.59 per kg in 1984 [82] and were ca. DM 3.60 per kg in the Federal Republic of Germany in 1986.

2005

The most important producers of ESBO are Argus, FMC, Ferro, Hall, Lankro, Rohm & Haas, and Viking (United States); and Air Liquide, AKZO, Ciba-Geigy, Henkel, PCUK, and VEB-Greiz (Europe).

The demand for other epoxides is difficult to estimate because much of the production never reaches the open market, but is used in-house to produce derived products.

# 7. Toxicology

Epoxides are not only extremely active in chemical syntheses, but are also equally active in the human body. By opening of the epoxide ring, they can bind in vivo to DNA, RNA, and proteins and are, thus, classified as alkylating agents. Low molecular mass linear epoxides are especially active and readily permeate to target areas in the cell where they can, for example, alkylate DNA [83].

Other chemicals can be metabolized to epoxides by the body. Epoxides of benzopyrenes and vinyl chloride formed in the body are considered to be the actual carcinogenic intermediates of these substances [84], [85].

In addition to the potential mutagenic properties of the epoxides [86], all but epoxidized fatty acids can cause skin inflammation [87] and allergic contact eczema.

**Linear Epoxides.** The mutagenicity of $\alpha$-epoxides decreases with increasing chain length. For instance, propylene oxide is 20 times more mutagenic than 1-hexene oxide. No further increase takes place beyond 1-octene oxide [88]. However, electrophilic substituents increase the mutagenicity.

**Cyclic Epoxides.** The mutagenicities of a few cyclic epoxides have the following order: vinylcyclohexene monoxide >> cyclopentene oxide > cyclohexene oxide > norbornene oxide [88], [89].

**Diepoxides.** The physiological effects of diepoxides were compared with those of nitrogen mustards [90]. Both 1,3-diepoxybutane and vinylcyclohexene dioxide gave rise to sarcoma in mammals following oral administration at 250 mg/kg [91]. The mutagenicities can be arranged in the following order: 1,3-diepoxybutane > 1,7-diepoxyoctane >>> 1,5-diepoxycyclooctane [88]. The last compound was shown to be neither carcinogenic [92] nor active in the mutagen assay [89].

**Literature References.**

*Mutagenic activity:*
cyclohexene oxide [93]; styrene oxide and 4-methylstyrene oxide [94]–[97]; vinylcyclohexene monoxide and dioxide [95]; cis-stilbene oxide and 2-methyl-1,3-diepoxybutane [93].

*Carcinogenic activity:*
1,3-diepoxybutane [86], [87], [93]; 1,6-diepoxyhexane [87], [98]; 1,7-diepoxyoctane [94]; styrene oxide [87], [92], [98]; vinylcyclohexene dioxide [87], [92]; limonene dioxide [84], [87].

**Specific Toxicity Data** (threshold limit values) [87].

1-Butene oxide: $LD_{50}$ 0.5 g/kg, $\leq 400$ ppm (in humans).
1,3-Diepoxybutane: $LD_{50}$ 0.078 g/kg (sarcoma in rats), $\leq 1$ ppm (in humans).
Styrene oxide: $LD_{50}$ 2 g/kg (lymphoma in rats).
Vinylcyclohexene monoxide: $LD_{50}$ 2.83 g/kg, $\leq 50$ ppm (in humans).
Vinylcyclohexene dioxide: $LD_{50}$ 2.83 g/kg (mutagenic, teratogenic in animal experiments), $\leq 1$ ppm (in humans).
Caryophyllene oxide: $LD_{50} > 5$ g/kg (negative in Ames test).
ESBO [99]: $LD_{50}$ 22.5 mL/kg.

# 8. References

[1] Peroxid-Chemie: *A 0.1.4., Langkettige Alpha-Epoxide,* Chapters 1–4, Höllriegelskreuth, 1981, pp. 1–29.
[2] J. G. Smith, *Synthesis* 1984, 629–656.
[3] G. H. Posner, D. Z. Rogers, *J. Am. Chem. Soc.* **99** (1977) 8208–8214.
[4] G. H. Posner, D. Z. Rogers, *J. Am. Chem. Soc.* **99** (1977) 8214–8218.
[5] M. Bischoff, U. Zeidler, H. Baumann, *Fette, Seifen, Anstrichm.* **79** (1977) 131–135.
[6] B. Werdelmann, *Fette, Seifen, Anstrichm.* **76** (1974) 1–8.
[7] R. E. Parker, N. S. Isaacs, *Chem. Rev.* **59** (1959) 737–799.
[8] G. H. Posner, M. Hulce, R. K. Rose, *Synth. Commun.* **11** (1981) 737–741.
[9] C. G. Moore, H. J. Porter, *J. Chem. Soc.* 1958, 2062–2064.
[10] M. O. Brimeyer, A. Merota, S. Quici, A. Nigan et al., *J. Org. Chem.* **54** (1980) 4254–4255.
[11] S. Torii, T. Okamoto, T. Oida, *J. Org. Chem.* **42** (1978) 2294–2296.
[12] H. Gross, B. Costisella DD 108 3051972.
[13] G. Kolaczinski, R. Mehren, W. Stein, *Fette, Seifen, Anstrichm.* **73** (1971) 553–557.
[14] K. Wiechert, P. Mohr, *Z. Chem.* **7** (1967) 229–230.
[15] H. Rutzen, *Fette, Seifen, Anstrichm.* **86** (1984) 109.
[16] H. Alper, D. Des Roches, T. Durst, R. Legault, *J. Org. Chem.* **41** (1976) 3611–3613.
[17] J. L. Eisenmann, *J. Org. Chem.* **27** (1962) 2706.
[18] S. Murata, M. Suzuki, R. Noyori, *J. Am. Chem. Soc.* **101** (1979) 2738–2739.
[19] T. Cohen, T. Tsuji, *J. Org. Chem.* **26** (1961) 1681.
[20] D. Braun, J. Weinert, *Justus Liebigs Ann. Chem.* 1979, 200–218.
[21] R. P. Hanzlik, M. Leinwetter, *J. Org. Chem.* **43** (1978) 438–440.
[22] H. Batzer, F. Lohse, R. Schmid, *Angew. Makromol. Chem.* **29/30** (1973) 349–351.
[23] Peroxid-Chemie: *A 0.1.2, Peroxygen Compounds in Organic Synthesis, Epoxidation,* Chapter 6, Höllriegelskreuth, 1981 pp. 11–17.

[24] J. Falbe: "C—O Verbindungen," in F. Korte: *Methodicum Chimicum,* **5** G. Thieme Verlag, Stuttgart 1975, pp. 131–201.

[25] P. D. Bartlett, *Rec. Chem. Prog.* **11** (1950) 47–50.

[26] D. Swern, *J. Am. Chem. Soc.* **69** (1947) 1692–1698.

[27] B. Plesnicar: "Oxidation with Peroxy Acids and Other Peroxides," in N. S. Trahanovsky: *Oxidation in Organic Chemistry,* **5,** Part C, Chapter 3, Academic Press, New York 1978, pp. 211–294.

[28] D. Swern: "Organic Peroxy Acids as Oxidizing Agent: Epoxidation,": in D. Swern: *Organic Peroxides,* **2,** 5, Wiley-Interscience, New York 1971, pp. 355–533.

[29] D. Swern: "Epoxidation and Hydroxylation of Ethylenic Compounds with Organic Peracids," in D. Swern: *Organic Reactions,* vol. **7,** J. Wiley & Sons, New York 1953, Chapter 7, pp. 378–433.

[30] D. J. Reif, H. O. House, *Org. Synth. Coll.* **4** (1963) 860.

[31] Hoechst, DE-OS 2 835 940, 1978 (H. Häberlein, F. Scheidl).

[32] Peroxid-Chemie, EP 55 387, 1981 (W. Edl., G. R. Sienel).

[33] Union Carbide, DE-AS 1 216 306, 1959 (B. Phillips, P. S. Starcher).

[34] Interox, GB-P 1 535 313, 1978.

[35] R. N. McDonald, R. N. Steppel, J. E. Dorsey, *Org. Synth.* **50** (1970) 15–18.

[36] V. I. Routledge, *Spec. Chem.* **6** (1986, May) 25–28.

[37] W. D. Emmons, G. B. Lucas, *J. Am. Chem. Soc.* **77** (1955) 2287–2288.

[38] Columbia-Southern, GB-P 837 464, 1957.

[39] W. M. Weigert, A. Kleemann, G. Schreyer, *Chem. Ztg.* **99** (1975) 19–25.

[40] Shell Development Co., US 2 833 787, 1955.

[41] H. J. Reich, F. Chow, S. L. Peake, *Synthesis* 1978, 299–301.

[42] M. Hori, K. B. Sharpless, *J. Org. Chem.* **43** (1978) 1689–1697.

[43] H. O. House, R. S. Ro, *J. Am. Chem. Soc.* **80** (1958) 2428–2433.

[44] G. B. Payne, P. H. Deming, P. H. Williams, *J. Org. Chem.* **26** (1961) 659–663. G. B. Payne, *Tetrahedron* **18** (1962) 763–765.

[45] R. L. Rowland, A. Rodgman, J. N. Schumacher, *J. Org. Chem.* **29** (1964) 16–21.

[46] Hoechst, DE-AS 2 557 655, 1975 (R. A. Sulzbach, F. Heller).

[47] R. Sheldon, *Bull. Soc. Chim. Belg.* **94** (1985) 651–670.

[48] C. Venturello, E. Alneri, M. Ricci, *J. Org. Chem.* **48** (1983) 3831–3833. Instituto Guido Donegani, GB-A 2 055 821, 1980.

[49] Interox, EP-P 68 564, 1982 (A. Lecloux, F. Legrand, C. Declerck).

[50] Ugine Kuhlmann, DE-OS 2 605 041, 1976 (M. Pralus, J.-P. Schirmann, S.-Y. Delavarenne); DE-OS 2 752 626, 1977 (J.-P. Schirmann, S.-Y. Delavarenne).

[51] M. Pralus, J. C. Lecoq, J. P Schirmann, *Fundam. Res. Homogeneous. Catal.* **3** (1979) 327–343.

[52] T. Katsuki, K. B. Sharpless, *J. Am. Chem. Soc.* **102** (1980) 5974–5976.K. B. Sharpless, *Chemtech.* **15** (1985) 692–700.

[53] K. B. Sharpless, S. S. Woodard, M. G. Finn, *Pure Appl. Chem.* **55** (1983) 1823–1836.

[54] K. B. Sharpless, T. R. Verhoeven, *Aldrichimica Acta* **12** (1979) 63–74. J. M. Klunder, M. Caron, M. Uchiyama, K. B. Sharpless, *J. Org. Chem.* **50** (1985) 912–915.

[55] J. G. Hill, B. E. Rossiter, K. B. Sharpless, *J. Org. Chem.* **48** (1983) 3607–3608. Massachusetts Institute of Technology, EP-A 197 766, 1986 (R. M. Hanson, S. Y. Ko, K. B. Sharpless).

[56] H. Waldman, *Compendium 75/76 Ergänzungsband der Zeitschrift Erdöl und Kohle, Erdgas, Petrochemie,* 1975 pp. 306–317.

[57] R. Landau, G. A. Sullivan, D. Brown, *Chem. Technol.* 1979, 602–607.

[58]   M. Bartok, K. L. Lang: "Oxiranes," in S. Patai: *The Chemistry of Ethers, Crown Ethers, Hydroxyl Groups and Their Sulphur Analogues,* J. Wiley & Sons, New York 1980, Part 2, Chapter 14, pp. 609–879.

[59]   A. Kleemann, R. Wagner: *Glycidol,* Dr. Alfred Hüthig Verlag, Heidelberg 1981, pp. 81–83.

[60]   K. Wheeler, T. Ruess, S. Takahashi: "Epichlorohydrin," *Chemical Economics Handbook Marketing Research Report,* Stanford Research Institute International, Menlo Park, Calif., 1987, pp. 642.3021 A–642.3021 W.

[61]   N. I. Sax: *Dangerous Properties of Industrial Materials,* 6th ed., Van Nostrand Reinhold Co., New York 1984, pp. 709–710.

[62]   C. Hine, V. K. Rowe, E. R. White, K. I. Darmer et al., in G. D. Glayton, F. E. Clayton (eds.): *Patty's Industrial Hygiene and Toxicology,* 3rd revised ed., vol. **2 A,** Wiley-Interscience, New York 1981, pp. 2242–2247.

[63]   ACGIH (ed.): *Threshold Limit Values (TLV) and Biological Exposure Indices,* ACGIH, Cincinnati, Ohio, 1986–1987.

[64]   Degussa, EP 32 989, 1980 (G. Käbisch, R. Trübe, H. Wittmann, S. Raupach et al.).

[65]   Hoechst, DE-OS 2 436 817, 1974 (H. Häberlein, H. Korbanka, G. Nowy).

[66]   *Beilstein,* **17,** 13; **17 (2),** 22; **17 (3), (4),** 71.

[67]   Solvay, EP 62 952, 1982 (M. Servais, R. Crochet).

[68]   Solvay, EP 44 111, 1981 (R. Walraevens, M. Servais).

[69]   Dragoco, DE-OS 2 827 957, 1978 (E. J. Brunke, E. Klein).

[70]   VEB-Miltitz, DE-OS 1 922 391, 1969 (M. Mühlstädt, W. Dollase, M. Herrmann, G. Feustel).

[71]   C. Corvi, BE 764 323, 1971.

[72]   Dow Chemical, Styrene Oxide, Technical Service & Development, Midland, Mich., 1958.

[73]   A. F. Chadwick, *J. Am. Oil Chem. Soc.* **35** (1958) 355.

[74]   Henkel, DE-OS 3 320 219, 1983 (G. Dieckelmann, K. Echwert, L. Jeromin, E. Peukert et al.).

[75]   H. Batzer, E. Nikles, *Chimia* **16** (1962) 57.

[76]   Union Carbide, US 2 716 123, 1953.

[77]   B. Dobinson, W. Hofmann, B. P. Stark: *The Determination of Epoxide Groups,* Pergamon Press, Oxford 1969, pp. 1–79.

[78]   J. L. Jungnickel, E. D. Peters, A. Polgar, F. T. Weiss: *Organic Analysis,* vol. **1,** Interscience, New York 1953, p. 127.

[79]   A. J. Durbetaki, *Anal. Chem.* **28** (1956) 2000–2001.

[80]   R. R. Jay, *Anal. Chem.* **36** (1964) 667–668.

[81]   W. B. Brown, *J. Soc. Chem. Ind.* **55** (1936) 321 T.

[82]   Predicast-Datenbank-Recherche, *Chem. Mark. Rep.* **225** (1984) 26.

[83]   L. Migliore, A. M. Rossi, N. Loprieno, *Mutat. Res.* **102** (1982) 425–437.

[84]   P. Kotin, H. L. Falk, *Radiat. Res.* 1963, Suppl. 3, 193–211.

[85]   W. K. Lutz: "Chemische Karzinogenese: Biochemische Aspekte," Blockkurs in Toxikologie, Institut für Toxikologie der Eidgenössischen Technischen Hoch-schule (ETH) und Univ. Zürich, Schwerzenbach, Mar. 1987.

[86]   L. Ehrenberg, S. Hussain, *Mutat. Res.* **86** (1981) 1–113.

[87]   C. Hine, V. R. Rowe, E. R. White, K. I. Darmer et al. in G. D. Clayton, F. E. Clayton (eds.): *Patty's Industrial Hygiene and Toxicology,* 3rd ed., vol. **2 A,** Wiley-Interscience, New York 1981, pp. 2141–2257.

[88]   C. E. Voogd, J. J. von der Stel, J. J. J A. A. Jacobs. *Mutat. Res.* **89** (1981) 269–282.

[89]   S. W. Frantz, J. E. Sinsheimer, *Mutat. Res.* **90** (1981) 67–78.

[90]   H. P. Kaufmann, G. Hauschild, R. Schickel, *Fette, Seifen, Anstrichm.* **63** (1961) 239–241.

[91] J. A. Hendry, R. F. Homer, F. L. Rose, A. L. Walpole, *Br. J. Pharmacol.* **6** (1951) 235–255.
[92] M. M. Manson, *Br. J. Ind. Med.* **37** (1980) 317–336.
[93] P. G. Gervasi, L. Citti, M. del Monte, V. Longo et al., *Mutat. Res.* **156** (1985) 77–82.
[94] G. Turchi, S. Bonatti, L. Citti, P. G. Gervasi, *Mutat. Res.* **83** (1981) 419–430.
[95] H. Norppa, K. Hemminki, M. Sorsa, H. Vainio, *Mutat. Res.* **91** (1981) 243–250.
[96] S. de Flora, R. Koch, K. Strobel, M. Nagel: *Toxico-logical Environmental Chemistry*, vol. **10**, Gordon & Breach Science Publ., London 1985, pp. 157–170.
[97] K. Sugiura, M. Goto, *Chem.-Biol. Interact.* **35** (1981) 71–91.
[98] B. L. van Duuren, *Ann. N.Y. Acad. Sci.* **163** (1969) 633–650.
[99] C. S. Weil, N. Condra, C. Haun, J. A. Striegel, *J. Am. Ind. Hyg. Assoc.* **24** (1963) 305–325.

# Esters, Organic

WILHELM RIEMENSCHNEIDER, from Hoechst AG, Frankfurt, Federal Republic of Germany

| | | | | | |
|---|---|---|---|---|---|
| 1. | Introduction | 2011 | 5.5. | Transesterification | 2028 |
| 2. | Physical Properties | 2013 | 5.6. | Acylation, of alcohols with ketenes | 2028 |
| 3. | Chemical Properties | 2015 | 5.7. | Carbonylation | 2029 |
| 3.1. | Hydrolysis | 2016 | 5.8. | Condensation of Aldehydes | 2029 |
| 3.2. | Transesterification | 2017 | 5.9. | Alcoholysis of Nitriles | 2030 |
| 3.3. | Ammonolysis | 2018 | 5.10. | Acylation of Olefins | 2031 |
| 3.4. | Reduction | 2019 | 5.11. | Ethoxylation of Carboxylic Acids | 2032 |
| 3.5. | Claisen Condensation | 2020 | 5.12. | Orthocarboxylic Acid Esters | 2033 |
| 3.6. | Pyrolysis | 2021 | 5.13. | Lactones | 2033 |
| 4. | Natural Sources | 2021 | 6. | Environmental Protection | 2034 |
| 5. | Production | 2022 | 7. | Quality Specifications and Analysis | 2034 |
| 5.1. | Esterification of Carboxylic Acids | 2022 | 8. | Storage and Transportation | 2035 |
| 5.1.1. | Catalysts | 2023 | 9. | Uses and Economic Aspects | 2036 |
| 5.1.2. | Industrial Processes | 2025 | 10. | Toxicology and Occupational Health | 2042 |
| 5.2. | Alkylation of Metal Carboxylates | 2026 | 11. | References | 2044 |
| 5.3. | Acylation with Acyl Halides | 2027 | | | |
| 5.4. | Acylation with Carboxylic Anhydrides | 2027 | | | |

# 1. Introduction

Organic esters are compounds with the general formula

$$R^1-C\underset{O-R^2}{\overset{O}{\diagup\!\!\!\!\diagdown}}$$

where $R^1$ and $R^2$ represent either the same or different aliphatic, aromatic, or heterocyclic groups. Esters are classified as carboxylic acid derivatives; they are prepared mainly by the reaction of a carboxylic acid and an alcohol (esterification).

Ortho esters are derived from ortho acids:

$$R^1-C{\begin{matrix}\nearrow O-R^2 \\ -O-R^2 \\ \searrow O-R^2\end{matrix}}$$

Ortho esters can be hydrolyzed to carboxylic acids and alcohols, but their properties are more similar to those of acetals. They are dealt with briefly in Section 5.12.

If a molecule possesses both a carboxyl and a hydroxyl group, it may form either polyesters:

$$\left[-(CH_2)_n-O-\overset{O}{\underset{\|}{C}}-(CH_2)_n-\overset{O}{\underset{\|}{C}}-O-\right]_x$$

or cyclic internal esters (lactones) (Section 5.13):

$$(\overset{\frown}{CR_2})_n\overset{O}{\underset{C=O}{|}}$$

Ring opening reactions of lactones lead to the same type of polyesters. Another type of polyesters with different characteristics is usually produced through condensation of dicarboxylic or polycarboxylic acids with di- or polyhydric alcohols.

$$\left[-(CH_2)_n-O-\overset{O}{\underset{\|}{C}}-(CH_2)_m-\overset{O}{\underset{\|}{C}}-O-\right]_x$$

Organic esters are of considerable economic importance. Esters of glycerol with fatty acids (glycerides) occur naturally in large quantities in fats and fatty oils (see Chap. 4 and → Fats and Fatty Oils); they are used predominantly in foods, but also as raw materials, especially in the production of surfactants.

Cellulose acetate is used on a large scale in fiber production. Synthetic esters are of increasing importance in many applications, e.g., fibers, films, adhesives, and plastics articles. Various synthetic esters have been developed for specific uses, e.g., solvents, extractants, plasticizers, lubricants, lubricant additives, and lacquer bases. A variety of volatile esters are used as aromatic materials in perfumes, cosmetics, and foods.

This article gives a general survey of production methods, properties, and uses of esters. The commercially most important esters are treated under separate keywords: → Formic Acid; Acetic Acid; Propionic Acid and Derivatives; Adipic Acid; Acrylic Acid and Derivatives; Methacrylic Acid and Derivatives; Benzoic Acid; Phthalic Acid and Derivatives; Terephthalic Acid, Dimethyl Terephthalate, and Isophthalic Acid; Fats and Fatty Oils; Waxes; Cellulose Esters;

**Nomenclature.** The term ester was coined by GMELIN (1850) from the German word **Essig-äther** (acetic ether, ethyl acetate). According to IUPAC rule C–463, esters of carboxylic acids are named by substituting the ending -ate for the -ic acid; the alkyl or aryl group is cited first. For example, the methyl ester of propionic acid is called methyl propionate [1]. Alternatively, the term propionic acid methyl ester is used.

Acid esters, such as the monoesters of dibasic acids, are named by inserting the word hydrogen between the names of the alcohol and acid (IUPAC rule C–463.2). Thus, methyl hydrogen succinate is the monomethyl ester of succinic acid.

Esters of glycerol such as fats and fatty oils are known as glycerides. Trivial names are usually preferred:

$$
\begin{array}{ll}
\text{CH}_2\text{-O-C(O)-CH}_3 & \text{CH}_2\text{-O-C(O)-(CH}_2)_{16}\text{-CH}_3 \\
\text{CH-O-C(O)-CH}_3 & \text{CH-OH} \\
\text{CH}_2\text{-O-C(O)-CH}_3 & \text{CH}_2\text{-O-C(O)-(CH}_2)_{16}\text{-CH}_3 \\
\text{Triacetin} & \text{1,3-Distearin}
\end{array}
$$

Ortho esters are named as derivatives of ortho acids, e.g., trimethyl orthoacetate (IUPAC rule C–464).

In substitutive nomenclature, different prefixes are used for the ester group, depending on the position of the principal group R:

$$
\begin{array}{ll}
\text{CH}_3\text{O-C(O)-R} & \text{CH}_3\text{-C(O)-OR} \\
\text{Methoxycarbonyl} & \text{Methylcarbonyloxy} \\
\text{derivative} & \text{derivative}
\end{array}
$$

## 2. Physical Properties

The physical properties of commercially important aliphatic and aromatic esters are listed in Table 1 [2]. The lower esters are colorless, mobile, and highly volatile liquids that usually have pleasant odors. As the molecular mass increases, volatility decreases and the consistency becomes waxy, then solid, and eventually even brittle, often with formation of lustrous crystals.

The melting point of an ester is generally lower than that of the corresponding carboxylic acid. The boiling point depends on the chain length of the alcohol component and may eventually exceed that of the acid. The lower esters are relatively stable when dry and can be distilled without decomposition.

With the exception of low molecular mass compounds formed by short-chain carboxylic acids, esters are insoluble in water (Table 2). By contrast, they are readily miscible with many organic solvents.

**Table 1.** Physical properties of the most important esters

| | CAS registry number | $M_r$ | $n_D^{20}$ (DIN 51 423) | $d_{20}^{20}$ (DIN 51 757) g/cm³ | bp at 101.3 kPa (DIN 53 171), °C | fp or mp, °C |
|---|---|---|---|---|---|---|
| *Esters of aliphatic acids* | | | | | | |
| Methyl formate | [107-31-3] | 60.05 | 1.344 | 0.975 | 32.0 | −99.8 |
| Ethyl formate | [109-94-4] | 74.08 | 1.3598 | 0.924 | 54.3 | −80 |
| Methyl acetate | [79-20-9] | 74.08 | 1.3594 | 0.933 | 57 | −98.1 |
| Ethyl acetate | [141-78-6] | 88.10 | 1.3723 | 0.900 | 77.06 | −83.6 |
| Propyl acetate | [109-60-4] | 102.13 | 1.3844 | 0.887 | 101.6 | −92.5 |
| Isopropyl acetate | [108-21-4] | 102.13 | 1.3773 | 0.872 | 90 | −73.4 |
| Butyl acetate | [123-86-4] | 116.16 | 1.3951 | 0.882 | 126 | −73.5 |
| Isobutyl acetate | [110-19-0] | 116.16 | 1.3902 | 0.871 | 117.2 | −98.6 |
| Pentyl acetate | [628-63-7] | 130.18 | 1.4023 | 0.876 | 149.25 | −70.8 |
| 2-Ethylhexyl acetate | [103-09-3] | 172.26 | 1.4204 | 0.873 | 199.3 | −93 |
| Vinyl acetate | [108-05-4] | 86.10 | 1.3959 | 0.932 | 72.2–72.3 | −93.2 |
| Ethylene glycol diacetate | [111-55-7] | 146.15 | 1.415 | 1.128⁴ | 191 | −31 |
| Methoxyethyl acetate | [110-49-6] | 118.13 | 1.4019 | 1.0067 | 145.0 | −65.1 |
| 2-Ethoxyethyl acetate | [111-15-9] | 132.16 | 1.4058 | 0.975 | 156.4 | −61.7 |
| 2-Butoxyethyl acetate | [112-07-2] | 160.21 | | 0.943 | 187.8 | |
| 2-(2-Ethoxyethoxy)ethyl acetate | [111-90-0] | 176.21 | | 1.011 | 217.4 | −25 |
| 2-(2-Butoxyethoxy)ethyl acetate | [112-34-5] | 204.27 | | 0.981 | 247 | −32.2 |
| Benzyl acetate | [140-11-4] | 148.15 | 1.5232 | 1.055 | 215.5 | −51.5 |
| Glyceryl triacetate | [102-76-1] | 218.23 | | 1.161 | 258 | −78 |
| Methyl propionate | [554-12-1] | 88.10 | 1.3775 | 0.915 | 79.8 | −87.5 |
| Ethyl propionate | [105-37-3] | 102.13 | 1.3839 | 0.892 | 99.10 | −72.6 |
| Glyceryl tripropionate | [139-45-7] | 260.3 | 1.4318 | 1.100 | 175–176 (2.67 kPa) | |
| Methyl butyrate | [623-42-7] | 102.13 | 1.3878 | 0.898 | 102.3 | −84.8 |
| Ethyl butyrate | [105-54-4] | 116.16 | 1.4000 | 0.878 | 121.6 | −100.8 |
| Butyl butyrate | [109-21-7] | 144.22 | 1.4075 | 0.871 | 166.6 | −91.5 |
| Methyl isobutyrate | [547-63-7] | 102.13 | 1.3840 | 0.891 | 92.6 | −84.7 |
| Ethyl isobutyrate | [97-62-1] | 116.16 | 1.3903 | 0.870 | 110–111 | −88 |
| Isobutyl isobutyrate | [97-85-8] | 144.22 | 1.3999 | 0.875 | 148.7 | −80.7 |
| Dimethyl adipate | [627-93-0] | 174.20 | 1.4283 | 1.060 | 115 (1.73 kPa) | 10.3 |
| Diethyl adipate | [141-28-6] | 202.25 | 1.4372 | 1.008 | 245 | −19.8 |
| Bis(2-ethylhexyl) adipate | [103-23-1] | 370.58 | | 0.927 | 214 (0.67 kPa) | −60 |
| Methyl stearate | [112-61-8] | 298.50 | 1.457⁴⁰ | 0.836⁶⁰ | 215 (2.0 kPa) | 40 |
| Ethyl stearate | [111-61-5] | 312.52 | 1.429⁴⁰ | 0.848³⁶ | 213–215 (2.0 kPa) | 33.7 |
| Butyl stearate | [123-95-5] | 340.58 | | 0.855 | 343 | 27.5 |
| Dodecyl stearate | [5303-25-3] | 440.80 | 1.433⁵⁰ | | | 28 |
| Hexadecyl stearate | [1190-63-2] | 496.9 | 1.441⁷⁰ | | | 57 |
| Methyl acrylate | [96-33-3] | 86.09 | 1.4040 | 0.953 | 80.5 | <−75 |
| Ethyl acrylate | [140-88-5] | 100.11 | 1.4068 | 0.923 | 99.8 | <−72 |
| Butyl acrylate | [141-32-2] | 128.17 | 1.4185 | 0.898 | 69 (6.7 kPa) | −64.6 |
| 2-Ethylhexyl acrylate | [103-11-7] | 184.28 | | 0.887 | 130 (6.7 kPa) | −90 |
| Methyl methacrylate | [80-62-6] | 100.13 | 1.4119²⁵ | 0.944 | 100–101 | −48 |

**Table 1.** (continued)

|  | CAS registry number | $M_r$ | $n_D^{20}$ (DIN 51 423) | $d_{20}^{20}$ (DIN 51 757), g/cm$^3$ | bp at 101.3 kPa (DIN 53 171), °C | fp or mp, °C |
|---|---|---|---|---|---|---|
| *Esters of aromatic acids* | | | | | | |
| Methyl benzoate | [93-58-3] | 136.14 | | 1.094$^{15}$ | 199.5 | − 12.5 |
| Ethyl benzoate | [93-89-0] | 150.17 | | 1.051$^{15}$ | 212.9 | − 34.2 |
| Methyl salicylate | [119-36-8] | 152.14 | | 1.184$^{15}$ | 223.3 | −8.6 |
| Ethyl salicylate | [118-61-6] | 166.17 | | 1.137$^{15}$ | 231.5 | 1.3 |
| Phenyl salicylate | [118-55-8] | 214.21 | | | 172.3 (1.6 kPa) | 43 |
| Dimethyl phthalate | [131-11-3] | 194.19 | | | 282 | 0 − 2 |
| Diethyl phthalate | [84-66-2] | 222.23 | 1.4990 | 1.118 | 295 | − 33 |
| Bis(2-ethylhexyl)phthalate | [117-81-7] | 390.57 | | 0.986 | 231 (0.66 kPa) | − 46 |
| Dimethyl isophthalate | [1459-93-4] | 194.18 | | | 124 (1.6 kPa) | 67 − 68 |
| Dimethyl terephthalate | [120-61-6] | 194.18 | | | 288 | 14 |
| Trimethyl trimellitate | [2459-10-1] | 252.30 | | | 194 (1.6 kPa) | − 13 |
| Methyl anthranilate | [134-20-3] | 151.16 | 1.584 | 1.168 | 132 (1.86 kPa) | 24 − 25 |
| Benzyl cinnamate | [103-41-3] | 238.27 | | | 240 − 244 (3.3 kPa) | 39 |

**Table 2.** Water solubility of lower esters

|  | CAS registry number | Solubility, g in 100 g of water | Temperature, °C |
|---|---|---|---|
| Methyl formate | [107-31-3] | 21.3 | 20 |
| Ethyl formate | [109-94-4] | 11.0 | 18 |
| Propyl formate | [110-74-7] | 2.2 | 22 |
| Butyl formate | [592-84-7] | 1.0 | 22 |
| Methyl acetate | [79-20-9] | 25 | 20 |
| Ethyl acetate | [141-78-6] | 7.7 | 20 |
| Propyl acetate | [109-60-4] | 1.5 | 16 |
| Isobutyl acetate | [110-19-0] | 0.8 | 20 |
| Butyl acetate | [123-86-4] | 0.7 | 20 |
| Methyl propionate | [554-12-1] | 6.3 | 20 |
| Ethyl propionate | [105-37-3] | 1.7 | 20 |
| Methyl butyrate | [623-42-7] | 1.6 | 21 |
| Ethyl butyrate | [105-54-4] | 0.6 | 22 |
| Methyl acetoacetate | [105-45-3] | 40 | 20 |
| Ethyl acetoacetate | [141-97-9] | 9 | 20 |

# 3. Chemical Properties

Due to the large number of possible acid and alcohol moieties, the chemical properties of esters may differ considerably. Only typical reactions applicable to the majority of esters are described in the following sections.

**Table 3.** Relative rates of acid-catalyzed hydrolysis of esters

|  | CAS registry number | $k_{ester}/k_{CH_3COOC_2H_5}$ |
|---|---|---|
| $CH_3COOC_2H_5$ | [141-78-6] | 1 |
| $C_2H_5COOC_2H_5$ | [105-37-3] | 0.47 |
| $(CH_3)_2CHCOOC_2H_5$ | [97-62-1] | 0.10 |
| $(CH_3)_3CCOOC_2H_5$ | [3938-95-2] | 0.01 |
| $CH_3COOCH_2CH(CH_3)_2$ | [110-19-0] | 0.70 |
| $CH_3COOCH_2C(CH_3)_3$ | [926-41-0] | 0.18 |
| $CH_3COOCH_2C(C_2H_5)_3$ | [10332-40-8] | 0.03 |
| $CH_3COOCH_3$ | [79-20-9] | 1.6 |
| $(COOCH_3)_2$ | [553-90-2] | 284 |
| $CH_2ClCOOCH_3$ | [96-34-4] | 1270 |
| $CHCl_2COOCH_3$ | [116-54-1] | 27000 |
| $CH_3COCOOC_2H_5$ | [617-35-6] | 17000 |

## 3.1. Hydrolysis

Esters are gradually split into the acid and alcohol through the action of humidity:

$$R^1-C(=O)OR^2 + H_2O \rightleftharpoons R^1-C(=O)OH + R^2OH$$

This hydrolysis is catalyzed by acids (protons) or bases (hydroxyl ions). The catalytic effect of bases is generally stronger. Industrially, the hydrolysis of esters is usually carried out with bases. However, acid hydrolysis is preferred for the production of pure acids and for alkali-sensitive compounds.

Hydrolysis of esters by use of water and a mineral acid leads to an equilibrium mixture of ester, alcohol, and free carboxylic acid. If bases are used, the salt of the carboxylic acid is formed; the carboxylic acid is thus removed from the equilibrium and the reaction becomes irreversible:

$$R^1-C(=O)OR^2 + NaOH \longrightarrow R^1-C(=O)ONa + R^2OH$$

Higher temperatures accelerate the reaction. Table 3 gives the relative rates of hydrolysis for some esters: hydrolysis is hindered by bulky acyl or alcohol groups, whereas acidifying groups increase the reaction rate.

Esters of secondary alcohols are hydrolyzed more slowly than those of primary alcohols. Esters of tertiary alcohols, e.g., *tert*-butyl esters, are exceptional in that they are readily hydrolyzed in the presence of acids and then often yield α-olefins instead of alcohols. Some esters are hydrolyzed only under extreme conditions; for example, esters of pivalic acid [75-98-9] (2,2-dimethylpropionic acid) and ortho-substituted benzoic acids react only in concentrated $H_2SO_4$.

**Basic Hydrolysis.** Basic hydrolysis (saponification) is one of the earliest known chemical reactions. By this method, soaps are obtained from natural fats and oils, with glycerol as a byproduct. Saponification based on wood ash as the alkali was described as early as 2500 BC; the principle is still employed today. The reaction was investigated by SCHEELE in 1779.

Basic hydrolysis requires stoichiometric amounts of alkali. The reaction is carried out by boiling the ester in an aqueous alkaline solution. Under these conditions, the alkali salt of the carboxylic acid is dissolved.

**Acidic Hydrolysis.** If the goal is to obtain free carboxylic acids rather than soaps, acid hydrolysis is the preferred method. Free acids are needed for various industrial purposes (→ Carboxylic Acids, Aliphatic; → Fatty Acids). Complete hydrolysis can only be achieved by removal of the alcohol from the equilibrium. The reaction rate in dilute acids is generally fairly low, mainly because of the poor solubility of esters in water. Therefore, in practice, emulsifiers such as sulfonated oleic acid or sulfonated aromatic compounds (Twitchell reagent) are added and the reaction is carried out at ca. 100 °C under pressure.

Another industrial process involves high-temperature hydrolysis with steam at 170 – 300 °C under pressure. Zinc, calcium, and magnesium compounds are used as catalysts [3]. Vinyl esters are readily hydrolyzed in the presence of palladium salts [4].

**Enzymatic Hydrolysis.** Specific lipolytic enzymes, the so-called esterases, effect hydrolysis at temperatures below 40 °C. Thus, lipase occurs in the digestive tracts of humans and most animals and in plants. Castor oil lipase is of commercial importance in hydrolysis of fats under mild conditions. The enzymes are especially effective in the weakly acidic range. Lipases may achieve importance as detergent additives; at low temperatures (20 °C) they facilitate the removal of fat-containing soils.

## 3.2. Transesterification

When esters are heated with alcohols, acids, or other esters, the alcohol or acid groups are exchanged more or less completely. This process is called transesterification. It is accelerated in the presence of a small amount of acid or alkali.

Three types of transesterification are known:

1) Exchange of alcohol groups (alcoholysis)

$$R^1-C(=O)OR^2 + R^3OH \rightleftharpoons R^1-C(=O)OR^3 + R^2OH$$

2) Exchange of acid groups (acidolysis)

$$R^1-CO-OR^2 + R^3-CO-OH \rightleftharpoons R^3-CO-OR^2 + R^1-CO-OH$$

3) Ester–ester interchange

$$R^1-CO-OR^2 + R^3-CO-OR^4 \rightleftharpoons R^1-CO-OR^4 + R^3-CO-OR^2$$

All three are equilibrium reactions and proceed to completion only if one component is removed from the reaction mixture.

Dispersed alkali metals, mainly sodium, are suitable catalysts for the transesterification of fats [5]. Tin salts have been used in the production of aryl esters, e.g., phenyl esters, at 190–250 °C. Methacrylic acid esters of higher alcohols, e.g., ethylhexanol, can be obtained by transesterification in the presence of tin compounds [6] or with magnesium alkoxides [7]. Other catalysts are discussed in Section 5.5.

Transesterification reactions play a major role in industry and are important in laboratory practice and in analytical chemistry. They can be used to reduce the boiling point of esters by exchanging a long-chain alcohol group for a short one such as methanol, e.g., in the analysis of fats, oils, and waxes (see Chap. 7) and in vapor-phase reactions. Transesterification reactions can also be useful if direct esterification with the desired alcohol is technically difficult because of physical reasons (e.g., high boiling point, low solu-bility, or high viscosity). Examples are to be found in polymer chemistry. Thus, poly(ethylene terephthalate) is produced mainly by transesterification of dimethyl terephthalate with ethylene glycol. Propanediols and 1,4-butanediol are used in a similar manner. The reaction is catalyzed with calcium, antimony, titanium, and germani-um salts. A further example is the reaction of diols (e.g., bisphenol A) with carbonic acid esters to produce polycarbonates.

An industrial example of acidolysis is the reaction of poly(vinyl acetate) with butyric acid to form poly(vinyl butyrate). Often a butyric acid–methanol mixture is used and methyl acetate is obtained as a coproduct.

## 3.3. Ammonolysis

Ammonia and esters react to form alcohols and amides:

$$R^1-CO-OR^2 + NH_3 \longrightarrow R^1-CO-NH_2 + R^2OH$$

This reaction is similar to hydrolysis. It is carried out in aqueous or alcoholic ammonia. Lower esters give good yields even at room temperature; higher esters require higher temperature and pressure.

If primary or secondary amines are used, N-substituted amides are formed. This reaction is called aminolysis. Hydrazines yield the corresponding hydrazides.

If esters are passed with ammonia over a catalyst such as alumina at 400–500 °C, dehydration of the intermediate amides occurs and the corresponding nitriles are obtained directly.

$$\text{R-C(=O)NH}_2 \longrightarrow \text{R-C} \equiv \text{N} + \text{H}_2\text{O}$$

Fatty nitriles can be produced in this manner [8] (see also → Amines, Aliphatic).

## 3.4. Reduction

Under certain conditions esters can be reduced to alcohols.

*Bouveault–Blanc Reduction.* This classical method developed in 1904 uses metallic sodium in alcohol (→ Fatty Alcohols). However, the Bouveault–Blanc reduction has been replaced nearly completely by catalytic hydrogenation and remains important only for laboratory purposes. The advantage of the Bouveault–Blanc process is that only the ester group is hydrogenated, while carbon–carbon double bonds are not reduced. Therefore, the reaction is suitable for the synthesis of unsaturated alcohols such as oleyl alcohol from unsaturated esters.

*Catalytic Hydrogenation.* A process of great industrial importance is the catalytic reduction of esters with hydrogen at 200–300 °C and 10–30 MPa (100–300 bar) in the presence of copper chromite catalysts [9]:

$$\text{R}^1\text{-C(=O)OR}^2 + 2\,\text{H}_2 \longrightarrow \text{R}^1\text{CH}_2\text{OH} + \text{R}^2\text{OH}$$

Sulfur and halogens are catalyst poisons. This method gives high yields and is used for the large-scale reduction of fats and fatty oils such as coconut oil and tallow, usually after transesterification to the methyl esters (→ Fatty Alcohols). The main products are dodecyl and decyl alcohols, which are important raw materials for the production of detergents. The method is also suitable for the conversion of 1,4-dimethyl terephthalate into 1,4-bis(hydroxymethyl) cyclohexane [105-08-8], an important intermediate in the production of polyesters.

Hardening of fats (hydrogenation of double bonds in unsaturated fats and oils) is carried out in the presence of nickel catalysts under slight hydrogen pressure and does not involve ester splitting.

*Grignard Reaction.* Esters react with alkylmagnesium halides under Grignard conditions. Formic acid esters give secondary alcohols; the other esters yield tertiary alcohols.

$$R^1-C{\overset{O}{\underset{OR^2}{\diagdown}}} + 2\ R^3MgX \longrightarrow R^1-\underset{R^3}{\overset{OMgX}{\underset{|}{C}}}-R^3 \xrightarrow{H_2O} R^1-\underset{R^3}{\overset{OH}{\underset{|}{C}}}-R^3$$

Yields are high, e.g., 95% in the production of triphenylmethanol from ethyl benzoate and phenylmagnesium bromide. The method is used mainly on a laboratory scale.

*Reduction with Complex Metal Hydrides.* The reduction of esters with complex metal hydrides, especially lithium aluminum hydride, has only preparative importance [10]. Its advantage is its broad applicability for all types of esters; double bonds are preserved. This method leads to very pure alcohols.

## 3.5. Claisen Condensation

The reaction of carboxylic acid esters with reactive methylene and methyl groups in the presence of alkaline compounds such as sodium alkoxides leads to the formation of $\beta$-oxocarboxylic acid esters ($\beta$-keto esters) [11]. The classical example is the condensation of two molecules of ethyl acetate to form acetoacetate:

$$2\ CH_3-C{\overset{O}{\diagdown}}_{OC_2H_5} \xrightarrow{NaOC_2H_5} CH_3-\overset{O}{\underset{||}{C}}-CH_2-C{\overset{O}{\diagdown}}_{OC_2H_5} + C_2H_5OH$$

Oxocarboxylic acid esters are among the most important intermediates in the laboratory and in industry. During hydrolysis with dilute alkali, these esters readily decompose into ketones, alcohols, and carbon dioxide:

$$CH_3-\overset{O}{\underset{||}{C}}-CH_2-C{\overset{O}{\diagdown}}_{OR} + H_2O \longrightarrow CH_3-\overset{O}{\underset{||}{C}}-CH_3 + CO_2 + ROH$$

Almost any ketone can be synthesized by using this process.

In industry, the Claisen condensation has been replaced by the synthesis of acetoacetic acid derivatives from diketene [*674-82-8*], see also Section 5.6 ($\rightarrow$ Oxocarboxylic Acids, Aliphatic; $\rightarrow$ Ketenes).

Condensation of esters in inert solvents in the presence of sodium rather than sodium ethoxide leads to formation of diketones and $\alpha$-hydroxy ketones (acyloins) [12]. This reaction is utilized in the synthesis of large-ring compounds.

## 3.6. Pyrolysis

When passed at 300–500 °C over an inert heat-transfer agent such as quartz glass, esters decompose into carboxylic acid and the unsaturated compound that corresponds to the alcohol. Esters of primary alcohols generally lead to α-olefins:

$$R^1-C(=O)-O-CH_2-CH_2-R^2 \xrightarrow{\text{Heat}} R^1-C(=O)-OH + H_2C=CH-R^2$$

The double bond does not migrate along the carbon chain. Methyl esters and other esters without hydrogen atoms in the β-position do not yield uniform pyrolysis products.

When heated to high temperatures in the presence of metal oxides such as thorium oxide, calcium oxide, manganese chromite, or zinc chromite, fatty acid esters or free fatty acids give high yields of ketones:

$$2\ R^1-C(=O)-OR^2 \xrightarrow[\text{Heat}]{\text{Cat.}} R^1-C(=O)-R^1 + CO_2 + \text{olefin}$$

The alcohol moiety is split off as olefin.

## 4. Natural Sources

Esters occur naturally in large quantities in fats and fatty oils, waxes, and fruit ethers.

**Fats and Fatty Oils.** Fats and fatty oils (as opposed to mineral oils) are almost exclusively esters of glycerol with long- and medium-chain fatty acids. About 1300 different fats and oils are known. They contain mainly even-numbered, straight-chain carboxylic acids (→ Fats and Fatty Oils). The most common are the triglycerides of stearic, palmitic, and oleic acid. The melting point of the natural products varies between 55 °C (mutton tallow) and −27 °C (nut oil) and decreases as the content of unsaturated and short-chain fatty acids increases.

Only a few of the animal fats and vegetable oils are suitable for commercial production. They are predominantly used as foods, but have also gained considerable industrial importance, e.g., in the production of lacquers, detergents, and solvents. The most frequently produced fats of animal origin are butter, beef and mutton tallow, lard, fish oil, and whale oil. Butter accounts for ca. 40 % of the world production of animal fats. The most important oils of plant origin are soybean oil, sunflowerseed oil, peanut oil, cottonseed oil, coconut oil, rapeseed oil, palm oil, and olive oil. Linseed oil, hempseed oil, and poppyseed oil contain a variable amount of unsaturated fatty acids, e.g., linoleic acid, linolenic acid, ricinoleic acid, and erucic acid. These oils gradually harden under the influence of oxygen in the air.

**Waxes.** Natural waxes (wax esters) are esters of monobasic carboxylic acids with higher monohydric and, less commonly, dihydric alcohols. Currently they are largely replaced by synthetic waxes, which are in most cases long-chain hydrocarbons, but not esters.

The most important sources of wax esters are beeswax, whose main constituent is myricyl palmitate [6027-71-0]; spermaceti from the head oil of the sperm whale, which contains cetyl palmitate [540-10-3]; and carnauba wax from the Brazilian palm, containing myricyl cerotate [84324-99-2]. Fossil waxes, such as montan waxes in bituminous lignite and peat waxes, also have major wax ester components. Lanolin, the neutral part of wool grease, contains higher fatty acid esters of cholesterol.

**Fruit Ethers.** Fruit ethers, i.e., volatile esters of short- and medium-chain carboxylic acids with short- and medium-chain alcohols, are major constituents of essential oils. Their name derives from their pleasant, usually fruity odor. They are used as aromatic materials in fragrances, foods, beverages, cosmetics, and soaps. Natural fruit ethers are obtained from plant materials by extraction, steam distillation, pressing, or a combination of these methods. The majority of simple, short-chain esters are currently produced synthetically.

# 5. Production

A great variety of production methods for carboxylic acid esters are known. However, many of these methods are valuable only for preparing special compounds [13]. In what follows, only those methods that have industrial significance are discussed.

## 5.1. Esterification of Carboxylic Acids

The simplest and most common method of esterification is the reaction of a concentrated alcohol with a concentrated carboxylic acid with the elimination of water:

$$R^1-C(=O)OH + R^2OH \xrightarrow{H^+} R^1-C(=O)OR^2 + H_2O$$

$R^1, R^2$ = alkyl or aryl

Esterification is the reverse of hydrolysis and leads to an equilibrium (see Section 3.1). With acetic acid and ethanol, the molar equilibrium ratio of ethyl acetate to acetic acid is ca. 2:1 (65 mol% of ethyl acetate). The equilibrium may be shifted toward the ester by increasing the concentration of one of the reactants, usually the alcohol. However, quantitative esterification is possible only by continuous removal of one of the products, i.e., ester or water. Usually the water is distilled off as an azeotrope (see Section 5.1.2).

Removal of the water by chemical or adsorptive binding is also possible. In preparative chemistry, calcium carbide, calcium hydride, and calcium and magnesium sulfate have proved successful; in industrial applications, activated bauxite is also used. The distilled, water-rich condensates are passed over the water-binding agents; alternatively, salts such as copper sulfate or calcium chloride can be added directly to the reaction mixture. Molecular sieves are also suitable [14]. Acetone dimethyl acetal, which binds water and simultaneously supplies methyl groups, may be used for the production of methyl esters [15].

Esterification is carried out at high temperature and in the presence of catalysts. The rate of ester formation depends on the carboxylic acid and the alcohol used. The lowest members, i.e., methanol and formic acid, react the most readily. Primary alcohols react faster than secondary alcohols, and the latter react faster than tertiary ones. Within each series, the reaction rate generally decreases with increasing molecular mass. Straight-chain acids react more readily than branched ones; the rate of esterification is lowered particularly by branching in the α-position. The relative rates of esterification are similar to those of hydrolysis (see Table 3). Esterification of aromatic acids, e.g., benzoic acid, is slow.

Experiments with primary and secondary alcohols containing $^{18}$O have shown that the oxygen in the water formed during acid-catalyzed esterification originates from the acid, not the alcohol [16]. The reaction of tertiary alcohols is more complex and does not proceed clearly via a tertiary carbonium ion. The mechanism is discussed in [17].

## 5.1.1. Catalysts

**Mineral Acids.** Only strong carboxylic acids react sufficiently quickly without a catalyst. Generally, a strong mineral acid must be added. Suitable catalysts include sulfuric acid, hydrogen chloride, arylsulfonic acids such as *p*-toluenesulfonic acid, and chlorosulfuric acid. Phosphoric acid, polyphosphoric acids, and mixtures of acids are also recommended [18]. If the acids are adsorbed on a solid support, esterification can be carried out as a continuous process.

**Ion Exchangers.** The acid groups can be bound chemically to a polymeric material. Cat-ion exchangers such as sulfonated polystyrene permit esterification under mild reaction conditions. The resulting esters are generally very pure because acid-catalyzed side reactions such as dehydration, etherification, and rearrangement are almost completely suppressed [19], [20]. This method gives high yields and is widely used in industry. After the reaction water has been removed, the ion exchanger is filtered off and the ester is purified by distillation. Zeolites are also used as esterification catalysts [21].

**Lewis Acids.** Lewis acids such as boron trifluoride increase the reactivity of the carboxyl group toward alcohols. Boron trifluoride reacts with alcohols under proton release:

$$BF_3 + ROH \longrightarrow [BF_3OR]^- + H^+$$

The proton acts as a catalyst. The complex with diethyl ether, boron trifluoride etherate [109-63-7], is generally preferred because it is easier to handle than gaseous boron trifluoride [22]. This method of esterification proceeds under mild conditions and is widely applicable. However, it is limited to the manufacture of high-value esters because of the relatively high price of $BF_3$. Each carboxyl group requires one equivalent of the boron trifluoride complex, which in most cases is destroyed during the further processing of the ester. All other groups that possess a free electron pair (e.g., hydroxyl, oxo, or amino groups) require an additional equivalent of $BF_3$.

Metal salts may also function as Lewis acids. For example, a complex acid is formed from the reaction of a zinc salt and an alcohol:

$$Zn(O-CO-R^1)_2 + R^2OH \longrightarrow [R^2O-Zn(O-CO-R^1)_2]^- + H^+$$

The actual catalysts are again the protons. Tin compounds have been used to prepare glyceryl esters. For the production of dimethyl ter-ephthalate, silica treated with aluminum, titanium, zinc, or tin compounds has proved successful [23]; this esterification proceeds also in the absence of catalysts (12.5 – 14.0 MPa, 270 –275 °C, 95 % yield).

**Amphoteric Hydroxides.** Esterification is catalyzed by amphoteric hydroxides of metals of groups I, II, III, and IV, especially aluminum hydroxide. The actual catalyst is probably the aluminate anion:

$$Al(OH)_3 + NaOH \longrightarrow [HO-Al(OH)_3]^- + Na^+$$

The aluminate may be generated in the reaction mixture from aluminum hydroxide and sodium hydroxide [24]. The esters produced by this method are light in color, and purification by distillation is generally unnecessary. Therefore, this process is especially suitable for preparing esters with high boiling points. Its main application is in the production of plasticizers, e.g., phthalates.

**Other Compounds.** A possible esterification catalyst is graphite hydrogen sulfate [12689-13-3], $(C_{24}HSO_4 \cdot 2\ H_2SO_4)_{57}$. It is obtained by electrolysis of 98% sulfuric acid with a graphite anode.

Silanes and phosphoric acid derivatives are utilized for syntheses under mild conditions, the latter especially for phenyl carboxylates.

Acid esterification of amino acids and peptides is difficult because protecting groups and peptide bonds are easily hydrolyzed. High yields of esters of both protected and free amino acids or peptides are obtained in the presence of a mild dehydration catalyst such as sulfuryl chloride or thionyl chloride at room temperature [25].

Trifluoroacetic anhydride is a suitable catalyst for the esterification of sterically hindered acids [26].

**Table 4.** Composition of common azeotropes

| Entrainer | bp, °C | | Water content, wt% |
|---|---|---|---|
| | Entrainer | Azeotrope | |
| Benzene | 80.2 | 69.2 | 8.8 |
| Toluene | 110.7 | 84.1 | 13.5 |
| Xylene | 139 | 92 | 35.8 |
| Cyclohexane | 80.7 | 68.9 | 9.0 |
| Tetrachloromethane | 76.7 | 66.0 | 4.1 |

## 5.1.2. Industrial Processes

Esterification is generally carried out by refluxing the reaction mixture until all the water has been split off. The water or the ester is removed from the equilibrium by distillation. The choice of method to achieve complete esterification depends on the boiling points of the alcohol, the acid, and the ester.

Basically, the following three possibilities exist:

1) The boiling point of the ester is lower than that of water. In this case the ester can be distilled off together with the alcohol. This method is used to produce methyl acetate, which forms an azeotrope with methanol.
2) Ester and water can be distilled off together, usually as an azeotropic mixture. On condensation, the mixture separates into an ester and a water phase. To achieve complete distillation of the ester, water or steam are often added to the reaction mixture. An example of this method is the production of *sec*-butyl acetate. Frequently alcohol, ester, and water form ternary azeotropes, e.g., during the esterification of acetic acid with ethanol.
3) The boiling point of the ester is higher than that of water. In this case, the water is distilled off, frequently as an azeotrope with the alcohol. Except for methanol, all alcohols of medium volatility form azeotropes with water. If the alcohol–water mixture separates into two phases during condensation, the alcohol can be recycled into the reaction mixture, as exemplified in the production of *n*-dibutyl phthalate. If water and alcohol do not separate, fresh alcohol must be continuously introduced into the reaction mixture.

Removal of water usually involves the addition of entrainers, which form azeotropes with relatively low boiling points and high water content. Table 4 lists some frequently used entrainers. After condensation, the azeotrope separates into an aqueous phase and an organic phase, and the entrainer is recycled into the reaction mixture.

With ethanol, *n*-propanol, 2-propanol, allyl alcohol, and *tert*-butyl alcohol, a cosolvent such as benzene or toluene is often added to the condensate to achieve separation of the organic phase [27].

With high-boiling esters, e.g., esters of polyhydric alcohols, the water of reaction may be removed by means of steam or inert gases. Steam distillation can also be used to remove excess alcohol after the esterification is complete.

Esterification is usually carried out batchwise. In commercial production, continuous processes are preferred. Special procedures have been devised in some cases.

Methyl and ethyl esters of mono- and dicarboxylic acids are produced continuously by pumping the alcohol and the carboxylic acid melt (e.g., lauric acid or adipic acid) into a pipe from below and drawing off the ester at the top [28]. In a continuous process for the production of volatile esters, the mixture of acid, alcohol, and catalyst is introduced into the upper part of a distillation column and an excess of the alcohol is introduced into the bottom. The column is heated so that the ester, water, and excess alcohol are distilled off [29].

Nonvolatile esters such as bis(2-ethylhexyl) phthalate [117-81-7] can be obtained continuously by the following procedure:

The mixture of acid or anhydride and alcohol together with sulfuric acid is introduced at the top of a plate column. An entrainer, e.g., benzene, is distilled through the column from below. The azeotrope of water and entrainer is drawn off the top and the ester accumulates in the sump.

Monochloroacetic acid esters of alcohols with three to five carbon atoms are obtained continuously by adding equimolar amounts of chloroacetic acid and alcohol to a chloroacetic acid melt and distilling off the ester and water.

Esterification can also be carried out continuously in the vapor phase by heating a mixture of acid, alcohol, ester, and catalyst to the reaction temperature. This process is called gas-phase esterification although one or more of the components may be present as liquids or solids.

Ion-exchange resins are especially suitable as catalysts in continuous processes. The reactants pass through or over the solid catalyst, and no separation or neutralization of the catalyst is necessary [20].

## 5.2. Alkylation of Metal Carboxylates

If metal carboxylates are treated with alky-lating agents, good yields of carboxylic acid esters result. A one-pot procedure is often used, in which the salt is first produced from the carboxylic acid and then directly alkylated to form the ester. Isolation of the salt is necessary only in the case of thermally unstable quaternary ammonium salts [30].

Alkali metal and alkaline-earth metal salts, as well as silver and copper salts, can be used. Dialkyl sulfates, aliphatic halides such as long-chain alkyl chlorides, and alicyclic halides are suitable alkylating agents. The method is used mainly in preparative chemistry.

## 5.3. Acylation with Acyl Halides

High yields of esters are obtained by acylation of alcohols with carboxylic acid halides:

$$R^1-COCl + R^2OH \longrightarrow R^1-COOR^2 + HCl$$

The acyl chloride is often produced in situ. The liberated hydrogen halide is generally bound to a base, e.g., pyridine or alkali hydroxide (Schotten–Baumann reaction).

This method is particularly important in preparative and analytical chemistry. A commercial application is the production of polycarbonates from phosgene and diols. Phenyl esters and esters of sterically hindered carboxylic acids can be produced from the corresponding acyl halides in the presence of Lewis acids, e.g., iron(III), aluminum, tin(IV), and titanium(IV) chlorides. Addition of phase-transfer agents such as tetrabutylammonium chloride has also proved effective [31]. An important application of this method is in the esterification of steroids.

## 5.4. Acylation with Carboxylic Anhydrides

Carboxylic anhydrides are suitable acylating agents for alcohols and phenols. With the exception of trifluoroacetic anhydride and mixed carboxylic anhydrides, they are less reactive than acyl halides.

The advantage of this method is that no water is formed; however, only one equivalent acyl per molecule of anhydride is available for esterification:

$$(R^1-CO)_2O + R^2OH \longrightarrow R^1-COOR^2 + R^1-COOH$$

Therefore, the method is especially suitable for the esterification of rare or expensive alcohols with cheap anhydrides such as acetic anhydride.

Acylation can be accelerated by addition of protonic acids (sulfuric acid, hydrochloric acid, perchloric acid, *p*-toluenesulfonic acid) or Lewis acids (zinc chloride). A large-scale commercial process is the production of cellulose acetate from cellulose and acetic anhydride (→ Cellulose Esters).

Another important product is acetylsalicylic acid (aspirin), which is obtained from salicylic acid and acetic anhydride.

Base-catalyzed acylation with anhydrides is used to synthesize esters from sterically hindered or acid-sensitive alcohols [32], [33]. Common bases are triethylamine and pyridine.

Reaction with anhydrides is the basis of a quantitative analytical method for hydroxyl groups.

## 5.5. Transesterification

This method has already been described in Section 3.2. Transesterifications take place under extremely mild conditions, which permits the reaction of components containing additional functional groups. Chiral centers can be preserved in many cases.

The most common transesterification catalysts are protonic acids (sulfuric acid, perchloric acid, p-toluenesulfonic acid, and molecular sieves). Alkaline catalysts such as alkoxides or anionic ion exchangers are especially effective in peptide chemistry. Potassium cyanide is recommended for unsaturated esters, which undergo cis–trans isomerization in the presence of acids. Thallium(I) salts are also employed. If titanium(IV) alkoxides or complexes of triphenylphosphine with trialkyltin alkoxides or with copper alkoxides are used, the alkoxide component need not be identical with the alcohol group that is exchanged.

## 5.6. Acylation, of alcohols with ketenes

Ketenes react with alcohols or phenols to form carboxylic acid esters:

$$\begin{array}{c} R^1 \\ R^2 \end{array} C=C=O + R^3OH \longrightarrow H-\underset{R^2}{\overset{R^1}{\underset{|}{C}}}-C\begin{array}{c} \diagup O \\ \diagdown OR^3 \end{array}$$

The addition is catalyzed by acids and bases (e.g., pyridine or sodium acetate). The use of ketenes is limited because they are difficult to handle and equally good results can be achieved with other acylating agents such as acyl halides and anhydrides.

The alcoholysis of diketene is an important method for the commercial production of β-oxocarboxylic acid esters (acetoacetates), especially ethyl acetoacetate [*141-97-9*] [34]:

$$\underset{O}{\overset{H_2C}{\square}}\!\!\!\!\diagdown_O + ROH \longrightarrow H_3C-\overset{O}{\overset{\|}{C}}-CH_2-C\begin{array}{c} \diagup O \\ \diagdown OR \end{array}$$

Acetoacetic acid anilides are produced in a similar way. These compounds are used in the synthesis of dyes, especially fast yellow dyes.

## 5.7. Carbonylation

The preparation of carboxylic acid esters by carbonylation of olefins in alcoholic solution (Reppe synthesis) is used on an industrial scale, e.g., in the production of methyl and ethyl propionate [35]:

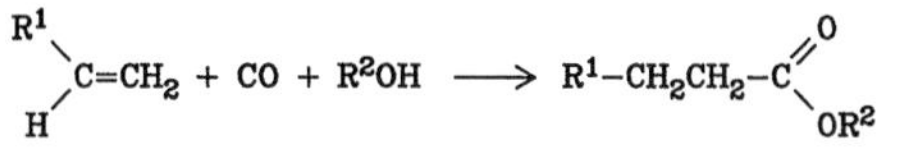

$$\underset{H}{\overset{R^1}{>}}C=CH_2 + CO + R^2OH \longrightarrow R^1-CH_2CH_2-C\overset{O}{\underset{OR^2}{<}}$$

The linear ester is obtained in 85% yield, with less than 10% branched ester of the following structure:

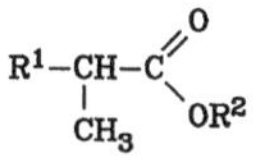

$$R^1-\underset{CH_3}{\overset{|}{C}}H-C\overset{O}{\underset{OR^2}{<}}$$

The reaction is carried out at a moderate temperature (ca. 100 °C) and a CO pressure of 20–25 MPa (200–250 bar). Iron, nickel, and cobalt salts, which are converted into the corresponding metal carbonyls during the reaction, and palladium or rhodium complexes are used as catalysts. A typical example is a bimetallic complex of bis(triphenylphosphine)palladium(II) chloride and tin(II) chloride [36]. Similarly, 1,3-dienes give $\gamma,\delta$-unsaturated esters.

Alkyl, benzyl, vinyl, aryl, and allyl halides can also be carbonylated in the presence of suitable catalysts such as Mn–Fe–Ni powders. These reactions require only a low CO pressure [37].

Carbonylation of alcohols at a pressure up to 70 MPa (700 bar) in the absence of olefins gives formates. Thus, methyl formate is obtained by carbonylation of methanol at 70 °C and 2–20 MPa (20–200 bar) in the presence of sodium methoxide [38]:

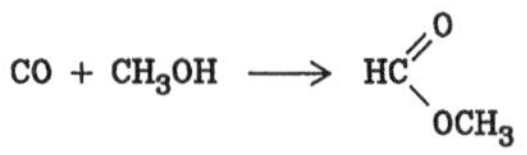

$$CO + CH_3OH \longrightarrow HC\overset{O}{\underset{OCH_3}{<}}$$

## 5.8. Condensation of Aldehydes

The condensation of two acetaldehyde molecules (Tishchenko reaction) is an important industrial process for the production of ethyl acetate: Aluminum ethoxide promoted with chloride ions serves as a catalyst. In practice, acetaldehyde is passed at 0–5 °C through a mixture of aluminum filings and traces of $AlCl_3$ in ethanol–ethyl acetate; the yield is 95%.

$$2\ CH_3CHO \longrightarrow CH_3C(=O)OC_2H_5$$

## 5.9. Alcoholysis of Nitriles

Nitriles react directly with alcohols to produce the corresponding esters:

$$R^1-C\equiv N + R^2OH + H_2O \longrightarrow R^1-C(=O)OR^2 + NH_3$$

The ammonia that is formed must be removed from the equilibrium, preferably by using an excess of a strong mineral acid, e.g., sulfuric acid or *p*-toluenesulfonic acid. The acid also acts as a catalyst [39].

This method is used widely in industry for the production of esters from unsaturated nitriles such as acrylonitrile and methacrylonitrile. Alcoholysis of acrylonitrile is carried out with concentrated sulfuric acid. This leads initially to formation of the acrylamide sulfate, which then reacts with the alcohol (preferably methanol or ethanol) to produce the ester:

$$CH_2=CH-C\equiv N + H_2SO_4 + H_2O \longrightarrow CH_2=CH-C(=O)NH_2 \cdot H_2SO_4$$

$$\xrightarrow{CH_3OH} CH_2=CH-C(=O)OCH_3 + NH_4HSO_4$$

Selectivity for methyl acrylate is > 90 %. A disadvantage of this process is the formation of ammonium hydrogen sulfate, which cannot be used economically. An alternative process, which avoids formation of this byproduct, is the vapor-phase alcoholysis of acrylonitrile in the presence of boric oxide [40].

Methyl methacrylate is obtained by alcohol-ysis of the methacrylonitrile precursor acetone cyanohydrin. In this case, methacrylamide sulfate is first produced by using 98 % $H_2SO_4$ at 80–140 °C, which is then reacted with methanol at 80 °C to produce methyl methacrylate [41].

$$CH_3-C(CH_3)(OH)-C\equiv N + H_2SO_4 + CH_3OH \longrightarrow$$

$$CH_2=C(CH_3)-C(=O)OCH_3 + NH_4HSO_4$$

Selectivity based on acetone is 77 %. Other methods, which avoid the formation of $NH_4HSO_4$, have been developed on a pilot-plant scale.

## 5.10. Acylation of Olefins

Direct addition of olefins to carboxylic acids is used on a large scale for the production of vinyl acetate, the ester of the hypothetical vinyl alcohol (→ Vinyl Esters).

$$CH_3-C(=O)OH + CH_2=CH_2 + 1/2\ O_2 \longrightarrow CH_3-C(=O)O-CH=CH_2 + H_2O$$

The presence of stoichiometric amounts of oxygen or air is necessary. Palladium or palladium salts are essential as catalysts; other group VIII metals are less effective.

The process can be carried out in the vapor phase or in the liquid phase. Currently, only the vapor-phase process is used on an industrial scale. The reaction conditions are 175 – 200 °C and 0.5 – 1 MPa (5 – 10 bar); the catalyst consists of palladium metal and alkali acetate on a carrier. The carrier is necessary for activation and higher selectivity. During the reaction, the alkali acetate is slowly removed from the catalyst and has to be continuously replaced [42]. The vapor-phase process avoids the corrosion problems of the liquid-phase process and gives better yields of vinyl acetate (94% based on ethylene or 98 – 99% based on acetic acid).

The liquid-phase process is no longer used on an industrial scale. This process is closely related to the Wacker – Hoechst acetaldehyde process. The catalyst contains copper salts as a redox system. The advantages of the liquid-phase process are a better control of the highly exothermic reaction and the generation of acetaldehyde as a byproduct, which can be oxidized to acetic acid. In this way the necessary acetic acid can be entirely produced from ethylene. The disadvantages are corrosion problems and a lower yield of vinyl acetate, based on ethylene.

The olefin process has replaced the older process based on acetic acid and acetylene, because ethylene is cheaper and more readily available in larger quantities.

Allyl esters are produced from propene, carboxylic acids, and oxygen by using a vapor-phase process. Supported catalysts impregnated with palladium salts and bismuth or cadmium compounds are used [43]. Methallyl acetate is obtained from isobutene and acetic acid in the presence of palladium(II) chloride [44]:

$$CH_3-C(=O)OH + (H_3C)_2C=CH_2 + 1/2\ O_2 \longrightarrow CH_3-C(=O)-O-CH_2-C(CH_3)=CH_2 + H_2O$$

Glycol mono- and diesters can be produced from olefins and carboxylic acids by adding stoichiometric amounts of oxygen:

$$2\ CH_3-C(=O)OH + CH_2=CH_2 + 1/2\ O_2 \longrightarrow$$

$$CH_3-C(=O)-O-CH_2-CH_2-O-C(=O)-CH_3 + H_2O$$

## 5.11. Ethoxylation of Carboxylic Acids

Addition of ethylene oxide to carboxylic acids generally proceeds smoothly and quickly. The low acidity of the carboxylic acids is sufficient to split the strained three-membered ring; no catalyst is required in the first addition step:

$$R-C(=O)OH + H_2C\underset{O}{-}CH_2 \longrightarrow R-C(=O)O-CH_2CH_2-OH$$

Further ethoxylation yields polyglycol esters; this reaction can be controlled or entirely suppressed by addition of bases such as alkali or amines.

$$R-C(=O)O-CH_2CH_2-OH + n\ H_2C\underset{O}{-}CH_2 \longrightarrow R-C(=O)O(CH_2CH_2O)_{n+1}H$$

Ethoxylation reactions are widely used in industry. An example is the production of bis(2-hydroxyethyl) terephthalate [6422-86-2], by controlled ethoxylation of terephthalic acid:

$$HO-C(=O)-C_6H_4-C(=O)-OH + 2\ H_2C\underset{O}{-}CH_2 \longrightarrow$$

$$HO-CH_2CH_2-O-C(=O)-C_6H_4-C(=O)-O-CH_2CH_2-OH$$

This ester is readily purified by crystallization from water, so that crude terephthalic acid can be used. The ethoxylation is carried out in the liquid phase at 2–3 MPa (20–30 bar) and 90–130 °C; no solvent is necessary. The presence of amines or quaternary alkylammonium salts is needed to avoid polyethoxylation. A considerable amount of poly(ethylene terephthalate) is produced by condensation of bis(2-hydroxyethyl) terephthalate.

Polyethoxylation of aliphatic carboxylic acids also is an industrial process. It is carried out under pressure at 120–220 °C in the presence of sodium hydroxide or sodium acetate. Addition of 10–30 mol of ethylene oxide per mole of acid causes fatty acids such as stearic acid to lose their hydrophobicity. The products have many applications, e.g., as surfactants. The degree of hydrophilicity depends on the number of glycol units. Ethoxylation of short-chain carboxylic acids yields hydrophilic lubricating oils.

## 5.12. Orthocarboxylic Acid Esters

Free orthocarboxylic acids are unstable. Their esters can be prepared by reaction of imido ethers with alcohols:

$$R-C(NH \cdot HCl)(OC_2H_5) + 2\ C_2H_5OH \longrightarrow R-C(OC_2H_5)_3 + NH_4Cl$$

Orthoformic acid ethyl ester [122-51-0] is obtained from chloroform and sodium ethoxide:

$$HCCl_3 + 3\ C_2H_5ONa \longrightarrow HC(OC_2H_5)_3 + 3\ NaCl$$

Orthoformates are relatively stable toward alkali; however, they are readily hydrolyzed by acids. In preparative chemistry, orthoformates are used to synthesize acetals and ketals. They also serve for the incorporation of methine groups into compounds with reactive methylene groups.

## 5.13. Lactones

Lactones are internal esters of hydroxycarboxylic acids. Of commercial interest are β-propiolactone [57-57-8] and γ-butyrolactone [96-48-0]. β-Propiolactone can be produced from ketene and formaldehyde in the presence of zinc chloride:

$$H_2C=C=O + CH_2O \longrightarrow \underset{H_2C-O}{\overset{H_2C-C(=O)}{|}}$$

γ-Butyrolactone is produced mainly by dehydrogenation of 1,4-butanediol (→ Butyrolactone):

(diol) → (lactone) + $H_2$

Lactones with five- and six-membered rings (γ- and δ-lactones) are the most stable. They readily form polyesters according to the following equation:

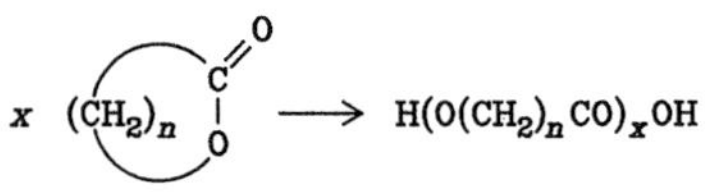

$$x \ (CH_2)_n \overset{O}{\underset{O}{C}} \longrightarrow H(O(CH_2)_n CO)_x OH$$

## 6. Environmental Protection

Most esters are synthesized by esterification of an acid with an alcohol, water being the only byproduct. Special environmental protection measures are usually unnecessary. Most esters possess low toxicity (Chap. 10). Because many of them are easily hydrolyzed on contact with water or moist air, the toxicological properties of the acid and alcohol component may need to be considered.

**Waste Gases.** Volatile esters with low carbon numbers are predominantly used as solvents and diluents for lacquers and plastics. Large-scale applications, e.g., in the auto and furniture industries, require excellent ventilation. The solvent vapors are suctioned off and burned in a flare or muffle furnace.

**Waste Liquors.** Short-chain esters and esters with hydrophilic groups are noticeably soluble in water; some results of biodegradability tests are listed in Table 5. The toxicity of some common esters for fishes and microorganisms is shown in Table 6.

## 7. Quality Specifications and Analysis

Table 7 shows some typical specifications of commercial products.

The most common analytical method is the determination of the saponification value, i.e., the number of milligrams of KOH necessary to hydrolyze 1.0 g of ester:

The ester (0.5–1 g) is refluxed for 30 min with 10 mL of 0.5 N alcoholic KOH. After cooling, the excess alkali is titrated with 0.5 N HCl using phenolphthalein as the indicator.

Before determination of the saponification value, determination of the free acid by titration with alkali in a water–alcohol mixture is often necessary.

The ester group can be readily identified by its characteristic absorption bands in the IR spectrum: the strong carbonyl absorption band at ca. 1740 cm$^{-1}$ and two C–O stretching bands between 1300 and 1100 cm$^{-1}$ [45].

Gas chromatography is suitable only for highly volatile esters. Esters with high boiling points must first be converted into low-boiling esters (usually methyl esters) either by transester-ification or by quantitative hydrolysis and sub-sequent reesterifica-

**Table 5.** Biodegradation of organic esters

|  | Degree of degradation,* % | BOD$_5$** |
|---|---|---|
| Methyl acetate | > 95 | ca. 500 |
| Ethyl acetate | > 90 | 770 |
| Vinyl acetate | > 90 | 810 |
| Butyl acetate | > 95 | 1000 |
| 2-Methoxyethyl acetate | 100 | 450 |
| 2-Butoxyethyl acetate | 100 | 260 |
| 2-(2-Butoxyethoxy)ethyl acetate | 100 | 380 |
| Methyl acetoacetate | 100 | 940 |
| Ethyl acetoacetate | > 90 | 780 |
| n-Butyl glycolate | 93 | 570 |
| Methyl crotonate | > 95 | 1050 |
| Dimethyl acetylsuccinate | > 95 | 1100 |
| Diethyl acetylsuccinate | > 95 | 1070 |
| Dimethyl maleate | 100 | 20 |
| Monomethyl maleate | > 95 | 150 |
| Diethyl maleate | > 90 | 200 |
| Dibutyl maleate | 99 | 630 |
| Di(2-ethylhexyl)maleate | 100 | 1450 |
| Methyl 3-aminobenzoate | 95 | 10 |
| Methyl 4-hydroxybenzoate | 100 | 1080 |
| Methyl 4-hydroxyphenylacetate | 98 | 320 |

\* According to DIN 38 412, Part 25.
\*\* Biochemical oxygen demand (milligrams of O$_2$ per liter of wastewater; sum over 5 d).

tion. Quantitative conversion into the methyl ester can be achieved with diazomethane, methyl iodide – calcium oxide in dimethyl sulfoxide (DMSO), or with the boron trifluoride – methanol complex [46].

Unsaturated esters, e.g., natural fats and oils, are often sufficiently characterized by their iodine value. This is defined as the number of grams of iodine bound by 100 g of an unsaturated ester and is a measure of the number of double bonds in the molecule.

An overview of the different analytical methods is given in [47]. Physical properties such as boiling range, melting point, density, refractive index, residue or ash content, and color are also used to identify esters or mixtures of esters. Previous separation of the different components, (e.g., by high-vacuum distillation) is often unnecessary.

# 8. Storage and Transportation

All organic esters are flammable. The flash points, autoignition temperatures, and flammable limits of representative commercial esters are given in Table 7. The ignition group according to the NFPA (National Fire Prevention Association) corresponds approximately to the danger class of the German VbF (Verordnung über brennbare Flüssigkeiten of 27. Feb. 1980).

**Table 6.** Toxicity for fishes and bacteria

|  | Fishes * $LC_0$, mg/L | Bacteria ** $EC_0$, mg/L |
|---|---|---|
| Methyl acetate |  | 1500 |
| Ethyl acetate | 135–310 | 2000 |
| Vinyl acetate | 9 | 400 |
| Butyl acetate | 44–70 | 1200 |
| 2-Methoxyethyl acetate | 150 | 1000 |
| 2-Butoxyethyl acetate | 25 | 800 |
| 2-Ethoxyethyl acetate | 90–116 |  |
| 3-Methoxybutyl acetate |  | > 2500 |
| 2-(2-Butoxyethoxy)ethyl acetate | 100 | 1000–2000 |
| Methyl acetoacetate | 250 | 500 |
| Ethyl acetoacetate | 173–412 | 3000 |
| N-Butyl glycolate | 50 | 100 |
| Methyl crotonate |  | 1250 |
| Butyl acrylate | 9 | 80 |
| Methyl dimethylacrylate |  | 200 |
| Dimethyl acetylsuccinate | 100 | 1500 |
| Diethyl acetylsuccinate | 50 | 2500 |
| Dimethyl maleate | 25 | 100 |
| Monomethyl maleate | 50 | 600 |
| Diethyl maleate | 25 | 800 |
| Dibutyl maleate | 12 | 40 |
| Di(2-ethylhexyl) maleate | 40 | 20 |
| Methyl 3-aminobenzoate | 100 | 1000 |
| Methyl 4-hydroxybenzoate | 100 | 1000 |
| Methyl 4-hydroxyphenylacetate |  | 400 |

\* DIN 38 412, Part 15.
\*\* Measurement of $CO_2$ evolution in fermentation of bacterial sludge; $EC_0$ is the concentration at which $CO_2$ evolution is not suppressed.

If esters are stored in tanks, a nitrogen blanket and a carbon filter is necessary. In some cases, the ventilation pipe must be linked to a flare.

Mild steel, aluminum, or other metallic materials can be used for storage and transportation. Plastics are unsuitable because the highly lipophilic esters may migrate into the container walls and cause them to soften or even dissolve. Esters with a high melting point, such as waxes, can also be transported in paper sacks or wooden barrels.

# 9. Uses and Economic Aspects

**Solvents.** Because of their highly lipophilic and hydrophobic nature and low polarity, esters are widely used as solvents, extractants, and diluents [48], [49]. Lower esters, especially the acetates of methanol, ethanol, propanols, and butanols, are good solvents for cellulose nitrate and other cellulose derivatives. Ethyl acetate is the most common technical solvent. Branched esters and esters of ethylene glycol also have good solvent properties. Considerable quantities of esters are used as diluents in paints, lacquers,

Table 7. Typical specifications of commercial esters

| | Color, max.[a] | $n_D^{20}$ | $d_{20}^{20}$ | Distillation range, °C (101.3 kPa) | Freezing point, °C | Flash point[b], °C | Autoignition temperature, °C | Lower flammable limit, vol % in air (t, °C) | Evaporation number (diethyl ether = 1.0) | Ignition group[d] |
|---|---|---|---|---|---|---|---|---|---|---|
| Methyl acetate | 10 | 1.3601 | 0.930 | 56–58 | | −13 | 455 | 3.1 | 2.2 | IA |
| Ethyl acetate | 10 | 1.371–1.373 | 0.889 | 76–78 | −84 | −4 | 460 | 2.1(38) | 2.9 | IA |
| Propyl acetate | 15 | 1.3844 | 0.885 | 99.0–103.0 | −93 | 14 | 457 | 1.71(38) | | IB |
| Isopropyl acetate | 10 | 1.3772 | 0.872 | 85.0–90.0 | −62 | 6 | 479 | 1.76(38) | | IB |
| Butyl acetate | 10 | 1.394 | 0.880 | 124–127 | −74 | 25 | 422 | 1.38(38) | 14 | IC |
| Isobutyl acetate | 10 | 1.390 | 0.870 | 115–118 | −99 | 18 | 478 | 1.27(93) | 8 | |
| 2-Ethylhexyl acetate | 15 | 1.4103 | 0.872 | 192.0–205.0 | −93 | 79 | 268 | 0.76(93) | | IIIA |
| Vinyl acetate | 5 | 1.369 | 0.93 | 72–74 | −93 | −8 | | | | IA |
| Ethylene glycol diacetate | 15 | 1.4159 | 1.107 | 187.9–193.0 | −42 | 99 | 482 | 1.6(135) | | IIIA |
| 2-Methoxyethyl acetate | 20 | 1.4602 | 1.003 | 142–148 | | 44 | 400 | 1.7 | 34 | II |
| 2-Ethoxyethyl acetate | 15 | 1.406 | 0.972 | 152–158 | −61 | 51 | 392 | 1.24(93) | 60 | II |
| 2-Butoxyethyl acetate | 15 | 1.4200 | 0.94 | 186.0–194.0 | −64 | 81 | 340 | 0.88(93) | | IIIA |
| 2-(2-Ethoxyethoxy)ethyl acetate | 15 | 1.4230 | 1.011 | 214.0–221.0 | −25 | 107 | 360 | 0.98(135) | | IIIB |
| 2-(2-Butoxyethoxy)ethyl acetate | 15 | 1.4265 | 0.978 | 244–250 | −32 | 116 | 290 | 0.76(135) | 1200 | IIIB |
| Glyceryl triacetate | 5 | 1.4296 | 1.160 | 258.0 | 3.2 | 153 | 432 | 1.05(189) | | IIIB |
| Glyceryl tripropionate | | 1.4314 | 1.09 | 285.0 | −58 | 167 | 421 | 0.8(186) | | IIIB |
| Isobutyl isobutyrate | 15 | 1.3990 | 0.855 | 144.0–151.0 | −80 | 38 | 432 | 0.96(93) | | II |
| Bis(2-ethylhexyl)adipate | 20 | 1.4472 | 0.927 | 417 | <−70 | 206 | 377 | 0.38(242) | | IIIB |
| Methyl acetoacetate | 20 | 1.411 | 1.076 | 166–173 | | 62 | 280 | 1.8 | 120 | IIIA |
| Ethyl acetoacetate | 20 | 1.420 | 1.030 | 175–184 | | 65 | 350 | 1 | 135 | IIIA |
| Butyl glycolate | 10 | 1.425 | 1.023 | 183–197 | | 75 | 404 | 0.6 | 460 | IIIA |
| Methyl acrylate | 10 | | 0.9568 | 79.8–80.3 | | 10 | | | | IA |
| Ethyl acrylate | 10 | | 0.9235 | 98.8–99.8 | | 10 | | | | IA |
| Butyl acrylate | 10 | | 0.9009 | 145.7–148.0 | | 49 | | | | II |
| 2-Ethylhexyl acrylate | 10 | | 0.8862 | 214.8–218.0 | | 91 | | | | IIIA |
| Dimethyl phthalate | 5 | 1.513 | 1.192 | 284 | −1 | 157 | 490 | 0.94(181) | | IIIB |
| Diethyl phthalate | 10 | 1.4990 | 1.120 | 298 | <−50 | 161 | 457 | 0.75(187) | | IIIB |

**Table 7.** (continued)

| | Color, max.[a] | $n_D^{20}$ | $d_{20}^{20}$ | Distillation range, °C (101.3 kPa) | Freezing point, °C | Flash point[b], °C | Autoignition temperature, °C | Lower flammable limit, vol % in air ($t$, °C) | Evaporation number (diethyl ether = 1.0) | Ignition group[d] |
|---|---|---|---|---|---|---|---|---|---|---|
| Dibutyl phthalate | 15 | 1.4905 | 1.048 | 340 | −35 | 190 | 404 | 0.47(236) | | IIIB |
| Bis(2-ethylhexyl) phthalate | 20–25 | 1.4836 | 0.9852 | 384 | −50 | 216 | 391 | 0.28(246) | | IIIB |
| Dimethyl terephthalate | < 15 | | | 288 | 140.6 | 153 | 519 | | | IIIB |
| Bis(2-ethylhexyl) terephthalate | 15 | 1.4867 | 0.9825 | 383 | −48 | 238 | 399 | | | IIIB |
| Bis(2-ethylhexyl) trimellitate | 40 | 1.4832 | 0.989 | 600 | −38 | 263 | 410 | 0.26 | 2.5 | IIIB |

[a] Pt–Co scale.
[b] According to ASTM D 56-52, similar to DIN 51 755 and 51 758.
[c] ASTM D-2155, similar to DIN 51 794.
[d] According to NFPA (National Fire Prevention Association).

**Table 8.** Production of miscellaneous esters in 1985, t

|  | CAS registry number | United States | Western Europe | Japan |
|---|---|---|---|---|
| Methyl acetate | [79-20-9] |  | 17 350* |  |
| Ethyl acetate | [141-78-6] | 87 270 | ca. 100 000* | 117 388 |
| Propyl acetate | [109-60-4] | 27 700 |  |  |
| n-Butyl acetate | [123-86-4] | 81 430 |  | 29 942 |
| Isobutyl acetate | [110-19-0] | 34 880 | ca. 50 000* |  |
| Vinyl acetate | [108-05-4] | 960 200 | > 200 000* | 402 930 |
| 2-Ethoxyethyl acetate | [111-15-9] | 46 840 |  |  |
| Ethyl acetoacetate | [141-97-9] |  | 19 450* |  |
| Acrylic acid esters, total |  |  | 250 000 | 103 880 |
| Butyl acrylate | [141-32-2] | 192 560 | 88 000 | 33 000 |
| Methyl acrylate | [96-33-3] | ca. 10 000 | 50 000 | 25 000 |
| Ethyl acrylate | [140-88-5] | 137 800 | 62 000 | 16 000 |
| 2-Ethylhexyl acrylate | [103-11-7] | 36 020 | 50 000 | 29 000 |
| Methyl methacrylate | [80-62-6] | 390 060 | ca. 200 000* | 236 690 |
| Dibutyl maleate | [105-76-0] | 1 540 |  |  |
| Dimethyl terephthalate | [120-61-6] | ca. $2 \times 10^6$ | ca. 500 000* | 327 870 |

* Federal Republic of Germany.

and coatings. Methyl acetate is employed in fast-drying lacquers. Due to the development of low-solvent and solvent-free paints, e.g., for auto paints, the use of esters as diluents is decreasing. Table 8 gives production data for some acetates.

**Plasticizers.** Large quantities of esters, especially phthalates, adipates, and fatty acid esters, are used as plasticizers [50], [51]. Production data for some important plasticizers are listed in Table 9. Isooctanoic acid esters and 2-ethylhexanoic acid esters of ethylene glycol are plasticizers for poly(vinyl choride).

**Surfactants.** Natural fats, oils, and waxes are used in considerable quantities as raw materials in the production of soap (alkali salts of fatty acids) and other detergents and surfactants (Table 10). Hydroxyethyl esters of long-chain fatty acids are sold as nonionic surfactants (nonionics), which are easily biodegradable. About 34 100 t of ethoxylated natural fats and oils were produced in the Federal Republic of Germany in 1985, as compared to only 1380 t in Japan. A defined number of ethylene oxide molecules can be added to the fatty acid by varying the process conditions; compounds with a specific water solubility — from moderate to excellent — can thus be obtained. Nonionics are important as low-foaming detergents and as emulsifiers. They are also used in the food, textile, cosmetics, and pharmaceutical industries. Because of their heat and salt resistance, they are suitable as lubricants in deep-well drilling.

**Polyesters.** Higher esters derived from natural products, e.g., cellulose acetate and cellulose nitrate, are of considerable economic importance (→ Cellulose Esters). However, they are being gradually replaced by synthetic polyesters produced from monomers such as acrylates, te-rephthalates, and vinyl acetate (Table 8).

**Table 9.** Production of plasticizers in 1985, t

|  | CAS registry number | United States | Western Europe | Japan |
|---|---|---|---|---|
| Phthalic acid esters, total |  | 516 350 | 960 000 | 356 910 |
| Dibutyl phthalate | [84-74-2] | 9 880 | 22 000* | 16 900 |
| Diethyl phthalate | [84-66-2] | 7 800 |  |  |
| Diisodecyl phthalate | [26761-40-0] | 66 500 | 70 000 | 10 510 |
|  |  |  | 39 850* |  |
| Dimethyl phthalate | [131-11-3] | 3 480 |  |  |
| Dioctyl phthalate | [117-81-7] | 125 180 | 525 000 | 240 720 |
|  |  |  | 222 320* |  |
| Ditridecyl phthalate | [119-06-2] | 9 910 |  |  |
| Trimellitic acid esters, total |  | 22 070 |  |  |
| Other cyclic esters, total |  | 23 060 |  |  |
| Adipic acid esters, total |  | 56 900 | 28 000 | 23 740 |
| Bis(2-ethylhexyl) adipate | [103-23-1] | 16 870 |  | 13 000 |
| Complex linear polyesters, total |  | 22 510 |  | 11 260 |
| Ethoxylated esters, total |  | 51 130 | 37 000 | 17 900 |
| Oleic esters, total |  | 5 240 |  |  |
| Butyl oleate | [142-77-8] | 760 |  |  |
| Palmitic acid esters |  | 1 550 |  |  |
| Sebacic acid esters |  | 1 650 | 7 000 |  |
| Stearic acid esters, total |  | 4 500 |  |  |
| Butyl stearate | [123-95-5] | 3 500 |  |  |

* Federal Republic of Germany.

**Table 10.** United States production of esters used as surfactants (1985)

|  | Production, t |
|---|---|
| Carboxylic acid esters, total | 117 800 |
| Anhydrosorbitol esters, total | 15 320 |
| Monostearate | 8 080 |
| Monooleate | 3 080 |
| Ethoxylated anhydrosorbitol esters, total | 14 020 |
| Monostearate | 5 040 |
| Monooleate | 3 480 |
| Glycerol esters, total | 27 600 |
| Monostearate | 5 070 |
| Monooleate | 2 790 |
| Glycerol esters of mixed acids | 14 260 |
| Diethylene glycol esters, total | 8 900 |
| Poly(ethylene glycol) esters, total | 21 170 |
| Monostearate | 2 700 |
| Monooleate | 1 620 |
| Monolaurate | 2 290 |
| Natural fats and oils, ethoxylated | 15 000 |
| Castor oil. ethoxylated | 6 260 |

Vinyl acetate, with a worldwide production of $2 \times 10^6$ t/a, is the most important aliphatic ester. Poly(vinyl acetate) is produced in large quantities for use in plastics, coatings, adhesives, and laminates. A considerable proportion is converted into poly(vinyl alcohol) [38]. Polymerization of acrylates such as methyl acrylate and methyl

**Table 11.** United States production of flavor and perfume esters (1985)

|  | CAS registry number | Production, t |
|---|---|---|
| Cedryl acetate | [77-54-3] | 90 |
| Geranyl acetate | [105-87-3] | 60 |
| Isoamyl acetate | [123-92-2] | 51 |
| Isoamyl butyrate | [106-27-4] | 45 |
| 1,3-Nonanediol diacetate | [63270-14-4] | 41 |
| Citronellyl acetate | [150-84-5] | 35 |
| Benzyl propionate | [122-63-4] | 20 |
| 2-Phenylethylphenyl acetate | [102-20-5] | 15 |
| Vetivenyl acetate | [117-98-6] | 13 |
| Isopentyl isovalerate | [659-70-1] | 11 |
| Allyl hexanoate | [123-68-2] | 11 |
| Necyl acetate | [141-12-8] | 10 |
| Citronellyl formate | [105-85-1] | 9 |

methacrylate yields homopolymers and mixed polymers used in paints, lacquers, and coatings. Polycondensation of dicarboxylic acids and diols gives polyester resins and plastics. Linear polyesters in particular are produced in large quantities.

Linear polyesters are used in the manufacture of synthetic fibers, packaging films, tapes, films for electric insulation, and a variety of plastics articles such as plastic bottles; they are also employed in hot-melt adhesives.

Polyesters are produced chiefly by transester-ification. The most important raw material is dimethyl terephthalate. This undergoes polycondensation with a diol (e.g., 1,2-ethanediol, 1,2-propanediol, 1,4-butanediol, 1,4-cyclohexanedimethanol), and methanol is released. However, direct esterification of terephthalic acid with ethylene oxide and polymerization of the resulting bis(2-hydroxyethyl) terephthalate is also used (see Section 5.11). Copolymerization with unsaturated acids such as maleic acid yields polymers that can be cross-linked (thermosetting polymers).

Alkyd resins are an important group of polyesters used mainly in lacquers. The stability of the lacquer, e.g., its weatherability, can be improved by adding esters that are highly resistant to hydrolysis, such as pivalic acid esters.

**Flavors and Fragrances.** Certain esters with a pleasant odor are used in fragrances, flavors, cosmetics, and soaps. Production data for the most important flavor and perfume esters are given in Table 11. Although the quantities that are used are small, these esters are of great commercial importance because they are expensive [52]. Terpene esters as well as simple esters such as butyl and isobutyl acetate (fruity flavor) and benzyl acetate (jasmine odor) are widely used. These esters can be obtained from plant extracts, but many are currently produced by synthetic processes.

**Other Uses.** Esters can be converted into various derivatives (see Chap. 3) and are useful *intermediates* in preparative chemistry. Ethyl formate is used in the synthesis of vitamin $B_1$. Methyl formate is converted on a large scale into formamide and formic acid; about 30 000 t/a of formic acid are produced by this route.

Natural oils, fats, and waxes are used as *lubricants* for high-speed engines. High-boiling synthetic esters that experience little change in viscosity when exposed to extremes of temperature, either high or low, have been developed for use in auto and aircraft turboengines. Examples are bis(2-ethylhexyl) sebacate [*122-62-3*] and glycol or polyglycol esters of branched-chain carboxylic acids. These esters are also used as hydraulic fluids.

An example of the use of esters in *pharmaceuticals* is the common analgesic aspirin [*50-78-2*] (2-acetoxybenzoic acid. The active salicylic acid is released only after hydrolysis in the stomach. Benzocaine [*94-09-7*], the ethyl ester of 4-aminobenzoic acid, is a topical anesthetic. Phenyl salicylate [*118-55-8*], Salol, has antipyretic, antirheumatic, and antiparasitic properties.

# 10. Toxicology and Occupational Health

With a few exceptions, esters have very low toxicity. Natural fats and oils are widely used in human nutrition. Table 12 lists toxicity data for commercially important esters [53], [54] – [56].

The ester group is nontoxic. However, all esters are hydrolyzed more or less rapidly on contact with water; therefore, the toxicity of their hydrolysis products must be taken into account [53].

Most esters readily penetrate skin and mucous membranes. The considerable absorption through the skin can cause health problems. Exposed skin should be thoroughly washed with soap and plenty of water; clothing should be changed.

Inhalation of highly volatile esters in high concentrations can have a narcotic effect. Secondary effects include breathlessness and fits of choking. Ester vapors should not be allowed to contact the eye. Good ventilation of the work area and wearing of breathing apparatus are generally recommended [53].

**Table 12.** Toxicity data for various esters

| | LD$_{50}$ (rat, oral), g/kg | Inhalation (rat, 4–6 h) | TLV, ppm | TLV, mg/m$^3$ | Comments |
|---|---|---|---|---|---|
| Methyl formate | 1.622[a] | 10 000 ppm: lethal[c] | 100 | 250 | irritation and burning of the eyes; narcotic; readily absorbed through the skin |
| Methyl acetate | 3.7[a] | 11 220 ppm: lethal[d] | 200 | 610 | irritation of mucous membranes |
| Ethyl acetate | 11.3 | 8 000 ppm: survival<br>16 000 ppm: lethal | 400 | 1400 | irritation of the eyes |
| Propyl acetate | > 3.2 | 5 300 ppm: survival | 200 | 840 | irritation of eyes, nose, and throat; defatting of skin |
| Isopropyl acetate | 3.0–6.5 | 32 000 ppm: lethal | 250 | 950 | irritation of eyes and respiratory tract |
| n-Butyl acetate | > 6.4 | 1 300 ppm: survival<br>14 000 ppm: lethal | 150 | 710 | |
| Isobutyl acetate | 3.2–6.4 | 3 000 ppm: survival<br>23 000 ppm: lethal | 150 | 700 | irritation of eyes and respiratory tract |
| Amyl acetate | 16.6 | 5 000 ppm: survival | 100 | 530 | narcotic |
| 2-Ethylhexyl acetate | 3.2–5.9 | > 1100 ppm: survival | – | – | slightly irritating to the eyes |
| 2-Methoxyethyl acetate | 5.1–11.9 | < 450 ppm: no effect | 5 | 24[e] | |
| 2-Ethoxyethyl acetate | 4.94[b] | | 5 | 27[e] | slight skin irritation |
| Ethylene glycol mono- and diacetate | 1.6–6.9 | | – | – | |
| Glyceryl triacetate | 6.4–12.8 | 8 200 ppm: no effect | – | – | |
| Glyceryl tripropionate | 1.6–3.2<br>6.4–12.8[b] | 600 ppm: no effect | – | – | |
| Isobutyl isobutyrate | 6.4–12.8 | 650 ppm: no effect<br>5 400 ppm: lethal | – | – | |

[a] Rabbit.
[b] Guinea pig.
[c] 5 % (20–30 min) or 2.5 % (30–60 min) are also lethal to guinea pigs.
[d] Mouse.
[e] Skin notation (considerable cutaneous absorption).

# 11. References

General References

S. Patai (ed.): *The Chemistry of Carboxylic Acids and Esters,* Wiley-Interscience, New York 1969. J. D'Ans, E. Lax: *Taschenbuch für Chemiker und Physiker,* 3rd ed., Springer Verlag, Berlin–Heidelberg–New York 1970.

Specific References

[1] J. D. Fletcher, O. C. Dermer, R. B. Fox (eds.): *Nomenclature of Organic Compounds,* Am. Chem. Soc., Washington D.C. 1974, pp. 137–145. IUPAC commission on nomenclature of organic chemistry:
*Nomenclature of Organic Chemistry,* 4th ed., Pergamon Press, Oxford-New York 1979.
[2] K. S. Markley (ed.): *Fatty Acids, Their Chemistry, Properties, Production and Uses,* 2nd ed., Interscience, New York 1960/61.
[3] D. Swern (ed.): *Bailey's Industrial Oil and Fat Products,* Wiley-Interscience, New York 1969, p. 937.
[4] J. Smidt et al., *Angew Chem.* **74** (1962) 93.
[5] Unilever, DE 2 327 729, 1973 (J. J. Muller, T. J. Kock).
[6] Rohm & Haas, DE 1 932 625, 1969 (W. R. Winslow).
[7] Degussa, DE 1 568 376, 1966 (G. Morlock, H. Trageser).
[8] A. M. Schwartz, J. W. Perry, J. Berch: *Surface-Active Agents and Detergents,* vol. **2,** Interscience, New York 1958, p. 104.
[9] H. Adkins, *Org. React. (N.Y.)* **8** (1954) 1–27.
[10] N. G. Gaylord: *Reduction with Complex Metal Hydrides,* New York 1956, pp. 391–543.
[11] C. B. Hauser, B. E. Hudson, Jr., *Org. React. (N.Y.)* **1** (1947) 266, 302.
[12] J. J. Bloomfield et al., *Org. React. (N.Y.)* **23** (1976) 259–403.
[13] *Houben-Weyl,* **E5,** 656–714.
[14] H. R. Harrison, W. M. Heynes et al., *Chem. Ind. (London)* 1968, 1568.
[15] J. R. Rachele, *J. Org. Chem.* **28** (1963) 2898.
[16] I. Roberts, H. C. Urey, *J. Am. Chem. Soc.* **60** (1938) 2391.
[17] H. Zimmermann, J. Rudolf, *Angew. Chem.* **77** (1965) 65–74; *Angew. Chem. Int. Ed. Engl.* **4** (1965) 40.
[18] R. Bader, A. D. Kontowicz, *J. Am. Chem. Soc.* **75** (1953) 5116.
[19] Knapsack, DE 2 226 829, 1972 (E. Lohmar, F. X. Werber).
[20] G. A. Olah et al., *Synthesis* 1978, 929 ff.
[21] H. J. Bergk, *Z. Chem.* **18** (1978) 22 ff.
[22] G. Hallas, *J. Chem. Soc.* 1965, 5770.
[23] B. F. Goodrich Comp. DE 1 005 947, 1955 (W. L. Bears, F. X. Werber). B. F. Goodrich Comp., DE 1 103 335, 1956 (F. X. Werber, S. J. Averill). Reichhold Chemie AG, DE 1 173 473, 1962 (H. Dalibar). Reichhold Chemie AG, DE-AS 1 185 611, 1963 (H. Dalibar). ICI, DE-AS 1 292 140, 1966 (D. K. Steel). Scholven Chemie AG, DE-OS 1 593 503, 1966 (K. S. Schmitt, W. Flakus et al.). Mobil Oil Comp., DE-AS 1 618 639, 1967 (A. B. Gainor, L. McMakin) British Titan Products Comp., DE-OS 1 807 103, 1968 (R. H. Stanley, D. W. Brook, et al.). Hüls AG, DE-AS 2 243 240, 1972 (F. List, K. Wember) F. Brill, L. Ferry, I. T. Baker, *High Polym.* **27** (1972) 506.

[24]  Hüls, DE 914 006, 1951 (A. Coenen, F. Broich).
[25]  E. Taschner, C. Wasielewski, *Liebigs Ann. Chem.* **640** (1961) 136–139.
[26]  R. C. Parish, C. M. Stock, *J. Org. Chem.* **30** (1965) 927.
[27]  S. Natelson, S. Gottfried, *Org. Synth., Coll. Vol.* **3** (1955) 381.
[28]  Ruhrchemie, DE 1 171 897, 1959 (H. Feichtinger, H. Noerke).
[29]  BASF, DE 878 348, 1942 (H. Dierichs, E. Braun, et al.).
[30]  W. William, B. Halper, *Synthesis* 1974, 727.
[31]  G. Schwabe, G. Westphal, H. G. Hennig, *Z. Chem.* **20** (1980) 184. S. Szeja, *Synthesis* 1980, 402.
[32]  J. Cason, *Org. Synth., Coll. Vol.* **3** (1955) 169.
[33]  W. Steglich, DE 1 958 954, 1969 (W. Steglich, G. Höfle).
[34]  Hoechst, DE 1 956 558, 1969 (E. Fischer, O. Mautz).
[35]  J. Falbe: *New Syntheses with Carbon Monoxide*, Springer Verlag, Heidelberg 1980.
[36]  F. Knifton, *J. Org. Chem.* **41** (1976) 2885.
[37]  C. Cassar, G. P. Chiuroli, F. Guerrieri, *Synthesis* 1973, 509.
[38]  K. Weissermel, H. J. Arpe: *Industrielle Organische Chemie*, 2nd ed., Verlag Chemie, Weinheim – New York 1978, pp. 42–43, pp. 216–221.
[39]  F. L. James, W. H. Bryan, *J. Org. Chem.* **23** (1958) 1225.
[40]  Mitsui Toatsu, JP 4 725 120, 1971.
[41]  Röhm & Haas, DE 1 468 939, 1965 (G. Schröder, H. Fink).
[42]  W. Schwerdtel, *Chem. Ing. Tech.* **40** (1968) 781. Hoechst AG, DE 1 191 366, 1961 (W. Riemenschneider, T. Quadflieg). Hoechst AG, DE-AS 1 618 391, 1967 (L. Hörnig, H. Fernholz, et al.) DE-AS 1 643 999, 1967 (H. Fernholz, H.-J. Schmidt). DE-AS 1 667 140, 1967 (H. Erpenbach, H. Glaser, et al.). Knapsack AG, DE-OS 1 668 352, 1967 (K. Sennewald, W. Vogt, et al.). Asahi Kasei Kogyo, DE 1 768 643, 1968; DE-AS 1 793 362, 1968 (N. Kominami, H. Nakajima). Bayer AG, DE 1 793 519, 1968 (W. Krönig, G. Scharfe). Knapsack AG, DE-AS 1 808 610, 1968 (K. Sennewald, W. Vogt). Hoechst AG, DE-OS 2 315 037, 1973 (H. Fernholz, H. Krekeler, et al.). Celanese Corp., DE-OS 2 361 098, 1973 (G. M. Severs, Jr.). General Electric Co., DE-OS 2 503 748, 1975; 2 503 926, 1975 (W. E. Smith, R. J. Gerhart). H. Krekeler, H. Schmitz, *Chem. Ing. Tech.* **40** (1968) 785. Hoechst AG, DE-AS 1 668 063, 1967 (G. Roscher, H. Schmitz).
[43]  Hoechst AG, DE 1 903 954, 1969 (H. Fernholz, H.-J. Schmidt). Air Products & Chemicals, DE-AS 1 768 770, 1968 (J. G. Schukys). General Electric Co., DE-OS 2 504 230, 1975; 2 504 231, 1975 (W. E. Smith, R. J. Gerhart).
[44]  ICI, DE 1 468 086, 1965 (D. Clark, D. Wright).
[45]  L. J. Bellamy, *The Infrared Spectra of Complex Molecules*, 3rd ed., J. Wiley & Sons, New York, 1975. L. J. Bellamy, *Advances in Infrared Group Frequencies*, Methuen & Co., London 1968, pp. 166–168.
[46]  L. D. Metcalfe, A. A. Schmitz, *Anal. Chem.* **33** (1961) 363.
[47]  T. S. Ma in S. Patai (ed.): *The Chemistry of Carboxylic Acids and Esters*, Wiley-Interscience, New York 1969, pp. 871–921.
[48]  E. H. Davies, T. H. Durrans: *Solvents*, 2nd ed., Chapman & Hall, London 1971, pp. 145–165.
[49]  C. Marsden, S. Mann (eds.): *Solvents Guide*, 2nd ed., Interscience Publishers, New York 1963.
[50]  D. N. Buttrey: *Plasticizers*, 2nd American ed., Franklin Publishing Co., Palisades, New York 1960.
[51]  I. Mellan: *Industrial Plasticizers*, Mac Millan, New York 1963.
[52]  D. L. J. Opdyke: "Monographs on Fragrance," *Food Cosmet. Toxicol.* **16** (1978) Suppl. 1, 839–841; **17** (1979) Suppl., 735–741, 841–843; **19** (1981) Suppl. 2, 237–245.

[53] N. I. Sax: *Dangerous Properties of Industrial Materials,* 4th ed., Reinhold Publishing Corp., New York 1975.
[54] *Threshold Limit Values for Chemical Substances in the Work Environment,* ACGIH Cincinnati, 1986.
[55] M. Windholz (ed.): *The Merck Index,* 10th ed., Merck & Co, Rahway, N.J., 1983.
[56] E. E. Sandmeyer, C. J. Kirwin in G. D. Clayton, F. E. Clayton (eds.): *Patty's Industrial Hygiene and Toxicology,* 3rd ed., vol. **2A,** Wiley-Interscience, New York 1981, pp. 2259 ff.
[57] D. L. J. Opdyke, *Food Cosmet. Toxicol.* **12** (1974) 719.
[58] Hoechst AG, unpublished results.

# Ethanol

Naim Kosaric, The University of Western Ontario, London, Ontario N6A 5B9, Canada (Chaps. 2, 3, 5.1, 5.3, 5.4, 8, and 9)

Zdravko Duvnjak, University of Ottawa, Ottawa, Ontario K1N 9B4, Canada (Chaps. 2, 3, 5.1, 5.3, 5.4, 8, and 9)

Adalbert Farkas, Consulting Chemist, Delray Beach, Florida 33483, United States (Chap. 4)

Hermann Sahm, Institut für Biotechnologie der KFA Jülich GmbH, Jülich, Federal Republic of Germany (Section 5.2)

Stefanie Bringer-Meyer, Institut für Biotechnologie der KFA Jülich GmbH, Jülich, Federal Republic of Germany (Section 5.2)

Otto Goebel, CORA Engineering, Chur, Switzerland (Chaps. 6, 7, and 10)

Dieter Mayer, Hoechst Aktiengesellschaft, Frankfurt, Federal Republic of Germany (Chap. 11)

| | | | | | |
|---|---|---|---|---|---|
| 1. | Introduction | 2048 | 5.1.4. | Direct Fermentation | 2074 |
| 2. | Physical Properties | 2049 | 5.2. | **Production by Bacteria** | 2075 |
| 3. | Chemical Properties | 2050 | 5.2.1. | Ethanol Tolerance | 2077 |
| 4. | Synthesis | 2052 | 5.2.2. | Fermentation, of saccharified Waste Starch | 2078 |
| 4.1. | Direct Catalytic Hydration of Ethylene | 2053 | 5.2.3. | Fermentation of Pentoses | 2080 |
| 4.1.1. | Chemistry | 2053 | 5.3. | **Fermentation Modes** | 2080 |
| 4.1.2. | Catalysts | 2054 | 5.3.1. | Batch Processes | 2081 |
| 4.1.3. | Production Process | 2055 | 5.3.2. | Continuous Processes | 2082 |
| 4.2. | Indirect Hydration of Ethylene | 2056 | 5.3.3. | Other Processes | 2084 |
| 4.2.1. | Chemistry | 2056 | 5.4. | **Raw Materials and Processes** | 2091 |
| 4.2.2. | Production Process | 2057 | 5.4.1. | Readily Fermentable Carbohydrates | 2091 |
| 4.3. | Other Methods | 2058 | 5.4.2. | Starch | 2095 |
| 4.3.1. | Homologation of Methanol | 2059 | 5.4.3. | Lignocellulosic Materials | 2101 |
| 4.3.2. | Carbonylation of Methanol and Methyl Acetate | 2059 | 5.4.4. | Waste Materials and Residues | 2105 |
| 4.3.3. | Conversion of Synthesis Gas | 2061 | 6. | **Recovery and Purification** | 2112 |
| 4.3.3.1. | Synthesis by Heterogeneous Catalysis | 2061 | 6.1. | **Distillation** | 2114 |
| | | | 6.1.1. | Distillation of Azeotropic Ethanol | 2114 |
| 4.3.3.2. | Synthesis by Homogeneous Catalysis | 2062 | 6.1.2. | Dehydration by Azeotropic Distillation | 2117 |
| | | | 6.1.3. | Motor Fuel Ethanol | 2118 |
| 5. | **Fermentation** | 2062 | 6.1.4. | Reduction of Energy Costs | 2120 |
| 5.1. | **Production by Yeast** | 2062 | 6.2. | **Nondistillative Methods** | 2121 |
| 5.1.1. | Nutrients | 2063 | 6.2.1. | Solvent Extraction | 2121 |
| 5.1.2. | Fermentation Pathways | 2063 | 6.2.2. | Carbon Dioxide Extraction | 2121 |
| 5.1.3. | Fermentation Variables | 2071 | 6.2.3. | Adsorptive Dehydration | 2122 |

| | | | | |
|---|---|---|---|---|
| 6.2.4. | Membrane Technology . . . . . . 2122 | 8. | Analysis . . . . . . . . . . . . . . . . | 2130 |
| 6.2.5. | Yarn-Filled column . . . . . . . . . . 2123 | 9. | Uses . . . . . . . . . . . . . . . . . . . . | 2131 |
| 6.3. | Storage and Transportation . . 2124 | 10. | Economic Aspects . . . . . . . . . | 2133 |
| 7. | Comparison of Process Economics for Synthetic and Fermentation Ethanol . . . . . . 2124 | 10.1. | Worldwide Production of Synthetic and Fermentation Ethanol . . . . . . . . . . . . . . . . . | 2133 |
| 7.1. | Summary of Cost Analysis for Synthetic and Fermentation Ethanol . . . . . . . . . . . . . . . . . 2124 | 10.2. | Major Producers of Fermentation Ethanol from Regenerable Resources . . . . . | 2135 |
| 7.2. | Production Costs of Synthetic Ethanol . . . . . . . . . . . . . . . . . 2125 | 11. | Toxicology . . . . . . . . . . . . . . . | 2137 |
| 7.3. | Production Costs of Fermentation Ethanol . . . . . . 2127 | 12. | References . . . . . . . . . . . . . . . | 2138 |

# 1. Introduction

Ethanol [64-17-5] or ethyl alcohol ($CH_3CH_2OH$), $M_r$ 46.7, is also referred to as alcohol spirit, spirit of wine, grain alcohol, absolute alcohol, and ethyl hydrate. Depending on its water content, preparation, and final use, several ethanol products exist on the market. The 99% alcohol (often referred to as absolute alcohol) is used extensively for tinctures and pharmaceutical preparations, as a solvent and preservative, as an antiseptic, and in perfume. Ethanol is an important functional component of alcoholic beverages, which are produced by fermentation of fermentable carbohydrates. The fermentation broth itself may constitute (after processing and aging) a beverage, e.g., in the case of beer or wine, or the alcohol can be concentrated from the broth to produce high-alcohol-containing spirits. If the alcohol is used for purposes other than as a beverage, it is denatured by the addition of substances such as methanol, pyridine, formaldehyde, or sublimate. The denatured alcohol is then used by industry and commerce, principally as a solvent, as a raw material for manufacturing chemicals, or as a fuel.

Chemically synthesized ethanol is usually derived from petroleum sources by the hydration of ethylene (see Chap. 4) and has found wide application as industrial alcohol. Various routes for the production of ethanol are depicted in Figure 1.

**History.** Ethanol can be considered to be one of the oldest human food products. Fermented beer was consumed in Babylon, and wine was produced as early as 3000 B.C. The distillation process probably orginated in the 10 – 14th centuries. At this time, the "spiritual" effect of ethanol was recognized; thus, the name *spiritus* was given to alcoholic drinks. The first wine distillates were used for medicinal rather than beverage purposes.

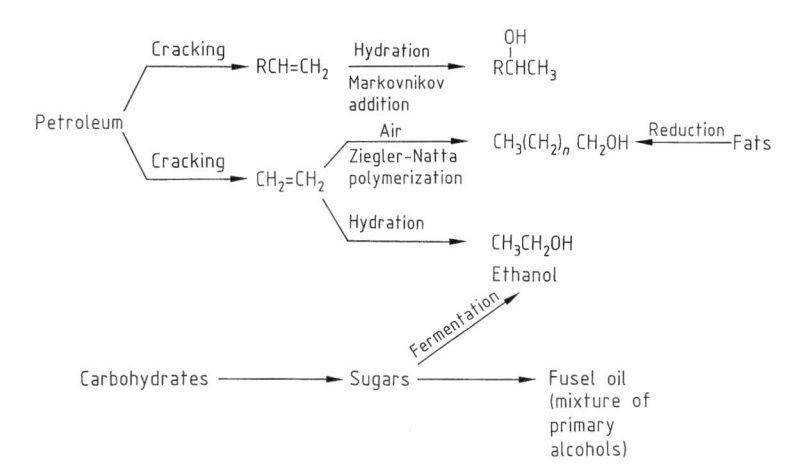

**Figure 1.** Industrial sources of ethanol

Until the 17th century, alcoholic fermentation was considered to be a spoilage process whereby the yeast produced was eliminated. The nature of fermentation was initially clarified in the 19th century with the discovery of the microscope, which showed that yeast cells were living organisms. However, recognition of the fact that these living organisms are responsible for the fermentation process took about 150 years.

In the 19th century, two theories were developed to explain the mechanism of fermentation: the "mechanistic" and the "vital" processes. LOUIS PASTEUR (1822–1895) promoted the vital theory, which stated that living organisms were responsible for the conversion of sugar to alcohol. The mechanistic theory was supported by JUSTUS FREIHERR VON LIEBIG (1803–1873) and by FRIEDRICH WÖHLER (1800–1882). A convincing proof of the mechanistic mechanism, by which physicochemical processes lead to chemical conversion of sugar to ethanol, came from EDWARD BUCHNER (1860–1917), who demonstrated that alcoholic fermentation is related not to the living cell but to a substance in the fermentation broth, which was later identified as an enzyme. As is now known, enzymes are ultimately responsible for the complex conversion of carbohydrates to ethanol.

## 2. Physical Properties

Ethanol in its pure form (absolute alcohol) is a colorless liquid. It is miscible in all proportions with water and also with ether, acetone, benzene, and some other organic solvents. Anhydrous alcohol is hygroscopic; at a water uptake of 0.3–0.4%, a certain stability does occur. Various physical properties of anhydrous ethanol are as follows:

| | |
|---|---|
| $bp$ | 78.39 °C |
| $fp$ | −114.15 °C |
| $n^{20}$ | 1.36048 |
| $d_4^{15}$ | 0.79356 |
| $d_4^{20}$ | 0.78942 |
| $d_{15}^{15}$ | 0.79425 |

| | |
|---|---|
| $d_{20}^{20}$ | 0.79044 |
| Surface tension at 20 °C | 22.03 mN/m |
| $C_p$ (16–21 °C) | 2.415 J g$^{-1}$ K$^{-1}$ |
| Heat of fusion | 4.64 kJ/mol |
| Heat of evaporation | |
|   At 70 °C | 855.66 kJ/kg |
|   At 80 °C | 900.83 kJ/kg |
|   At 100 °C | 799.05 kJ/kg |
| Heat of combustion (at constant volume) | 1370.82 kJ/mol |
| Thermal conductivity at 20 °C | 18 µW m$^{-1}$ K$^{-1}$ |
| Dynamic viscosity $\eta$ | 1.19 mPa · s |
| Volumetric expansion coefficient | $1.1 \times 10^{-3}$ K$^{-1}$ * |
| Heat of mixing 30 wt% ethanol and 70 wt% water at 17.33 °C | 39.32 J/g |
| Flash point (in a closed vessel) | 13 °C |
| Autoignition temperature | 425 °C |
| Explosion limit (amount of ethanol in a mixture with air) | |
|   Lower, 3.5 vol% | 67 g/m$^3$ |
|   Upper, 15 vol% | 290 g/m$^3$ |
| Maximum explosion pressure | 736 kN/m$^2$ |
| Specific conductivity | $135 \times 10^{-11}$ $\Omega^{-1}$ cm$^{-1}$ |
| Dilution number at 20 °C (diethyl ether = 1) | 8.2 |
| Diffusion coefficient for vapors at 20 °C and 101.3 kPa | 0.12 cm$^{-1}$ |
| Heating value | |
|   Upper | 29 895 kJ/kg |
|   Lower | 29 964 kJ/kg |

* In practice, the volume increase for 1000 L is taken as 1 L/K.

Tables 1 and 2 show the freezing and flash points of ethanol–water mixtures. Table 3 shows the vapor–liquid equilibria of ethanol–water mixtures; the azeotropic mixture contains 95.57 wt% ethanol and 4.43 wt% water. Therefore, the highest concentration of ethanol, obtained by distillation from an ethanol–water mixture, is 95.57 wt%. Azeotropic distillation, with the help of a tertiary solvent (e.g., benzene), must be introduced to produce absolute (anhydrous) ethanol.

When ethanol is mixed with water, the volume contracts slightly, as shown in Table 4. For example, when 52 volumes of absolute ethanol and 48 volumes of water are mixed, 96.3 volumes of diluted ethanol result.

# 3. Chemical Properties

The chemical properties of ethanol are dominated by the functional —OH group, which can undergo many industrially important chemical reactions, e.g., dehydration, halogenation, ester formation, and oxidation (→ Alcohols, Aliphatic).

Because ethanol can be produced efficiently not only by chemical synthesis from petroleum and coal-based feedstock, but also by fermentation of abundantly available organic materials, its commercial role as a raw material for various chemicals is of increasing importance.

**Table 1.** Freezing points of ethanol–water mixtures

| Ethanol, vol% | fp, °C | Ethanol, vol% | fp, °C |
|---|---|---|---|
| 50 | −36.9 | 30 | −15.3 |
| 45 | −28.1 | 25 | −11.3 |
| 40 | −24 | 20 | −7.6 |
| 38 | −22.3 | 15 | −5.1 |
| 32 | −16.8 | 10 | −3 |

**Table 2.** Flash points of ethanol–water mixtures

| Ethanol, wt% | Flash point, °C | Ethanol, wt% | Flash point, °C |
|---|---|---|---|
| 100 | 13 | 50 | 24.5 |
| 94.5 | 16 | 40 | 26.5 |
| 80 | 19.5 | 30 | 30 |
| 70 | 21.5 | 10 | 46 |
| 60 | 22.6 | 5.5 | 56 |

**Table 3.** Vapor–liquid equilibria and boiling points of the ethanol–water system at 101.3 kPa

| Ethanol in liquid, wt% | Ethanol in vapor, wt% | bp, °C |
|---|---|---|
| 0 | 0.00 | 100.00 |
| 1 | 6.5 | 98.90 |
| 3 | 20.5 | 96.75 |
| 5 | 38.0 | 94.95 |
| 10 | 52.0 | 91.45 |
| 15 | 59.5 | 88.95 |
| 20 | 64.8 | 87.15 |
| 25 | 68.6 | 85.75 |
| 30 | 71.4 | 84.65 |
| 35 | 73.3 | 83.75 |
| 40 | 74.7 | 83.10 |
| 45 | 75.9 | 82.45 |
| 50 | 77.1 | 81.90 |
| 55 | 78.2 | 81.45 |
| 60 | 79.4 | 81.00 |
| 65 | 80.7 | 80.60 |
| 70 | 82.2 | 80.20 |
| 75 | 83.9 | 79.80 |
| 80 | 85.9 | 79.35 |
| 85 | 88.3 | 78.95 |
| 90 | 91.3 | 78.50 |
| 95 | 95.04 | 78.15 |
| 97 | 96.86 | 78.20 |
| 99 | 98.93 | 78.25 |
| 100 | 100.00 | 78.30 |

**Table 4.** Volume contraction for 1 mol of ethanol–water mixture at 15°C

| $x$ *$C_2H_5OH$ | $d_4^{15}$ | $\Delta V$** |
|---|---|---|
| 1.00000 | 0.79354 | 0.0000 |
| 0.97483 | 0.79666 | −0.1063 |
| 0.92674 | 0.80270 | −0.2903 |
| 0.88143 | 0.80851 | −0.4429 |
| 0.81817 | 0.81688 | −0.6250 |
| 0.74155 | 0.82756 | −0.8028 |
| 0.65639 | 0.84032 | −0.9495 |
| 0.56708 | 0.85510 | −1.0540 |
| 0.47734 | 0.87186 | −1.1123 |
| 0.35078 | 0.89980 | −1.1116 |
| 0.22077 | 0.93482 | −0.9695 |
| 0.15550 | 0.95355 | −0.7864 |
| 0.07909 | 0.97311 | −0.4030 |
| 0.05987 | 0.97790 | −0.2869 |
| 0.03726 | 0.98435 | −0.1586 |
| 0.01605 | 0.99196 | −0.0582 |
| 0.00792 | 0.99542 | −0.0267 |
| 0.00000 | 0.99913 | 0.0000 |

\* Mole fraction of ethanol.
\*\* Change in volume.

Conversion of ethanol to "alkochemicals" is an entirely new approach to producing familiar petrochemicals (Fig. 2). Some of these routes are already being used industrially in large alcohol-producing countries.

# 4. Synthesis

The formation of ethanol from ethylene [74-85-1] by reaction with sulfuric acid was demonstrated over a hundred years ago [1], [2]. Economic production of synthetic ethanol on an industrial scale was first accomplished in 1930 by the Carbide and Chemical Corporation (now Union Carbide). The process used involved absorption of ethylene in sulfuric acid and subsequent hydrolytic cleavage of the resulting ethyl sulfates. This indirect process has been gradually superseded by direct catalytic hydration of ethylene, introduced by the Shell Chemical Company in 1948 [3].

More recent routes for ethanol synthesis depend on coal-based starting materials such as methanol [67-56-1] or synthesis gas (a mixture of carbon monoxide and hydrogen). They include homologation and carbonylation processes and the direct conversion of synthesis gas to ethanol.

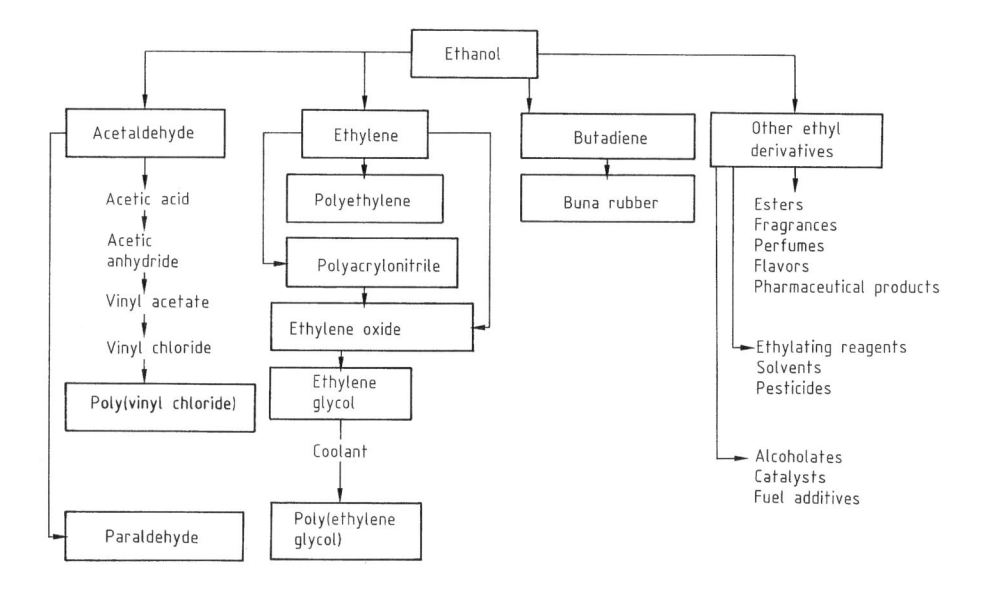

**Figure 2.** Chemicals produced from ethanol

## 4.1. Direct Catalytic Hydration of Ethylene

### 4.1.1. Chemistry

In the temperature range at which industrial catalysts operate, the hydration of ethylene to ethanol is a reversible reaction controlled by the equilibrium

$$C_2H_4 \text{ (g)} + H_2O \text{ (g)} \rightleftharpoons C_2H_5OH \text{ (g)} \qquad \Delta H = -43.4 \text{ kJ} \tag{1}$$

Calculation of the equilibrium is described in [4]–[6]. The kinetics of Reaction (1) have been studied on a phosphoric acid–silica gel catalyst [7], [8] and on a blue tungsten oxide–silica gel catalyst (energy of activation = 125 kJ/mol) [9].

A nomogram for the correlation between ethylene conversion in this reaction and several process variables (pressure, temperature, and the water–ethylene molar ratio) has been developed [10]. Under usual reactor conditions (approximately equimolar ethylene–water feed, 250–300 °C, 5–8 MPa), the equilibrium conversion of ethylene is 7–22%.

A lower temperature favors higher ethylene conversion, but Reaction (1) is then accompanied by reversible formation of diethyl ether [*60-29-7*] according to Reaction (2):

$$C_2H_5OH + C_2H_4 \rightleftharpoons C_2H_5OC_2H_5 \tag{2}$$

At high pressures, ethylene polymerizes to yield butenes and higher olefins, which are converted to higher alcohols by hydration.

The mechanism of ethylene hydration is assumed to consist of four steps [11].

1) Formation of a π-complex by the addition of a proton to the ethylene molecule:

$$\text{C=C} + H_3O^+ \underset{}{\overset{Fast}{\rightleftharpoons}} \left[ \text{C} \overset{H}{\underset{}{\vdots}} \text{C} \right]^+ + H_2O \quad (3)$$
$$\pi\text{-Complex}$$

2) Conversion of the π-complex to a carbocation:

$$\left[ \text{C} \overset{H}{\underset{}{\vdots}} \text{C} \right]^+ \underset{}{\overset{\text{Rate determining}}{\rightleftharpoons}} \left[ -\overset{}{\underset{}{C}} - \overset{}{\underset{+}{C}} - \right] \quad (4)$$

3) Addition of water to the carbocation:

$$\left[ -\overset{}{\underset{}{C}} - \overset{}{\underset{+}{C}} - \right] + H_2O \underset{}{\overset{Fast}{\rightleftharpoons}} -\overset{}{\underset{}{C}} - \overset{}{\underset{}{C}} - \overset{+}{O} \overset{H}{\underset{H}{\diagdown}} \quad (5)$$

4) Removal of a proton from the protonated ethanol:

$$-\overset{}{\underset{}{C}} - \overset{}{\underset{}{C}} - \overset{+}{O} \overset{H}{\underset{H}{\diagdown}} + H_2O \underset{}{\overset{Fast}{\rightleftharpoons}} -\overset{}{\underset{}{C}} - \overset{}{\underset{}{C}} - OH + H_3O^+ \quad (6)$$

## 4.1.2. Catalysts

The technical and patent literature describes numerous catalysts for ethylene hydration [12]. Most of them are acidic because the reaction involves carbocations. However, only phosphoric acid catalysts supported by diatomaceous earth (kieselguhr, Celite) [13]–[15], montmorillonite [16], bentonite [17], silica gel [18]–[21], or opoka (Volga sandstone) [22] are of industrial importance.

Improved mechanical properties are claimed for catalysts on a diatomaceous earth support with added silica gel [23] or bentonite [24]. Catalyst performance depends on the porosity of the support [21] which can be increased by incorporation of combustible material [24].

Under operating conditions, the support must be resistant to phosphoric acid but able to retain it in large amounts; it must also have adequate mechanical strength and a sufficiently long life and, ultimately, must provide an active, selective catalyst for the hydration [13]. The support is impregnated with 50–77% phosphoric acid but is first treated with hydrochloric or sulfuric acid to remove most of its iron and aluminum. Otherwise, these materials are entrained on contact with phosphoric acid and tend to clog the heat exchangers. Heating the supported catalyst to 700–1100 °C has been recommended because even though this treatment impairs catalyst activity and selectivity, it ensures that the phosphoric acid binds firmly to the support [25]–[27].

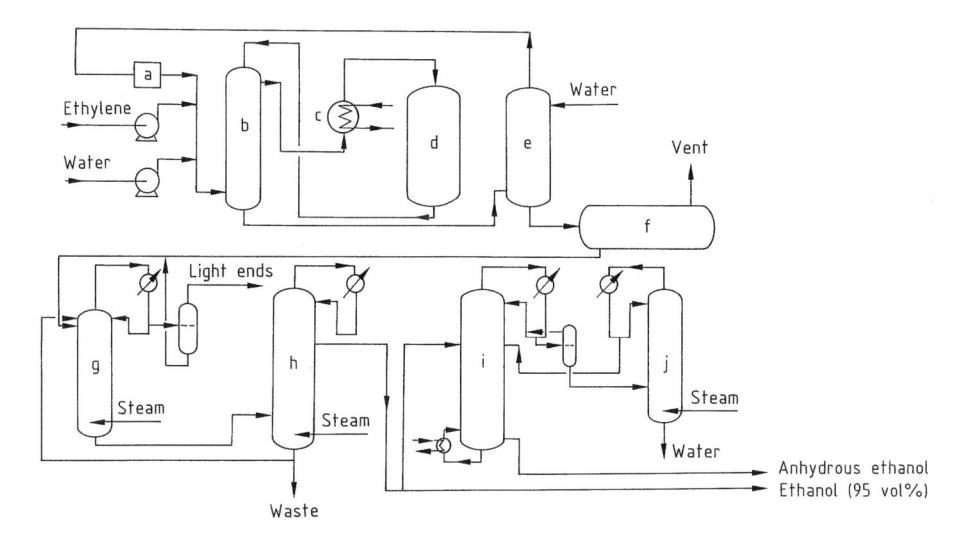

**Figure 3.** Synthesis of ethanol by direct hydration
a) Circulation compressor; b) Heat exchanger; c) Superheater; d) Reactor; e) Washer; f) Crude ethanol tank; g) Extractive distillation column; h) Concentration column; i) Dehydration; j) Concentration column

The following compounds have also been suggested as catalysts for ethylene hydration: acidic oxides (such as tungsten trioxide) on silica [11], silicotungstic acid [28], [29], silicoborotungstic acid [30], or phosphoric acid on silica–alumina [31]; high-silica HZSM-5, dealuminated Y [32], HVSN-5 [33], or H-form [34] zeolites, as well as zeolites containing metal ions [35]; and salts such as boron phosphate [36]–[40], zirconium phosphate [41], other phosphates [40], cobalt sulfate [42], or magnesium sulfate [43].

## 4.1.3. Production Process

Ethanol production based on the catalytic hydration of ethylene is shown schematically in Figure 3. Ethylene and deionized water (molar ratio range 1:0.3–1:0.8) are heated to 250–300 °C at 6–8 MPa by passage through a heat exchanger and a superheater. Since hydration is exothermic, the gaseous reaction products leave the reactor at a temperature 10–20 °C higher than when they entered and thus are used as a source of heat in the heat exchanger. Some phosphoric acid is entrained by the gas stream and is neutralized by injecting a dilute solution of sodium hydroxide.

After condensation and separation of the liquid reaction products, the gas is freed from residual ethanol by water washing and then recompressed and recycled to the reactor. The phosphoric acid lost by entrainment and evaporation is replaced continuously or periodically by spraying it on the catalyst bed. Crude product collects in the sump of the washer and contains ca. 10–25 wt% ethanol. It is decompressed to recover the dissolved ethylene, which is recycled. The ethanol is then purified, ideally by extractive distillation followed by rectification, to obtain a 95 vol% ethanol–water azeotrope. The azeotrope can be dehydrated by azeotropic distillation to give anhydrous ethanol. For further details of distillation see Chapter 6.

Prior to distillation, the crude ethanol is often catalytically hydrogenated to convert acetaldehyde [75-07-0] and higher aldehyde impurities to the corresponding alcohols. Acetaldehyde is a highly undesirable impurity because it tends to form crotonaldehyde [4170-30-3].

Approximately 2% diethyl ether is formed as a byproduct; it can be easily purified to a marketable product from the light ends fraction (→ Ethers, Aliphatic) or recycled to the reactor where it is quantitatively converted to ethanol.

Ethylene is currently available in high purity (> 99.9 vol%); therefore, only small amounts need to be vented even if an ethylene level of 85% must be maintained in the reactor stream.

Cylindrical steel containers with a diameter up to 4 m and an internal volume of up to 150 m$^3$ are used as reactors. They are lined with copper or carbon bricks for protection against attack by phosphoric acid. The heat exchangers and the connecting piping exposed to phosphoric acid are made of, or lined with, copper or copper alloys. The remaining synthesis and distillation equipment is made of steel. Waste gas does not present any problems. Vented ethylene is either returned to the ethylene plant or burned. The synthesis step does not produce any liquid waste. Wastewater from distillation does not contain any organic compounds, and its low phosphate content (0.3 kg of $Na_2HPO_4$ per cubic meter) permits ready disposal.

Optimal reaction conditions depend on the correlation of activity and selectivity of the catalyst with temperature, pressure, residence time, and molar ratio of water to ethylene. These relationships have been studied extensively [44]–[48]. A rate equation for the decrease in catalyst activity with time has been derived and used to develop a mathematical model for computing optimum reaction parameters [49], [50].

## 4.2. Indirect Hydration of Ethylene

The indirect hydration, esterification–hydrolysis, or sulfuric acid process is based on the absorption of a large volume of ethylene in concentrated sulfuric acid; ethanol and some diethyl ether are formed when the sulfuric acid solution is diluted with water.

### 4.2.1. Chemistry

Ethylene is absorbed in two steps:

$$C_2H_4 + H_2SO_4 \rightleftharpoons C_2H_5OSO_3H \qquad \Delta H = -60 \text{ kJ} \qquad (7)$$

$$C_2H_4 + C_2H_5OSO_3H \rightleftharpoons C_2H_5OSO_2OC_2H_5 \qquad (8)$$

Kinetic studies have shown that this reaction also proceeds via the mechanism in Reactions (3)–(6) [51] but is complicated by the fact that only part of the ethylene is physically dissolved in sulfuric acid.

Hydrolysis involves three steps:

$$C_2H_5OSO_3H + H_2O \rightleftharpoons C_2H_5OH + H_2SO_4 \tag{9}$$

$$C_2H_5OSO_3C_2H_5 + H_2O \rightleftharpoons C_2H_5OH + C_2H_5OSO_3H \tag{10}$$

$$C_2H_5OSO_3C_2H_5 + C_2H_5OH \rightleftharpoons C_2H_5OC_2H_5 + C_2H_5OSO_3H \tag{11}$$

The products are ethanol and 5–10% diethyl ether. The ether yield can be controlled by varying the reaction conditions, especially the ratio of ethylene to sulfuric acid, and the hydrolysis conditions (→ Ethers, Aliphatic).

*Transesterification.* The ethyl sulfates formed in Reactions (7) and (8) can also be transesterified with acetic acid [64-19-7] at 104 °C to yield ethyl acetate [141-78-6], which is then recovered by distillation and hydrolyzed to ethanol and acetic acid [52]. This process avoids problems associated with reconcentration of sulfuric acid.

## 4.2.2. Production Process

Figure 4 shows the flow scheme for the production of ethanol by the sulfuric acid process. The feed gas must contain a minimum of 35 vol% ethylene and only inert gases such as methane and ethane. Higher homologues of ethylene cause resin formation and, therefore, must be removed from the feed.

The absorption of ethylene increases almost linearly with pressure. Operating pressure is generally 1–3.5 MPa, the higher pressure being used when the ethylene level in the feed is low. Each mole of sulfuric acid absorbs up to 1.4 mol of ethylene. The absorption is carried out with 94–98 wt% sulfuric acid in wash towers at 65–85 °C. Temperatures above 90 °C lead to formation of resins and oils and, therefore, must be avoided. The resulting liquid is agitated for several hours under pressure to complete the reaction.

Hydrolysis is usually performed in two stages. First, diethyl sulfate is hydrolyzed at a low temperature (70 °C) in the presence of less than the equivalent amount of water. Then more water is added gradually, and the temperature is raised to ca. 100 °C. Hydrolysis is complete within 1 h, and the sulfuric acid is then diluted to 40–55 wt%.

The ethanol formed is recovered, together with the diethyl ether byproduct, in a stripping column. The product mixture is then washed with sodium hydroxide solution to neutralize any acid, and the diethyl ether is removed in the ether distillation column. The ethanol is purified by distillation and concentrated to give a 95 vol% ethanol–water azeotrope. Anhydrous ethanol can be obtained as described in Chapter 6.

Reconcentration of the dilute sulfuric acid is the most expensive part of this process. It is encumbered by high energy consumption, corrosion problems, and oxidation of

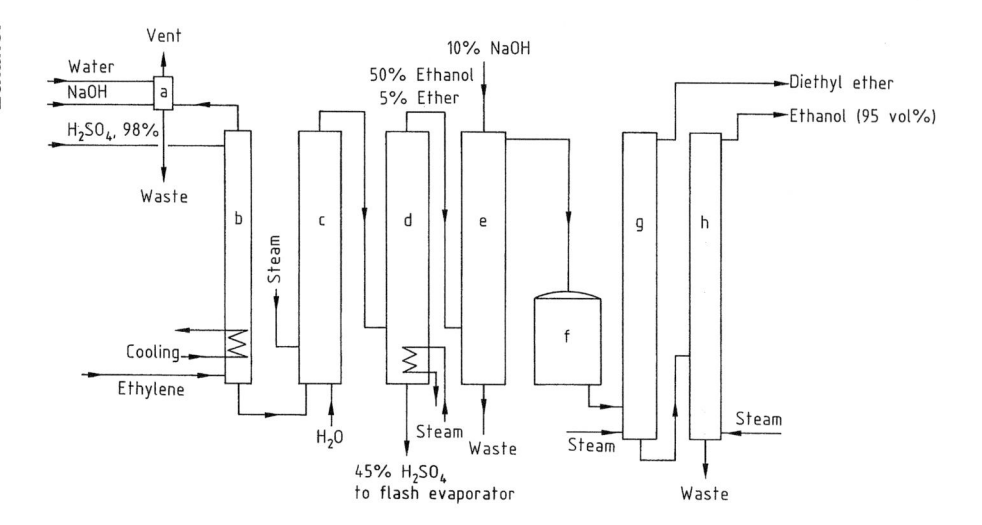

**Figure 4.** Synthesis of ethanol by the sulfuric acid method
a) Wash column for residual gas; b) Absorber; c) Hydrolyzer; d) Stripping column; e) Wash column for crude ethanol; f) Crude product tank; g) Diethyl ether distillation column; h) Ethanol distillation column

the organic compounds remaining in the sulfuric acid. Although tar formed in the absorption is separated as completely as possible, the sulfuric acid loss amounts to more than 5% based on the ethanol formed.

To avoid these disadvantages, an operating pressure below 0.5 MPa has been recommended [53]; less than 0.5 mol of ethylene is absorbed per mole of sulfuric acid, which contains up to 7% silver sulfate to accelerate the reaction. The sulfuric acid can then be reconcentrated by simple flash evaporation.

The following construction materials are used in this process: steel for absorption and distillation; lead and acid-resistant masonry for hydrolysis; silumin (an aluminum alloy), tantalum, cast iron, and lead for reconcentration of sulfuric acid.

## 4.3. Other Methods

Efforts have been made to develop methods for synthesizing ethanol that are not based on raw materials derived from petroleum and are likely to be of industrial importance in the future. Methanol [67-56-1] and synthesis gas, which are available from coal, can be used as starting materials.

## 4.3.1. Homologation of Methanol

The homologation (hydrocarbonylation) of alcohols discovered in 1949 [54], [55] allows the chain length of an alcohol to be increased by one —$CH_2$— group by reaction with synthesis gas (CO + $H_2$). The reaction occurs in the presence of a cobalt carbonyl catalyst at elevated temperature and pressure:

$$ROH + CO + 2H_2 \xrightarrow{Cat.} RCH_2OH + H_2O \quad (12)$$

When homologation was originally applied to methanol, the ethanol yield was relatively low and the reaction product contained higher alcohols, formate and acetate esters, and other oxygenated products. The complexity of the mixture formed was a result of further homologation of ethanol and other alcohols and of additional carbonylation reactions.

The selectivity for ethanol has been improved to up to 89 % by reaction in tetrahydrofuran [109-99-9] solution [56] or by addition of iodine or iodide together with organic phosphines [57]–[61]. High ethanol selectivities have also been claimed for combinations of cobalt catalysts with copper, iron, nickel, ruthenium, or rhodium compounds [62]–[70]. An alumina-supported rhodium–iron catalyst has been described as well [71]. Results of these studies are summarized in Table 5.

## 4.3.2. Carbonylation of Methanol and Methyl Acetate

The first step of another method for converting methanol to ethanol is carbonylation of methanol to acetic acid [64-19-7] [72].

$$CH_3OH + CO \rightleftharpoons CH_3COOH \quad (13)$$

This reaction is the basis of a commercial process developed by Monsanto for synthesis of acetic acid [73] by using a rhodium carbonyl–iodine complex as catalyst [74] (→ Acetic Acid).

Acetic acid can be hydrogenated directly to ethanol, but this requires expensive highpressure equipment, and the process is corrosive [72]. According to a Davy McKee patent, carbonylation is carried out in a methanolic solution containing rhodium trichloride [10049-07-7], methyl iodide [74-88-4], and acetic acid at 175 °C and 7 MPa [75]. The acetic acid formed is esterified with ethanol in the presence of sulfuric acid to yield ethyl acetate [141-78-6]. Ethyl acetate is then converted to ethanol by hydrogenolysis in a tubular reactor at 200 °C over a copper–zinc oxide catalyst. Thus, the overall reaction sequence is:

$$CH_3OH + CO \longrightarrow CH_3COOH \quad (13)$$

$$CH_3COOH + C_2H_5OH \longrightarrow CH_3COOC_2H_5 + H_2O \quad (14)$$

**Table 5.** Homologation of methanol to ethanol[a]

| Catalyst | Additive | Ethanol selectivity, % | Reference |
|---|---|---|---|
| $Co(OAc)_3$ | tetrahydrofuran | 72 | [56] |
| $Bu_3PCo(CO)_3$ | $I_2$ | 65 | [57] |
| $Co(OAc)_3$ | $I_2$, $Bu_3P$ | 70 | [58] |
| $Co(OAc)_3$ | $I_2$, $Ph_3P$ | 80 | [59] |
| $Co(OAc)_3$ | $I_2$, $[Ph_2P(CH_2)_3]_2$ | 65 | [60] |
| $CoI_3$ | | 89 | [61] |
| $Co_2(CO)_8$, $RuCl_3$ | $I_2$ | 79 | [62] |
| $Bu_3PCo(CO)_3$, $Ru(acac)_3$ | $I_2$ | 72 | [63] |
| $Co(acac)_3$, $Ru(acac)_3$ | $I_2$, $(Ph_2PCH_2)_2$ | 82[c] | [64] |
| $CoI_3$, $FeI_3$, $RuCl_3$ | $I_2$, MeOEt, MeOAc | 86 | [65] |
| $CoI_3$, $RhI_3$ | N-methyl-2-pyrrolidone | 62 | [66] |
| $Co_{20}B_{10}RhH_6$ | | 78 | [67] |
| $Co_2(CO)_8$, $NiCl_2$ | triphenylamine | 82 | [68] |
| CoS, Cu | $Bu_3P$, N-methyl-2-pyrrolidone | 96 | [69] |
| $Co(acac)_3$, $W(CO)_6$ | HI, $(Ph_2PCH_2)_2$ | 89[d] | [70] |
| $Co(acac)_3$, $Fe(acac)_3$ | HI, $(Ph_2PCH_2)_2$ | 83[d] | [70] |
| Rh–Fe/$\gamma$–alumina | | 37–87[e] | [71] |

[a] Except where otherwise specified, homologation was performed at 150–250 °C and 20–35 MPa.
[b] Abbreviations: OAc = acetate; acac = acetylacetonate; Bu = butyl; Et = ethyl; Me = methyl; Ph = phenyl.
[c] Ethanol content of product.
[d] Selectivity for ethanol and acetaldehyde.
[e] Yield of ethanol and methyl acetate; conditions for homologation as described in [71].

$$CH_3COOC_2H_5 + 2\,H_2 \longrightarrow 2\,C_2H_5OH \tag{15}$$

The net reaction is:

$$CH_3OH + CO + 2\,H_2 \longrightarrow C_2H_5OH + H_2O \tag{16}$$

HALCON has shown that carbonyls of non-noble metals, such as molybdenum, tungsten, and chromium, catalyze the carbonylation step [76], [77] and has developed an improved esterification procedure that is used commercially [72]. In a modification of this process, methyl acetate [*79-20-9*] rather than methanol is carbonylated [76], [78]:

$$CH_3COOCH_3 + CO \longrightarrow (CH_3CO)_2O \tag{17}$$

Reaction of the resulting acetic anhydride [*108-24-7*] with ethanol and methanol then yields ethyl and methyl acetates:

$$(CH_3CO)_2O + CH_3OH + C_2H_5OH \longrightarrow CH_3COOCH_3 + CH_3COOC_2H_5 + H_2O \tag{18}$$

After separation of the acetates, the ethyl acetate is hydrogenolyzed to ethanol (Reaction 15) and the methyl acetate is carbonylated (Reaction 17). The net reaction is given by Reaction (16).

Methyl acetate is carbonylated on a commercial scale in Eastman Kodak's acetic anhydride plant. Various techniques are available for both the esterification (Reaction 18) and the hydrogenolysis (Reaction 15) steps [72].

## 4.3.3. Conversion of Synthesis Gas

Soon after methanol had been synthesized from carbon monoxide and hydrogen (→ Methanol), and the Fischer–Tropsch synthesis had been discovered, a search started for related routes to ethanol and higher alcohols, based on synthesis gas and the use of heterogeneous [79] or homogeneous [80] catalysis.

Highly specific methods for producing ethanol from synthesis gas are not yet available, but some promising leads have been found.

### 4.3.3.1. Synthesis by Heterogeneous Catalysis

Attempts to improve the conversion of synthesis gas to ethanol have been based on modifying industrial catalysts containing oxides of zinc, chromium, or copper that are normally used for the synthesis of methanol. Early work showed that the incorporation of alkali [81] and cobalt [82] improved yields of ethanol and higher alcohols.

The Institut Français du Pétrole [83] has developed complex catalysts containing copper, cobalt, chromium, potassium, lanthanum, and other elements for conversion of synthesis gas to alcohols to be blended with gasoline. The product contains 50–70% methanol, 16–23% ethanol, 8–14% propanol, and some higher alcohols. Institut patents report on two catalysts that can convert synthesis gas to an alcohol mixture containing 39% ethanol, 24% methanol, and 36% propanols + butanols at 250 °C and 6 MPa [84], [85].

Use of a Cu–Ni–Ba catalyst is reported to give a product containing 46% ethanol, 38% methanol and 9% propanol [86]. A Cu–Zn–Ni–Na catalyst showed a 55% selectivity for alcohols, but the alcohol fraction contained only 29% ethanol [87]. Iron and potassium have also been added to copper–zinc catalysts normally used for methanol synthesis; their effects on ethanol synthesis and the reaction mechanisms involved have been discussed [88], [89].

On certain supports, platinum-group metal catalysts prepared from carbonyl clusters convert synthesis gas to methanol–ethanol mixtures at 200–240 °C and atmospheric pressure [90]. Thus, a supported rhodium–lanthanum trioxide catalyst gives ethanol as the predominant product, with a total alcohol selectivity up to 80%. More recent studies using supported rhodium catalysts promoted with cerium or europium report a similarly high selectivity for ethanol [91], [92].

The only catalysts derived from rhodium carbonyl clusters that produce ethanol as the major product are those which contain added cobalt [93]. A number of formulations containing rhodium and a variety of cocatalysts and promotors are claimed by the patent literature to be selective catalysts for the synthesis of ethanol, e.g., [94][95][96], acetaldehyde, and acetic acid, e.g., [97]. According to a Hoechst patent, a composition of cobalt, rhenium, gold, and barium oxides on a silica support produces ethanol from synthesis gas with 60% selectivity [98].

### 4.3.3.2. Synthesis by Homogeneous Catalysis

The homogeneous catalytic conversion of synthesis gas to ethanol and ethylene glycol [*107-21-1*] has been reviewed with particular reference to the reaction mechanism, the nature of the active catalyst species, and their interaction with solvents, cocatalysts, and other additives [80].

Ruthenium is the only group 8 metal to show a high selectivity for the conversion of synthesis gas to ethanol when used in conjunction with organic phosphorus compounds. Synthesis gas can be converted to ethanol with a selectivity >50% by using an iodine-promoted ruthenium carbonyl catalyst dissolved in tripropylphosphine oxide [99], [100]. Lower ethanol selectivities (20–30%) have been reported for ruthenium catalysts in a melt of tetrabutylphosphonium bromide [101]–[103] or alkyltriphenylphosphonium bromides [104].

A combination of ruthenium and cobalt carbonyls dissolved in toluene is reported to convert synthesis gas to a mixture of ethanol, methanol, and acetic acid [105]. Ethanol is also reported to be the major product in a conversion catalyzed by rhodium carbonyl, potassium chloroplatinate, and stannous chloride [106].

# 5. Fermentation

## 5.1. Production by Yeast

Yeasts are unicellular, uninucleate fungi that can reproduce by budding, fission, or both. They have been used for centuries to make alcoholic beverages.

Yeasts are the most commonly used organisms in the industrial production of ethanol. Some widely used, high-productivity strains are *Saccharomyces cerevisiae*, *S. uvarum* (formerly *S. carlsbergensis*), and *Candida utilis*. *Saccharomyces anamensis* and *Schizosaccharomyces pombe* are also used in some instances. *Kluyveromyces* species, which ferment lactose, are good producers of ethanol from whey.

Ethanol production by yeast is characterized by high selectivity, low accumulation of byproducts, high ethanol yield, high fermentation rate, good tolerance toward both increased ethanol and substrate concentrations, and lower pH value. Viability and genetic stability of yeast cells under process conditions and at high temperature are also desirable. Although finding a strain that has all these characteristics is difficult, some yeast strains can fulfill them to a great extent [107], [108].

**Table 6.** Ability of and species to ferment sugars

| Carbon number of basic subunit | Type of basic subunit | Sugar | Basic unit | Yeast * | | |
|---|---|---|---|---|---|---|
| | | | | S. cerevisiae | S. uvarum (carlsbergensis) | K. fragilis |
| C$_6$ sugars | aldose sugars | glucose | glucose | + | + | + |
| | | maltose | glucose | + | + | − |
| | | maltotitrirose | glucose | + | + | − |
| | | cellobiose | glucose | − | − | − |
| | | trehalose | glucose | +/− | +/− | − |
| | | galactose | galactose | + | + | + |
| | | mannose | mannose | + | + | + |
| | | lactose | glucose, galactose | − | − | + |
| | | melibiose | glucose, galactose | − | + | |
| | ketose sugars | fructose | fructose | + | + | + |
| | | sorbose | sorbose | − | − | − |
| | aldoses and ketoses | sucrose | glucose, fructose | + | + | + |
| | | raffinose | glucose, fructose, galactose | +/− | + | +/− |
| | deoxy sugars | rhamnose | 6-deoxymannose | − | − | − |
| | | deoxyribose | 2-deoxyribose | +/− | +/− | +/− |
| C$_5$ sugars | aldose sugars | arabinose | arabinose | − | − | − |
| | | xylose | xylose | − | − | − |

* + Indicates good ability to ferment sugars; − cannot ferment sugars; +/− low ability to ferment sugars.

## 5.1.1. Nutrients

Yeasts require the following for growth: carbon, nitrogen, phosphorus, sulfur, oxygen, hydrogen, minor quantities of potassium, magnesium, calcium, trace minerals, and some organic growth factors (vitamins, nucleic acids, and amino acids).

Various carbon compounds can serve as a carbon source for yeast (Table 6). Nitrogen can be supplied as ammonia, ammonium salts, urea, or amino acids. Orthophosphate salts and phosphoric acid are good sources of phosphorus. Sulfur, potassium, magnesium, and calcium can be supplied in the form of their salts. Table 7 shows the amount of these elements and trace minerals required by yeasts.

Yeast extract is a good source of trace minerals and organic growth factors. Many raw materials used for industrial ethanol production supply all the nutrients necessary for yeast growth; additional nutrient supplements are required in some cases.

## 5.1.2. Fermentation Pathways

**Hexoses.** Yeasts are able to metabolize various carbon compounds. Metabolic pathways differ under aerobic and under anaerobic conditions. Conversion of glucose by *Saccharomyces cerevisiae* under both conditions is shown in Figure 5 [111].

**Table 7.** Growth optimum and calculated inlet concentrations of inorganic ions required for optimum growth of yeast in a CSTR

| Ion | Growth optimum, mmol/L | Cellular concentration, mmol/100 g dry mass | CSTR inlet concentration for a cell concentration of 10 g/L, mmol/L |
|---|---|---|---|
| $B^+$ | 0.0004 – 0.001 | $0.5 \times 10^{-3}$ | 0.0004 – 0.001 |
| $Ca^{2+}$ | 0.5 – 5.0 [b] | 0 – 1.5 | 0.5 – 5.0 |
| $Co^{2+}$ | 0.0001 – 0.001 [c] | $(0.03 – 1) \times 10^{-3}$ | 0.0001 – 0.001 |
| $Cu^{2+}$ | 0.001 – 0.01 [c] | 0.008 – 0.20 | 0.002 – 0.03 |
| $Fe^{2+}$ | 0.001 – 0.01 | 0.036 – 0.18 | 0.005 – 0.02 |
| $K^+$ | 2.0 – 10.0 [b] | 8 – 56 | 3 – 15 |
| $Mg^{2+}$ | 2.0 – 4.0 | 4 – 17 | 2 – 6 |
| $Mn^{2+}$ | 0.002 – 0.01 [b] | 0.007 – 0.055 | 0.003 – 0.01 |
| $Mo^{2+}$ | 0.001 – 0.01 | $(0.04 – 0.08) \times 10^{-3}$ | 0.001 – 0.01 |
| $Ni^+$ | 0.001 – 0.05 | $(0.03 – 2) \times 10^{-3}$ | 0.001 – 0.05 |
| $Zn^{2+}$ | 0.005 – 0.015 [b, c] | 0.08 – 0.3 | 0.006 – 0.04 |
| $Cl^-$ | $\simeq 1$ | 11 – 140 | 2 – 15 |
| $I^-$ | $\simeq 0.001$ | $0.04 – 1 \times 10^{-3}$ | 0.001 |
| $SO_4^{2-}$ | $\simeq 1$ | 0.6 – 15.0 | 2 – 10 |
| $H_2PO_4^-$ | 2 – 4 | 40 – 65 | 6 – 10 |
| $NH_4^+$ | $\simeq 1$ | $\simeq 700$ | $\simeq 100$ |

[a] CSTR = continuous stirred tank reactor.
[b] Optimum concentration may be higher under conditions where specific inhibitory ions are present or any ion is present at an inhibitory level.
[c] These ions are most likely to be deficient in a complex industrial medium.

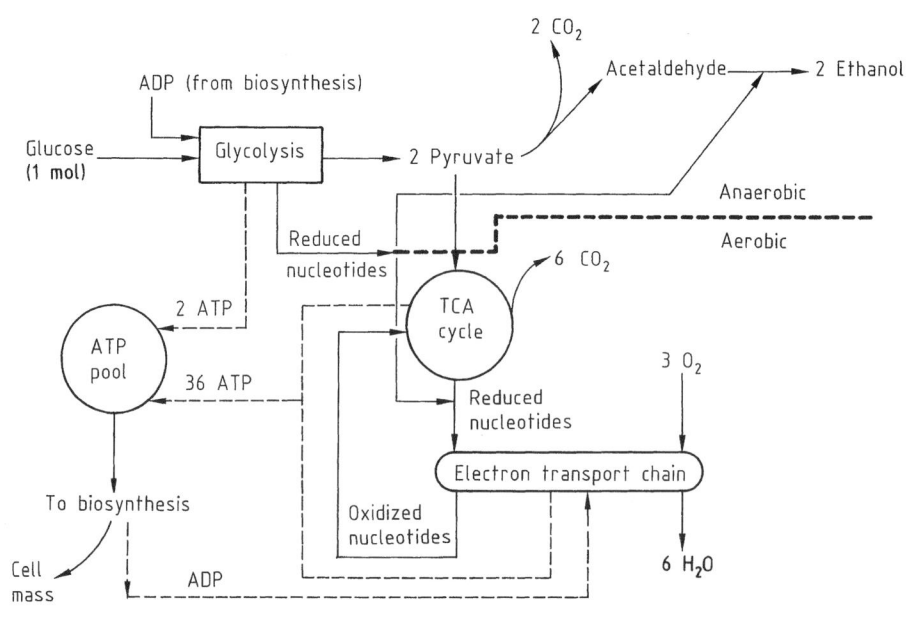

**Figure 5.** Simplified chart of anaerobic and aerobic catabolism of *Saccharomyces cerevisiae* [111]
ADP = adenosine diphosphate; ATP = adenosine triphosphate; TCA = tricarboxylic acid (citric acid).

Under *anaerobic conditions*, the yeast produces ethanol, according to the Gay–Lussac equation:

$$C_6H_{12}O_6 \longrightarrow 2\ C_2H_5OH + 2\ CO_2$$

Therefore, from each gram of glucose consumed, 0.51 g of ethanol can be produced. However, some of the carbon source is used for biomass generation so that the actual ethanol yield is ca. 90–95% theoretical. In addition to ethanol, carbon dioxide, and biomass, 2 mol of adenosine triphosphate (ATP) are produced per mole of glucose. A lower than theoretical yield of ethanol is also caused by the formation of small amounts of byproducts, such as glycerol and succinate, at the expense of the carbon source. If the carbon source is not used for the production of biomass and byproducts, the ethanol yield based on sugar would increase by 1.6 and 2.7%, respectively [112]. In ethanol production, some higher alcohols (fusel oils) are formed partly from the carbon source and partly from deamination and subsequent conversion of certain amino acids.

Under *aerobic conditions* glucose is converted to carbon dioxide and water according to the equation:

$$C_6H_{12}O_6 + 6\ O_2 \longrightarrow 6\ CO_2 + 6\ H_2O$$

In addition, biomass and energy are also generated.

The sequence of intermediates and reactions in the transformation of glucose (or a polysaccharide) to ethanol and $CO_2$ is shown in Figure 6. The Embden–Meyerhof–Parnas or glycolytic pathway represents the major route of glucose catabolism in most cells that convert hexoses to pyruvate. This pathway, which must provide all the intermediates and most of the energy required for cell growth, is also used in microbial fermentation of glucose to ethanol, lactate, glycerol, glycols, and a variety of other products.

| | | |
|---|---|---|
| Glycolysis: | glucose + 2 $P_i$ + 2 ADP | → 2 lactate + 2 ATP (classical mammalian muscle or brain) |
| | glucose + $P_i$ | → α-glycerol phosphate + pyruvate (insect flight muscle, striated muscle) |
| Fermentation: | glucose + 2 $P_i$ + 2 ADP | → 2 ethanol + 2$CO_2$ + 2 ATP + 2 $H_2O$ (first form) |
| | glucose + $HSO_3^-$ | → glycerol + acetaldehyde · $HSO_3$ + $CO_2$(second form, no net ATP) |
| | 2 glucose | → 2 glycerol + ethanol + acetate + 2 $CO_2$ (third form, no net ATP) |
| | glucose + ($P_i$) | → α-glycerol phosphate + acetaldehyde + $CO_2$ |
| | | ↓ |
| | | glycerol + $P_i$ (third form, no net ATP) |

Enzymes, characteristic inhibitors, and standard free energy changes for each of the reactions are listed in Table 8. The reactions do not require oxygen and provide the cellular energy supply under completely anaerobic conditions.

As shown in Table 8 and Figure 6, conversion of D-glucose to ethanol starts with the formation of D-glucose 6-phosphate (Robinson ester) (Reaction 1). This reaction is catalyzed by hexokinase (ATP:D-hexose 6-phosphotransferase). The enzyme uses ATP as a phosphate donor and requires a divalent cation (usually $Mg^{2+}$, but $Mn^{2+}$ or frequently $Ca^{2+}$ and $Co^{2+}$ can replace the cofactor). This enzyme is relatively nonspecific and can catalyze the phosphorylation of a variety of hexoses of appropriate

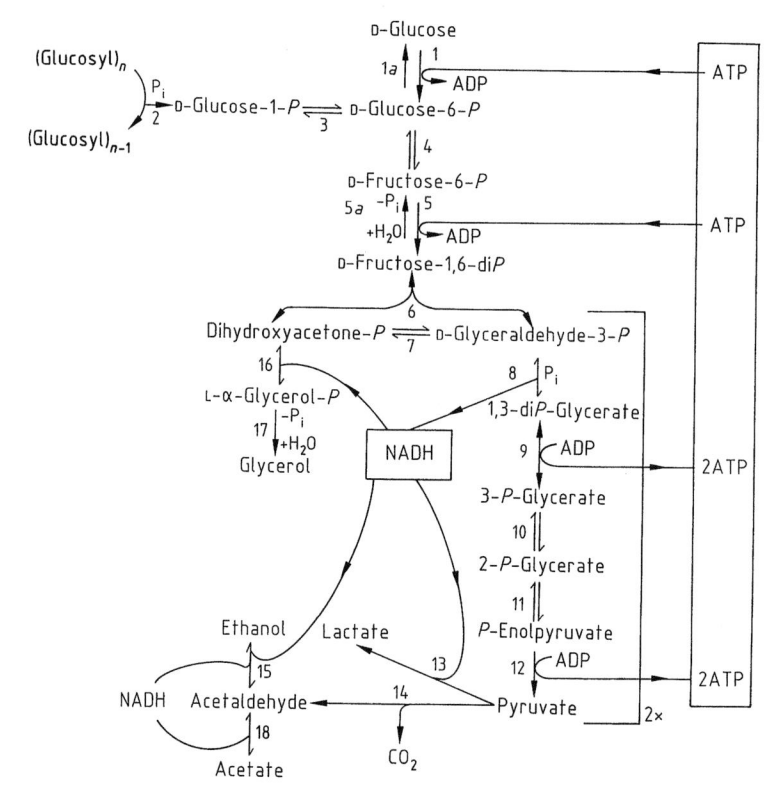

**Figure 6.** Embden–Meyerhof–Parnas scheme of glycolysis [113]

configuration including D-fructose, D-mannose, 2-deoxy-D-glucose, and D-glucosamine. The equilibrium of the reaction lies so far in the direction of product formation as to render the reaction virtually irreversible thermodynamically ($K_{eq} \simeq 6.5 \times 10^3$ at pH 7.4 and 25 °C) and of no practical importance in gluconeogenesis.

The conversion of starch and other starchlike glucose-containing polysaccharides is catalyzed by α-1,4-glucan phosphorylases (α-1,4-glucose:orthophosphate glycosyltransferases), which are widely distributed in a variety of organisms (Fig. 6, Reaction 2). This phosphorolysis commences at the free, nonreducing end of an amylose chain and removes one glucose unit at a time, yielding a total of (n + 1) glucose 1-phosphate molecules until the reducing end is reached. If the substrate is amylopectin, phosphorolysis continues only until the branch points are reached and the product is a limit dextrin. The 1,6-bond at the branch may be cleaved by an amylo-1,6-glucosidase, yielding free glucose, after which phosphorylase is able to act again until the next branch is reached, and so on.

Phosphoglucomutase (α-D-glucose 1,6-diphosphate:α-D-glucose 1-phosphate phosphotransferase) is responsible for the interconversion of D-glucose 1-phosphate (G1P) (Cori ester) and D-glucose 6-phosphate (G6P) (Fig. 6, Reaction 3). The enzyme is specific for α-D-pyranose phosphate and the $K_{eq}$ is 19 at pH 7 and 25 °C for $c_{G6P}/c_{G1P}$. The reaction requires $Mg^{2+}$ and is inhibited by fluoride.

The interconversion of D-glucose 6-phosphate and D-fructose 6-phosphate (Neuberg ester) is catalyzed by phosphoglucoisomerase (D-glucose 6-phosphate ketol-isomerase) (Fig. 6, Reaction 4). The enzyme is quite specific. It is competitively and specifically inhibited by 2-deoxy-D-glucose 6-

**Table 8.** Reactions and thermodynamics of glycolysis and alcoholic fermentation

| Reaction number | Equation[a] | Name of enzyme | Characteristic inhibitor | $\Delta G^{0'\,b}$, kJ/mol |
|---|---|---|---|---|
| 1 | Glucose + ATP $\xrightarrow{Mg^{2+}}$ glucose 6-$P$ + ADP | hexokinase glucokinase | | −14.32 |
| 1a | Glucose 6-$P$ + H$_2$O $\xrightarrow{Mg^{2+}}$ glucose + P$_i$ | glucose 6-phosphatase Nonspecific phosphatases | glucose, orinase | −16.83 |
| 2 | Glycogen + $n$ P$_i$ $\rightleftharpoons$ $n$ glucose 1-$P$ | α-1,4-glucan phosphorylase | | 3.06 |
| 3 | Glucose 1-$P$ $\xrightarrow{\text{glucose 1,6-di}P}$ glucose 6-$P$ | phosphoglucomutase | F, organophosphorus inhibitors | −7.29 |
| 4 | Glucose 6-$P$ $\rightleftharpoons$ fructose 6-$P$ | phosphoglucose (glucose phosphate) isomerase | 2-deoxyglucose 6-phosphate | 2.09 |
| 5 | Fructose 6-$P$ + ATP $\xrightarrow[\text{(ADP, AMP)K}^+]{Mg^{2+}}$ fructose-1,6-di$P$ + ADP + H$^+$ | phosphofructokinase | ATP, citrate | −14.24 |
| 5a | Fructose 1,6-di$P$ + H$_2$O $\xrightarrow{Mg^{2+}}$ fructose 6-$P$ + P$_i$ | fructose diphosphatase nonspecific phosphatases | AMP, fructose 1,6-diphosphate, Zn$^{2+}$, Fe$^{2+}$ | −16.75 |
| 6 | Fructose 1,6-di$P$ $\rightleftharpoons$ dihydroxyacetone $P$ + glyceraldehyde 3-$P$ | (fructose phosphate) aldolase | chelating agents (microbial enzymes only) | 23.99 |
| 7 | Dihydroxyacetone $P$ $\rightleftharpoons$ glyceraldehyde 3-$P$ | triose phosphate isomerase | | 7.66 |
| 8 | 2 × (Glyceraldehyde 3-$P$ + P$_i$ + NAD$^+$ $\rightleftharpoons$ 1,3-diphosphoglycerate + NADH + H$^+$) | glyceraldehyde phosphate dehydrogenase; triose phosphate dehydrogenase | ICH$_2$COR D-threose 2,4-diphosphate | 2 × (6.28) |
| 9 | 2 × (1,3-Diphosphoglycerate + ADP + H$^+$ $\xrightarrow{Mg^{2+}}$ 3 phosphoglycerate + ATP) | phosphoglycerate kinase | | 2 × (−28.39) |
| 10 | 2 × (3-Phosphoglycerate $\xrightarrow{\text{glycerate 2,3-di}P}$ 2-phosphoglycerate | phosphoglyceromutase | | 2 × (4.43) |
| 11 | 2 × (2-Phosphoglycerate $\xrightarrow{Mg^{2+} \text{ or } Mn^{2+}}$ phosphoenolpyruvate) | enolase (phosphopyruvate hydratase) | Ca$^{2+}$ F$^-$ plus P$_i$ | 2 × (1.84) |
| 12 | 2 × (Phosphoenolpyruvate + ADP + H$^+$ $\xrightarrow[K^+(Rb^+, Cs^+)]{Mg^{2+}}$ pyruvate + ATP) | pyruvate kinase | Ca$^{2+}$ vs. Mg$^{2+}$ Na$^+$ vs. K$^+$ | 2 × (−23.95) |
| 13 | 2 × (Pyruvate + NADH + H$^+$ $\rightleftharpoons$ lactate + NAD$^+$) | lactate dehydrogenase | oxamate | 2 × (−25.12) |
| 14 | 2 × (Pyruvate + H$^+$ $\rightarrow$ acetaldehyde + CO$_2$) | pyruvate (de)carboxylase | | 2 × (−19.76) |
| 15 | 2 × (Acetaldehyde + NADH + H$^+$ $\rightleftharpoons$ ethanol + NAD$^+$) | alcohol dehydrogenase | (HSO$_3^-$) | 2 × (−21.56) |
| Sums: | (glucose)$_n$ + H$_2$O $\rightarrow$ 2 lactate + 2 H$^+$ + (glucose)$_{n-1}$ | | | −219.40 |
| | (glucose)$_n$ + 3 P$_i$ + 3 ADP $\rightarrow$ 2 lactate + 3 ATP + (glucose)$_{n-1}$ | glycolysis (muscle) | | −114.38[c] |
| | glucose $\rightarrow$ 2 lactate + 2 H$^+$ | | | −198.45 |
| | glucose + 2 P$_i$ + 2 ADP $\rightarrow$ 2 lactate + 2 ATP | glycolysis or lactate fermentation | | −124.52[c] |
| | glucose $\rightarrow$ 2 ethanol + CO$_2$ | | | −234.88 |
| | glucose + 2 P$_i$ + 2 ADP $\rightarrow$ 2 ethanol + 2 CO$_2$ + 2 ATP | alcoholic fermentation | | −156.92[c] |

[a] Cosubstrates or coenzymes shown above, activators below arrow; P = phosphate. [b] $\Delta G^{0'}$ values refer to pH 7.0 with all other reactants including H$_2$O at unit activity; the free energy of formation of glucose in aqueous solution equals 910.88 kJ/mol, its $\Delta G^0$ of combustion to CO$_2$ + H$_2$O is 2872 kJ/mol, and $\Delta G^{0'}$ for (glucose)$_n$ + H$_2$O $\rightarrow$ (glucose)$_{n-1}$ equals −21.06 kJ/mol. [c] From this table.

phosphate, which is readily formed from the corresponding free deoxysugar in the presence of hexokinase.

Phosphofructokinase (ATP:D-fructose 6-phosphate 1-phosphotransferase) is a highly specific enzyme that controls the formation of D-fructose 1,6-diphosphate (Harden–Young ester) from fructose 6-phosphate. It requires ATP and $Mg^{2+}$. The enzyme can use phosphate from uridine triphosphate or from inosine triphosphate instead of from ATP. Its activity can be inhibited by high ATP concentrations, which occurs when lactate and pyruvate are oxidized aerobically to $CO_2$ via the Krebs cycle. In that case, glycolysis is blocked and glucose synthesis is favored. At very low ATP concentration, glycolysis is required for energy generation, and carbohydrate synthesis is shut off.

Fructose 1,6-diphosphatases (D-fructose 1,6-diphosphate 1-phosphohydrolases) are required for hydrolytic conversion of the diphosphate to the 6-monophosphate (Fig. 6, Reaction 5 a). The enzymes are specific and can be inhibited by high concentrations of their substrate and also by adenosine monophosphate (AMP).

The conversion of fructose 1,6-diphosphate to two molecules of triosephosphate [3-phosphoglyceraldehyde (Ficher–Bauer ester) and phosphodihydroxyacetone] (Fig. 6, Reaction 6) is carried out by aldolase (D-fructose 1,6-diphosphate:D-glyceraldehyde 3-phosphate-lyase). These two triosephosphate molecules are in equilibrium, supported by triosephosphate isomerase (D-glyceraldehyde 3-phosphate ketol-isomerase) (Fig. 6, Reaction 7). Only 3-phosphoglyceraldehyde undergoes further conversion to ethanol. In the course of cleavage of the hexose, triose products must be removed because the combined equilibrium lies in favor of hexose. The enzyme from yeast contains $Zn^{2+}$ and, after removal or complexation of the metal, is activated equally well by $Zn^{2+}$, $Fe^{2+}$, and $Co^{2+}$. The activity of the enzyme is enhanced about 35 times in the presence of $K^+$ ions [112].

Glyceraldehyde 3-phosphate dehydrogenase [D-glyceraldehyde 3-phosphate:nicotinamide–adenine dinucleotide (NAD) oxidoreductase (phosphorylating)] catalyzes Reaction 8. Free sulfhydryl groups are required for its activity. The presence of $NAD^+$ and $P_i$ (inorganic phosphate) is necessary. In the reaction, $NAD^+$ is reduced to NADH, which is then reoxidized in the reduction of acetaldehyde to ethanol or pyruvate to lactate. As a dehydrogenase, the enzyme is specific for D-glyceraldehyde 3-phosphate, the free dephosphorylated aldehyde, or acetaldehyde, and is strongly inhibited by D-triose 2,4-diphosphate.

The enzyme phosphoglycerate kinase (ATP:3-phospho-D-glycerate 1-phosphotransferase) catalyzes the conversion of 1,3-diphosphoglycerate to 3-phosphoglycerate, with the generation of 1 mol of ATP (Fig. 6, Reaction 9). It requires a divalent metal cofactor ($Mg^{2+}$, $Mn^{2+}$, or $Co^{2+}$).

3-Phosphoglycerate, which is formed in Reaction (9) (Fig. 6), is isomerized to 2-phospho-D-glycerate by phosphoglyceromutase (2,3-diphospho-D-glycerate:2-phospho-D-glycerate phosphotransferase) (Fig. 6, Reaction 10). The diester 2,3-diphosphoglycerate and $Mg^{2+}$ are obligatory cofactors for the reaction.

The conversion of 2-phosphoglycerate to phosphoenolpyruvate (Fig. 6, Reaction 11) can be considered to be a dehydration reaction that is catalyzed by enolase (phosphopyruvate hydratase). This enzyme is widely distributed in living organisms. Divalent cations ($Mn^{2+}$, $Mg^{2+}$, $Zn^{2+}$, or $Cd^{2+}$) that can be antagonized by $Ca^{2+}$ or $Sr^{2+}$ are essential. Fluoride ion is an effective inhibitor, especially in the presence of $P_i$.

Pyruvate kinase (ATP:pyruvate phosphotransferase) catalyzes Reaction (12), (Fig. 6). For this reaction, $Mg^{2+}$ or $Mn^{2+}$ is required; $Co^{2+}$ ions are competitive antagonists. Some enzymes of this type require $K^+$, $Rb^+$, or $Cs^+$ for full activity. The ions can be antagonized by $Na^+$ or $Li^+$. The reaction equilibrium is very favorable for the formation of pyruvate, and the turnover number is $6 \times 10^{13}$ compared to $12 \times 10^{13}$ for the opposite direction [113].

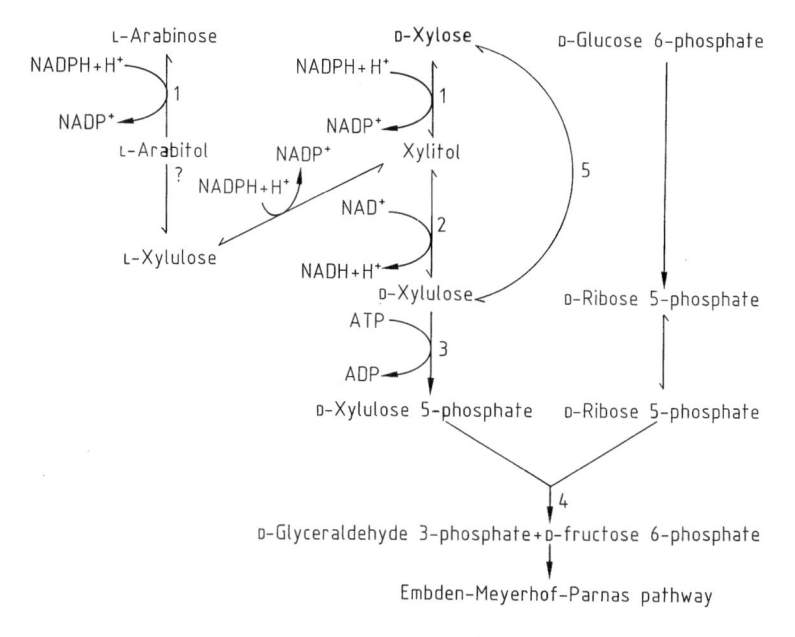

**Figure 7.** D-Xylose and L-arabinose metabolism in yeast [114]
1) Aldose reductase; 2) D-Xylulose reductase; 3) D-Xylulokinase; 4) Transaldolase and transketolase; 5) D-Xylose isomerase

To produce ethanol from pyruvate, yeasts that operate anaerobically irreversibly decarboxylate pyruvate with the aid of pyruvate decarboxylase (2-oxo acid carboxylase) to yield acetaldehyde (Fig. 6, Reaction 14). Alcohol dehydrogenase (alcohol:NAD oxidoreductase) then catalyzes the reduction of acetaldehyde to ethanol (Fig. 6, Reaction 15). The enzyme responsible for the decarboxylation can satisfy its metal requirements by using either certain divalent ions ($Mg^{2+}$, $Mn^{2+}$, $Co^{2+}$, $Ca^{2+}$, $Cd^{2+}$, or $Zn^{2+}$) or trivalent ions ($Al^{3+}$ or $Fe^{3+}$). Alcohol dehydrogenase from yeast contains $Zn^{2+}$. The enzyme has four catalytic sites and a molecular mass of 151 000. Both enzymes have been isolated from yeast and produced in crystalline form.

**Pentoses.** Yeasts metabolize aldopentoses by an oxidation–reduction reaction. They reduce D-xylose to xylitol, which is subsequently oxidized to D-xylulose (Fig. 7). The reduction is catalyzed by aldose reductase (alditol:NADP 1-oxidoreductase) [115]. The enzyme has a specificity for NADPH (reduced nicotinamide–adenine dinucleotide phosphate).

D-Xylulose reductase (xylitol:NAD 2-oxidoreductase) is involved in the oxidation of xylitol to D-xylulose. The reaction is readily reversible, and NADH oxidation occurs with some ketoses including D-xylulose, D-fructose, and D-ribulose.

D-Xylulokinase catalyzes the phosphorylation of D-xylulose to D-xylulose 5-phosphate. The enzyme is found in bacteria [116], [117]. The presence of D-xylulokinase in yeast is indicated by the ability of many yeasts to use D-xylulose under both aerobic and anaerobic conditions [115], [118]. D-Xylulose 5-phosphate can be converted to pyruvate via both the pentose–phosphate and the Embden–Meyerhof–Parnas pathways.

**Table 9.** Pentose (xylose and xylulose) fermenting yeasts *

| Yeast | Process | Advantage | Disadvantage |
|---|---|---|---|
| Candida sp. XF 217 | batch | yield (0.42 g per gram of xylose), low xylitol formation | low ethanol tolerance |
| Kluyveromyces cellobiovorus | batch | cellobiose also | xylitol formation, low ethanol tolerance |
| Pachysolen tannophilus | batch | none | xylitol formation, low ethanol tolerance |
| Pichia stipitis | batch | yield (0.42 g per gram of xylose) low xylitol formation | low ethanol tolerance |
| Saccharomyces cerevisiae | simultaneous isomerization and xylulose fermentation | high ethanol tolerance, fast, yield (0.45 g per gram of xylulose) | isomerization, cost (xylitol formation) |
| Schizosaccharomyces pombe | simultaneous isomerization and xylulose fermentation | high ethanol tolerance, fast, yield (0.48 g per gram of xylulose) | isomerization, cost |

* Reprinted with permission from [126].

Yeasts also metabolize L-arabinose by oxidation–reduction. The aldose is reduced to L-arabitol, which is converted to D-xylulose 5-phosphate, the key intermediate of the pentose phosphate pathway (Fig. 7) [114].

In the metabolism of D-xylose, the actual conversion of the sugar occurs after a lag period of adaption for the production of D-xylose-metabolizing enzymes [119], [120].

Some yeasts, such as *Candida utilis* and *Rhodotorula gracilis*, have inducible D-xylose isomerase in addition to the initial oxidation–reduction reaction and can isomerize D-xylose directly to D-xylulose [119], [121].

Ethanol can be produced readily from D-xylulose in high yield by yeasts that normally produce ethanol from glucose [118], [122].

About half of all yeasts can assimilate D-xylose under aerobic conditions [123]. Yeasts that could consume xylose or other pentoses under anaerobic conditions were unknown. But anaerobic conversion of D-xylose to ethanol has been recently demonstrated in the yeasts *Pachysolen tannophilus* [124] and *Pichia stipitis* [125]. Many yeasts can convert D-xylulose (which is a keto isomer of D-xylose) to ethanol under anaerobic conditions. The growth of yeasts on D-xylose under anaerobiosis has not been reported.

Certain yeasts can produce a high yield of ethanol directly from D-xylose (*Candida* sp. XF 217, *Pichia stipitis*), while others (*Saccharomyces cerevisiae*, *Schizosaccharomyces pombe*) can do so after D-xylose is isomerized to xylulose by a bacterial isomerase. Table 9 shows some advantages and disadvantages of these yeasts in the production of ethanol from xylose and xylulose.

## 5.1.3. Fermentation Variables

**Effect of Substrate Concentration.** The carbon substrate concentration has a significant effect on ethanol production. Previously, batch fermentation was carried out with an initial concentration of fermentable substrate of 14–18 wt%, which resulted in a final ethanol concentration of ca. 6–9% (wt/vol) [127]. Increasing sugar concentration was an advantage because it led to higher final ethanol concentration and, subsequently, to reduction of distillation costs. At the same time, the growth of osmosensitive contaminants was suppressed. However, at sugar concentrations >14%, plasmolysis of yeast cells begins. In addition, the initial rate of fermentation starts to decline before the ethanol concentration reaches a significant value [109].

Under anaerobic conditions and at low-to-moderate glucose concentrations, the rate of ethanol production may be represented by a Monod-type relationship [111]:

$$V = V_{max} S/(K_s + S)$$

where $V$ is the ethanol production rate, g $g_{cell}^{-1}$ $h^{-1}$; $S$ is the glucose substrate concentration, g/L; and $K_s$ is a constant with a low value of 0.35 g/L.

This relationship shows that under anaerobic conditions, maximum ethanol production per cell is achieved when sugar concentration is considerably $>3.5$ g/L and $<150$ g/L. Inhibition of the ethanol production rate above 150 g/L of glucose is significant. If the inhibition caused by ethanol is considered, it appears that it limits the concentration of fermentable sugars.

**Effect of Oxygen.** Although biomass is produced under aerobic conditions and anaerobic conditions favor production of ethanol, oxygen was found to be essential for good fermentation. In addition to being a terminal acceptor of electrons from the respiratory chain, it also acts as a yeast growth factor and is involved in the synthesis of unsaturated fatty acids and ergosterol, which stimulate growth of yeast under anaerobic conditions and increase viability of cells [109], [128].

Oxygen is especially necessary when batch fermentation is carried out at high sugar levels requiring prolonged growth of yeast, or in continuous processes, because the yeast is unable to grow for more than four to five generations under fully anaerobic conditions. When oxygen disappears from the broth, the yeast continues to grow anaerobically. Cell division leads to a redistribution of the sterols and unsaturated fatty acids that have accumulated during aeration. After several generations the sterol and unsaturated fatty acid level is too low to enable normal functioning of the membrane causing a change in the physiology of the yeast [129], [130].

The necessary level of oxygen depends on the strain of yeast. Usually an oxygen tension of 6.7–13.3 Pa is required in the broth. If the concentration exceeds this value, cell growth is enhanced at the expense of ethanol production by promoting the Pasteur effect, i.e., the complete oxidation of glucose to $CO_2$ and $H_2O$ via the tricarboxylic acid

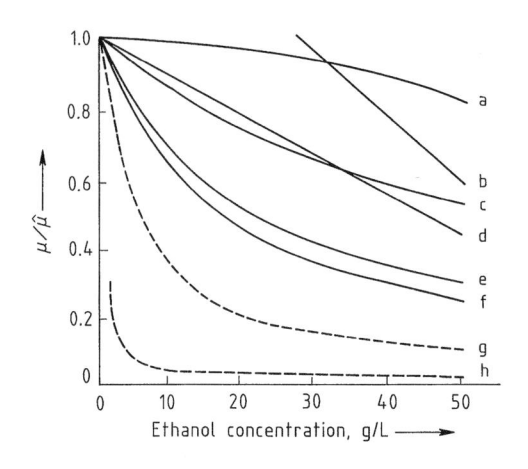

**Figure 8.** Comparison of the effect of various ethanol inhibition functions [137]
$\mu$ = specific growth rate; $\hat{\mu}$ = maximum specific growth rate; Source of ethanol: —— added, – – – autogenous
a) Continuous [138]; b) Batch [139]; c) Continuous [140]; d) Continuous [136]; e) Batch [141]; f) Batch [140]; g) Continuous [142]; h) Batch [143]

(TCA) cycle and the respiratory chain. In some yeasts, cell mass productivity increases drastically with a corresponding decrease in ethanol production at an oxygen tension > 0.23 kPa [131]. This process is controlled by the concentration of fermentable sugar in the growth medium. If the concentration is high and conditions are aerobic, the Pasteur effect is likely to be diminished and glucose will be degraded by the aerobic pathway. This phenomenon, known as "aerobic" fermentation, results from the operation of the so-called reverse Pasteur or Crabtree effect, which represses both the synthesis and the activity of respiratory enzymes at glucose or sucrose concentrations greater than 0.02–0.1% [109], [132].

Air sparging through the growth medium can be omitted if certain supplements, such as oleic acid, linolenic acid, tween 80, or ergosterol are present in the medium [129], [133]. Cultures enriched with these supplements can attain a high ethanol concentration (15.5 wt%) with a high substrate conversion efficiency (95%) under anaerobic conditions.

Various sterols and fatty acids increase the viability of resting cells and prolong their fermentative activity [133], [134]. Incorporation of these compounds into the cell membrane appears to increase its permeability to ethanol and permits a high rate of ethanol transport out of the cells.

**Effect of Ethanol.** Ethanol is toxic to yeast. The general effect is most noticeable on the cell membrane; the major toxic effect has been postulated as membrane damage or a change in membrane properties [135]. Ethanol inhibits both growth and ethanol production in a noncompetitive manner [136]. When ethanol is present in concentrations of up to 2%, the observed inhibition is almost negligible for most yeasts. With a higher concentration, the effect of ethanol becomes more evident (Figs. 8 and 9).

Concentrations of ethanol >110 g/L stop both growth and ethanol production in most yeast strains. However, with most tolerant yeasts, ethanol production (not growth) is possible even in the presence of up to 20% ethanol [109].

As shown in Figure 8, the effect of externally added ethanol on cell growth is much lower than the effect of autogenous ethanol. The effect of ethanol on cell viability

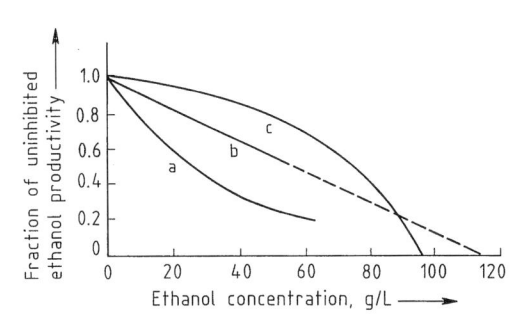

**Figure 9.** Effect of ethanol concentration on ethanol productivity [136]
a) [144]; b) [136]; c) [138]

increases at higher temperature [145] or if the cells are exposed to heat stress for a short period of time before coming into contact with ethanol [146].

Prolonged, continuous exposure of yeast to high ethanol concentration results in a loss of viability so that recycled cells must be continually replenished or discarded completely at intervals [147].

**Effect of pH.** The concentration of hydrogen ions in a fermentation broth affects yeast growth, ethanol production rate, byproduct formation, and bacterial contamination control.

Usually, industrial ethanol fermentation by yeast has an initial pH value of ca. 4–6, depending on the buffering capacity of the medium. In lightly buffered media, the initial pH is ca. 5.5–6 and in more highly buffered media, ca. 4.5–4.7 [127], [148]. If the pH value is <5 during fermentation, bacterial growth is severely repressed; the pH values for growth of most strains of *Saccharomyces cerevisiae* are 2.4–8.6, with an optimum of 4.5. Yeast sugar fermentation rates are relatively insensitive to pH values between 3.5 and 6 [109], [138].

**Effect of Temperature.** Most brewer's yeasts have a maximum growth temperature around 39–40 °C [126]. The maximum growth temperature reported for any species of yeast was 49 °C for *Kluyveromyces marxianus* [149]. Mesophilic strains of *Saccharomyces* have optimum cell yields and growth rates between 28 and 35 °C. The optimum and maximum temperatures for growth of thermophilic yeast are ca. 40 and 50 °C, respectively; these strains have a high maintenance requirement and more complicated nutritional requirements [109].

In batch processing, the optimum temperature for the complete utilization of glucose and the highest final ethanol concentration is generally slightly below the optimum growth temperature. This is attributed to enhanced ethanol inhibition at higher temperature [109], [150]. At higher temperature, the ethanol production rate is higher than its transport rate through the cell membrane. The difference in these rates results in an increase of ethanol concentration in the cells, a subsequent inhibition of some enzymes, and cell death. For continuous processes, the maximum process temperature (35–40 °C) in the absence of ethanol must be reduced by ca. 1 °C for each percentage increase in ethanol concentration [109].

Some yeasts have an optimum fermentation temperature of ca. 40–42 °C. They produce up to 12% of ethanol with yields >90% of theoretical [126], [151], [152]. Because sugar fermentation is exothermic (586 J of heat produced per gram of glucose consumed) [127], using yeasts that ferment at higher temperature substantially reduces cooling costs of fermentors.

## 5.1.4. Direct Fermentation

When polysaccharides (cellulose and starch) are used for ethanol production, they first must be hydrolyzed by means of different physicochemical or enzymatic methods and then converted to ethanol. Combining this two-step process into a single operation would be an advantage. This can be achieved by simultaneous saccharification and fermentation (SSF) processes using cocultures of microbes that are able to hydrolyze polysaccharides and convert sugars to ethanol, or by direct fermentation of polysaccharides to ethanol with monocultures.

In these cases the rate of enzymatic hydrolysis is increased by constant removal of the produced sugars through their conversion to the product. This is particularly important in the case of cellulose hydrolysis where glucose and cellobiose severely inhibit enzymes participating in the hydrolysis.

The coupled enzymatic hydrolysis of cellulose and simultaneous fermentation of the resulting sugars to ethanol by yeast have been investigated [153]–[157]. Table 10 shows simultaneous hydrolysis of ball-milled cellulosic substrates by cellulases of *Trichoderma reesei* QM 9414 and conversion of sugars to ethanol by *Candida brassicae*. Use of a thermotolerant yeast and cellulolytic enzymes in a combined saccharification fermentation process is an advantage because the optimum temperature for the hydrolysis of cellulose is 45–50 °C, and cooling problems can be simplified during large-scale fermentation [158].

A simultaneous saccharification–fermentation process was also applied in the production of ethanol from starchy materials (cassava and corn). Enzymes from *Aspergillus niger, A. awamori,* and *Rhizopus* sp. were used for hydrolysis of starch with simultaneous yeast fermentation [159]–[161]. Ethanol yields in these processes ranged from 82 to 99% of theoretical. In certain cases ethanol concentration reached a level of 20 vol% in five days [159].

The yeast *Schwanniomyces* is able to hydrolyze starch directly and partially ferment it to ethanol [162]. Amylolytic systems of *S. castellii* and *S. alluvis* have been reported [163]–[165]. Economic use of these and other amylolytic species is probably not feasible because of their limited ethanol tolerance [162], [166].

**Table 10.** Simultaneous saccharification and fermentation (SSF) of various substrates with and without milling pretreatment ( QM 9414 cellulase with )

| Feedstock | Conversion to ethanol | | | |
|---|---|---|---|---|
| | Unpretreated | | Pretreated | |
| | 24 h | 48 h | 24 h | 48 h |
| *Eucalyptus grandis* sawmill waste | 0 | 0 | 33.44 | 38.64 |
| *Acacia mearnsii* sawmill waste | 0 | 0 | 36.73 | 45.71 |
| Southern pine sawmill waste | 0 | 0 | 38.01 | 42.59 |
| Digester rejects | 41.76 | 54.55 | 40.31 | 72.86 |
| Primary clarifier sludge | 51.22 | 66.93 | 57.49 | 84.24 |
| Newsprint | 34.61 | 45.40 | 54.25 | 60.34 |
| Cardboard | 35.12 | 42.96 | 53.45 | 77.93 |
| Air classified sawmill waste | 49.58 | 53.00 | 59.05 | 75.13 |
| Cotton linters | 10.41 | 8.74 | 18.37 | 16.59 |
| Rice straw | 81.49 | 110.06 | 57.66 | 85.47 |
| Rice hulls | 45.99 | 50.19 | 65.57 | 96.05 |
| Corn stillage | 14.37 | 16.69 | 45.57 | 65.92 |

## 5.2. Production by Bacteria

In recent years, a number of facultative and obligatory anaerobes have become of increasing interest for the production of ethanol, as shown in Table 11 [167].

Thermophilic bacteria grow optimally at 60–70 °C. In the production of fermentation ethanol, their chief advantages over yeasts are

1) cooling costs are lower because less of the heat given off during fermentation has to be removed;
2) less energy is required for stirring the medium because the viscosity of the culture medium decreases with increasing temperature; and
3) the ethanol produced can be continuously separated from the culture medium by using a slightly reduced pressure.

Thermophilic bacteria, however, have a low ethanol tolerance [176], [177] and, in addition to ethanol, produce considerable amounts of lactic and acetic acids during fermentation [178], [179].

The mesophilic bacterium *Zymomonas mobilis*, isolated at the beginning of the 20th century from palm wine and from the Mexican beverage pulque, appears to be an excellent ethanol producer [180], [181]. This bacterium has several advantages over yeast (Table 12):

1) its growth rate is approximately double that of yeast;

**Table 11.** Bacteria of recent interest for the production of ethanol

| Organsim | Optimal temperature required for growth, °C | Hexose fermentation | Pentose fermentation | Reference |
|---|---|---|---|---|
| **Facultative anaerobes** | | | | |
| Bacillus macerans | 37 | + | + | [168] |
| Klebsiella planticola | 30 | + | + | [169] |
| **Obligatory anaerobes** | | | | |
| Zymomonas mobilis | 30 | + | − | [170] |
| Sarcina ventriculi | 37 | + | +* | [171], [172] |
| Clostridium saccharolyticum | 37 | + | + | [173] |
| thermocellum | 55–62 | + | + | [174] |
| thermosaccharolyticum | 58–64 | + | + | [174] |
| thermohydrosulfuricum | 69 | + | + | [174] |
| Thermoanaerobacter ethanolicus | 69 | + | + | [174] |
| finnii | 64–66 | + | + | [175] |
| brockii | 65–70 | + | +* | [174] |

* Only arabinose.

2) the rate of ethanol production is six to seven times faster than in conventional yeast fermentation, probably because sugar uptake in this small bacterium (size 1–2 µm) is more efficient than in yeast (size 5–10 µm); and

3) the yield of ethanol is ca. 5% higher than in yeast fermentation because less sugar is incorporated into the bacterial cell mass.

*Zymomonas mobilis* uses the 2-keto-3-deoxy-6-phosphogluconate or Entner–Doudoroff pathway (and not, like yeast, the fructose-1,6-diphosphate or Embden–Meyerhof–Parnas pathway) to metabolize sugar [170], pyruvate being formed as an intermediate. Details of these pathways can be found in [182] (see also Section 5.1.2). The formation of acetaldehyde by decarboxylation of pyruvic acid and its subsequent reduction to ethanol are identical to the reactions in yeast: 2 mol of ethanol and 2 mol of $CO_2$ are produced per mole of glucose. The energy (ATP) yield of this bacterium, based on the amount of glucose metabolized, is, however, only half that obtained with yeast. Therefore, *Z.* mobilis can synthesize only half as much new cell mass per mole of glucose degraded as yeast does, but this results in a higher ethanol yield.

*Zymomonas mobilis*, unlike yeast, does not require oxygen for growth; this greatly simplifies fermentative ethanol production. In contrast to a number of strictly anaerobic bacteria, *Z. mobilis* is relatively insensitive to oxygen. Indeed, it possesses an electron-transport system similar to the respiratory chain of aerobic organisms [183] and can transfer hydrogen, formed in the conversion of glucose to pyruvate, to molecular oxygen. These reducing equivalents are then not available for reduction of acetaldehyde to ethanol; acetaldehyde thus accumulates and inhibits the growth of *Z. mobilis* [184].

**Table 12.** Fermentation properties of the bacterium and of the yeast

| Property | Zymomonas mobilis | Saccharomyces carlsbergensis |
|---|---|---|
| Doubling time, h | 2.51 | 5.64 |
| Ethanol production rate $q_{P/X}$,* g g$^{-1}$ h$^{-1}$ | 5.44 | 0.82 |
| Cell yield $Y_{X/S}$,* g/g | 0.028 | 0.043 |
| Product yield $Y_{P/S}$,* g/g | 0.465 ** | 0.460 ** |

\* Explanation of symbols:

$$q_{P/X} = \frac{\text{g ethanol formed}}{\text{g cells formed} \cdot \text{h}}$$

$$Y_{X/S} = \frac{\text{g cells formed}}{\text{g substrate used}}$$

$$Y_{P/S} = \frac{\text{g ethanol formed}}{\text{g substrate used}}$$

\*\* Max. theoretical product yield = 0.511 g/g.

## 5.2.1. Ethanol Tolerance

The growth of most bacteria is inhibited by ethanol concentrations of 10–20 g/L. However, both *Z. mobilis* and yeast can tolerate high ethanol concentrations. In fact, alcohol concentrations of 120 g/L have been attained by fermenting sufficiently concentrated sugar solutions with *Z. mobilis* [180]. High ethanol concentrations generally destroy the structure and function of the cell membrane [185]. Large amounts of pentacyclic triterpenoids (hopanoids), e.g., hopene, diplopterol, and tetrahydroxybacteriohopane, have recently been found in *Z. mobilis*. These substances were first found a few years ago in various mineral oil fractions [186], [187].

Hopene [1615-91-4]
(hop-22(29)-ene diploptene)

Diplopterol [1721-59-1]
(hopan-22-ol)

Tetrahydroxybacteriohopane [51024-98-7]

The tetrahydroxybacteriohopane content of bacterial cells increases considerably with increasing ethanol concentration (Fig. 10). This indicates that ethanol tolerance is probably linked to hopanoid content. *Zymomonas mobilis* appears to counteract the ethanol-induced changes in fluidity of the membrane by increasing the biosynthesis and incorporation of hopanoids. The function of hopanoids in bacteria is apparently similar

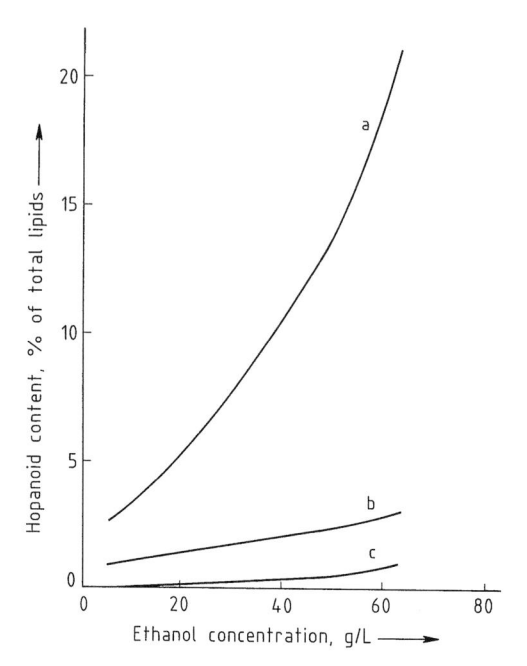

**Figure 10.** Dependence of the hopanoid content of *Zymomonas mobilis* on ethanol concentration
a) Tetrahydroxybacteriohopane; b) Diplopterol; c) Hopene

to that of sterols in yeast. They stabilize the membranes, making the microorganism resistant to ethanol [188]. Hopanoids are synthesized in bacterial cells under both aerobic and anaerobic conditions, whereas sterol biosynthesis in yeast requires oxygen.

## 5.2.2. Fermentation, of saccharified Waste Starch

A major consideration in achieving economic ethanol production is the cost of the sugar used as a raw material. For this reason, the fermentation of enzymatically saccharified (hydrolyzed) waste starch by *Z. mobilis* has been carefully investigated [189]. Pfeifer & Langen has recently developed a process for isolating glucose syrup from wheat flour (Fig. 11). The wheat flour is first separated into three fractions — pure starch (A-starch), protein (gluten), and a residue (B-starch) [190]. The pure A-starch fraction is subsequently converted enzymatically into a liquid sugar product, which is used in the food industry. The residue fraction contains not only starch, but also fibrous material, proteins, and lipids, making it unsuitable for the production of pure glucose syrup. However, after enzymatic liquefaction and saccharification of the waste starch in this fraction, the resulting glucose can be efficiently fermented to ethanol by *Z. mobilis*. Indeed, an ethanol production rate of 3.5–4.5 g $L^{-1}$ $h^{-1}$ and 99% glucose conversion have been achieved in a continuous culture, with dilution rates of 0.06–0.08 $h^{-1}$ and an initial glucose concentration of 120 g/L.

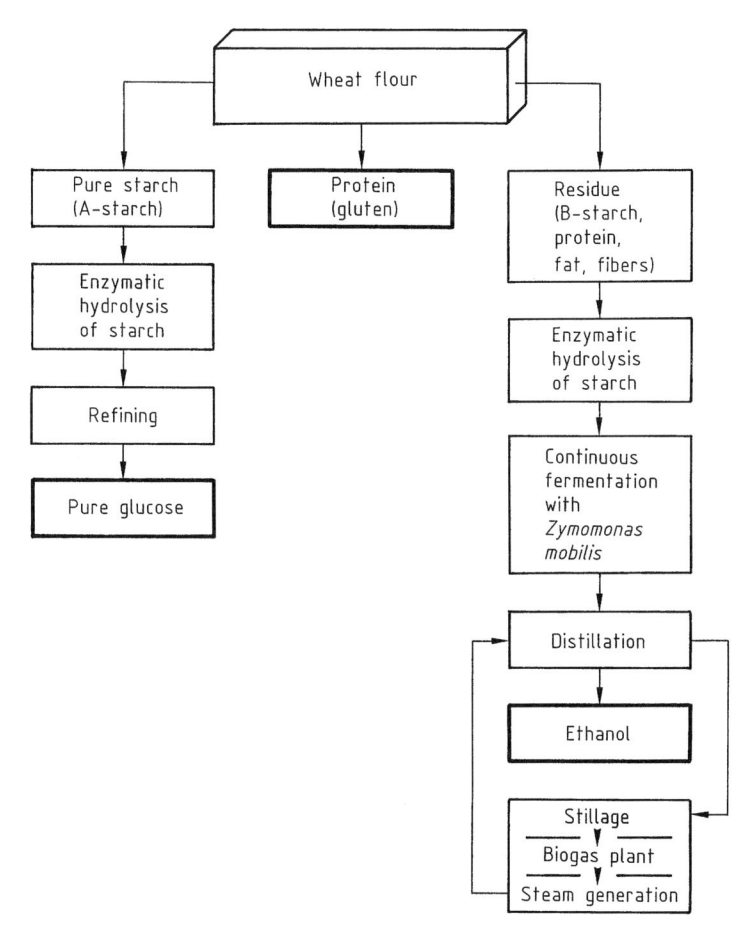

**Figure 11.** Flow diagram of the industrial production of glucose, gluten, and ethanol from wheat flour

In 1984 Pfeifer & Langen constructed a plant for the continuous production of ethanol by *Z. mobilis* from waste starch fractions. Fermentation carried out in two 70-m$^3$ fermentors could give a daily production of ca. 10 000 L of 96% alcohol. Ethanol production and the processing of wheat flour are summarized in Figure 11.

After fermentation of glucose to ethanol and its subsequent removal by distillation, stillage is left, which contains the bacterial biomass, fibers, proteins, and fat. These organic substances cannot be degraded by *Z. mobilis*. The wastewater is then fed into a "biogas" plant where a mixed culture of anaerobic bacteria degrades the organic components to biogas, a mixture of methane and $CO_2$. The methane is used to heat the distillation units.

## 5.2.3. Fermentation of Pentoses

Yeast and *Z. mobilis* can produce ethanol from glucose, fructose, or sucrose in yields of up to 95%, based on the maximum theoretical yield. However, no yeast or bacterial strains are currently available that can ferment pentoses (e.g., xylose or arabinose) to ethanol as efficiently.

The price of the sugar substrate represents as much as 70% of the total cost of fermentation ethanol. Efficiency could, however, be improved considerably by using both the hexoses and the pentoses of the plant biomass as substrates. A number of facultative and obligatory anaerobes are capable of fermenting pentoses (Table 11), but in addition to ethanol and $CO_2$, they also synthesize large amounts of other fermentation products (e.g., acetic acid, formic acid, lactic acid, acetone, and hydrogen). Metabolic studies show that these bacteria use both the pentose phosphate pathway and the Embden–Meyerhof pathway to metabolize pentoses to pyruvic acid (Fig. 12). In *Z. mobilis*, pyruvic acid is decarboxylated to acetaldehyde and $CO_2$ by the key fermentation enzyme pyruvate decarboxylase (E.C. 4.1.1.1) [*9001-04-1*], but facultative and obligatory anaerobic bacteria do not possess this enzyme. Facultative anaerobic bacteria contain pyruvate formate-lyase (formate acetyltransferase E.C. 2.3.1.54) [*9068-08-0*], which cleaves pyruvic acid to formic acid and acetyl-CoA. Some of the formic acid is then converted to hydrogen and $CO_2$ by formate hydrogen lyase. Strict anaerobes (e.g., clostridia) contain a pyruvic-ferredoxin oxidoreductase (E.C. 1.2.7.1) [*9082-51-3*] and a hydrogenase, which convert pyruvate to acetyl-CoA, hydrogen, and $CO_2$. The acetyl-CoA formed by these two routes participates in further enzymatic reactions, and some of it is reduced to ethanol. Bacterial pentose metabolism is summarized in Figure 12.

The gene for pyruvate decarboxylase from *Z. mobilis* has been inserted into *Escherichia coli* and *Klebsiella planticola*. In comparison with the wild types, these manipulated strains produce more ethanol and fewer organic acids during the fermentation of sugar [169], [190], [191]. Preliminary results indicate that this technique will prove useful in developing bacteria that can efficiently produce ethanol from pentoses.

## 5.3. Fermentation Modes

The two basic modes for production of ethanol by fermentation are the batch and continuous processes. A semicontinuous mode, which is a combination of the two, is also used.

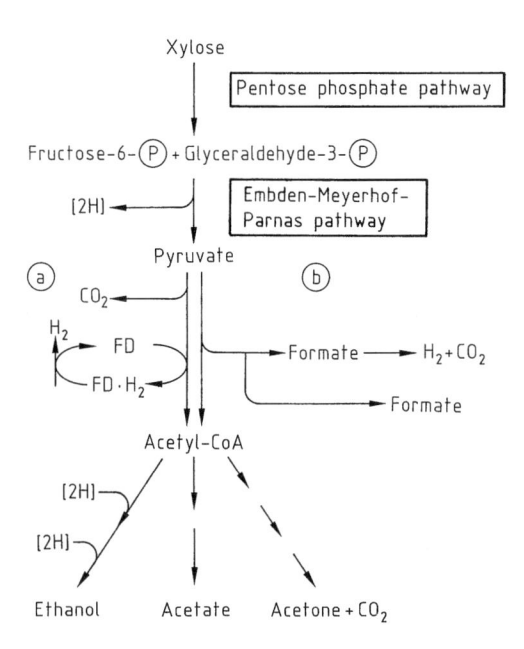

**Figure 12.** Pentose fermentation pathways in bacteria
a) Pyruvate: ferredoxin oxidoreductase (E.C. 1.2.7.1) [9082-51-3] (clostridia); b) Pyruvate formate-lyase (E.C. 2.3.1.54) [9068-08-0] (enterobacteria)
FD = ferredoxin; ⓟ = phosphate

## 5.3.1. Batch Processes

Most fermentation ethanol is produced by batch processes [111], [192]. A batchwise process for the fermentation of corn is shown in Figure 24. Typical kinetics in a batch process are shown in Figure 13.

Conversion of sugars in a simple batch system is ca. 75–95 % of theoretical, with a final ethanol concentration of 10–16 vol % [192], [194]. The productivity of simple, conventional batch processes is usually 1.8–2.5 g of ethanol per liter of fermentor volume per hour [111].

To increase productivity, a batch process with recycling of the yeast was developed; this is known as the "Melle Boinot" process. Recycling of yeast cells creates a high biomass concentration at the beginning of the process, which reduces the time for the conversion of substrate to ethanol (Table 13) [195]. This rapid fermentation also increases productivity. Increasing the initial concentration of yeast cells decreases yield. With a yeast cell concentration of $\geq 21$ g/L, no growth was noticed; the existing cells used nutrients for both ethanol production and their own maintenance [195].

Under rapid fermentation conditions, a high ethanol concentration is attained quickly and the death rate of yeast cells is quite high. The death rate can be decreased by fermenting at lower temperature and by maintaining the dissolved oxygen concentration in the medium at 10–20 % of oxygen saturation [145], [196]. The tolerance of some yeasts to ethanol is also increased in the presence of yeast vitamins [195], [197].

**Table 13.** Effect of initial cell concentration on time of fermentation during batch culture

| Initial cell concentration, g/L | Time to develop 12 vol% ethanol, h |
|---|---|
| 21.4 | 7 |
| 23.6 | 6 |
| 27.2 | 5 |
| 26.13 * | 4.5 |

* With yeast vitamins.

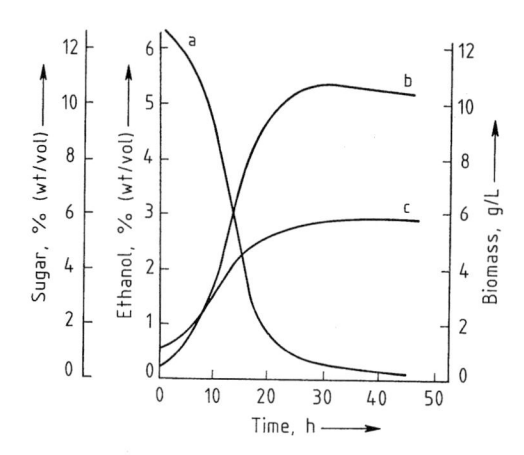

**Figure 13.** Ethanol production by *Kluyveromyces marxianus* 10606 [193]
a) Sugar; b) Ethanol; c) Biomass

Although classic batchwise processes for ethanol production are attractive because of their simplicity, they have many disadvantages such as low productivity, difficulty in automating, long and frequent downtime, and significant labor cost.

## 5.3.2. Continuous Processes

**Without Cell Recycle.** Generally, continuous processes eliminate most of the disadvantages that are inherent in the batch processes. They may be carried out for a long time period without shutdown, thus eliminating downtime between batches, increasing overall productivity. Because of increased productivity, smaller volume reactors are required than in batch production. Continuous processes can be fully automated and operated under conditions that give a uniform product. Continuous stirred tank reactors are used for these processes (Fig. 14).

Simple continuous processes for ethanol production are characterized by continuous addition of feed (including oxygen) to a reactor in which a desirable steady state has been established. Beer is also removed continuously from the reactor. The beer contains ethanol, biomass, and unconsumed nutrients. The productivity of such a process (up to 6 g L$^{-1}$ h$^{-1}$ with highly productive strains) is about three times higher than that of a batchwise process. Therefore, the reactor volume need be only one-third that used to produce the same amount of ethanol in a batchwise manner. In this process, both specific ethanol productivity and total productivity of the reactor are limited by ethanol

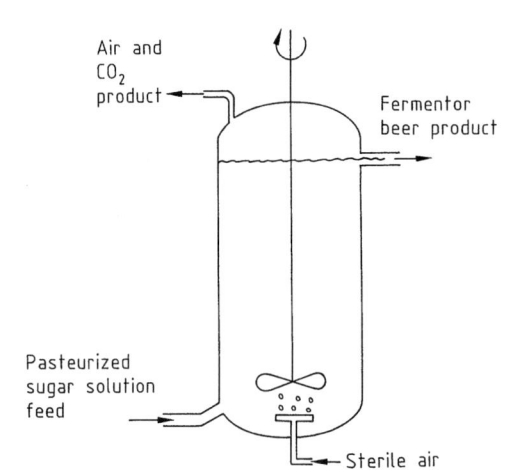

**Figure 14.** Continuous stirred tank reactor [111]

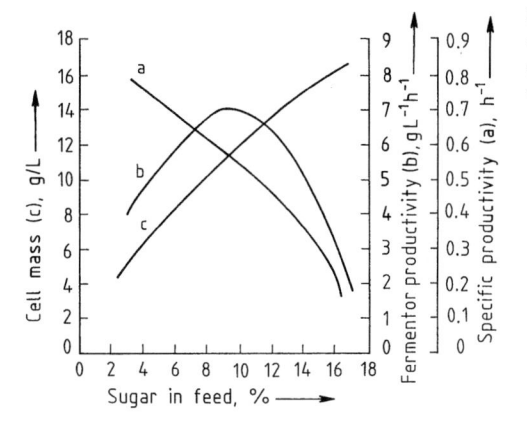

**Figure 15.** Effect of glucose concentration on continuous fermentation [198]
a) Specific productivity; b) Fermentor productivity; c) Cell mass

inhibition. In addition, total productivity is also limited by low biomass concentration (10–12 g/L) [111]. A sugar concentration of ca. 10% in the feed gives the highest reactor productivity (Fig. 15). Economic analysis showed a 53% reduction in operating costs and 50% in capital costs for this process compared to classic batch processes [198].

Sometimes two or more continuous stirred tank reactors can be joined in series to increase productivity. With two reactors, fresh medium is pumped to the first reactor; residence time is adjusted so that conversion of sugar is incomplete. This results in a lower ethanol concentration in the first reactor compared to the final concentration and, therefore, less ethanol inhibition. The outflow from the first reactor is fed to the second, where fermentation is completed. The productivity of the second reactor is lower than that of the first, but the overall productivity is increased compared to the system with one reactor [136], [199].

**With Cell Recycle.** Continuous systems with cell recycle were developed in order to increase productivity over systems that did not recycle biomass. The biomass from the stream leaving the reactor is separated by centrifugation and returned to the reactor. Yeast concentration as high as 83 g/L can be maintained during ethanol production [196]. Such high concentrations permit rapid and complete fermentation of concentrated sugar solutions. Fermentor ethanol productivities of 30–51 g L$^{-1}$ h$^{-1}$ have been achieved, which represents a more than tenfold increase over continuous fermentation without cell recycle [195], [196], [200], [201].

Both investment and operating costs of these processes are higher because of the required separation of cell mass from beer by centrifugation. Simpler and less expensive systems for cell separation are needed. Because of the low settling rate of yeast (<1 cm/h), shallow-depth sedimentation of biomass with the use of plate or tube settlers has been proposed [202]. The tube settler was found to be slower than the plate settler in reaching maximum effluent clarity. The plates and tubes are inclined to allow self-cleaning by the yeast sliding down the collection surfaces. The sedimentation rate of the yeast cells increases when evolution of $CO_2$ is repressed. This repression can be obtained by heat treatment for 20 min at 50 °C, which does not seriously impair cell viability. These settlers help reduce bacterial contamination because of their lower settling velocity.

Other separators, such as the whirlpool separator in which yeast cells are deposited in a central cone when the overflow is pumped tangentially into a vertical cylindrical vessel, have also been developed [203]. Another method involves a partial recycle fermentor that has a separate settling zone, which is not disturbed by agitation. This allows partial separation of cells [204].

### 5.3.3. Other Processes

**Flocculating Cells (Internal Recycle).** High-ethanol-producing and flocculating yeasts are of particular commercial interest. These cells can readily be concentrated and separated without the use of mechanical devices (e.g., centrifuges and conventional settlers). This simplifies the process and makes it more economic [201], [205]–[207]. A highly flocculating yeast, *Saccharomyces diastaticus,* was investigated for continous ethanol production from various substrates [201]. The process involved one fermentor with internal cell settling and no cell recycle or cell concentration by centrifuges or other such mechanical devices. A high specific ethanol productivity rate of up to 50 g L$^{-1}$ h$^{-1}$, with an inflowing sucrose concentration of 112 g/L (using fodder beet juice), was obtained at a dilution rate of 1.033 h$^{-1}$.

A semicontinuous (fill-and-draw) process was also developed using the flocculating yeast, *S. diastaticus* [208].

> The yeast is allowed to grow to a high biomass concentration in an agitated bioreactor; this assures good mixing and mass transfer during fermentation. The agitation is stopped after almost all the sugar

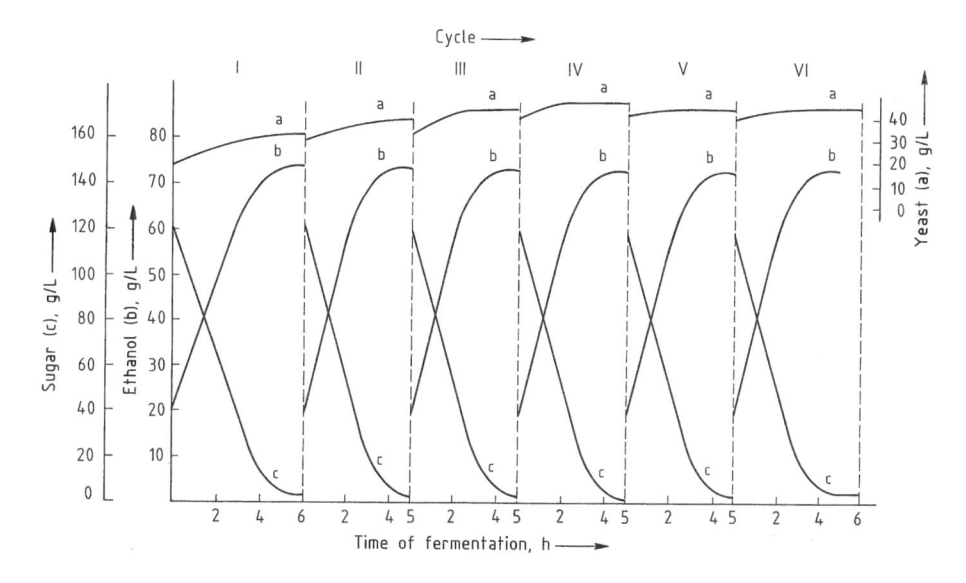

**Figure 16.** Semicontinuous fermentation of fodder beet juice with *Saccharomyces diastaticus* [208]
a) Yeast; b) Ethanol; c) Sugar

has been consumed, and the yeast is allowed to settle rapidly ($\leq 1$ min) in the bioreactor. The clear supernatant, which is the alcohol-containing liquid broth, is decanted and sent to distillation. Next, fresh medium is added to the bioreactor, which starts a new fermentation batch. These cycles can be repeated (ten times or more) without loss in productivity and cell viability. High ethanol productivity is achieved with a very short fermentation time (Fig. 16).

**Tower Reactor.** The tower reactor is convenient for internal settling of flocculating yeast cells [209]. It consists of a vertical cylindrical tower with a conical bottom (Fig. 17). Above the cylindrical section is a large, baffled settling zone (d), which has the form of a yeast separator. The separator can have various forms and provides a space free from rising gas, which enables yeast cells to settle and be returned to the main part of the reactor. An overall aspect ratio of 10:1, with an aspect of 6:1 on the cylindrical section, is used. The reactor does not have a mechanical agitator. High cell concentrations in excess of 100 g/L are achieved without any auxiliary mechanical separator. Compared to simple batch processes, productivities are 32–80 times higher in the tower reactor [111], [206], [207]. Hydraulic retention time is less than 0.4 h, and conversion efficiency is up to 95% of theoretical [207].

Flocculation and sedimentation characteristics remain unchanged over several months of continuous operation. Selective yeast retention is maintained even after moderate contamination, and any infecting organisms can be quickly washed out without any subsequent effect on reactor performance. Cell viability of 90–95% is attainable at biomass concentrations in excess of 100 g/L and remains at this level

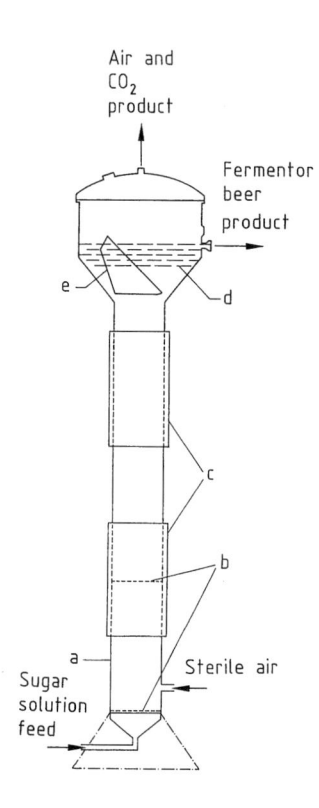

**Figure 17.** Tower fermentor (APV Co.) [111]
a) Flocculating yeast plug; b) Baffles; c) Attemperator jackets; d) Yeast settling zone; e) Clarifying tube

when the substrate is limited. The viability drops to 70–80 % when the ethanol concentration is allowed to exceed 75–85 g/L [207].

A major disadvantage of this type of reactor is the long time required (ca. two weeks) for the initial start-up.

**Membrane Bioreactor.** Buildup of a high yeast concentration can be achieved in a simple continuous dialysis fermentor. The reactor has fermentation and nutrient zones to which fresh medium is fed and from which beer is removed. These two zones are separated by a membrane, which prevents biomass escaping from the fermentation zone and allows transfer of nutrients. At the same time ethanol passes through the membrane to the nutrient zone where it is recovered in an overflow [111], [210].

The rate of ethanol production in this bioreactor is limited by the rate of nutrient diffusion across the membrane. This problem can be overcome by using a pressure dialysis reactor (Fig. 18) in which a pressure differential is applied across the membrane (c) to force a flow of medium through the fermentation zone [111].

Membrane fouling by proteins, which results from cell lysis and can affect flow, is a problem with this type of bioreactor. The problem can be alleviated by replacing the fixed membrane with a rotating microporous membrane cylinder (rotofermentor) enclosed in a stationary fermentor vessel. The rotofermentor allows continuous removal of metabolic products (including ethanol) in the beer by filtration through the rotating

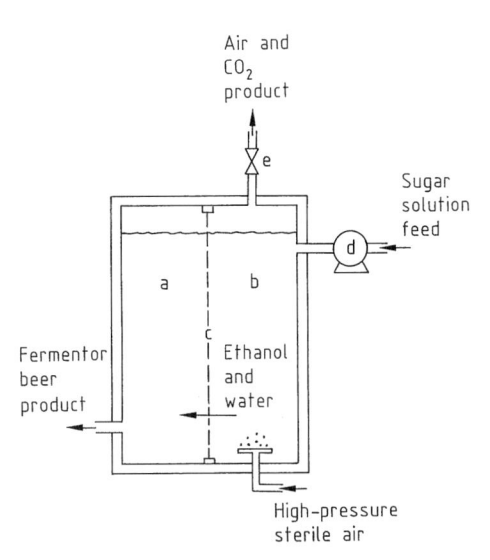

**Figure 18.** Pressure dialysis fermentor [111]
a) Low-pressure product recovery zone; b) High-pressure fermentation zone; c) Permeable dialysis membrane; d) High-pressure feed pump; e) Pressure-regulating valve

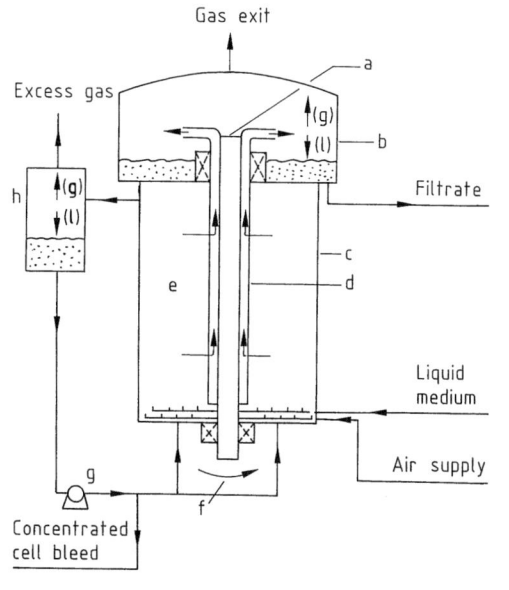

**Figure 19.** Schematic diagram of the rotofermentor assembly [211]
a) Rotor; b) Filtrate chamber; c) Fermentor; d) Rotating microporous membrane; e) Concentrated cell growth; f) Motor drive; g) Recycle pump; h) Gas–liquid separator

membrane under a pressure of 115–170 kPa (Fig. 19). The cells are retained in the annular part of the reactor outside the membrane [211]. Large molecules that tend to accumulate on the membrane surface are sent back into the annular zone by the centrifugal force generated by rapid rotation of the membrane. This cleans the membrane and allows diffusion through it. In the rotofermentor, the ethanol production rate reaches 26.8 g $L^{-1}$ $h^{-1}$ at concentrations of 24.8 and 50.4 g/L of yeast cells and ethanol, respectively [212]. The reactor has several serious mechanical and operating disadvantages related to the complexity of the system and durability of certain parts of the reactor (membrane and seals).

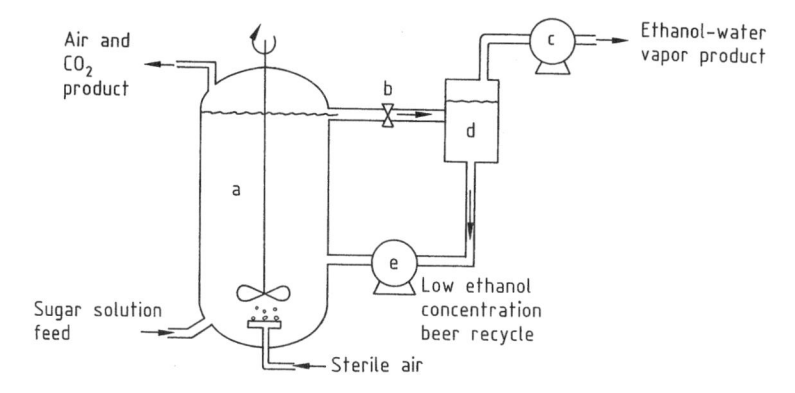

**Figure 20.** Continuous flash fermentation [111]
a) Atmospheric pressure fermentor; b) Flow-regulating valve; c) Vacuum compressor; d) Flash vessel; e) High-compression pump

Membrane bioreactors in which the fermentation vessel is coupled in a semi-closed loop configuration to a membrane filtration unit are available [213], [214]. The fermentation broth is pumped through the filtration unit. Inhibitory products permeate through the membrane, while retained yeast cells are recycled to the fermentor. High cell concentrations (up to 100 g/L) and ethanol production rates (up to 100 g $L^{-1}$ $h^{-1}$) are achieved [213].

**Vacuum Bioreactor.** Because high concentrations of ethanol are toxic to yeast, a vacuum process for continuous removal of ethanol from the beer has been developed [200], [215].

The bioreactor operates at a total pressure of 6.7 kPa. At that pressure, water and ethanol boil at a temperature that is compatible with the yeast. The liquid level in the bioreactor remains constant during operation of the bioreactor. An ethanol production rate of 82 g $L^{-1}$ $h^{-1}$ is achieved with a yeast density of 120 g/L. In the reactor, the ethanol concentration is kept at 35 g/L, while the ethanol concentration in the vapor product is >200 g/L. Oxygen is sparged instead of air in order to satisfy vacuum operation requirements [216]. Oxygen sparging can be omitted if the medium is supplemented with sterol and fatty acids [215].

The compressors used in the process must be large and must operate at low pressure. Capital cost of the compressors is quite high. Energy requirements exceed those of the normal ethanol fermentation process [195]. The probability of contamination is increased in this process.

A modification of the process exists (Fig. 20) in which the reactor operates at atmospheric pressure and $CO_2$ is released without compression from the reactor [216]. Beer is circulated continuously through a flash vessel (d) where ethanol–water solution is stripped under vacuum.

**Bioreactor with Solvent Extraction.** The concentration of ethanol in beer can be reduced during fermentation by solvent extraction. This reduces inhibition caused by the product, increases productivity of the reactor, and allows a higher concentration of sugar to be fermented in a shorter period.

In some cases, after separation of the cells, the clear beer is treated with an immiscible solvent that removes most of the ethanol. The resulting beer containing only small amounts of ethanol is returned to the reactor, and the saturated solution undergoes ethanol separation [217].

In another system, both extractant and nutrients can be continuously fed to a reactor that contains immobilized cells. The ethanol produced is extracted, which reduces its concentration in the aqueous phase, and is later recovered from the extractant [218]. By using the technique of ethanol extraction from beer with a mixture of *n*-dodecanol (60%) and *n*-tetradecanol (40%), the ethanol production is five times greater than in the same system without extraction. A solution of 407 g/L of glucose can be totally fermented by a yeast that cannot normally transform more than 200 g/L glucose [218].

The extractant used in this technology must have the following properties [219]:

1) it must be nontoxic to yeast;
2) it should be selective for ethanol in comparison with water and secondary fermentation products;
3) it must have a high distribution coefficient for ethanol; and
4) it should not form emulsions with fermentation broth.

Further development of this extractive technology is necessary to avoid inhibitory effects of ethanol and attain high rates of ethanol production. If a good extractant is found, the cost of ethanol separation by means of a membrane technique will be considerably decreased compared to the classic distillation of ethanol from beer.

**Immobilized Cells.** A high yeast concentration ($5.4 \times 10^{10}$ cells per milliliter) [220] in the fermentor can be obtained by various cell immobilization techniques, e.g., by entrapment in a gel matrix, covalent binding to surfaces of various support materials, or adsorption on a support. These systems do not require agitation. The immobilized cells are retained in the reactor; therefore, cell separation devices and recycle are not needed. High dilution rates without cell washout can be achieved.

Immobilized cells can be used in fixed- and fluidized-bed reactors. In these cases, the substrate solution flows continuously through the reactor, and the immobilized cells convert available sugar to ethanol. Calcium alginate [*9005-35-0*] can be used to entrap yeast cells. In a system with calcium alginate beads, a maximum ethanol productivity of 53.8 g/L was achieved at a dilution rate of 4.6 h$^{-1}$ and an initial glucose concentration of 127 g/L [221].

Carrageenan [*9000-07-1*] can also be used to immobilize yeast cells, giving ethanol production rates of 43 g L$^{-1}$ h$^{-1}$ at a dilution rate of 1.0 h$^{-1}$ and sugar utilization of 86% from the feed containing 100 g/L glucose [222].

**Table 14.** Ethanol productivity from immobilized systems (*Saccharomyces cerevisiae* and *Kluyveromyces fragilis*)

| System | Feed sugar | Feed sugar concentration, g/L | Feed sugar used, % | Dilution rate, h$^{-1}$ | Maximum ethanol productivity, g L$^{-1}$ h$^{-1}$ | Reference |
|---|---|---|---|---|---|---|
| S. cerevisiae carrageenan | glucose | 100 | 86 | 1.0 | 43 | [222] |
| S. cerevisiae calcium alginate | glucose | 127 | 63 | 4.6 | 53.8 | [221] |
| S. cerevisiae calcium alginate | molasses | 175 | 83 | 0.3 | 21.3 | [223] |
| S. cerevisiae carrier A. | molasses | 197 | 74 | 0.35 | 25 | [224] |
| S. cerevisiae glass beads coated with gelatin and glutaraldehyde | glucose | 150 | 99 | | 43 | [225] |
| S. cerevisiae anionic exchange resin XE-352 | glucose | 120 | 94 | | 53.1 | [227] |
| K. fragilis ceramic Intalox saddles coated with gelatin and glutaraldehyde (two fixed-bed reactors) | lactose | 109 | 99 | | 50.6 | [226] |

**Table 15.** Ethanol productivity in free and immobilized systems (with *Saccharomyces cerevisiae*)

| Process | Substrate | Volumetric ethanol production rate, g L$^{-1}$ h$^{-1}$ |
|---|---|---|
| Batch | molasses | 2.0* |
| Continuous (free cell) | molasses | 3.35 |
| Continuous (immobilized) | molasses | 28.6 |

\* Excluding downtime of fermentor.

Table 14 shows the ethanol production rates of *Saccharomyces cerevisiae* and *Kluyveromyces fragilis* cells immobilized by entrapment, covalent binding, and adsorption.

A comparison between batch and continuous (free and immobilized) processes for ethanol production is shown in Table 15. The ethanol production rate in an immobilized cell system had a lower optimum temperature (30 °C) than a system with free cells (37 °C) [221]. The downshift of the temperature optimum was possible because of the effect of ethanol inhibition (which increases with increased temperature) within the gel matrix.

A diffusion limitation is imposed by the gel matrix, which produces a substrate and product gradient within the bead. Therefore, the immobilized system should be operated at lower temperature than the free cell system to minimize inhibition and maximize ethanol production.

The inhibition caused by high substrate concentration is excluded if the rate of sugar conversion to ethanol is higher or equal to the rate of sugar diffusion into the matrix.

A difference between the optimum pH values for the production of ethanol by free and immobilized cells was also found. Systems with free *Saccharomyces cerevisiae* cells have an optimum pH of 4.5, but the immobilized system showed a broad optimum spectrum with a maximum activity between pH 3.0 and 7.5. The reason for this difference is a pH gradient within the beads [221], [229].

## 5.4. Raw Materials and Processes

Raw materials for the production of ethanol by fermentation can be classified as:

1) readily fermentable carbohydrates that can be used directly, and
2) starch and other organic materials that must be converted to a fermentable form prior to fermentation.

The raw materials come from three major sources:

1) agricultural crops,
2) forest products, and
3) industrial and agricultural byproducts and residues.

Depending on need, end use, and availability, the choice of raw material varies for different regions, countries, and industries.

### 5.4.1. Readily Fermentable Carbohydrates

Various sugar crops, such as sugarcane, sugar and fodder beet, fruit crops, and crops based on crassulacean acid metabolism (CAM), are in this category.

**Sugarcane.** Sucrose [57-50-1] ($\alpha$-D-glucopyranosyl-$\beta$-D-fructofuranoside) is the sugar obtained from cane or beet. Sugarcane is a tropical crop whose successful cultivation is limited to an area spanning 37 °N to 31 °S.

Although sugarcane is grown mainly for production of table sugar and molasses, it is also an excellent raw material for the production of ethanol. The fermentable carbohydrates from sugarcane can be used either as the cane juice directly or as blackstrap molasses (a sugar byproduct). A material balance shows that 160 kg of fermentable solids can be obtained from 1 t of cane [230].

The cane juice is prepared by crushing raw cane and extracting the sugar with water, followed by clarification using milk of lime and $H_2SO_4$ to precipitate the inorganic materials [231]. The resulting extract is a green, sticky fluid, slightly more viscous than water, with an average sucrose content of 12–13 % [232].

Blackstrap molasses is the residue remaining after sucrose has been crystallized from cane juice. Molasses is a heavy viscous material, which contains sucrose, fructose, and glucose (invert sugar) at a total concentration of ca. 50–60 % (wt/vol) [233]. In

**Table 16.** Yields in production of ethanol from sugarcane

| | Alcohol, indirectly from molasses | Alcohol, directly from sugarcane juice |
|---|---|---|
| Sugarcane yield in 1.5–2-year cycle (south-central region), t/ha* | 63 | 63 |
| Average sucrose yield (13.2 wt%), t/ha | 8.32 | 8.32 |
| Crystal sugar production, t/ha | 7.0 | |
| Final molasses or cane juice production, t/ha | 2.21 | 66.2 |
| Fermentable sugar, molasses, or juice, t/ha | 1.32 | 8.73 |
| Alcohol yield at 100% global efficiency, kg/ha | 675 | 4460 |
| Alcohol yield at reasonable 85% global efficiency, L per ton of cane or in L/ha | 11.5<br>730 | 75<br>4800 |

* Hectare (ha) = $10^4$ $m^2$.

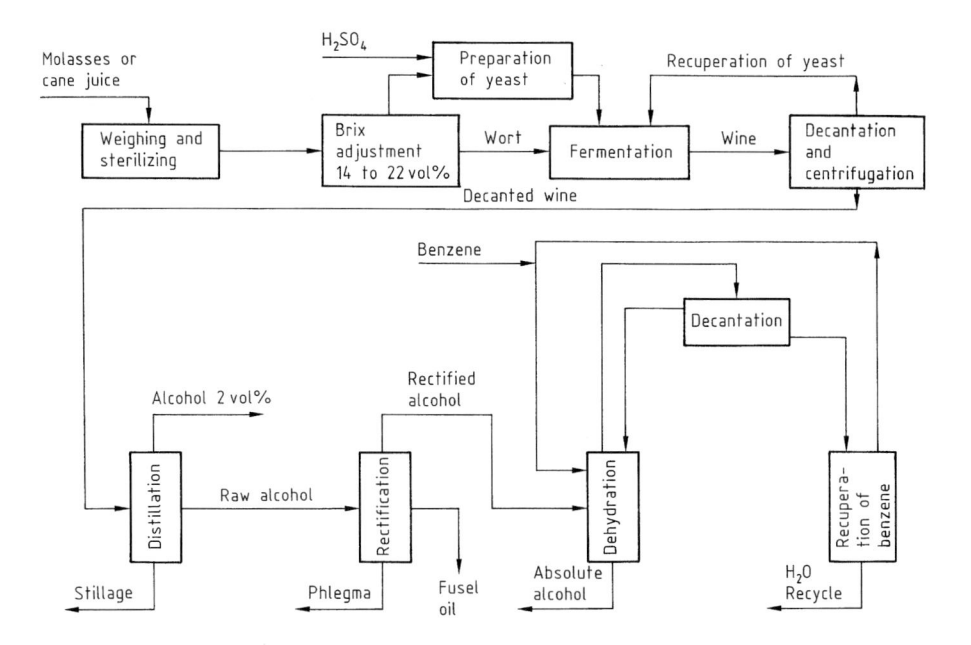

**Figure 21.** Typical process for the production of ethanol from sugarcane [234]

contrast to cane juice, molasses is stable on storage and is usually diluted to the desired concentration just prior to fermentation.

A typical process for the production of ethanol from sugarcane is depicted in Figure 21. Production data are listed in Table 16. Ethanol production reaches a maximum after 14–20 h and then decreases until ca. 95% of the available sugar is consumed. The process is usually batchwise, but some semicontinuous [199] and continuous operations [209], [234]–[236] are also used.

**Table 17.** Performance data for the Danish Distilleries process *

|  | Fermentor 1 ($f_1$)** | Fermentor 2 ($f_2$)** |
|---|---|---|
| Amount of yeast and dry matter per liter, g | 10 | 10 |
| pH | 4.7 | 4.8 |
| Alcohol, vol% | 6.1 | 8.4 |
| Residual sugar, % | 1.0 | 0.1 |
| Temperature, °C | 35 | 35 |

\* Residence time in each fermentor: 10.5 h; influx: 600 kg of molasses diluted in $22 \times 10^3$ L/h.
\*\* See Figure 22.

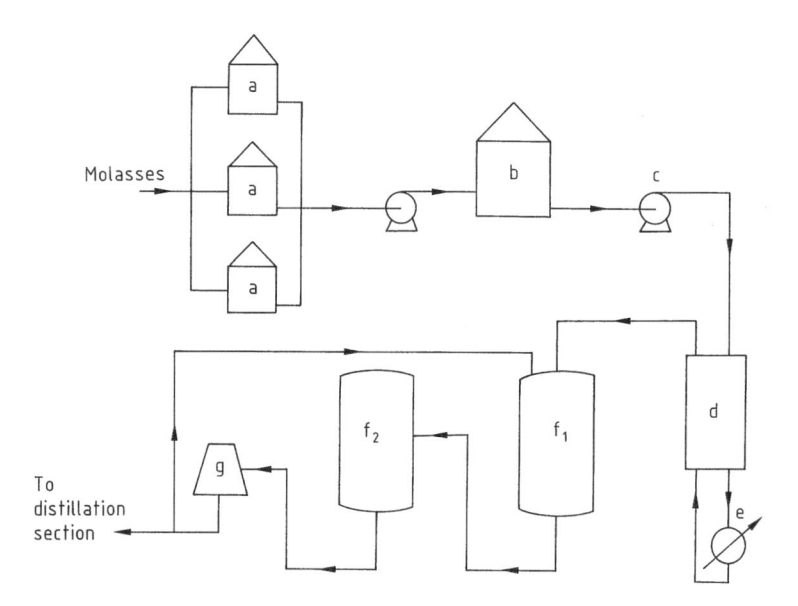

**Figure 22.** Continuous production of ethanol by Danish Distilleries [235]
a) Storage tank; b) Intermediate container; c) Metering pump; d) Regeneration section; e) Plate heat exchanger; f) Fermentor; g) Yeast separator

In the *batch process*, several fermentors are usually operating at staggered intervals to provide a continuous feed to the distillation columns. Overall productivity is ca. 18–25 kg of ethanol per cubic meter of fermentor volume per hour [192]. The "Melle Boinot" process is used in most Brazilian distilleries (see Section 5.3.1).

Ethanol has been produced in a *continuous process* (using continuously stirred tank reactors) from molasses by Danish Distilleries [235]. The process is shown in Figure 22, and performance data are given in Table 17.

According to this process, the molasses is stored in two or three 1500-m³ tanks from which it is pumped to intermediate containers. The material is adjusted for pH and nutrients (nitrogen and phosphorus), sterilized at 100 °C by using plate heat exchangers, and then introduced to three fermentors with a total volume of 170 m³. The fermented wort is centrifuged after fermentation, and the live yeast returned to the first fermentor. At the start, sufficient yeast propagation must be

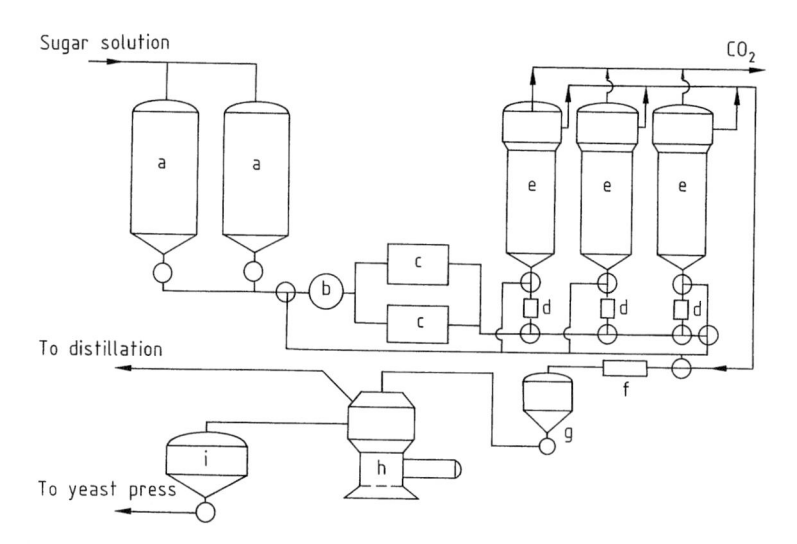

**Figure 23.** Commercial ethanol tower fermentation system (APV Company) [209]
a) Buffer storage tank; b) Divert line; c) Pasteurizer; d) Flow controller; e) Vertical cylindrical tower; f) Chiller; g) Yeast green beer buffer storage; h) Centrifuge; i) Separator

accomplished by aeration (0.02 – 0.03 L of air per liter of liquid per minute). The yield is ca. 28.29 L of alcohol per 100 kg of molasses, or a maximum of ca. 65 L of alcohol per 100 kg of fermentable sugar.

A continuous process for production of beer from sugar by use of *tower fermentors* is shown in Figure 23 [237], [238].

The key to the process is a vertical cylindrical tower fermentor with a conical bottom. A baffled yeast settling zone constitutes the upper part of the fermentor. The fermentor uses a flocculent yeast, which is pumped into the base of the tower. As the reaction proceeds, the beer rises, and the flocculating yeast settles and is retained in the reactor. High cell densities of 50–60 g/L are achieved without the use of mechanical cell concentration or separation devices. Short residence time (<4 h) with a sugar concentration of up to 12% (wt/vol) sucrose, 90% sugar utilization, and 90% conversion to ethanol, produce up to 5% ethanol in the final broth. The overall productivity of this system can be up to 80 times higher than that of the simple batch system.

**Sugar Beet.** Like sugarcane, sugar beet produces carbohydrates that consist primarily of sucrose. Sugar beet is a more versatile crop than sugarcane. It can tolerate a wide range of soil and climatic conditions, and is grown throughout nearly half of the United States, Europe, Africa, Australia, and New Zealand.

In addition to sucrose, sugar beet contains sufficient nitrogen and other organic and microorganic nutrients [239] so that little, if any, fortification is required prior to fermentation. Another benefit is the high yield of coproducts such as beet tops and extracted pulp. The pulp has a high feed value, and the tops may be returned to the soil for erosion control and nutrient replacement. The yield of fodder beets is high (ca. 50 – 150 t/ha); their composition is described in [240].

A new fodder beet crop, produced in New Zealand through a genetic cross between sugar beets and marigolds, gives greater yields of fermentable carbohydrates per hectare than does sugar beet [241]. In addition, the sugar from fodder beet is reported to be more resistant to degradation over long storage.

Processes for the production of alcohol from sugar and fodder beets are basically the same as from sugarcane.

**Fruit Crops.** Many crops (grapes, plums, peaches, apricots, pineapples, etc.) contain variable proportions of sugars (sucrose plus fructose, usually 6–12%). The fruit sugars can be readily fermented to alcohol, and this is done on a large scale for production of alcoholic beverages. The alcohol content of the product, which basically is the liquid after fermentation, separation of yeast, further treatment, and aging, depends on use and fermentation conditions. Table wines have <14% alcohol, whereas wines with >14% alcohol fall in the category of desert wines and aperitifs. Higher concentrations of alcohol are achieved by means of distillation to produce "strong" alcoholic beverages (e.g., brandy, whiskey, gin, vodka).

Alcohol for industrial or fuel use is seldom produced from fruit and vegetable crops. However, some fruit from tropical and semiarid climates, such as dates [242], mohwa flowers [243], and rain tree fruit [244], have been investigated for fuel alcohol production.

**Crops Based on Crassulacean Acid Metabolism.** Interest in using the agriculturally semi- or nonproductive regions of the world to grow alcohol-producing crops has increased [245]. These regions could be used to grow plants that utilize crassulacean acid metabolism (CAM) because their photosynthetic metabolism is extremely efficient with respect to irrigation requirements. These plants exhibit above-average productivity (expressed as a function of biomass production per unit of existing biomass) compared to other agricultural crops.

The CAM plants that are high in fermentable carbohydrates include various cacti (e.g., *Opuntia* sp.) and other plants such as *Euphorbia lathyrus* and *Agave* sp. Few data are available on potential ethanol production from these crops and its economic feasibility; however, an estimated 50 t/ha of these crops could be produced annually in subagricultural areas [246].

## 5.4.2. Starch

A variety of starch materials, such as grains, cassava, sweet potatoes, sweet sorghum, and Jerusalem artichoke, can be used for fermentation to ethanol. Selection depends on various factors, the major ones being climate and availability for large-scale production. Corn, wheat, potatoes, and Jerusalem artichokes are the most common raw materials in Europe and North America, whereas rice, cassava, sweet potato, and sweet sorghum are important in tropical countries.

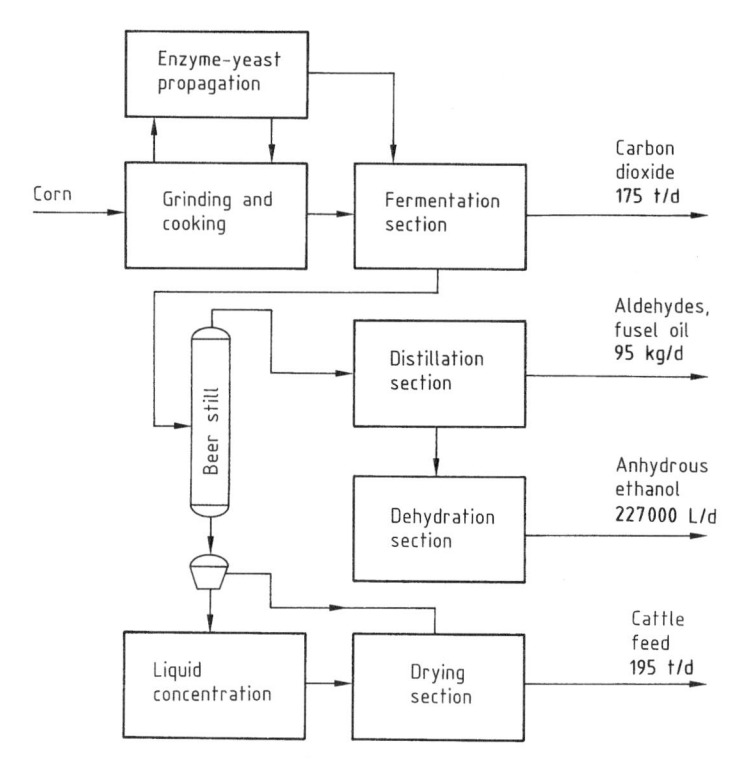

**Figure 24.** Flow diagram for a conventional fermentation plant producing anhydrous ethanol from corn [247]

**Corn.** Corn is the preferred raw material for conversion to alcohol in the United States and parts of Europe. It is available in large quantity, and its price (especially for low-grade or distressed corn) is thus acceptable for conversion to ethanol. Conversion to ethanol is efficient, and byproducts, such as corncobs, stalks, and leaves, are valuable as animal feed, energy source, or fertilizer. About 66 % of corn production is used for food and feed, and ca. 5 % is used to make alcohol.

A number of batch and continuous processes have been developed for production of ethanol from corn. A conventional fermentation plant producing $76 \times 10^3$ m$^3$ of anhydrous ethanol per year from $816.5 \times 10^3$ kg of corn per day is shown in Figure 24.

In this process, corn is ground and cooked to dissolve and gelatinize the starch. The enzymes α-amylase and glucoamylase are then added to hydrolyze the starch to fermentable monosaccharides. After yeast fermentation for ca. 48 h at 32 °C, about 90 % of the starch is converted to ethanol. The fermentation broth is fed to the beer still where alcohol (ca. 50 vol %) is distilled. Subsequent distillation produces 95 % alcohol, which can be further concentrated by azeotropic distillation using benzene. After centrifugation, the stillage is concentrated to ca. 50 % solids in a multiple-effect evaporator, further concentrated in a fluidized-bed, transport-type dryer to ca. 10 % moisture, and then used as such for animal feed. This feed contains all the protein originally present in the grain, plus the additional protein from the yeast, resulting in a product containing 28 – 36 wt % protein.

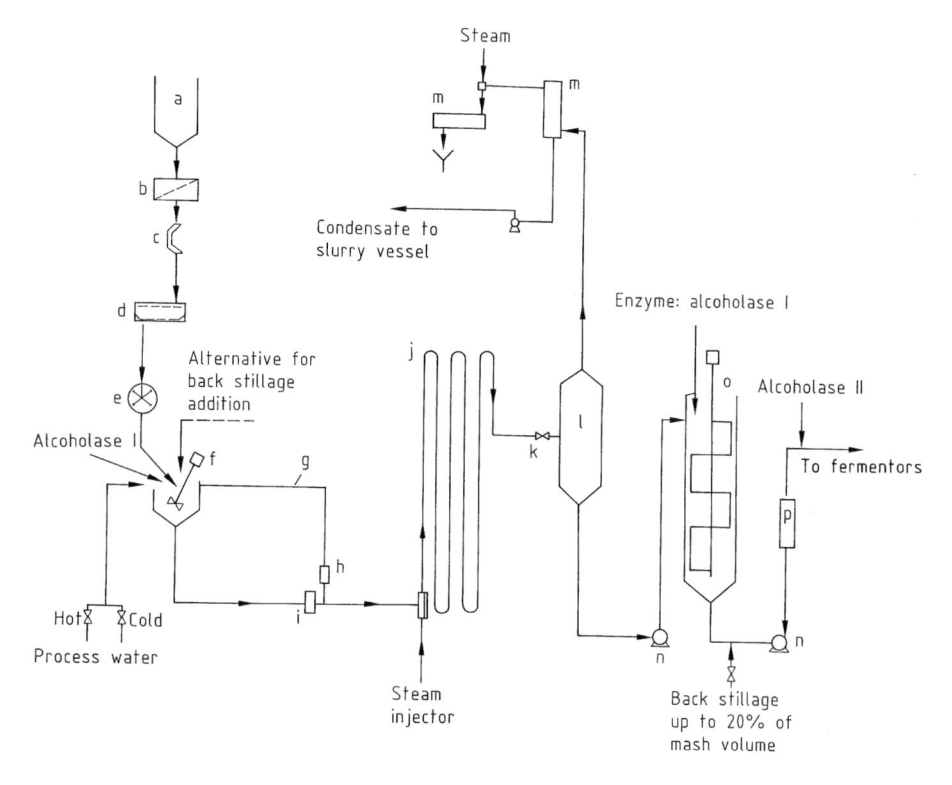

**Figure 25.** Alltech process for continuous whole mash cooking [248]
a) Grain hopper; b) Screen; c) Magnets; d) Continuous weigher; e) Hammer mill; f) Slurry vessel with agitator, temperature 50–70 °C; g) Rupture disk; h) Expansion vessel; i) Positive displacement pump; j) Continuous cooker tube, residence time 5 min, temperature variable up to 150 °C; k) Pressure valve; l) Flash vacuum cooler to 66–76 °C; m) Condenser; n) Open impeller pump; o) Converter, residence time 20 min, agitator 1 rpm; p) Wort cooler

In addition to alcohol and cattle feed, the original $816.5 \times 10^3$ kg of corn yields $175 \times 10^3$ kg of $CO_2$ and 95 kg of byproduct aldehydes, ketones, and fusel oils.

Alltech developed a method for an integrated grain-processing–fermentation route [248]. The grain pretreatment step prior to fermentation is shown in Figure 25.

Two enzymes, alcoholase I (from *Bacillus subtilis*) and alcoholase II (from *Aspergillus niger* and *Rhizopus niveus*) are used to hydrolyze the starch to fermentable sugars. Continuous whole mash cooking is applied. The ground starch is first mixed with water and alcoholase I at 60 °C, and then cooked at 93–165 °C in a batch or continuous cooker. The cooked mash is then cooled to ca. 66–76 °C, and a second portion of alcoholase I is added; 20 min is allowed for conversion. After this first hydrolysis step, the temperature is adjusted to 32 °C and the mash, supplemented with alcoholase II, is fermented with yeast.

**Cassava.** Cassava, also known as manioc, mandioc, aipum, yuca, cassada, and tapioca, is second in importance only to the sweet potato as a root crop throughout the tropics and in parts of South America where the plant originated. It was taken to West Africa by the Portuguese around 1914, where it now seems to have replaced yams and cocoyams because it adapts easily and requires less labor than other crops. Cassava is one of the highest yielding plants of the vegetable kingdom (10 – 30 t/ha); it requires little cultivation and the tubers can be left in the ground until required without serious deterioration.

Cassava (genus *Manihot*) is in the family Euphorbiaceae, which belongs to the subdivision Angiospermae, class Dicotyledoneae, order Geraniales. This large, widely spread family comprises 283 genera including 7300 species, with an almost worldwide distribution [249].

*Manihot esculenta, M. utilissima,* and *M. dulcis* are some economically important members of a genus which includes over 150 species that are distributed throughout tropical countries. The species include herbs, shrubs, and trees, many of them producing latex and some yielding rubber. Brazil, Indonesia, and Zaire are the largest producers of cassava.

Roots are generally of interest for alcohol production. They contain 20 – 35 wt% starch and 1 – 2 wt% protein, although strains with up to 38% starch have been developed [250]. The advantages of cassava for fuel ethanol production (which can amount to 7600 L/ha) have been assessed [234], [251], [252]. The process used to obtain ethanol from cassava is shown in Figure 26.

Fresh roots are washed, peeled, and ground into a mash. Part of this mash is dried; it can be stored in this form up to a year and is used for animal feed. For fermentation to ethanol, the starch is hydrolyzed with α-amylase, which is added in two steps. The first addition decreases the vicosity of the mash and facilitates cooking. In the second addition, the enzyme completes liquefaction of the starch. After that glucoamylase is added, which converts the liquefied starch to glucose and prepares the mash for fermentation. The fermentation process is the same as the one used for production of alcohol from sugarcane.

Alcohol yield from cassava is 165 – 180 L/t, which, on a mass basis, is higher than that obtained from sugarcane [252]. However, because sugarcane production can be as high as 90 t/ha, the alcohol yield per unit area is greater from cane under present cultivation conditions. Another advantage of cane is its dry fiber content, which equals the amount of total sugar present. This amount of fiber (bagasse) is sufficient to maintain the energy requirements of the plant; this is not the case with cassava, which only contains ca. 3.5% fiber. Another disadvantage of cassava is that it does not contain readily fermentable sugars and, therefore, requires considerable processing of the roots prior to fermentation.

**Sweet Sorghum.** Sweet sorghum (*Sorghum sacchartum*) contains both starch and sugar. Its yield of ethanol from fermentable sugars is ca. 3500 – 4000 L/ha; an additional 1600 – 1900 L/ha can be produced from stalk fibers. There are more than 17 000

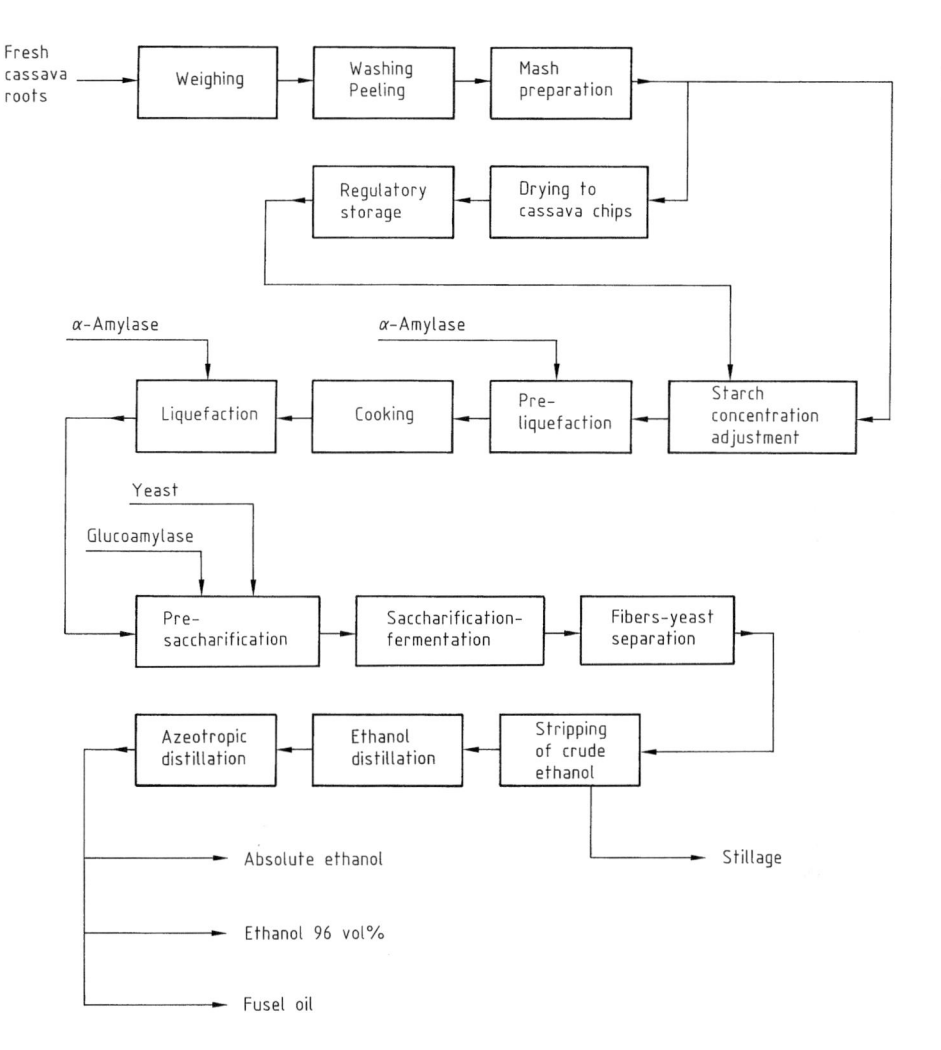

**Figure 26.** Production of ethanol from cassava root [234]

varieties of sorghum, and the yield is anticipated to increase by 30% with some new hybrids [253]. The plant is adaptable to most of the world's agricultural regions; it is resistant to drought, and its nutrients are efficiently utilized by animals.

The fermentable sugars and starches are treated conventionally for ethanol production. The free sugars are fermented directly, whereas the starches are hydrolyzed by use of amylases, as is the case with cassava.

**Potato.** The potato is a common starch crop worldwide. The potato originated in South America (Chile and Peru) and came to Europe through Spain at the end of the 16th century. It is now grown in almost all climates and almost all types of soil, including dry and sandy soil [254].

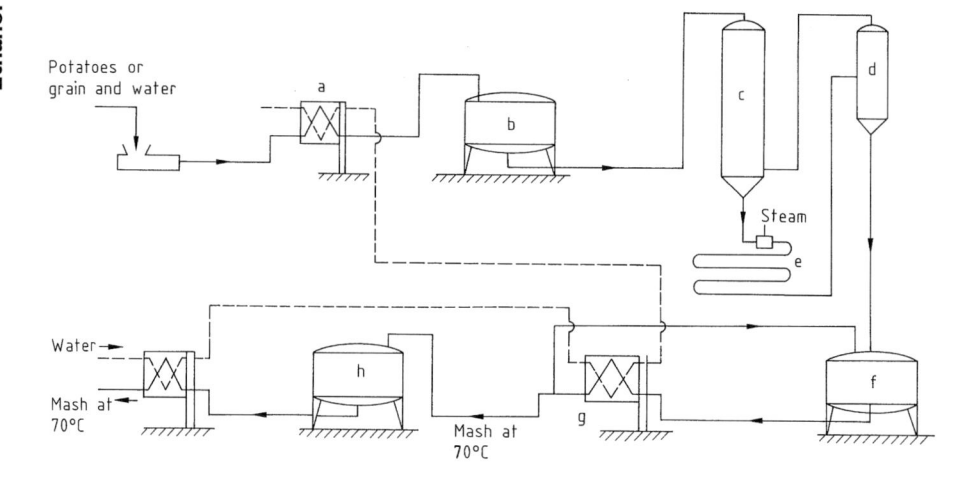

**Figure 27.** Danish Distilleries semicontinuous production of alcohol from potatoes or grain [235]
a) Preheater; b) Pulper; c) Enzyme treatment vessel; d) Flash cooler; e) Boiler tube; f) Holding tank; g) Condenser; h) Liquefaction vessel

Starch is the main carbohydrate component of potato (ca. 68–80%). Depending on cultivation and variety of potato, starch content can vary between 12 and 21% in raw potatoes. Only small quantities of soluble sugars are present (0.07–1.5% sucrose, glucose, and fructose), as well as some rubber and dextrins (0.2–1.6%) and pentosans (0.75–1.00%).

The production of ethanol is based on fermentation of the available starch. A process developed by Danish Distilleries is shown in Figure 27 [235]. The process is semi-continuous and is applicable to both potatoes and grain.

Potatoes are mashed and then treated with amylases to hydrolyze the starch. The treatment section involves rapid steam treatment at 150 °C for ca. 3 min. The mash is cooled to 70 °C for liquefaction with commercial amylase preparations of bacterial origin; it is then cooled further to 30 °C and used for alcohol fermentation in the customary manner.

**Jerusalem Artichoke.** The Jerusalem artichoke (*Helianthus tuberosus*) is a member of the Compositae family and is closely related to the sunflower (*Helianthus annus*), earning it the nickname "wild sunflower." About 102 different names are synonymous with the name *H. tuberosus*.

The plant is native in North America. It was originally grown by the Cree and Huron Indians who called it askipaw and skibwan, respectively. The plant was introduced to Europe at the beginning of the 16th century where it rapidly spread through the Mediterranean countries. The addition of "Jerusalem" to the name is most likely the result of an English version of *Girasole*, the Italian name for this plant [255].

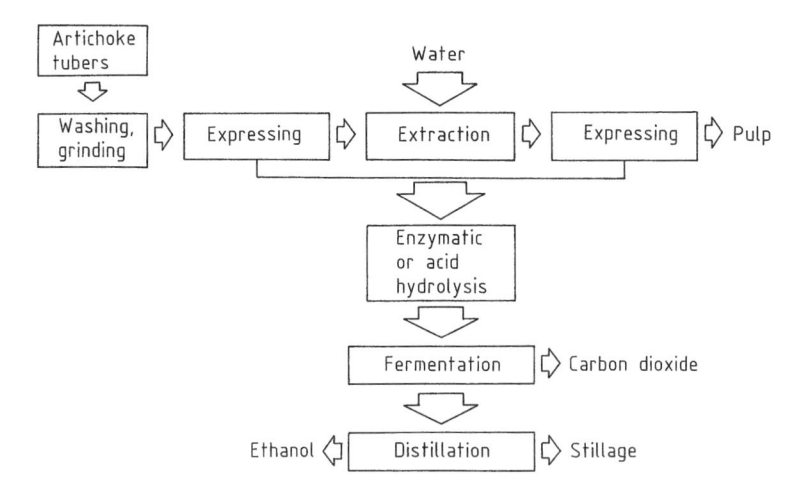

**Figure 28.** Production of ethanol from Jerusalem artichoke tubers [257]

The plant grows 1.5–2.5 m tall for wild strains and up to 3.7 m under cultivation. Top growth accounts for 40–56% of the total plant biomass. The tubers are of greatest interest as a raw material for fermentation to ethanol.

The main soluble carbohydrate in the Jerusalem artichoke is inulin, which is composed of a homologous series of polyfructofuranose units. These units consist of linear chains of D-fructose molecules joined by $\beta$-2,1-linkages. The chains are terminated by a D-glucose molecule linked to fructose by an $\alpha$-1,2-bond as in sucrose [256].

A process to produce $360 \times 10^3$ kg/a of ethanol from Jerusalem artichoke tubers is presented in Figure 28 [257].

In this process, the juice is expressed from the tubers and extracted with water to obtain a carbohydrate concentration of about 20%. The carbohydrates (predominantly inulin) are then hydrolyzed enzymatically by activating the endogenous inulinases at ca. 50–60 °C; acid hydrolysis (pH ca. 1) of inulin is also effective. The resulting fermentable sugars are then converted to ethanol by a conventional route.

Novel routes for conversion of the juice to ethanol have also been explored; the flocculating yeast *Saccharomyces diastaticus* has been used in semicontinous and continuous modes [201].

## 5.4.3. Lignocellulosic Materials

Lignocellulose is the largest terrestrial source of biomass that is renewably produced through photosynthesis. The solar energy reaching the earth surface is $3.67 \times 10^{21}$ kJ/a [258]. Gobal photosynthesis (with an efficiency of 0.07%) could convert $2.57 \times 10^{18}$ kJ of that energy to cellulose-containing biomass. This would result in a net production of

$1.8 \times 10^{11}$ t/a of biodegradable material, 40% of which is cellulose [259]. Estimates are that $1-1.25 \times 10^{11}$ t/a of terrestrial dry mass is produced together with $0.44-0.55 \times 10^{11}$ t/a in the oceans [260]. Present removal of this potential energy source is ca. 0.5% of the total growing stock on a global basis [261].

The fermentation potential of lignocellulose is based mainly on the cellulose content of the biomass. Chemically, cellulose is similar to starch. It is a polymer of glucose in which the glucose units are linked by β-1,4-glucosidic bonds, whereas the bonds in starch are predominantly α-1,4-linkages. The degree of polymerization (DP) varies for different sources of cellulose; for example, newsprint cellulose has a DP of 1000, whereas cotton has a DP of ca. 10 000 [262].

The cellulose molecule is more resistant to hydrolysis compared to starch. This resistance is due not only to the primary structure based on glucosidic bonds, but also, to a great extent, to the secondary and tertiary configuration of the cellulose chain, as well as its close association with other protective polymeric structures such as lignin, starch, pectin, hemicellulose, proteins, and mineral elements.

The lignin molecule seems to be primarily responsible for difficulties in hydrolyzing the lignocellulosic material, because it forms a protective sheath around the cellulose microfibrils. Lignin is a macromolecule of phenolic character and can be viewed as a dehydration product of three monomeric alcohols: *trans*-4-coumaryl alcohol, *trans*-coniferyl alcohol, and *trans*-sinapyl alcohol. The relative amount of each varies with the source [263].

When cotton cellulose is treated with dilute acid, partial hydrolysis occurs rapidly, and ca. 15% of the cellulose chain is degraded to glucose. The remaining 85% is more resistant to hydrolysis, possibly because this portion of the cellulose exists in a highly crystalline order [262].

In order to use lignocellulosic materials for fermentation to ethanol, they must be pretreated and then hydrolyzed to fermentable sugars. Pretreatment may be physical or chemical, e.g., milling, steam explosion, or use of solvents and various swelling agents.

In recent years, *steam explosion* has become a popular pretreatment method. In this process, green wood chips are heated to ca. 180–200 °C for 5–30 min in a continuous operation (Stake process), or to a temperature of 245 °C for 0.5–2 min (Iotech process) [264]. The acids formed from hemicellulose under these high-temperature and high-pressure conditions start to "autohydrolyze" the cellulose and intact lignin. Lignin is sufficiently softened at the end of the steaming period, so that when the vessel is suddenly depressurized to atmospheric pressure, an explosion occurs within the woody cells. This partially disrupts the close association of cellulose with lignin and consequently increases the surface area available for further hydrolysis. The effect of steam pretreatment on the enzymatic hydrolysis of various cellulose-containing materials is shown in Table 18.

The pretreated lignocellulosic material is then subjected to further hydrolysis, which can be acidic or enzymatic. A comparison of enzymatic and acid hydrolysis for cellulose degradation is given in Table 19. Cellulose and its degradation products are the only materials considered for fermentative purposes.

**Table 18.** Effect of steam pretreatment on the enzymatic hydrolysis * of cellulosic substrates

| Substrate | Pretreatment | Total reducing sugars, mg/mL | |
|---|---|---|---|
| | | 4 h | 24 h |
| Hardwoods | | | |
| Poplar | none | 1.4 | 2.4 |
| | steam | 15.3 | 25.8 |
| Aspen | none | 1.8 | 3.0 |
| | steam | 12.8 | 24.8 |
| Agriculture residues | | | |
| Corn stover | none | 4.9 | 7.8 |
| | steam | 15.7 | 22.5 |
| Sugarcane bagasse | none | 1.7 | 2.5 |
| | steam | 9.5 | 16.1 |
| Urban waste | none | 10.5 | 18.0 |
| | steam | 6.2 | 10.8 |
| Softwoods | | | |
| Eastern spruce | none | 2.0 | 3.8 |
| | steam | 3.5 | 6.4 |
| Douglas fir | none | 1.6 | 3.2 |
| | steam | 2.8 | 4.3 |

* *Trichoderma reesei* cellulase (QM9414), 19 IU (International Units) per gram of substrate; 5% substrate slurries, pH 4.8, 50 °C; steamed substrates washed prior to enzymatic hydrolysis.

**Table 19.** Comparison between enzymatic and acid hydrolysis of cellulosic materials

| Acid | Enzyme |
|---|---|
| Nonspecific catalyst; therefore, will delignify material as well as hydrolyze cellulose | Specific macromolecular catalyst; therefore, extensive physical and chemical pretreatment of material necessary to make cellulose available for degradation |
| Decomposes hemicellulose to inhibitory compounds (i.e., furfural) | Produces clear sugar syrup ready for subsequent anaerobic fermentation |
| Harsh reaction conditions necessary; therefore, increased cost of heat- and corrosion-resistant equipment | Run under mild conditions (50 °C, atmospheric pressure, pH 4.8) |
| High chemical cost requires catalyst recovery and reuse | Cost of producing cellulases is the most expensive process step; therefore, recycle is necessary |
| High rate of hydrolysis | Lower rate of hydrolysis |
| Low overall yield of glucose because of degradation | High glucose yield depending on system and pretreatment |

An example of a semicontinuous process for ethanol production from wood is shown schematically in Figure 29; this process uses dilute sulfuric acid for cellulose prehydrolysis.

The hydrolysate percolates through a bed of wood chips. Optimum conditions for the process are described in [266]. After acid digestion (a), the effluent passes through a flash evaporator (b), which separates the vapors containing furfural and methanol from the underflow condensate containing the

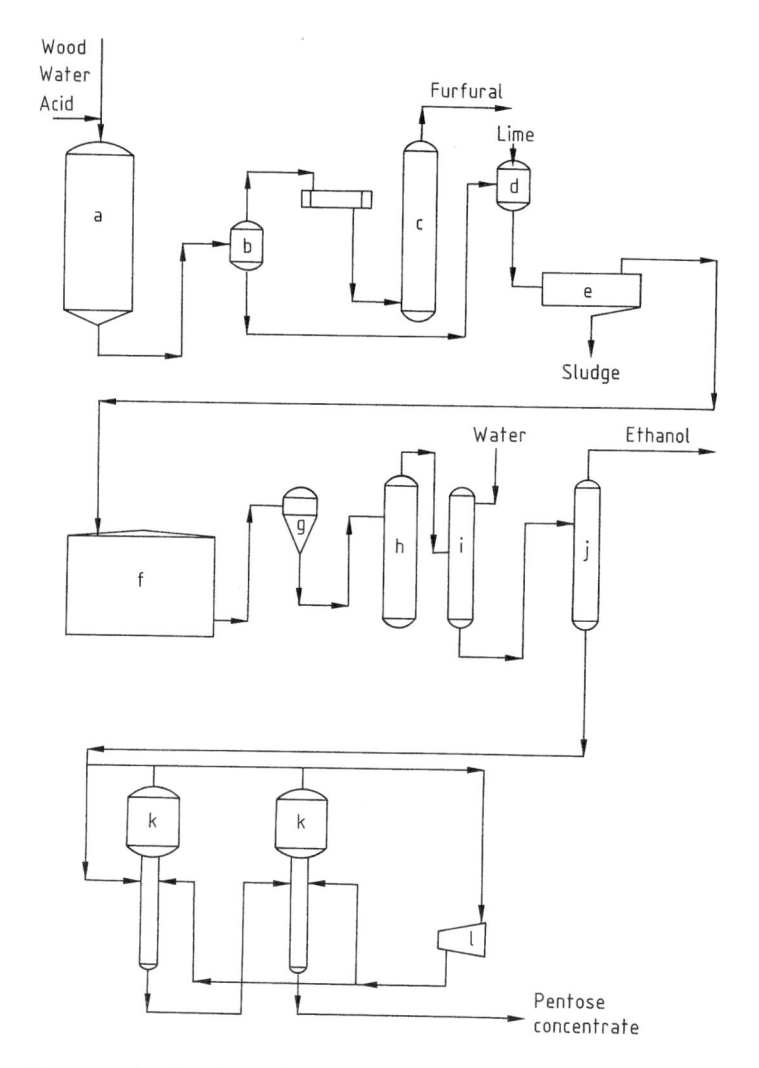

**Figure 29.** Ethanol production from wood [266]
Optimum conditions for this process are: acid concentration in total water, 0.53%; maximum temperature of percolation, 196 °C; rate of temperature rise, 4 °C/min; percolation time, 145–190 min; ratio of total water to oven-dried wood, 10; percolation rate, 8.69–14.44 L min$^{-1}$ m$^{-3}$
a) Digester; b) Flash evaporator; c) Furfural tower; d) Neutralization vessel; e) Clarifier; f) Fermentor; g) Yeast separator; h) Alcohol stripper; i) Extraction tower; j) Rectifying tower; k) Evaporator; l) Vapor compressor

sugar solution. The acid hydrolysate solution is further neutralized with a lime slurry, and the precipitated calcium sulfate is separated in a clarifier (d) as a 50% solids sludge. The neutralized liquor is blended with recovered yeast (*Saccharomyces cerevisiae*) from previous fermentation and is fermented to ethanol (e), which is further concentrated to 95% by distillation (i).

The bottom material from the alcohol stripping column (g), which contains pentose sugars, is further concentrated in multiple-effect evaporators (j) to a 65% solution, that can be used as animal feed or for chemical conversion to furfural.

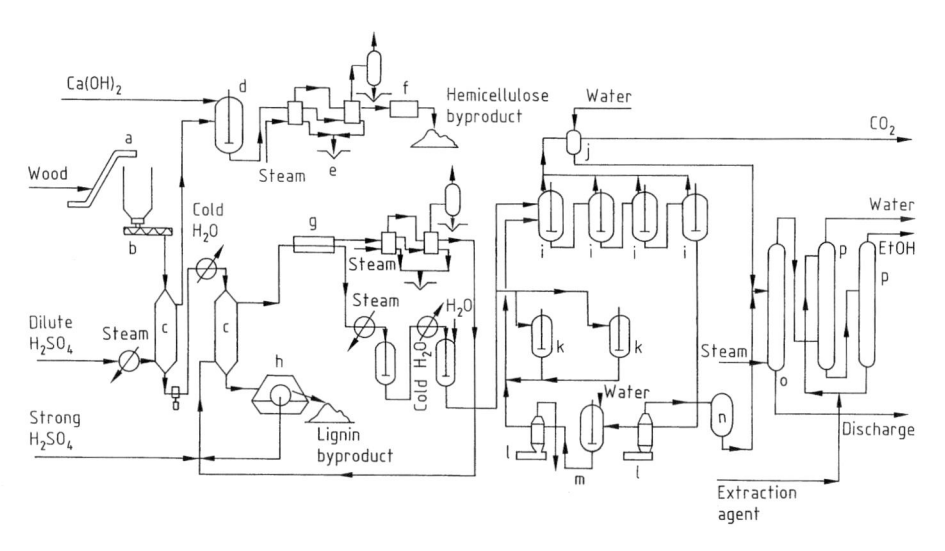

**Figure 30.** Ethanol production from wood by use of strong acid hydrolysis [267]
a) Feed hopper; b) Feeder; c) Digester; d) Neutralization vessel; e) Multiple-effect evaporators; f) Dryer; g) Electrodialysis membrane; h) Filter; i) Fermentor; j) Carbon dioxide scrubber; k) Seed fermentor; l) Centrifuge; m) Yeast wash vessel; n) Surge vessel; o) Beer still; p) Alcohol column

Figure 30 shows a strong acid hydrolysis process.

The air-dry wood is first pretreated with dilute sulfuric acid (c). Complete hydrolysis is accomplished in a subsequent strong acid cycle in which cellulose is hydrolyzed at room temperature with 70–80% $H_2SO_4$. The glucose, retained by the dialysis membrane (g), is neutralized, deionized, and then sent to fermentation (i). The sulfuric acid permeate from the dialysis unit is evaporated and reconcentrated for recycle. Lignin is separated from the concentrated acid by filtration (h) and washing.

To illustrate enzymatic hydrolysis of cellulose for alcohol production, a process used by the Natick Development Center is shown in Figure 31. A large part of this process involves the preparation of the cellulase enzyme. Newspaper is the cellulose-containing substrate.

## 5.4.4. Waste Materials and Residues

Various types of agricultural, industrial, or municipal refuse and waste can be used as substrates for ethanol fermentation. The fermentation is based on available sugar, starch, or cellulose in the waste material. The major advantage of this route lies in coupling waste treatment with the production of a higher value product. Both environmental pollution abatement and process economics are thus improved.

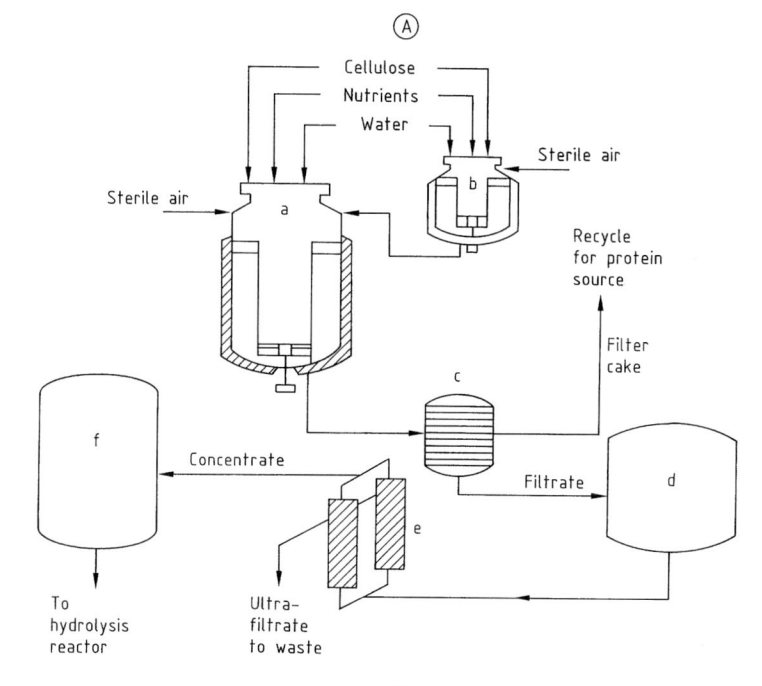

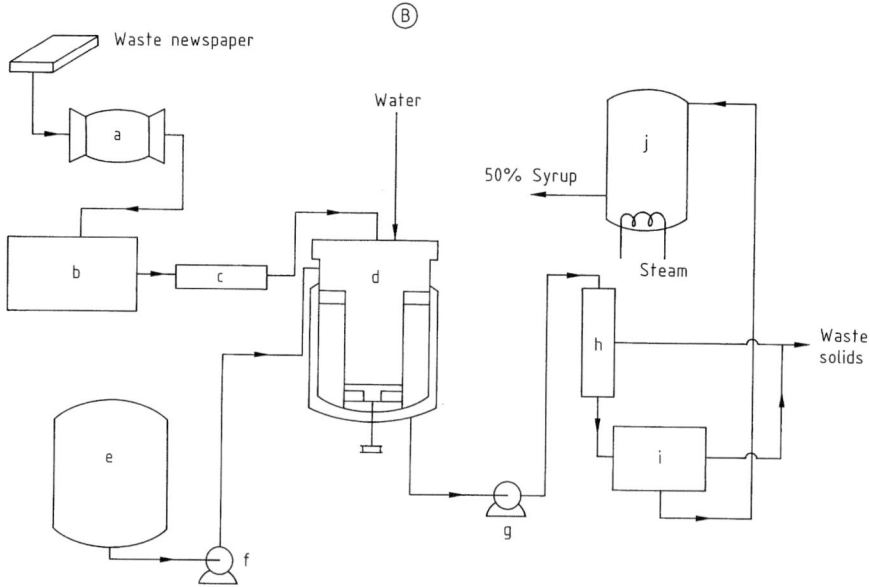

**Figure 31.** Enzymatic hydrolysis of newsprint by Natick Development Center (NDC) [265]
A) Pilot plant process for cellulase production:
a) Production vessel (vertical filters); b) Inoculum vessel; c) Filter; d) Harvest storage; e) Ultrafilter; f) Concentrate storage
B) Pilot-plant process for newspaper hydrolysis: a) Ball mill; b) Solids metering; c) Solids transfer; d) Bioreactor; e) Enzyme storage; f) Metering pump; g) Harvest pump; h) Crude filter; i) Polish filter; j) Evaporator

**Cornstalks.** Cornstalks are available in large quantities as a byproduct of corn agriculture. This material is predominantly composed of lignocellulose. A two-stage process using dilute acid treatment followed by concentrated acid impregnation of the lignocellulosic material is shown in Figure 32.

In this process, ground corn stover (841 nm, 20 mesh) is treated with 4.4% $H_2SO_4$ at 100 °C for 50 min (a). The mixture is then filtered (b) and the xylose-rich liquid is processed by electrodialysis (g) for acid recovery. The solids are dried further (c) and impregnated with 85% $H_2SO_4$ (d), followed by dilution with water to give a $H_2SO_4$ concentration of 8% (e). Subsequent hydrolysis is carried out at 110 °C for ca. 10 min, and acid is again recovered by electrodialysis. The combined yield of xylose is 94%, and the yield of glucose is 89%. Glucose is converted to ethanol by *Saccharomyces cerevisiae* (j) and xylose by *Fusarium oxysporum* (h), both in immobilized cell reactors. The overall annual capacity of the plant is $17 \times 10^3$ $m^3$.

**Domestic Refuse.** Domestic refuse contains a complex variety of materials that come mainly from cellulosic-type residues. A large quantity of this material, ca. 1.3 – 2.2 kg per person, is generated daily. This waste is currently disposed of by incineration or landfill. A process for the production of 36.5 t/d of ethanol from domestic refuse is illustrated in Figure 33 [269].

The refuse is separated into dense and light fractions by the use of a flotation separator or a special pulper. The pulped fraction, which contains cellulose, is first subjected to removal of fines and plastics, and then introduced into a reactor where it is hydrolyzed with 0.4% $H_2SO_4$ for ca. 1.2 min at 230 °C. This process is followed by flash cooling, neutralization with $CaCO_3$, and filtration. Fermentation of the sugar solution takes ca. 20 h at 40 °C and yields ca. 1.7% aqueous ethanol solution, which is further concentrated by distillation to ca. 95% ethanol.

**Waste Liquor from the Pulp and Paper Industry.** Two chemical pulping methods are predominant in the pulp and paper industry: the sulfate (Kraft) and the sulfite processes. The basis of the these operations is treatment of the lignocellulosic material (wood, straw, etc.) with highly concentrated acid or base, which should dissolve the lignin portion of the wood and leave cellulose fibers that are processed into the final paper product. Depending on conditions (temperature, pressure, concentration of chemicals, chemical to wood ratio, and time of digestion), more or less delignification and breakdown of the original cellulose occur. As a result, a product pulp is produced as well as a waste chemical liquor, which basically consists of spent cooking chemicals. The more drastic the delignification conditions (low yield process), the better is the quality of the paper obtained. The high-yield process refers to milder delignification and a pulp that still contains a considerable amount of lignin. Low-yield processes are characterized by waste liquors with a high concentration of chemicals and a higher organic content.

The organic content of these liquors is primarily sulfonated lignin (e.g., 43% of organic dry substance in a spent spruce sulfite liquor). However, cellulose and hemicellulose are also partially hydrolyzed during digestion so that waste liquors contain a

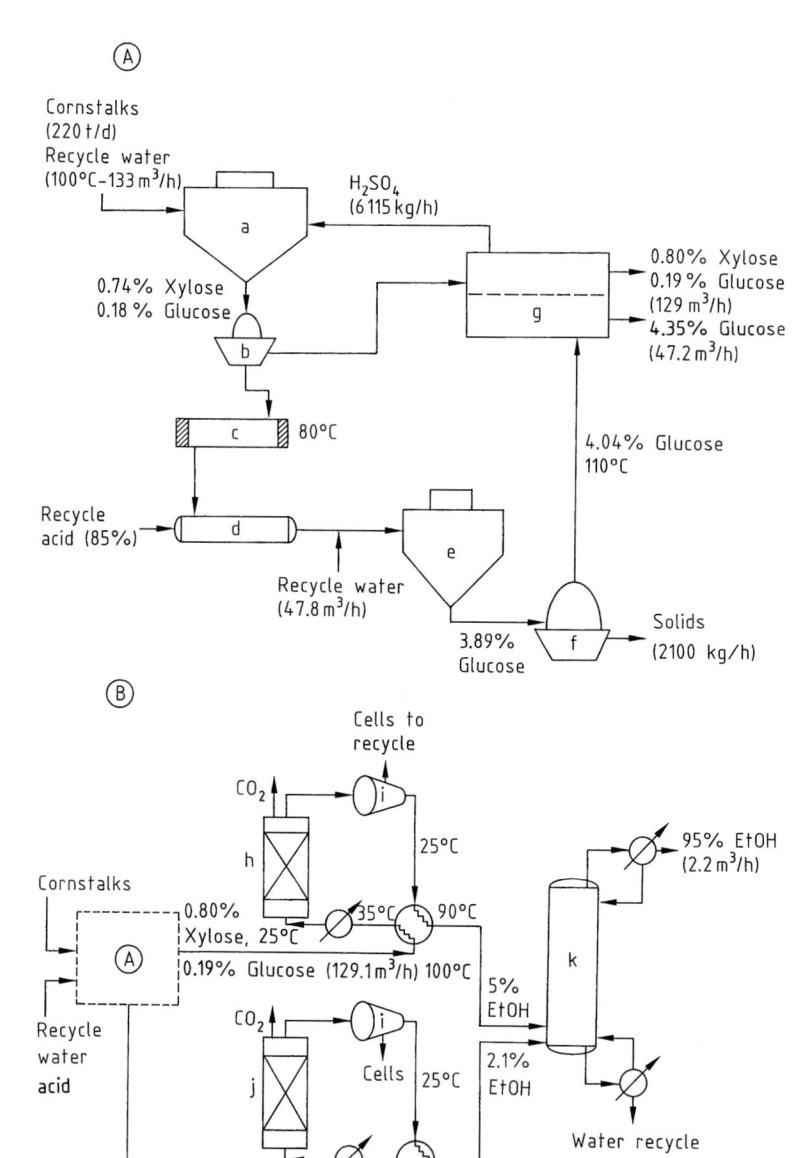

**Figure 32.** Production of ethanol from cornstalks [268]
A) Acid hydrolysis: a) Prehydrolysis tank, 4.4% $H_2SO_4$; b) Filter; c) Rotary dryer; d) Impregnator; e) Hydrolysis tank, 8.0% $H_2SO_4$; f) Filter; g) Electrodialysis unit
B) Fermentation: A) Acid hydrolysis process; h) Fixed film of *Fusarium oxysporum*; i) Centrifuge; j) Fixed film of *Saccharomyces cerevisiae*; k) Distillation column

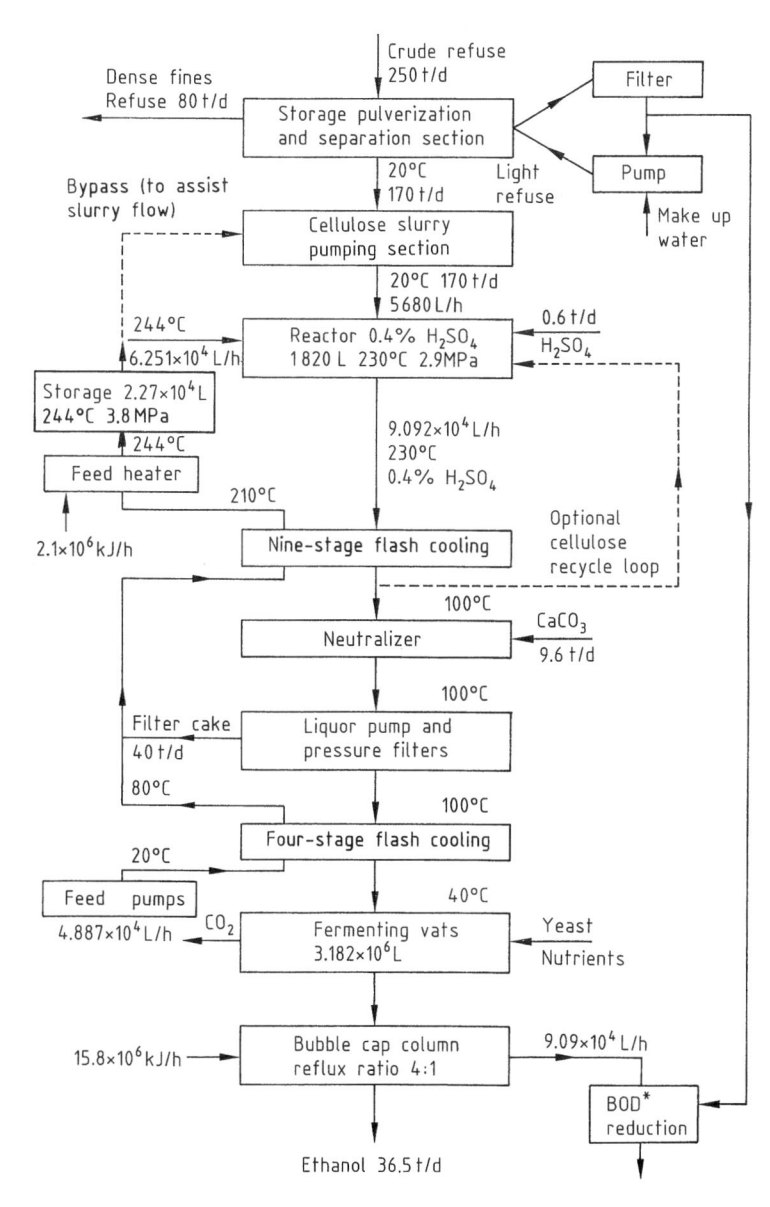

**Figure 33.** Flow diagram for continuous production of ethanol from refuse with 60% content of cellulose [269]
* Biochemical oxygen demand.

certain proportion of free sugars (hexoses and pentoses in ca. 2–4% concentration and ca. 14% total solids) [270].

Tremendous quantities of waste liquor are generated in a pulp and paper mill; they amount to ca. 9180 L per ton of pulp produced [271]. Consequently, a chemical pulping process with a medium capacity of 500 t/d of pulp produces $4.59 \times 10^6$ L/d of waste

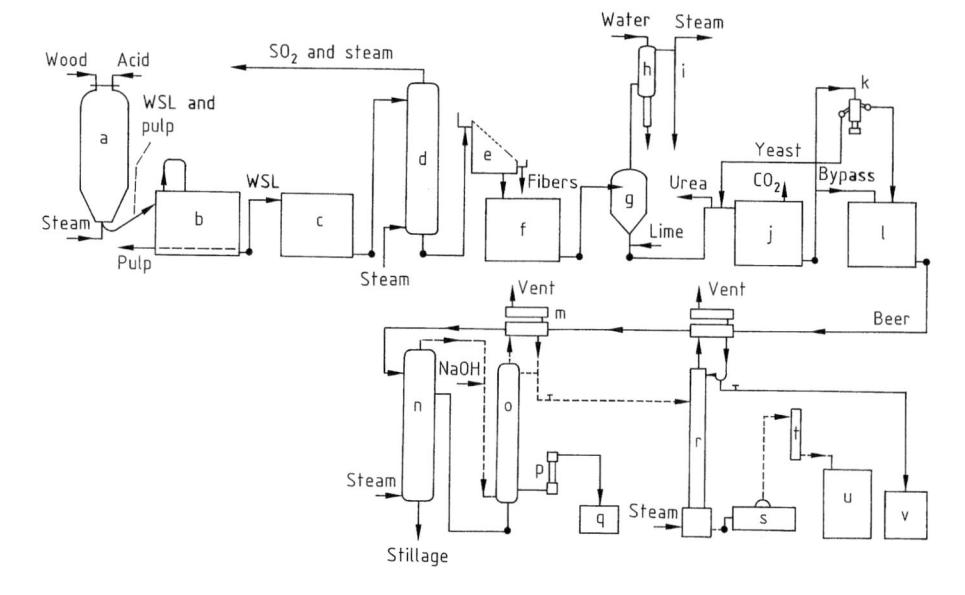

**Figure 34.** Production of ethanol from waste sulfite liquors (WSL) [272]*
a) Digester; b) Blowpits; c) Storage; d) Stripper; e) Screen; f) Storage; g) Flash cooler; h) Barometric condenser; i) Ejectors; j) Fermentor; k) Yeast separator; l) Storage; m) Preheaters; n) Beer still; o) Rectifying column; p) Oil washer; q) Fusel oil; r) Purifying column; s) Vaporizer; t) Condenser; u) Alcohol; v) Heads
* Reprinted with permission of American Institute of Chemical Engineers.

liquors. Release of this liquor into natural waters is prohibited because both organic and toxic pollution result.

The sulfate (Kraft) process is designed so that the majority of the waste liquor can be recycled and its organic value converted to energy by combustion in a specially designed steam boiler (recovery furnace).

Previously, the majority of pulping mills worldwide were sulfite mills. Because of economic and environmental problems, sulfite pulping is gradually being phased out and the process converted to sulfate pulping or modified in other ways. Recovery of chemicals in the sulfite process is not as feasible as in the Kraft process, so large quantities of waste sulfite liquors (WSL) are discharged to the environment.

Because WSL contain fermentable sugars, these liquors have been used efficiently as fermentation substrates for alcohol production. The process is relatively old (1908 in Sweden) but is still in operation in some mills (e.g., the Ontario Paper Company, Canada). A typical process for fermentation of WSL is shown schematically in Figure 34.

The waste liquor is first stripped of $SO_2$ with a conventional steam stripper. This is necessary because $SO_2$ would inhibit subsequent fermentation. The liquor is adjusted to give a ca. 10–12% concentration of sugars, the pH is adjusted to 4.5, and nitrogen and phosphorus nutrient sources are added (e.g., urea and phosphate). The fermentation is conventionally carried out with yeast (*Saccha-*

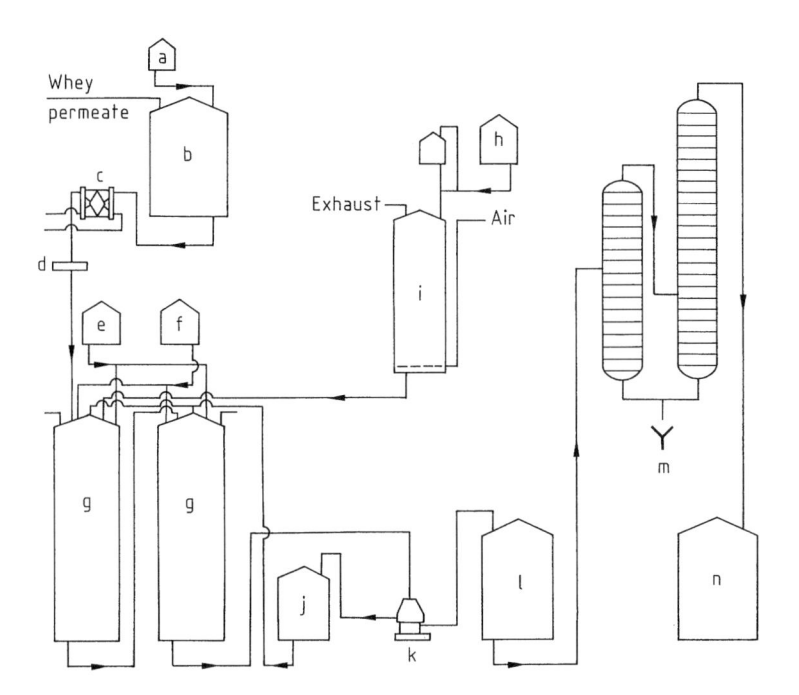

**Figure 35.** Continuous production of ethanol from whey [278]
a) Acid; b) Storage tank; c) Heat exchanger; d) Control; e) Antifoam; f) Chemicals; g) Fermentor; h) Substrate; i) Propagation plant; j) Storage; k) Separator; l) Buffer tank; m) Distillation; n) Alcohol storage

*romyces cerevisiae*) at 30 °C for ca. 20 h. The yeast is usually concentrated and recycled, and the broth containing ethanol is sent to the distillation section.

**Cheese Whey.** Whey is a byproduct of cheese production. An estimated $74 \times 10^6$ t of whey is produced annually worldwide. This amount of whey contains ca. $0.7 \times 10^6$ t of milk protein and 3.2 t of lactose [273].

Whey with its protein, carbohydrate, and vitamin content is a valuable, nutritious material; its composition is described in [274]. Whey is used in various forms as a component of either animal or human food. However, the utilization and recycling of whey nutrients depend on many factors, one of which is the size of the cheese factory. The smaller the factory, the less recycling of whey is practiced.

The alcohol produced from whey is derived mainly from the fermentation of available lactose. However, only a few microorganisms can convert lactose to ethanol, and conventional yeast (*Saccharomyces cerevisiae*) is not among them. The most efficient lactose-utilizing organism is reported to be *S. fragilis* [275].

A process using *Kluyveromyces fragilis* yeast was developed in Denmark (Fig. 35) [276]–[278]. The whey is first concentrated by reverse osmosis and ultrafiltration, and then introduced into fermentation vessels. The yield based on lactose is about 80 % of theoretical. About 42 L of whey, containing 4.4 % lactose, is required to produce 1 L of absolute ethanol.

A better substrate for industrial fermentation of whey is enzymatically hydrolyzed lactose. $\beta$-Galactosidase-treated whey yields a mixture of monosaccharides, glucose, and galactose, which can be efficiently fermented by high-alcohol-producing yeasts [275], [279].

# 6. Recovery and Purification

Over the past 20 years a series of distillation systems have been developed for the efficient recovery of ethanol from synthetic and fermentation feedstocks. These units produce high-grade industrial alcohol, anhydrous alcohol, alcoholic spirits, and ethanol for motor fuels. Ethanol quality and recovery have been improved; energy consumption has decreased.

**Distillation.** Synthetic ethanol is purified in a simple three-column distillation unit (see Sections 4.1.3 and 4.2.2). Recovery is 98%, and the high-grade product contains less than 20 mg/kg of total impurities and has a permanganate time of over 60 min.

The following are key features of the efficient recovery of high-grade ethanol from fermentation feedstocks:

1) *Extractive distillation* results in a higher degree of purity than is possible in conventional purification columns; both investment and operating costs are reduced.
2) *Pressure-cascading installations* and *heat pumps* permit substantial heat recovery and recycling, thus minimizing heat loss and steam consumption. Virtually all (95–99%) the ethanol in the crude feed is recovered as high-grade product.
3) *Advanced control systems* ensure stable operating conditions. Product quality can be maintained with a total impurity content of less than 50 mg/kg and a permanganate time of over 45 min.
4) *Energy requirements* are minimized. The flash heat recovered from the grain-cooking system is used to heat the ethanol distillation unit, thus reducing the energy consumption for ethanol production by ca. 10%. Use of a vapor recompression technique can reduce the energy required for the evaporation of stillage to as little as one-tenth of that required in a triple- or quadruple-effect evaporator.

**Dehydration.** To produce anhydrous ethanol, the water–ethanol azeotrope obtained from distillation of the crude synthetic or fermentation feedstock must be dehydrated. For economic reasons, large distilleries rely mostly on azeotropic distillation for ethanol dehydration. Benzene has been used as an azeotropic dehydrating (entraining) agent in many plants, but some concern exists about its carcinogenicity and toxicity ($\rightarrow$ Benzene). However, proper design and control minimize benzene loss and exposure of operating personnel to this substance. Cyclohexane and ethylene glycol are used in some distilleries; they are also effective dehydrating agents.

Some smaller ethanol plants use molecular sieve adsorption techniques to dry the ethanol azeotrope. Pervaporation through semipermeable membranes or use of a solid dehydrating agent may reduce energy and equipment costs.

**Concentration Units.** Several systems are used to define ethanol concentration in addition to those based on conventional units (e.g., weight percent, volume percent).

*Proof.* In Canada, Great Britain, and the United States, the ethanol concentration of beverage spirits is expressed in terms of "proof." United States law defines proof as follows: "Proof spirits shall be held to be that alcoholic liquor which contains one-half of its volume of alcohol of a specific gravity of 0.7939 at 15.5 °C." More simply, the figure for proof is twice the percentage of ethanol content by volume. For example, 100 proof means 50 vol% alcohol.

In Great Britain as well as Canada, "proof spirit is such as at 10 °C weighs exactly twelve-thirteenths of the weight of an equal bulk of distilled water." A proof of 87.7 therefore signifies an ethanol concentration of 50 vol%. British proof can be converted into U.S. proof by multiplying by 1.11.

*Degree Gay-Lussac.* The degree Gay–Lussac (°GL) is measured with a hydrometer, which reads the volume percentage of ethanol in a mixture of ethanol and water at 15 °C.

**Grades of Ethanol.** Several grades of ethanol are available.

*Industrial ethanol* (96.5 vol%) is used for industrial and technical purposes as a solvent and a fuel, and is also converted into many other products. Industrial ethanol is usually denatured with 0.5 – 1 wt% of crude pyridine and is sometimes colored with $5 \times 10^{-4}$ wt% methyl violet for easy recognition.

*Denatured spirit* (88 vol%) is a term used in some countries to describe industrial 88 vol% ethanol, which has been denatured and colored, and is generally used for heating and lighting.

*Fine alcohol* (96.0 – 96.5 vol%) is a purer type of ethanol used mainly for pharmaceutical and cosmetic preparations and for human consumption.

*Absolute or anhydrous ethanol* (99.7 – 99.8 vol%) is the term given to very pure, extremely dry ethanol, which is used in the food and pharmaceutical industries and also in the manufacture of aerosols. Anhydrous denatured ethanol mixed with 70 – 80 vol% of gasoline is used as a fuel for internal combustion engines. The specifications of anhydrous ethanol obtained by azeotropic distillation are given in Table 20.

*Motor fuel ethanol* refers to fermentation ethanol that is used as an anhydrous or hydrous fuel or as a blending agent to improve the octane number of gasoline.

**Table 20.** Specifications for anhydrous ethanol (99.6 vol%)

| Parameter | Specification | Typical value | Test method |
|---|---|---|---|
| Purity, vol% | 99.8 min. | >99.9 | ETM 240.03* |
| $d_{20}^{20}$ | 0.7911 max. | 0.790 | ASTM D 268 |
| Acidity, wt% acetic acid | 0.002 max. | 0.001 | ASTM D 1613 |
| Color, Pt–Co | 10 max. | 5 | ASTM D 1209 |
| Nonvolatile matter, mg/100 mL | 2.0 max. | 1.0 | ASTM D 1353 |
| Distillation temperature, °C | | | |
|   initial | 77.5 min. | 78.2 | |
|   dry point | 79.0 max. | 78.4 | |
| Permanganate time, min | 25 min. | 45 | ETM 110.77* |
| Residual odor | nil | nil | ASTM D 1296 |
| Appearance | clear and free of suspended matter | clear and free of suspended matter | ETM 80.65* |

* ETM = Ethanol testing method (CORA Engineering).

## 6.1. Distillation

Ethanol is recovered as an azeotrope from ethanol–water mixtures by means of distillation [280], [281]. The boiling point diagram for this system is shown in Figure 36; the pure water azeotrope at the azeotropic point (a) contains 95.57 wt% (97.3 vol%) ethanol and has a boiling point of 78.15 °C.

To obtain anhydrous ethanol, the pure ethanol–water azeotrope must be dehydrated. This is generally accomplished by azeotropic distillation with an entraining agent, usually benzene. The water is thus removed in the form of an overhead ternary benzene–ethanol–water azeotrope.

### 6.1.1. Distillation of Azeotropic Ethanol

The crude ethanol synthesized by the direct or indirect hydration of ethylene contains ca. 50 vol% ethanol, whereas the crude product obtained after fermentation contains no more than 10 vol% ethanol. Consequently, the distillation of synthetic ethanol requires less energy and is less expensive than that of fermentation ethanol.

The distillation of synthetic azeotropic ethanol obtained by the direct and indirect hydration of ethylene is described in Sections 4.1.3 and 4.2.2, respectively. The principles on which it is based are the same as those used in the distillation of fermentation ethanol described in this section.

The low-energy distillation of ethanol from a fermentation feedstock is shown schematically in Figure 37 [281]–[283]. Fermented mash containing ca. 10 wt% ethanol and ca. 10 wt% total solids is preheated to near saturation temperature, degassed to remove $CO_2$, and fed to the stripping column. The overhead product leaving the stripping column contains 75–85 vol% ethanol. The bottom liquid (stillage) contains less than 0.02 wt% ethanol and is sent either for disposal or for animal feed production.

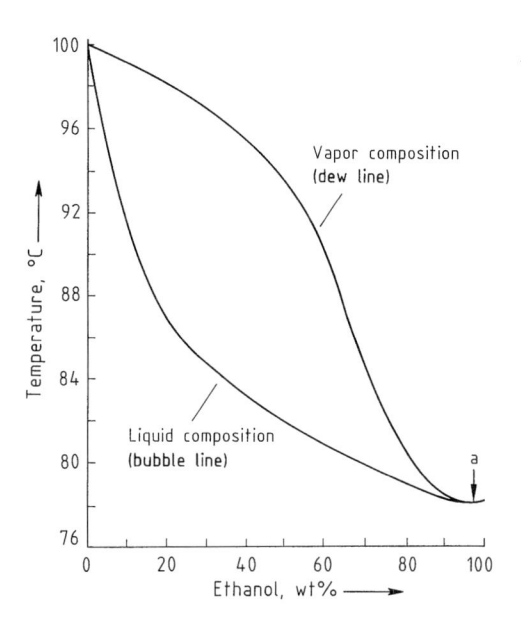

**Figure 36.** Boiling point diagram of ethanol–water mixtures
a) Azeotropic point

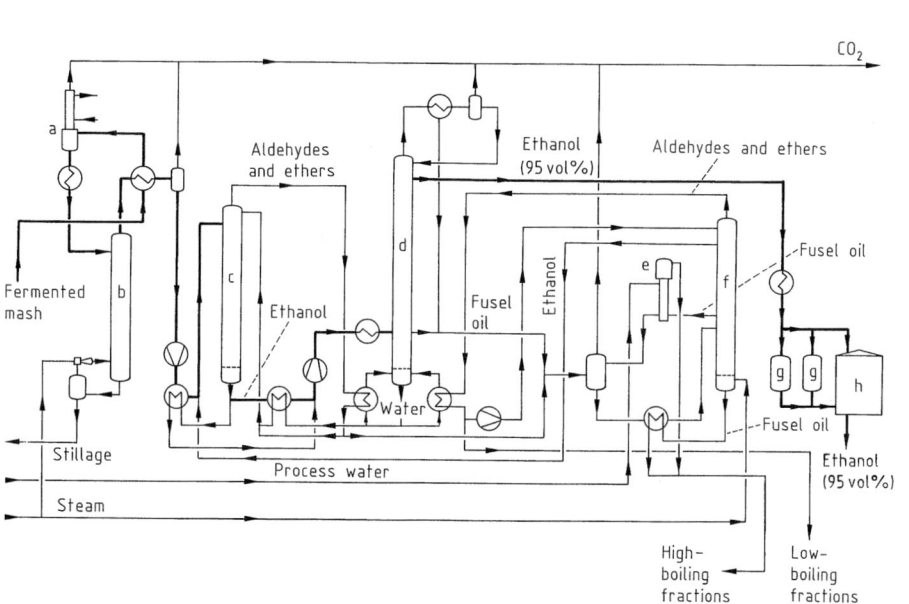

**Figure 37.** Distillation of 95 vol% ethanol from a fermentation feedstock
a) Degasser; b) Stripping column; c) Extractive distillation column; d) Rectifying column; e) Fusel oil washer; f) Concentrating column; g) Charcoal filtration; h) Storage tank

**Table 21.** Average composition of fusel oils (in wt%)

| Fusel oil | 1-Propanol [71-23-8] | 1-Butanol (n-butyl alcohol) [71-36-8] | 2-Methyl-1-propanol (isobutyl alcohol) [78-83-1] | 2-Methyl-1-butanol (active amyl alcohol) [137-32-6] | 3-Methyl-1-butanol (isoamyl alcohol) [123-51-3] |
|---|---|---|---|---|---|
| Molasses | 13.2 | 0.2–0.7 | 15.8 | 28.4 | 37.4 |
| Wheat–cereals | 9.1 | 0.2–0.7 | 19.0 | 20.0 | 51.2 |
| Potatoes | 14.0 | 0.5 | 15.5 | 15.0 | 55.0 |
| Sulfite waste liquor | 7.0 | | 22.0 | 13.0 | 55.0 |
| Fruit | 8.0 | 2.0 | 19.0 | 14.0 | 57.0 |

A vapor recompression system may be used to heat the reboiler of the stripping column for steam economy.

The overhead distillate from the stripping column is mixed with recycled ethanol from the concentrating column and fed into the extractive distillation column, which operates at a pressure of 0.6–0.7 MPa. This column removes essentially all fermentation byproducts, mainly aldehydes, ethers, methanol, and higher alcohols, from the ethanol. The aldehydes, ethers, and methanol are more volatile than ethanol and leave the top of the column. However, the higher alcohol byproducts, known collectively as fusel oil, are normally less volatile than ethanol. The average composition of fusel oil from fermentation ethanol derived from different raw materials is shown in Table 21 [284]. The technique used to extract fusel oil from ethanol exploits the fact that higher alcohols are more volatile than ethanol in solutions containing a high concentration of water. They can, therefore, also be steam distilled and removed in the overhead steam to leave a virtually pure ethanol–water mixture.

Dilute ethanol is then sent from the bottom of the stripping column to the rectifying column where it is brought up to strength. The rectifying column is heated by overhead vapors from the extractive distillation column and the concentrating column. The ethanol (95 vol%) is withdrawn as a side stream from one of the upper trays; it is then filtered through charcoal and stored. The water that collects at the bottom of the rectifying column contains traces of ethanol and is recycled to the extractive distillation column. Fusel oil is withdrawn as a side stream from one of the lower trays of the rectifying column and fed to the fusel oil washer.

In the concentrating column, overhead vapor from the extractive distillation column, which contains aldehyde, ether, and alcohol impurities, is separated into low-boiling and high-boiling fractions. Ethanol is also recovered and recycled to the extractive distillation column. A small stream is taken from the overhead condensate (low-boiling fraction), which contains acetaldehyde and a small amount of ethanol. It may be sold as a byproduct or burned as a fuel. A side stream containing a high concentration of fusel oil is sent through a cooler to the fusel oil washer. In the washer, ethanol is extracted from the fusel oil with water, and the washings are recycled to the concentrating column. High-boiling fusel oil is also run off from the bottom of the concentrating column. The combined fusel oil (high-boiling) fractions may be sold as a byproduct.

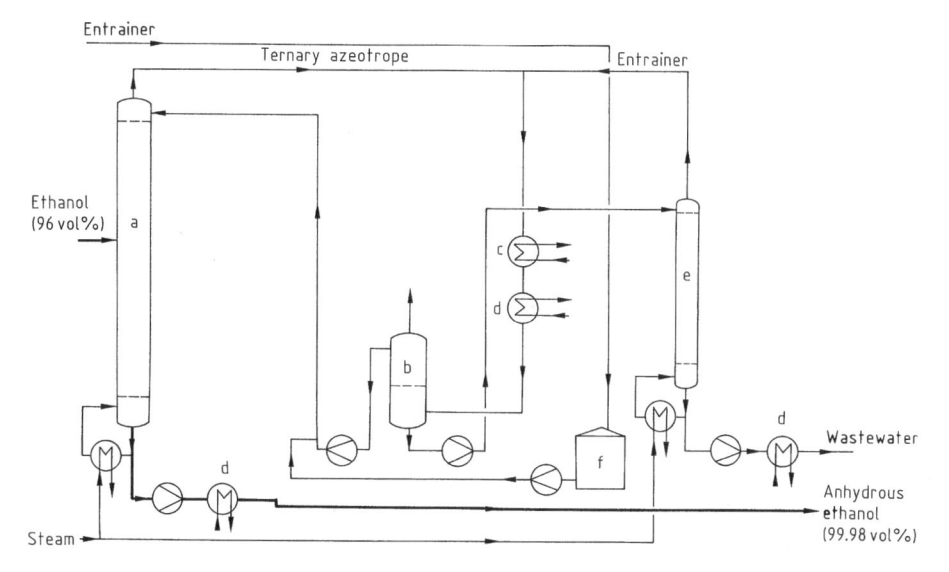

**Figure 38.** Production of anhydrous ethanol by azeotropic distillation
a) Dehydrating column; b) Decanter; c) Condenser; d) Cooler; e) Hydrocarbon stripping column; f) Entrainer tank

## 6.1.2. Dehydration by Azeotropic Distillation

Azeotropic distillation systems are designed for the production of pure, anhydrous ethanol (99.98 vol%), which contains <200 mg/kg of water and <20 mg/kg of total impurities (for other specifications, see Table 20) [281]–[283]. A flow scheme for the two-column azeotropic distillation of anhydrous ethanol is shown in Figure 38. The dehydrating column and the hydrocarbon stripping column operate at atmospheric pressure. Therefore, they may be heated with low-pressure steam, hot condensate, or hot waste streams from other parts of the ethanol processing plant, thus minimizing steam consumption.

Water is removed from the ethanol–water azeotrope in the form of a ternary azeotrope which is produced by adding an entraining agent such as benzene, heptane ($C_6$–$C_8$ cut), or cyclohexane. The 95 vol % ethanol feed enters the dehydrating column near the midpoint. The anhydrous ethanol product collects at the bottom of the tower and is sent through a cooler prior to storage. The ternary azeotrope leaves the column as an overhead product which is condensed and then separated into an organic and an aqueous phase in the decanter. The upper organic layer containing the entrainer is returned to the top of the dehydrating column. The lower aqueous layer is pumped to the hydrocarbon stripper, where the hydrocarbon entrainer, the ethanol, and some water vapor are recovered overhead and sent to the condenser–decanter system. Water from the stripper is pumped off as waste; if it contains a substantial amount of ethanol, it may be recycled to the ethanol distillation unit.

The overall efficiency and reliability of the anhydrous ethanol system are the result of the following special features:

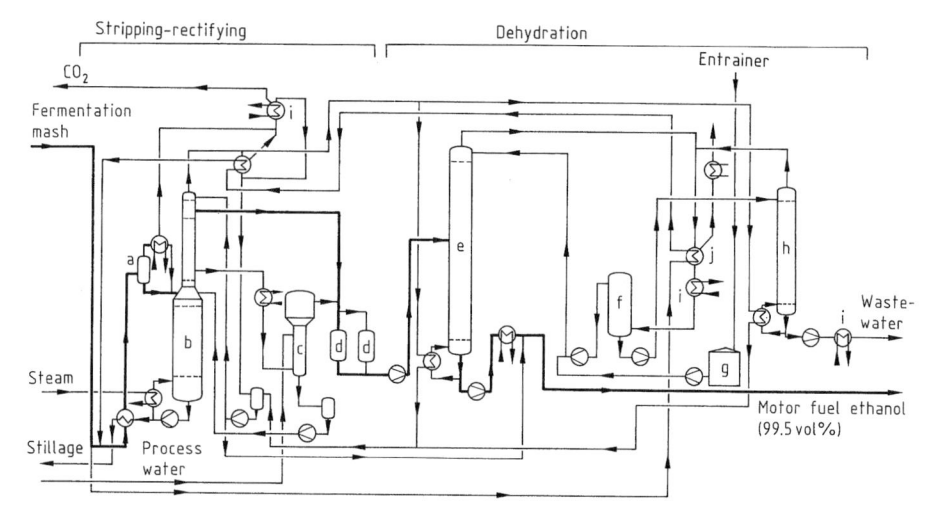

**Figure 39.** Production of anhydrous motor fuel ethanol
a) Degasser; b) Stripping–rectifying column; c) Fusel oil washer; d) Filter; e) Dehydrating column; f) Decanter; g) Entrainer tank; h) Entrainer stripping column; i) Cooler; j) Condenser

1) The use of a common condenser and decanter for the two columns reduces capital costs.
2) The use of very efficient column trays which can operate at a low throughput.
3) The BOD of the wastewater is low.
4) Consumption of the entraining agent is low.
5) Steam consumption is low: 1–1.5 kg is required per liter of anhydrous ethanol depending on the quality of the end product.

In some plants that produce motor fuel ethanol, gasoline is substituted for benzene as the entrainer. Since gasoline is required in the end product, the hydrocarbon stripping step is eliminated. However, gasoline dehydrating agents give products of varying composition because the gasoline contains a number of different components.

## 6.1.3. Motor Fuel Ethanol

Two types of ethanol for motor fuel are produced industrially [281]–[283], namely, anhydrous and hydrous motor fuel ethanol.

**Anhydrous Motor Fuel Ethanol.** The distillation and dehydration of motor fuel ethanol from a fermentation feedstock are shown schematically in Figure 39.

*Stripping–Rectifying.* Ethanol is distilled in a single stripping–rectifying column where the fermented mash is separated into an overhead, ca. 95-vol% ethanol fraction and a bottom liquid fraction (stillage) containing <0.02 wt% ethanol. The mash is preheated at its saturation temperature and degassed to remove residual $CO_2$ before it enters the stripping–rectifying column. The column operates at a pressure of ca. 0.3 MPa and is heated with steam by a forced circulation reboiler. This is the only part of the system that uses steam. The stillage leaving the bottom of the column is cooled to its boiling point at atmospheric pressure by heat exchange, the extracted heat being used to preheat the mash feed. The temperature of the resulting stillage is such that it can be subjected to vaporrecompression evaporation without preheating or flashing in an evaporator.

The pressurized overhead vapors from the stripping–rectifying column are used to preheat the mash feed and to heat the reboilers of the dehydration column and of the entraining stripping column. The condensed vapors are returned to the top of the stripping–rectifying column. A small overhead stream is drawn off to remove volatile acetaldehyde from the system; it contains <1% of the product ethanol and is either added to the purified anhydrous ethanol product or burned as a fuel in the plant boiler.

A side stream containing fusel oil is also taken from the rectifying section of the column and sent to a fusel oil washer. The aqueous washings are returned to the stripping section of the column; the decanted, washed oil is combined with the ethanol product stream and fed to the dehydration column. Fusel oil has a higher energy value than ethanol and acts as an agent for blending ethanol with gasoline.

*Dehydration.* After leaving the stripper–rectifier, the 95 vol% ethanol and fusel oil are filtered and fed to a dehydrating column, which operates at atmospheric pressure. Here, water is removed from the feed by use of benzene, heptane ($C_6$–$C_9$ cut), cyclohexane, or some other entraining agent. The bottom stream from the dehydrating column consists of anhydrous, 99.5 vol% ethanol and is cooled prior to storage. The overhead fraction from the dehydrating column is a ternary azeotrope, which is combined with overhead vapors from the entrainer stripping column. The combined vapors are condensed and cooled, forming two phases that are separated in a decanter. The upper entrainer layer from the decanter is pumped to the top of the dehydrating tower, while the lower aqueous layer is fed to the entrainer stripping column for recovery of the entrainer and ethanol.

Steam consumption in this system is 1.8–2.5 kg per liter of ethanol produced, depending on the alcohol concentration of the mash.

**Hydrous Motor Fuel Ethanol.** A low-energy system has been developed for distillation of the 85–95 vol% motor fuel ethanol used in engines that require neat alcohol and not gasoline. This distillation process is used mainly in Brazil; a flow scheme is shown in Figure 40. For maximum steam economy, the fermentation feed is fed into two stripper–rectifiers, the first of which operates at a pressure of ca. 0.4 MPa, the second at atmospheric pressure. Steam is used only in the high-pressure column, which takes 55–60% of the fermentation feed. The steam consumption of this system is 1.2–1.5 kg per liter of 85–95 vol% motor fuel ethanol. The overhead and fusel oil byproducts are removed and processed in a fashion similar to that described for production of anhydrous motor fuel ethanol.

2119

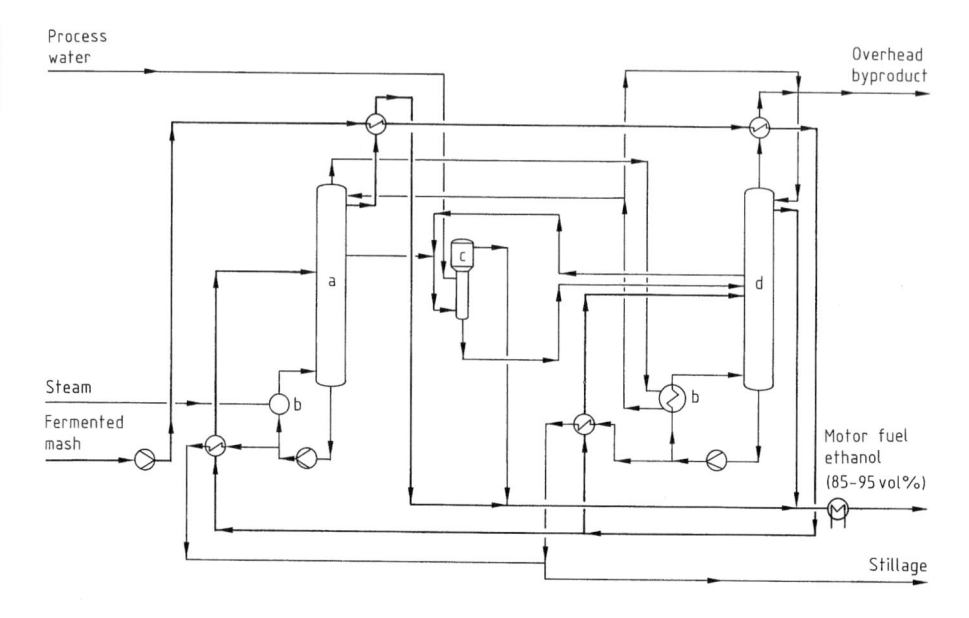

**Figure 40.** Distillation of hydrous motor fuel ethanol
a) High-pressure stripping–rectifying column; b) Reboiler; c) Fusel oil washer; d) Stripping–rectifying column (operating at atmospheric pressure)

## 6.1.4. Reduction of Energy Costs

Distillation of fermentation ethanol requires large amounts of energy. Energy costs have been cut by reducing steam consumption and by using vapor recompression systems.

Earlier columns for the distillation of ethanol from corn, wheat, or molasses were operated at atmospheric pressure. The development of new multiple-stage, high-pressure systems has reduced steam consumption by 40 % compared to previous systems. The new commercial installations are based on a pressure-cascading technique and consume 3.0–4.2 kg of steam for every liter of 96 vol% ethanol produced. Earlier distillation systems required 6 kg of steam per liter of ethanol.

A modern motor fuel ethanol plant has a total energy consumption of 1.1–1.4 MJ per liter of ethanol. Steam consumption figures for the systems described in Sections 6.1.2 and 6.1.3 are summarized in Table 22 [285]–[287].

*Vapor Recompression Systems.* Energy costs can be reduced by up to 80 % in low-pressure distillation columns by using a vapor recompression system [288].

In this system, compressed overhead vapor is used as the heat source for the reboiler, instead of expensive steam. The temperature of the column overhead vapor is increased by compression.

Vapor recompression can be applied to distillation columns operating at or below atmospheric pressure. It is used mainly in the distillation of motor fuel ethanol.

**Table 22.** Steam consumption and ethanol concentration in the low-energy distillation of ethanol

| Parameter | Distillation product | | | |
|---|---|---|---|---|
| | Ethanol–water azeotrope | Anhydrous ethanol | Anhydrous motor fuel ethanol | Hydrous motor fuel ethanol |
| Maximum ethanol concentration in the feed, vol% | 10 | 96 | 10 | 10 |
| Ethanol concentration of the product, | | | | |
| vol% | 96 | 100 | 99.5 | 95 |
| proof (U.S.) | 192 | 200 | 199 | 190 |
| Steam consumption per liter of ethanol, kg | 4.1 | 1.4 | 2.2 | 1.2 |

Investment costs for ethanol distillation units equipped with a vapor recompression system are almost 50 % higher than those for conventional distillation.

## 6.2. Nondistillative Methods

The energy required to remove water from ethanol can be reduced significantly by using methods that do not rely on distillation.

### 6.2.1. Solvent Extraction

Ethanol dissolves in some liquids that are virtually immiscible with water. These solubility differences can be exploited to recover ethanol from an aqueous solution by means of solvent extraction [289].

In the United States, the Energol Corporation employs liquid–liquid extraction with a proprietary solvent to separate ethanol from water. The solvent is then removed by distillation. This method does not require energy-intensive azeotropic distillation and thus has a low energy consumption. The energy budget for the entire plant is 3500 – 3700 kJ per kilogram of ethanol produced. In 1987, Energol's technique was used in four 40 000 to 45 000-L/d plants which came into operation in the mid 1980s.

The University of Pennsylvania and General Electric have developed a process that uses dibutyl phthalate [84-74-2] as a water-immiscible solvent for purifying ethanol. This solvent has a much higher boiling point than ethanol, and ethanol can therefore be separated in a single distillation step; solvent losses are low.

### 6.2.2. Carbon Dioxide Extraction

Another type of solvent extraction makes use of a critical fluid, i.e., a gas compressed to the point at which the distinction between gas and liquid disappears.

Ethanol is selectively extracted from grain mash with carbon dioxide close to its critical point of 7.3 MPa and 31 °C [288]. The ethanol-rich stream is flashed at ca.

4.8 MPa to remove the carbon dioxide, leaving ethanol separated in the liquid phase. This route requires a third to a half of the energy needed for conventional ethanol distillation, but the capital cost is 20% higher. A further advantage of this method is that the carbon dioxide used as solvent is obtained as a low-cost byproduct of fermentation. This is important because some solvent inevitably escapes during ethanol recovery and has to be replaced.

### 6.2.3. Adsorptive Dehydration

**Molecular Sieves.** Ethanol azeotropes are dehydrated industrially by adsorption with molecular sieves whose pores are permeable to water but not to ethanol [290]–[292]. The molecular sieve may be a synthetic or naturally occurring zeolite (e.g., clinoptilolite) or a proprietary resin. The 95 vol% ethanol is dehydrated in molecular sieve columns; 75% of the adsorbed material is water and 25% ethanol. When the column is saturated, the stream is directed to a fresh column and the saturated column is regenerated. The regeneration stream containing 25 vol% ethanol is fed back to the ethanol distillation system.

**Solid Agents.** Ethanol can also be dehydrated by adsorption with solid agents [293]. Less energy is required to vaporize water from cellulose or corn-starch than from calcium hydroxide, because of their low heats of adsorption. Therefore, ground cornmeal is used as a dehydrating agent for removing water from an 85% ethanol feed stream. The cornmeal adsorbent can be recycled ca. 20 times before being used as animal feed.

### 6.2.4. Membrane Technology

**Pervaporation.** A new method of ethanol purification based on pervaporation has been developed [294], [295] and is shown schematically in Figure 41. The pervaporator consists of a number of semipermeable membrane modules made of poly(vinyl alcohol) resins.

The 94 vol% ethanol feed is preheated to 60 °C and pumped to the semipermeable membrane modules of the pervaporator. Water permeates the membrane down its concentration gradient; a phase change occurs from the liquid phase at the membrane inlet to the vapor phase in the permeate. Water is thus separated without azeotrope formation. The driving force for permeate flow is provided by a vacuum of less than 1 kPa at the permeate condenser inlet. The total energy consumption is the sum of the evaporation and the condensation enthalpies.

The condensed permeate contains a small amount of ethanol and can be recycled to a rectifying or distillation tower for recovery of ethanol. Pervaporation of 1096 L of a 91 wt% (94 vol%) ethanol feed yields 1000 L of anhydrous ethanol (99.85 wt%,

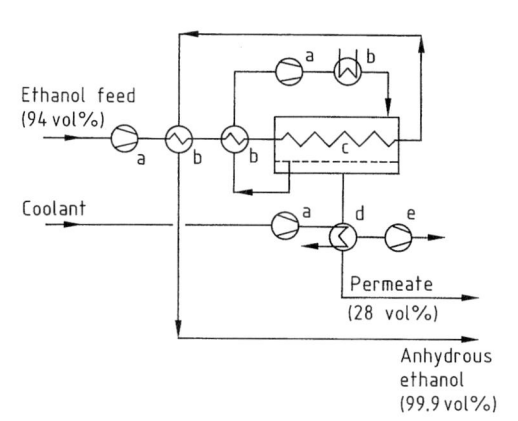

**Figure 41.** Production of anhydrous ethanol by pervaporation
a) Pump; b) Heater; c) Pervaporator; d) Condenser; e) Vacuum pump

99.9 vol%) and 107.5 L of permeate byproduct containing 23 wt% ethanol. The production of 1000 L of anhydrous ethanol requires 135 kg of steam (200 kPa), 10 m$^3$ of cooling water (20 °C), and 15 kW · h of electricity.

**Reverse Osmosis.** Purification of ethanol by reverse osmosis employs membranes that are relatively impermeable to ethanol but permeable to water [296]. A pressure of 4–7 MPa is usually applied to remove the water by forcing it across the membrane. The ethanol retention of new noncellulosic membranes is much higher than that of the cellulose acetate membranes used earlier (80% compared to 50%). Reverse osmosis may prove useful for savings in energy costs by concentrating ethanol to about 10% prior to distillation.

## 6.2.5. Yarn-Filled column

Textile yarns, such as rayon, retard the movement of water vapor but allow ethanol vapor to pass. This phenomenon has been exploited to develop a technique for separating ethanol–water mixtures [297].

Water containing 12 wt% ethanol is vaporized by injecting it into a stream of air; the ethanol–water–air mixture is then sent through a yarn-filled column equipped with heating elements. A zone of water builds up at the beginning of the column; air and ethanol pass through and are cooled to condense the ethanol. Before the water zone moves too far into the column, the direction of flow is reversed and the water is flushed out with air. The cycle takes about 25 s; flow occurs for 10 s in the forward direction and 15 s in the reverse direction.

A continuous version of the process is being developed in which the yarn is made into a continuous belt by tying the ends together. The belt moves slowly through the column, in the opposite direction to the flow of the ethanol–water–air vapor. Ethanol is then produced continuously at one end of the column. The yarn leaves the other end of the column and passes through a heating zone where water is driven off.

## 6.3. Storage and Transportation

Ethanol is shipped in railroad tank cars, 200-L tank trucks, 20-L drums, and smaller glass or metal containers. In the case of special quality requirements, the drums may be lined with phenolic resin.

The hazard classifications of ethanol are as follows:

GGVE/GGVS   3/3 b
IMDG Code   3.2, 3.3, UN No. 1170
United States   CFR 49:172.101, flammable liquid

# 7. Comparison of Process Economics for Synthetic and Fermentation Ethanol

The figures and calculations in this chapter are based on data taken from [298]–[300].

## 7.1. Summary of Cost Analysis for Synthetic and Fermentation Ethanol

The cost of producing ethanol by ethylene hydration (synthetic ethanol) or fermentation (fermentation ethanol) depends primarily on the price of the raw material used. Fermentation ethanol became economically competitive with synthetic ethanol when crude-oil prices rose in the late 1970s and early 1980s, because ethylene is derived from refinery products. Production costs for fermentation ethanol may be further reduced by introduction of new processes and improved technologies that use raw materials such as wood, waste, and other low-cost feedstocks.

The feedstock price for synthetic ethanol production is high, but the capital cost requirement is low. Even if fermentation ethanol is produced from an inexpensive feedstock, the operating and capital costs are higher than those for synthetic ethanol.

The following analysis shows that the production costs for synthetic ethanol based on 1986 ethylene prices of $ 350/t are in the same range as those for fermentation ethanol produced from molasses costing $ 45/t.

Production figures are given in liters and metric tons. To convert liters to gallons, divide by 3.785; 1 t = 1260 L.

## 7.2. Production Costs of Synthetic Ethanol

The synthesis of ethanol from ethylene is divided into three processing steps: hydration, purification, and distillation (for details, see Chaps. 4 and 6). Approximately 87 major pieces of processing equipment are required.

**Investment Costs.** To calculate the production costs of synthetic ethanol, the number of installed pieces of major equipment and their costs were estimated in U.S. dollars based on 1986 cost levels. Figure 42 shows the estimated total investment costs for hypothetical plants and offsites (auxiliary equipment, primarily storage tanks) with capacities of 25 000, 50 000, 100 000, and 150 000 t/a. The breakdown of the investment for the main processing plant — the total inside-battery-limit investment — is as follows:

1) Equipment and machinery, including piping, insulation, painting, and electrical installations, 50%
2) Construction and erection, 25%
3) Civil engineering work, 10%
4) Engineering, costs for equipment purchase and start-up, 15%

License fees are considered separately.

**Production Costs.** The annual costs (in $10^6$ dollars) of synthetic ethanol produced in a 50 000-t/a plant from ethylene costing \$ 350/t are as follows (with the assumption that 0.625 t of ethanol can be obtained from 1 t of ethylene):

| | | |
|---|---|---|
| Raw materials | 11.26 | (59%) |
| Utilities | 4.18 | (22%) |
| Labor | 0.24 | (1%) |
| Depreciation, interest, maintenance | 3.59 | (18%) |
| Total production costs | 19.27 | (100%) |

**Payback Time.** The payback time (in years) for a 50 000-t/a plant can be calculated by using the following figures (given in $10^6$ \$/a, with the assumption that ethanol sells at \$ 450/t and that ethylene costs \$ 350/t):

| | |
|---|---|
| Annual turnover | 22.50 |
| Total production costs | 19.27 |
| Difference | 3.23 |
| Depreciation | 1.65 |
| Cash flow | 4.88 |

**Ethanol**

```
Water                                Diethyl ether
                                           │
                                     Anhydrous
Ethylene                             ethanol
  ──→ [ a ] ──→ [ b ] ──→ [ c ] ──→
```

| | | | Offsite costs 10⁶ $ | Total investment costs 10⁶ $ | Ethanol-plant capacity t/a | L/d |
|---|---|---|---|---|---|---|
| Number of major pieces of equipment | | | | | | |
| Investment costs, 10⁶ $ | | | | | | |
| 28 | 25 | 34 | 1.5 | 16.5 | 50 000 | 193 000 |
| 4.0 | 3.0 | 8.0 | 0.9 | 9.9 | 25 000 | 97 000 |
| | | | 2.5 | 27.5 | 100 000 | 386 000 |
| | | | 3.3 | 36.3 | 150 000 | 579 000 |

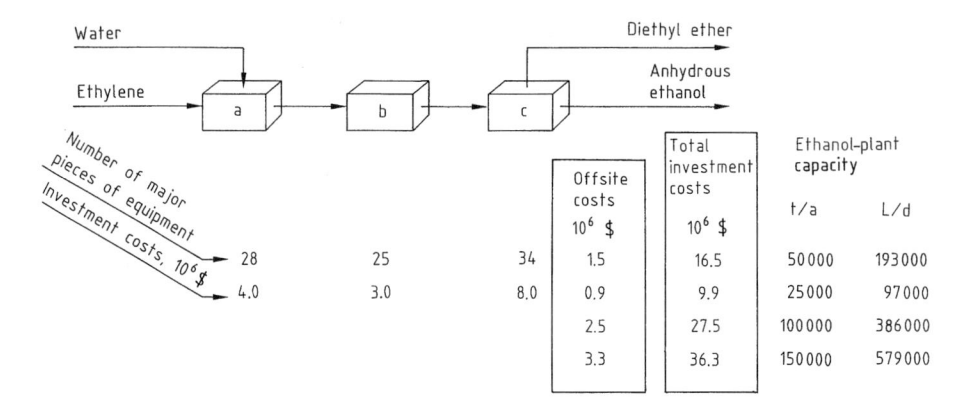

**Figure 42.** Investment costs for a plant producing synthetic ethanol
a) Hydration; b) Purification; c) Distillation

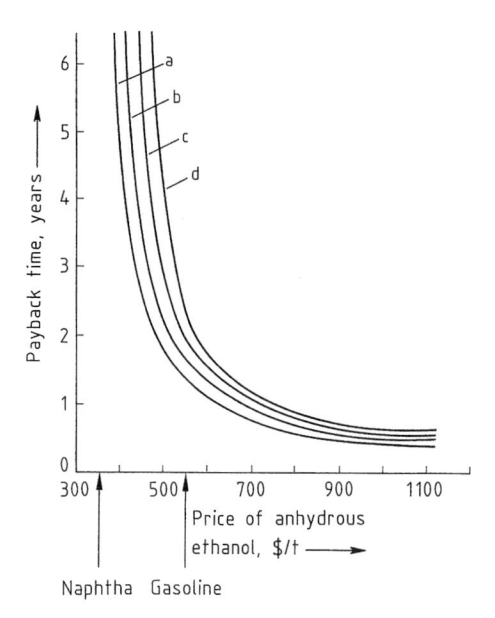

**Figure 43.** Price of anhydrous ethanol as a function of the payback time in a synthetic alcohol plant
Cost of ethylene (per ton): a) $ 300; b) $ 350; c) $ 400; d) $ 450 Production capacity: 50 000 t/a (19 300 L/d)

$$\text{Payback time} = \frac{\text{Investment}}{\text{Cash flow}} = 3.4 \text{ years}$$

For a plant with double this capacity (100 000 t/a), the payback time is reduced from 3.4 to 2.5 years. Figure 43 shows the payback times for synthetic ethanol production in a 50 000-t/a plant as a function of the selling price for ethanol at various ethylene costs. The 1986 prices for naphtha and gasoline are indicated for comparison.

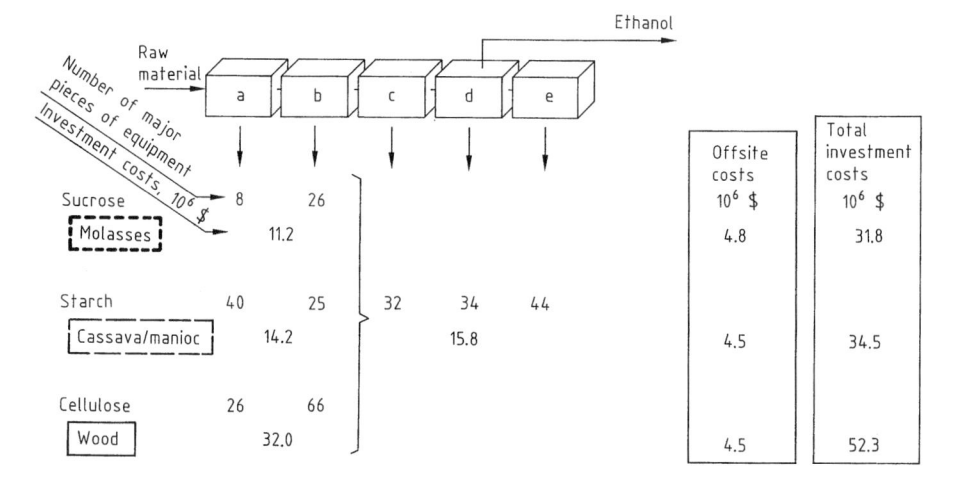

|  | Number of major pieces of equipment | Investment costs, 10⁶ $ | | | | | Offsite costs 10⁶ $ | Total investment costs 10⁶ $ |
|---|---|---|---|---|---|---|---|---|
| Sucrose | | 8 | 26 | | | | 4.8 | 31.8 |
| Molasses | | | 11.2 | | | | | |
| Starch | | 40 | 25 | 32 | 34 | 44 | 4.5 | 34.5 |
| Cassava/manioc | | | 14.2 | | 15.8 | | | |
| Cellulose | | 26 | 66 | | | | 4.5 | 52.3 |
| Wood | | | 32.0 | | | | | |

**Figure 44.** Investment costs for a plant producing fermentation ethanol from molasses, cassava, or wood *
a) Handling of raw materials; b) Hydrolysis pretreatment; c) Fermentation; d) Distillation and dehydration; e) Stillage processing
* Production capacity: 52 000 t/a (200 000 L/d).

## 7.3. Production Costs of Fermentation Ethanol

The processing scheme of ethanol fermentation from regenerable resources is divided into five major steps (for details, see Chaps. 5 and 6):

1) Handling of raw materials
2) Hydrolysis pretreatment
3) Fermentation
4) Distillation and dehydration
5) Processing of stillage and production of optional byproducts

Approximately 140 major pieces of process equipment are required for a molasses plant and almost 200 for a wood saccharification plant.

**Investment Costs.** Figure 44 shows the estimated investment costs for a fermentation-based ethanol plant with a capacity of 52 000 t/a (200 000 L/d); costs are given for each processing step and for the entire processing plant and offsites. The total investment costs are indicated for plants using sucrose, starch, or cellulose-containing raw materials such as molasses, cassava/manioc, or wood.

The breakdown of the inside-battery-limit investment is similar to that for synthetic ethanol plants.

**Table 23.** Production costs of fermentation ethanol obtained from sugarcane molasses, cassava, and eucalyptus wood

|  | Cost, $10^6$ \$/a | | |
|---|---|---|---|
|  | Molasses | Cassava | Eucalyptus wood |
| Raw materials | 9.27 | 14.77 | 7.19 |
| Utilities | 3.77 | 7.66 | 12.32 |
| Labor | 0.76 | 1.10 | 1.18 |
| Depreciation, interest, maintenance | 7.06 | 7.75 | 11.58 |
| Total production costs | 20.86 | 31.28 | 32.27 |

**Production Costs for Three Different Raw Materials.** The annual production costs of fermentation ethanol produced in a 52 000-t/a (200 000-L/d) plant from three basic raw materials (sugarcane molasses, cassava, and eucalyptus wood) are listed in Table 23.

Both wheat and corn (maize) are used by large ethanol-producing companies in the corn-belt region of the United States. The economics of ethanol production from these raw materials are similar to those of molasses-based ethanol production, but are strongly influenced by byproduct utilization and market values. The wet-milling process facilitates the production of high-quality ethanol, and retains the corn oil and corn protein in the food chain. In this process, the corn kernel is separated into starch, corn oil, gluten, and fiber, which are converted into a variety of other products. Utility costs are reduced by the use of coal-based energy. In some plants, the enzymatic hydrolysis is designed so as to permit production to be switched seasonally from ethanol to high-fructose corn syrup.

**Payback Time.** The payback time (in years) can be calculated by using the following figures (given in $10^6$ \$/a, with the assumption that ethanol sells at \$ 450/t and that molasses cost \$ 45/t):

| | |
|---|---|
| Annual turnover | 24.90 |
| Total production costs | 20.86 |
| Difference | 4.04 |
| Depreciation | 3.18 |
| Cash flow | 7.22 |

$$\text{Payback time} = \frac{\text{Investment}}{\text{Cash flow}} = 4.4 \text{ years}$$

If the selling price of ethanol is higher, e.g., \$ 550/t, the payback time will be reduced to 2.8 years.

Figure 45 shows the payback times for fermentation ethanol production as a function of the selling price for ethanol. Sugarcane molasses, cassava, and eucalyptus wood have been considered as raw materials, their costs being \$ 45, \$ 30, and \$ 15 per ton, respectively. In the early 1970s, synthetic ethanol was less expensive than fermentation ethanol because crude-oil prices, and thus ethylene prices, were low. Increases in the price of oil and, therefore, ethylene were accompanied by drops in the price of corn and

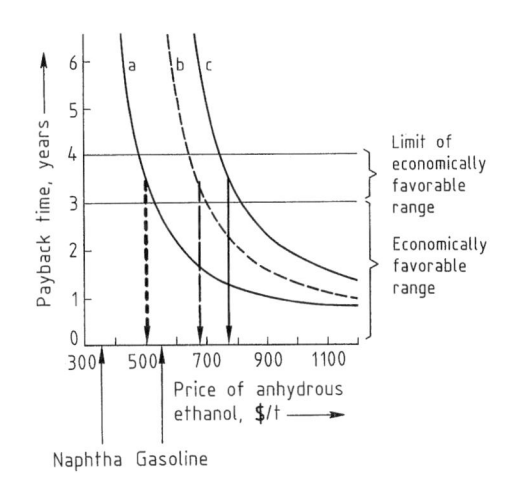

**Figure 45.** Price of anhydrous ethanol as a function of the payback time in a plant* producing fermentation alcohol

Raw material and price (per ton): a) Sugarcane molasses, $ 45; b) Cassava, $ 30; c) Eucalyptus wood, $ 15

* Production capacity: 52 000 t/a (200 000 L/d).

sugarcane molasses; this meant that fermentation ethanol was less expensive to produce than synthetic ethanol. Fermentation ethanol can be expected to maintain a slight price advantage over synthetic ethanol as long as the raw material prices remain low. If ethanol is blended with gasoline for use as a motor fuel, i.e., gasohol, a payback time of 3–4 years is acceptable. The minimum selling prices for ethanol which would economically justify its production from molasses, cassava, and eucalyptus wood would therefore be $ 500, $ 675, and $ 775 per ton, respectively (see Fig. 45). In 1986, Western European prices were $ 300–400 per ton for raw naphtha and $ 500–600 per ton for gasoline from the refinery (tax not included).

**Stillage Processing Options.** If stillage is used directly as a fertilizer without further processing (as in developing countries), the economic feasibility of fermentation ethanol production is increased by about 15%.

Stillage can also be processed further in four main ways:

1) Multiple-effect evaporation, drying of the concentrated stillage, and production of distillers' dry grain (a byproduct, which is used as an animal feed)
2) Mechanical vapor recompression, syrup evaporation, drying of the concentrated stillage, and production of distillers' dry grain
3) Evaporation and syrup combustion
4) Anaerobic digestion with generation of biogas

Analysis of the economic merits and drawbacks of these four schemes with regard to the production of fermentation ethanol reveals that system (2) is the most advantageous; systems (1), (3), and (4) are less favorable, but equal. The disadvantage of schemes (1) and (2) is the unstable selling price of distillers' dry grain. Scheme (3) is expensive because an off-gas purification system is required. The advantage of scheme (4) is the production of energy in the form of biogas.

# 8. Analysis

The classic *physical methods* for the determination of ethanol concentration in a sample are based on measurements of relative density (with a calibrated pycnometer, hydrometer, or density meter) [301]; boiling-point depression of alcohol–water mixtures (using an ebulliometer) [302]; and refractive index [301], [303]. In addition, gas chromatography is quick and reliable. Various internal standards, such as acetone, ethyl acetate, *n*-propanol, 2-propanol, and ethylene glycol monoethyl ether, have been used [301], [304], [305].

*Chemical methods* include oxidation to acetic acid by dichromate in the presence of sulfuric acid and titration of unreacted dichromate with ferrous ammonium sulfate (sodium diphenylamine 4-sulfonate or 1,10-*o*-phenanthroline are added as indicators):

$$2\ Cr_2O_7^{2-} + 3\ C_2H_5OH + 16\ H^+ \longrightarrow 4\ Cr^{3+} + 3\ CH_3COOH + 11\ H_2O$$
$$Cr_2O_7^{2-} + 6\ Fe^{2+} + 14\ H^+ \longrightarrow 2\ Cr^{3+} + 6\ Fe^{3+} + 7\ H_2O$$

The pH of the reaction is critical because ethanol could be oxidized either to acetaldehyde or to a mixture of acetaldehyde and acetic acid [302], [306], [307]. If the ethanol–water mixture is denatured, other compounds can interfere.

The chemical purity of ethanol is often determined using acetylation or phthalation reactions in which ethanol reacts with a defined amount of either acetic or phthalic anhydride in pyridine solution. The decrease in acidity of the anhydride solution (after hydrolysis and comparison with a standard solution) corresponds to the reaction of the hydroxyl groups. Any reactive hydroxyl group can take part in these reactions, which are not specific for ethanol. Other functional groups do not interfere [308].

Trace amounts of ethanol can be determined by colorimetric methods using compounds such as 8-hydroxyquinoline or vanadic acid, which form colored complexes with alcohols [309].

In addition to physical and chemical methods, a *biochemical method* is used to determine ethanol concentration [310]. This method is based on the oxidation of ethanol to acetaldehyde by nicotinamide–adenine dinucleotide ($NAD^+$). The enzyme alcohol dehydrogenase catalyzes the reaction:

$$CH_3CH_2OH + NAD^+ \longrightarrow CH_3CHO + NADH + H^+$$

To bring the reaction to completion, semicarbazide, (aminooxy)acetic acid, or the enzyme aldehyde dehydrogenase is added to remove acetaldehyde from the reaction mixture. The reaction is followed spectrophotometrically at a wavelength of 340 nm where the absorbance is proportional to the increase in NADH concentration [302]. Despite the fact that some other alcohols can also react, the method is good and can be used to determine very small ethanol concentrations.

Some enzymatic methods use enzymatic probes with immobilized enzymes [311]. Recently, electrode probes with whole microbial cells have also been developed for alcohol determination [311].

# 9. Uses

Ethanol is an organic chemical with many applications, e.g.,

1) alcoholic beverage,
2) solvent,
3) raw material in chemical synthesis, and
4) fuel.

In most countries, ethanol produced by fermentation has been used for beverages and specialty chemicals, and ethanol produced by chemical synthesis has been used for industrial purposes. Certain countries, such as Brazil and India, use fermentation ethanol for industrial purposes.

For many years, alcoholic beverages have been taxed worldwide. However, the price of taxed ethanol was too high for its use as an industrial raw material. To enable industry to obtain alcohol at a lower price, the Tax-Free Industrial and Denatured Alcohol Act of 1906 was passed in the United States. Similar laws have been enacted in other countries.

The U.S. government implemented financial, administrative, and chemical controls to prevent the utilization of tax-free ethanol in beverages. These regulations established the following four distinct categories of industrial ethanol:

1) Completely denatured alcohol
2) Proprietary solvents and special industrial solvents
3) Specially denatured alcohol
4) Pure (absolute) ethanol

Pure ethanol and slightly denatured ethanol are under strict control, whereas highly denatured ethanol has the fewest financial and administrative controls. A wide range of chemicals can be used for ethanol denaturation [312].

**Solvent.** Ethanol is the most important solvent after water. Its major commercial applications are in the manufacture of toiletries and cosmetics, detergents and disinfectants, pharmaceuticals, surface coatings, and in food and drug processing. Both synthetic and fermentation ethanol can be used for these purposes; however, fermentation ethanol is preferred (particularly in Europe) for applications involving human consumption (or body use) such as cosmetics, toiletries, and pharmaceuticals.

The amout of denatured ethanol used as a solvent is increasing. For example, in the United States 197 000 and 340 000 t of ethanol were used in 1960 and 1979, respectively. In Japan, between 1974 and 1978 the amount of ethanol consumed as a solvent increased by 28% [313], [314]

**Raw Material.** Ethanol is also used to produce various chemicals (see Fig. 2):

1) Acetaldehyde (→ Acetaldehyde)

2) Butadiene (→ Butadiene)
3) Diethyl ether (→ Ethers, Aliphatic)
4) Ethyl acetate (→ Acetic Acid)
5) Ethylamines (→ Amines, Aliphatic)
6) Ethylene (→ Ethylene)
7) Glycol ethers and other products formed by reaction with ethylene oxide or epoxides
   (→ Epoxides; → Ethylene Oxide; → Propylene Oxide)
8) Vinegar

Table 24 shows the most important products derived from ethanol in the United States. The amount of specially denatured ethanol used as a raw material in the manufacture of chemicals was 37.6% of the total amount of denatured ethanol used in the country during the specified period of time. In the United States, consumption of ethanol as a raw material reached a peak in 1960 (627 000 t) and then decreased substantially. The amount of ethanol used to manufacture chemicals in 1979 was only one-third of the amount used in 1960. The main reasons for this decline are the use of ethylene glycol rather than ethanol in antifreeze, replacement of ethanol by ethylene and ethane in the production of acetaldehyde, and production of ethylhexanol and butyraldehyde from other raw materials. Certain trends indicate that further decline in the use of ethanol to manufacture chemicals may be reversed [313].

In some countries, such as Brazil, the production of a variety of chemical products is based on ethanol. Among these products are those, such as acetaldehyde, that are made from petrochemicals in countries richer in crude oil. In the United States and elsewhere, ethylene is used to produce synthetic ethanol, whereas in Brazil, fermentation ethanol is used for ethylene production.

The example of Brazil is stressed because it is the world's largest producer of ethanol from sugar. Table 25 shows the amount of sugar used for ethanol production in the period from 1981–1982 to 1985–1986 in Brazil and in all other countries, except the United States in which production of ethanol is based on corn. In 1985–1986 Brazil's production represented 95% of world production. Brazil almost doubled its ethanol production in the tabulated five year period [315]. Most of the ethanol is used as a fuel; only a small fraction is employed as a chemical feedstock for the production of acetaldehyde, acetic acid, butanol, octanol, chlorinated ethylenes, glycols, polyethylene, styrene, vinyl acetate, and other chemicals [313], [316].

In 1977, $158 \times 10^6$ L of ethanol was used as a raw material for chemical manufacturing in the EEC. A much larger amount was used as a solvent. In these countries, ca. 55% of the ethanol was derived from fermentation [313].

More than 50% of the total ethanol produced in India in 1978 was used to produce chemicals. The total production of acetaldehyde, acetic acid, acetic anhydride, and DDT (dichlorodiphenyltrichloroethane) was based on ethanol, as was a significant fraction of the production of organic acetates, acetone, butanol, and a certain amount of polyethylene and poly(vinyl chloride) and styrene [313].

**Table 24.** Ethanol used as a raw material in chemical manufacturing in the United States (from July 1, 1978 to June 30, 1979)

| Product | Ethanol consumption, L |
| --- | --- |
| Vinegar | 73 853 912 |
| Acetic acid | 1 570 148 |
| Ethyl acetate | 33 190 215 |
| Ethyl chloride | NA* |
| Other ethyl esters | 84 679 089 |
| Sodium ethoxide | 4 180 679 |
| Ethylenamines for rubber processing | 35 208 990 |
| Dyes and intermediates | 1 357 936 |
| Acetaldehyde | NA* |
| Ether, diethyl | 327 524 |
| Ether, glycol, and others (excluding diethyl ether) | 11 419 306 |
| Xanthanes | 752 425 |
| Drugs and medicinal chemicals | 2 518 511 |
| Organosilicon products | 126 683 |
| Other chemicals | 9 733 531 |
| Synthetic resins | 12 609 456 |
| **Total** | 271 528 405 |

\* Not available.

**Table 25.** Sugar consumed to produce ethanol

| Crop years | Sugar consumption, $10^6$ t | | |
| --- | --- | --- | --- |
| | Brazil | Other countries* | Worldwide |
| 1981–1982 | 4.8 | 0** | 4.8 |
| 1982–1983 | 4.9 | 0** | 4.9 |
| 1983–1984 | 5.9 | 0.1 | 6.0 |
| 1984–1985 | 9.1 | 0.2 | 9.3 |
| 1985–1986 | 8.5 | 0.5 | 9.0 |

\* Does not include the United States.
\*\* Negligible amount.

# 10. Economic Aspects

The data given in this chapter are taken from [317]–[319].

## 10.1. Worldwide Production of Synthetic and Fermentation Ethanol

The capacities and annual production figures for already installed plants producing synthetic and fermentation ethanol in the developed and developing countries are listed in Table 26. Only industrial installations have been taken into consideration; small-scale production units with capacities of less than 5000 L/d (e.g., small units on farms and in local communities) have been ignored.

Table 26. Ethanol plant capacities and annual production

| | Synthesis | | | Fermentation | | |
|---|---|---|---|---|---|---|
| | Installed plant capacity, $10^6$ L/d | Number of plants | Production, $10^6$ L/a | Installed plant capacity, $10^6$ L/d | Number of plants | Production, $10^6$ L/a |
| **Developed (industrialized) countries** | | | | | | |
| Australia, New Zealand | | | | 0.270 | 9 | 40 |
| Eastern Europe | | | | 1.100 | 28 | 165 |
| Western Europe | 2.283 | 8 | 610 | 4.327 | 123 | 974 |
| Israel, Japan, South Africa | 0.231 | 4 | 69 | 0.918 | 31 | 138 |
| North America | 3.262 | 10 | 880 | 9.569 | 74 | 2153 |
| Soviet Union | 0.200 | 2 | 55 | 0.850 | 75 | 128 |
| Total | 5.976 | 24 | 1614 | 17.034 | 340 | 3598 |
| **Developing countries** | | | | | | |
| Africa | | | | 1.035 | 35 | 155 |
| Asia | 0.230 | 4 | 62 | 4.608 | 163 | 691 |
| Asia (centrally planned economies) | 2.240 | | 605 | | | |
| Central America | | | | 1.931 | 67 | 290 |
| South America | | | | 7.704 | 116 | 1156 |
| Oceania | | | | 0.036 | 2 | 5 |
| Total | 2.470 | 4 | 667 | 15.314 | 383 | 2297 |
| **Total worldwide** | 8.446 | 28 | 2282 | 32.348 | 723 | 5895 |

The total worldwide production capacity for synthetic ethanol is $8.446 \times 10^6$ L/d and annual production is $2282 \times 10^6$ L. Most of the 28 plants producing synthetic ethanol are located in industrialized countries; the developing countries produce mainly fermentation ethanol. In Italy, the Federal Republic of Germany, the United Kingdom, and Canada, ca. 50 % of the ethanol is produced by ethylene hydration; the corresponding figure for the United States is ca. 35 %. The developed and developing countries produce approximately equal amounts of fermentation ethanol in a total of ca. 720 plants.

The total worldwide production capacity for fermentation ethanol far exceeds that for synthetic ethanol and amounts to more than $32 \times 10^6$ L/d; the average production is, however, less than $20 \times 10^6$ L/d. The reason is that fermentation plants in most of the developing countries are only supplied with raw materials derived from sugarcane for part of the year.

Fermentation ethanol producers in the industrialized nations import some of their raw materials in the form of molasses from the sugar-producing industries of developing countries.

## 10.2. Major Producers of Fermentation Ethanol from Regenerable Resources

More than 60% of the total world production capacity for fermentation ethanol is shared by Brazil, the United States, and India. These countries all have favorable climates, good soil conditions, and enough land to produce the necessary raw materials.

Figure 46 shows the development of production capacities for fermentation ethanol in these countries and worldwide from 1970 to 1986. Since the first large increase in price of crude oil in the early 1970s, worldwide capacities of fermentation ethanol plants have more than doubled. The rate of capacity increase is, however, slowing down. Capacities of fermentation ethanol plants can be expected to decrease because of reduced investments and closure of obsolete plants. Drastic increases in the price of oil may, however, change this situation.

**Brazil.** Brazil's 65 industrial fermentation ethanol plants have a total capacity of over $6.7 \times 10^6$ L/d but produce only $3.4 \times 10^6$ L/d. Raw materials for the production of fermentation ethanol are molasses and sugarcane juice; cassava and roots (manioc) have also been tested. Ethanol, therefore, originates from sugar factories and distilleries. The sugar industry has traditionally played a major role in the economic and social development of Brazil.

In 1975, the following two factors encouraged Brazil to set up a national program, called "Proalcool," for the production of fermentation alcohol:

1) the fluctuations and sudden decline in international sugar prices, and
2) the increase in price of crude oil.

The establishment of new agroindustrial complexes requiring large investments and close coordination between agricultural and industrial planning is perhaps the major challenge facing the Proalcool program. The Brazilian automobile industry has already spent $ (30-40) \times 10^6$ in development costs for the "alcohol engine."

**India.** The production of ethanol from agricultural raw materials (primarily molasses) in India has a long tradition. The present "green revolution" is in favor of establishing more sugarcane plantations.

Until recently, India was the only country to use ethanol as a raw material for producing chemicals customarily derived from ethylene, but Brazil is now also adopting this approach.

The total capacity of India's approximately 80 industrial ethanol plants is over $3 \times 10^6$ L/d. Unfortunately, most of these plants operate at less than full capacity and for only five to six months of the year, so that actual production is ca. $1.2 \times 10^6$ L/d.

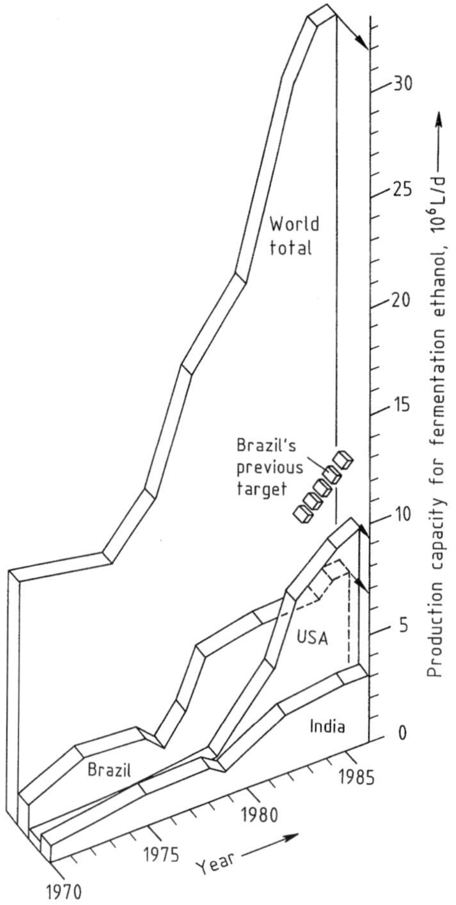

**Figure 46.** Development of production capacities for fermentation ethanol

**United States.** Because of its vast cornfields, the United States uses corn as the primary raw material for production of fermentation ethanol. Some companies use corn to produce both ethanol and high-fructose corn syrup. Most fermentation ethanol producers are located in the corn-belt states of Iowa, Illinois, and Indiana.

Sugarcane molasses is used as a raw material for ethanol production primarily in the Southern States and the coastal areas. A few paper-manufacturing companies also use sulfite waste liquor as a raw material.

Tax reductions offered by the U.S. government as an incentive for gasohol utilization have led to the creation of more than 70 fermentation ethanol projects with an estimated total capacity of $13 \times 10^6$ L/d. The number of government-supported projects has, however, been reduced to about 45 because of budget cuts.

The total production capacity for fermentation ethanol in the United States is $9.5 \times 10^6$ L/d, and annual production is about $2 \times 10^9$ L.

# 11. Toxicology

The acute oral toxicity of ethanol in rats ($LD_{50}$) is 11.5 [10.1] – 13.7 g/kg [10.2]. In mice [10.3], guinea pigs [10.1], rabbits [10.4] and dogs [10.3] the $LD_{50}$ values are 9.5, 9.6, 9.9, and 6 g/kg, respectively.

Ethanol can cause mild to severe irritation in the rabbit eye depending on the concentration and amount used [10.5], but it is not appreciably irritating to the intact skin [10.5], [10.6].

Inhalation of ethanol causes irritation of the mucous membranes, excitation, ataxia, drowsiness, narcosis, and death due to respiratory failure. Lethal concentrations are in the range of 20 000 – 40 000 ppm in the mouse, guinea pig, and rat after inhalation for several hours. Concentrations of less than 6000 ppm can be tolerated without symptoms of intoxication [10.7], [10.8].

Repeated oral administration (12 weeks) of 10 g $kg^{-1}$ $d^{-1}$ to juvenile rats resulted in low weight gains and fatty liver degeneration [10.9]. Triglyceride, cholesterol, and phospholipid metabolism were disturbed in monkeys fed a diet in which 40% of the total calories were replaced by ethanol. Fatty degeneration occurred in the liver and the myocardium [10.10].

Ethanol increased embryolethality in pregnant rats and retarded fetal development when administered at levels of 5 mL/kg [10.11], [10.12]. Major malformations were not induced in the fetus.

Ethanol doses of 1 – 1.5 g/kg cause dominant lethal mutations in male mice [10.13], but not in female mice [10.14]. In the Ames test, ethanol does not exhibit mutagenic activity [10.15]. An increase of chromatid breakage was observed in human fibroblast cultures, but this is considered to be a cytotoxic effect rather than evidence of mutagenicity [10.16].

Ethanol is oxidized to acetaldehyde by means of the following enzyme systems [10.17]:

1) Alcohol dehydrogenase [10.18]
2) Catalase acting as a peroxidase and coupled to a system supplying oxygenated water [10.19]
3) A microsomal ethanol oxidizing system requiring NADPH as a coenzyme [10.20].

In the liver 80% of the absorbed ethanol is metabolized first to acetaldehyde [10.19] and then to acetic acid. Acetic acid is finally degraded to $CO_2$ and water. The oxidation to acetaldehyde seems to be much slower than that of acetaldehyde to acetic acid [10.21].

General symptoms produced by intake of ethanol in humans are listed in Table 27 [10.7]; however, considerable variations are observed between individual subjects.

In humans suffering from alcoholism, fatty liver degeneration is thought to be due to increased fatty acid synthesis from acetate, increased transportation of lipids from peripheral fat depots to the liver, and decreased oxidation of fatty acids [10.22]. Fatty

**Table 27.** Symptoms produced after ethanol intake in humans

| Symptom | Blood ethanol concentration, % |
|---|---|
| Beginning of uncertainty | 0.06 – 0.08 |
| Slow comprehension | 0.10 |
| Stupor | 0.11 – 0.15 |
| Drunkenness | 0.16 |
| Severe intoxication | 0.2 – 0.4 |
| Death | 0.4 – 0.5 |

infiltration of the myocardium and chronic leptomeningitis are also reported in chronic alcoholism and are responsible for the well defined symptomatology of this condition [10.23].

The ACGIH and OSHA exposure limits for ethanol are both 1000 ppm; its MAK value is also 1000 ppm (1900 mg/m$^3$).

# 12. References

[1]   E. A. Cotelle, US 41 685, 1861.
[2]   W. Gorianoff, A. Butleroff, *Justus Liebigs Ann. Chem.* **180** (1876) 245.
[3]   C. R. Nelson, M. L. Courter, *Chem. Eng. Prog.* **50** (1954) 526.
[4]   Yu. M. Bakshi, A. I. Gelbshtein, M. I. Temkin, *Dokl. Akad. Nauk SSSR* **126** (1959) 314; *Chem. Abstr.* **55** (1961) 1944 f.
[5]   D. S. Tsiklis, A. I. Kulikova, *Zh. Fiz. Khim.* **35** (1961) 954; *Chem. Abstr.* **58** (1963) 6660g.
[6]   Z. Novosad, *Chem. Prum.* **5** (1955) no. 30, 72.
[7]   A. I. Gelbshtein, Yu. M. Bakshi, M. I. Temkin, *Dokl. Akad. Nauk SSSR* **132** (1960) 384; *Chem. Abstr.* **57** (1962) 6663d.
[8]   B. S. Bouden, A. I. Gelbshtein et al., *Neftepererab. Neftekhim.* (Moscow) 1968, no. 5, 33.
[9]   C. V. Mace, C. F. Bomilla, *Chem. Eng. Prog.* **50** (1954) 385.
[10]  Ch.-K. Feng, *Hua Hsueh Tung Pao* 1978, no. 6, 367; *Chem. Abstr.* **90** (1979) 6558 f.
[11]  R. W. Taft, P. J. Riesz, C. A. Fazio, *J. Am. Chem. Soc.* **74** (1952) 5372.
[12]  F. Vachez, *Rev. Inst. Fr. Pet. Ann. Combust. Liq.* **18** (1963) 724.
[13]  De Bataafsche Petroleum Maatschappij, DE 1 042 561, 1955.
[14]  Eastman Kodak Co., GB 1 144 947, 1966.
[15]  *Hydrocarbon Process.* **46** (1967) no. 11, 168.
[16]  Veba Chemie, GB 1 201 181, 1967.
[17]  Hibernia-Chemie, BE 715 907, 1968.
[18]  National Distillers & Chemical Co., DE-OS 2 015 536, 1970.
[19]  BP Chemicals International, DE-OS 2 237 015, 1972.
[20]  C. L. Thomas: *Catalytic Processes and Proven Catalysts,* Academic Press, New York-London 1970, p. 230.
[21]  G. G. Eremeeva, V. M. Dronkin, G. I. Gagarina, *Neftepererab. Neftakhim* (Moscow) 1979, no. 16, 19 – 20; *Chem. Abstr.* **91** (1979) 107 638.
[22]  K. W. Toptschijewa, S. M. Rachowskaja, I. K. Kutschkajewa, *Neftekhimiya* **3** (1963) 271.

[23] Chemopetrol Koncernovy Podnik Chemicke Zavody Ceskoslovensko-Sovetskeho Praptelstvi, BE 888 479, 1981 (V. Kadlec, V. Grosser, J. Rosenthal); *Chem. Abstr.* **96** (1982) 37 267.
[24] Union Carbide, BE 854 606, 1977.
[25] Kurashiki Rayon Co., DE-OS 1 280 833, 1964.
[26] Kurashiki Rayon Co., DE-OS 1 543 130, 1965.
[27] Esso Res. & Eng. Co., US 3 686 334, 1969.
[28] Tokuyama Soda K. K., DE-OS 2 022 568, 1970.
[29] Tokuyama Soda K. K., DE-OS 2 215 380, 1972.
[30] G. K. Boreskow, W. A. Dsisjko et al., *Khim. Promst. (Moscow)* 1961, no. 2, 97.
[31] Eastman Kodak Co., US 3 554 926, 1968.
[32] L. N. Tolkacheva, L. A. Novikova, V. A. Plakhotnik, V. Y. Danyushevski et al., *Neftekhimiya* **23** (1983) no. 6, 803; *Chem. Abstr.* **100** (1984) 105 486.
[33] Mobil Oil Co., US 4 214 107, 1980 (C. D. Cheng, N. J. Morgan).
[34] Toray Industries, JP 7 245 323, 1968.
[35] A. M. Tsybulevskii, L. A. Novikova, V. A. Kondrat'ev, L. N. Tolkacheva et al., *Neftekhimiya* **19** (1979) no. 5, 771; *Chem. Abstr.* **92** (1980) 110 468 r.
[36] G. F. Nencetti, R. Tartarelli, *Chim. Ind. (Milan)* **48** (1966) 30.
[37] M. Baccaredda, G. F. Nencetti et al., *Riv. Combust.* **22** (1968) 65.
[38] R. Tartarelli, M. Giorgini et al., *J. Catal.* **17** (1970) 41.
[39] A. A. Kubashov, K. V. Topchieva, M. G. Mitichenko, G. G. Eremeeva et al., *Zh. Prikl. Khim. (Leningrad)* **53** (1980) no. 5, 1090; *Chem. Abstr.* **93** (1980) 132 011 j.
[40] M. Giorgini, R. Tartarelli, *Chim. Ind. (Milan)* **58** (1976) no. 9, 611; *Chem. Abstr.* **86** (1977) 11 804.
[41] UOP, US 4 465 874, 1984 (J. A. Koca).
[42] Mitsui Toatsu Chemicals, JP-Kokai 79 27 505, 1979 (F. Matsuda, T. Kato).
[43] Mitsui Toatsu Chemicals, JP-Kokai 79 16 4 143, 1979 (F. Matsuda, T. Kato).
[44] T. C. Carle, D. M. Stewart, *Chem. Ind. (London)* 1962, 830.
[45] B. S. Bouden, G. G. Goryacheva, G. I. Gagarina, *Neftepererab. Neftekhim. (Moscow)* 1967, no. 12, 24.
[46] Zh. A. Bril, V. M. Platonov, M. Ya. Klimenko, *Khim. Promst. (Moscow)* **45** (1969) 332.
[47] B. S. Bouden, A. I. Gelbshtein et al., *Neftepererab. Neftekhim. (Moscow)* 1968, no. 4, 17.
[48] G. G. Goryacheva, B. S. Bouden, A. I. Gelbshtein, *Sov. Chem. Ind. (Engl. Transl.)* 1970, no. 6, 15.
[49] J. Koutensky, *Sb. Pr. Vyzk. Chem. Vyuziti Uhli Dehtu Ropy* **15** (1978) 211; *Chem. Abstr.* **89** (1978) 75 211.
[50] J. Koutensky, *Sb. Pr. Vyzk. Chem. Vyuziti Uhli Dehtu Ropy* **16** (1979) 183; *Chem. Abstr.* **93** (1980) 167 145.
[51] H. G. Harris, D. M. Himmelblau, *J. Chem. Eng. Data* **9** (1964) 61.
[52] UOP, US 4 374 286, 1983 (R. J. Schmidt).
[53] Haldor F. A. Topsøe, DE-OS 2 158 795, 1971.
[54] I. Wender, R. Levine, M. Orchin, *J. Am. Chem. Soc.* **71** (1949) 4160.
[55] I. Wender, H. Greenfield, M. Orchin, *J. Am. Chem. Soc.* **73** (1951) 2656.
[56] British Petroleum Co., EP 3 876, 1979 (B. R. Gane, D. E Stewart).
[57] Gulf Res. & Dev. Co., US 4 239 924, 1980 (J. E. Bozik, T. P. Kobylanski, W. R. Pretzer).
[58] British Petroleum Co., EP 1 937, 1979 (B. R. Gane, D. C. Stewart).
[59] British Petroleum Co., GB 2 053 915, 1981 (W. J. Ball, D. G. Stewart).
[60] British Petroleum Co., JP-Kokai 8 049 326, 1980.
[61] Y. Sugi, K. Brando, Y. Takami, *Chem. Lett.* 1981, no. 1, 63.

[62] Catalytica Associates, Review of PERC Program no. FE-7093; *Homogeneous Reactions-*, "The Homologation of Methanol," Dept. of Energy Contract EF-77-C-01-2536, May 1978.
[63] Gulf Res. & Dev. Co., US 4 239 924, 1980 (J. E. Bozic, T. P. Kobylinsky, W. R. Pretzer).
[64] Gulf Res. & Dev. Co., DE 3 228 769, 1983 (M. M. Habib, W. R. Pretzer).
[65] Agency of Industrial Sciences and Technology, JP-Kokai 59 110 637, 1984.
[66] Ruhrchemie AG, DE 3 330 507, 1985 (W. Lipps, H. Bahrmann, B. Cornils, W. Konkal).
[67] Air Products & Chemicals, US 4 171 461, 1979 (G. M. Bartish).
[68] Agency for Industrial Sciences and Technology, JP-Kokai 59 110 637, 1982.
[69] Mitsubishi Gas Chem. Co., JP-Kokai 82 80 334, 1982.
[70] Union Chemische Braunkohlen Kraftstoff, DE 3 045 891, 1982 (K. H. Keim, J. Koff).
[71] Ethyl Corp., US 4 309 314, 1982 (D. C. Hargis, M. Dubek).
[72] B. Juran, A. V. Porcelli, *Hydrocarbon Process. Int. Ed.* **64** (1985) no. 10, Section 1, 85.
[73] Monsanto Chemical Co., BE 713 296, 1968 (P. E. Paulik, J. F. Roth et al.).
[74] J. F. Roth, J. H. Cradock, A. Hershman, P. E. Paulik, *Chem. Technol.* 1971, 600.
[75] Davy McKee (London), WO 8 303 409, 1983 (M. W. Bradley, N. Harris, K. Turner).
[76] Halcon SD Group, DE 3 335 594, 1984 (N. Rizkalla).
[77] Halcon SD Group, DE 3 335 694, 1984 (N. Rizkalla).
[78] Halcon SD Group, US 4 497 967, 1985 (C. G. Wan).
[79] G. Natta, U. Colombo, I. Pasquon in P. H. Emmett (ed.): *Catalysis,* vol. **5**, Reinhold Publ. Co., New York 1957, p. 131.
[80] B. D. Dombek, *Adv. Catal.* **32** (1983) 325.
[81] D. A. Popeckov, A. A. Shokol, *Zap. Inst. Khim. Akad. Nauk Ukr. RSR* **4** (1937) 205.
[82] R. Taylor, *J. Chem. Soc.* 1934, 1429.
[83] P. Courty, J. P. Arlie, A. Convers, P. Mitchenko et al., *Actual. Chim.* 1983, no. 9, 19.
[84] Inst. Français du Pétrole, DE 2 748 097, 1978 (A. Suguir, E. Freund).
[85] Inst. Français du Pétrole, DE 2 949 952, 1980 (A. Suguir, E. Freund).
[86] New Fuel Oil Dev. Technology Res. Assoc., JP-Kokai 8 545 537, 1985.
[87] Res. Assoc. for Petrol Alternative Dev., EP 110 357, 1984 (M. Shibata, Y. Aoki, S. Uchiyama).
[88] I. A. Sibilia, J. M. Dominguez, R. G. Herman, K. Klier, *Prepr. Pap. Am. Chem. Soc., Div. Fuel Chem.* **29** (1984) no. 5, 261.
[89] K. J. Smith, R. B. Anderson, *Prepr. Pap. Am. Chem. Soc., Div. Fuel Chem.* **29** (1984) no. 5, 269.
[90] M. Ichikawa, *J. Chem. Soc. Chem. Commun.* 1978, 566; *Bull. Chem. Soc. Jpn.* **51** (1978) no. 8, 2268, 2273.
[91] R. Bardet, J. Thivolle-Cazat, Y. Trambouze, *C. R. Seances Acad. Sci. Ser. 2* **292** (1981) no. 12, 883.
[92] Societé Chimique de Grand Paroisse, Azote et Prod. Chim., FR 2 530 159, 1984 (R. Bardet, J. Thivolle-Cazat, Y. Trambouze, C. Harmon).
[93] A. Cariotti, S. Martingo, L. Sanderighi, C. Tonelli et al., *J. Chem. Soc. Faraday Trans. 1* **80** (1984) no. 6, 1605.
[94] Sagami Chemical Research Center, JP-Kokai 82 108 026, 1982.
[95] Agency for Industrial Sciences and Technology, JP-Kokai 8 532 730, 8 532 734, 8 532 736, 1985.
[96] Agency for Industrial Sciences and Technology, JP-Kokai 8 525 943, 8 532 729, 8 532 731, 8 532 732, 8 532 733, 8 532 735, 1985.
[97] Showa Denko K. K., JP-Kokai 8 262 231, 8 262 232, 8 262 233, 82 109 728, 82 109 729, 8 262 730, 82 126 432, 1982.
[98] Hoechst, EP 21 241, 1981 (H. Hachenberg, F. Wunder, E. I. Leupold, H. J. Schmidt).

[99]   Union Carbide, US 4 301 253, 1981 (B. K. Warren).
[100]  B. K. Warren, P. D. Dombek, *J. Catal.* **79** (1983) 334.
[101]  J. F. Knifton, *Prepr. Pap. Am. Chem. Soc. Div. Petr. Chem.* **29** (1984) no. 2, 586.
[102]  J. F. Knifton, *J. Am. Chem. Soc.* **103** (1981) 3959.
[103]  Texaco Dev. Corp., US 4 265 828, 1981 (J. K. Knifton).
[104]  Texaco Dev. Corp., US 2 041 924, 1980 (J. F. Knifton).
[105]  Agency for Industrial Sciences and Technology, JP-Kokai 84 164 738, 1984.
[106]  Agency for Industrial Sciences and Technology, JP-Kokai 83 172 333, 1983.
[107]  S. E. Ferrari, J. J. C. Lopes, J. R. A. Leme, E. R. de Oliveira: *Proc. 4th Int. Symp. Alcohol Fuels Technology,* Sao Paulo, Brazil, Oct. 1980, p. 139.
[108]  C. S. Gong, L. F. Chen, *Biotechnol. Bioeng. Symp.* **14** (1984) 257–268.
[109]  R. P. Jones, N. Pamment, P. F. Greenfield, *Process Biochem.* **16** (1981) no. 3, 42–49.
[110]  R. P. Jones, P. F. Greenfield, *Process Biochem.* **19** (1984) no. 2, 48–58.
[111]  B. Maiorella, Ch. R. Wilke, H. W. Blanck: "Alcohol Production and Recovery," in A. Fiechter (ed.): *Advances in Biochemical Engineering,* **20** Springer-Verlag, New York 1981 pp. 43–92.
[112]  E. Oura, *Process Biochem.* **12** (1977) no. 3, 19–21, 35.
[113]  H. R. Mahler, E. H. Cordes: *Biological Chemistry,* Int. ed., Harper & Row, New York 1968, pp. 404–435.
[114]  C. S. Gong, L. F. Chen, G. T. Tsao, M. C. Flickinger: "Conversion of Hemicellulose Carbohydrates," in A. Fiechter (ed.): *Advances in Biochemical Engineering,* **20** Springer-Verlag, New York 1981 pp. 93–118
[115]  T. W. Jeffries: "Utilization of Xylose by Bacteria, Yeasts, and Fungi," in A. Fiechter (ed.): *Advances in Biochemical Engineering/Biotechnology,* vol. **27,** Springer-Verlag, Berlin-New York 1983, pp. 1–32.
[116]  S. Mitsuhashi, J. O. Lampen, *J. Biol. Chem.* **204** (1953) 1011–1018.
[117]  B. L. Wilson, R. P. Mortlock, *J. Bacteriol.* **113** (1973) 1404–1411.
[118]  P. Y. Wang, H. Schneider, *Can. J. Microbiol.* **26** (1980) 1165–1171.
[119]  M. Hoefer, A. Betz, A. Kotyk, *Biochim. Biophys. Acta* **252** (1971) 1–12.
[120]  R. W. Detroy, R. L. Cunningham, A. I. Herman, *Biotechnol. Bioeng. Symp.* **12** (1982) 81–89.
[121]  M. Tomoyeda, H. Horitsu, *Agric. Biol. Chem.* **28** (1964) 139.
[122]  C. S. Gong, L. F. Chen, M. C. Flickinger, L. C. Chiang et al., *Appl. Environ. Microbiol.* **41** (1981) 430–436.
[123]  J. A. Barnett, *Adv. Carbohydr. Chem. Biochem.* **32** (1976) 125–234.
[124]  T. W. Jeffries, *Biotechnol. Bioeng. Symp.* **12** (1982) 103–110.
[125]  H. Dellweg, M. Rizzi, H. Methner, D. Debus, *Biotechnol. Lett.* **6** (1984) 395–400.
[126]  M. Korhola, I. Suomalainen, E. Vaisanen, H. Tuompo: "Distiller's Yeasts," in M. Korhola, H. Tuompo, V. Kauppinen (eds.): *Proceedings of the 7th Conference on Global Impacts of Applied Microbiology: Symposia on Alcohol Fermentation and Plant Cell Culture,* vol. 4, Foundation for Biotechnical and Industrial Fermentation Research. Helsinki 1986, pp. 29–61.
[127]  A. H. Rose, J. S. Harrison (eds.): *The Yeasts,* vol. **3,** Academic Press, London 1970, p. 306.
[128]  G. K. Hoppe, G. Hansford, *Biotechnol. Lett.* **6** (1984) 681–686.
[129]  A. A. Andreasen, T. J. B. Stier, *J. Cell. Comp. Physiol.* **43** (1954) 271–281.
[130]  R. D. Tyagi, *Process Biochem.* **19** (1984) no. 4, 136–141.
[131]  G. R. Cysewski, C. R. Wilke, *Biotechnol. Bioeng.* **18** (1976) 1297–1313.
[132]  C. Beck, H. K. von Meyenburg, *J. Bacteriol.* **96** (1968) 479–486.
[133]  K. Watson, *Biotechnol. Lett.* **4** (1982) 397–402.

[134] S. Laffon-Lafourcade, F. Larue, P. Ribereau-Gayon, *Appl. Environ. Microbiol.* **38** (1979) 1069–1073.
[135] F. H. White, *Proc. Int. Brewing Conv. (Australia-New Zealand Section)* **15** (1978) 133.
[136] T. K. Ghose, R. D. Tyagi, *Biotechnol. Bioeng.* **21** (1979) 1401–1420.
[137] G. K. Hoppe, G. S. Hansford, *Biotechnol. Lett.* **4** (1982) 39–44.
[138] C. D. Bazua, C. R. Wilke, *Biotechnol. Bioeng. Symp.* **7** (1977) 105–118.
[139] I. Holzberg, R. K. Finn, K. H. Steinkraus, *Biotechnol. Bioeng.* **9** (1967) 413–427.
[140] S. Aiba, M. Shoda, *J. Ferment. Technol.* **47** (1969) 790–794.
[141] N. B. Egamberdiev, N. D. Ierusalimskii, *Microbiologiya* **37** (1968) 687.
[142] G. K. Hoppe (1981), as cited in[137]
[143] P. Strehaiano, M. Moreno, G. Goma, *C. R. Acad. Sci. Ser. D.* **286** (1978) 225.
[144] S. Aiba, M. Shoda, M. Nagatani, *Biotechnol. Bioeng.* **10** (1968) 845–864.
[145] T. W. Nagodawithana, C. Castellano, K. H. Steinkraus, *Appl. Microbiol.* **28** (1974) 383–391.
[146] K. Watson, R. Cavicchioli, *Biotechnol. Lett.* **5** (1983) 683–688.
[147] R. Espinosa, V. Cojulun, F. Maroquin, *Biotechnol. Bioeng. Symp.* **8** (1978) 69–74.
[148] G. Reed (ed.): *Prescott and Dunns's Industrial Microbiology*, 4th ed., AVI Publishing Co., Westport, Conn., 1984, pp. 835–859.
[149] D. B. Hughes, N. J. Tudroszen, C. J. Moye, *Biotechnol. Lett.* **6** (1984) 1–6.
[150] H. Suomalainen, O. Kauppila, R. J. Peltonen, L. Nykanen: *Handbuch der Lebensmittelchemie*, vol. **7**, Springer-Verlag, Berlin-Heidelberg 1968, pp. 496–653.
[151] T. Seki, S. Myoga, S. Limtong, S. Vedono et al., *Biotechnol. Lett.* **5** (1983) 351–356.
[152] A. J. Hacking, I. W. F. Taylor, C. M. Hanas, *Appl. Microbiol. Biotechnol.* **19** (1984) 361–363.
[153] P. J. Blotkamp, M. Takagi, M. S. Pemberton, G. H. Emert: *Proc. 84th Natl. Meeting AIChE*, Atlanta, Ga., Feb. 1978.
[154] Gulf Oil Corp., GB 2 036 074 A, 1980 (M. S. Pemberton, S. D. Crawford).
[155] J. J. Savarese, S. D. Young, *Biotechnol. Bioeng.* **20** (1978) 1291–1293.
[156] M. Mes-Hartree, C. Hogan, R. D. Hayes, J. N. Saddler, *Biotechnol. Lett.* **5** (1983) 101–106.
[157] D. B. Rivers, G. H. Emert: *Proc. Bioenergy '80 World Congress and Exposition*, Atlanta, Ga., Apr. 1980, p. 157.
[158] L. D. McCracken, C. S. Gong, *Biotechnol. Bioeng. Symp.* **12** (1982) 91–102.
[159] S. Hayashida, K. Ohta, P. Q. Flor, N. Nanri et al., *Agric. Biol. Chem.* **46** (1982) 1947–1950.
[160] N. Matsumoto, A. Fukushi, M. Miyanaga, K. Kakihara et al., *Agric. Biol. Chem.* **46** (1982) 1549–1558.
[161] C. L. Weller, M. P. Steinberg, E. D. Rodda, *Biotechnol. Bioeng. Symp.* **13** (1983) 437–447.
[162] J. J. Wilson, G. G. Khachatourians, W. M. Ingledew, *Biotechnol. Lett.* **4** (1982) 333–338.
[163] A. M. Sills, G. G. Stewart, *J. Inst. Brew.* **88** (1982) 313–316.
[164] K. Oteng-Gyang, G. Moulin, P. Galzi, *Z. Microbiol.* **21** (1981) 537–544.
[165] J. J. Wilson, W. M. Ingledew, *Appl. Environ. Microbiol.* **44** (1982) 301–307.
[166] R. De Mot, E. van Oudendijck, H. Verachtert, *Biotechnol. Lett.* **6** (1984) 581–586.
[167] H. Giesel, *Nachr. Chem. Tech. Lab.* **32** (1984) 316.
[168] H.-J. Schepers, S. Bringer-Meyer, H. Sahm, *Z. Naturforsch. C: Biosci.* **42 C** (1987) 401.
[169] J. S. Tolan, R. K. Finn, *Appl. Environ. Microbiol.* (1987) in press.
[170] J. Swings, J. De Ley, *Bacteriol. Rev.* **41** (1977) 1.
[171] R. K. Finn, S. Bringer, H. Sahm, *Appl. Microbiol. Biotechnol.* **19** (1984) 161.
[172] S. Bringer, E. Durst, H. Sahm, R. K. Finn, *Biotechnol. Bioeng. Symp.* **14** (1984) 269.
[173] W. D. Murray, A. W. Khan, L. van den Berg, *Int. J. Syst. Bacteriol.* **32** (1982) 132.

[174] L. H. Carreira, L. G. Ljungdahl in D. L. Wise (ed.): *Liquid Fuel Developments*, CRC Press Inc., Boca Raton, Fla., 1983, p. 1.
[175] U. Schmid, H. Giesel, S. M. Schoberth, H. Sahm, *Syst. Appl. Microbiol.* **8** (1986) 80.
[176] J. G. Zeikus: "Biology of the Spore-forming Anaerobes," in A. L. Demain, N. A. Solomon (eds.): *Biology of Industrial Microorganisms*, The Benjamin/Cummings Publ. Company Inc., London-Amsterdam-Don Mills-Sidney-Tokyo 1985, p. 79.
[177] B. Sonnleitner, A. Fiechter, *Trends Biotechnol.* **1** (1983) 74.
[178] J. Wiegel, *Experientia* **38** (1982) 151.
[179] M. C. Flickinger, *Biotechnol. Bioeng.* **22** (1980) Suppl. 1, 27.
[180] P. L. Rogers, K. J. Lee, M. L. Skotnicki, D. E. Tribe, *Adv. Biochem. Eng.* **23** (1982) 37.
[181] P. L. Rogers, A. E. Goodman, R. H. Heyes, *Microbiol. Sci.* **1** (1984) 133.
[182] G. Gottschalk: *Bacterial Metabolism*, Springer-Verlag, Berlin 1979.
[183] J. P. Belaich, J. C. Senez, *J. Bacteriol.* **89** (1965) 1195.
[184] S. Bringer, R. K. Finn, H. Sahm, *Arch. Microbiol.* **139** (1984) 376.
[185] L. O. Ingram, *Tibtech* 1986 (Feb.), 40.
[186] G. Ourisson, P. Albrecht, M. Rohmer, *Spektrum Wissensch.* 1984, no. 10, 54.
[187] S Bringer, T. Härtner, K. Poralla, H. Sahm, *Arch. Microbiol.* **140** (1985) 312.
[188] A. Schmidt, S. Bringer-Meyer, K. Poralla, H. Sahm, *Appl. Microbiol. Biotechnol.* **25** (1986) 32. H. Sahm, S. Bringer-Meyer, *Chem. Ing. Tech.* **59** (1987) 695–700.
[189] S. Bringer, H. Sahm, W. Swyzen, *Biotechnol. Bioeng. Symp.* **14** (1984) 311.
[190] S. Bringer-Meyer, K.-L. Schimz, H. Sahm, *Arch. Microbiol.* **146** (1986) 105.
[191] B. Bräu, H. Sahm, *Arch. Microbiol.* **144** (1986) 296.
[192] D. Rose, *Process Biochem.* **3** (1976) 10–12, 36.
[193] A. Duvnjak, N. Kosaric, S. Kliza, *Biotechnol. Bioeng.* **24** (1982) 2297–2308.
[194] N. Kosaric, A. Wieczorek, G. P. Cosentino, R. J. Magee et al.: "Ethanol Fermentation," in H. J. Rehm, G. Reed (eds.): *Biotechnology*, vol. 3, Verlag-Chemie, Weinheim 1983, pp. 257–385.
[195] T. K. Ghose, R. D. Tyagi, *Biotechnol. Bioeng.* **21** (1979) 1387–1400.
[196] E. J. Del Rosario, K. J. Lee, P. L. Rogers, *Biotechnol. Bioeng.* **21** (1979) 1447–1482.
[197] O. Rahn, *Growth* **16** (1952) 59.
[198] G. R. Cysewski, C. R. Wilke, *Biotechnol. Bioeng.* **20** (1978) 1421–1444.
[199] V. L. Yarovenko: "Theory and Practice of Continuous Cultivation of Microorganisms in Alcoholic Production," in A. Fiechter, T. K. Ghose, N. Glakenbrough (eds.): *Advances in Biochemical Engineering*, vol. **9**, Springer-Verlag, Berlin-Heidelberg-New York 1978, pp. 1–30.
[200] G. R. Cysewski, C. R. Wilke, *Biotechn. Bioeng.* **19** (1977) pp. 1125–1143.
[201] N. Kosaric, Z. Duvnjak, A. Wieczorek in, Foundation for Biotechnological and Industrial Fermentation Research, [126] vol. **4**, pp. 101–120.
[202] T. J. Walsh, H. R. Bungay, *Biotechnol. Bioeng.* **21** (1979) 1081–1084.
[203] F. Jirmann, U. D. Runkel, *Brauwelt* **107** (1967) 1453.
[204] J. Hough et al., *J. Inst. Brew.* **68** (1962) 478.
[205] C. B. Netto, A. Destrauhaut, G. Goma, *Biotechnol. Lett.* **7** (1985) no. 5, 355–360.
[206] I. G. Prince, J. P. Barford, *Biotechnol. Lett.* **4** (1982) 263–268.
[207] D. M. Comberbach, J. D. Bu'Lock, *Biotechnol. Lett.* (1984) no. 2, 129–134.
[208] N. Kosaric, A. Wieczorek, Z. Duvnjak: *Proceedings of the First Bioenergy Specialists' Meeting on Biotechnology*, Waterloo, Ontario, Oct. 14–18, 1984, pp. 262–272.
[209] R. N. Greenshields, E. L. Smith, *Process Biochem.* **9** (1974) no. 3, 11–17, 28.
[210] S. J. Pirt: *Principles of Microbe and Cell Cultivation*, Blackwell Scientific Publications, Oxford, 1975, pp. 218–221.

[211]  A. Margaritis, C. R. Wilke, *Biotechnol. Bioeng.* **20** (1978) 709–726.
[212]  A. Margaritis, C. R. Wilke, *Biotechnol. Bioeng.* **20** (1978) 727–753.
[213]  M. Cheryvan, M. Mehaia, *Process Biochem.* **12** (1984) 204–208.
[214]  M. H. V. Mulder, C. A. Smolders, *Process Biochem.* **4** (1986) 35–39.
[215]  A. Ramalingham, R. K. Finn, *Biotechnol. Bioeng.* **19** (1977) 583–589.
[216]  B. Maiorella, H. W. Blanck, C. R. Wilke: "Rapid Ethanol Production via Fermentation," Univ. Calif., Lawrence Berkley Lab. Report 10 219, *AIChE 72nd National Meeting*, San Francisco, Calif., Nov. 19, 1979.
[217]  E. K. Pye, A. E. Humphrey, "The Biological Production of Liquid Fuels from Biomass," Univ. Penn. Interim Report to DOE, Washington D.C., Jun.–Aug. 1979, p. 79, Task 7.
[218]  M. Minier, G. Goma, *Biotechnol. Lett.* **3** (1981) 405–408.
[219]  E. K. Pye, A. E. Humphrey: "Production of Liquid Fuels from Cellulosic Biomass," *Proc. 3rd. Ann. Biomass Energy Systems Conf.*, DOE, Solar Energy Res. Inst. Golden, Colo., Jun. 5, 1980, p. 69.
[220]  F. B. Kolot, *Process Biochem.* **2** (1984) 7–13.
[221]  D. Williams, D. M. Munnecke, *Biotechnol. Bioeng.* **23** (1981) 1813–1825.
[222]  M. Wada, J. Kato, I. Chibata, *Eur. J. Appl. Microbiol. Biotechnol.* **11** (1979) 67.
[223]  Y. Linko, P. Linko, *Biotechnol. Lett.* **3** (1981) 21–26.
[224]  T. K. Ghose, K. K. Bandyopadhyay, *Biotechnol. Bioeng.* **22** (1980) 1489–1496.
[225]  J. L. Vega, J. D. O'Malley, E. C. Clausen, J. L. Gaddy, *Biotechnol. Bioeng. Symp.* **15** (1985) 263.
[226]  M. C. Dale, M. R. Okos, P. C. Wankat, *Biotechnol. Bioeng.* **27** (1985) 932–942.
[227]  A. J. Dauglis, N. H. Brown, W. R. Cluett, D. B. Dunlop, *Biotechnol. Lett.* **3** (1981) 651–656.
[228]  R. D. Tyagi, T. K. Ghose, *Biotechnol. Bioeng.* **24** (1982) 781–795.
[229]  W. Halwacks, C. Wandrey, K. Schugerl, *Biotechnol. Bioeng.* **20** (1978) 541–554.
[230]  R. A. Nathan, "Fuels from Sugar Crops," DOE Technical Information Center, Oak Ridge, Tenn., 1978.
[231]  J. L. Prouty in [107] p. 57.
[232]  H. W. Ockerman: *Source Book of Food Scientists*, AVI Publishing Co., Westport, Conn., 1978.
[233]  B. P. Baker: "The Availability, Composition and Properties of Cane Molasses," in E. Sinda, E. Parkkinen (eds.): *Problems with Molasses in the Yeast Industry, Symposium*, Aug. 31–Sept. 1, 1979, Helsinki, Finland, p. 126.
[234]  L. R. Lindeman, C. Rocchicciolo, *Biotechnol. Bioeng.* **21** (1979) 1107.
[235]  K. Rosen, *Process Biochem.* **13** (1978) no. 5, 25.
[236]  Alcon Biotechnology, London, 1980.
[237]  *Chem. Eng.* **25** (1980) Aug. 11.
[238]  W. M. Ingledew: *Proc. Canpac '81 Conf.*, Winnipeg, Manitoba, Oct. 1981.
[239]  R. A. M. Ginuis: *Beet Sugar Technology*, 3rd ed., The Beet Sugar Development Foundation, Fort Collins, Colo., 1982.
[240]  J. K. Paul, *Ethyl Alcohol Production and Use as a Motor Fuel*, Noyes Data Corp., Park Ridge, N.J., 1979.
[241]  W. B. Earl, W. A. Brown: *3rd Int. Symp. Alcohol Fuels Technology*, Asilomar, Calif., May 1979.
[242]  A. A. Al-Talibi, N. D. Benjamin, A. R. Abboud, *Nahrung* **19** (1975) 335.
[243]  P. K. Agrawal: *Proc. 44th Am. Conv. Sugar Technologists Assn.*, Kanpur, India, 1980.
[244]  K. Nand, S. Srikanta, V. Screenivasa-Purthy, *J. Food Sci. Technol.* **14** (1977) 80.
[245]  J. A. Bassham: *2nd Int. Symp. Nonconventional Energy*, Trieste, Italy, Jun. 1981.
[246]  P. S. Nobel in [107] p. 13.

[247] W. A. Scheller: Fermentation in Cereal Processing," *61st National Meeting American Assn. Cereal Chemists*, New Orleans, Oct. 1976.
[248] Alltech, Lexington, Ky. 1981.
[249] E. L. Core: *Plant Taxonomy*, Prentice Hall, Englewood Cliffs, N. J., 1955, p. 346.
[250] S. K. Chan, Investigation at the Federal Experiment Station, Serdang, Malaysia, 1969.
[251] T. J. B. De Menezes, T. Arakaki, P. De Lamo, A. Sales, *Biotechnol. Bioeng.* **20** (1978) 555.
[252] E. A. Jackson, *Proc. Biochem.* **11** (1976) no. 5, 29.
[253] T. A. McClure, M. F. Arthur, S. Kresovich, D. A. Scantland in [107] p. 123.
[254] T. Schormüller: *Lehrbuch der Lebensmittelchemie*, Springer-Verlag, Berlin-Heidelberg 1974, p. 506.
[255] N. Kosaric, G. P. Cosentino, A. Wieczorek, Z. Duvnjak, *Biomass* **5** (1984) 1–36.
[256] F. Schneider, F. W. Conti, *Stärke* **8** (1956) 269.
[257] N. Kosaric, A. Wieczorek, Z. Duvnjak, S. Kliza: *Bioenergy Research and Development Seminar*, Winnipeg, Canada, Mar. 29–31, 1982.
[258] J. P. Holdren, P. R. Ehrlich, *Am. Sci.* **62** (1974) 282.
[259] R. H. Whittaker: *Communities and Ecosystems*, MacMillan, New York 1970.
[260] M. Slesser, C. Lewis: "Biological Energy Resources," E & FN Spon Ltd., London, 1979.
[261] T. K. Ghose, P. Ghosh, *J. App. Chem. Biotechnol.* **28** (1978) 309.
[262] C. D. Callihan, J. E. Clemmer in A. H. Rose (ed.): *Microbial Biomass*, Academic Press, New York 1979, p. 271.
[263] E. Adler, *Wood Sci. Technol.* **11** (1977) 169.
[264] M. Wayman in [107]
[265] L. Spano, T. Tassinari, D. D.Y. Ryu, A. Allen et al., *29th Canadian Chem. Eng. Conf.*, Sarnia, Canada, Oct. 1979.
[266] A. E. Hokanson, R. Katzen, *Chem. Eng. Prog.* **74** (1978) no. 1, 67.
[267] J. Yu, S. F. Miller, *Ind. Eng. Chem. Prod. Res. Dev.* **19** (1980) 237.
[268] G. L. Foutch, G. L. Margruder, J. L. Gaddy: "Agricultural Energy," *Biomass Energy Crop Production*, vol. **2**, American Society of Agricultural Engineers, St. Joseph, Mich., 1980, p. 299.
[269] A. Porteous, *Pap. Trade J.* **156** (1972) no. 6, 30.
[270] R. W. Detroy, C. W. Hesseltine, *Process Biochem.* **13** (1978) no. 9, 2.
[271] A. J. Forage, R. C. Righelato, in [262] p. 289.
[272] E. O. Ericsson, *Chem. Eng. Prog.* **43** (1947) no. 4, 165.
[273] N. Kosaric, Y. J. Asher: "The Utilization of Cheese Whey and Its Components," in A. Fiechter (ed.): *Advances in Biochemical Engineering Biotechnology*, **32**, Springer-Verlag, Berlin-Heidelberg 1985, pp. 25–60.
[274] S. M. Drews, *Ber. Landwirtsch. Sonderh.* **192** (1975) 599.
[275] V. S. O'Leary, R. Green, B. C. Sullivan, V. H. Holsinger, *Biotechnol. Bioeng.* **19** (1977) 1019.
[276] *Chem. Eng. News* **56** (1978) no. 17, 21.
[277] S. Elias, *Food Eng.* **51** (1979) no. 1, 99.
[278] L. Reesen, *Dairy Ind. Int.* **43** (1978) no. 1, 9.
[279] E. T. Reese, *Biotechnol. Bioeng. Symp.* **5** (1975) 77.
[280] G. T. Austin: *Shreve's Chemical Process Industries*, 5th ed., McGraw-Hill, New York 1984, pp. 581–590.
[281] CORA Engineering: *Production of Ethanol from Renewable Resources*, Conceptual Engineering Manual, Chur, Switzerland, 1981, rev. 1986.
[282] R. Katzen: "Low Energy Distillation Systems," *Bio-Energy Conference*, Atlanta 1985.

[283] R. Katzen: "Large Scale Ethanol Production Facilities," *Bio-Energy '84 World Conference*, Gothenburg, Sweden, 1984.

[284] J. M. Paturau: *By-Products of the Cane Sugar Industry*, Elsevier, Amsterdam 1975, p. 183.

[285] N. Kosaric et al. in H. J. Rehm, G. Reed (eds.): *Biotechnology*, vol. **3,** Verlag Chemie, Weinheim 1983, pp. 345–358.

[286] Technipetrol, "Development and Progress in Ethylalcohol Technology," Rome 1982.

[287] U. Tegtmeier, *Biotechnology Letters* **7** (1985) no. 2, 129–134.

[288] *Chem. Eng.* **88** (1981) no. 11, 29.

[289] *Chem. Eng.* **81** (1980) no. 18, 19.

[290] *Chem. Eng.* **92** (1985) no. 10, 14.

[291] M. A. Scharf: "Prism Separators Remove Water from Ethanol," *Monsanto Newsletter*, Jun. 1986, Monsanto Europe, Brussels.

[292] *Chem. Eng.* **88** (1981) no. 19, 17.

[293] *Chem. Eng.* **87** (1980) no. 23, 103.

[294] J. Kaschemekat, J. Barbknecht, K. W. Böddeker, *Chem. Ing. Tech.* **58** (1986) no. 9, 740–742.

[295] Vogelbusch Co.: *Alcohol Pervaporation*, Information 2/84, Vienna 1984.

[296] Paterson Candy International, "Reverse Osmosis Concentration and Ultrafiltration Separation Applications in Biochemical Processes," Whitechurch, United Kingdom, 1985.

[297] *Chem. Week* (1979) Jun. 6, 42–44.

[298] A. Chauvel et al.: *"Procédés de pétrochimie – Charactéristiques techniques et économiques,"* vol. **2,** Editions Technip, Paris 1980.

[299] J. A. Roels et al.: "Biotechnology and Base Chemicals," *Int. Conf. on Biomass*, Venice, Mar. 1985.

[300] CORA Engineering: *Production of Ethanol from Renewable Resources*, Conceptual Engineering Manual, Chur, Switzerland, 1981, rev. 1986.

[301] S. Williams (ed.): *Official Methods of Analysis of the Assn. of Official Analytical Chemists*, 14th ed., Assn. of Official Analytical Chemists, Arlington, Va., 1984.

[302] M. A. Amerine, S. C. Ough: *Methods for Analysis of Musts and Wines*, J. Wiley & Sons, New York 1980.

[303] *Pure Appl. Chem.* **17** (1968) 273–312.

[304] "Petroleum Products, Lubricants and Fossil Fuels," *1985 Annual Book of ASTM Standards*, vol. 05.01 (1), ASTM, Philadelphia 1985, D56–D1660.

[305] T. S. Ma, R. E. Lang: "General Principles," *Quantitative Analysis of Organic Mixtures,* J. Wiley & Sons, New York 1979, Part 1, pp. 264–268.

[306] H. W. Zimmermann, *Am. J. Enol. Vitic.* **14** (1963) 205–213.

[307] H. Rebelein, *Allg. Dtsch. Weinfachztg.* **107** (1971) 590–594.

[308] V. C. Mehlenbacher: "Determination of Hydroxyl Groups," in J. Mitchell, Jr., I. M. Kolthoff, E. S. Proskaner, A. Weissberger (eds.): *Organic Analysis*, vol. 1, Interscience, New York 1953, pp. 1–65.

[309] M. Stiller, *Anal. Chim. Acta* **25** (1961) 85–89.

[310] S. P. Colowichk, N. O. Kaplan, in D. B. McCormick, L. D. Wright (eds.): *Methods in Enzymology*, vol. **66**, Academic Press, New York 1980, pp. 41–42.

[311] G. G. Guilbault: *Analytical Uses of Immobilized Enzymes*, Marcel Dekker, New York 1984, pp. 112–230.

[312] *Chem. Eng. News* **55** (1977) no. 2, 12.

[313] Alcohol Production from Biomass in the Developing Countries, World Bank, Washington, D.C., Sept. 1980.

[314] Statistical Release, Dept. of Treasury, Bureau of Alcohol, Tobacco and Firearms, Washington, D.C., Jun. 16, 1982.
[315] United States Department of Agriculture, Foreign Agricultural Service, Foreign Agricultural Circular FS 1–86, May 1986.
[316] H. Rothman, R. Greenshield, R. F. Calte: *Energy from Alcohol: The Brazilian Experience*, Univ. Press of Kentucky, Lexington, Ky., 1983, pp. 75–85.
[317] "Ethanol – Chemical Profile," *Chem. Mark. Rep.* (1985), Feb. 25.
[318] R. Katzen: "Large Scale Ethanol Production Facilities,"*Bio-Energy '84 World Conference*, Gothenburg, Sweden 1984.
[319] CORA Engineering: *Production of Ethanol from Renewable Resources*, Conceptual Engineering Manual, Chur, Switzerland, 1981, rev. 1986.
[320] H. F. Smyth Jr., *J. Ind Hyg. Toxicol.* **23** (1941) 253.
[321] E. T. Kimura, D. M. Ebert, P. W. Dodge, *Toxicol. Appl. Pharmacol.* **19** (1971) 699.
[322] W. S. Spector (ed.): "Acute Toxicities," *Handbook of Toxicol.*, vol. 1, Saunders, Philadelphia-London 1956, p. 128.
[323] J. C. Munch, E. W. Schwartze, *J. Lab Clin. Med.* **10** (1925) 985.
[324] W. L. Guess, *Toxicol. Appl. Pharmacol.* **16** (1970) 382.
[325] L. Phillips, M. Steinberg, H. Maibach, W. Akers *Toxicol. Appl. Pharmacol.* **21** (1972) 369.
[326] K. B. Lehmann, F. Flury: *Toxicologie und Hygiene der technischen Lösungsmittel*, Springer-Verlag, Berlin 1938, p. 152.
[327] A. Loewy, R. van der Heyde, *Biochem. Z.* **86** (1918) 125.
[328] J. L. Hall, D. T. Rowlands, *Am. J. Pathol.* **60** (1970) 153.
[329] S. C. Vasder, R. N. Chakravarti, D. Subrahmanyam, A. C. Jain et al., *Cardiovasc. Res.* **9** (1975) 134.
[330] A. M. Skosyreva, *Akush. Ginekol. (Sofia)* **4** (1973) 15.
[331] W. J. Tze, M. Lee, *Nature (London)* **257** (1975) 479.
[332] F. M. Badr, R. S. Badr, *Nature (London)* **253** (1975) 134.
[333] L. Machemer, D. Lorke, *Mutat. Res.* **29** (1975) 209.
[334] J. McCann, E. Choi, E. Yamasaki, B. Ames, *Proc. Nat. Acad. Sci. USA* **72** (1975) 5135.
[335] L. F. Meisner, S. L. Juhorn, *Acta Cytologica* **16** (1972) 41.
[336] V. K. Rowe, S. B. McCollister in G. D. Clayton, F. E. Clayton (eds.): *Patty's Industrial Hygiene and Toxicology*, 3rd ed., vol. **2 C,** Wiley-Interscience, New York 1982, p. 4549.
[337] R. Derache, *Int. Encycl. Pharmacol. Ther.* **20** (1970) 507.
[338] M. K. Roach, W. N. Reese, P. J. Creaven, *Biochem. Biophys. Res. Commun.* **36** (1969) 596.
[339] E. Umdagard, *C. R. Trav. Lab. Carlsberg* **22** (1938) 333.
[340] E. Jacobsen, *Pharmacol. Rev.* **4** (1952) 107.
[341] O. Farsander, *Int. Encycl. Pharmacol. Ther.* **20** (1967) 117.
[342] E. Petri in F. Hencke, O. Lubarsch (eds.): *Handbuch der Speziellen Pathologischen Anatomie und Histologie*, vol. **10,** Springer-Verlag, Berlin 1930, p. 226.

# Ethanolamines and Propanolamines

Hans Hammer, BASF Aktiengesellschaft, Ludwigshafen/Rhein, Federal Republic of Germany (Chap. 2)

Wolfgang Körnig, BASF Aktiengesellschaft, Ludwigshafen/Rhein, Federal Republic of Germany (Chaps. 3 and 5–7)

Theodor Weber, BASF Aktiengesellschaft, Ludwigshafen/Rhein, Federal Republic of Germany (Chap. 4)

Heinz Kieczka, BASF Aktiengesellschaft, Ludwigshafen/Rhein, Federal Republic of Germany (Chap. 8)

| | | | | | |
|---|---|---|---|---|---|
| 1. | Introduction | 2149 | 4.1.1. | Physical Properties | 2164 |
| 2. | Ethanolamines | 2150 | 4.1.2. | Chemical Properties | 2165 |
| 2.1. | Properties | 2150 | 4.2. | Production | 2166 |
| 2.1.1. | Physical Properties | 2150 | 4.3. | Quality Specifications | 2168 |
| 2.1.2. | Chemical Properties | 2151 | 4.4. | Uses | 2169 |
| 2.2. | Production | 2154 | 5. | N-Alkylated Propanolamines and 3-Alkoxypropylamines | 2170 |
| 2.3. | Quality Specifications | 2156 | 5.1. | Properties | 2170 |
| 2.4. | Uses | 2156 | 5.1.1. | Physical Properties | 2170 |
| 2.5. | Economic Aspects | 2158 | 5.1.2. | Chemical Properties | 2174 |
| 3. | N-Alkylated Ethanolamines | 2158 | 5.2. | Production | 2174 |
| 3.1. | Properties | 2159 | 5.3. | Quality Specifications | 2175 |
| 3.1.1. | Physical Properties | 2159 | 5.4. | Uses and Economic Aspects | 2175 |
| 3.1.2. | Chemical Properties | 2159 | 6. | Storage and Transportation | 2176 |
| 3.2. | Production | 2161 | 7. | Environmental Protection | 2176 |
| 3.3. | Quality Specifications | 2162 | 8. | Toxicology and Occupational Health | 2177 |
| 3.4. | Uses and Economic Aspects | 2162 | 9. | References | 2180 |
| 4. | Isopropanolamines | 2163 | | | |
| 4.1. | Properties | 2164 | | | |

# 1. Introduction

Amino alcohols have been prepared industrially since the 1930s. However, large-scale production started only after 1945, when alkoxylation with ethylene oxide and propylene oxide replaced the older chlorohydrin route. In industry, amino alcohols are usually designated as alkanolamines. Ethanolamines (aminoethanols) and propanolamines (aminopropanols) are by far the most important compounds of this group. Both are used widely in the manufacture of surfactants and in gas purification [1]–[6].

# 2. Ethanolamines

Monoethanolamine [141-43-5] (**1**) (MEA; 2-aminoethanol), diethanolamine [111-42-2] (**2**) (DEA; 2,2′-iminodiethanol), and triethanolamine [102-71-6] (**3**)(TEA; 2,2′,2″-nitrilotriethanol) can be regarded as derivatives of ammonia in which one, two, or three hydrogen atoms have been replaced by a –CH$_2$CH$_2$–OH group.

$$H_2N-CH_2CH_2-OH \quad\quad HN\!\!\begin{array}{l}{-CH_2CH_2-OH}\\{-CH_2CH_2-OH}\end{array}$$
$$\mathbf{1} \quad\quad\quad\quad \mathbf{2}$$

$$N\!\!\begin{array}{l}{-CH_2CH_2-OH}\\{-CH_2CH_2-OH}\\{-CH_2CH_2-OH}\end{array}$$
$$\mathbf{3}$$

Ethanolamines were prepared in 1860 by WURTZ from ethylene chlorohydrin and aqueous ammonia [7]. It was only toward the end of the 19th century that an ethanolamine mixture was separated into its mono-, di-, and triethanolamine components; this was achieved by fractional distillation.

Ethanolamines were not available commercially before the early 1930s; they assumed steadily growing commercial importance as intermediates only after 1945, because of the large-scale production of ethylene oxide. Since the mid-1970s, production of very pure, colorless triethanolamine in industrial quantities has been possible. All ethanolamines can now be obtained economically in very pure form.

The most important uses of ethanolamines are in the production of emulsifiers, detergent raw materials, and textile chemicals; in gas purification processes; and in cement production, as milling additives. Monoethanolamine is an important feedstock for the production of ethylenediamine and ethylenimine.

## 2.1. Properties

### 2.1.1. Physical Properties

Monoethanolamine and triethanolamine are viscous, colorless, clear, hygroscopic liquids at room temperature; diethanolamine is a crystalline solid. All ethanolamines absorb water and carbon dioxide from the air, and are infinitely miscible with water and alcohols. The freezing points of all ethanolamines can be lowered considerably by the addition of water. Some physical properties of ethanolamines are listed in Table 1.

## 2.1.2. Chemical Properties

Because of their basic nitrogen atom and the hydroxyl group, ethanolamines have chemical properties resembling those of both amines and alcohols. They form salts with acids, and the hydroxyl group permits ester formation. When mono- and diethanolamine react with organic acids, salt formation always takes place in preference to ester formation. With weak inorganic acids, e.g., $H_2S$ and $CO_2$, thermally unstable salts are formed in aqueous solution. This reaction of ethanolamines is the basis for their application in the purification of acidic natural gas, refinery gas, and synthesis gas [1], [8]. In the absence of water, monoethanolamine and diethanolamine react with $CO_2$ to form carbamates:

$$2\ HO\text{-}CH_2CH_2\text{-}NHR + CO_2 \longrightarrow HO\text{-}CH_2CH_2\text{-}NR\text{-}COOH \cdot RHN\text{-}CH_2CH_2\text{-}OH$$
$$R = H \text{ or } -CH_2CH_2\text{-}OH$$

Triethanolamine does not form a carbamate.

By reaction with ammonia in the presence of hydrogen, the hydroxyl group of monoethanolamine can be replaced by an amine group to form ethylenediamine [*107-15-3*] and piperazine [*110-85-0*] [9], [10]:

$$NH_2\text{-}CH_2CH_2\text{-}OH + NH_3 \longrightarrow NH_2\text{-}CH_2CH_2\text{-}NH_2 + H_2O$$

$$2\ NH_2\text{-}CH_2CH_2\text{-}OH \longrightarrow HN\underset{}{\bigcirc}NH + 2\ H_2O$$

Considerable quantities of monoethanolamine are converted into ethylenimine (aziridine [*151-56-4*]) by adding sulfuric acid and cyclizing the hydrogen sulfate with sodium hydroxide ($\rightarrow$ Aziridines) [11], [12]:

$$NH_2\text{-}CH_2CH_2\text{-}OH \xrightarrow{H_2SO_4}$$
$$NH_2\text{-}CH_2CH_2\text{-}OSO_3H \xrightarrow{NaOH} H_2C\text{---}CH_2$$
$$\underset{H}{\overset{}{N}}$$

In a similar manner, morpholine [*110-91-8*] is formed from diethanolamine ($\rightarrow$ Amines, Aliphatic) [13]:

$$HN\begin{array}{c}CH_2CH_2\text{-}OH\\CH_2CH_2\text{-}OH\end{array} + 2\ H_2SO_4 \longrightarrow$$

$$HN\begin{array}{c}CH_2CH_2\text{-}OSO_3H\\CH_2CH_2\text{-}OSO_3H\end{array} \xrightarrow{NaOH} HN\underset{}{\bigcirc}O$$

As primary and secondary amines, monoethanolamine and diethanolamine also react with acids or acid chlorides to form amides. The amine first reacts with the acid, e.g., stearic acid, to form a salt, which can be dehydrated to the amide by heating:

# Ethanolamines and Propanolamines

**Table 1.** Physical properties of ethanolamines

| Compound | $M_r$ | mp, °C | bp (101.3 kPa), °C | ϱ (20 °C), g/cm³ | Heat of vaporization (101.3 kPa), kJ/kg | Specific heat $c_p$, kJ kg⁻¹ K⁻¹ | Cubic expansion coefficient, K⁻¹ | Viscosity (20 °C), mPa·s | $n_D^{20}$ | Surface tension (20 °C), N/m | Flash point[a], °C | Ignition temperature[b], °C | Temperature class[c] |
|---|---|---|---|---|---|---|---|---|---|---|---|---|---|
| Monoethanolamine | 61.08 | 10.53 | 170.3 | 1.0157 | 848.1 | 2.72 | $7.78 \times 10^{-4}$ | 23.2 | 1.4544 | 0.049 | 94.5 | 410 | T2 |
| Diethanolamine | 105.1 | 27.4 | 268.5 | 1.0912 (30 °C) | 638.4 | 2.73 | $5.86 \times 10^{-4}$ | 389 (30 °C) | 1.4747 | 0.0477 | 176.0 | 365 | T2 |
| Triethanolamine | 149.2 | 21.6 | 336.1 | 1.1248 | 517.8 | 2.33 | $4.82 \times 10^{-4}$ | 930 | 1.4852 | 0.0484 | 192 | 325 | T2 |

[a] According to DIN 51 758.
[b] According to DIN 51 794.
[c] According to VDE 0165. Ethanolamines do not belong to any hazard class according to VbF (regulation governing flammable liquids).

$$NH_2-CH_2CH_2-OH + C_{17}H_{35}-COOH \longrightarrow$$
$$HO-CH_2CH_2-NH_2 \cdot C_{17}H_{35}-COOH \xrightarrow{-H_2O}$$
$$HO-CH_2CH_2-NH-CO-C_{17}H_{35}$$

The action of aromatic aldehydes on monoethanolamine yields oxazolidines in accordance with the following reaction [14]:

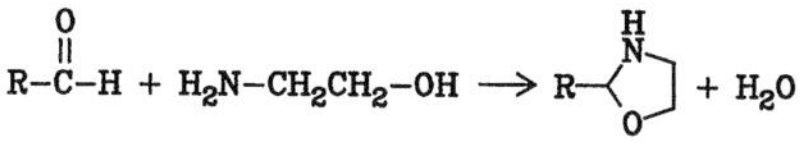

$$R-\underset{\underset{O}{\|}}{C}-H + H_2N-CH_2CH_2-OH \longrightarrow R-\overset{H}{\underset{O}{N}}\!\!\diagdown\!\!\diagup + H_2O$$

On the other hand, oxazolines are formed when ethanolamides of fatty acids are heated at fairly high temperature with simultaneous removal of water [15]:

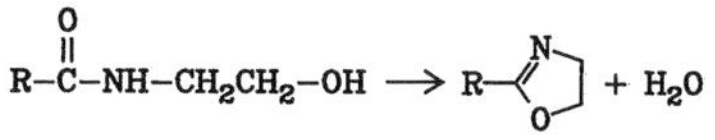

$$R-\underset{\underset{O}{\|}}{C}-NH-CH_2CH_2-OH \longrightarrow R-\overset{N}{\underset{O}{\diagdown}}\!\!\diagup + H_2O$$

Formaldehyde reacts with monoethanolamine and diethanolamine to form hydroxymethyl compounds, which can be reduced to the *N*-methyl derivatives [16] (see Section 3.2).

Monoethanolamine reacts with carbon disulfide to form 2-mercaptothiazoline [96-53-7] [17]:

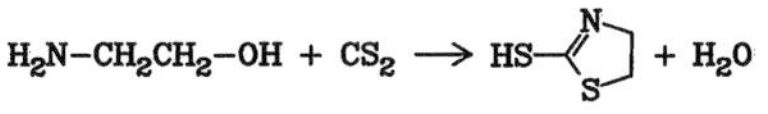

$$H_2N-CH_2CH_2-OH + CS_2 \longrightarrow HS-\overset{N}{\underset{S}{\diagdown}}\!\!\diagup + H_2O$$

The hydroxyl groups of ethanolamines can be replaced with chlorine by reaction with thionyl chloride or phosphorus pentachloride. The chloroethylamines formed are hazardous because of their skin toxicity. For example, tris(2-chloroethyl)amine has been used as a gas in warfare. Bis(2-chloroethyl)amine is obtained in good yield by reaction of diethanolamine with thionyl chloride:

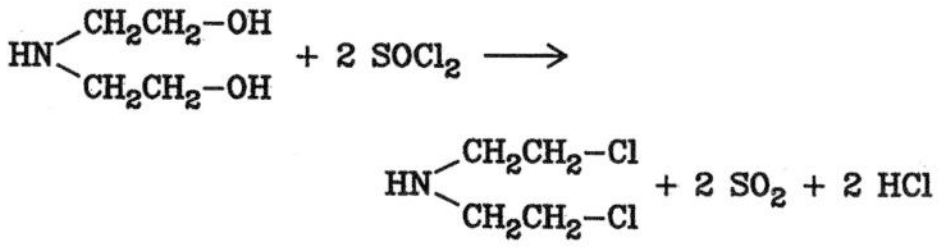

$$HN\!\!<\!\!\begin{matrix}CH_2CH_2-OH\\CH_2CH_2-OH\end{matrix} + 2\ SOCl_2 \longrightarrow$$
$$HN\!\!<\!\!\begin{matrix}CH_2CH_2-Cl\\CH_2CH_2-Cl\end{matrix} + 2\ SO_2 + 2\ HCl$$

Reaction of triethanolamine with ethylene oxide gives unstable alkaline quaternary compounds and the corresponding stable ethers. For example, undistilled triethanolamine prepared from mono- or diethanolamine always contains some triethanolamine monoglycol ether [18].

Mono- or diethanolamine can also be used as the amine component in aminoalkylation, the so-called Mannich reaction, which is very important in the biosynthesis of many alkaloids [19].

$$\text{C}_6\text{H}_5\text{OH} + \text{CH}_2\text{O} + \text{H}_2\text{N}-\text{CH}_2\text{CH}_2-\text{OH} \longrightarrow$$

$$\text{(HO-C}_6\text{H}_4\text{)}-\text{CH}_2-\text{NH}-\text{CH}_2\text{CH}_2-\text{OH} + \text{H}_2\text{O}$$

N-Nitrosamines can be formed by reacting diethanolamine or, under suitable conditions, triethanolamine with nitrous acid, nitrites, or oxides of nitrogen. Animal experiments have shown that N-nitrosamines are carcinogenic [20].

$$\text{HN}(\text{CH}_2\text{CH}_2\text{-OH})_2 + \text{HONO} \longrightarrow \text{O=N-N}(\text{CH}_2\text{CH}_2\text{-OH})_2 + \text{H}_2\text{O}$$

Monoethanolamine can form complexes with heavy-metal ions (e.g., copper, nickel, and iron); some of these complexes are water-soluble.

## 2.2. Production

Today, ethanolamines are produced on an industrial scale exclusively by reaction of ethylene oxide with excess ammonia, this excess being considerable in some cases [21].

The reaction of ethylene oxide with ammonia takes place slowly and is accelerated by water. An anhydrous procedure employs a fixed-bed catalyst consisting of an ion-exchange resin [22].

In all conventional processes, reaction takes place in the liquid phase, and the reactor pressure must be sufficiently large to prevent vaporization of ammonia at the reaction temperature. In current procedures, ammonia concentrations in water between 50 and 100%, pressures up to 16 MPa (160 bar), reaction temperatures up to 150 °C, and an excess up to 40 mol of ammonia per mole of ethylene oxide are used. The reaction is highly exothermic; the enthalpy of reaction is about 125 kJ per mole of ethylene oxide [23]. The following competing reactions occur:

$$\text{NH}_3 + \text{H}_2\text{C}\overset{\text{O}}{-\!-}\text{CH}_2 \xrightarrow{k_1} \text{H}_2\text{N}-\text{CH}_2\text{CH}_2-\text{OH}$$

$$\text{H}_2\text{N}-\text{CH}_2\text{CH}_2-\text{OH} + \text{H}_2\text{C}\overset{\text{O}}{-\!-}\text{CH}_2 \xrightarrow{k_2} \text{HN}(\text{CH}_2\text{CH}_2\text{-OH})_2$$

$$\text{HN}(\text{CH}_2\text{CH}_2\text{-OH})_2 + \text{H}_2\text{C}\overset{\text{O}}{-\!-}\text{CH}_2 \xrightarrow{k_3} \text{N}(\text{CH}_2\text{CH}_2\text{-OH})_3$$

Further possible reactions to tetraethanolammonium hydroxide or the monoglycol ether of triethanolamine do not play any role in the synthesis. The kinetics of these reactions have been studied [24].

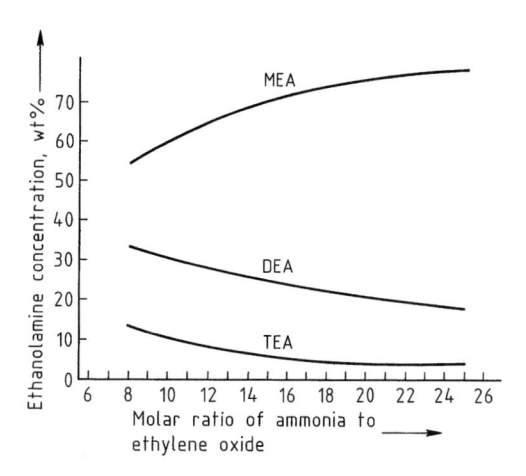

**Figure 1.** Product distribution of monoethanolamine (MEA), diethanolamine (DEA), and triethanolamine (TEA) as a function of the molar ratio of ammonia to ethylene oxide (concentration of the aqueous ammonia solution ca. 85%)

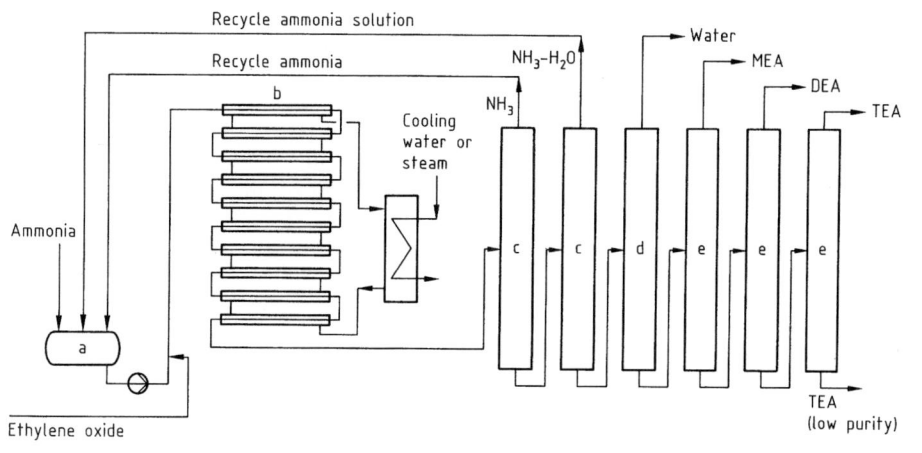

**Figure 2.** Flow sheet for the production of ethanolamines
a) Aqueous ammonia tank; b) Tubular reactor; c) Ammonia and ammonia solution columns; d) Dehydration column; e) Vacuum fractionation columns

All the reaction steps have about the same activation energy and show a roughly quadratic dependence of the reaction rate on the water content of the ammonia–water mixture used. Therefore, product composition depends solely on the molar excess of ammonia and not on water content, reaction temperature, or pressure [25]. The product distribution as a function of the molar ratio of the reactants is shown in Figure 1 [26].

Unconsumed ammonia and water fed in as a catalyst are separated from the end products in a distillation line downstream of the reactor and are recycled. In large-scale continuous single-line plants, the requirement for low energy use (i.e., operation with minimum steam consumption) determines both the transport of heat from the reactor and the design of the thermally integrated distillation line. The ethanolamine mixture is worked up in vacuum distillation columns to give the pure products (Fig. 2).

**Table 2.** Specifications of commercial ethanolamines

| Compound | Minimum purity, wt% | $\varrho$ (DIN 51 757), g/cm$^3$ | Boiling range 5–95 mL (DIN 51 751), °C | Color (DIN 53 409), APHA |
|---|---|---|---|---|
| Monoethanolamine | 99 (25 °C) | 1.01 (25 °C) | 170–172 | ≤ 10 |
| Diethanolamine | ≥98 | 1.09 (30 °C) | | ≤ 20 |
| Triethanolamine | 98 | 1.12–1.13 | | |
| Pure triethanolamine | 99 | 1.12–1.13 (25 °C) | | ≤ 30 |

Product distribution of the three ethanolamines can be controlled by appropriate choice of the ammonia:ethylene oxide ratio. A higher diethanolamine or triethanolamine content can also be obtained by recycling monoethanolamine or diethanolamine to the reactor or by reacting them with ethylene oxide in a separate unit.

For safety reasons, ethylene oxide must be metered into the ammonia stream; in the reverse procedure, ammonia or amines may cause ethylene oxide to undergo an explosive polymerization reaction.

In all large-scale processes, virtually complete conversion to the three ethanolamines, without significant formation of byproducts, is achieved. Therefore, feedstock costs are independent of the type of production process. On the other hand, manufacturing costs and, in particular, energy costs depend to a large extent on the product composition desired, plant design, and degree of thermal integration.

## 2.3. Quality Specifications

Today, all ethanolamines are prepared with > 99% purity. Water, the two other ethanolamines, and a small amount of triethanolamine glycol ether are minor components; all other impurities are in the ppm range. Table 2 gives specifications for commercial ethanolamines.

Purity is determined by gas chromatography, and the residual water content by the Karl Fischer method. After pretreatment, the N-nitrosamine content can also be determined by gas chromatography using a thermal analyzer.

## 2.4. Uses

**Surfactants.** Ethanolamines are used widely as intermediates in the production of surfactants, which have become commercially important as detergents, textile and leather chemicals, and emulsifiers [2]–[6]. Their uses range from drilling and cutting oils to medicinal soaps and high-quality toiletries [27].

Properties of ethanolamine derivatives can be varied within wide limits by appropriate choice of the ethanolamine, the acid component, and their ratio. Synergistic effects in complex emulsifier systems can be obtained by simultaneous use of different combinations of amines and acids. Ethanolamines, particularly triethanolamine which had presented problems with regard to color, are now produced in sufficient purity that no color problems arise when they are reacted with acids or heated [28].

Ethanolamine-based surfactants can be formulated as weakly basic or neutral products and are, therefore, particularly well tolerated by the skin; this is especially true for triethanolamine soaps. Moreover, they are noncorrosive and can be used on virtually all textiles without damage. Ethanolamine soaps prepared from oleic acid, stearic acid, lauric acid, or caprylic acid are constituents of many toiletries and medicinal soaps.

Ethanolamine soaps produced from fatty acids are among the most industrially important emulsifiers. They are used in cosmetics [29], polishes, shoe creams, car care products, drilling and cutting oils, and pharmaceutical ointments. Ethanolamine soaps combined with wax and resins are used as impregnating materials, protective coatings, and products for the care of textile and leather goods.

Ethanolamine soaps obtained from alkylarylsulfonic acids, preferably alkylbenzenesulfonic acids, or from alcohol sulfates are growing steadily in importance and dominate the market for household cleansers [30]. The use of linear alkyl groups instead of branched chains in these products has resulted in greater biodegradability.

Fatty acid ethanolamides and products obtained by further ethoxylation have foam-stabilizing ability and are, therefore, important additives for detergents [31], [32]. Diethanolamides obtained from coconut fatty acids and from oleic acid are used industrially. These amides were first described in 1937 [33]. Liquid detergents are based predominantly on triethanolamine [34]–[36].

In the manufacture of leather, ethanolamine-based chemicals are used for dressing, dyeing, and finishing. For paints and coatings, ethanolamines are employed both in production and in softeners and paint removers.

**Corrosion Inhibitors.** Diethanolamine and triethanolamine are important components of corrosion inhibitors, particularly in coolants for automobile engines, as well as in drilling and cutting oils. They are also employed as additives in lubricants.

When the corrosion inhibitor sodium nitrite and diethanolamine are used together, $N$-nitrosamines may form. These nitrosamines are carcinogenic; therefore, sodium nitrite and diethanolamine or triethanolamine should not be used simultaneously.

**Gas Purification.** Large amounts of ethanolamines, principally diethanolamine, are used in absorptive gas purification to remove weakly acidic components. For example, hydrogen sulfide and carbon dioxide are removed from natural gas, refinery gas, and synthesis gas [1].

**Intermediates.** A substantial portion of the monoethanolamine produced in Europe is used as a starting material in the preparation of ethylenimine and ethylenediamine.

**Table 3.** Approximate ethanolamine capacity and production in 1985

|  | Capacity, t/a | Estimated production, t/a |
| --- | --- | --- |
| United States | 300 000 | 220 000 |
| Western Europe | 220 000 | 145 000 |
| FRG | 110 000 | 80 000 |
| Southeast Asia | 55 000 | 40 000 |
| South America | 20 000 | 18 000 |
| Eastern Europe | 5 000 | 4 000 |
| Total | 600 000 | 427 000 |

Ethylenimine is converted to polyethylenimine, an important chemical in paper technology. Diethanolamine is used in the production of morpholine.

**Cement Additives.** Since the 1960s, aqueous solutions of triethanolamine and triethanolamine acetate have been used as milling additives in cement production [37]. During grinding of the clinker in ball mills, the small amount of triethanolamine added prevents agglomeration and cushioning of the grinding medium by saturating the surfaces of the freshly broken particles. Like other milling additives (e.g., various glycols), they reduce the power required for milling the clinker. Today, the milling of virtually all high-quality portland cement involves milling additives. In addition to assisting the milling process, triethanolamine also improves the flow properties and setting behavior of the cement [38].

## 2.5. Economic Aspects

Large-scale, economical production of ethanolamines from ethylene oxide and ammonia in large, single-line plants has greatly promoted their use in many industrial sectors. In 1985, world ethanolamine capacity was about 600 000 t/a (Table 3). About 50 % of world production is monoethanolamine; 30 – 35 %, diethanolamine; and 15 – 20 %, triethanolamine.

## 3. N-Alkylated Ethanolamines

N-Alkylated ethanolamines are amino alcohols that contain a secondary or tertiary nitrogen and one or two hydroxyl groups.

$$R^1R^2N-CH_2CH_2-OH$$

where $R^1$ = alkyl or H
$R^2$ = alkyl

$$R-N(CH_2CH_2-OH)_2$$

where R = alkyl

Numerous compounds with different properties and uses can be synthesized by varying the alkyl radical.

## 3.1. Properties

### 3.1.1. Physical Properties

N-Alkylated ethanolamines are generally liquid and have a faint amine-like odor. They tend to discolor on prolonged storage, especially if exposed to air, light, and temperatures above 40 °C. This tendency to discoloration can be reduced substantially by the use of additives when purifying the products by distillation, or by pretreating the crude products before distillation [39]–[41]. N-Alkylated ethanolamines are partially to readily soluble in alcohol, water, acetone, glycols, glycerol, and glycol ethers. They are only sparingly soluble or even insoluble in nonpolar solvents, e.g., aliphatic hydrocarbons. 2-Dimethylaminoethanol [108-01-0] (N,N-dimethylethanolamine) and 2-diethylaminoethanol [100-37-8] (N,N-diethylethanolamine) form azeotropes with water; N-methyldiethanolamine [105-59-9] does not form an azeotrope. All N-alkylethanolamines are hygroscopic and also absorb $CO_2$ from air. Physical properties of typical compounds are listed in Table 4.

### 3.1.2. Chemical Properties

The chemical properties of N-alkylated ethanolamines are very similar to those of unsubstituted ethanolamines (see Section 2.1.2). They form unstable salts with inorganic acids such as $CO_2$ and $H_2S$. Salts with organic acids give a neutral to weakly basic reaction. N-Alkylated ethanolamines also react with acids or acid derivatives to give the corresponding esters. Important products are dimethylethanolamine acrylate [2439-35-2] and diethylethanolamine acrylate [2426-54-2], which are used as intermediates in the synthesis of flocculating agents.

N-Alkylated diethanolamines can be converted to the corresponding morpholine derivatives by reaction with sulfuric acid at an elevated temperature [42]. For example, N-methyldiethanolamine reacts with sulfuric acid to give N-methylmorpholine [109-02-4] in good yield [43]:

$$CH_3-N(CH_2CH_2-OH)_2 \xrightarrow{H_2SO_4} CH_3-N\underset{}{\bigcirc}O$$

Secondary N-alkylethanolamines also react with acids to form acid amides. The reaction of N-alkylethanolamines with ethylene oxide to give longer chain alkoxy derivatives has low yields.

# Ethanolamines and Propanolamines

Table 4. Physical properties of -alkylethanolamines

| Compound | CAS registry number | $M_r$ | mp, °C | bp, °C | $\varrho$ (20 °C), g/cm³ | Cubic expansion coefficient, K⁻¹ | Heat of vaporization, kJ/kg | Specific heat, kJ kg⁻¹ K⁻¹ | Specific electrical conductivity, $\Omega^{-1}$ cm⁻¹ | Viscosity (20 °C), mPa·s | $n_D^{20}$ | Surface tension (20 °C), mN/m | Flash point, °C | Ignition temperature, °C | Temperature class |
|---|---|---|---|---|---|---|---|---|---|---|---|---|---|---|---|
| Monomethylethanolamine $CH_3NH-CH_2CH_2OH$ | [109-83-1] | 75.1 | ca. −3 | 159 | 0.939–0.942 | $7.8 \times 10^{-4}$ | 564.4 | 2.55 | $3.33 \times 10^{-6}$ (25 °C) | 13.0 | 1.4389 | 34.4 (22 °C) | ca. 74 | 350 | T2 |
| Methyldiethanolamine $CH_3N(CH_2CH_2OH)_2$ | [105-59-9] | 119.2 | ca. −21 | 247 | 1.038–1.041 | $7.5 \times 10^{-4}$ | 418.7 | 1.72 | $8.1 \times 10^{-7}$ | 101 | 1.4694 | 40.9 (20.3 °C) | 137 | 265 | T3 |
| Dimethylethanolamine $(CH_3)_2N-CH_2CH_2OH$ | [108-01-0] | 89.1 | −70[a] | 134 | 0.887 | $1.35 \times 10^{-3}$ | 454.3 | 2.30 | $3.8 \times 10^{-7}$ | 3.85 | 1.4296 | 27.1 (24.5 °C) | ca. 38 | 235 | T3 |
| Diethylethanolamine $(C_2H_5)_2N-CH_2CH_2OH$ | [100-37-8] | 117.2 | −70[a] | 162 | 0.883–0.889 | $1.07 \times 10^{-3}$ | 383.9 | 2.42 | $1.1 \times 10^{-3}$ | 5 | 1.4417 | 29.2 | ca. 46 | 260 | T3[b] |
| n-Butylethanolamine $C_4H_9NH-CH_2CH_2OH$ | [111-75-1] | 117.2 | −2.1 | 199 | 0.8917 | $7.6 \times 10^{-4}$ | 397.7 | 2.45 | $2.9 \times 10^{-7}$ | 19.7 | 1.4435 | 29.9 (22 °C) | 85 | 265 | T3 |
| n-Butyldiethanolamine $C_4H_9N(CH_2CH_2OH)_2$ | [102-79-4] | 161.2 | −45[a] | ca. 270 | 0.970 | $7.7 \times 10^{-4}$ | 329.5 | 2.26 | $6.3 \times 10^{-7}$ | 75.9 | 1.462 | 33.9 (22 °C) | 130 | 260 | T3 |
| Dibutylethanolamine $(C_4H_9)_2N-CH_2CH_2OH$ | [102-81-8] | 173.3 | −75[a] | 222–234 | 0.860 | $8.1 \times 10^{-4}$ | 277.6 | 2.32 | $9.5 \times 10^{-7}$ | 7.4 | 1.4444 | 26.3 (22 °C) | 96 | 165 | T4[c] |
| Cyclohexylethanolamine $C_6H_{11}NH-CH_2CH_2OH$ | [2842-38-8] | 143.2 | 37.5 | 235 | 0.9797 | $7.6 \times 10^{-4}$ | 327.2 | 2.45 | $2.0 \times 10^{-8}$ | 317 | 1.4860 | 39.5 (25.1 °C) | 122 | 270 | T3 |
| Cyclohexyldiethanolamine $C_6H_{11}N(CH_2CH_2OH)_2$ | [4500-29-2] | 187.3 | | 297 | 1.0339 | $6.6 \times 10^{-4}$ | 319.0 | 2.30 | $6.6 \times 10^{-4}$ | 815 | 1.4930 | 40.1 | 172 | 255 | T3 |
| 1-(2-Hydroxyethyl)-piperazine (4) | [103-76-4] | 130.2 | −24 | 112 (1.33 kPa) | 1.0592 | $7.5 \times 10^{-4}$ | 411.1 | 2.00 | $6.1 \times 10^{-7}$ | 1040 | 1.110 (40 °C) | 38.3 | 135 | 280 | T3 |
| 4-(2-Hydroxyethyl)-morpholine (5) | [622-40-2] | 131.2 | 1.5–2 | 224 | 1.0714 | $8.1 \times 10^{-4}$ | 342.7 | 1.91 | $4.7 \times 10^{-5}$ | 26.6 | 1.4783 | 40.8 | 113 | 205 | T3 |
| Hydroxyethylaniline (6) | [122-98-5] | 137.2 | −30[a] | 282–287 | 1.0954 | $1.14 \times 10^{-3}$ | 439.2 | 2.08 | | 102 | 1.5796 | | ca. 148 | 410 | T2 |
| Ethylhydroxyethyl-aniline (7) | [92-50-2] | 165.2 | 37 | 270 | 1.030 (40 °C) | $7.3 \times 10^{-4}$ | | 2.14 | | 86 | | | 139 | 335 | T2 |

[a] Pour point. [b] Hazard class A II (VbF). [c] Hazard class A III (VbF).

$HN\!\!\begin{array}{c}\diagup\!\!\diagdown\\ \diagdown\!\!\diagup\end{array}\!\!N-CH_2CH_2-OH$  $O\!\!\begin{array}{c}\diagup\!\!\diagdown\\ \diagdown\!\!\diagup\end{array}\!\!N-CH_2CH_2-OH$  $\langle\!\!\!\bigcirc\!\!\!\rangle-NH-CH_2CH_2-OH$  $\langle\!\!\!\bigcirc\!\!\!\rangle-N\!\!\begin{array}{c}-CH_2CH_3\\ -CH_2CH_2-OH\end{array}$

4  5  6  7

## 3.2. Production

Industrially, N-alkylated ethanolamines are produced almost exclusively by batchwise or continuous reaction of primary, secondary, or tertiary amines with ethylene oxide [44]–[46]. For safety reasons, ethylene oxide must be added to the amine, never vice versa (see Section 2.2). Depending on the product involved, reaction temperature varies from 50 to 170 °C and pressure from 0.2 to 4 MPa (2 to 40 bar). Water accelerates the reaction.

When *primary amines* are used, a mixture of N-alkylethanolamines and N-alkyldiethanolamines is always formed:

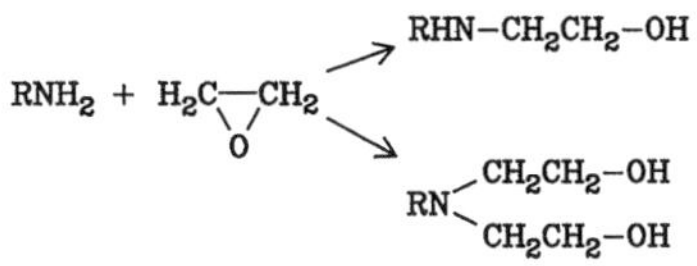

$$RNH_2 + H_2C\overset{O}{-}CH_2 \longrightarrow RHN-CH_2CH_2-OH$$
$$\longrightarrow RN(CH_2CH_2-OH)_2$$

where R = alkyl

The larger the excess of amine, the lower is the proportion of N-alkyldiethanolamines. If N-alkylethanolamines are desired, a continuous process is usually preferred for economic reasons. Excess amine is distilled from the reaction product and recycled.

With *secondary amines*, only one product is formed:

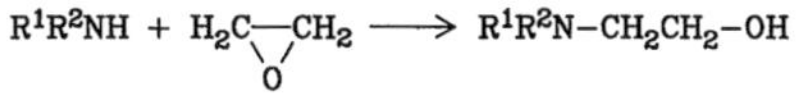

$$R^1R^2NH + H_2C\overset{O}{-}CH_2 \longrightarrow R^1R^2N-CH_2CH_2-OH$$

where $R^1$, $R^2$ = alkyl

*Trialkylammonium chlorides* yield the corresponding 2-hydroxyethylammonium chlorides. One of the commercially most important salts is choline chloride [67-48-1] (→ Choline):

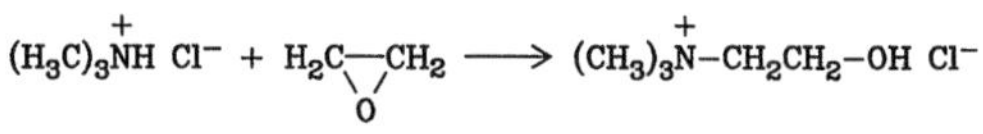

$$(H_3C)_3\overset{+}{N}H\ Cl^- + H_2C\overset{O}{-}CH_2 \longrightarrow (CH_3)_3\overset{+}{N}-CH_2CH_2-OH\ Cl^-$$

Choline chloride can also be prepared from trimethylamine and ethylene chlorohydrin, but this route has no commercial significance.

Another common method of preparing N-alkylethanolamines is by N-alkylation of ethanolamines [47]. Methyl derivatives can also be obtained by reacting monoethanolamine or diethanolamine, or their homologues, with formaldehyde and then reducing the hydroxymethyl compound:

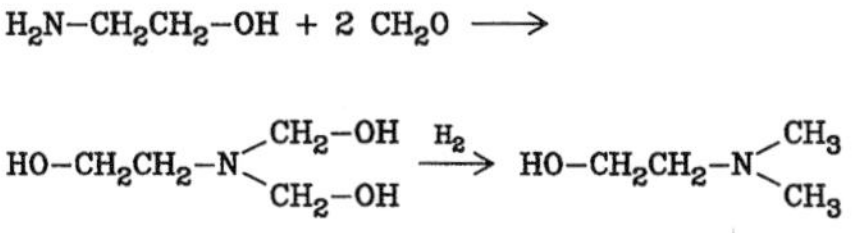

$$H_2N-CH_2CH_2-OH + 2\ CH_2O \longrightarrow$$

$$HO-CH_2CH_2-N(CH_2-OH)_2 \xrightarrow{H_2} HO-CH_2CH_2-N(CH_3)_2$$

Pure components are obtained from the crude products by distillation.

**Table 5.** Specifications for commercial N-alkylethanolamines

| Compound | Boiling range 5–95 mL$^a$, °C | Purity$^c$ (min.), % | Water content$^d$, % | Color$^e$, APHA |
|---|---|---|---|---|
| Dimethylethanolamine | 133.5 – 135.5 | 99.0 | 0.5 | 15 |
| Diethylethanolamine | 161.5 – 163 | 99.0 | 0.5 | 15 |
| Methyldiethanolamine | 115.5 – 118 | 98.0 | 0.5 | 50 |
| Methylethanolamine | 158.5 – 160.0 | 99.0 | 0.5 | 10 |
| 4-(2-Hydroxyethyl)morpholine | 223 – 225 | 99.0 | 0.2 | 40 |
| 1-(2-Hydroxyethyl)piperazine | 121 – 123$^b$ | 98.5 – 99.0 | 0.2 | 40 |
| 1-(2-Hydroxyethyl)piperidine | 92 – 96$^b$ | 99.0 | 0.1 | 2 |
| Hydroxyethylaniline | 282 – 287 | 99.0 | 0.3 | 100 – 150 |
| Ethylethanolamine |  | 99.6 | 0.4 | 5 |
| Dibutylethanolamine | 226 – 228 | 99.6 | 0.1 | 10 |
| Diisopropylethanolamine | 190 – 192 | 99.6 | 0.2 | 5 |
| Butylethanolamine | 196 – 198 | 99.1 | 0.1 | 5 |

$^a$ DIN 51 751.
$^b$ DIN 53 406.
$^c$ Determined by gas chromatography.
$^d$ Determined by Karl Fischer method.
$^e$ According to DIN 53 409.

The reaction of epichlorohydrin with amines to give N-alkylethanolamines is today of only minor commercial importance.

## 3.3. Quality Specifications

Table 5 lists specifications of commercial products. Purity is almost always measured by gas chromatography.

## 3.4. Uses and Economic Aspects

N-Alkylethanolamines are used mainly as intermediates, especially in the production of pharmaceuticals such as the local anesthetics procaine (diethylethanolamine 4-aminobenzoate [59-46-1]) and tetracaine (dimethylethanolamine 4-butylaminobenzoate [136-47-0]), crop protection agents, and flocculants (dimethylethanolamine acrylate and diethylethanolamine acrylate) [48], [49]. They are also important in the preparation of chemicals for the paper and leather industries. Use of N-alkylethanolamines in the production of plastics has risen substantially in recent years.

Direct uses of N-alkylethanolamines, especially methyldiethanolamine [50], include gas purification methods for removing acidic gases ($CO_2$, $H_2S$). Choline chloride is very important in the animal feedstuff industry (→ Choline) [51]. World production was ca. 85 000 t/a in 1984.

Annual worldwide demand for N-alkylethanolamines (excluding choline chloride) ranges from 2 to 10 000 t, depending on the product and application. Some products in the category of more than 1000 t/a are N,N-dimethylethanolamine, N,N-diethylethanolamine, 1-(2-hydroxyethyl)piperazine derivatives, N-methyldiethanolamine, and N-methylethanolamine.

# 4. Isopropanolamines

The propanolamines described here are propanol derivatives containing one amino group. They have the following general formulas:

$$\underset{\mathbf{8}}{\text{CH}_3-\underset{\underset{\text{NR}_2}{|}}{\text{CH}}-\text{CH}_2-\text{OH}} \qquad \underset{\mathbf{9}}{\text{CH}_3-\underset{\underset{\text{OH}}{|}}{\text{CH}}-\text{CH}_2-\text{NR}_2}$$

where R = H or $C_3H_6OH$

2-Amino-1-propanols (**8**) and 1-amino-2-propanols (**9**) have been prepared industrially since 1937 by reacting propylene oxide with the appropriate amine. The products are mixtures, which are processed further; the isomer mixtures are not usually separated on an industrial scale. The most important members of this class of compounds are the isopropanolamines, which are used in the form of mixtures of various isomers.

$$\underset{\text{Isopropanolamine}}{\text{H}_2\text{N}-\text{CH}_2-\underset{\underset{\text{OH}}{|}}{\text{CH}}-\text{CH}_3} \qquad \underset{\text{Diisopropanolamine}}{\text{HN}\underset{\diagdown \text{CH}_2-\underset{\underset{\text{OH}}{|}}{\text{CH}}-\text{CH}_3}{\diagup \text{CH}_2-\underset{\overset{\overset{\text{OH}}{|}}{}}{\text{CH}}-\text{CH}_3}}$$

$$\underset{\text{Triisopropanolamine}}{\text{CH}_3-\underset{\underset{\text{OH}}{|}}{\text{CH}}-\text{CH}_2-\text{N}\underset{\diagdown \text{CH}_2-\underset{\underset{\text{OH}}{|}}{\text{CH}}-\text{CH}_3}{\diagup \text{CH}_2-\underset{\overset{\overset{\text{OH}}{|}}{}}{\text{CH}}-\text{CH}_3}}$$

3-Amino-1-propanol, $H_2N-CH_2CH_2CH_2-OH$, which is obtained not from propylene oxide but from hydrogenation of ethylene cyanohydrin, is discussed in Chapter 5.

**Table 6.** Physical properties of propanolamines

| Compound | CAS registry number | $M_r$ | $fp$, °C | $bp$ (101.3 kPa), °C | $\varrho$, g/cm³ | Cubic expansion coefficient (40–100 °C), K⁻¹ | Heat of vaporization, kJ/kg | Specific heat, kJ kg⁻¹ K⁻¹ | Viscosity, mPa · s | Refractive index, $n_D$ |
|---|---|---|---|---|---|---|---|---|---|---|
| 1-Amino-2-propanol (d,l) | [78-96-6] | 75.11 | −0.3 | 159.7 | | | | | | 1.4466 (25 °C) |
| 2-Amino-1-propanol (d,l) | [6168-72-5] | 75.11 | +7.5 | 168.5 | | | | | | 1.4481 (25 °C) |
| Monoisopropanolamine [a] | | 75.11 | +1.9 | 159 | 0.962 (20 °C) | $8.2 \times 10^{-4}$ | 604 | 2.73 | 30 (20 °C) | 1.4463 (20 °C) |
| Diisopropanolamine [b] | [110-97-4] | 133.19 | +38.9 | 244 | 0.9842 (50 °C) | $8.2 \times 10^{-4}$ | 375 | 2.49 | 58 (60 °C) | 1.4542 (45 °C) |
| Triisopropanolamine [b] | [122-20-3] | 191.27 | ca. 55 | 291 | 0.9904 (60 °C) | $6.9 \times 10^{-4}$ | 309 | 2.47 | 141 (60 °C) | 1.4555 (60 °C) |

[a] Commercial isomer mixture consisting of ca. 94 % 1-amino-2-propanol and 6 % 2-amino-1-propanol.
[b] Commercial isomer mixture.

## 4.1. Properties

### 4.1.1. Physical Properties

Isopropanolamines are hygroscopic, virtually colorless substances with a slight amine-like odor. They are readily soluble in water, ethanol, glycol, and acetone but only slightly soluble in hydrocarbons and diethyl ether. Monoisopropanolamine is more soluble than monoethanolamine in organic solvents; it forms a miscibility gap with heptane and toluene.

Diisopropanolamine and triisopropanolamine undergo thermal decomposition above ca. 170 °C and can therefore be distilled only at reduced pressure. The freezing points of diisopropanolamine and triisopropanolamine can be greatly depressed by adding a small amount of water.

All isopropanolamines have at least one asymmetrically substituted carbon atom and thus can be optically active. The commercial products are racemic mixtures.

Important physical data are listed in Table 6 and safety data in Table 7.

Table 7. Safety data for isopropranolamines

|  | Monoisopropanolamine | Diisopropanolamine | Triisopropanolamine |
|---|---|---|---|
| Flash point, °C | 71 | 135 | 160 |
| Ignition temperature, °C | 335 | 290 | 275 |
| Explosion limits, vol % | 1.9–10.4 | 1.6–8.0 | 0.8–5.8 |
| Vapor pressure, hPa | 1.9 (20 °C) | 2.06 (100 °C) | 1 (100 °C) |

## 4.1.2. Chemical Properties

The properties of amines and alcohols are combined in isopropanolamines, and these compounds exhibit typical reactions of both functional groups; the amino group can be primary, secondary, or tertiary. In chemical behavior, isopropanolamines differ only slightly from the ethanolamines described in Section 2.1.2.

Isopropanolamines react with acidic gases ($H_2S$, $CO_2$) in aqueous solution to form salts, which decompose to the starting materials when heated. This effect is utilized in the purification of natural gas and synthesis gas. The salts of strong acids, such as HCl or $H_2SO_4$, are crystalline compounds.

Diisopropanolamine sulfate eliminates water at elevated temperature to form 2,6-dimethylmorpholine sulfate [53].

The sodium salt of the sulfuric acid half-ester of monoisopropanolamine undergoes cyclization when heated, and propylenimine is formed [54]:

$$H_2N-CH_2-\underset{\underset{CH_3}{|}}{CH}-OSO_3Na \xrightarrow{Heat} H_3C-\underset{\underset{\underset{H}{N}}{\diagdown\diagup}}{CH-CH_2} + H_2O + NaHSO_4$$

Fatty acids, such as stearic acid, react with isopropanolamine at room temperature to form neutral, waxlike isopropanolamine soaps. In reactions above 140 °C and with elimination of water, amides are formed preferentially, and minor amounts of fatty esters [55] are produced in a side reaction. Complete esterification can be effected by reaction with acyl chlorides in the presence of pyridine [56]. The reaction of esters with isopropanolamines above 100 °C gives amides [57]:

$$H_3C-\underset{\underset{OH}{|}}{CH}-CH_2-NH_2 + R^1-\underset{\underset{O}{\|}}{C}-O-R^2 \longrightarrow H_3C-\underset{\underset{OH}{|}}{CH}-CH_2-NH-\underset{\underset{O}{\|}}{C}-R^1 + R^2OH$$

where $R^1$, $R^2$ = alkyl

When thionyl chloride is used as a reactant and chloroform as the solvent, the hydroxyl group can be exchanged for chlorine. Isopropanolamine is converted to 2-chloropropylamine [58], which is obtained in good yield.

Isopropanolamine and diisopropanolamine react with epoxides to give mixed isopropanolamines. The amino group is several times more reactive than the hydroxyl group [59]. In the presence of suitable catalysts such as sodium hydroxide or sodium alkoxides, the epoxide reacts preferentially with the hydroxyl group to form polyethers [60]. Alkylating substances, e.g., alkyl halides or dimethyl sulfate, give N-alkylated derivatives [61].

Isopropanolamine can be converted to 1,2-diaminopropane by catalytic amination under pressure (→ Amines, Aliphatic).

Aldehydes and ketones react with the primary nitrogen atom of isopropanolamine to form a Schiff base.

$$CH_3-\overset{OH}{\underset{|}{CH}}-CH_2NH_2 + R-CHO \longrightarrow$$

$$CH_3-\overset{OH}{\underset{|}{CH}}-CH_2-N=CH-R + H_2O$$

Formaldehyde reacts with isopropanolamines to form the corresponding hydroxymethyl compounds, which can be converted to the methyl derivatives by catalytic hydrogenation.

$$(CH_3-\overset{OH}{\underset{|}{CH}}-CH_2)_2NH + CH_2O \longrightarrow (CH_3-\overset{OH}{\underset{|}{CH}}-CH_2)_2N-CH_2-OH$$

$$\xrightarrow[H_2]{Cat.} (CH_3-\overset{OH}{\underset{|}{CH}}-CH_2)_2N-CH_3 + H_2O$$

Compounds containing an active hydrogen atom undergo a Mannich reaction with formaldehyde and an isopropanolamine that has a primary or secondary amino group.

$$H_3C-\overset{O}{\underset{\|}{C}}-CH_3 + CH_2O + HNR^1R^2 \longrightarrow$$

$$CH_3-\overset{O}{\underset{\|}{C}}-CH_2CH_2-NR^1R^2 + H_2O$$

where $R^1$ = H or $C_3H_6OH$; $R^2$ = $C_3H_6OH$

Diisopropanolamine reacts with nitrous acid or its salts to form N-nitrosamines. Isopropanolamines react with optically active acids, such as tartaric acid, to form diastereomeric pairs of salts; this enables the separation of isopropanolamines into their enantiomeric forms [62].

The use of triisopropanolamine as an esterification catalyst has been suggested [63].

## 4.2. Production

Isopropanolamines are produced like ethanolamines, by reacting propylene oxide with $NH_3$ in the liquid phase via a continuous or batchwise procedure [64], [65]. The product is always a mixture of 1-amino-2-propanols and 2-amino-1-propanols, with the 1-amino isomers being the major component.

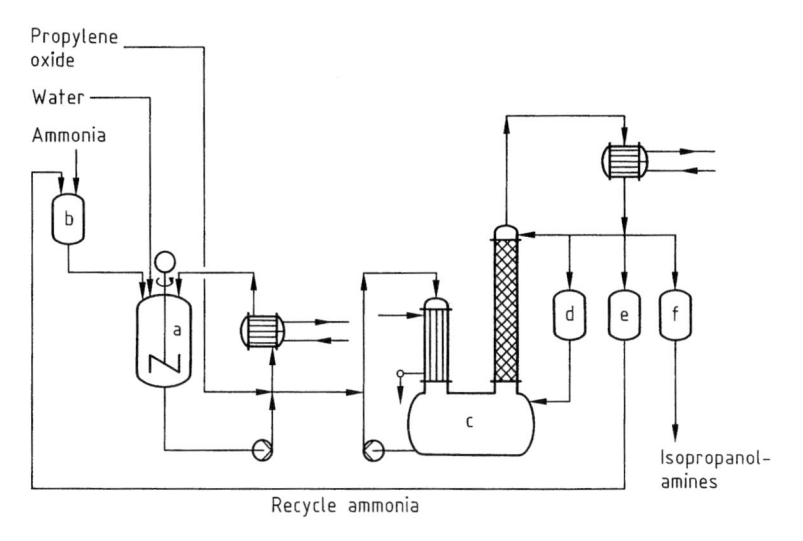

**Figure 3.** Flow sheet for the batchproduction of iso-propanolamines
a) Reaction kettle; b) Ammonia tank; c) Vessel; d, e, f) Storage tanks

$$H_3C-CH-CH_2 + NH_3 \rightarrow \begin{array}{l} H_3C-CH-CH_2-NH_2 \\ \phantom{H_3C-CH-}| \\ \phantom{H_3C-CH-}HO \end{array}$$
$$\phantom{H_3C-CH-CH_2 + NH_3 \rightarrow} \begin{array}{l} H_3C-CH-CH_2-OH \\ \phantom{H_3C-CH-}| \\ \phantom{H_3C-CH-}H_2N \end{array}$$

Industrially, the manner in which the propylene oxide ring is cleaved cannot yet be controlled, hence the isomer ratio of isopropanolamines.

Excess ammonia (about 2–20 mol per mole of propylene oxide) facilitates removal of the reaction heat (about 93 kJ/mol) and reduces the amount of polypropoxylated byproducts. This amount also depends on the relative reaction rates of propylene oxide with any amines present in the mixture.

Reaction temperatures can vary from 50 to 150 °C but are preferably around 100 °C. The resulting pressure can be reduced by adding water (ca. 10–60%). Water and hydroxylcontaining compounds such as alcohols, phenols, and alkanolamines (autocatalysis) accelerate the reaction [66].

A *continuous procedure* is used for large-scale production, e.g., of monoisopropanolamine and diisopropanolamine [67], [68]. This procedure is similar to that used to produce ethanolamines (see Section 2.2).

Figure 3 shows a flow sheet for the *batchwise production* of isopropanolamines. The procedure is also suitable for preparing N-alkylated derivatives (see Chap. 5); in this case, alkylamines are used instead of ammonia.

The oxygen-free reaction kettle (a) is filled with the amine raw material from the stock tank (b); water is added in an amount of 10–60%, based on the amine; and the kettle is heated to a minimum

**Table 8.** Minimum quality requirements for commercial isopropanolamines (as of 1985)

| Quality criterion | Method | Monoisopropanolamine | Diisopropanolamine | Diisopropanolamine containing 10% H$_2$O | Triisopropanolamine | Triisopropanolamine containing 15% H$_2$O |
|---|---|---|---|---|---|---|
| Water content, wt % | DIN 51 777 | 0.5 | 0.5 | 10 ± 1 | 0.5 | 15 ± 1 |
| Purity (min.), %* | GC | 98 | 98 | 88 | 97 | 83 |
| Low boilers (max.), %* (without H$_2$O) | GC | 0.3 | 0.3 | 0.3 | 1.5 | 1.5 |
| High boilers (max.) %* | GC | 0.5 | 0.5 | 0.5 | 1.0 | 1.0 |
| Color (max.), APHA | DIN 53 409 | 20 | 100 | 100 | 100 | 100 |
| Acid consumption, mL of 1 N HCl per gram | | 13.15–13.45 | 7.35–7.65 | 6.60–6.90 | 5.10–5.40 | 4.30–4.60 |
| Freezing point, °C | BS 523:1964 | ca. 2 | ca. 39 | ca. −10 | ca. 58 | < −10 |

* Determined as area under GC peak.

temperature around 50 °C. Metering of propylene oxide is then begun, the feed rate depending on the removal of heat and the operating pressure of the reaction kettle.

The optimum amount of propylene oxide in relation to starting amine is determined by cost-effectiveness (byproducts, work-up costs). When addition of propylene oxide is complete, the minimum reaction temperature is maintained for some time to ensure that the product discharged from the reactor is free of propylene oxide. The reaction mixture is separated by fractional distillation. Unreacted amine and water are removed as the first fraction and recycled to the subsequent batch. Pure isopropanolamines are obtained by distillation at reduced pressure.

The most important manufacturers of isopropanolamines are BASF, Bayer, Hüls, and ICI in Western Europe, and Dow Chemical and Union Carbide in the United States.

## 4.3. Quality Specifications

Because isopropanolamines are manufactured, sold, and used worldwide, the commercial products differ only slightly in delivery specifications. Typical specifications are listed in Table 8.

Purity is usually determined by gas chromatography. Depending on analytical conditions, isomeric and enantiomeric forms of isopropanolamines exhibit different numbers of peaks, which can be identified by test analyses. Derivatives obtained by reaction with trifluoroacetic anhydride can be analyzed more easily and analysis time can be reduced.

## 4.4. Uses

Possible uses of isopropanolamines are similar to those of ethanolamines; as a rule, the isopropanolamine derivatives exhibit better solubility in hydrophobic media. For example, neutralization with fatty acids, such as stearic or lauric acid, gives the corresponding soaps which are used in lubricant additives, textile finishing, and coating paper and wood [69]. Salts of isopropanolamines with fatty acids are also used as ionic emulsifiers in emulsion paints and as dispersants for pigments. Salts of isopropanolamines with alkylarylsulfonic acids are employed as degreasers and cleaning promoters [70]. The fatty acid amides of isopropanolamines and mixtures of these with esters are frequently used in the cosmetics industry as shampoo thickeners and foam regulators [71].

Isopropanolamines are effective corrosion inhibitors in antifreezes [72] and cutting oils. Simultaneous use of nitrites should be avoided, since carcinogenic nitrosamines may form under certain circumstances. Isopropanolamines are important catalysts for urethane foams [73]. In epoxy resin systems, they have proved useful additives to curing agents [74].

Sulfur-containing isopropanolamine derivatives are used to extract metal ions, such as mercury, gold, and platinum, from salt solutions and wastewater [75].

A very important use of aqueous diisopropanolamine solutions is as a gas wash in the Adip and Sulfinol processes. Here, $CO_2$ and $H_2S$, in particular, are washed out of natural gas, synthesis gas, or coke oven gas and, after desorption, can be obtained in some cases in their pure form [76].

Monoisopropanolamine is employed extensively as an intermediate in the production of 1,2-diaminopropane (→ Amines, Aliphatic), while diisopropanolamine is used for the synthesis of 2,6-dimethylmorpholine [53].

# 5. N-Alkylated Propanolamines and 3-Alkoxypropylamines

N-Alkylated propanolamines can be divided into N-alkylated 1-amino-2-propanols, 2-amino-1-propanols, and 3-amino-1-propanols.

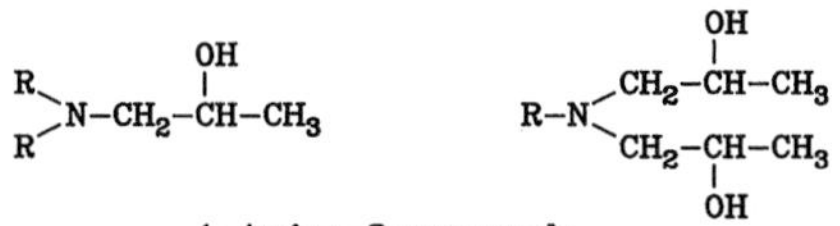

$$R_2N-CH_2-CH(OH)-CH_3 \qquad R-N(CH_2-CH(OH)-CH_3)_2$$

1-Amino-2-propanols

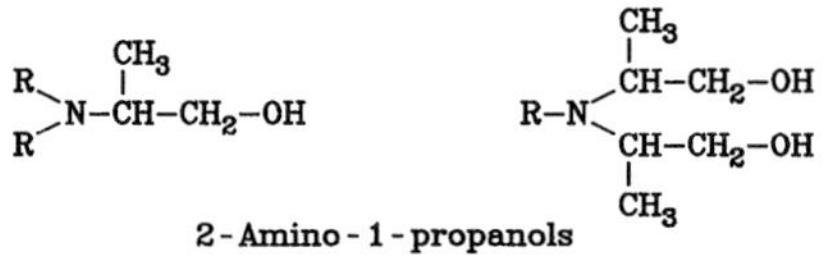

$$R_2N-CH(CH_3)-CH_2-OH \qquad R-N(CH(CH_3)-CH_2-OH)_2$$

2-Amino-1-propanols

$$R_2N-CH_2CH_2CH_2-OH$$

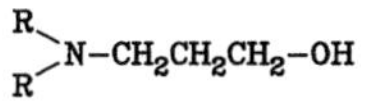

3-Amino-1-propanols

2-Amino-1-propanols and 1-amino-2-propanols are referred to industrially as isopropanolamines (Chap. 4). N-Alkylated isopropanolamines are prepared by reacting propylene oxide with the corresponding amines, which yields almost exclusively 1-amino-2-propanols [77]. Alkylated 2-amino-1-propanols have no industrial importance and are not discussed in this chapter.

## 5.1. Properties

### 5.1.1. Physical Properties

With a few exceptions, N-alkylated isopropanolamines, 3-amino-1-propanols, and 3-alkoxypropylamines are colorless liquids at room temperature. Some of them, such as cyclohexylisopropanolamine, are crystalline solids. The compounds tend to discolor on prolonged storage, especially if exposed to light, air, and heat (>40 °C). They have an amine-like odor, and absorb water and carbon dioxide. They are soluble in water, alcohol, acetone, glycol, glycerol, and glycol ethers, but only sparingly soluble in aliphatic or aromatic hydrocarbons and diethyl ether. Methyldiisopropanolamine does not form an azeotrope with water, but an azeotrope is formed upon distillation of aqueous dimethylisopropanolamine and 3-methoxypropylamine. Physical properties of N-alkylated isopropanolamines are given in Table 9 and those of 3-amino-1-propanols and 3-alkoxypropylamines in Table 10.

Table 9. Physical properties of N-alkylated isopropanolamines

| Compound | CAS registry number | $M_r$ | mp, °C | bp, °C | $\varrho$ (20 °C), g/cm³ | Cubic expansion coefficient, K⁻¹ | Heat of vaporization, kJ/kg | Specific heat, kJ kg⁻¹ K⁻¹ | Specific electrical conductivity, Ω⁻¹ cm⁻¹ | Viscosity (20 °C), mPa·s | $n_D^{20}$ | Surface tension (20 °C), mN/m | Flash point, °C | Ignition temperature, °C | Temperature class |
|---|---|---|---|---|---|---|---|---|---|---|---|---|---|---|---|
| Monomethylisopropanolamine $CH_3-CH-CH_2-NHCH_3$ $\quad\quad\quad \mid$ $\quad\quad\quad OH$ | [16667-45-1] | 89.14 | 12.5 | 149 | 0.9062 | | 455.6 | | $4.8 \times 10^{-7}$ | | 1.4340 | 28.7 (22 °C) | 58.7 | 305 | T2 |
| Methyldiisopropanolamine $(CH_3-CH-CH_2)_2NCH_3$ $\quad\quad\quad \mid$ $\quad\quad\quad OH$ | [4402-30-6] | 147.22 | −32 | 226 | 0.954 | | 329.3 | | $1.4 \times 10^{-7}$ | 47.3 | 1.4472 (25 °C) | 31.0 (22 °C) | 110.5 | 300 | T3 |
| Dimethylisopropanolamine $CH_3-CH-CH_2-N(CH_3)_2$ $\quad\quad \mid$ $\quad\quad OH$ | [108-16-7] | 103.17 | −85* | 125.8 | 0.849 0.852 | $7.0 \times 10^{-4}$ | 393.6 | 2.45 (25 °C) | $2 \times 10^{-6}$ | 1.5 | 1.4189 | 24.5 (25 °C) | 26 | 225 | T3 |
| Diethylisopropanolamine $CH_3-CH-CH_2-N(C_2H_5)_2$ $\quad\quad \mid$ $\quad\quad OH$ | | 131.22 | | 157.5 159 | 0.8570 | | | | | | 1.4265 | | | | |
| Dibutylisopropanolamine $CH_3-CH-CH_2-N(n-C_4H_9)_2$ $\quad\quad \mid$ $\quad\quad OH$ | | 187.33 | −80* | 229.1 | 0.8419 | $7.8 \times 10^{-4}$ | 246.6 | | | 3.5 | 1.4361 | | 205 | | |
| Cyclohexylisopropanolamine $CH_3-CH-CH_2-NH(C_6H_{11})$ $\quad\quad \mid$ $\quad\quad OH$ | | 157.26 | 45.1 | 238 | 0.9365 (40 °C) | | 329.3 | 2.47 (60 °C) | $9.0 \times 10^{-4}$ | 11.5 (60 °C) | 1.4732 | | 111 | 270 | T3 |
| Cyclohexyldiisopropanolamine $(CH_3-CH-CH_2)_2-N(C_6H_{11})$ $\quad\quad\quad \mid$ $\quad\quad\quad OH$ | | 215.34 | 2.0 | | | | | | | | | | | | |

## N-Alkylated Propanolamines and 3-Alkoxypropylamines

## Table 9. continued

| Compound | CAS registry number | $M_r$ | mp, °C | bp, °C | $\varrho$ (20 °C), g/cm³ | Cubic expansion coefficient, K$^{-1}$ | Heat of vaporization, kJ/kg | Specific heat, kJ kg$^{-1}$ K$^{-1}$ | Specific electrical conductivity, $\Omega^{-1}$ cm$^{-1}$ | Viscosity (20 °C), mPa·s | $n_D^{20}$ | Surface tension (20 °C), mN/m | Flash point, °C | Ignition temperature, °C | Temperature class |
|---|---|---|---|---|---|---|---|---|---|---|---|---|---|---|---|
| Cyclohexyldiisopropanolamine $(CH_3-CH-CH_2)_2-N(C_6H_{11})$ $\quad\quad\;\,$OH | | 215.34 | 2.0 | | | | | | | | | | | | |
| Cyclooctyliisopropanolamine $CH_3-CH-CH_2-NH(C_8H_{15})$ $\quad\quad\;\,$OH | | 185.31 | −12.85* | 275 | 0.953 | 7.6 × 10$^{-4}$ | 234.5 | 2.24 (20 °C) | 5.1 × 10$^{-4}$ | 665 | 1.4902 | 37.0 (20.3 °C) | 141 | 245 | T3 |
| Cyclooctyldiisopropanolamine $(CH_3-CH-CH_2)_2-N(C_8H_{15})$ $\quad\quad\;\,$OH | | 243.39 | 29.6 | 318 | 0.977 | 7.4 × 10$^{-4}$ | 233.95 | 1.98 (20 °C) | 5.7 × 10$^{-11}$ | 10870 | 1.4890 | 38.0 (20.3 °C) | 176 | 245 | T3 |
| Aminoethylisopropanolamine $CH_3-CH-CH_2-NH-CH_2-CH_2NH_2$ $\quad\quad\;\,$OH | [123-84-2] | 119.19 | −38* | | 0.9868 | | | | | 145 | 1.4750 | | 123 | 345 | T2 |
| 1-(2-Hydroxypropyl)-piperazine (**10**) | | 144.22 | −17* | | 1.0108 | | | | | 423 | 1.4910 | | 115 | 290 | T3 |
| 1,4-Bis(2-hydroxypropyl)-piperazine (**11**) | | 202.30 | 97−98 | | | | | | | | | | 157 | 250 | T3 |
| 4-(2-Hydroxypropyl)-morpholine (**12**) | | 145.21 | −47* | 218 | 1.0163 | | 314.3 | 1.92 (20 °C) | | 11.0 | 1.4642 | 34.5 | 92 | 215 | T3 |
| N,N,N′,N′-Tetrakis-2-hydroxypropylethylenediamine $(CH_3-CH-CH_2)_2-N-CH_2-CH_2-N(CH_2-CH-CH_3)_2$ $\quad\quad\;\,$OH $\quad\quad\quad\quad\quad\quad\quad\quad\quad\quad\quad\quad\quad\,$OH | | 292.43 | +11* | 190 (133 Pa) | 1.03 (25 °C) | 6.7 × 10$^{-4}$ | | 2.297 (50 °C) | 2.4 × 10$^{-4}$ | 45 (100 °C) | 1.478 (25 °C) | 35.1 | 201 | 285 | T3 |

* Pour point.

$CH_3-CH-CH_2-N\bigcirc NH$
$\quad\;\,$OH
**10**

$CH_3-CH-CH_2-N\bigcirc N-CH_2-CH-CH_3$
$\quad\;\,$OH $\quad\quad\quad\quad\quad\quad\quad\quad\,$OH
**11**

$CH_3-CH-CH_2-N\bigcirc O$
$\quad\;\,$OH
**12**

Table 10. Physical properties of 3-amino-1-propanols and 3-alkoxypropylamines

| Compound | CAS registry number | $M_r$ | mp, °C | bp, °C | ϱ (20 °C), g/cm³ | Cubic expansion coefficient, $K^{-1}$ | Heat of vaporization, kJ/kg | Specific heat, $kJ\,kg^{-1}\,K^{-1}$ | Viscosity, mPa·s | $n_D$ | Flash point, °C | Ignition temperature, °C | Temperature class |
|---|---|---|---|---|---|---|---|---|---|---|---|---|---|
| 3-Amino-1-propanol $HO-CH_2CH_2CH_2-NH_2$ | [156-87-6] | 75.11 | 11.5 | 189 | 0.98–0.99 | $8.3 \times 10^{-4}$ | 641 | 2.71 | 36 (20 °C) | 1.460–1.462 | 98–100 | 385 | T2 |
| 3-Dimethylamino-1-propanol $HO-CH_2CH_2CH_2-N(CH_3)_2$ | [3179-63-3] | 103.17 | −32.7 | 162–163 | 0.882–0.885 | | | | | | 57 | | |
| 3-Methoxypropylamine $CH_3O-CH_2CH_2CH_2-NH_2$ | [5332-73-0] | 89.14 | <−70 | 118 | 0.8729 | $1.2 \times 10^{-3}$ | 422.9 | | 0.9 (25 °C) | 1.4159 (25 °C) | 27 | 270 | T3 |
| 3-Ethoxypropylamine $C_2H_5O-CH_2CH_2CH_2-NH_2$ | [6291-85-6] | 103.21 | <−70 | 134.7 | 0.861 | | | | | | 33.7 | 245 | T3 |
| 3-(2-Ethylhexyloxy)-propylamine (13) | [5397-31-9] | 187.30 | <−70 | 237 | 0.8466 | | 258.85 | 2.27 | | | 104 | 220 | T3 |

$H_2N-CH_2CH_2CH_2-O-CH_2-CH\begin{smallmatrix}CH_2CH_2CH_2-CH_3\\CH_2-CH_3\end{smallmatrix}$

13

# N-Alkylated Propanolamines and 3-Alkoxypropylamines

## 5.1.2. Chemical Properties

Because N-alkylated propanolamines contain hydroxyl groups and basic nitrogen, they possess the properties of both amines and alcohols. They form salts with organic and inorganic acids [78]. The sulfuric acid esters of propanolamines can be cyclized with sodium hydroxide to give propylenimines (see Section 4.1.2) [79]. Propanolamines with secondary amino groups form alkanolamides with fatty acids. Reaction of the amides with ethylene oxide gives water-soluble or easily dispersible nonionic surfactants [80]. 1-Amino-2-propanols are chiral and can be separated into enantiomers by means of appropriate reagents [62], [81]. N-Alkylpropanolamines with tertiary amino groups can be converted by means of benzoyl chloride, without the assistance of a base, into the amino alcohol ester hydrochlorides; with thionyl chloride, they give 2-chloropropylammonium chlorides.

The condensation of 3-amino-1-propanol with carbonyl compounds yields tetrahydro-1,3-oxazines [82], Schiff bases, or mixtures of both compounds. 3-Amino-1-propanol and 2-hydroxy-3,3-dimethylbutyrolactone react almost quantitatively to give panthenol [83].

Alkoxypropylamines react analogously to primary aliphatic amines. They form salts with inorganic or organic acids, and can be converted to acid amides, sulfonamides, urethanes, and Schiff bases. They can be alkylated with formaldehyde or formic acid to give dimethylamino compounds. 3-Methoxypropylamine catalyzes the addition of hydrogen cyanide to vinyl esters to give 1-cyanoethyl esters [84]. 3-Methoxypropylsulfanilylurea obtained from the 3-methoxypropylammonium salt of N,N-di(acetylsulfanilyl)urea is an oral antidiabetic agent [85].

## 5.2. Production

Industrially, N-alkylated isopropanolamines are prepared primarily from amines and propylene oxide according to the following reactions [66], [86]:

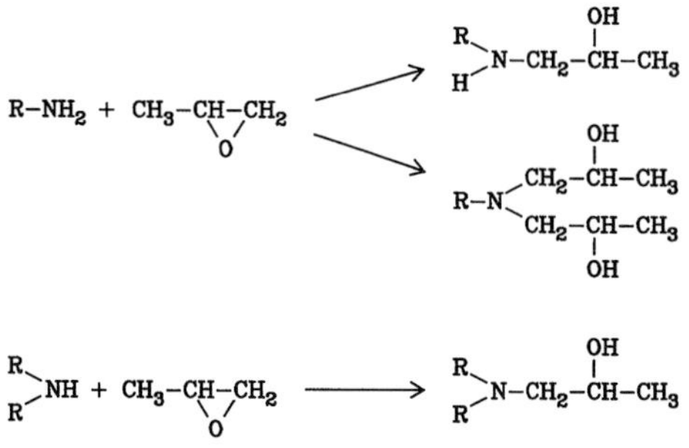

$$R-NH_2 + CH_3-CH-CH_2 \longrightarrow \begin{matrix} R \\ H \end{matrix}N-CH_2-\underset{OH}{CH}-CH_3$$

$$\longrightarrow R-N\begin{matrix} CH_2-\underset{OH}{CH}-CH_3 \\ CH_2-\underset{OH}{CH}-CH_3 \end{matrix}$$

$$\begin{matrix} R \\ R \end{matrix}NH + CH_3-CH-CH_2 \longrightarrow \begin{matrix} R \\ R \end{matrix}N-CH_2-\underset{OH}{CH}-CH_3$$

N-Alkylated 1-amino-2-propanols are formed predominantly [65], [87], [88]. The isomeric 2-amino-1-propanols are obtained only as byproducts (0.2–5%, depending

Table 11. Specifications for commercial N-alkylated propanolamines and 3-alkoxypropylamines

| Compound | Purity, % | $\varrho$ (20 °C) (DIN 51 757), g/cm$^3$ | Boiling range 5–95 mL (DIN 51 751, DIN 53 406), °C | Color (DIN 53 409), APHA |
|---|---|---|---|---|
| Monomethylisopropanolamine | 98 | 0.906 | 146–152 | $\leq 50$ |
| Methyldiisopropanolamine | 98 | 0.954 | 226–230 | $\leq 50$ |
| Dimethylisopropanolamine | $\geq 99$ | 0.849–0.852 | 125–127 | $\leq 50$ |
| Aminoethylisopropanolamine | $\geq 98$ | 0.9868 | 232.5–235.5 | $\leq 20$ |
| $N,N,N',N'$-Tetrakis(2-hydroxypropyl)ethylenediamine | $\geq 99$ | 1.009–1.013 | | 20 |
| 3-Dimethylamino-1-propanol | 99 | 0.882–0.885 | 162–163 | |
| 3-Methoxypropylamine | 99 | 0.871–0.873 | 117–119 | 20 |
| 3-Ethoxypropylamine | 99 | 0.855–0.861 | 130–135 | 5 |
| 3-(2-Ethylhexyloxy)propylamine | $\geq 98$ | 0.847–0.848 | 233–238 | $\leq 40$ |

on the nature of the amino component). The reactions are carried out at 80–160 °C and pressures up to 50 bar, batchwise or continuously. A detailed description can be found in Section 4.2. As in ethanolamine production, the oxide must be added to the amine, not vice versa (see Section 2.2). The crude reaction products are purified by continuous or batchwise distillation.

3-Amino-1-propanol is obtained by hydrogenating ethylene cyanohydrin. N-Methylated 3-amino-1-propanols are prepared from 3-amino-1-propanol and formaldehyde, with subsequent hydrogenation of the adducts [89]. 3-Alkoxypropylamines are prepared from acrylonitrile by addition to alcohols and subsequent hydrogenation [90].

## 5.3. Quality Specifications

Table 11 gives the specifications for commercial N-alkylated propanolamines and 3-alkoxypropylamines. Purity is generally determined by gas chromatography.

## 5.4. Uses and Economic Aspects

N-Alkylated propanolamines and 3-alkoxypropylamines are used mainly as intermediates, especially in the preparation of crop-protection agents and pharmaceuticals. 3-Alkoxypropylamines have considerable commercial importance in the preparation of dyes. N-Alkylated propanolamines and their derivatives have been employed, in the form of soaps, as lubricating oil additives, mold release agents [91], [92], textile finishes, and coatings for paper and wood [93]. The propanolammonium salts of alkylarylsulfonic acids are used as cleansing boosters and degreasing agents [94]. 3-

Alkoxypropylamines are utilized as catalysts in the production of polyurethane foams and epoxy resins.

3-Alkoxypropylamines and *N*-alkylpropanolamines are used as corrosion inhibitors and in the preparation of aqueous polymer solutions [95]–[99]. The pentachlorophenol salt of dimethylisopropanolamine has been described as a herbicide [100]. Anion exchangers can be prepared by reacting cellulose with dimethylisopropanolamine and epichlorohydrin [101].

World demand for *N*-alkylpropanolamines and 3-alkoxypropylamines is estimated at 5000 t/a. More than 100 t/a of 3-methoxypropylamine and dimethylisopropanolamine are used.

# 6. Storage and Transportation

Alkanolamines should be stored in stainless steel containers with exclusion of air ($O_2$, $CO_2$) and moisture, preferably under dry nitrogen. Storage temperature should not exceed 40 °C (50 °C for ethanolamines). Steel tanks may be used if absorption of iron (up to 10 ppm) is not important. Alkanolamines turn yellow on prolonged storage, especially in the presence of oxygen.

Depending on quality requirements and sensitivity of the products, steel, stainless steel, or polyethylene containers can be used for transportation. The containers must possess airtight closures to prevent absorption of water and carbon dioxide. Zinc and other nonferrous metals are attacked by alkanolamines. For transportation, particular national regulations, such as GGVS in the Federal Republic of Germany, should be observed [102].

Rubber gloves and safety goggles should be worn when handling alkanolamines. In some cases, especially with 3-alkoxypropylamines, a full face shield and rubber clothing are advisable.

For safety data of alkanolamines, see the sections on physical properties.

# 7. Environmental Protection

Ammonia- or amine-containing off-gases from alkanolamine production are either burned or purified by acid scrubbing. Very low amine concentrations ($10^{-3}$ ppm) can be determined by modern analytical methods [103].

Wastewater from plant cleaning and acid scrubbing is treated in a sewage plant. When properly fed into a biological treatment plant, alkanolamines are readily degraded by appropriate bacteria.

Spilled product should be removed with an absorptive material such as urea resin foam (Hygromull) or peat dust, which is then incinerated.

# 8. Toxicology and Occupational Health

**Ethanolamines.** Prominent among the toxic effects of ethanolamines is the irritating effect on skin and mucous membranes. This effect decreases from primary to tertiary alkanolamines and is also weaker in their salts.

*Monoethanolamine* has an oral $LD_{50}$ of 1500 mg/kg in rats [104] and a dermal $LD_{50}$ of 1000 mg/kg in rabbits [105]; it is classified as harmful. Rats survive 8-h inhalation of the saturated vapor at 20 °C without any symptoms. In rabbits, skin exposure for 1–5 min leads to corrosion; comparable effects are observed on the eyes [104]. Oral administration of 1.28 g kg$^{-1}$ d$^{-1}$ to rats for 90 d is lethal (mortality rates not given) [106], whereas a dose of 0.64 g kg$^{-1}$ d$^{-1}$ leads to changes in the liver and kidney weights; a dose of 0.32 g kg$^{-1}$ d$^{-1}$ is tolerated without any toxic effects. Inhalation at a concentration of 12–25 ppm is not lethal to dogs, guinea pigs, or rats, but inhalation of 66–75 ppm is lethal to guinea pigs and rats and 100 ppm to dogs (details of the test are not given) [106]. In the Ames test, no mutagenic effect is found [107]. The MAK and TLV are established at 3 ppm.

*Diethanolamine* is not irritating to the skin of rabbits after short exposure up to 15 min; longer exposure leads to erythema of the skin, edema, and surface necrosis. Severe irritation of the eyes is observed upon exposure to diethanolamine. The oral $LD_{50}$ is rats is about 1600 mg/kg [104]. The dermal $LD_{50}$ in rabbits is reported to be 12 200 mg/kg [105]. Rats survive an 8-h inhalation of saturated vapor at 20 °C without any symptoms [104]. Inhalation of 1471 ppm as vapor/aerosol over a period up to 2 h is lethal to rats [108]. A single inhalation (duration of exposure not given) of 200 ppm (vapor) and 1400 ppm (aerosol) is lethal to some of the animals exposed [109]. Liver and kidneys are particularly damaged after repeated administration. This is observed after repeated intraperitoneal administration of 250 mg/kg (frequency of administration not given) as well as after continuous inhalation of 25 ppm for 216 h or 6 ppm for 13 weeks (5 d/week), which is lethal to some of the exposed rats, and after a seven-week administration of neutralized diethanolamine at a concentration of 4 mg/mL in drinking water (species not given) [109]. Oral administration of 0.02 g kg$^{-1}$ d$^{-1}$ to rats for 90 d is tolerated without any signs of toxicity, whereas a dose of 0.09 g kg$^{-1}$ d$^{-1}$ causes an increase in the liver and kidney weights. A dose of 0.17 g kg$^{-1}$ d$^{-1}$ is lethal [106]. In the Ames test, no mutagenic effect is found [110]. The TLV is established at 3 ppm (15 mg/m$^3$).

*Triethanolamine* is not irritating to the skin of rabbits after exposure times of up to 4 h. Longer applications (up to 20 h) cause irritation. This substance does not irritate the eyes. The $LD_{50}$ (rat, oral) for the technical product is ca. 7.2 g/kg. Rats survive an 8-h inhalation of saturated vapor at 20 °C without any symptoms [111]. Triethanolamine can be absorbed through the skin [112]. Occasional cases of allergic contact dermatitis are reported [113], [114]; however, no sensitizing effect could be found in guinea pigs [115]. Oral administration of 0.08 g kg$^{-1}$ d$^{-1}$ to rats for 90 d is tolerated without toxic effects, but a dose of 0.17 g kg$^{-1}$ d$^{-1}$ causes changes in the liver and kidney weights. A dose of 0.73 g kg$^{-1}$ d$^{-1}$ leads to histopathological changes (which have not been characterized) and deaths (mortality rates not given) [106]. Triethanolamine is reported to be carcinogenic in mice following lifelong administration at concentrations of 0.3 and 0.03% in the feed; however, a dose-related increase in malignant tumors, mainly of the lymphatic system, is observed only in females [116]. This effect may be due to contaminants, such as nitrosamines, which are possibly formed in the course of the study. On the other hand, in a 2-year carcinogenicity study, in which rats were administered 1 and 2% triethanolamine in drinking water, only kidney lesions were found in both dose groups, but no carcinogenic effect was observed [117]. In a battery of mutagenicity tests, no genotoxic effects of triethanolamine could be determined in the Rec-assay with *Bacillus subtilis;* the Ames test with *Salmonella typhimurium* TA 98 and TA 100 as well as *Escherichia coli* WP2 try$^-$, with and without metabolic activation; chromosome aberration test in CHL cells; cell transformation test using hamster embryo cells [118]; chromosome aberration test in CHO cells; SCE test in CHO cells; and sex-linked recessive lethal test on *Drosophila* [119]. A cell transformation test with and without metabolic activation as well as a UDS test on rat hepatocytes did not show any mutagenic effects [120].

*Diethylethanolamine* causes dermal corrosion in rabbits after exposures up to 1 h [121]. Comparable effects are also observed on the eyes. The oral LD$_{50}$ in rats is about 1330 mg/kg. Exposure to a saturated vapor at 20 °C for more than 1 h is lethal for rats [122]. The dermal LD$_{50}$ for rabbits is reported to be 1260 mg/kg [123]. The MAK and TLV are established at 10 ppm (50 mg/m$^3$), with a warning of the danger of skin resorption.

*Dibutylethanolamine* is not irritating to the skin of rabbits after exposure up to 5 min. Longer exposure leads to irritation. Severe irritation is observed in the eyes, with corneal opacity. The oral LD$_{50}$ (rat) is about 600 mg/kg. Rats survive 8-h inhalation of saturated vapor at 20 °C without any symptoms [124]. The dermal LD$_{50}$ (rabbit) is reported to be 1680 mg/kg [125]. Five-week administration of neutralized dibutylethanolamine in drinking water in doses of 0.43, 0.20, and 0.13 g kg$^{-1}$ d$^{-1}$ (male rats) and 0.33, 0.24, and 0.14 g kg$^{-1}$ d$^{-1}$ (female rats) leads to temporary loss in body weight in the two upper dose groups for males and in all groups for females. Increased relative kidney weights are observed in the highest dose groups of both sexes, whereas no changes are observed in the lowest dose groups. Histopathological investigation of the kidneys reveals no substance-related effects [126]. Irritation of the eyes and nose is observed after inhalative exposure of rats (70 ppm, 5 d, 6 h/d). This exposure is lethal to one of five animals and leads to an increase in the relative liver and kidney weights and to liver damage. If a concentration of 33 ppm is administered over the same period, the relative kidney weights increase only slightly. A concentration of 22 ppm for 6 h/d over a period of 6 months is tolerated by rats without any signs of toxicity [126]. The substance probably acts on the central nervous system through inhibition of acetylcholinesterase, an effect which was measured in vitro [127]. The TLV is established at 2 ppm (14 mg/m$^3$).

**Table 12.** Acute toxicity of propanolamines and 3-alkoxypropylamines

| Compound | LD$_{50}$, g/kg | | AIH[a] (rats) | Skin irritation[b,c] (rabbits) | Eye irritation[c] (rabbits) | Reference |
|---|---|---|---|---|---|---|
| | Oral (rats) | Dermal (rabbits) | | | | |
| Monoisopropanolamine | 4.26 | 1.64 mL/kg | | m i | se i | [105], [106] |
| | 2.7 | | >8 h | 15 min: c | c | [128] |
| Diisopropanolamine | 6.72 | | | si | se i | [105] |
| | 6 | | >8 h | 15 min: n i; 20 h: m i | m i | [128] |
| Triisopropanolamine | 6.5 | >10 | | | se i | [105], [106] |
| | 4 | | >8 h | 15 min: n i | m i | [128] |
| | | | | 20 h: sl i | se i | [105], [128] |
| Monomethylisopropanolamine | 1 | | >7 h | 15 min: c | c | [128] |
| Cyclohexylisopropanolamine | 0.8 | | >8 h | 15 min: m i | c | [128] |
| | | | | 20 h: c | | |
| Dimethylisopropanolamine | 1.36 | | 10 min | 5 min: c | c | [128] |
| | 1.89 | | | | | [105] |
| Butyldiisopropanolamine | 3.5 mL/kg | | >8 h | 15 min: sl i; 20 h: se i | c | [128] |
| Cyclohexyldiisopropanolamine | 6.4 | | >8 h | 15 min: n i | se i/c | [128] |
| | | | | 20 h: c | | |
| 3-Amino-1-propanol | 2.83 | 1.25 | | hi | | [105], [106] |
| | 1.3 | >2 (rats) | >8 h | 5 min: se i; 15 min: c | c | [128] |
| 3-Dimethylamino-1-propanol | 1.86 | | 3 h | 15 min: se i/c | se i/c | [128] |
| 3-Methoxypropylamine | 6.26 | | 1 h | 5 min: c | c | [128] |
| 3-Ethoxypropylamine | 1.1 | | 1 h | 1 min: c | c | [128] |
| 3-(2-Ethylhexyloxy)propylamine | 0.85 | | 8 h | 5 min: c | c | [128] |
| | 0.32 | 0.36 | | | | [105] |

[a] AIH = acute inhalation hazard: rats inhale an atmosphere enriched with vapor at 20 °C; duration of exposure required to cause the first death is given.
[b] The skin irritation for the specific time of exposure to the substance is given.
[c] Symbols: i = irritating (n = not; sl = slightly; m = moderately; se = severely); c = corrosive.

**Propanolamines.** Like ethanolamines, propanolamines cause irritaions and corrosions to the skin. However, they are less pronounced than those of the corresponding alkylamines. Furthermore, skin irritation decreases with increased number of hydroxyl groups. Although very few propanolamines have been investigated for acute dermal toxicity, they are presumably absorbed through the skin. The low vapor pressure of these compounds at room temperature explains the usually low acute inhalation hazard. Acute toxicity data are summarized in Table 12.

*Diisopropanolamine* is nonmutagenic in the Ames test [129]. *Isopropanolamine* has been tested in the Ames test with questionable results [129]. The acute inhalative LC$_{50}$ in rats is > 16.4 mg L$^{-1}$ h$^{-1}$ for *3-amino-1-propanol* administered as an aerosol [130].

When *diisopropanolamine* was administered to rats in doses of 100, 300, 600, 1200, and 3000 mg kg$^{-1}$ d$^{-1}$ in the drinking water for 2 weeks, no symptoms were observed up to 600 mg kg$^{-1}$ d$^{-1}$. At 1200 mg kg$^{-1}$ d$^{-1}$, apart from a slight decrease in feed and water consumption in both sexes, a slight decrease in body weight and a slight increase in the relative kidney weights were observed in male rats. A dose of 3000 mg kg$^{-1}$ d$^{-1}$ is lethal to two out of five male animals; the prominent effects for this dose group include a significant loss in body and organ weight, acute inflammation and degeneration of kidneys and urinary bladder, and general liver atrophy [131].

*Triisopropanolamine* is nonmutagenic in the Ames test [119]. When triisopropanolamine is administered to rats in doses of 100, 300, 600, 1200, and 2000 mg kg$^{-1}$ d$^{-1}$ in the drinking water for 2 weeks, no changes are observed at 100 and 300 mg kg$^{-1}$ d$^{-1}$. At the higher doses, there is a dose-related increase in the relative kidney weights. Feed and drinking water consumption as well as total serum protein and albumin values are reduced [131].

# 9.  References

General References

[1]  A. L. Kohl, F. C. Riesenfeld: *Gas Purification*, 3rd ed., Gulf Publishing Co., Houston, Texas, 1979.

[2]  K. Lindner: *Tenside-Textilhilfsmittel-Waschrohstoffe*, 2nd ed., vol. **1**,Wissenschaftliche Verlagsgesellschaft, Stuttgart 1964.

[3]  E. J. Fischer: *Triethanolamine und andere Alkanolamine*, 4th ed., Straßenbau, Chemie und Technik Verlagsgesellschaft, Heidelberg 1953.

[4]  N. Schönfeldt: *Grenzflächenaktive Alkylenoxid-Addukte*, Wissenschaftliche Verlagsgesellschaft, Stuttgart 1976.

[5]  A. N. Schwartz, J. W. Perry, J. Berch: *Surface Active Agents and Detergents*, vol. **II**, Interscience, New York 1958.

[6]  John W. McCutcheon, Inc.: *Detergents and Emulsifiers*, Morristown, N.J., published yearly.

Specific References

[7]  A. Wurtz, *Justus Liebigs Ann. Chem.* **114** (1860) 51–54; **122** (1862) 226–232.

[8]  C. M. Yon, G. R. Atwood, C. D. Swaim, *Hydrocarbon Process.* 1979, July, 197–200.

[9]  BASF, DE-OS 1 953 263, 1972;US 4 014 933, 1977 (H. Corr, G. Boettger).

[10]  M. Arné: *Alkyl-Amines*, Report 138, Stanford Research Institute, Menlo Park, Calif., March 1981.

[11]  H. Kindler et al., *Chem.-Ing.-Tech.* **37** (1965) 400–402.

[12]  H. Wenker: "Preparation of Ethyleneimine from Ethanolamine," *J. Am. Chem. Soc.* **57** (1935) 2328.

[13]  L. W. Jones, G. R. Burns, *J. Am. Chem. Soc.* **47** (1925) 2966.

[14]  L. Knorr, H. Mathes, *Ber. Dtsch. Chem. Ges.* **34** (1901) 3484-3489.

[15]  W. Seeliger, H. Hellmann et al., *Angew. Chem.* **78** (1966) 913–927; *Angew. Chem. Int. Ed. Engl.* **5** (1966) 875.

[16]  BASF, DE-OS 2 618 580, 1977 (H. Hoffmann, K. Merkel, H. Toussaint, D. Voges).

[17]  L. Knorr, P. Rössler: "Zur Kenntnis des Ethanolamins," *Ber. Dtsch. Chem. Ges.* **36** (1903) 1281.

[18]  BASF, FR 2 307 902. 1976 (H. Bosche, H. Lüdemann).

[19] H. A. Bruson, *J. Am. Chem. Soc.* **58** (1936) 1741-45.
[20] H. Druckrey, R. Preussmann, S. Ivankovic, D. Schmähl, *Z. Krebsforsch.* **69** (1967) 103–201.
[21] BASF, DE-OS 1 768 335, 1972 (R. Dahlinger, W. Goetze, G. Schulz, U. Sönksen).
[22] Mo och Domsjö AB, US 3 697 598, 1972 (B. J. G. Weibull, L. U. F. Folke, S. O. Lindstrom).
[23] BASF, unpublished measurements.
[24] *Yukagaku* **15** (1966) no. 5, 215–220.
[25] G. Schulz, W. Goetze, P. Wolf: "Kinetische Untersuchungen zur Oxethylierung von $NH_3$ in konz. wässrigen Lösungen," *Chem.-Ing.-Tech.* **40** (1968) no. 9/10, 446–448.
[26] BASF, unpublished measurements.
[27] W. Wirth: "Nichtionische Tenside: Bedeutung, Herstellungsverfahren, Eigenschaften und Einsatzmöglichkeiten," *Tenside Deterg.* **5** (1975) 245–247.
[28] BASF, DE-OS 2 810 135, (H. Bosche, H. Hammer, G. Jeschek).
[29] *Seifen, Öle, Fette, Wachse, Kosmet. Aerosole* **17** (1977) 487–491.
[30] S. Scholz-Weigl, S. Holzmann, M. Schechter: *Tenside in unserer Welt – heute und morgen; Berichte vom Welt-Tensid-Kongress München 1984 (Proceedings of the World Congress on Surfactants, Munich 1984)*, Kürle Druck und Verlag, Gelnhausen (FRG) 1984, pp. 89–105 and 167–174.
[31] H. Grossmann, *Fette, Seifen, Anstrichm.* **74** (1972) 58–63.
[32] H. Manneck, *Seifen, Öle, Fette, Wachse* **4** (1959) 77–79.
[33] W. Kritchevsky, US 089 212, 1937; US 2 094 608, 1937.
[34] Procter & Gamble, DE-OS 2 527 793, 1976 (C. H. Nicol).
[35] Procter & Gamble, GB 1527–141, 1975 (J. L. Arnan, C. R. Barrot, P. N. Crisp, I. Siklossy).
[36] Procter & Gamble, EP 8–142, 1978 (P. Gosset, R. Ludewick).
[37] H. Schneider: "Über die Verwendung von Mahlhilfen bei der Zementmahlung," *Zem.-Kalk-Gips* **22** (1969) 193–201.
[38] W. Lieber, W. Richartz: "Einfluß von Triethanolamin, Zucker und Borsäure auf das Erstarren und Erhärten von Zement," *Zem.-Kalk-Gips* **25** (1972) 403–409.
[39] Dow Chemical, US 146 489, 1971; DE 2 225 015, 1972 (W. Dowd).
[40] Jefferson Chemical, DE 1 808 941, 1969 (V. A. Currier, A. Lichaa).
[41] Yokkaichi Gosei, JP 35 515, 1969.
[42] BASF, US 4 504 363, 1985 (B. Girgensohn, N. Goetz, F. Zanker).
[43] *Houben-Weyl,* **6/4**, pp. 510–520.
[44] Hori Todashi: "Alkylamines and Derivatives V – Ethoxylated Chemicals," *Bosei Kanri* **20** (1976) no. 12, 37.
[45] IG-Farbenindustrie, DE 650 574, 1928.
[46] Carbides and Carbon Chem. Corp., US 1 904 013, 1927 (E. Reid, D. C. Lewis).
[47] Du Pont, US 2 163 099, 1938 (R. Maxwell).
[48] Texaco Development, US 99 794 T 30, 1970.
[49] Rohm & Haas Co., DE-OS 1 929 581, 1970 (S. N. Lewis, I. F. Levy).
[50] Dow Chemical, US 4289 738, 1980 (C. Martin, R. L. Pearce).
[51] J. Gropp: *Chemie und Ernährung – BASF Forum Tierernährung am 28./29. 10. 1982*, Verlag Wissenschaft und Politik, p. 111.
[52] BASF, *Safety data sheets,* Ludwigshafen 1985.
[53] *Houben Weyl,* **6/4**, p. 511. CS 146 401 (1972).
[54] Hoechst, DE 832 152, 1952 (H. Bestian, W. Schuhmacher).
[55] A. Davison, B. Milwidisky: *Synthetic Detergents*, 4th ed., C.R.C. Press, Cleveland, Ohio, 1968.
[56] G. D. Jones, *J. Org. Chem.* **9** (1944) 491.
[57] E. Jungermann, D. Taber: *Nonionic Surfactants*, Marcel Dekker, New York 1967.

[58] US 3 025 325, 1959 (R. Velgos).
[59] Dow Chemical, US 2 649 483, 1953 (M. Huscher, M. Long et al.).
[60] Jefferson Chemical, US 3 317 609, 1967 (L. Sherman).
[61] FR 2 103 968, 1972 (B. S. Polanskyi).
[62] Merck & Co., US 2 189 808, 1940 (R. T. Major, H. T. Bonnett).
[63] S. Hünig, *J. Prakt. Chem.* **5** (1958) 224–232.
[64] Oxirane, DE 1 004 620, 1957 (A. Lowe, D. Butler et al.).
[65] Progressive Products Co., US 3 454 647, 1966 (P. Kersnar).
[66] Texaco, US 3 761 523, 1972 (R. E. Reid, T. P. Chen).
[67] Shell, BE 834 514, 1975 (W. A. Gleich).
[68] Mo och Domsjö AB, DE-OS 1 941 859, 1969 (B. J. Weibull, L. U. Thorsell et al.).
[69] Chemische Fabrik Pfersee, GB 827 469, 1956.
[70] US 2 673 215, 1952(V. J. Keenan, F. J. Gozlow).
[71] D. V. Beio, J. McCarthy: "The Fatty Alkanolamines," *Cosmet. Technol.* 1980, Mar.
[72] Asahi Denka Kogyo, JP 56 032 581, 1979.
[73] L. Thiele, *Acta Polym.* **30** (1979) 323–342.
[74] H. John: *Epoxidharze*, VEB Deutscher Verlag für Grundstoffindustrie, Leipzig 1969.
[75] Chemische Werke Buna, VEB, DD 240 648, 1984.
[76] *Hydrocarbon Process.* 1975, no. 4, 79–111. E. E. Isaals, F. D. Otto et al., *Can. J. Chem. Eng.* **55** (1977) 210.
[77] J. Rosen, US 2 993 934, 1956.
[78] UCC, US 2 129 805, 1934 (A. Lamendin, B. Matel, J. Dhenin). Société des Produits Chimiques Marles-Kuhlmann, DE-OS 1 918 388, 1969 (A. L. Wilson).
[79] H. W. Heine, R. W. Greiner, M. A. Boote, B. A. Brown, *J. Am. Chem. Soc.* **75** (1953) 2505–2506. Y. Minoura, T. Takebayashi, C. Price, *J. Am. Chem. Soc.* **81** (1959) 4589–4692.
[80] A. Tettamanzi, *Atti Accad. Sci. Torino Cl. Sci. Fis. Mat. Nat.* **69** (1934) 369–377. B. Kirson, *Bull. Soc. Chim. Fr.* 1958, 223–226. ICI, GB 899 948, 1959 (R. Hulse, H. J. Twitchett). British Titan Products, BE 665 840, 1965.
[81] Merck & Co., US 2 118 054, 1935 (R. Major, H. T. Bonnett).
[82] E. D. Bergmann, A. Kaluszyner, *Recl. Trav. Chim. Pays-Bas* **78** (1957) 315–326.
[83] Hoffmann – La Roche, GB 582 156, 1945.
[84] American Cyanamid, GB 591 489, 1947.
[85] C. H. Boehringer Sohn, DE 1 012 598, 1957 (E. Haak, A. Hagedorn).
[86] H. Tiltscher, *Angew. Makromol. Chem.* **25** (1972) 1–14.
[87] Oxirane, GB 763 434, 1951 (A. J. Lowe, D. Butler, M. Megde).
[88] Carbon Chemicals Corp., US 2 137 627, 1938 (J. N. Wickert).
[89] Atlantic Richfield, US 3 457 313, 1966 (T. Baker).
[90] Union Carbide, US 2 813 877, 1954 (J. Lambrach).
[91] Atlas Powder Corp., US 2 775 604, 1953.
[92] Esso Res. & Eng. Corp., US 2 764 551, 1954 (S. Lippncott, C. Muessig, P. U. Smith).
[93] Chemische Fabrik Pfersee, GB 827 469, 1956.
[94] The Atlantic Refining Co., US 2 673 215, 1952 (F. Gozlow).
[95] Hermann Wülfing Wings, BE 574 022, 1966.
[96] Nox-Rust Chem. Corp., US 2 512 949, 1945 (E. Lieber).
[97] Montclair Res. Corp., US 2744 883, 1956 (I. Hamer, J. Rust).
[98] Carbide and Carbon Chemicals Corp., US 2 137 627, 1938 (M. Reed).
[99] BASF, DE 1 052 685, 1956 (F. Ebel, E. Meyer).

[100] Diamond Alkali Corp., US 2 993 934, 1956 (I. Rosen).
[101] Ciba, GB 770 570, 1955.
[102] *Gefahrgutverordnung Straße – GGVS 1985 (regulation governing the transport of hazardous materials on roads)*, Bundesgesetzblatt I, 1985, p. 1550.
[103] P. W. Langvardt, R. G. Melcher, *Anal. Chem.*, **52** (1980) 669–671.
[104] BASF, unpublished results, 1966.
[105] *NIOSH Registry of Toxic Effects of Chemical Substances*, US Department of Health, Education, and Welfare, Washington, D.C., 1986.
[106] G. J. Clayton et al. (eds.): *Patty's Industrial Hygiene and Toxicology*, 3rd ed., vol. **2B**, John Wiley & Sons, New York 1981.
[107] *NTP Annual Plan for Fiscal Year 1985*, March 1985.
[108] G. v. Foster, *Diss. Abstr. Int. B* **32** (1972) no. 11, 6549.
[109] R. Hartung et al., *Toxicol. Appl. Pharmacol.* **17** (1970) 308.
[110] S. Haworth et al., *Environ. Mutagen.* **5** (1983) Suppl. no. I, 3–142.
[111] BASF, unpublished results, 1966–1983.
[112] G. M. Kostrodymova et al., *Gig. Sanit.* **3** (1976) no. 3, 20.
[113] G. Angelini et al., *Contact Dermatitis* **12** (1985) 263.
[114] B. Scheurer, *Hautarzt* **34** (1983) 126.
[115] Cited in *J. Am. Coll. Toxicol.* **2** (1983) no. 7, 183.
[116] H. Hoshino et al., *Cancer Res.* **38** (1978) 3918.
[117] A. Maekawa et al., *J. Toxicol. Environ. Health* **19** (1986) 345.
[118] K. Inoue et al., *Mut. Res.* **101** (1982) 305.
[119] *NTP Annual Plan for Fiscal Year 1983*, Jan. 1983.
[120] BASF, unpublished results of Litton Bionetics, 1982/83.
[121] BASF, unpublished results, 1982.
[122] BASF, unpublished results, 1969.
[123] ACGIH: *Documentation of the Threshold Limit Values*, 1986, p. 198.
[124] BASF, unpublished results, 1967.
[125] H. F. Smyth et al., *Arch. Ind. Hyg. Occup. Med.* **10** (1954) 61.
[126] H. H. Cornish et al., *Am. Ind. Hyg. Assoc. J.* **30** (1969) 46.
[127] R. Hartung et al., *Toxicol. Appl. Pharmacol.* **12** (1968) 486.
[128] BASF, unpublished results, 1963–1978.
[129] *NTP Annual Plan for Fiscal Year 1984, Feb. 1984*.
[130] BASF, unpublished results, 1979.
[131] Dow Chemical, unpublished result, cited in *Final Report of the Safety Assessment for Diisopropanolamine, Triisopropanolamine, Isopropanolamine, Mixed Isopropanolamines*, The Expert Panel of the Cosmetic Ingredient Review, Washington, D.C., Sept. 26, 1986.

# Ethers, Aliphatic

*Individual keywords:* → *Cellulose Ethers;* → *Crown Ethers;* → *Dimethyl Ether;* → *Dioxane;* → *Epoxides;* → *Ethylene Oxide;* → *Furan and Derivatives;* → *Methyl Tert-Butyl Ether;* → *Propylene Oxide;* → *Tetrahydrofuran;* → *Vinyl Ethers. For allyl ethers* → *Allyl Compounds; for aromatic ethers* → *Naphthalene Derivatives,* → *Phenol Derivatives; for glycol ethers* → *Ethylene Glycol*

WILHELM HEITMANN, Hüls AG, Herne, Federal Republic of Germany (Chaps. 2 and 3, Sections 4.1, 4.3, and 4.4)

GÜNTHER STREHLKE, Deutsche Texaco, Moers, Federal Republic of Germany (Section 4.2)

DIETER MAYER, Hoechst Aktiengesellschaft, Frankfurt, Federal Republic of Germany (Chap. 5)

| | | | | | |
|---|---|---|---|---|---|
| 1. | **Introduction** | 2185 | 4.3. | **Other Aliphatic Ethers** | 2195 |
| 2. | **Properties** | 2186 | 4.3.1. | Di-*n*-propyl Ether | 2195 |
| | | | 4.3.2. | Di-*n*-butyl Ether | 2196 |
| 2.1. | **Physical Properties** | 2186 | 4.3.3. | Diamyl Ether | 2197 |
| 2.2. | **Chemical Properties** | 2186 | 4.4. | **Chloroalkyl Ethers** | 2198 |
| 3. | **Synthesis** | 2188 | 4.4.1. | Bis(chloromethyl) Ether | 2199 |
| | | | 4.4.2. | Bis(2-chloroethyl) Ether | 2199 |
| 4. | **Individual Aliphatic Ethers** | 2189 | 4.4.3. | Bis(2-chloroisopropyl) Ether | 2200 |
| 4.1. | **Diethyl Ether** | 2189 | 5. | **Toxicology** | 2201 |
| 4.2. | **Diisopropyl Ether** | 2193 | 6. | **References** | 2203 |

# 1. Introduction

Aliphatic ethers are compounds of the type R–O–R', where R and R' are alkyl radicals that may or may not be identical and may be substituted. Structurally, ethers may be regarded as oxides. However, as far as their chemical stability is concerned, ethers have little in common with organic oxides (basicity, see Section 2.1). They are generally regarded as anhydrides of alcohols since they can be prepared by dehydration of alcohols. Ethers are classified as either symmetrical or unsymmetrical, the latter being the anhydrides of two different alcohols. Ethers are named according to IUPAC rules as hydrocarbon derivatives, e.g., ethoxyethane.

Only the low molecular mass ethers are discussed in this article; for a list of separate articles dealing with other ethers, see under the main title of this article.

## 2. Properties

### 2.1. Physical Properties

The lower molecular mass aliphatic ethers have a pleasant and characteristic odor. They are highly volatile and highly mobile. Their boiling points are lower than those of the corresponding alcohols and are similar to those of hydrocarbons of equivalent molecular mass and structure. This indicates that ether molecules do not associate in the liquid phase. Ethers are insoluble or only poorly soluble in water but they are readily miscible with almost all organic liquids. Their solvent power for many organic compounds is considerable and is often enhanced in the presence of the corresponding alcohol. Some physical properties of commercially important ethers are summarized in Table 1 and the most important properties of others in Table 2.

### 2.2. Chemical Properties

Unsubstituted ethers are chemically very stable. At room temperature, they are neither attacked by alkali metals nor undergo base hydrolysis. They are also resistant to acids, with the exception of hydroiodic acid which cleaves ethers with formation of alkyl iodide; this reaction is the basis of the Zeisel method for the determination of alkoxy groups.

At higher temperature, ethers are also hydrolyzed by other acids, especially nitric and hydrobromic acids, acyl halides, and phosphorus pentachloride [10]. The rate of hydrolysis increases rapidly with the degree of branching of the alkyl group. Therefore, breakdown or removal of nonvolatile methoxy compounds or methyl-protecting groups by boiling with hydrobromic acid is advantageous. At higher temperature, ethers can be cleaved to produce olefins, water, and alcohol, especially in the presence of catalysts such as aluminum oxide.

Ethers react more quickly with chlorine or bromine than do the corresponding alkanes, substitution taking place mainly in the $\alpha$-position. Mono- as well as bis(1-haloalkyl) ethers can be obtained by appropriate control of the reaction.

Ethers form peroxides in the presence of atmospheric oxygen, especially when exposed to light. This has long been known because violent explosions can occur toward the end of ether distillations due to the peroxides present in the distillation residue [11]–[14]. Peroxides can be detected with vanadium(V) sulfate or with iron(II) sulfate and potassium rhodanide [15]. They can be destroyed by treatment with an iron(II) salt solution [12], potassium pyrosulfite, or triethylenetetramine [16].

One of the important properties of ethers is their ability to form oxonium compounds with protons and Lewis acids, e.g., boron trifluoride, zinc chloride, or aluminum chloride. This is a result of the nucleophilic character of the ether oxygen:

**Table 1.** Physical properties of commercially important ethers

| Ether | CAS registry number | mp, °C | bp, °C | $d_4^{20}$ | $n_D^{20}$ | Heat of vaporization, kJ/kg | Flash point, °C | Ignition temperature, °C | Explosive limits in air, vol% ether |
|---|---|---|---|---|---|---|---|---|---|
| Dimethyl | [115-10-6] | −138.5 | −24.8 | | 1.3527 | | −42 | 240 | 3.0 – 18.6 |
| Diethyl | [60-29-7] | −116.3 | 34.5 | 0.7135 | 1.3807 | 362 | −40 | 180 | 1.7 – 48 |
| Di-n-propyl | [111-43-3] | −122 | 90.5 | 0.7360 | 1.3682 | | −21 | | |
| Diisopropyl | [108-20-3] | −86.2 | 68.5 | 0.7258 | 1.3981 | 285 | −28 | 405 | 1.0 – 21 |
| Di-n-butyl | [142-96-1] | −95.2 | 142.4 | 0.7704 | 1.4119 | 288 | 25 | 175 | 0.9 – 8.5 |
| Di-n-amyl | [693-65-2] | −69.3 | 187.5 | 0.7751 | 1.4085 | 276* | 57* | 171* | |
| Diisoamyl | [544-01-4] | | 173.2 | 0.7777 | 1.4346 | | | | |
| Bis(chloromethyl) | [542-88-1] | | 105 | 1.315 | 1.4572 | | | | |
| Bis(2-chloroethyl) | [111-44-4] | −50 | 178 | 1.2199 | 1.4505 | 268.4 | 55 | 365 | |
| Bis(2-chloroisopropyl) | [39638-32-9] | <−20 | 187.1 | 1.1127 | | | 85 | | |

* For technical-grade amyl ether

**Table 2.** Physical properties of some representative ethers

| Ether | CAS registry number | Formula | bp, °C | $d_4^{20}$ | $n_D^{20}$ |
|---|---|---|---|---|---|
| Di-sec-butyl | [6863-58-7] | [(CH₃)(C₂H₅)CH]₂O | 122 | 0.7590* | 1.3931 |
| Diisobutyl | [628-55-7] | [(CH₃)₂CHCH₂]₂O | 123 | 0.7600 | 1.4107 |
| Di-tert-butyl | [6163-66-2] | (CH₃)₃C-O-C(CH₃)₃ | 108 | 0.7622 | 1.3946 |
| Di-n-hexyl | [112-58-3] | C₆H₁₃-O-C₆H₁₃ | 226 | 0.7936 | 1.4204 |
| Methyl n-propyl | [557-17-5] | CH₃-O-C₃H₇ | 38.9 | 0.738 | 1.3579 |
| Methyl isopropyl | [598-53-8] | CH₃-O-CH(CH₃)₂ | 32 | 0.7347** | 1.3579 |
| Methyl n-butyl | [628-28-4] | CH₃-O-C₄H₉ | 70 | 0.7443 | 1.3736 |
| Methyl tert-butyl | [1634-04-4] | CH₃-O-C(CH₃)₃ | 55.1 | 0.7406 | 1.3690 |
| Ethyl n-butyl | [628-81-9] | C₂H₅-O-C₄H₉ | 91.4 | 0.7490 | 1.3818 |
| Bis(1-chloroethyl) | [6986-48-7] | (CH₃CHCl)₂O | 113 | 1.1285 | 1.4250 |
| 1-Chloroethyl ethyl | [7081-78-9] | CH₃CHCl-O-C₂H₅ | 98 (decomp.) | 0.9495 | 1.4040 |
| 2-Chloroethyl ethyl | [628-34-2] | ClCH₂CH₂-O-C₂H₅ | 106 | 0.9945 | 1.412 |

* $d_4^{25}$ ; ** $d_{20}^{20}$

$$R-\ddot{\underset{R}{O}}: + HX \rightleftharpoons \left[R-\underset{R}{\overset{..}{O}}-H\right]^+ + X^-$$

Therefore, ethers also dissolve in cold, concentrated sulfuric acid to form oxonium sulfates. This reaction is exploited to separate ethers from hydrocarbons, whereby the ether is recovered by diluting the sulfuric acid. Organometallic compounds (in particular, organomagnesium Grignard reagents) dissolve in ethers to form complexes. The tertiary oxonium salts $[R_3O]^+X^-$ described by MEERWEIN are effective alkylating agents [1].

## 3. Synthesis

Ethers are generally prepared by catalytic dehydration of alcohols or by reaction of alkyl halides with alkoxides:

1) *Catalytic dehydration of alcohols in the liquid phase.* Acids (particularly sulfuric acid) and electrophilic metal and nonmetal halides are effective catalysts for dehydrating alcohols in the liquid phase to produce ethers. Moderately strong acids and salts of weak bases and strong acids are most effective [17] but favorable results are obtained only for the lower alcohols (1 – 3 carbon atoms). The reaction proceeds via intermediate formation of an oxonium compound:

$$\underset{R}{\overset{H}{>}}O^+-H + \underset{H}{\overset{R}{>}}O \longrightarrow H-O-H + R-\underset{H}{\overset{R}{\overset{+}{O}}}$$

2) *Catalytic dehydration of alcohols in the gas phase* over, e.g., aluminum oxide, anhydrous alumina, or bauxite at a temperature of 180 – 250 °C gives ether yields of up to 75%.
3) *Reaction of alkyl halides with alkali-metal alkoxides* by the Williamson synthesis [18]:
   RONa + R'X ⟶ R-O-R' + NaX
   This method is used for preparing most unsymmetrical and symmetrical aliphatic ethers. Symmetrical ethers can be prepared similarly from an alkyl halide and silver oxide. Thus, di-*tert*-butyl ether [6163-66-2], whose preparation is difficult, can be obtained by reacting *tert*-butyl chloride with silver carbonate [19].
   Introduction of dimethyl sulfoxide as a solvent considerably improved yields in the Williamson ether synthesis [20]. A mixture of sodium hydroxide and alcohol can be used in place of the alkoxide.
4) *Other methods.* Ethers can also be prepared from secondary or tertiary alcohols [21] or by indirect addition of an alcohol to an olefin [22].

# 4. Individual Aliphatic Ethers

Dimethyl ether and Methyl Tert-Butyl Ether are separate keywords.

## 4.1. Diethyl Ether

Diethyl ether [60-29-7], also known as ethoxyethane or simply ether, $CH_3CH_2$–O–$CH_2CH_3$, $M_r$ 74.12, is one of the most important ethers. It is a clear, mobile liquid with a sweetish, slightly pungent, characteristic odor.

**Physical Properties** (see also Table 1). The vapor–liquid equilibria of the diethyl ether–ethanol and diethyl ether–ethanol–water systems at different pressures are described in [23]. Binary azeotropes of diethyl ether are listed in Table 3. Diethyl ether is completely miscible with common organic solvents, but only partially miscible with water.

Vapor pressure of diethyl ether:

| $t$, °C | 0 | 10 | 20 | 30 | 40 | 60 | 80 | 100 |
|---|---|---|---|---|---|---|---|---|
| $p$, kPa | 18.89 | 38.88 | 56.28 | 86.29 | 122.8 | 231.1 | 396.4 | 647.2 |

Critical data $t_{crit}$ 194 °C; $p_{crit}$ 3.63 MPa; $\varrho_{crit}$ 262.5 kg/m³
Specific heat capacity $c_p$:

| $t$, °C | 0 | 30 | 120 | 180 |
|---|---|---|---|---|
| $c_p$, kJ kg$^{-1}$ K$^{-1}$ | 2.21 | 2.29 | 3.36 | 4.36 |

Heat of formation (25 °C): –273 kJ/mol
Heat of combustion: 36.85 MJ/kg

Density:

| $t$, °C | 0 | 10 | 15 | 20 | 30 |
|---|---|---|---|---|---|
| $p$, kg/m³ | 736.3 | 725.0 | 719.3 | 713.5 | 701.8 |

Electrical conductivity: $3.7 \times 10^{-13}$ S
Dielectric constant (20 °C): 4.34; ether vapor
  (103.3 kPa, 100 °C): 1.0049
Viscosity
  (20 °C): 0.245 mPa·s;
  (25 °C): 0.223 mPa·s
Surface tension (20 °C): 17.1 mN/m

Solubility limits in the system diethyl ether–water:

| $t$, °C | 0 | 10 | 20 | 30 | 60 | 100 |
|---|---|---|---|---|---|---|
| Water in diethyl ether, wt% | 1.00 | 1.09 | 1.19 | 1.30 | 1.80 | |
| Diethyl ether in water, wt% | 11.7 | 8.7 | 6.5 | 5.2 | 3.6 | 2.1 |

**Table 3.** Binary azeotropes of diethyl ether

| Component | Component concentration, wt% | Azeotrope bp, °C |
|---|---|---|
| Water (98 kPa) | 1.26 | 34.15 |
| Water (231 kPa) | 2.0 | 60 |
| Water (1080 kPa) | 4.5 | 114 |
| Acetaldehyde | 76.5 | 18.9 |
| Methyl formate | 50 | 28.2 |
| Pentane | 32 | 33.4 |
| Propylene oxide | 49.6 | 32.6 |

**Chemical Properties** (see also Section 2.2). Due to its high chemical stability, diethyl ether has found no applications as an intermediate. Although it can be used to prepare 1-chloroethyl ethyl ether and bis(1-chloroethyl) ether by chlorination at low temperature [24], the reaction is not exploited commercially.

Diethyl ether is oxidized to acetic acid by strong oxidants such as chromium trioxide or concentrated nitric acid. If nitric acid is used, a risk of explosion exists, especially if concentrated sulfuric acid is also present.

It must once again be stressed that peroxides may be formed when diethyl ether comes into contact with air. On distillation of the ether, the peroxides may be concentrated and then decompose violently [11]–[14]. Therefore, diethyl ether should always be checked for the presence of peroxides before use [15]. The peroxides can be removed by shaking the diethyl ether with reducing agents such as 5% aqueous iron(II) sulfate solution [11], [25]. Peroxide formation can be prevented by the addition of radical scavengers such as phenols or amines. 2,6-Di-*tert*-butyl-*p*-cresol [*128-37-0*] is often added at a concentration of 1–30 mg/kg, but other stabilizers (e.g., sodium diethylthiocarbamate) are also used, depending on the purpose for which the ether is employed.

**Production.** In countries where ethanol is produced synthetically, diethyl ether is produced in sufficient quantities as a byproduct to make its synthesis unnecessary (→ Ethanol). Otherwise, ether is produced from ethanol by dehydration, either with sulfuric acid or catalytically in the vapor phase over alumn. The latter process is simpler, but gives lower yields.

In the *Sulfuric acid diethyl ether process*, ethanol and sulfuric acid (96%) are heated to ca. 125 °C in a ratio of 1:3. This is done in a lead-lined steel kettle that is fitted with a heating coil and further protected by an acid-resistant brick lining. The ether is distilled with part of the unreacted alcohol. Any sulfuric acid and sulfur dioxide (produced via the reduction of the sulfuric acid by ethanol with formation of acetaldehyde) present in the distillate are removed by washing with dilute sodium hydroxide solution.

Ethanol is introduced into the sulfuric acid either in vapor or in liquid form depending on the degree of conversion. The temperature in the reactor must be carefully controlled because the reaction should not proceed below 125 °C or over

130 °C. The process runs continuously for several months and is then stopped to replace the sulfuric acid, which becomes heavily contaminated with tarlike polymers. The yield is ca. 95%.

*Catalytic dehydration over alumn* is carried out from 180 to 230 °C, the higher temperatures being required toward the end of the life of the catalyst. The reaction is carried out in a steel tube furnace, which contains the alumn catalyst. The ethanol (94.5 wt%) is first passed through a vaporizer and then over the catalyst. The reaction heat (ca. 25 kJ/mol) is removed by pumping water under pressure to generate steam. The drum pressure is also utilized to control the temperature in the reactor. Ethanol conversion per pass is 60–80%, depending on the catalyst activity. The vapor leaving the reactor is condensed and washed with dilute sodium hydroxide solution to remove traces of sulfuric acid. The maximum yield is 90% due to a side reaction, which results in the formation of ethylene.

Water is removed from the alumn catalyst by heating crystalline potassium aluminum sulfate in flat iron pans until a viscous melt is obtained. After cooling, the resulting mass is ground to give pieces with a diameter of ca. 0.5–1 cm; more water is then removed by gentle heating to give a residual water content of 15–20%.

During *ethanol synthesis* by the direct catalytic hydration of ethylene (→ Ethanol), about 2% of the ethylene is converted to diethyl ether as a byproduct. Indirect hydration of ethylene with sulfuric acid also yields 10% ether. The amount of ether thus obtained exceeds the demand. Therefore, the ether yield is matched to the demand by recycling it into the process (for reaction equilibrium, see [26]) or by appropriate adjustment of the reaction conditions.

*Purification.* Irrespective of the process used, the diethyl ether must be purified to eliminate small amounts of acetaldehyde, hydrocarbons, and varying amounts of ethanol and water. Removal of the acetaldehyde by distillation leads to loss of ether due to the formation of a binary azeotrope. Removal is also incomplete because part of the acetaldehyde is bound by ethanol as an acetal and is liberated during further processing. Therefore, high-quality ether cannot be obtained by this method.

Acetaldehyde can be removed by hydrogenation or by washing the ether vapor with sodium hydrogen sulfite solution; subsequent washing with sodium hydroxide solution is required to remove any sulfur dioxide present. Water can be removed from the ether azeotrope by passing the vapor over a desiccant such as sodium hydroxide flakes or through a molecular sieve or by high-pressure distillation. At a pressure of ca. 1.1 MPa, a binary azeotrope containing 4.5 wt% water leaves the top of the fractionating column and, after condensation and cooling, separates into two phases. The upper phase is returned to the column, whereas the aqueous phase is recycled into the low-pressure ether column.

For environmental reasons, hydrogenation is preferred for purification and high-pressure distillation for dehydration.

**Quality Specifications and Analysis.** Two grades of diethyl ether are commercially available: *technical-grade ether* and *anesthetic ether*. There are no universally accepted specifications for technical-grade ether, but it is conventionally free of peroxides. Anhydrous technical-grade ether has a maximum water content of 500 mg/kg. Anesthetic ether has a neutral reaction, is free from foreign odors, passes the tests for peroxides and aldehydes, and has a maximum water content of 0.2% [27].

Potassium iodide–starch solution is used to test for traces of peroxide: no color should develop after 30 min in the dark.

Nessler's reagent is used to test for aldehydes and acetone: no more than a weak opalescence is permissible after 5 min in the dark. Water is determined by the Karl Fischer method.

**Storage and Transportation.** Diethyl ether is a hazardous, flammable liquid, it is subject to the Gefahrstoffverordnung of the Federal Republic of Germany [28] and must be appropriately labeled for transportation. Its hazard classifications are as follows:

| | |
|---|---|
| VbF | A I |
| GGVE/GGVS | 3/2 a |
| IMDG code | 3.1, UN No. 1155 |
| Ignition group (VDE 0171) | G 4 |
| United States | CFR 49: 172.02, flammable liquid |

Diethyl ether is stored in mild steel containers. Blanketing with nitrogen is recommended but not essential. Tin-plate cans, mild steel drums, and aluminum containers are suitable for transportation, but zinc-coated vessels are not.

Being heavier than air (ca. 2.5-fold), ether vapor creeps along the ground Therefore, good ventilation is necessary when ether is handled to prevent the formation of ignitable mixtures. Because of its low autoignition temperature (180 °C), diethyl ether vapor can ignite even if it comes into contact with a noninsulated 1.0-MPa steam pipeline.

The fact that ether is a poor conductor of electricity presents a further hazard because static electrical charge can be generated in flowing ether. Therefore, the flow rate of ether must be restricted; a limit of 1–1.5 m/s, for example, has been set for pipelines with a diameter of 12 mm [8], [10]. Pipelines and pumps must be grounded (earthed), and flanged joints must be bridged by a conductor.

When diethyl ether is poured or transferred from one container to another, both containers must be grounded to prevent the buildup of static charge and possible explosions. If glass or plastic containers are used, filling should be performed by using a metallic pipeline that is below the surface of the ether or by placing grounded chains or other metallic conductors in the free-flowing ether to remove static electrical charge.

**Uses.** Diethyl ether is a good solvent for many oils, fats, resins, alkaloids, odorants, and dyes and, therefore, is widely used as a solvent and as an extractant. When mixed with ethanol, diethyl ether is used to gel nitrocellulose and to dissolve collodion wool.

Ether is used as a reaction medium in the laboratory and in industry due to its chemical stability, low boiling point, and solvent properties for organometallic compounds (Grignard reagents). It is also employed as an anesthetic in medicine, but this use is declining.

**Economic Aspects.** Of the ethers discussed in this article, diethyl ether is the only one to be produced in quantities that warrant discussion. In Western Europe current production is 13 000 – 14 000 t/a, 2000 t/a being produced in the Federal Republic of Germany. Demand remains stagnant.

## 4.2. Diisopropyl Ether

Diisopropyl ether [108-20-3], also known as 2-isopropoxyisopropane or isopropyl ether, $(CH_3)_2$ CH–O–CH$(CH_3)_2$, $M_r$ 102.18, is a clear, colorless, mobile, highly flammable liquid with a characteristic ethereal odor. Its solubility in water is low, but it is soluble in organic solvents and readily forms explosive peroxides. It has a higher boiling point, a lower vapor pressure, and a lower water solubility than diethyl ether, which means that diisopropyl ether can be recovered without significant loss after use as a solvent or as an extractant.

**Physical Properties** [31] (see also Table 1). Azeotropes of diisopropyl ether are listed in Table 4. Diisopropyl ether is miscible with most organic solvents, including acetic acid, but it is not miscible with glycols, glycerol, glycerol monoacetate, or ethanolamines. Its mutual solubility with water is lower than that of diethyl ether.

Saturation concentration (20 °C): 734 g/m$^3$
Vapor density (air = 1): 3.5
Evaporation rate (ether = 1): 1.6
Critical data: $t_{crit}$ = 226.9 °C; $p_{crit}$ = 2.875 MPa; $\varrho_{crit}$ = 0.265 kg/m$^3$
Specific heat capacity (30 °C): 2.21 kJ kg$^{-1}$ K$^{-1}$
Heat of formation: –3443 kJ/kg
Heat of combustion (25 °C): 39.279 MJ/kg
Vapor pressure:

| $t$, °C | 0 | 10 | 20 | 30 | 40 | 68.47 |
|---|---|---|---|---|---|---|
| $p$, kPa | 5.53 | 9.93 | 17.5 | 24.9 | 37.7 | 101.3 |

Coefficient of cubic expansion: 0.00144 K$^{-1}$
Surface tension (23 °C): 32 mN/m
Solubility limits in the system water – diisopropyl ether:

| $t$, °C | 10 | 20 | 30 | 50 |
|---|---|---|---|---|
| Water in diisopropyl ether, wt% | 0.50 | 0.55 | 0.61 | 0.84 |
| Diisopropyl ether in water, wt% | 1.35 | 0.88 | 0.70 | 0.59 |

**Chemical Properties** (see also Section 2.2). The chemistry of diisopropyl ether differs little from that of diethyl ether. However, it has a significantly greater tendency to form peroxides and must, therefore, be tested for the presence of peroxides before use. This can be done by adding potassium iodide solution in 1 mol/L sulfuric acid and a few drops of ammonium molybdate solution and then titrating the free iodine with sodium thiosulfate.

As with diethyl ether, peroxide formation can be prevented during long storage periods largely by adding stabilizers such as 1-naphthol or 2,6-di-*tert*-butyl-*p*-cresol. Methods for removing peroxides are described in [30].

**Production.** Diisopropyl ether is obtained as a byproduct in the production of isopropyl alcohol from propylene and water (→ Propanols). The crude ether contains water, isopropyl alcohol, $C_6$ hydrocarbons, and other byproducts.

Purification of the ether involves extraction of isopropyl alcohol with water and drying with the binary azeotrope, diisopropyl ether–water. Further distillation stages afford diisopropyl ether as an overhead product with a purity >99%.

**Uses.** Diisopropyl ether is used as a solvent for chemical reactions. Its volatility and water solubility are lower than those of diethyl ether, thus, diisopropyl ether is used for extracting substances from aqueous solutions, e.g., for extracting acetic acid, lactic acid, and phenols [31] and for removing phenols from wastewater [32]–[34]. Diisopropyl ether is also used as an extractant for essential oils, nicotine, and vitamins. As with diethyl ether, diisopropyl ether dissolves nitrocellulose when mixed with ethanol and, therefore, can be used in the explosives industry. Due to its high solvent power and volatility, diisopropyl ether is used as a stain remover, since it tends to leave no solvent marks.

Diisopropyl ether is employed for the dealkylation of mineral oil because the solubility of alkane wax in diisopropyl ether (unlike that of the other components of mineral oil) decreases greatly when it is cooled or when, for example, acetone is added. Large quantities of diisopropyl ether, usually still containing 2-propanol, are used in the fuel industry; diisopropyl ether has good antiknock properties and helps to prevent the carburetor from icing up and deposits from forming in the engine. Although diisopropyl ether is less effective than benzene in removing water from ethanol (→ Ethanol) or 2-propanol (→ Propanols), it can be used for this purpose when the treated alcohol must be free from benzene (<0.5 mg/kg) for toxicological reasons.

**Storage and Transportation.** The requirements for the storage and transportation of diisopropyl ether are similar to those described for diethyl ether. The hazard classifications for diisopropyl ether are as follows:

| | |
|---|---|
| GGVE/GGVS | 3/3b |
| IMDG code | 3.1, UN No. 1159 |
| United States | CFR 49: 172.101, flammable liquid |
| VbF | AI |

**Table 4.** Azeotropes of diisopropyl ether

| Component | Component concentration, wt% | Azeotrope bp, °C |
|---|---|---|
| **Binary azeotropes** | | |
| Water | 4.5 | 62.2* |
| Water | 7.6 | 92.0* (203 kPa) |
| Ethanol | 17.1 | 64.0 |
| 2-Propanol | 14.1 | 65.2 |
| Acetonitrile | 17.0 | 61.7 |
| Propionitrile | >4.0 | <67.5 |
| Acetone | 61.0 | 54.2 |
| Hexane | 47.0 | 67.5 |
| Methylcyclopentane | <20.0 | <68.0 |
| **Ternary azeotropes** | | |
| Water | 1.9 | 53.8 |
| Acetone | 53.4 | |
| Water | 5.0 | 59.0* |
| Acetonitrile | 13.0 | |
| Water | 4.0 | 61.0* |
| Ethanol | 6.5 | |
| Water | 9.1 | 128.5* (700 kPa) |
| Ethanol | 14.2 | |
| Water | 4.1 | 61.6* |
| 2-Propanol | 7.1 | |

\* Condensate separates into two phases

## 4.3. Other Aliphatic Ethers

Only a limited market exists for the remaining aliphatic ethers.

## 4.3.1. Di-n-propyl Ether

Di-n-propyl ether [*111-43-3*], $M_r$ 102.17, $CH_3(CH_2)_2-O-(CH_2)_2CH_3$, is a colorless, mobile liquid with an ethereal odor.

**Properties** (see also Table 1). Azeotropes of di-n-propyl ether are listed in Table 5. The solvent properties of di-n-propyl ether are similar to those of diisopropyl ether (see Section 4.2). At 25 °C the solubility of water in di-n-propyl ether is 0.68%, and that of the ether in water is 0.25%.

The danger of peroxide formation is considerably less for di-n-propyl ether than for diisopropyl ether.

**Table 5.** Azeotropes of di-n-propyl ether

| Component | Component concentration, wt% | Azeotrope bp, °C |
|---|---|---|
| **Binary azeotropes** | | |
| Water | | 75 |
| Methanol | 72 | 63.8 |
| Ethanol | 44 | 74.5 |
| 1-Propanol | 32 | 85.8 |
| 2-Propanol | 52 | 78.2 |
| Isobutanol | 10 | 89.5 |
| 2-Butanol | 22 | 87.0 |
| tert-Butanol | 52 | 79.0 |
| **Ternary azeotropes** | | |
| Water | 11.7 | 74.8 |
| 1-Propanol | 20.2 | |

Vapor pressure:

| $t$, °C | 0 | 21.6 | 50.4 | 70.0 |
|---|---|---|---|---|
| $p$, kPa | 2.67 | 8.00 | 26.7 | 53.3 |

**Production and Uses.** Di-n-propyl ether can be prepared from 1-propanol by dehydration with sulfuric acid. It is used for the same purposes as diethyl ether, advantages are offered by its considerably higher boiling point.

## 4.3.2. Di-n-butyl Ether

Di-n-butyl ether, also known as 1-butoxybutane, is the only butyl ether that is commercially available. Di-n-butyl ether [142-96-1], $CH_3(CH_2)_3$–O–$(CH_2)_3CH_3$, $M_r$ 130.22, is a colorless liquid with a characteristic, fruity odor.

**Properties** (see also Table 1). Azeotropes of di-n-butyl ether are listed in Table 6.

Vapor pressure:

| $t$, °C | 11.3 | 29.7 | 66.8 | 127.7 |
|---|---|---|---|---|
| $p$, kPa | 1.0 | 2.0 | 7.6 | 76.0 |

Viscosity (30 °C): 1.6 mPa·s
Surface tension (20 °C): 22.9 mN/m

**Production.** Di-n-butyl ether is produced by dehydrating 1-butanol with sulfuric acid. It is also obtained as a byproduct during the synthesis of the butyl esters of higher acids; it occurs in the recovered alcohol and can be obtained by extractive distillation with water, drying, and rectification.

**Table 6.** Azeotropes of di-*n*-butyl ether

| Component | Component concentration, wt% | Azeotrope bp, °C |
|---|---|---|
| **Binary azeotropes** | | |
| Water | 33 | 92.9 |
| 1-Butanol | 82.5 | 117.6 |
| 1-Pentanol | 50 | 134.5 |
| Isoamyl alcohol | 65 | 129.8 |
| Ethylene glycol | 6.4 | 139.5 |
| Butyl acetate | 95 | 125.9 |
| 2-Methoxyethanol | 68 | 122 |
| 2-Ethoxymethanol | 50 | 127 |
| 2-Propoxymethanol | 37 | 138.5 |
| **Ternary azeotropes** | | |
| Water | 29.9 | 90.6 |
| 1-Butanol | 34.6 | |
| Water | 31.2 | 45 (1.3 kPa) |
| 1-Butanol | 24.6 | |
| Water | 24.7 | 86.6 |
| 2-Butanol | 56.1 | |
| **Quarternary azeotrope** | | |
| Water | 30 | 90.6 |
| 1-Butanol | 13 | |
| *n*-Butyl acetate | 51 | |

Typical *specifications* are: boiling range, 139–144 °C; $d_{20}^{20}$, 0.7690–0.7715; water content, max. 0.1%; acid content as butyric acid, max. 0.02%; peroxide content as dibutyl peroxide, max. 0.05%.

**Uses.** Di-*n*-butyl ether is an excellent solvent for many natural and synthetic resins, fats, oils, organic acids, and alkaloids, especially at high temperature. When mixed with ethanol or 1-butanol, di-*n*-butyl ether also dissolves ethyl cellulose. The mutual solubility of di-*n*-butyl ether with water is low: at 20 °C, the solubility of the ether in water is 0.03% and that of water in the ether is 0.19%. Di-*n*-butyl ether is consequently used as an extractant for aqueous solutions. It is also employed as a reaction medium (Grignard reactions).

## 4.3.3. Diamyl Ether

Technical-grade diamyl ether is a mixture of di-*n*-amyl ether (di-*n*-pentyl ether) [693-65-2], $CH_3(CH_2)_4-O-(CH_2)_4CH_3$, $M_r$ 158.28, and diisoamyl ether (diisopentyl ether) [544-01-4], $(CH_3)_2CH(CH_2)_2-O-(CH_2)_2CH(CH_3)_2$, with traces of other amyl ethers and amyl compounds.

**Table 7.** Binary azeotropes of diamyl ethers

| Component | Di-n-amyl ether azeotrope | | Diisoamyl ether azeotrope | |
| --- | --- | --- | --- | --- |
| | bp, °C | Amyl ether content, wt% | bp, °C | Amyl ether content, wt% |
| Water | 98.4 | | 97.4 | 46 |
| Ethylene glycol | 168.8 | | 162.8 | 81 |
| Ethylene glycol monoacetate | 180.8 | 52 | 170.2 | 72 |
| 2-Butoxyethanol | 169.0 | 33 | 164.95 | 46 |
| Bis(2-chloroethyl) ether | <176.5 | | 169.35 | 61 |

**Properties** (see also Table 1). Di-n-amyl ether and diisoamyl ether are colorless to pale yellow liquids with a pearlike odor. Binary azeotropes of diamyl ethers are listed in Table 7.

Ternary azeotropes:
Water – n-amyl alcohol – di-n-amyl ether: 95.94 °C
Water – isoamyl alcohol – diisoamyl ether: bp, 94.4 °C

*Technical-grade diamyl ether*

Vapor pressure:

| $t$, °C | 18.6 | 57 | 96 | 129 | 150 |
| --- | --- | --- | --- | --- | --- |
| $p$, kPa | 0.133 | 1.33 | 8.00 | 26.6 | 53.3 |

Specific heat capacity (20 °C): 2.15 kJ kg$^{-1}$ K$^{-1}$
Surface tension (20 °C): 24.8 mN/m

**Production and Uses.** The production and uses of diamyl ether are similar to those of di-n-butyl ether.

## 4.4. Chloroalkyl Ethers

A few symmetrical chloroalkyl ethers are commercially important. They can be synthesized by direct chlorination of the appropriate ethers or by the sulfuric acid process (see Chapter 3); however, the former method is not used on a commercial scale. Chloroalkyl ethers can also be produced by reacting aldehydes with hydrogen chloride [35]:

RCHO + HCl $\rightleftharpoons$ RCHClOH
2 RCHClOH $\rightleftharpoons$ RClCH–O–CHClR + H$_2$O

The yield of the bis(1-chloroalkyl) ether can be improved by removing the water formed, e.g., by using chlorosulfonic acid.

## 4.4.1. Bis(chloromethyl) Ether

Bis(chloromethyl) ether [542-88-1], also known as dichlorodimethyl ether, $M_r$ 114.96, $ClH_2C-O-CH_2Cl$, is a clear liquid and a strong irritant to the eyes and the respiratory system. Its physical properties are listed in Table 1.

**Production and Uses.** Bis(chloromethyl) ether is produced by reacting paraformaldehyde and chlorosulfonic acid in sulfuric acid (65%) under cooling at a temperature of 26–28 °C. The reaction is controlled to prevent any hydrochloric acid escaping.

Bis(chloromethyl) ether is employed for chloromethylating compounds that cannot be chloromethylated with formaldehyde and hydrochloric acid, e.g., aromatic compounds and anilides.

## 4.4.2. Bis(2-chloroethyl) Ether

Bis(2-chloroethyl) ether [111-44-4], also known as dichloroethyl ether, $M_r$ 143.01, $ClCH_2CH_2-O-CH_2CH_2Cl$, is a colorless liquid with a chloroform-like odor that irritates the eyes and mucous membranes.

**Properties** (see also Table 1). Bis(2-chloroethyl) ether is miscible with most organic solvents except alkanes. The solubility of the ether in water at 40 °C is 1.1%, and that of water in the ether is 0.5%. Azeotropes of bis(2-chloroethyl) ether are listed in Table 8.

Vapor pressure:

| $t$, °C | 23.5 | 49.3 | 101.5 | 134.0 |
|---|---|---|---|---|
| $p$, kPa | 0.133 | 0.667 | 8.00 | 26.7 |

Specific heat capacity (20–30 °C): 1.55 kJ kg$^{-1}$ K$^{-1}$
Viscosity (25 °C): 2.065 mPa · s
Surface tension (25 °C): 41.8 mN/m

Unlike bis(chloromethyl) ether, bis(2-chloroethyl) ether is somewhat resistant to hydrolysis. Heating with solid sodium hydroxide to 200–220 °C is necessary to produce 2-chloroethyl vinyl ether [110-75-8], and with solid potassium hydroxide to 200–240 °C in the presence of ammonia to produce divinyl ether [109-93-3](→ Vinyl Ethers). Reaction of bis(2-chloroethyl) ether with primary aromatic amines leads to N-aryl morpholines; reaction with hydrazine leads to 4-aminomorpholine.

**Production.** Bis(2-chloroethyl) ether is obtained as a byproduct in the production of ethylene chlorohydrin (→ Chlorohydrins), where it collects in the distillation sump and is purified by vacuum distillation. Bis(chloromethyl) ether is also produced by reaction of ethylene chlorohydrin with sulfuric acid at 90–100 °C or by saturation of an aqueous ethylene chlorohydrin solution with ethylene and chlorine below 85 °C.

**Table 8.** Azeotropes of bis(2-chloroethyl) ether

| Component | Component concentration, wt% | Azeotrope bp, °C |
|---|---|---|
| **Binary azeotropes** | | |
| Water | 65.5 | 98 |
| Monoethylene glycol | 12.5 | 170.5 |
| Diethylene glycol | 8 | 174.6 |
| 2-Ethylhexanol | 10 | 96 (6.7 kPa) |
| Diisoamyl ether | 61 | 169.35 |
| **Ternary azeotrope** | | |
| Water | 53 | 97.5 |
| 2-Chloroethanol | 25 | |

Typical *specifications* of commercially available bis(2-chloroethyl) ether are as follows: $d_{20}^{20}$, 1.219–1.224; *bp*, 175–181 °C; water content, max. 0.1%; acid content as hydrochloric acid, max. 0.05%; ethylene dichloride content, max. 1.0%.

**Uses.** Bis(2-chloroethyl) ether is a good solvent for many natural and synthetic resins. It dissolves ethyl cellulose, but 10–30% ethanol is also required to dissolve other cellulose esters.

Bis(2-chloroethyl) ether is used in the textile industry in dyeing, washing, and mercerizing solutions. It can be used for the isolation of butadiene from $C_4$ cuts by extractive distillation. Bis(2-chloroethyl) ether is also employed as an insecticide and an acaricide.

### 4.4.3. Bis(2-chloroisopropyl) Ether

Bis(2-chloroisopropyl) ether [39638-32-9], also known as dichloroisopropyl ether, $(CH_2Cl)(CH_3)CH–O–CH(CH_3)(CH_2Cl)$, $M_r$ 171.07, is a colorless liquid.

**Properties** (see also Table 1). Except for its low volatility, the physical and chemical properties of bis(2-chloroisopropyl) ether are very similar to those of bis(2-chloroethyl) ether. Its solubility in water is slightly less: 0.17% at 20 °C.

Bis(2-chloroisopropyl) ether does not form an azeotrope with propylene chlorohydrin, but does form an azeotrope with 62.6 wt% water, *bp* 98.5 °C.

**Production and Uses.** Bis(2-chloroisopropyl) ether can be prepared by dehydration of 2-chloroisopropyl alcohol with sulfuric acid.

Typical *specifications* are: Color, 40 APHA max.; $d_{20}^{20}$, 1.113–1.119; boiling range, 180–190 °C; water content, max. 0.1%; acid content as hydrochloric acid, max. 0.01%.

The uses of bis(2-chloroisopropyl) ether are the same as those of bis(2-chloroethyl) ether.

# 5. Toxicology

**Diethyl Ether.** The most important characteristic of diethyl ether is its anesthetic effect. Diethyl ether is rapidly absorbed into the blood when it is inhaled and is equally rapidly exhaled when inhalation is stopped. Very little diethyl ether is absorbed through the skin, but contact with the skin and mucous membranes causes irritation.

A vapor concentration of ca. 10 vol% is necessary to induce anesthesia, whereas a concentration of ca. 5 vol% is required to maintain this state [36]. All the classical stages of narcosis are observed. Prolonged inhalation of over 10 vol% can cause death. In human beings, after intoxication has been overcome, reversible functional changes are observed in the kidneys and liver in the form of an increase in serum glutamic oxaloacetic transaminase and alkaline phosphatase [37].

The acute lethal dose has been found to be 6.4 vol% for rats, 10.6 vol% for dogs, and 7.16 – 19.25 vol% for monkeys [38]. The oral $LD_{50}$ for rats is ca. 2 g/kg [39]. A small proportion of diethyl ether administered to rats is metabolized to $CO_2$ [40].

MAK: 400 ppm (1200 mg/m$^3$)
TLV: 400 ppm (1200 mg/m$^3$)

**Diisopropyl Ether.** The anesthetic effects of diisopropyl ether are less pronounced than those of diethyl ether; the local irritating effects on the skin and eyes are also milder.

In experiments on monkeys, rats, and guinea pigs, a vapor concentration of 6 vol% in the inhaled air was found to be lethal as a result of respiratory paralysis. A concentration of 3 vol% is tolerated without causing death, but anesthetic effects are then observed. Repeated inhalation (ten times) of diisopropyl ether vapor at a concentration of 3 vol% is tolerated by the animals without any permanent detrimental effects; this result implies that apparently no cumulative toxicity is to be found [41].

The acute oral toxicity ($LD_{50}$) of diisopropyl ether in adult rats is 16.0 mL/kg [39].

MAK: 500 ppm (2100 mg/m$^3$)
TLV-TWA: 250 ppm (1050 mg/m$^3$)
TLV-STEL: 310 ppm (1320 mg/m$^3$)

**Di-n-butyl Ether.** Dibutyl ether's threshold concentration for producing irritation of the eyes and nose in humans is ca. 200 ppm [42].

In rats, the acute oral toxicity ($LD_{50}$) is ca. 7.4 mL/kg. The dermal $LD_{50}$ in rabbits is ca. 10 mL/kg. Inhalation of 4000 ppm for 30 min is lethal to rats; thus, the inhalation toxicity of dibutyl ether is higher than that of the other alkane ethers [43].

**Bis(chloromethyl) Ether.** Halogenated ethers generally exert stronger anesthetic effects than alkyl ethers.

Bis(chloromethyl) ether is known to display carcinogenic properties. Chronic dermal treatment of mice with a dose of 2 mg, three times a week, produces papillomas and squamous epithelium carcinomas within one year [44]. In chronic inhalation tests on mice employing nominal concentrations of ca. 0.005 mg/L, pulmonary tumors were found in almost all of the exposed animals [45]. Similar observations were made in tests on rats, where concentrations of 0.5 ppm were used [46]. Subcutaneous administration induced tumors in mice [47] and rats [48]. In the mice, an increase in pulmonary tumors was once again observed, but in the rats only fibromas and fibrosarcomas were formed at the site of injection.

Evidence also indicates that bis(chloromethyl) ether has carcinogenic effects in humans [49], [50]. Therefore, this ether should neither be manufactured as such nor used as an intermediate. However, if this is unavoidable, special precautions must be taken to prevent any form of contact such as inhalation or skin absorption. The same applies if chloromethylation is performed with formaldehyde and hydrochloric acid, since it has been reported that bis(chloromethyl) ether is then formed within less than 1 min at room temperature [51]. However, subsequent studies have failed to confirm this finding [52].

MAK: class III A 1 carcinogen
TLV-TWA: 0.001 ppm (0.005 mg/m$^3$), class A1 a carcinogen

Monochlorodimethyl ether [107-30-2] is also classified as carcinogenic in humans (MAK, class III A carcinogen; TLV-TWA, class A 2 carcinogen) because it contains up to 7% bis(chloromethyl) ether [50]. Monochlorodimethyl ether and bis(chloromethyl) ether can be chemically detoxified [53], [54]. Reaction with methanolic ammonia, secondary amines, sodium phenolate, or sodium methoxide is particularly recommended and destroys more than 99% of these ethers.

**Bis(2-chloroethyl) Ether.** The position of the halogen atom is of considerable importance in determining the properties of halogenated ethers. Substitution at the α-position leads to higher reactivity than at the β- or γ-position. This obviously also affects the toxicological properties of these ethers.

In humans, bis(2-chloroethyl) ether vapor concentrations of 550 ppm cause significant irritation of the mucous membranes of the respiratory tract. The threshold value required to evoke a response appears to be between 260 and 100 ppm [55]. Acute oral toxicity is high, the oral LD$_{50}$ values being 75 mg/kg for rats [56] and 136 mg/kg for mice [57]. Inhalation of 250 ppm for 4 h is lethal to rats [58]. In chronic inhalation studies on rats and guinea pigs, a concentration of 69 ppm is tolerated without producing any symptoms [59].

Studies on mice [44], [60] and rats [61] do not reveal any carcinogenic effects. Mutagenic effects cannot be demonstrated in *Drosophila* [62] or in the Ames test [63]. Alkylation of DNA does not occur [64].

MAK: 10 ppm (60 mg/m$^3$)
TLV: not specified

**Bis-(2-chloroisopropyl) Ether.** Bis(2-chloroiso-propyl) ether is less toxic than bis(2-chloroethyl) ether. In guinea pigs, the oral LD$_{50}$ is ca. 450 mg/kg [57]. When subacute doses were given to rats, a concentration of 200 mg/kg of body mass caused changes in the liver and kidneys [65].

In acute inhalation experiments on rats [65], a vapor concentration of 700 ppm is lethal; lethal effects start to appear at a concentration of 175 ppm. Morphological investigations reveal hepatic cell necrosis and stagnation of blood in the lungs. Irritation of the skin and mucous membranes is low [65].

MAK and TLV values are not specified for bis(2-chloroisopropyl) ether.

# 6. References

General References

[1] "Methoden zur Herstellung und Umwandlung von Ether" in *Houben-Weyl* **IV/3**.
[2] "Eigenschaften und Reinigung der wichtigsten organischen Lösemittel" in *Houben-Weyl* **I/2**.
[3] A. Weissberger: *Techniques of Organic Chemistry*, 3rd ed., vol. **VII**, Wiley Interscience, New York 1970.
[4] F. Korte (ed.): *Methodicum Chimicum*, vol. **5**, Thieme Verlag, Stuttgart, Academic Press, New York-San Francisco-London 1975.
[5] C. Mardsen, S. Mann: *Solvents Guide*, 2nd ed., Cleaver-Hume Press Ltd., London 1963.
[6] E. W. Flick (ed.): *Industrial Solvents Handbook*, Noyes Data Corporation, Parke Ridge, N.J. 1985.
[7] A. K. Doolittle: *The Technology of Solvents and Plasticizers*, J. Wiley & Sons, New York, Chapman & Hall Ltd., London 1954.
[8] Hauptverband der gewerblichen Berufsgenossenschaften: *Richtlinien für die Vermeidung von Zündgefahren infolge elektrostatischer Aufladungen*, Carl Heymanns Verlag, Köln 1980.
[9] N. I. Sax: *Dangerous Properties of Industrial Materials*, 6th ed., van Nostrand Reinhold Comp., New York-Cincinnati-Toronto-London-Melbourne 1984.

Specific References

[10] L. R. Burwell, *Chem. Rev.* **54** (1954) 615–685.
[11] A. Rieche, *Z. Angew. Chem.* **44** (1931) 896–899.
[12] R. Neu, *Z. Angew. Chem.* **45** (1932) 519–520.
[13] A. Rieche, K. Koch, *Ber. Deutsch Chem. Ges.* **75** (1942) 1016–1028.
[14] V. J. Karnojitzky, *Chim. Ind. (Paris)* **88** (1962) 233.
[15] *DAB 7*, p. 293.
[16] A. C. Hamstead, *Ind. Eng. Chem.* **56** (1964) no. 6, 37–42.
[17] J. van Elphen, *Recl. Trav. Chim. Pays-Bas* **49** (1930) 754.
[18] A. Williamson, *Justus Liebigs Ann. Chem.* **77** (1851) 37; **81** (1852) 373.
[19] J. L. Erickson, W. H. Ashton, *J. Am. Chem. Soc.* **63** (1941) 1769.
[20] R. G. Smith, A. Vanterpool, H. J. Kulak, *Can. J. Chem.* **47** (1969) 2015–2019.
[21] B. Sjöberg, K. Sjöberg, *Acta Chem. Scand.* **26** (1972) 275.

[22] G. L. Grady, S. K. Chokshi, *Synthesis* 1972, 483–84.
[23] W. P. Moeller, S. W. Englund, Tsu Kan Tsui, D. F. Othmer, *Ind. Eng. Chem.* **43** (1951) 711–717.
[24] G. E. Hall, F. M. Umbertini, *J. Org. Chem.* **15** (1950) 715–719.
[25] *Houben-Weyl*, **I/2**, p. 811.
[26] C. S. Cope, B. F. Dodge, *AIChE J.* **5** (1959) 10–16.
[27] Europäisches Arzneibuch, Band III, Deutscher Apotheker Verlag Stuttgart, GOVI Verlag GmbH Frankfurt 1978, pp. 173–174.
[28] Gefahrstoffverordnung vom 26.08.1986, Bundesgesetzbl. Teil 1, Nr. 47 vom 05.09.1986.
[29] R. Kühn, K. Birett, *Merkblätter gefährlicher Arbeitsstoffe*, 22. Erg. Lieferung II/83-P 38-1, Ecomed Verlagsgesellschaft mbH.
[30] A. C. Hamstead, *Ind. Eng. Chem.* **56** (1964) no. 6, 37.
[31] V. I. Sushchenya, G. P. Solomakka, L. I. Blyakhman, *Zh. Prikl. Kleim (Leningrad)* **51** (1968) 437.
[32] I. W. Filippow, S. S. Kagan, M. I. Kondratjewa, *Koks Khim.* **1963**, no. 12, 46; **1965**, no. 14, 2063.
[33] E. Yu. Nyarep, V. V. Sheloumov, *Tr. Nauch.-Issled, Inst. Slantser* 1968 no. 17, 108.
[34] R. G. Edmonds, G. F. Jenkins, *Chem. Eng. Progr.* **50** (1954) 111.
[35] R. Summers, *Chem. Rev.* **55** (1955) 301–353.
[36] H. L. Price, R. D. Dripps, L. S. Goodman, A. Gilman, eds.: *The Pharmacological Basis of Therapeutics*, 4th ed., Macmillan Publ. Co., New York 1970.
[37] B. Dawson et al., *Mayo Clin. Proc.* **41** (1966) 599.
[38] W. S. Spector, ed.: *Handbook of Toxicology*, vol. **1**, Saunders, Philadelphia 1956.
[39] E. T. Kimura, D. M. Ebert, P. W. Dodge, *Toxicol. Appl. Pharmacol.* **19** (1971) 699.
[40] J. C. Krantz, jr., C. J. Carr: *The Pharmacologic Principles of Medical Practice*, 7th ed., Williams and Wikins, Baltimore 1969, pp. 97–112.
[41] W. Machle, E. W. Scott, J. Treon, *J. Ind. Hyg. Toxicol.* **21** (1939) 72.
[42] L. Silverman, H. F. Schulte, W. M. First, *J. Ind. Hyg. Toxicol.* **28** (1946) 262.
[43] H. F. Smyth, jr., C. P. Carpenter, C. W. Weil, U. C. Pozzani, *AMA Arch. Ind. Hyg. Occup. Med.* **10** (1954) 61.
[44] B. L. van Duuren, C. Katz, M. Goldschmidt, K. Frenkel, A. Sivak, *J. Nat. Cancer Inst.* **48** (1972) 1431.
[45] B. K. J. Leong, H. N. MacFarland, W. H. Reese, jr., *Arch. Environ. Health* **22** (1971) 663.
[46] S. Laskin, M. Kuschner, R. T. Drew, V. P. Cappiello, N. Nelson, *Arch. Environm. Health* **23** (1971) 135.
[47] J. L. Gargus, W. H. Reese, jr., H. A. Rutter, *Toxicol. Appl. Pharmacol.* **15** (1969) 92.
[48] B. L. van Duuren, A. Sivak, B. M. Goldschmidt, C. Katz, S. Melchionne, *J. Nat. Cancer Inst.* **43** (1969) 481.
[49] A. M. Thiess, W. Hey, H. Zeller, *Zbl. Arbeitsmed.* **23** (1973) 97.
[50] D. Heuschler: *Gesundheitsschädliche Arbeitsstoffe*, Verlag Chemie, Weinheim 1973.
[51] *Am. Ind. Hyg. Ass. J.* **33** (1972) 381.
[52] G. J. Kallos et al., *Am. Ind. Hyg. Ass. J.* **34** (1973) 469.
[53] D. Martinez, *Z. Chem.* **26** (1986) 309–315.
[54] M. Iovu, V. Radulescu, M. Iqbal, *Rev. Roum. Chim.* **30** (1985) 713–717; *Chem. Abstr.* **105** (1986) 114584.
[55] H. H. Schrenk, F. A. Patty, W. P. Yant, *Public Health Rep.* **48** (1933) 1389.
[56] H. F. Smyth, jr., C. P. Carpenter, *J. Ind. Hyg. Toxicol.* **30** (1948) 63.
[57] H. F. Smyth, in W. S. Spector (ed.): *Handbook of Toxicology*, "Unveröffentlichte Untersuchung," vol. I, Saunders, Philadelphia 1956.

[58] C. P. Carpenter, H. F. Smyth, jr., U. C. Pozzani, *J. Ind. Hyg. Toxicol.* **31** (1949) 343.
[59] C. L. Hake, V. K. Rowe in F. A. Patty (ed.): *Industrial Hygiene and Toxicology*, "Ethers" 2nd rev. ed., vol. **II**, New York 1963, p. 1673.
[60] J. R. M. Innes et al., *J. Nat. Cancer Inst.* **42** (1969) 1101.
[61] B. Ulland, E. K. Weisburger, J. H. Weisburger, *Toxicol. Appl. Pharmacol.* **25** (1973) 446.
[62] C. Auerbach, J. M. Robson, J. G. Carr, *Science* **105** (1947) 243.
[63] J. Obermeier, H. Frohberg, unpublished results, Institut für Toxikologie, E. Merck, Darmstadt 1976.
[64] F. P. Guengerich, P. S. Mason, W. T. Stoff, T. R. Fox, P. G. Watanabe, *Cancer Res.* **41** (1981) 4391.
[65] Biochemical Research Laboratory, Dow Chemical Co., unpublished results.

# Ethylbenzene

*Individual keywords:* → *Benzene;* → *Styrene;* → *Toluene;* → *Xylenes*

ROBERT R. COTY, Badger Engineers, Cambridge, Massachusetts 02142, United States
VINCENT A. WELCH, Badger Engineers, Cambridge, Massachusetts 02142, United States
SANJEEV RAM, Badger Engineers, Cambridge, Massachusetts 02142, United States
JASBIR SINGH, Badger Engineers, Cambridge, Massachusetts 02142, United States

| | | | | |
|---|---|---|---|---|
| 1. | Introduction ............. 2207 | 5. | Environmental Protection ... 2215 |
| 2. | Physical Properties ........ 2208 | 6. | Quality Specifications....... 2215 |
| 3. | Chemical Properties........ 2209 | 7. | Handling, Storage, and |
| 4. | Production .............. 2209 | | Transportation............ 2216 |
| 4.1. | Production by Liquid-Phase Alkylation.............. 2209 | 8. | Uses ................... 2217 |
| | | 9. | Economic Aspects ......... 2217 |
| 4.2. | Production by Vapor-Phase Alkylation............... 2212 | 10. | Toxicology and Occupational Health................... 2218 |
| 4.3. | Separation from Mixed $C_8$ Streams................. 2214 | 11. | References............... 2219 |

# 1. Introduction

Ethylbenzene [*100-41-4*], also known as phenylethane and ethylbenzol, $C_6H_5CH_2CH_3$, $M_r$ 106.168, is a single-ring, alkylaromatic compound. It is almost exclusively (>99%) used as an intermediate for the manufacture of *styrene monomer* [*100-42-5*], $C_6H_5CH = CH_2$, one of the most important large-volume chemicals. Styrene production, which uses ethylbenzene as a starting raw material, consumes ca. 50% of the world's benzene production. Less than 1% of the ethylbenzene produced is used as a paint solvent or as an intermediate for the production of diethylbenzene and acetophenone.

Currently, almost all ethylbenzene is produced commercially by alkylating benzene with ethylene, primarily via two routes: in the liquid phase with aluminum chloride catalyst, or in the vapor phase with a fixed bed of either a Lewis acid or a synthetic zeolite catalyst developed by Mobil Corporation.

The alkylation of aromatic hydrocarbons with olefins in the presence of aluminum chloride catalyst was first practiced by M. BALSOHN in 1879 [1]. However, CHARLES FRIEDEL and JAMES M. CRAFTS pioneered much of the early research on alkylation and aluminum chloride catalysis. Over a century later, processes that employ the classic Friedel–Crafts reaction chemistry remain a dominant source of ethylbenzene.

In 1965 ca. 10% of the United States ethylbenzene production was from the superfractionation of the mixed xylenes stream produced by the catalytic reforming of naphtha [2]. In 1986, the amount of ethylbenzene derived from this source was insignificant because of the escalating cost of energy.

Ethylbenzene was first produced on a commercial scale in the 1930s by Dow Chemical in the United States and by BASF in the Federal Republic of Germany. The ethylbenzene–styrene industry remained relatively insignificant until World War II. The tremendous demand for synthetic styrene butadiene rubber (SBR) during the war prompted accelerated technology improvements and tremendous capacity expansion. This enormous wartime effort led to the construction of several large-scale factories, turning styrene production quickly into a giant industry. In 1986 the world annual production capacity of ethylbenzene exceeded $14 \times 10^6$ t, of which ca. 40% was produced in North America.

## 2. Physical Properties

Under ordinary conditions, ethylbenzene is a clear, colorless liquid with a characteristic aromatic odor. Ethylbenzene is an irritant to the skin and eyes and is moderately toxic by ingestion, inhalation, and skin adsorption. The physical properties of ethylbenzene are as follows [3]:

| | | |
|---|---|---|
| Density | at 15 °C | 0.87139 g/cm$^3$ |
| | at 20 °C | 0.8669 g/cm$^3$ |
| | at 25 °C | 0.86262 g/cm$^3$ |
| mp | | −94.949 °C |
| bp | at 101.3 kPa | 136.186 °C |
| Refractive index | at 20 °C | 1.49588 |
| | at 25 °C | 1.49320 |
| Critical pressure | | 3609 kPa (36.09 bar) |
| Critical temperature | | 344.02 °C |
| Flash point (closed cup) | | 15 °C |
| Autoignition temperature | | 460 °C |
| Flammability limit | lower | 1.0% |
| | upper | – |
| Latent heat | fusion | 86.3 J/g |
| | vaporization | 335 J/g |
| Heating value, | gross | 42 999 J/g |
| | net | 40 928 J/g |
| Kinematic viscosity | at 37.8 °C | 0.6428 × 10$^{-6}$ m$^2$/S |
| | at 98.9 °C | 0.390 × 10$^{-6}$ m$^2$/S |
| Surface tension | | 28.48 mN/m |
| Specific heat capacity | | |
|   ideal gas, 25 °C | | 1169 J kg$^{-1}$ K$^{-1}$ |
|   liquid, 25 °C | | 1752 J kg$^{-1}$ K$^{-1}$ |
| Acentric factor | | 0.3011 |
| Critical compressibility | | 0.264 |

# 3. Chemical Properties

The most important commercial reaction of ethylbenzene is its dehydrogenation to *styrene*. The reaction is carried out at high temperature (600–660 °C), usually over an iron oxide catalyst. Steam is used as a diluent. Commercially, selectivities to styrene range from 89 to 96% with per-pass conversions of 65–70%. Side reactions involve mainly the dealkylation of ethylbenzene to benzene and toluene.

Another reaction of commercial importance is the oxidation of ethylbenzene by air to the *hydroperoxide*, $C_6H_5CH(OOH)CH_3$ [*3071-32-7*]. The reaction takes place in the liquid phase, with no catalyst required. However, because hydroperoxides are unstable compounds, exposure to high temperature must be minimized to reduce the rate of decomposition. The production of byproducts is reduced if the temperature is gradually lowered during the course of the reaction. The hydroperoxide is subsequently reacted with propene in a process that yields *styrene* and *propylene oxide* as coproducts.

With a suitable catalyst, ethylbenzene can be converted to *xylenes*. Commercial processes for isomerizing xylenes usually involve the catalytic isomerization or dealkylation of ethylbenzene.

Like toluene, ethylbenzene may be dealkylated catalytically or thermally to *benzene*. Ethylbenzene also undergoes other reactions typical of alkylaromatic compounds [4].

# 4. Production

Currently, the primary source of ethylbenzene is the alkylation of benzene with ethylene. The only other source, the superfractionation of mixed $C_8$ aromatic streams, supplies only a small portion of the ethylbenzene produced. Two distinct types of ethylbenzene alkylation processes are currently used commercially: liquid-phase alkylation and vapor-phase alkylation.

## 4.1. Production by Liquid-Phase Alkylation

Liquid-phase aluminum chloride processes have been the dominant source of ethylbenzene since the 1930s. Several companies have developed variations of this technology. Processes currently in use include those of Dow Chemical, BASF, Shell Chemical, Monsanto, Société Chimique des Charbonnages, and Union Carbide/Badger. The Monsanto process is currently the most modern commercially licensed aluminum chloride alkylation technology.

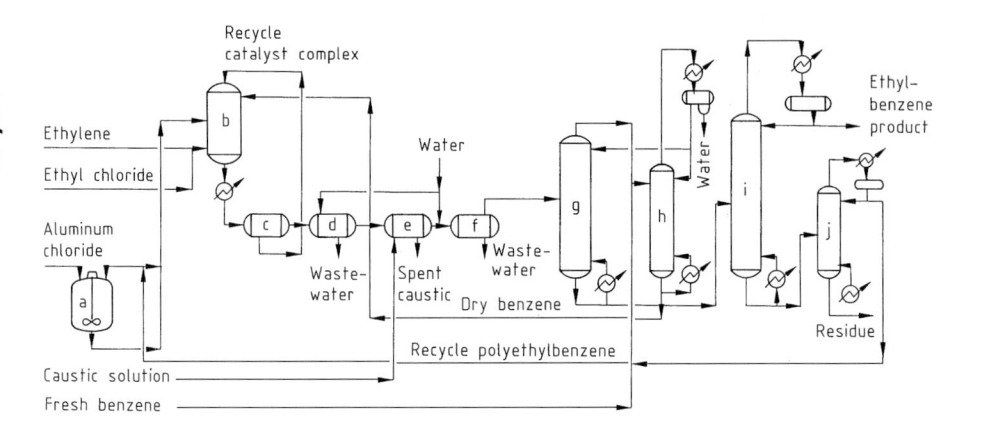

**Figure 1.** Aluminum chloride ethylbenzene process for ethylbenzene production
a) Catalyst mix tank; b) Alkylation reactor; c) Settling tank; d) Acid separator; e) Caustic separator; f) Water separator; g) Benzene recovery column; h) Benzene dehydrator column; i) Ethylbenzene recovery column; j) Polyethylbenzene column

Alkylation of benzene with ethylene in the presence of an aluminum chloride catalyst complex is exothermic ($\Delta H - 114$ kJ/mol); the reaction is very fast and produces almost stoichiometric yields of ethylbenzene. In addition to $AlCl_3$, a wide range of Lewis acid catalysts, including $AlBr_3$, $FeCl_3$, $ZrCl_4$, and $BF_3$, have been used. Aluminum chloride processes generally use ethyl chloride or hydrogen chloride as a catalyst promoter. These halide promoters reduce the amount of $AlCl_3$ required. The reaction mechanism has been studied in detail [5].

*Alkylation.* In the *conventional $AlCl_3$ process* (see Fig. 1), three phases are present in the reactor: aromatic liquid, ethylene gas, and a liquid catalyst complex phase (a reddish-brown material called red oil). A mixture of catalyst complex, dry benzene, and recycled polyalkylbenzenes is continuously fed to the reactor and agitated to disperse the catalyst complex phase in the aromatic phase. Ethylene and the catalyst promoter are injected into the reaction mixture through spargers, and essentially 100% of the ethylene is converted. Low ethylene:benzene ratios are used to give optimum overall yield of ethylbenzene. Commercial plants typically operate at ethylene:benzene molar ratios of ca. 0.3–0.35. As this ratio is increased, more side reactions, such as transalkylation and isomeric rearrangement, occur. Further alkylation of ethylbenzene leads to the reversible formation of lower molecular mass polyalkylbenzenes. The loss in net yield due to residue is minimized by recycling this material to the alkylation reactor. In addition, because the reaction occurs close to thermodynamic equilibrium, the traditional processes use a single reactor to alkylate benzene and transalkylate polyalkylbenzenes.

The reaction temperature is generally limited to 130 °C; a higher temperature rapidly deactivates the catalyst and favors formation of nonaromatics and polyalkylbenzenes, which are preferentially absorbed by the highly acidic catalyst complex, resulting in byproduct formation. Sufficient pressure is maintained to keep the reactants in the liquid phase. Because the reaction mixture is highly corrosive, the alkylation reactors are lined with either brick or glass. High-alloy materials of construction are also required for the piping and handling systems.

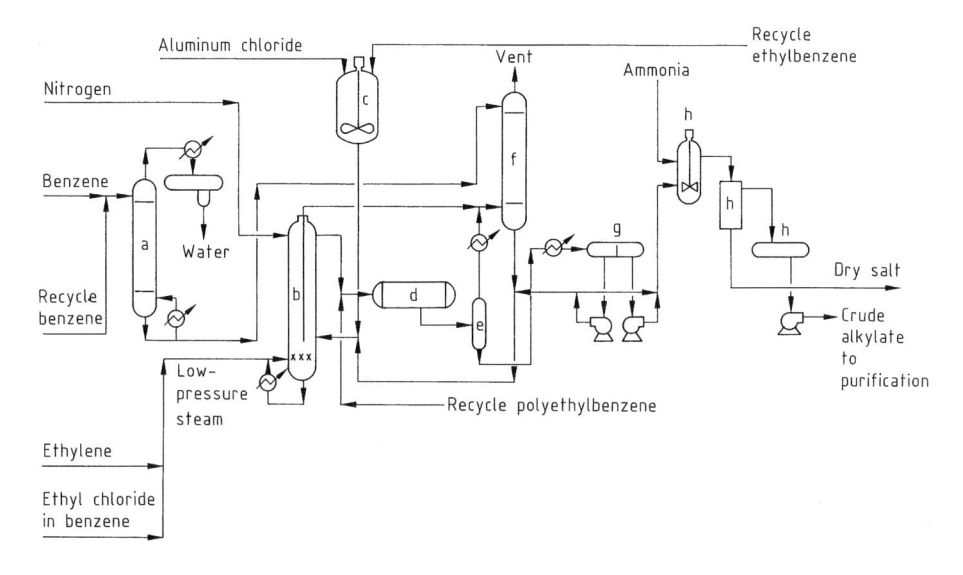

**Figure 2.** Homogeneous liquid-phase alkylation process for ethylbenzene production
a) Benzene drying column; b) Alkylation reactor; c) Catalyst preparation tank; d) Transalkylator; e) Flash drum;
f) Vent gas scrubbing system; g) Decantor; h) Neutralization system

The liquid reactor effluent is cooled and discharged into a settler, where the heavy catalyst phase is decanted from the organic liquid-phase and recycled. The organic phase is washed with water and caustic to remove dissolved AlCl$_3$ and promoter. The aqueous phase from these treatment steps is first neutralized and then recovered as a saturated aluminum chloride solution and a wet aluminum hydroxide sludge.

Removal of dissolved catalyst from the organic stream has long been a problem for ethylbenzene producers. Recently CdF Chimie found that more complete recovery of AlCl$_3$ could be achieved by first contacting the organic phase with ammonia instead of sodium hydroxide [6].

*Separation.* Purification of the ethylbenzene product is usually accomplished in a series of three distillation columns. The unreacted benzene is recovered by the first column as an overhead distillate. The second column separates the ethylbenzene product from the heavier polyalkylated components. The bottoms product of the second column is fed to a final column, where the recyclable polyalkylbenzenes are stripped from nonrecyclable high molecular mass residue compounds. The residue, or flux oil, consisting primarily of polycyclic aromatics, is burned as fuel.

Because the alkylation mixture can tolerate only minor amounts of water, the recycled benzene and fresh benzene must be dried thoroughly prior to entering the reactor. Water not only increases corrosion, but also decreases catalyst activity. Benzene dehydration is accomplished in a separate column.

The improved *Monsanto process* (see Fig. 2) has distinct advantages compared to conventional AlCl$_3$ processes. The most important of these is a significant reduction in the AlCl$_3$ catalyst use, thus lessening the problem of waste catalyst disposal. Monsanto found that by an increase in temperature and by careful control of ethylene addition, the required AlCl$_3$ concentration could be reduced to the solubility limit, thereby eliminating the separate catalyst complex phase [7]. Therefore, the alkylation occurs in a single homogeneous liquid phase instead of in the two liquid phases of earlier processes. Monsanto claims that a separate catalyst complex phase may actually prevent the attainment of maximum reactor yields. With a few exceptions, the flow scheme of the Monsanto process is nearly the same as that of more traditional processes. The process is also capable of operating with low-concentration ethylene feed. Typically, the alkylation temperature is maintained at 160–180 °C. This higher operating temperature enhances catalyst activity, with the additional benefit that the heat of reaction can be recovered as low-pressure steam.

Whereas the traditional process accomplishes alkylation and transalkylation in a single reactor, the homogeneous catalyst system must employ a separate transalkylation reactor. At the lower catalyst concentrations, the recycle of substantial amounts of polyalkylbenzenes terminates the alkylation reaction. Therefore, only dry benzene, ethylene, and catalyst are fed to the alkylation reactor. The recycle polyethylbenzene stream is mixed with the alkylation reactor effluent prior to entering the transalkylation reactor. The transalkylation reactor is operated at much lower temperature than the primary alkylation reactor.

After transalkylation, the reaction products are washed and neutralized to remove residual $AlCl_3$. With the homogeneous process, all of the catalyst remains in solution. The catalyst-free organic reaction mixture is then purified using the sequence described previously for the conventional $AlCl_3$ process. As with other $AlCl_3$ processes, the organic residue is used as fuel and the aluminum chloride waste streams are usually sold, or sent to treatment facilities.

## 4.2. Production by Vapor-Phase Alkylation

Vapor-phase alkylation has been practiced since the early 1940s, but at that time processes were unable to compete with liquid-phase aluminum chloride based technology. The *Alkar process* developed by UOP, based on boron trifluoride catalyst, had modest success in the 1960s, but fell from favor because of high maintenance costs resulting from the severe corrosion caused by small quantities of water. Nevertheless, some ethylbenzene units continue to use this process.

The *Mobil–Badger ethylbenzene process* represents the latest and most successful vapor-phase technology to be introduced. The process was developed in the 1970s around Mobil's versatile ZSM-5 synthetic zeolite catalyst. Earlier attempts at using zeolites or molecular sieves for benzene alkylation had suffered from rapid catalyst deactivation because of coke formation and poor transalkylation capabilities. The Mobil catalyst combines superior resistance to coke formation with high catalytic activity for both alkylation and transalkylation. In 1980 the process was first commercialized by American Hoechst Corp. at their $408 \times 10^3$ t/a Bayport, Texas plant. Currently nine commercial plants have been licensed, representing ca. $3 \times 10^6$ t/a of production capacity.

**Alkar process.** This process produces a high-purity ethylbenzene product and can use dilute ethylene feedstock. If the entry of water into the process is strictly prevented, the corrosion problems associated with aluminum chloride processes are avoided. However, even small amounts of water (<1 mg/kg) hydrolyze the $BF_3$ catalyst.

The alkylation reaction takes place at high pressure (2.5–3.5 MPa; 25–35 bar) and low temperature (100–150°C). Dehydrated benzene, ethylene, and makeup $BF_3$ catalyst are fed to the reactor. Typically, ethylene:benzene molar ratios between 0.15 and 0.2 are used. The reactor inlet temperature is controlled by recycling a small portion of the reactor effluent.

Transalkylation takes place in a separate reactor. Dry benzene, $BF_3$ catalyst, and recycled polyethylbenzenes are fed to the transalkylation reactor, which operates at higher temperature (180–230°C) than the alkylation reactor. The effluent streams from the two reactors are combined and passed to a benzene recovery column, where benzene is separated for recycle to the reactors. Boron trifluoride and light hydrocarbons are taken overhead as a vapor stream from which the $BF_3$ is recovered for recycle. The bottoms from the benzene recovery column is sent to a product column, where ethylbenzene of >99.9% purity is taken overhead. A final column serves to recover polyethylbenzenes for recycle to the transalkylation reactor.

The Alkar process can operate with an ethylene feed containing as low as 8–10 mol% ethylene, enabling a variety of refinery and coke-oven gas streams to be used. However, purification of these streams is necessary to remove components that poison the $BF_3$ catalyst, e.g., trace amounts of water, sulfur compounds, and oxygenates.

**Mobil–Badger Process.** The fixed-bed ZSM-5 catalyst promotes the same overall alkylation chemistry as those used in the other processes; however, the *reaction mechanism* is different. Ethylene molecules are adsorbed onto the Brønsted acid sites within the catalyst, which activates the ethylene molecule and allows bonding with benzene molecules to occur. Hence, the range of higher alkylated aromatic byproducts formed by the Mobil–Badger process is somewhat different than that for the Friedel–Crafts processes. These components do not affect the ethylbenzene product purity and are recycled to the reactor for transalkylation or dealkylation.

The Mobil–Badger heterogeneous catalyst system offers several *advantages* when compared to the other commercially available processes. The most important are that it is noncorrosive and nonpolluting. The catalyst is essentially silica–alumina, which is environmentally inert. Because no aqueous waste streams are produced by the process, the equipment for waste treatment and for catalyst recovery is eliminated. In addition, carbon steel is the primary material of construction; high-alloy materials and brick linings are not required.

The reactor (see Fig. 3) typically operates at 400–450°C and 2–3 MPa (20–30 bar). At this temperature >99% of the net process heat input and exothermic heat of reaction can be recovered as steam. The reaction section includes two parallel multibed reactors, a fired heater, and heat recovery equipment. The high-activity catalyst allows transalkylation and alkylation to occur simultaneously in a single reactor.

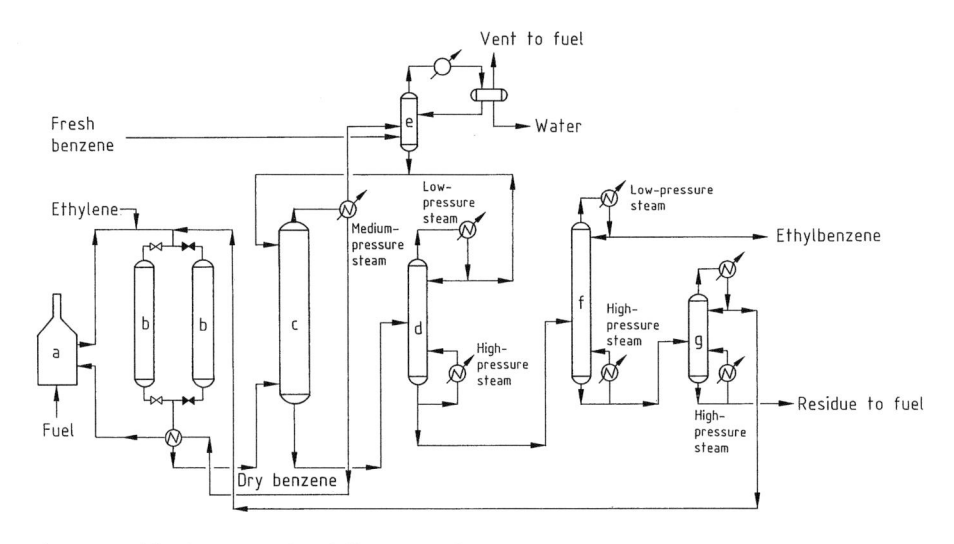

**Figure 3.** Mobil-Badger process for ethylbenzene production
a) Heater; b) Reactor; c) Prefractionator; d) Benzene recovery column; e) Vent gas scrubber; f) Ethylbenzene recovery column; g) Diethylbenzene recovery column

Because the catalyst slowly deactivates as a result of coke formation and requires periodic regeneration, two reactors are included to allow uninterrupted production; one is on stream while the other is regenerated. Regeneration takes ca. 36 h and is necessary after 6–8 weeks of operation. The catalyst is less sensitive to water, sulfur, and other poisons than the Lewis acid catalysts.

The reactor effluent passes to the purification section as a hot vapor. This stream is used as the heat source for the first distillation column, which recovers the bulk of the unreacted benzene for recycle to the reactor. The remaining benzene is recovered from a second distillation column. The ethylbenzene product is taken as the overhead product from the third column. The bottoms product from this column is sent to the last column, where the recyclable alkylbenzenes and polyalkylbenzenes are separated from heavy nonrecyclable residue. The low-viscosity residue stream, consisting mainly of diphenylmethane and diphenylethane, is burned as fuel.

The Mobil–Badger process also can use dilute ethylene feedstocks. In semicommercial applications, the process has operated on streams containing as little as 15 mol % ethylene.

## 4.3. Separation from Mixed C$_8$ Streams

Less than 4 % of worldwide ethylbenzene production is recovered from mixed xylene streams, usually in conjunction with xylene production from reformate. Although adsorption processes have been developed, most notably the Ebex process of UOP, ethylbenzene production from these sources has been by distillation. Because of the difficulty of the separation, the process is generally termed *superfractionation*. It was first undertaken by Cosden Oil & Chemical Co. in 1957, using technology jointly developed with the Badger Co. The separation generally requires three distillation columns in series, each having over 100 stages. Several units were built during the 1960s in the United States, Europe, and Japan. However, the increased cost of energy has made this route uncompetitive.

# 5. Environmental Protection

The U.S. EPA considered imposing a national emission standard for benzene emissions from ethylbenzene-styrene plants, but concluded that the health risks were too small to warrant federal regulatory action under Clean Air Act provisions [8]. Nevertheless, producers of ethylbenzene have taken steps to reduce benzene emissions, mainly by incinerating vent streams and installing improved pump seals. The health hazards of benzene are discussed elsewhere ($\rightarrow$ Benzene).

Alkylation plants that use aluminum chloride alkylation technology produce an aqueous waste stream from the reactor effluent wash section. In the mid 1970 s, plants produced a wet aluminum hydroxide sludge, which was deposited in a landfill (Class I). More modern plants recover a concentrated aluminum chloride solution that has found use in municipal water treatment or industrial floc applications. However, where demand from such applications does not exist, disposal can present a problem.

Studies have shown ethylbenzene to be toxic to aquatic life in relatively low concentration (10 – 100 mg/kg). Therefore, runoff from spills, fire control, etc. should be diked to prevent it from entering streams or water supplies.

# 6. Quality Specifications

The product specification on ethylbenzene is set to provide a satisfactory feedstock to the associated styrene unit. Objectionable impurities in the ethylbenzene can be grouped into two categories: those that are detrimental to the operation of the styrene unit and those that affect the purity of the styrene product. Impurities in product ethylbenzene that pose an operating problem in the conventional dehydrogenative styrene process are

1) *halides*, which deactivate the dehydrogenation catalyst and contribute to downstream equipment corrosion. Usually these are chlorides from an $AlCl_3$ alkylation section or fluorides from an Alkar unit.
2) *diethylbenzenes*, which dehydrogenate to divinylbenzenes in the styrene reactor section. The divinylbenzenes form insoluble cross-linked polymers in the downstream process equipment. A limit of 5 – 40 mg/kg of diethylbenzenes in the ethylbenzene product is usually imposed.

Ethylbenzene contaminants that can affect styrene purity are components having a boiling range between ethylbenzene and styrene. These include xylenes, propylbenzenes, and ethyltoluenes. The levels of cumene, $n$-propylbenzene, ethyltoluenes, and xylenes in the ethylbenzene is controlled to meet the required styrene purity specification.

A typical sales specification for a United States manufacturer is a follows:

| | |
|---|---|
| Purity | 99.5 wt% min. |
| Benzene | 0.1 – 0.3 wt% |
| Toluene | 0.1 – 0.3 wt% |
| $o$-Xylene + cumene | 0.02 wt% max. |
| $m,p$-Xylenes | 0.2 wt% max. |
| Allylbenzene + $n$-propylbenzene + ethyltoluene | 0.2 wt% max. |
| Diethylbenzene | 20 mg/kg max. |
| Total chlorides as Cl$^-$ | 1 – 3 mg/kg max. |
| Total organic sulfur | 4 mg/kg max. |
| Relative density at 15 °C | 0.869 – 0.872 |
| APHA* Color | 15 max. |

*American Public Health Assoc.

# 7. Handling, Storage, and Transportation

Ethylbenzene is a flammable liquid. It is stored and transported in steel containers and is subject to the control of the appropriate regulatory agencies. The U.S. DOT identification number is UN1175, and the reportable quantity is 454 kg. Details on regulations concerning the transport of ethylbenzene can be found in the CFR or from DOT's Material Transportation Bureau. Other countries have regulations and safety practices similar to those of the United States.

Foam, carbon dioxide, dry chemical, halon, and water (fog pattern) extinguishing media are used in fighting ethylbenzene fires.

Adequate ventilation is necessary in handling and storage areas. The use of NIOSH-approved respirators is recommended at high concentration. Skin contact should be avoided. Chemical gloves and safety glasses should be worn if contact is possible.

Exposure of ethylbenzene to heat, ignition sources, and strong oxidizing agents should be avoided.

# 8. Uses

Essentially all commercial ethylbenzene production is captively consumed for the manufacture of *styrene monomer*. Styrene is used in the production of polystyrene and a wide variety of other plastics (→ Styrene).

Of the minor uses, the most significant is in the paint industry as a *solvent*, which accounts for <1% of production capacity. Even smaller volumes go toward the production of acetophenone, diethylbenzene, and ethyl anthraquinone [2].

# 9. Economic Aspects

Ethylbenzene production is linked directly to the styrene monomer market. A total of 99% of the ethylbenzene produced worldwide is used to make styrene monomer. Through the 1960s and into the early 1970s, annual growth rates for styrene and ethylbenzene averaged 10%. During this period sustained growth was powered by the expanding polystyrene market. Subsequent production growth from 1972 to 1985, while averaging 2–3%, has been erratic with large swings in demand. In the 1980s, driven by high energy costs, extensive styrene unit modernizations were undertaken. Ethylbenzene–styrene plants were either upgraded for energy efficiency or closed. Production capacity remained approximately unchanged during this period, with 1800 t/a of nameplate capacity replacing the shutdown capacity. Current growth rate projections through 1990–1995 range from 3.0 to 3.5% per year.

The 1985 *worldwide capacity* (in $10^3$ t/a) by region is as follows:

| | |
|---|---:|
| North America | 5 900 |
| South America | 380 |
| Western Europe | 3 820 |
| Eastern Europe | 1 630 |
| Asia, Oceania, and Far East | 640 |
| Japan | 1 470 |
| Middle East and Africa | 390 |
| *Total* | 14 230 |

The prominent operating *ethylbenzene producers* in the United States and their 1986 capacities (in $10^3$ t/a) were as follows [9]:

| | |
|---|---:|
| American Hoechst (Huntsman), Bayport, Tex. | 520 |
| Amoco, Texas City, Tex. | 395 |
| ARCO, Channelview, Tex. | 575 |
| CosMar, Carville, La. | 770 |
| Dow, Freeport, Tex. | 860 |
| El Paso Products, Odessa, Tex. | 140 |
| Chevron, Donaldsonville, La. | 310 |
| Koch, Corpus Christi, Tex. | 50 |

| | |
|---|---|
| Monsanto (Sterling), Texas City, Tex. | 770 |
| Total | 4390 |

Note: Charter Co., Du Pont, and Tenneco, all of which produce ethylbenzene by superfractionation of mixed xylene streams, were not operating in 1986.

Similar to the direct link of ethylbenzene production to that of styrene, ethylbenzene production cost is tied to feedstock cost. Modern processes all have raw material yields >99%. Integration of the ethylbenzene and styrene processes enables efficient energy recovery of the exothermic alkylation heat of reaction. With 90–99% energy recovery of the heat of reaction plus the heat input to the process, production costs are directly related to benzene and ethylene feedstock prices. The United States unit sales values of ethylbenzene in $/kg from 1960 to 1986 are as follows [10]:

| Year | Sales ($/kg) |
|---|---|
| 1960 | 0.13 |
| 1965 | 0.09 |
| 1970 | 0.09 |
| 1973 | 0.11 |
| 1974 | 0.37 |
| 1975 | 0.20 |
| 1978 | 0.24 |
| 1979 | 0.35 |
| 1980 | 0.51 |
| 1983 | 0.50 |
| 1986 [11] | 0.48–0.51 |

Price increases in 1974 and 1980 reflect the radical change in oil prices experienced during these periods.

Stand-alone ethylbenzene production is limited and usually associated with special situations. The high energy cost for ethylbenzene produced by superfractionation from mixed xylenes has resulted in idling most of this capacity. At Koch Industries' plant at Corpus Christi, a dilute ethylene stream valued as fuel provides this facility with a low-cost feedstock, which offsets the higher utility cost.

# 10. Toxicology and Occupational Health

Ethylbenzene is a moderately toxic compound that is also an irritant to the skin, eyes, and upper respiratory tract. Systemic absorption can cause depression of the central nervous system, with narcosis at very high concentrations.

Current knowledge of the toxic effects puts this compound in a category quite separate from related compounds, such as benzene, and therefore, processing requirements are not nearly as stringent. However, it is sufficiently hazardous to warrant extreme vigilance in its handling and proper medical treatment in the event of exposure.

Direct contact with liquid ethylbenzene causes irritation to the skin and mucous membranes. This may be followed by erythema and inflammation of the skin unless treated immediately. Skin rash may develop after prolonged or repeated exposure.

Acute exposure to vapor concentrations of up to 0.1% in air causes noticeable irritation to the eyes; increasing the concentration to 0.2% produces extreme irritation and lachrymation of the eyes, in addition to irritation of the nose and throat followed by dizziness and constriction of the chest. Exposures of up to 1% are sufficient to cause death (through respiratory failure) in guinea pigs [12].

Although many of the effects of acute exposure are documented, no specific data on the symptoms of chronic exposure are available.

Chronic poisoning can result through inhalation, ingestion, and by rapid absorption through the skin. The $LD_{50}$ is sufficiently high (3.5 g/kg) for ingestion to be of no practical concern [12], but TLV–TWAs of 100 ppm, 435 mg/m$^3$, determined by the ACIGH, dictate strict monitoring of vapor emissions [13]. The short-term exposure limits (TWA–STEL) for vapors, specified by the ACIGH are only slightly higher at 125 ppm, 1100 mg/m$^3$. The MAK values for ethylbenzene quoted by the DFG are 100 mL/m$^3$, 440 mg/m$^3$ [14].

Ethylbenzene absorbed in the blood, either by inhalation or absorption through the skin, is readily metabolized. It is excreted mainly as mandelic acid and phenylglyoxylic acid in the urine [15].

There are no reports of ethylbenzene being carcinogenic; both the ACGIH and the DFG exclude it from their list of identified carcinogens. Similarly, no specific kidney, liver, respiratory, or skin diseases have been found to be associated with ethylbenzene [13]. In terms of other long-term illnesses, reference has been made to possible teratogenic effects in the fetus [12].

# 11. References

[1] R. H. Boundy, R. F. Boyer (eds.): *Styrene, Its Polymers, Copolymers, and Derivatives*, Reinhold Publ. Co., New York 1952, p. 16.
[2] H. Tobin, J. Bakker, K. Tsuchiya: "Xylenes" in *Chemical Economics Handbook*, SRI International, Menlo Park, Calif., May 1985, p. 300. 7300.
[3] American Petroleum Institute (ed.): *Technical Data Book – Petroleum Refining*, 2nd ed., vol. **1**, American Petroleum Institute, Washington, D.C., 1970, pp. 1.46–47.
[4] *Beilstein* **5**, 776–786.
[5] G. A. Olah (ed.): *Friedel-Crafts and Related Reactions*, vol. **2**, Wiley-Interscience, New York 1964, Part 1.
[6] CdF Chimie, US 4 117 023, 1978 (P. J. Gillet, G. Henrich).
[7] Monsanto, US 3 848 012, 1974 (F. Applegath, L. E. DuPree Jr., A. C. MacFarlane, J. D. Robinson).

[8]  *Fed. Reg.* **49** (1984, Jun. 6) no. 110, 23 478. *Fed. Reg.* **50** (1985, Aug. 23) no. 164, 34 144.
[9]  *Chem. Mark. Rep.* **230** (1986, Jul. 28) no. 4, 50.
[10] H. Tobin, J. Bakker, K. Tsuchiya: "Ethylbenzene," in*Chemical Economics Handbook,*SRI International, Menlo Park, Calif., May 1985, p. 645.3000.
[11] *Chem. Mark. Rep.* **230** (1986, Oct. 13) no. 15, 47.
[12] N. I. Sax: *Dangerous Properties of Industrial Materials,* 6th ed., Van Nostrand Reinhold, New York 1984, p. 1322.
[13] ACGIH (ed.): *Threshold Limit Values (TLV) and Biological Exposure Indices,* ACGIH, Cincinnati, Ohio, 1986–1987.
[14] DFG (ed.): *Maximum Concentration at the Workplace (MAK),* VCH Verlagsgesellschaft, Weinheim 1986.
[15] F. W. Mackison, R. S. Stricoff, L. J. Partridge (eds.): "NIOSH/OSHA Occupational Health Guidelines for Chemical Hazards," U.S. Dept. of Health & Human Services (National Institute for Occupational Safety & Health) Publication no. 81–123, Washington, D.C., Jan. 1981.

# Ethylene

HEINZ ZIMMERMANN, Linde AG, Hoellriegelskreuth, Federal Republic of Germany

ROLAND WALZL, Linde AG, Hoellriegelskreuth, Federal Republic of Germany

| | | | | |
|---|---|---|---|---|
| 1. | Introduction | 2221 | 5.3. | Recovery Section ... 2264 |
| 2. | Physical Properties | 2222 | 5.3.1. | Products ... 2265 |
| | | | 5.3.2. | Cracked Gas Processing ... 2266 |
| 3. | Chemical Properties | 2223 | 5.3.2.1. | Front-End Section ... 2266 |
| 4. | Raw Materials | 2224 | 5.3.2.2. | Hydrocarbon Fractionation Section ... 2270 |
| 5. | Production | 2226 | 5.3.3. | Utilities ... 2281 |
| 5.1. | Ethylene from Pyrolysis of Hydrocarbons | 2226 | 5.3.4. | Process Advances ... 2283 |
| 5.1.1. | Cracking Conditions | 2227 | 5.4. | Other Processes and Feedstocks ... 2283 |
| 5.1.2. | Heat Requirements for Hydrocarbon Pyrolysis | 2232 | 6. | Environmental Protection ... 2285 |
| 5.1.3. | Commercial Cracking Yields | 2234 | 7. | Quality Specifications ... 2286 |
| 5.1.4. | Commercial Cracking Furnaces | 2239 | 8. | Chemical Analysis ... 2286 |
| 5.1.5. | Tube Metallurgy | 2250 | 9. | Storage and Transportation ... 2288 |
| 5.1.6. | Thermal Efficiency of Ethylene Furnaces | 2252 | 10. | Uses and Economic Aspects ... 2289 |
| 5.1.7. | Coking and Decoking of Furnaces and Quench Coolers | 2254 | 11. | Toxicology and Occupational Health ... 2290 |
| 5.2. | Quenching of Hot Cracked Gas | 2256 | 12. | References ... 2291 |

# 1. Introduction

Ethylene [74-85-1], ethene, $H_2C=CH_2$, $M_r$ 28.52, is the largest-volume petrochemical produced worldwide. Ethylene, however, has no direct end uses, being used almost exclusively as a chemical building block. It has been recovered from coke-oven gas and other sources in Europe since 1930 [1]. Ethylene emerged as a large-volume intermediate in the 1940s when U.S. oil and chemical companies began separating it from refinery waste gas and producing it from ethane obtained from refinery byproduct streams and from natural gas. Since then, ethylene has almost completely replaced acetylene for many syntheses. Ethylene is produced mainly by thermal cracking of hydrocarbons in the presence of steam, and by recovery from refinery cracked gas.

In 1996 total worldwide ethylene production capacity was $79.3 \times 10^6$ t, with an actual demand of ca. $71 \times 10^6$ t/a [2], which has growth projections of 4.5 % per year worldwide for the period of 1996 to 2005 [3], [4][5].

## 2. Physical Properties

Ethylene is a colorless flammable gas with a sweet odor. The physical properties of ethylene are as follows:

| | |
|---|---|
| *mp* | −169.15 °C |
| *bp* | −103.71 °C |
| Critical temperature, $T_c$ | 9.90 °C |
| Critical pressure, $P_c$ | 5.117 MPa |
| Critical density | 0.21 g/cm$^3$ |
| Density | |
|   at *bp* | 0.57 g/cm$^3$ |
|   at 0 °C | 0.34 g/cm$^3$ |
| Gas density at STP | 1.2603 g/L |
| Density relative to air | 0.9686 |
| Molar volume at STP | 22.258 L |
| Surface tension | |
|   at *bp* | 16.5 mN/m |
|   at 0 °C | 1.1 mN/m |
| Heat of fusion | 119.5 kJ/kg |
| Heat of combustion | 47.183 MJ/kg |
| Heat of vaporization | |
|   at *bp* | 488 kJ/kg |
|   at 0 °C | 191 kJ/kg |
| Specific heat | |
|   of liquid at *bp* | 2.63 kJ kg$^{-1}$ K$^{-1}$ |
|   of gas at $T_c$ | 1.55 kJ kg$^{-1}$ K$^{-1}$ |
| Enthalpy of formation | 52.32 kJ/mol |
| Entropy | 0.220 kJ mol$^{-1}$ K$^{-1}$ |
| Thermal conductivity | |
|   at 0 °C | $177 \times 10^{-4}$ W m$^{-1}$ K$^{-1}$ |
|   at 100 °C | $294 \times 10^{-4}$ W m$^{-1}$ K$^{-1}$ |
|   at 400 °C | $805 \times 10^{-4}$ W m$^{-1}$ K$^{-1}$ |
| Viscosity of liquid | |
|   at *mp* | 0.73 mPa · s |
|   at *bp* | 0.17 mPa · s |
|   at 0 °C | 0.07 mPa · s |
| of gas | |
|   at *mp* | $36 \times 10^{-4}$ mPa · s |
|   at 0 °C | $93 \times 10^{-4}$ mPa · s |
|   at 150 °C | $143 \times 10^{-4}$ mPa · s |

Vapor pressure
   at −150 °C                               0.002 MPa
   at *bp*                                      0.102 MPa
   at −50 °C                                 1.10 MPa
   at 0°                                         4.27 MPa
Explosive limits in air at 0.1 MPa and 20 °C
   lower (LEL)                         2.75 vol% or 34.6 g/cm$^3$
   upper (UEL)                        28.6 vol% or 360.1 g/cm$^3$
Ignition temperature                     425 – 527 °C

# 3. Chemical Properties

The chemical properties of ethylene result from the carbon–carbon double bond, with a bond length of 0.134 nm and a planar structure. Ethylene is a very reactive intermediate, which can undergo all typical reactions of a short-chain olefin. Due to its reactivity ethylene gained importance as a chemical building block. The complex product mixtures that have to be separated during the production of ethylene are also due to the reactivity of ethylene.

Ethylene can be converted to saturated hydrocarbons, oligomers, polymers, and derivatives thereof. Chemical reactions of ethylene with commercial importance are: addition, alkylation, halogenation, hydroformylation, hydration, oligomerization, oxidation, and polymerization.

The following industrial processes are listed in order of their 1993 worldwide ethylene consumption [6]:

1) Polymerization to low-density polyethylene (LDPE) and linear low-density polyethylene (LLDPE)
2) Polymerization to high-density polyethylene (HDPE)
3) Addition of chlorine to form 1,2-dichloroethane
4) Oxidation to oxirane [75-21-8] (ethylene oxide) over a silver catalyst
5) Reaction with benzene to form ethylbenzene [100-41-4], which is dehydrogenated to styrene [100-42-5]
6) Oxidation to acetaldehyde
7) Hydration to ethanol
8) Reaction with acetic acid and oxygen to form vinyl acetate
9) Other uses, including production of linear alcohols, linear olefins, and ethylchloride [75-00-3], and copolymerization with propene to make ethylene–propylene (EP) and ethylene–propylene–diene (EPDM) rubber

**Table 1.** Raw materials for ethylene production (as a percentage of total ethylene produced)

| Raw materials | USA | | W. Europe | | Japan | | World | |
|---|---|---|---|---|---|---|---|---|
| | 1979 | 1991 | 1981 | 1991 | 1981 | 1991 | 1981 | 1991 |
| Refinery gas | 1 | 3 | | 2 | | | | 17 |
| LPG, NGL | 65 | 73 | 4* | 14 | 10* | 2* | 31* | 27 |
| Naphtha | 14 | 18 | 80 | 72 | 90 | 98 | 58 | 48 |
| Gas oil | 20 | 6 | 16 | 12 | 0 | 0 | 11 | 8 |

*Including refinery gas

# 4. Raw Materials

Table 1 lists the percentage of ethylene produced worldwide from various feedstocks for 1981 and 1992 [7]. In Western Europe and Japan, over 80 % of ethylene is produced from naphthas — the principal ethylene raw materials.

A shift in feedstocks occurred for the period from 1980 to 1991. In the United States and Europe larger amounts of light feedstocks (LPG: propane + butane) and NGL (ethane, propane, butane) are used for ethylene production, whereas in Japan more naphtha was used in 1991 compared to 1981. The use of gas oils for ethylene production decreased slightly during the 1980s.

Ethane [74-84-0] is obtained from wet natural gases and refinery waste gases. It may be cracked alone or as a mixture with propane. Propane [74-98-6] is obtained from wet natural gases, natural gasolines, and refinery waste gases. Butanes are obtained from natural gasolines and refinery waste gases. A mixture of light hydrocarbons such as propane, isobutane [75-28-5], and n-butane [106-97-81], commonly called liquefied petroleum gas (LPG) and obtained from natural gasolines and refinery gases, is also used as a feedstock.

Naphthas, which are the most important feedstocks for ethylene production, are mixtures of hydrocarbons in the boiling range of 30–200 °C. Processing of light naphthas (boiling range 30–90 °C, full range naphthas (30–200 °C) and special cuts ($C_6$–$C_8$ raffinates) as feedstocks is typical for naphtha crackers.

A natural-cut full-range naphtha contains more than 100 individual components, which can be detected individually by gas chromatography (GC). Depending on the origin naphtha quality can vary over a wide range, which necessitates quality control of the complex feed mixtures. Characterization is typically based on boiling range; density; and content of paraffins (n-alkanes), isoalkanes, olefins, naphthenes, and aromatics (PIONA analysis) by carbon number. This characterization can be carried out by GC analysis or by a newly developed infrared method [8]. Full characterization of feedstocks is even more important when production is based on varying feedstocks, e.g. feedstocks of different origins purchased on spot markets.

The quality of a feedstock is depending on the potential to produce the target products (ethylene and propylene). Simple yield correlations for these products can be used to express the quality of a feedstock in a simple figure, the quality factor, which indicates wether yields of the target products are high or low, with aromatic feedstocks being poor and saturated feedstocks being good feedstocks.

Quality characterization factors for naphthas have been developed, which indicate the aromatics content by empirical correlation. Since aromatics contribute little to ethylene yields in naphtha cracking, a rough quality estimate can be made for naphthas with a typical weight ratio of n- to isoparaffins of 1–1.1. The K factor is defined as [9]:

$$K = \frac{(1.8\, T_k)^{1/3}}{d}$$

where $T_k$ is the molal average boiling point in K. Naphthas with a $K$ factor of 12 or higher are considered saturated; those below 12 are considered naphthenic or aromatic. The $K$ factor does not differentiate between iso- and n-alkanes. The U.S. Bureau of Mines Correlation Index (BMCI) [10] can also be used as a rough quality measure of naphthas:

$$\text{BMCI} = 48\,640/T + 473.7\,d - 456.8$$

where $T$ is the molal average boiling point in K and $d$ is the relative density $d_{15.6}^{15.6}$. A high value of BMCI indicates a highly aromatic naphtha; a low value, a highly saturated naphtha.

Gas oils are feedstocks that are gaining importance in several areas of the world. Gas oils used for ethylene production are crude oil fractions in the boiling range of 180–350 °C (atmospheric gas oils, AGO) and 350–600 °C (vacuum gas oils, VGO). In contrast to naphtha and lighter gas feeds, these feedstocks can not be characterized by individual components.

Gas chromatography coupled with mass spectrometry (GC–MS) or high performance liquid chromatography (HPLC) allow the analysis of structural groups, i.e., the percentage of paraffins, naphthenes, olefins, monoaromatics, and polyaromatics in the gas oil, and can be used to determine the quality of the hydrocarbon fraction. If this information is used together with data such as hydrogen content, boiling range, refractive index, etc., the quality can be determined quite accurately. A rough estimate of feed quality can be made by using the BMCI or the calculated cetane number of a gas oil. The cetane number, normally used to calculate the performance of diesel fuels, is an excellent quality measure, since it is very sensitive to the n-paraffin content, which is one of the key parameters for the ethylene yield. The cetane number CN is calculated as follows [11]:

$$\text{CN} = 12.822 + 0.1164\,\text{CI} + 0.012976\,\text{CI}^2$$

where

$$\text{CI} = 0.9187\,(T_{50}/10)^{1.26687} \times (n_D^{20}/100)^{1.44227},$$

where $T_{50}$ is the volume average boiling point in °C and $n_D^{20}$ the refractive index at 20 °C.

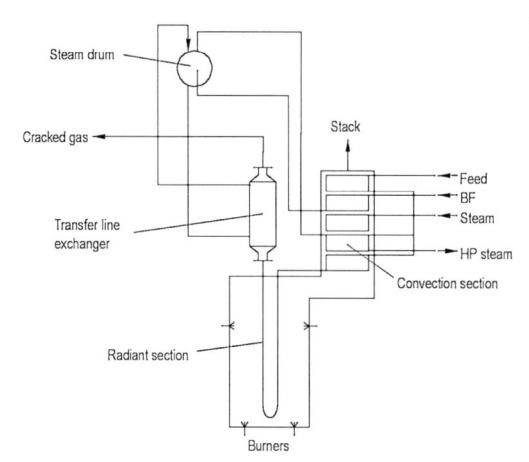

**Figure 1.** Principal arrangement of a cracking furnace

# 5. Production

## 5.1. Ethylene from Pyrolysis of Hydrocarbons

The bulk of the worldwide annual commercial production of ethylene is based on thermal cracking of petroleum hydrocarbons with steam; the process is commonly called pyrolysis or steam cracking. The principal arrangement of such a cracking reactor is shown in Figure 1.

A hydrocarbon stream is heated by heat exchange against flue gas in the convection section, mixed with steam, and further heated to incipient cracking temperature (500–680 °C, depending on the feedstock). The stream then enters a fired tubular reactor (radiant tube or radiant coil) where, under controlled residence time, temperature profile, and partial pressure, it is heated from 500–650 to 750–875 °C for 0.1–0.5 s. During this short reaction time hydrocarbons in the feedstock are cracked into smaller molecules; ethylene, other olefins, and diolefins are the major products. Since the conversion of saturated hydrocarbons to olefins in the radiant tube is highly endothermic, high energy input rates are needed. The reaction products leaving the radiant tube at 800–850 °C are cooled to 550–650 °C within 0.02–0.1 s to prevent degradation of the highly reactive products by secondary reactions.

The resulting product mixtures, which can vary widely, depending on feedstock and severity of the cracking operation, are then separated into the desired products by using a complex sequence of separation and chemical-treatment steps.

The cooling of the cracked gas in the transfer-line exchanger is carried out by vaporization of high-pressure boiler feed water (BFW, $p = 6-12$ MPa), which is separated in the steam drum and subsequently superheated in the convection section to high-pressure superheated steam (HPSS, 6–12 MPa).

## 5.1.1. Cracking Conditions

Commercial pyrolysis of hydrocarbons to ethylene is performed almost exclusively in fired tubular reactors, as shown schematically in Figure 1. These furnaces can be used for all feedstocks from ethane to gas oil, with a limitation in the end point of the feedstock of 600 °C. Higher boiling materials can not be vaporized under the operating condition of a cracking furnace.

Increasing availability of heavy gas oil fractions, due to a shift in demand to lighter fractions, offers cost advantages for processing heavy feedstocks in some areas of the world. Furthermore, the availability of large quantities of residual oil have led some companies to investigate crude oil and residual oils as ethylene sources. Such feedstocks cannot be cracked in conventional tubular reactors. Various techniques employing fluidized beds, molten salts, recuperators, and high-temperature steam have been investigated, but none of these have attained commercial significance [12].

Pyrolysis of hydrocarbons has been studied for years. Much effort has been devoted to mathematical models of pyrolysis reactions for use in designing furnaces and predicting the products obtained from various feedstocks under different furnace conditions. Three major types of model are used: empirical or regression, molecular, and mechanistic models [13].

Today, mechanistic computer models, which are available from various companies, are used for design, optimization and operation of modern olefin plants. Sophisticated regression models are also used, mainly by operators, and offer the advantage of a much lower computer performance requirements than mechanistic models.

The regression models are based on a data set, which can consist of historical data or calculated data. Depending on the quality of the data base the empirical regression models can be of sufficient accuracy for most operating problems, within the range of the data field. These models can be run on small computers and are well suited for process computer control and optimization.

Molecular kinetic models that use only apparent global molecular reactions and thus describe the main products as a function of feedstock consumption have been applied with some success to the pyrolysis of simple compounds such as ethane, propane, and butanes.

For example, cracking of propane can be described as

$$C_3H_8 \longrightarrow a\,H_2 + b\,CH_4 + c\,C_2H_4 + d\,C_3H_6 + e\,C_4H_8 + f\,C_{5+}$$

where $a, b, c, d, e, f$ are empirical factors depending on the conversion of propane.

Gross oversimplification is required if these models are applied to complex mixtures such as naphthas or gas oils, but some success has been attained even with these materials.

In recent years, advances have been made in mechanistic modeling of pyrolysis, facilitated by the availability of more accurate thermochemical kinetic and pyrolysis data and of high-speed computers. The major breakthrough in this area, however, has been the development of methods to integrate large systems of differential equations [14]–[16].

Mechanistic models need less experimental data and can be extrapolated. The accuracy of these models is very good for most components, but they require permanent tuning of the kinetic parameters, especially for computing the cracked-gas composition for ultrashort residence times. The main application for mechanistic models is the design of cracking furnaces and complete ethylene plants. The accuracy of the models has been improved, driven by the competition between the contractors for ethylene plants. A number of mechanistic models are used today in the ethylene industry, describing the very complex kinetics with hundreds of kinetic equations [17]–[19].

To demonstrate the complexity of the chemical reactions, the cracking of ethane to ethylene is discussed here in detail. A simple reaction equation for ethane cracking is:

$$C_2H_6 \longrightarrow C_2H_4 + H_2 \tag{1}$$

If this were the only reaction, the product at 100 % conversion would consist solely of ethylene and hydrogen; at lower conversion, ethylene, hydrogen and ethane would be present. In fact, the cracked gas also contains methane, acetylene, propene, propane, butanes, butenes, benzene, toluene, and heavier components. This reaction (Eq. 1) is clearly not the only reaction occurring.

In the 1930s, the free-radical mechanism for the decomposition of hydrocarbons was established [20]. Although the free-radical treatment does not explain the complete product distribution, even for a compound as simple as ethane, it has been extremely useful. Ethane cracking represents the simplest application of the free-radical mechanism. Ethane is split into two methyl radicals in the chain initiation step (Eq. 2). The methyl radical reacts with an ethane molecule to produce an ethyl radical (Eq. 3), which decomposes to ethylene and a hydrogen atom (Eq. 4). The hydrogen atom reacts with another ethane molecule to give a molecule of hydrogen and a new ethyl radical (Eq. 5).

*Initiation*

$$C_2H_6 \longrightarrow CH_3\cdot + CH_3\cdot \tag{2}$$

*Propagation*

$$CH_3\cdot + C_2H_6 \longrightarrow CH_4 + C_2H_5\cdot \tag{3}$$

$$C_2H_5\cdot \longrightarrow C_2H_4 + H\cdot \tag{4}$$

$$H\cdot + C_2H_6 \longrightarrow H_2 + C_2H_5\cdot \tag{5}$$

If reactions (4) and (5) proceed uninterrupted, the molecular reaction in Equation (1) results. If only reactions (3)–(5) occurred, the cracked gas would contain traces of methane (Eq. 3) and equimolar quantities of ethylene and hydrogen with unreacted ethane. This is not observed.

Reactions (3) and (4) terminate if either an ethyl radical or a hydrogen atom reacts with another radical or atom by reactions such as:

*Termination*

$$H\cdot + H\cdot \longrightarrow H_2 \tag{6}$$

$$CH_3\cdot + H\cdot \longrightarrow CH_4 \tag{7}$$

$$H\cdot + C_2H_5\cdot \longrightarrow C_2H_6 \tag{8}$$

$$C_2H_5\cdot + CH_3\cdot \longrightarrow C_3H_8 \tag{9}$$

$$C_2H_5\cdot + C_2H_5\cdot \longrightarrow C_4H_{10} \tag{10}$$

On termination of chain propagation, new methyl or ethyl radicals or a new hydrogen atom must be generated (Eqs. 2–4) to start a new chain. Thus, every time a new chain is initiated, a molecule of methane is formed (Eq. 3) and a molecule of ethylene is produced (Eq. 4). Other normal and branched-chain alkanes decompose by a similar, but more complex, free-radical mechanism [21]. The number of possible free radicals and reactions increases rapidly as chain length increases.

The free-radical mechanism is generally accepted to explain hydrocarbon pyrolysis at low conversion [20]. As conversion and concentrations of olefins and other products increase, secondary reactions become more significant. Partial pressures of olefins and diolefins increase, favoring condensation reactions to produce cyclodiolefins and aromatics. The cracking of heavy feed, such as naphthas or gas oils, often proceeds far enough to exhaust most of the crackable material in the feedstock.

The reaction scheme with heavier feeds is much more complex than with gaseous feedstocks, due to the fact the hundreds of reactants (feed components) react in parallel and some of those components are formed as reaction products during the reaction. Since the radicals involved are relatively short lived, their concentration in the reaction products is rather low.

Radical decomposition is one of the most important types of reaction and it directly produces ethylene according to the following scheme:

*Radical decomposition*

$$RCH_2CH_2CH_2\cdot \longrightarrow RCH_2\cdot + C_2H_4 \tag{11}$$

This β-scission reaction produces a shorter radical ($RCH_2\cdot$) and ethylene. Radicals normally decompose in the β-position, where the C–C bond is weaker due to electronic effects. Large radicals are more stable than smaller ones and can therefore undergo isomerization.

*Radical isomerization*

$$RCH_2CH_2CH_2\cdot \longrightarrow RCH_2\dot{C}HCH_3 \tag{12}$$

The free-radical decomposition of *n*-butane (Eqs. 12–14) results in the molecular equation (Eq. 15):

$$n\text{-}C_4H_{10} + H\cdot \longrightarrow n\text{-}C_4H_9\cdot + H_2 \tag{14}$$

$$n\text{-}C_4H_9\cdot \longrightarrow C_2H_4 + C_2H_5\cdot \tag{15}$$

$$C_2H_5\cdot \longrightarrow C_2H_4 + H\cdot \tag{16}$$

$$n\text{-}C_4H_{10} \longrightarrow 2\,C_2H_4 + H_2 \tag{17}$$

Reactions like (1) and (15) are highly endothermic. Reported values of $\Delta H$ at 827 °C are + 144.53 kJ/mol for Equation (1) and + 232.244 kJ/mol for Equation (15).

The mathematical description of these complex systems requires special integration algorithms [22]. Based on the pseudo steady state approximation, the chemical reactions can be integrated and the concentration of all components at each location of the reactor (cracking coil) can be computed [23], [24].

In a generalized and very simplified form the complex kinetics of cracking of hydrocarbons (ethane to gas oil) in steam crackers can be summarized as follows:

| | Primary reactions | | Secondary reactions | |
|---|---|---|---|---|
| Feedstock/steam | $\longrightarrow$ | ethylene | $\longrightarrow$ | $C_4$ products |
| | | propylene | | $C_5$ products |
| | | acetylene | | $C_6$ products |
| | | hydrogen | | aromatics |
| | | methane | | $C_7$ products |
| | | etc. | | heavier products |

The fundamentals of furnace design and the main influences of the different parameters can be understood even with this simplified mechanism:

- Residence time: From the above scheme it is clear that a long residence time favors the secondary reactions, whereas a short residence time increases the yields of the primary products, such as ethylene and propylene.
- Partial pressure: Since most of the secondary products result from reactions in which the number of molecules decreases, increasing the pressure favors the secondary products. One function of the steam present in the system is to reduce the hydrocarbon partial pressure and thus favor the formation of primary products.
- Temperature and temperature profiles: The oligomerization reactions involved in the formation of secondary products are favored by lower temperatures; therefore, special temperature profiles are applied along the cracking coil to avoid long residence times at low temperatures.

Most commercial pyrolysis to produce ethylene is carried out in fired tubular reactors in which the temperature of the reactant increases continuously from the inlet to the outlet. Typical inlet temperatures are 500 – 680 °C, depending on the material being processed. Typical outlet temperatures are 775 – 875 °C.

Modern cracking furnaces are designed for rapid heating at the radiant coil inlet, where reaction rate constants are low because of the low temperature. Most of the heat transferred simply raises the reactant from the inlet temperature to the necessary reaction temperature. In the middle of the coil, the rate of temperature rise is lower, but cracking rates are appreciable. In this section, the endothermic reaction absorbs

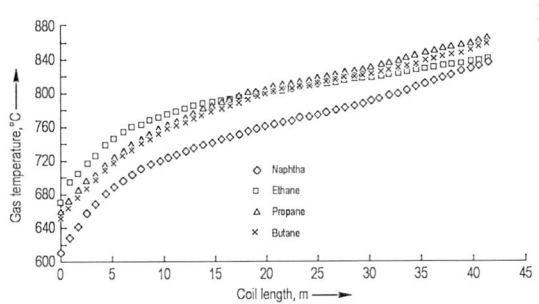

Figure 2. Process gas temperatures along radiant coils

most of the heat transferred to the mixture. At the coil outlet, the rate of temperature rise again increases but never becomes as rapid as at the inlet.

The designers of cracking coils try to optimize the temperature and pressure profiles along the radiant coils to maximize the yield of valuable products yields by special coil design that allows rapid temperature increase in the inlet section and low pressure drops in the outlet section of the cracking coils.

Typical process gas temperature profiles along the radiant coil of modern ethylene furnaces are shown in Figure 2 for ethane, propane, butane, and naphtha cracking.

The quantity of steam used, generally expressed as steam ratio (kilograms of steam per kilogram of hydrocarbon), varies with feedstock, cracking severity, and design of the cracking coil. Typical steam ratios used at a coil outlet pressure of 165–225 kPa (1.65–2.25 bar) for various feedstocks are:

| | |
|---|---|
| Ethane | 0.25–0.35 |
| Propane | 0.30–0.40 |
| Naphthas | 0.4–0.50 |
| Atmospheric gas oils (cut: 180–350 °C) | 0.60–0.70 |
| Hydrocracker bottoms (cut: 350–600 °C) | 0.70–0.85 |

Steam dilution lowers the hydrocarbon partial pressure, thereby enhancing olefin yield. It also reduces the partial pressure of high-boiling, high-molecular-mass aromatics and heavy tarry materials, reducing their tendency to deposit and form coke in the radiant tubes and to foul quench-exchanger tubes. Steam reduces the fouling of the radiant tubes by reacting with deposited coke to form carbon monoxide, carbon dioxide, and hydrogen in a water gas reaction, but as can be seen from the concentration of carbon monoxide in the cracked gas (normally ca. 100 ppm), only to a limited extent.

The sulfur content of the feed is important, since sulfur passivates active Ni sites of the cracking coil material by forming nickel sulfides, which do not catalyze coke gasification, in contrast to nickel metal and nickel oxides. This explains why CO production is high (up to 1 % in the cracked gas) if the feedstock is free of sulfur. To prevent this effect, ca. 20 ppm of sulfur (e.g., as dimethyl sulfide) are added to sulfur-free feedstocks.

Feedstock composition is another important consideration in commercial production of ethylene. Ethylene-plant feedstocks generally contain straight- and branched-chain alkanes, olefins, naphthenes, and aromatics. Ethylene and other olefins are formed primarily from alkanes and naphthenes in the feed. *n*-Alkanes are the preferred component for high ethylene yields, and those containing an even number of carbon atoms give slightly better ethylene yields than odd-numbered ones. *n*-Alkanes also produce propene, whose yield decreases with increasing chain length.

Isoalkanes, in general, produce much smaller yields of ethylene and propene than *n*-alkanes. They give higher yields of hydrogen and methane, and of $C_4$ and higher olefins. Isoalkanes tend to produce more propene than ethylene compared to *n*-alkanes.

Olefin and diolefin yields from cyclopentane, methylcyclopentane, and cyclohexane have been reported [21]; methylcyclopentane and cyclohexane give substantial amounts of butadiene.

Simple and condensed ring aromatics produce no ethylene. Benzene is stable under normal cracking conditions. It is formed during cracking and remains unchanged in the feed. Other aromatics yield primarily higher molecular mass components. Aromatics with long side chains, such as dodecylbenzene, yield ethylene and other olefins by side-chain cracking. The aromatic nucleus still, however, generally produces pyrolysis fuel oil and tar.

In summary, maximum ethylene production requires:

– A highly saturated feedstock
– High coil outlet temperature
– Low hydrocarbon partial pressure
– Short residence time in the radiant coil
– Rapid quenching of the cracked gases

These conditions maximize the yield of olefins and minimize the yield of methane and high molecular mass aromatic components.

Cracking conditions and yield structure are often optimized for economic reasons in plant design and operation; even small changes in cracked gas composition influence the overall plant economics dramatically.

The dilution steam ratio is in most cases at the minimum of the range for the specified feedstock, due to the contribution of the dilution steam to the production costs.

## 5.1.2. Heat Requirements for Hydrocarbon Pyrolysis

Primary hydrocarbon pyrolysis reactions are endothermic; heat requirements are determined by feedstock and cracking conditions. The temperature required to attain a given reaction rate decreases as the carbon chain length in the feedstock increases. Thus, the enthalpy required to reach cracking temperature depends on feedstock and cracking conditions.

Heat requirements for hydrocarbon pyrolysis can be divided into three components:

1) The enthalpy required to heat the feedstock, including the latent heat of vaporization of liquids
2) The endothermic heat of cracking
3) The enthalpy required to heat the cracked gas from the radiant-coil inlet temperature to the radiant-coil outlet temperature

The first is accomplished in the convection section of the furnace; the last two, in the radiant section. Enthalpies (1) and (3) can be calculated from standard heat-capacity data; they represent 25–60 % of total process heat input in the furnace as the feedstock changes from ethane to gas oils [21].

A typical heat balance for a modern naphtha furnace including the TLE is shown in Figure 3.

The endothermic heat of cracking is most simply calculated from heats of formation by the following equation:

$$\Delta H_r = \Delta H_p - \Delta H_f$$

where $\Delta H_r$ = endothermic heat of cracking, kJ/kg; $\Delta H_p$ = heat of formation of the cracked products, kJ/kg; and $\Delta H_f$ = heat of formation of the feed, kJ/kg.

Heats of formation for some alkanes, olefins, naphthenes, and aromatics are as follows (in kJ/kg) [25]:

| | |
|---|---|
| Hydrogen | 0.0 |
| Methane | −5646.3 |
| Ethane | −3548.1 |
| Ethylene | 1345.3 |
| Acetylene | 8540.7 |
| Propane | −2949.9 |
| Propene | −17.9 |
| Methylacetylene | 4317.1 |
| Propadiene | 4496.8 |
| n-Butane | −2694.4 |
| Isobutane | −2827.5 |
| 1-Butene | −459.4 |
| cis-2-Butene | −649.5 |
| trans-2-Butene | −668.2 |
| Isobutylene | −746.5 |
| Ethylacetylene | 2720.4 |
| 1,3-Butadiene | 1721.8 |
| Benzene | 783.6 |
| Toluene | 243.4 |
| Cyclopentane | −1611.3 |
| Methylcyclopentane | −1722.6 |
| Cyclohexane | −1846.4 |
| Methylcyclohexane | −1890.3 |

As the alkane chain length increases, the heat of formation approaches 1800 kJ/kg asymptotically. Ethane, the most refractory alkane besides methane, has the most

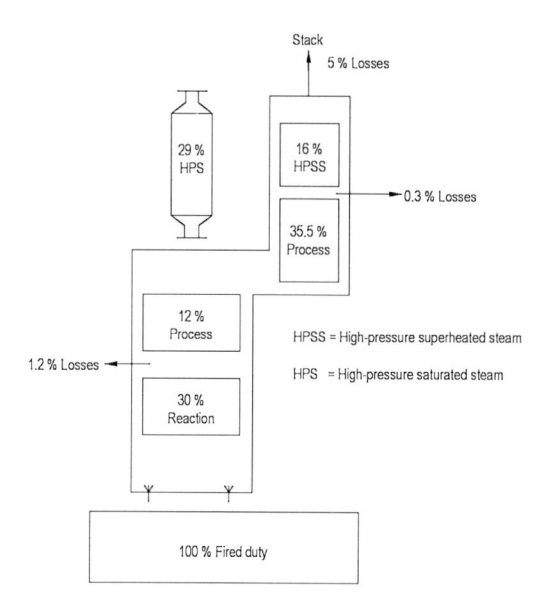

**Figure 3.** Heat balance of a naphtha cracking furnace

endothermic heat of cracking, + 4893 kJ/kg. The corresponding figure for propane is + 4295 kJ/kg and for a high-molecular-mass $n$-alkane, ca. 1364 kJ/kg.

Endothermic heat of cracking increases with degree of saturation of the feed; as conversion increases, secondary condensation products increase. These products have less negative heats of formation than alkanes, or positive heats of formation.

## 5.1.3. Commercial Cracking Yields

Determining the complete yield patterns (gas and liquid products) for a commercial cracking furnace requires a standardized procedure, especially if the feedstock is changed from ethane to naphtha and gas oil. A homogeneous sample is withdrawn from the furnace effluent, quenched, and separated into gaseous and liquid fractions. These are analyzed separately and then combined to give a complete cracked-gas analysis.

The material balance is obtained by matching the hydrogen, carbon, and sulfur content of the feedstock with that of the cracked gas. This type of analysis is time-consuming. In a faster system, gas–liquid separation is eliminated [26] or only a gas-phase analysis is taken. Due to many sources of errors, series of samples have to be taken and analyzed. Statistical methods are used to check the quality of the results.

For gas feedstocks, conversion is used as a measure of the severity of cracking:

*Conversion* $= (C_{n,\text{in}} - C_{n,\text{out}})/C_{n,\text{in}}$

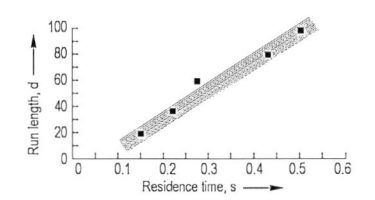

**Figure 4.** Run length versus residence time for ethane cracking

where $C_{n,\text{in}}$ = quantity of component $n$ at inlet in kg/h and $C_{n,\text{out}}$ = quantity of component $n$ at outlet in kg/h.

For liquid feedstocks the severity is defined as the ratio of propylene to ethylene (on weight basis) P/E or as ratio of methane to propylene M/P. Both ratios P/E and M/P are a linear function of temperature over a wide range and can be measured with much higher accuracy than the apparent bulk coil outlet temperature of a cracking furnace. The use of P/E and M/P ratios for severity definition is accepted in industry.

**Ethane.** Ethane is cracked commercially in all types of furnaces, from long residence time to short residence time. However, since ethane is a very stable paraffin, cracking temperatures at the coil inlet typically are higher than for other hydrocarbons. For ethane the coil inlet temperatures are in the range of 650–680 °C. Typical conversion ranges of commercial furnaces are 60–70 %, with 67 % conversion being a typical figure for modern designs.

Table 2 lists yields for ethane cracking at various residence times. There is a slight increase in yield at short residence times. However, as the run length of short-residence-time ethane furnaces is very low (Fig. 4) and yield improvement is moderate, application of short-residence-time furnaces for ethane cracking is not feasible.

In naphtha plants the ethane contained in the cracked gas is separated and recycled to a segregated ethane-cracking furnace, in which cracking is carried out under typical ethane-cracking conditions. In case the capacity of the ethane furnace is not sufficient to crack all the ethane, cocracking of ethane with naphtha is performed, with ethane conversions of 40–50 %. Due to the build up of large recycle streams in such operations, cocracking should be avoided wherever possible.

**Propane** is cracked in all types of furnaces at typical conversions of 90–93 %. The application of short-residence-time furnaces for propane cracking has to be evaluated carefully, due to the dramatic reduction in run length between successive decokings. In many naphtha plants with propane recycle, propane is co-cracked together with naphtha, without a dramatic drop in propane conversion.

Typical yields for propane cracking for various residence times are listed in Table 3. For very mild propane cracking conditions (70 % conversion) yields of propylene show a maximum at 18–19 wt % based on propane feed.

**Butane.** All types of furnaces can be used for butane cracking, which typically is performed at 94–96 % conversion. Yields for $n$-butane and isobutane differ widely. For

**Table 2.** Yields from ethane cracking with various residence times

| | | | | | |
|---|---|---|---|---|---|
| Conversion, kg/kg | | 65.01 | 64.97 | 65.01 | 65.01 |
| Steam dilution, kg/kg | | 0.3 | 0.3 | 0.3 | 0.3 |
| Residence time, s | | 0.4607 | 0.3451 | 0.186 | 0.1133 |
| $H_2$ | wt% | 4.04 | 4.05 | 4.09 | 4.12 |
| CO | wt% | 0.04 | 0.04 | 0.03 | 0.03 |
| $CO_2$ | wt% | 0.01 | 0.01 | 0.01 | 0.01 |
| $H_2S$ | wt% | 0.01 | 0.01 | 0.01 | 0.01 |
| $CH_4$ | wt% | 3.75 | 3.52 | 3.19 | 2.84 |
| $C_2H_2$ | wt% | 0.44 | 0.47 | 0.54 | 0.75 |
| $C_2H_4$ | wt% | 51.88 | 52.31 | 52.85 | 53.43 |
| $C_2H_6$ | wt% | 34.99 | 35.03 | 34.99 | 34.99 |
| $C_3H_4$ | wt% | 0.02 | 0.02 | 0.02 | 0.02 |
| $C_3H_6$ | wt% | 1.22 | 1.13 | 1.06 | 0.97 |
| $C_3H_8$ | wt% | 0.12 | 0.12 | 0.12 | 0.13 |
| $C_4H_4$ | wt% | 0.05 | 0.05 | 0.05 | 0.06 |
| $C_4H_6$ | wt% | 1.80 | 1.80 | 1.79 | 1.65 |
| $C_4H_8$ | wt% | 0.19 | 0.19 | 0.18 | 0.16 |
| $C_4H_{10}$ | wt% | 0.21 | 0.21 | 0.21 | 0.22 |
| Benzene | wt% | 0.55 | 0.47 | 0.38 | 0.26 |
| Toluene | wt% | 0.08 | 0.07 | 0.06 | 0.04 |
| Xylenes | wt% | 0.00 | 0.00 | 0.00 | 0.00 |
| Ethylbenzene | wt% | 0.01 | 0.00 | 0.00 | 0.00 |
| Styrene | wt% | 0.03 | 0.02 | 0.02 | 0.01 |
| Pyrolysis gasoline | wt% | 0.35 | 0.32 | 0.29 | 0.24 |
| Pyrolysis fuel oil | wt% | 0.21 | 0.16 | 0.11 | 0.06 |
| **Sum** | **wt%** | 100.00 | 100.00 | 100.00 | 100.00 |

n-butane, ethylene yields in the range of 40% are typical (Table 4), whereas for isobutane only 15 wt% ethylene yield is achieved. Due to the poor ethylene yield from isobutane, other applications such as dehydrogenation, oligomerization, and alkylation seem to be more feasible than cracking isobutane to produce ethylene. However, mild cracking at 70% conversion and high pressures has been claimed to yield isobutene as primary product [27].

**Naphtha.** Naphtha, the refinery hydrocarbon fraction with the boiling range of 35–180 °C, can vary in composition and boiling range, depending on source and refinery conditions. Today naphtha cuts from 35 to 90 °C (light naphtha), 90–180 °C (heavy naphtha) and 35–180 °C (full-range naphtha) are processed. In Europe and Asia/Pacific naphtha is the primary raw marerial for ethylene cracking, with full range naphthas being processed most.

The properties of the naphtha feed stock used to generate the yield patterns given in Tables 5, 6, and 7 are as follows:

**Table 3.** Yields from propane cracking with various residence times

|  |  |  |  |  |  |
|---|---|---|---|---|---|
| Conversion, kg/kg |  | 90.020 | 90.035 | 89.926 | 89.983 |
| Steam dilution, kg/kg |  | 0.3 | 0.3 | 0.3 | 0.3 |
| Residence time, s |  | 0.4450 | 0.3337 | 0.1761 | 0.1099 |
| $H_2$ | wt% | 1.51 | 1.55 | 1.61 | 1.68 |
| CO | wt% | 0.04 | 0.04 | 0.03 | 0.04 |
| $CO_2$ | wt% | 0.01 | 0.01 | 0.01 | 0.01 |
| $H_2S$ | wt% | 0.01 | 0.01 | 0.01 | 0.01 |
| $CH_4$ | wt% | 23.43 | 23.27 | 22.82 | 22.40 |
| $C_2H_2$ | wt% | 0.46 | 0.51 | 0.59 | 0.82 |
| $C_2H_4$ | wt% | 37.15 | 37.51 | 38.05 | 38.59 |
| $C_2H_6$ | wt% | 3.06 | 2.80 | 2.37 | 1.96 |
| $C_3H_4$ | wt% | 0.52 | 0.57 | 0.65 | 0.89 |
| $C_3H_6$ | wt% | 14.81 | 14.82 | 15.01 | 15.27 |
| $C_3H_8$ | wt% | 9.97 | 9.96 | 10.07 | 10.01 |
| $C_4H_4$ | wt% | 0.08 | 0.08 | 0.09 | 0.11 |
| $C_4H_6$ | wt% | 2.85 | 2.9 | 2.98 | 2.99 |
| $C_4H_8$ | wt% | 1.00 | 1 | 1.02 | 1.09 |
| $C_4H_{10}$ | wt% | 0.04 | 0.04 | 0.05 | 0.05 |
| Benzene | wt% | 2.15 | 2.12 | 2.02 | 1.80 |
| Toluene | wt% | 0.43 | 0.4 | 0.36 | 0.28 |
| Xylenes | wt% | 0.05 | 0.05 | 0.04 | 0.03 |
| Ethylbenzene | wt% | 0.01 | 0.01 | 0.01 | 0.00 |
| Styrene | wt% | 0.21 | 0.2 | 0.18 | 0.15 |
| Pyrolysis gasoline | wt% | 1.27 | 1.26 | 1.27 | 1.24 |
| Pyrolysis fuel oil | wt% | 0.94 | 0.89 | 0.76 | 0.58 |
| **Sum** | **wt%** | 100.00 | 100.00 | 100.00 | 100.00 |

| | |
|---|---|
| Density at 20 °C | 0.692 g/mL |
| S content | 55 mg/kg |
| H content | 15.17 wt% |
| C content | 84.80 wt% |
| Average molar mass | 92 g/mol |
| *Boiling curve (ASTM D 86)* | |
| *initial boiling point* | 26 °C |
| 5 vol% | 49 °C |
| 10 vol% | 53 °C |
| 20 vol% | 58 °C |
| 30 vol% | 64 °C |
| 40 vol% | 70 °C |
| 50 vol% | 77 °C |
| 60 vol% | 86 °C |
| 70 vol% | 99 °C |
| 80 vol% | 116 °C |
| 90 vol% | 138 °C |
| 95 vol% | 152 °C |
| *final boiling point* | 183 °C |
| *PIONA analysis* | |
| *n*-Paraffins | 36.13 wt% |
| Isoparaffins | 36.62 wt% |
| Olefins | 0.21 wt% |
| Napthenes | 21.06 wt% |
| Aromatics | 5.98 wt% |

**Table 4.** Yields from butane cracking with various residence times

|  |  |  |  |  |  |
|---|---|---|---|---|---|
| Conversion, kg/kg |  | 93.95 | 93.43 | 93.44 | 93.92 |
| Steam dilution, kg/kg |  | 0.35 | 0.35 | 0.35 | 0.35 |
| Residence time, s |  | 0.4775 | 0.3440 | 0.1781 | 0.1117 |
| $H_2$ | wt % | 1.05 | 1.09 | 1.16 | 1.22 |
| CO | wt % | 0.04 | 0.04 | 0.03 | 0.04 |
| $CO_2$ | wt % | 0.01 | 0.01 | 0.01 | 0.01 |
| $H_2S$ | wt % | 0.01 | 0.01 | 0.01 | 0.01 |
| $CH_4$ | wt % | 20.48 | 20.29 | 19.85 | 19.33 |
| $C_2H_2$ | wt % | 0.37 | 0.42 | 0.51 | 0.71 |
| $C_2H_4$ | wt % | 35.03 | 35.81 | 37.25 | 38.36 |
| $C_2H_6$ | wt % | 4.57 | 4.16 | 3.47 | 3.02 |
| $C_3H_4$ | wt % | 0.87 | 0.93 | 1.04 | 1.32 |
| $C_3H_6$ | wt % | 17.37 | 17.24 | 17.03 | 16.99 |
| $C_3H_8$ | wt % | 0.37 | 0.35 | 0.33 | 0.33 |
| $C_4H_4$ | wt % | 0.09 | 0.1 | 0.11 | 0.13 |
| $C_4H_6$ | wt % | 3.98 | 4.08 | 4.27 | 4.38 |
| $C_4H_8$ | wt % | 2.85 | 2.83 | 2.78 | 2.85 |
| $C_4H_{10}$ | wt % | 6.05 | 6.07 | 6.06 | 6.08 |
| Benzene | wt % | 2.78 | 2.69 | 2.52 | 2.18 |
| Toluene | wt % | 0.83 | 0.77 | 0.66 | 0.49 |
| Xylenes | wt % | 0.12 | 0.11 | 0.09 | 0.06 |
| Ethylbenzene | wt % | 0.01 | 0.01 | 0.01 | 0.01 |
| Styrene | wt % | 0.25 | 0.24 | 0.22 | 0.18 |
| Pyrolysis gasoline | wt % | 1.85 | 1.81 | 1.80 | 1.72 |
| Pyrolysis fuel oil | wt % | 1.02 | 0.94 | 0.79 | 0.58 |
| **Sum** | **wt %** | 100.00 | 100.00 | 100.00 | 100.00 |

The yield increases with decreasing residence time. The maximum benefit of a selective cracking coil is achieved in high-severity cracking, in which a 5 % feedstock saving can be achieved by using short residence times. However, careful analysis is required since, similar to gas cracking, shorter run lengths are typical for short-residence-time cracking. Maximum annual furnace productivity is achieved with a combination of short residence time and long run length.

**Gas Oil (AGO; Boiling Range 180–350 °C).** Compositions and boiling ranges of gas oils vary greatly with source and refinery conditions. Saturated gas oils may have BMCI values as low as 20, whereas aromatic gas-oil BMCI values can exceed 45. Typical yield distributions are shown in Table 7 for an AGO with properties as listed in Table 9.

Gas oils produce much more $C_{5+}$ material than the above feedstocks because they contain large amounts of condensed polynuclear aromatics. These components resist cracking and may remain unchanged or condense to higher molecular mass materials.

**Hydrocracker Residue (HVGO or HCR; Boiling Range 350–600 °C).** Hydrocracker residues are the unconverted product fraction from hydrocrackers. Due to the severe operating conditions of the hydrocrackers, HCR is a highly saturated product, with only a limited content of aromatics and low contents of polyaromatics. HCR or HVGO is a feedstock that can result in ethylene yields as high as in naphtha

**Table 5.** Yields from low-severity naphtha cracking

| | | | | | |
|---|---|---|---|---|---|
| P/E *, kg/kg | | 0.65 | 0.65 | 0.65 | 0.65 |
| Steam dilution, kg/kg | | 0.4 | 0.4 | 0.4 | 0.4 |
| Residence time, s | | 0.4836 | 0.3526 | 0.1784 | 0.1096 |
| $H_2$ | wt % | 0.82 | 0.83 | 0.85 | 0.86 |
| CO | wt % | 0.03 | 0.02 | 0.02 | 0.02 |
| $CO_2$ | wt % | 0.00 | 0.00 | 0.00 | 0.00 |
| $H_2S$ | wt % | 0.00 | 0.00 | 0.00 | 0.00 |
| $CH_4$ | wt % | 13.56 | 13.2 | 12.44 | 11.83 |
| $C_2H_2$ | wt % | 0.32 | 0.34 | 0.36 | 0.45 |
| $C_2H_4$ | wt % | 25.23 | 25.6 | 26.08 | 26.43 |
| $C_2H_6$ | wt % | 4.76 | 4.36 | 3.62 | 3.16 |
| $C_3H_4$ | wt % | 0.59 | 0.63 | 0.70 | 0.92 |
| $C_3H_6$ | wt % | 16.42 | 16.63 | 16.97 | 17.18 |
| $C_3H_8$ | wt % | 0.68 | 0.65 | 0.61 | 0.60 |
| $C_4H_4$ | wt % | 0.09 | 0.09 | 0.10 | 0.12 |
| $C_4H_6$ | wt % | 4.63 | 4.86 | 5.27 | 5.68 |
| $C_4H_8$ | wt % | 5.76 | 5.94 | 6.34 | 6.69 |
| $C_4H_{10}$ | wt % | 0.72 | 0.75 | 0.80 | 0.85 |
| Benzene | wt % | 6.37 | 5.98 | 5.18 | 4.47 |
| Toluene | wt % | 3.08 | 2.9 | 2.60 | 2.26 |
| Xylenes | wt % | 1.17 | 1.15 | 1.09 | 0.96 |
| Ethylbenzene | wt % | 0.79 | 0.8 | 0.83 | 0.85 |
| Styrene | wt % | 1.08 | 1.01 | 0.87 | 0.77 |
| Pyrolysis gasoline | wt % | 11.25 | 11.81 | 13.19 | 14.09 |
| Pyrolysis fuel oil | wt % | 2.65 | 2.45 | 2.08 | 1.81 |
| **Sum** | **wt %** | 100.00 | 100.00 | 100.00 | 100.00 |

* Propylene to ethylene ratio, a measure of severity.

cracking. Typical yields for a HCR are shown in Table 8 for an AGO with properties as listed in Table 9.

A comparison of the feedstock requirements as a function of residence time for a 500 000 t/a plant for ethane, propane, butane and naphtha is shown in Figure 5. It can be seen that a naphtha cracker has about three times the feedstock throughput of an ethane cracker, and that naphtha cracking shows some sensitivity to short residence times, whereas ethane crackers are not very sensitive with respect to feedstock savings as short residence times are applied. A careful economic analysis is required to select the optimum residence time range for a given feedstock, since shorter residence times lead to reduced availability due to shorter run length, as shown for ethane cracking in Figure 4.

## 5.1.4. Commercial Cracking Furnaces

**Furnace Design.** Modern cracking furnaces typically have one or two rectangular fireboxes with vertical radiant coils located centrally between two radiant refractory walls. Firebox heights of up to 15 m and firebox widths of 2–3 m are standard design practice in the industry. Firing can be performed with wall- or floor-mounted radiant burners, or a combination of both, which use gaseous or combined gaseous and liquid

**Table 6.** Yields from medium-severity naphtha cracking

| | | | | | |
|---|---|---|---|---|---|
| P/E*, kg/kg | | 0.55 | 0.55 | 0.55 | 0.55 |
| Steam dilution, kg/kg | | 0.45 | 0.45 | 0.45 | 0.45 |
| Residence time, s | | 0.4840 | 0.3572 | 0.1828 | 0.1132 |
| $H_2$ | wt% | 0.90 | 0.92 | 0.94 | 0.96 |
| CO | wt% | 0.04 | 0.04 | 0.03 | 0.04 |
| $CO_2$ | wt% | 0.00 | 0.00 | 0.00 | 0.00 |
| $H_2S$ | wt% | 0.00 | 0.00 | 0.00 | 0.00 |
| $CH_4$ | wt% | 15.23 | 14.91 | 14.31 | 13.82 |
| $C_2H_2$ | wt% | 0.44 | 0.47 | 0.53 | 0.71 |
| $C_2H_4$ | wt% | 27.95 | 28.45 | 29.24 | 29.87 |
| $C_2H_6$ | wt% | 4.59 | 4.23 | 3.57 | 3.14 |
| $C_3H_4$ | wt% | 0.71 | 0.76 | 0.87 | 1.21 |
| $C_3H_6$ | wt% | 15.38 | 15.64 | 16.09 | 16.43 |
| $C_3H_8$ | wt% | 0.53 | 0.51 | 0.48 | 0.47 |
| $C_4H_4$ | wt% | 0.12 | 0.13 | 0.16 | 0.20 |
| $C_4H_6$ | wt% | 4.54 | 4.79 | 5.28 | 5.79 |
| $C_4H_8$ | wt% | 4.41 | 4.52 | 4.75 | 4.95 |
| $C_4H_{10}$ | wt% | 0.42 | 0.44 | 0.47 | 0.49 |
| Benzene | wt% | 7.45 | 7.12 | 6.5 | 5.85 |
| Toluene | wt% | 3.26 | 3.10 | 2.82 | 2.46 |
| Xylenes | wt% | 1.18 | 1.16 | 1.12 | 0.97 |
| Ethylbenzene | wt% | 0.62 | 0.62 | 0.64 | 0.63 |
| Styrene | wt% | 1.21 | 1.14 | 1.00 | 0.88 |
| Pyrolysis gasoline | wt% | 7.70 | 7.96 | 8.52 | 8.79 |
| Pyrolysis fuel oil | wt% | 3.32 | 3.09 | 2.68 | 2.34 |
| **Sum** | **wt%** | 100.00 | 100.00 | 100.00 | 100.00 |

* Propylene to ethylene ratio, a measure of severity.

fuels. Fireboxes are under slight negative pressures with upward flow of flue gas. Flue gas flow and draft are established by induced-draft fans. In some furnace designs, burners are mounted on terraced walls.

Firebox length is determined by the total ethylene production rate desired from each furnace and the residence time of the cracking operation. This also determines the number of individual radiant coils needed, with short residence times requiring many more individual coils than longer residence times for the same production capacity. This is due to the shorter lengths of the short-residence-time coils, which can be as short as 10–16 m per coil. Long-residence-time coils can have lengths of 60–100 m per coil. The number of coils required for a given ethylene capacity is determined by the radiant coil surface, which is in the range of 10–15 $m^2$ per tonne of feedstock for liquid feedstocks.

Production rate for each coil is determined by its length, diameter, and charge rate, which translates into a certain heat flux on the radiant coil. Values of 85 kW/$m^2$ for the average heat flux of a coil should be the maximum.

Radiant coils are usually hung in a single plane down the center of the firebox. They have also been nested in a single plane or placed parallel in a staggered, double-row tube arrangement. Staggered tubes have nonuniform heat-flux distributions because of the shadowing effect of one coil on its companion. If maximum capacity and full

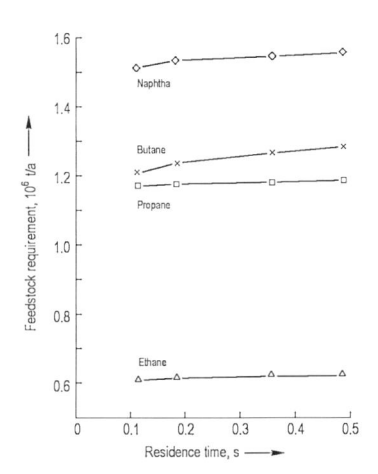

**Figure 5.** Feedstock requirements for a 500 000 t/a ethylene plant

**Table 7.** Yields from high-severity naphtha cracking

| | | | | | |
|---|---|---|---|---|---|
| P/E *, kg/kg | | 0.45 | 0.45 | 0.45 | 0.45 |
| Steam dilution, kg/kg | | 0.5 | 0.5 | 0.5 | 0.5 |
| Residence time, s | | 0.4930 | 0.3640 | 0.1897 | 0.1170 |
| $H_2$ | wt % | 0.9 | 1.00 | 1.03 | 1.06 |
| CO | wt % | 0.06 | 0.06 | 0.05 | 0.06 |
| $CO_2$ | wt % | 0.00 | 0.00 | 0.00 | 0.00 |
| $H_2S$ | wt % | 0.00 | 0.00 | 0.00 | 0.00 |
| $CH_4$ | wt % | 16.9 | 16.61 | 16.10 | 15.74 |
| $C_2H_2$ | wt % | 0.64 | 0.69 | 0.80 | 1.14 |
| $C_2H_4$ | wt % | 30.25 | 30.81 | 31.74 | 32.38 |
| $C_2H_6$ | wt % | 4.32 | 4.00 | 3.43 | 3.00 |
| $C_3H_4$ | wt % | 0.81 | 0.88 | 1.02 | 1.48 |
| $C_3H_6$ | wt % | 13.63 | 13.88 | 14.29 | 14.55 |
| $C_3H_8$ | wt % | 0.37 | 0.35 | 0.33 | 0.32 |
| $C_4H_4$ | wt % | 0.18 | 0.2 | 0.24 | 0.32 |
| $C_4H_6$ | wt % | 4.20 | 4.45 | 4.90 | 5.39 |
| $C_4H_8$ | wt % | 3.11 | 3.17 | 3.26 | 3.28 |
| $C_4H_{10}$ | wt % | 0.17 | 0.18 | 0.19 | 0.18 |
| Benzene | wt % | 8.32 | 8.06 | 7.63 | 7.15 |
| Toluene | wt % | 3.40 | 3.26 | 3.00 | 2.63 |
| Xylenes | wt % | 1.14 | 1.12 | 1.09 | 0.95 |
| Ethylbenzene | wt % | 0.44 | 0.44 | 0.44 | 0.40 |
| Styrene | wt % | 1.44 | 1.37 | 1.24 | 1.12 |
| Pyrolysis gasoline | wt % | 5.25 | 5.35 | 5.55 | 5.57 |
| Pyrolysis fuel oil | wt % | 4.38 | 4.12 | 3.67 | 3.28 |
| **Sum** | **wt %** | 100.00 | 100.00 | 100.00 | 100.00 |

* Propylene to ethylene ratio, a measure of severity.

**Table 8.** Yields for AGO and HCR cracking

|  |  | AGO | HCR |
|---|---|---|---|
| P/E*, kg/kg |  | 0.54 | 0.53 |
| Steam dilution, kg/kg |  | 0.8 | 1 |
| Residence time, s |  | 0.32 | 0.3 |
| $H_2$ | wt% | 0.71 | 0.68 |
| CO | wt% | 0.01 | 0.01 |
| $CO_2$ | wt% | 0.01 | 0.01 |
| $H_2S$ | wt% | 0.00 | 0.00 |
| $CH_4$ | wt% | 10.58 | 11.08 |
| $C_2H_2$ | wt% | 0.38 | 0.60 |
| $C_2H_4$ | wt% | 25.93 | 25.98 |
| $C_2H_6$ | wt% | 2.82 | 2.81 |
| $C_3H_4$ | wt% | 0.63 | 0.97 |
| $C_3H_6$ | wt% | 14.07 | 16.02 |
| $C_3H_8$ | wt% | 0.37 | 0.42 |
| $C_4H_4$ | wt% | 0.12 | 0.04 |
| $C_4H_6$ | wt% | 5.73 | 8.03 |
| $C_4H_8$ | wt% | 3.61 | 4.14 |
| $C_4H_{10}$ | wt% | 0.05 | 0.06 |
| Benzene | wt% | 5.44 | 5.64 |
| Toluene | wt% | 3.49 | 2.73 |
| Xylenes | wt% | 0.92 | 0.46 |
| Ethylbenzene | wt% | 0.37 | 0.34 |
| Styrene | wt% | 1.18 | 1.12 |
| Pyrolysis gasoline | wt% | 7.28 | 6.91 |
| Pyrolysis fuel oil | wt% | 16.30 | 7.95 |
| **Sum** | **wt%** | 100 | 100 |

* Propylene to ethylene ratio, a measure of severity.

**Table 9.** Properties of AGO and HCR used for yield calculations in Table 4

|  | AGO | HCR |
|---|---|---|
| Density at 20 °C, g/mL | 0.8174 | 0.8280 |
| C, wt% | 86.20 | 85.28 |
| H, wt% | 13.64 | 14.34 |
| S, wt% | 0.18 | 0.003 |
| BMCI | 24.8 | 9.37 |
| Paraffins + naphthalenes, wt% | 80.38 | 89.4 |
| Monoaromatics, wt% | 12.45 | 7.3 |
| Polyaroamtics, wt% | 7.17 | 3.3 |
| Boiling point analysis, °C |  |  |
| ibp | 170 | 313 |
| 10% | 203 | 351 |
| 30% | 227 | 393 |
| 50% | 248 | 405 |
| 70% | 273 | 422 |
| 90% | 316 | 451 |
| 95% | 340 | 460 |
| fbp | 352 | 475 |

utilization of the firebox are desired, tube staggering can be adjusted so that the shadowing effect is only marginally greater than with a single-row arrangement.

Heat transfer to the radiant tubes occurs largely by radiation, with only a small contribution from convection. Firebox temperature is typically 1000–1200 °C.

The radiant coil of an ethylene furnace is a fired tubular chemical reactor. Inlet conditions must meet the required temperature, pressure, and flow rate. Hydrocarbon feed is heated from its convection-section entry temperature to the temperature needed for entry into the radiant coils. Dilution steam is introduced into the hydrocarbon stream in the convection section. The combined hydrocarbon and steam stream is then superheated to the target temperature at the radiant coil inlet (crossover temperature, XOT). Gaseous feeds require only the energy necessary to heat the feed and steam to the radiant coil inlet temperature. Liquid feeds require additional energy to heat the liquid to its vaporization temperature plus its latent heat of vaporization.

Energy is saved in the convection section by preheating the cracking charge; by using utility preheating, i.e., boiler-feedwater preheating, dilution-steam superheating, high-pressure steam superheating or by preheating the combustion air.

A typical arrangement of furnace elements is shown in Figure 6. In this case, the convection section contains six separate zones. As an example, the feed could be preheated in the upper zone, the boiler feedwater in the next, and mixed feed and dilution steam in the third zone. The process steam is superheated in the fourth zone, the HP steam is superheated in the fifth zone and the final feed and dilution steam are preheated in the bottom zone. The number of zones and their functions can differ according to furnace design. For example, the upper zone could be used to recover heat for air preheating.

With the trend to higher capacities per individual furnace a new furnace set-up has been introduced to the industry with great commercial success. In this type two fireboxes are connected to one convection section. This design allows liquid feedstock cracking furnaces with capacities of 130 000–150 000 t/a to be installed. This twin radiant cell concept has been applied more than 60 industrial furnaces since the late 1980s.

The induced-draft fans allow complex convection sections to be designed with very high gas velocities to give optimum heat transfer in the convection section banks. Modern designs include integrated catalyst systems for $NO_x$ removal by SCR (selective catalytic reduction) technology, whereby the $NO_x$ present in the flue gas reacts with ammonia:

$$4\,NO_x + 4\,NH_3 + O_2 \longrightarrow 4\,N_2 + 6\,H_2O$$

The temperature of the cracked gas leaving the radiant coils can range from 750 to 900 °C. Rapid reduction of gas temperature to 500–650 °C, depending on the feedstock, is necessary to avoid losses of valuable products by secondary reactions. This is accomplished by a transfer-line exchanger (TLE) (see Fig. 6), which cools the cracked gases and recovers much of the heat contained in the cracked gas as high-pressure steam.

**Figure 6.** Typical arrangement of furnace elements

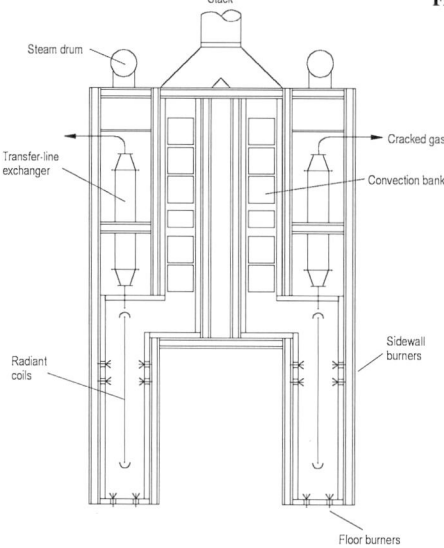

In all modern furnaces the radiant coils exit the firebox at the top; the TLEs are top-mounted. In older designs coils also leave the bottom of the firebox, with TLEs mounted vertically along the side of the furnace or horizontally underneath it. Radiant-coil designs are generally considered proprietary information. All modern designs strive for short residence time, high temperature, and low hydrocarbon partial pressure. Most coil configurations have two features in common: coils are hung vertically, usually suspended on spring or constant hangers above the top of the radiant firebox, and they are fired from both sides of the radiant coil. Horizontal tubes are no longer used in ethylene furnaces.

Coils can range from a single, small-diameter tube with low feed rate and many coils per furnace, to long, large-diameter tubes, with high feed rate and few coils per furnace. Longer coils consist of lengths of tubing connected with return bends. Individual tubes can be the same diameter or swaged to larger diameters one or more times along the complete coil; two or more individual tubes may be combined in parallel.

An example of a so-called split radiant coil arrangement is shown in Figure 7; here parallel small-diameter coils are combined into a larger diameter outlet coil. This arrangement allows advantageous temperature profiles to be applied, with a rapid temperature increase in the inlet section, and offers slightly higher yields than a uniform diameter coil of the same residence time.

Coil lengths and diameters vary widely as can be seen from the following design characteristics:

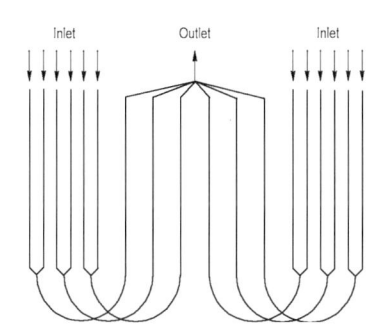

**Figure 7.** Short-residence-time coil arrangement

| | |
|---|---|
| Ethylene capacity (per furnace) | $(50-130) \times 10^6$ t/a |
| Number of coils (per furnace) | 4–200 |
| Tube diameter | 25–180 mm |
| Coil length | 10–100 m |
| Coil outlet temperature | 750–890 °C |
| Tube wall temperature | |
| clean | 950–1040 °C |
| maximum | 1040–1120 °C |
| Average heat flux | 50–90 kW/m$^2$ |
| Residence time | 0.08–1.0 s |

Cracking coils with internal fins have been offered to the industry, to increase the inner surface of a coil. These coils should offer the advantage of processing more feedstock in a coil of identical external dimensions. However, the fins result in a higher pressure drop of the coil, with disadvantages in yields, and coke spalling problems occur when the furnace is cooled down rapidly. Furthermore, the surface of a finned coil is of lower quality than that of a bare tube, and this increases the coking rate of the finned tube.

**Lummus Furnace** (Lummus Crest, Bloomfield, N.J.). The coil design of Lummus furnaces has been modified in the light of pilot-plant investigations. Six generations of coils SRT-I (short residence time - first model) through SRT-VI, have been produced.

The SRT-I model radiant coil is uniform in diameter. Many SRT-I heaters are still operating, primarily for cracking ethane, propane, or butane. All other coils are so-called split coils, with smaller inlet coils combined into larger outlet coils. The latest designs — SRT-V and SRT-VI — are short residence time coils with residence times in the range of 0.15–0.20 s. In the SRT-V coil, 8–12 small coils of ca. 40 mm internal diameter are headered into a single outlet coil of large diameter (150–180 mm). In SRT-VI designs a smaller number of parallel coils is headered into a larger outlet tube, which is connected to a linear quench exchanger.

Lummus furnaces can be paired, as shown in Figure 6, or can operate alone. Stacks may be shared, with a common induced-draft fan; the flue gas flows upward. Some convection sections contain boiler feedwater preheating coils and steam superheating coils, along with steam desuperheat stations and process-heating coils. In some designs, steam is superheated in separate heaters. Lummus heaters can be fired by wall-mounted burners or by combinations of wall- and floor-mounted burners. Oil-fired floor burners are available.

Overall fuel efficiencies of 92–95% net heating value (NHV) can be obtained, depending on feedstock, fuel sulfur content, firing control, and convection section type. Lummus furnaces can use any available transfer-line exchanger. Radiant-coil outlets may be paired, leading to a common exchanger. Capacities per furnace for full-range naphtha or atmospheric gas oil vary from 25 000 to 100 000 t/a.

**Millisecond Furnace** (M. W. Kellogg Co., Houston, Texas). The Millisecond Furnace is designed for shortest residence times and low hydrocarbon partial pressures. It is recommended for gas and liquid feedstocks, offering the greatest advantages for the latter. This heater, developed in the early 1970s from a prototype installed by Idemitsu Petrochemical in Japan, reduces residence time to ca. 0.10 s. Kellogg successfully marketed the Millisecond concept in the mid-1980s, but with the trend of the industry to robust furnaces rather than maximum selectivity, which started in the 1990s, Kellogg lost market shares.

The Millisecond Furnace uses Kellogg's proprietary primary quench exchanger (PQE; see Section 5.2). The furnace is decoked on-line by using steam through the PQE and into the downstream equipment. Coil outlet temperature is ca. 50 °C higher than for competitive designs. The Millisecond Furnace, which has a greater number of smaller diameter and shorter tubes than other furnaces designed for short residence time, is of modular design. Its capacity is increased by adding more units in series. Internal coil diameters are in the range of 30–40 mm. A single small PQE is fed by combining two or four of the small-diameter radiant tubes. Individual radiant tubes can number up to 200 per furnace, with 50 or 100 PQEs per furnace. Several PQEs are mounted in a single enclosure, for ease of installation.

Furnaces can be paired to use a common steel structure or they can stand alone. Paired furnaces can have a common stack and induced draft fan. Burners are located on opposite walls of the firebox at floor level; they use gas, oil, or combinations thereof with forced or induced draft. The design can be adapted easily for preheating combustion air. A Millisecond furnace is shown in Figure 8.

Flow is distributed to the many small-diameter radiant coils by pigtails connecting the individual radiant coils to their common feed header. Due to the short residence time and the resulting higher process temperatures, run lengths of Millisecond furnaces are shorter than those of competing furnaces and they thus require more decokes per year.

**Stone & Webster Furnace** (Stone & Webster Engineering Corp., Boston, Mass.). Stone & Webster supplies an ultraselective cracking (USC) furnace coupled with an ultraselective quench exchanger (see Section 5.2). Radiant coils have internal diameters of 50–90 mm. In contrast to many other contractors, Stone & Webster is not offering split coils.

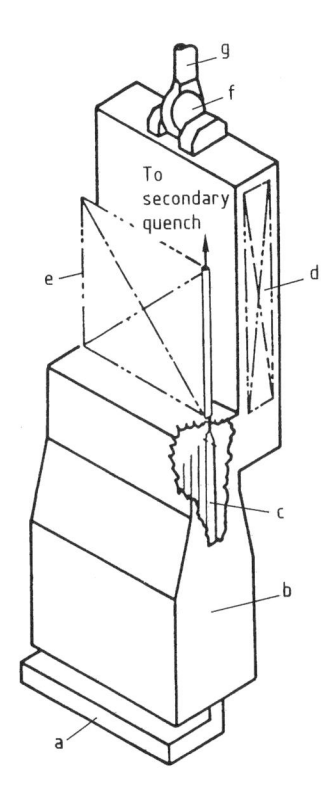

**Figure 8.** Kellogg Millisecond Furnace module*
a) Preheated process fluid inlet; b) Radiant section; c) Radiant tubes; d) Convection section; e) Primary quench exchangers; f) Induced-draft fan; g) Stack

* Reprinted with permission of M. W. Kellogg Co.

Their swaged USC-U (Super U Coil) coil configuration consists of two tubes connected with a return bend. The diameter of the outlet tube is larger than that of the inlet tube, allowing for thermal, molecular, and pressure expansion. Generally, outlet tubes from two USC-U coils are combined to feed a single ultraselective exchanger (USX), but as many as four coils have been fed to a single USX.

Feed rates per radiant coil are low; individual furnace capacity is increased by using as many as 48 individual USC-U coils per furnace. Today Stone & Webster is offering Linear Exchangers (LinEx) instead of USX exchangers, for cost reasons.

The swaged USC-W coil configuration consists of four thin tubes, which differ in diameter. Individual diameters for each coil increase from inlet to outlet. A coil outlet can go directly to a USX or LinEx quench system or two coils can share a quench system. The individual furnace capacity is obtained by installing 12 – 24 USC-W coils in the same firebox. Spring hangers support radiant coils from the top in the center plane of the firebox. Feed to the coils is equalized by critical-flow venturi nozzles.

The USC furnaces may be paired as shown in Figure 5, with a common stack and induced-draft fan, or they can stand alone. Burners are mounted on the wall, or on the wall and floor. Floor burners are gas- or oil-fired.

For furnaces using gas-turbine integration (GTI) or preheating of combustion air, a floor-fired, short-residence-time (FFS) design is available. This is completely floor fired, which reduces large-volume ductwork to a minimum. Coil geometry is similar to USC-U or USC-W designs, but to maintain a uniform firebox temperature, more and shorter coil legs are used. Overall fuel efficiencies of Stone & Webster furnaces are competitive.

**KTI Furnace** (KTI Corp., The Hague, the Netherlands). A number of radiant-coil designs are available, ranging from a simple serpentine coil of constant diameter to GK-1 (gradient kinetics, model 1) and GK-5 coils.

Gaseous feed, naphtha, and gas oil can be cracked with all coils. Residence times range from 0.15 to 0.5 s. Capacities range up to 120 000 t/a per furnace.

The GK radiant coil has two thin, parallel tubes from the coil inlet to ca. 66 % of the total coil length, where they combine to a single, larger tube. The thin tubes can be staggered or in-line; the larger diameter tubes must be in-line.

The KTI furnaces can be paired with a common stack and induced-draft fan, or they can stand alone. Top outlet coils are preferred. Burners are wall-mounted or wall- and floor-mounted.

**Linde-Pyrocrack Furnace** (Linde AG, Munich, Germany). Linde-Pyrocrack furnaces provide capacities of 20 000 – 130 000 t/a. Feedstocks can be all straight-run or pretreated hydrocarbons from ethane, propane, butane, natural gas liquids, naphthas, raffinates, gas oils, gas condensates and hydrocracked vacuum gas oils. Furnaces are designed for segregated cracking or cocracking and for maximum feedstock flexibility. Linde and its U.S. subsidiary Selas Fluid Processing have installed furnaces with a total ethylene capacity $> 13 \times 10^6$ t/a.

Pyrocrack coils are all split coils with small inlet coils headered into a larger outlet coil, with residence times from 0.15 to 0.5 s. The three basic Pyrocrack coil types are shown in Figure 9. Coil diameters range from ca. 40/55 mm (inlet/outlet) for Pyrocrack 1-1 coils to 100/ 150 mm for Pyrocrack 4-2 coils. Critical Venturi nozzles are used to ensure equal distribution and residence times in all parallel coils of a furnace.

Today almost exclusively a combination of sidewall and floor firing is used, which offers the best distribution of the released heat over the whole firebox and a excellent radiant efficiency. Fuel for the radiant burners can be gas or oil or both. Oil firing offers the opportunity to use a fraction of the pyrolysis fuel oil (i. e., the liquid product of the pyrolysis boiling above 180 °C) as fuel and thus of increasing the amount of the pyrolysis methane fuel fraction available for export to other comsumers.

The convection section of Linde Pyrocrack furnaces contains process heating bundles for feed, dilution steam, boiler feedwater, and high-pressure steam superheater bundles. Overall efficiencies of up to 94 – 95 % (on lower heating value, LHV) are typical for Linde furnaces. High-pressure steam systems for heat recovery have been developed by Linde for up to 14 MPa. Linde has built furnaces with SCR-based DeNOx technology integrated into the convection section. The design of the side and front walls of the convection section provides uniform heat distribution and prevents flue gas bypassing. Furnace draft is maintained by an induced-draft fan mounted over the convection section.

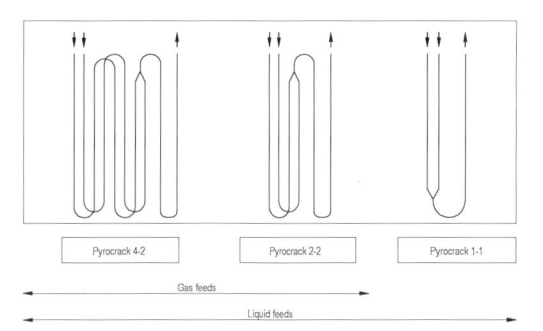

**Figure 9.** Linde-Pyrocrack coils

Linde has introduced with great commercial success the Twin Radiant Cell concept, in which a single high-capacity furnace has two radiant sections, which are connected to a common convection section (Fig. 10). Other new features are a proprietary system for processing gas condensates, which have an extremely high and variable final boiling point. This system was developed from the well-established Linde technology for heavy hydrocracker bottoms and adapted to the requirements for gas condensates.

Linde is offering optimizers (OPTISIM) for the optimization of the operation of furnaces and complete olefin plants.

**Other Furnaces.** The furnaces listed below show some interesting features, but none is of commercial importance in today's worldwide olefin business. Only the furnace designs of the contractors listed above (Lummus, Kellogg, Stone & Webster, KTI, and Linde) are considered to be competitive.

The *terraced-wall furnace* (Foster & Wheeler Energy Corp., Livingston, N.J.) uses burners firing upward from shelves or ledges built into the walls of the firebox (Fig. 11). A similar firebox design is used for reformer heaters. This arrangement apparently gives uniform heating of the radiant coil. Furnaces are all-gas or all-liquid fuel fired.

The *Mitsui advanced cracker* (Mitsui Engineering & Shipbuilding Co., Tokyo, Japan) offers lower construction costs, smaller land area requirements, high thermal efficiency, and low $NO_x$ emissions. The mass of steel used in a 400 000 t/a ethylene plant is 30 % lower than that used in a 300 000 t/a plant with conventional furnaces. This is achieved by a patented installation of three radiant coils in the firebox space normally occupied by one or two coils [28].

The *Mitsubishi Furnace* jointly developed by Mitsubishi Petrochemical Co. and Mitsubishi Heavy Industries of Japan is quite different in design from other furnaces (Fig. 12). Known as the Mitsubishi thermal cracking furnace (M-TCF), it is a vertical, downdraft furnace with the convection section located below the radiant firebox. It is fired with gas or oil burners located in two rows mounted on the ceiling and two rows mounted on terraced walls. All burners are down-firing, with long flat flames. According to Mitsubishi, up to 70 % of the total heat can be supplied with any type of distillate fuel, including the cracked residue from an ethylene plant. Flue gas flows to a common

**Figure 10.** Twin radiant cell furnace

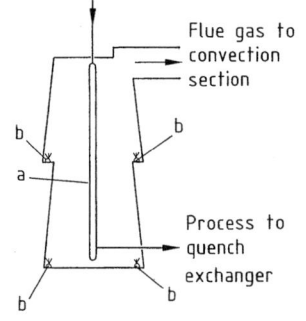

**Figure 11.** Foster – Wheeler terraced wall furnace
a) Radiant coil; b) Burners

stack by means of an induced-draft fan, which produces suction at the bottom of the convection section.

## 5.1.5. Tube Metallurgy

Ample evidence indicates that maximum yields of ethylene from tubular cracking of hydrocarbons require high temperature, short residence time, and low hydrocarbon pressure in the radiant coil. Modified coils can help achieve these conditions, but coil design is limited by metallurgy. Radiant coils must be selected in light of the following considerations:

- Operating temperature
- Tube service life
- Cost
- Carburization resistance

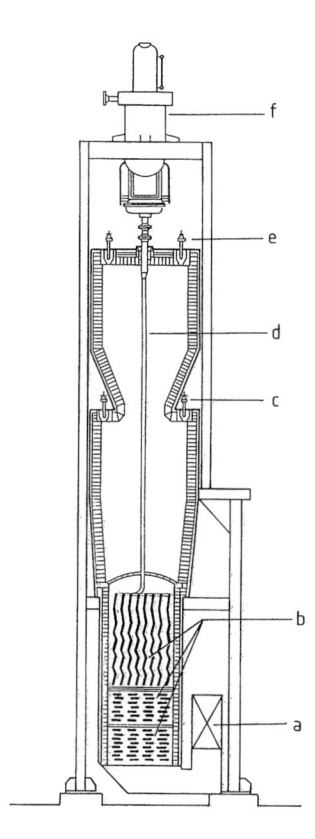

**Figure 12.** Section view of Mitsubishi M-TCF furnace *
a) Flue-gas duct; b) Convection tubes; c) Side burners; d) Cracking coil; e) Roof burner; f) TLE
* Reprinted with permission of Mitsubishi Heavy Industries.

- Creep-rupture strength
- Ductility
- Weldability

An outside tube wall operating temperature up to and even exceeding 1100 °C is now possible. Coils are designed to operate at 975 – 1000 °C with an internally clean radiant coil. Tube wall temperature rises with time because of coke deposition inside the coil. Coke acts as a thermal insulator inside the coil, requiring increased tube wall temperature for a given furnace loading. Localized overheating must be avoided during removal of coke from the coil (when steam or a mixture of steam and air is used). With steam-air decoking, combustion occurs inside the coils.

Tube service life is economically important in plant operation. Furnace capital costs represent ca. 20 % of the total cost of an ethylene plant. About one-third of this is for radiant coils. With current metallurgy, a five-year service life in the hottest section and a seven-year service life of the inlet section of a furnace are typical. Service life is shortened by carburization of the inner tube surface, which is favored by the environment inside a cracking coil. Temperature is high, and feeds are usually highly carburizing; reaction occurs between the gaseous phase in the tube and its base metal.

Coil alloys are normally protected by an oxide layer on the inner surface. If this layer is destroyed and cannot be regenerated, carbon diffuses into the base metal, which itself diffuses toward the inside surface. Carburization may be more severe during decoking and can be reduced by lowering the percentage of air mixed with steam during this cycle. Aluminized coatings on the inner surface of the tube may reduce carburization [29]–[34]. The extent of carburization can be measured by magnetic tests [32].

The materials used for cracking coils are high-alloy steels, e.g. with 35% Ni, 25% Cr and several important trace metals (Si, Mn, Nb, Mo, W, Ti) [35]. The technical designation of this most common radiant coil material is HP 40 modified. Only a limited number of manufacturers worldwide are able to produce these materials according to very strict quality standards.

## 5.1.6. Thermal Efficiency of Ethylene Furnaces

Cracking furnaces represent the largest energy consumer in an ethylene plant; their thermal efficiency is a major factor in operating economics. New plants are designed for 93–95% thermal efficiency, and revamping of older ones can increase efficiencies to 89–92%. Only a limited number of heat sinks are available for heat released in the radiant firebox of a cracking furnace:

1) Process heat duty and duty for utility heating
2) Wall or radiation loss
3) Stack heat loss

Process heat duty is the energy required to heat the feedstock and dilution steam from the temperature at which they enter the furnace convection section to the temperature at which they leave the radiant coil. Process heat duty varies with feedstock composition, feed rate, and cracking severity. The duty for heating the utilities is the heat requirement to heat boiler feedwater in the convection section (from 110 °C to ca. 180 °C), before it is used for the HP steam production in the TLEs, and for superheating the saturated HP steam produced in the TLEs in the convection section.

Wall or radiation loss is that portion of the heat released in the radiant firebox which is lost to the atmosphere through the refractory and outer steel shells of the radiant and convection sections. Wall loss is initially determined by the type and thickness of thermal insulation in the furnace. Typical wall losses in modern furnaces are 1.2% of the fired duty for the radiant section and 0.3% of the fired duty for the convection section.

Stack heat loss is that portion of the heat released in the radiant firebox which is discharged to the atmosphere in the flue gas. Stack heat losses are around 5% of the fired duty. This is the only heat loss that furnace operators can control, given a fixed furnace design, quality of maintenance, feed rate, and coil outlet temperature. Stack heat loss is sensitive to excess combustion air in the furnace and, thus, to draft, fuel composition, and air leakage.

**Radiative Heat Transfer and Furnace Thermal Efficiency.** The heat absorbed by the radiant coils and lost through the walls of a radiant-section firebox must equal the heat given up by the flue gas as it drops from the adiabatic flame temperature to the temperature at which it leaves the firebox [36]. The heat absorbed by the radiant coils of a furnace is a function of coil geometry, burner location, flame emissivity, temperature of the flue gas leaving the radiant firebox, and tube wall temperature of the radiant coil [37]. Typical radiant efficiencies are in the range of 38–42 % for side wall firing, 40–45 % for a combination of sidewall and floor firing and 42–47 % for 100 % floor firing. The differences result from the direct radiation loss to the convection section if the burners are mounted in the upper section of the radiant box.

The effects of excess combustion air and combustion air preheating on radiative heat transfer have been determined [38], [39]. Excess air rates in modern furnaces are ca.10 %, but some plants are operated at rates as low as 5 % excess air, resulting in reduced fuel consumption.

At 20 % excess combustion air (for a given furnace configuration and a radiant-section process duty), a 21 % reduction in fuel consumption is possible if combustion air is heated to 270 °C. Fuel usage increases rapidly as excess combustion air increases. At 49 % excess air, fuel consumption is the same as if combustion air were not preheated.

Combustion air preheating may be used in a new ethylene plant or in a modernized furnace to increase thermal efficiency. Various heat-exchange techniques can be used, including direct exchange between flue-gas discharge and inlet combustion air by use of rotating refractory vane elements, and indirect exchange by means of heat pipes, high-pressure water, or circulating high-temperature thermal fluids.

The most common technique for combustion air preheating is the use of hot exhaust gas from gas turbines (GTs). Such plants co-produce electricity (in the range of 30–60 mW$_{el}$), and this has a major impact on the plant economy. The GTs are exclusively used for electricity production. Direct driving of large compressors by the GTs is not feasible, due to the maintenance requirements for GTs, which require an annual shutdown. In modern plants the ethylene plant can run with and without the GT [40]. The economy of operation with an intregated GT is highly dependent on the credit for the electricity and the increased HP steam production. Scenarios that favor GT integration exist in Japan, Korea, and Brazil.

Radiant-coil inlet and outlet process temperatures in an ethylene furnace must be the same after application of preheated combustion air as they were before, to obtain the same cracking severity.

**Convective Heat Transfer and Furnace Thermal Efficiency.** Convective heat transfer is widely used in isolated green-fields ethylene plants to improve thermal efficiency by preheating boiler feedwater, generating steam, or superheating HP steam in the convection section of a furnace. Recovered energy must be considered in the overall energy balance for a plant of this type. Such techniques can be employed in older plants if boiler feedwater, generated steam, or superheated steam can be used. Otherwise, preheating the combustion air is an alternative.

## 5.1.7. Coking and Decoking of Furnaces and Quench Coolers

**Radiant-Coil Coking.** Hydrocarbon pyrolysis produces acetylenic, diolefinic, and aromatic compounds, which are known to deposit coke on the inside surface of the radiant coil [41], [42]. This coke layer inhibits heat transfer from the tube to the process gas, raises tube wall temperature, and reduces the cross-sectional flow area of the tube, thus increasing the pressure drop along the radiant coil. Cracking yield is thus lowered because of increasing hydrocarbon partial pressure. When tube temperature limits are reached, or pressure limitations of the sonic venturi nozzles used for flow distribution in short-residence-time furnaces, the furnace must be shut down to remove the coke. During the course of one on-stream cycle, deposited coke can reduce the heat-transfer efficiency of the firebox by $1-2\%$, resulting in a 5% increase in fuel consumption [43]. As radiant efficiency decreases, convection section heat transfer increases. This results in an unwanted increase in process crossover temperature and a lower demand for radiant-coil heat input. After furnace firing equilibration, actual fuel demand increases by $1-3\%$.

Gaseous feedstocks such as ethane yield a harder, denser coke with higher thermal conductivity than naphthas or gas oils. Highly aromatic feeds produce the most coke.

The mechanism of coil coking is not fully understood [44]. The possibility that tube metals such as iron and nickel catalyze coke formation is supported by the observation that sulfur in the feed inhibits coking. A sulfur content above 400 ppm, however, may increase coking.

**Quench-Cooler Coking.** Coke deposited in transfer-line exchangers (TLEs) reduces both heat transfer and the amount of steam generated to such an extent that the equipment must be cleaned. Different mechanisms have been proposed for TLE coking or fouling [45]:

1) Spalled coke produced in the radiant coil is carried into the TLE inlet cone where it blocks some of the tubes on the TLE entry tube sheet. Such fouling is most serious in ethane cracking.
2) Poor flow distribution in the entry cone and at the tube sheet causes eddies and backmixing. Gas so trapped experiences long residence time at high temperature, which increases tar and coke production.
3) Heavy polynuclear aromatics and other high-boiling components in the cracked gas condense on the cool TLE tube walls. This is particularly true for gas oil and heavy naphtha feedstocks. The condensed high-boiling substances are gradually converted to cokelike materials. This mechanism is supported by the fact that fouling is fastest at the start of the run (SOR). Typically, TLE outlet temperatures increase by $50-100$ °C during the first day or two in gas oil cracking. Deposits during this period reduce the heat-transfer rate, and inner surface temperature at a given location in a tube becomes higher than the original clean wall temperature. The

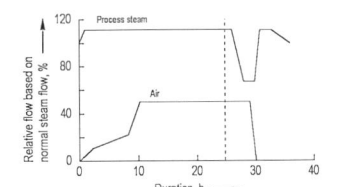

**Figure 13.** Steam–air decoking

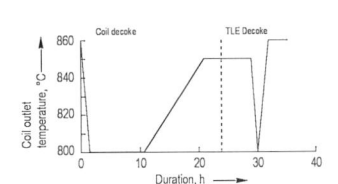

fouling rate drops as the surface temperature increases; with time process gas temperature slowly increases.

**Decoking.** At some time, furnace operation that is affected by coke deposits must be stopped so that the coke can be removed. Mechanical techniques are not feasible. Usually, the coke is burned out with a mixture of steam and air (Fig. 13).

The furnace is taken off-line, the residual hydrocarbons are purged downstream with steam, and the process flow is rerouted to a special decoking system. A steam–air mixture is then introduced into the radiant coils to burn out the coke at coil outlet temperatures of ca. 800 °C. The air concentration is increased carefully to avoid overheating the radiant coil. In modern furnaces the $CO_2$ content is measured continuously during decoking and the air flow rate is adjusted accordingly. After ca. 20 h the decoking of the radiant coils is complete and the decoking conditions are adjusted to decoke the TLE. Total decoking of radiant coil and TLE takes ca. 36 h. Mechanical cleaning of the TLE is not required in modern plant employing appropriate decoking procedures.

Steam-only decoking makes use of the water gas reaction, which is endothermic and less likely to burn out the radiant coils. This method does not require the furnace to be disconnected from downstream processing equipment, if that equipment is designed to accept the required amount of steam and handle the carbon monoxide produced. Temperatures as high as 950–1000 °C are required for steam-only decoking. The decoking gas from the radiant coils may be routed through the transfer-line exchangers, to partially decoke them.

Decoking time and radiant-tube wall temperature decrease as the amount of air mixed with steam increases [46]. The result of a steam-only decoking operation is poorer than that of a steam–air decoke.

Modern furnace quench systems are designed so that off-gas from radiant-coil decoking passes directly through the transfer-line exchangers, which are decoked as a result. About once a year a modern TLE must be opened for inspection and cleaned mechanically. This is carried out by shutting down the furnace, disconnecting the TLE, and removing the coke from the tubes mechanically or hydraulically.

The furnace should never be shut down and cooled without radiant-coil decoking. Differences between the coefficients of thermal expansion of the coke and of the metal tube wall are such that the radiant coils may rupture.

Off-gas from decoking is usually diverted to a decoking system containing a knockout pot equipped with a quench water inlet or a high-performance cyclone separator with dry or wet separation of the coke particles in the effluent. The cleaned, cooled gases are discharged to the atmosphere. Alternatively, the decoke off-gas can be rerouted to the firebox, where the coke particles are incinerated.

## 5.2. Quenching of Hot Cracked Gas

Cracked gases leave the radiant coil of an ethylene furnace at 750 – 875 °C. The gases should be cooled instantaneously to preserve their composition. However, this is not practical, so furnaces are designed to minimize residence time of the hot gas in the adiabatic section between furnace outlet and quench-system cooling zone. Typically, residence time in this adiabatic zone should not exceed 10 % of the residence time in the radiant coil.

The cracked gas is generally cooled by indirect quenching in the TLEs, which permits heat recovery at a temperature high enough to generate valuable high-pressure steam. Direct quenching by injection of oil has been displaced almost totally by indirect quenching.

Design objectives for transfer-line exchangers are as follows:

1) Minimum residence time in the adiabatic section between furnace outlet and TLE cooling surfaces, limiting secondary reactions that reduce the value of the cracked gas
2) Uniform flow to all tubes across the surface of the tube sheet, preventing eddies in the entry cone, which prolong residence time of the cracked gas and favor coke deposition.
3) Low pressure drop across the exchanger; pressure drop across the TLE is directly reflected by the outlet pressure of the radiant coil; increasing pressure at the outlet of the radiant coil degrades the value of the cracking pattern.
4) High heat recovery in the cracked gas, which directly affects operating economics. The pressure for steam generation is affected by the cracking feedstock. For light hydrocarbon cracking, in which little TLE fouling from condensation results, steam of much lower pressure (60 bar) can be generated. With such feeds, two TLEs can be used in series, one for generating high-pressure steam, the other for low-pressure steam or other heat recovery. The TLE outlet temperature must be higher for liquid feeds (360 °C) because of condensation of heavy components in the cracked gas; therefore, the pressure of the generated steam can be higher (8 – 12 MPa). In these

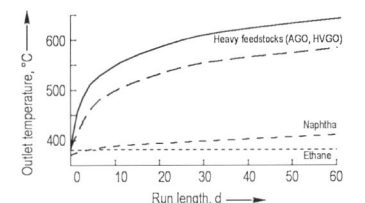

**Figure 14.** Transfer-line exchanger outlet temperatures

cases, because of the high outlet temperature, TLEs are generally followed by a direct oil quench to lower the gas temperature for entry into the downstream portion of the plant.

5) Reasonable TLE run lengths. Fouling of TLE tubes by condensation and blockage by coke increase outlet temperature and pressure drop. Ultimately, the system may have to be shut down and cleaned. The TLE outlet temperature usually increases rapidly for several hours from the start-of-run (SOR) value; the increase then falls to a few °C per day.

Figure 14 shows the TLE outlet temperature over the run length for ethane, naphtha, and heavy feedstock (AGO, HVGO) cracking in a modern highly selective furnace.

The length of run for a TLE in gas cracking service is limited by coke blockage of the TLE inlet tubes [44]. For liquid feeds, TLE run length is governed by fouling at the outlet end of the tubes resulting from condensation of heavy components in the cracked gas. Fouling rate increases as boiling range and relative density of the feed increases.

The transfer-line exchangers with commercial importance are the Borsig exchangers (shell-and-tube and a double pipe closed coupled linear exchanger), Schmidtsche (multi double pipe and a double pipe linear exchanger), the Stone & Webster USX exchanger and the Kellogg exchanger (similar to the USX).

**Borsig Transfer-Line Exchangers** (Borsig, Berlin, Germany). The Borsig TLE design has gained a large share of the world market (over $50 \times 10^6$ t/a ethylene capacity) for use with feedstocks ranging from ethane to atmospheric gas oils. Borsig offers two alternatives: a conventional shell-and-tube design and a linear exchanger. The latest series of the shell-and-tube exchanger, the tunnel-flow exchanger (Fig. 15), which was introduced in 1989, are specially designed to operate with large temperature and pressure differences across the hot-end tube sheet. This sheet is thin (ca. 15 mm) but is supported by a studded stiffening system. The main feature of the tunnel-flow exchanger is the controlled water flow pattern in the inlet section to avoid particle deposition on the tube sheet causing hot spots.

Water flows from downcomers into a circumferential water-jacket annulus. From here water enters several parallel tunnels, which distribute the water perfectly over the bottom tubesheet and via annular gaps in the stiffening plate to the individual tubes) containing the hot cracked gas, which is cooled down by the water, producing high-pressure saturated steam.

**Figure 15.** Borsig tunnel-flow exchanger

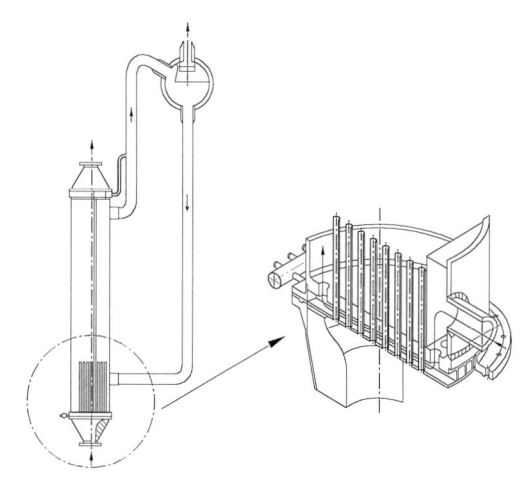

Water flows through the tunnel system toward the opposite side at high velocity in intimate contact with the tube sheet, thereby preventing deposition of solids. On the opposite side, water flows upward through the circumferential annulus and enters the main shell volume. Debris may be deposited on the outer edges of the tube sheet as water flows from it to the main shell, where blowdown nozzles are present. Minor deposits here cause no problems because no heat load is present in this area.

Borsig exchangers can be mounted horizontally or vertically; the latter is the design for all modern furnaces with a top-out concept. Outlet tube sheets are generally thick and uncooled. Thin, studded tube sheets may be needed for TLEs with high outlet temperatures, e.g., in gas-oil cracking [47].

A new development is the Borsig Linear Quencher, which is a double-pipe exchanger, connected directly to the coil outlet. This concept offers a minimum adiabatic volume of the cracking system. However, due to the larger tube diameter of this exchanger (up to 85 mm od), the heat transfer is not as efficient as in the conventional system with smaller tubes. This translates into tube lengths of ca. 20 m, which lead to the characteristic U shape of the linear exchangers, to keep the elevation of the steam drum within a reasonable range. Due to the length of 20 m, the pressure drop is higher than in a conventional system.

Figure 16 shows a Borsig Linear Quencher (BLQ) The main features of the BLQ are the turboflow inlet chamber, with high water velocities to avoid solids deposition and a stepped bend at the end of the primary leg to resist erosion in the U-shaped return bend.

**Schmidt'sche Transfer-Line Exchangers** (Schmidt'sche Heissdampf Gesellschaft, Kassel-Bettenhausen, Germany). SHG has been designing transfer-line exchangers for ethylene crackers since 1960, with more than 3000 conventional and 400 linear exchangers supplied to the industry up to 1996.

The SHG transfer-line exchanger shown in Figure 17 has separate exchangers, steam drums, risers, and downcomers. The basic heat-exchange element consists of concentric

**Figure 16.** Borsig Linear Quencher

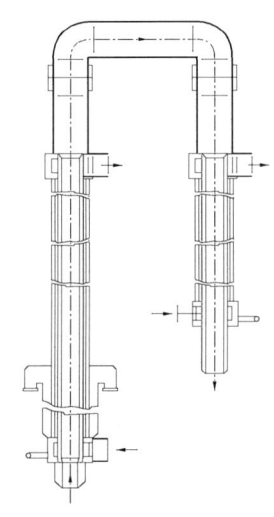

**Figure 17.** SHG double-pipe transfer-line exchanger

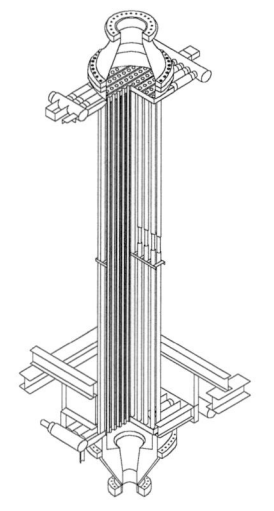

tubes welded at each end to oval tubes. Process gas flows through the inner tube, and water circulates through the annulus between the tubes. Many double-tube elements are combined into a single exchanger by welding the oval headers to form a gastight tube sheet at each end. Process gas is delivered to the inner tubes of the basic tube element by a specially designed entry cone, which provides uniform process gas flow through all tubes and reduces the probability of eddy formation in the cone. Process gas leaving the exchanger is collected in a conical chamber and delivered to downstream equipment. Water flow to and from the oval tubes is provided by downcomers and risers connected to the oval tubes by transition pieces.

The SHG transfer-line exchangers can be mounted vertically with upward or downward flow of process gases. They may also be mounted horizontally with the gas-discharge end slightly elevated with respect to the gas-inlet end. They are normally cleaned by steam – air decoking and from time to time cleaned by hydrojetting.

Schmidt'sche has also developed a closed coupled linear exchanger, which offers a reduced adiabatic reaction zone, but due to the increased length has a higher pressure drop than the conventional exchanger. Figure 18 shows a Schmidtsche close coupled linear exchanger (SLE) in the typical U-shape. Feeds for all Schmidtsche exchangers range from ethane to gas oils [48].

**Stone & Webster Quench System** (Stone & Webster Engineering Corp., Boston, Mass.). The Stone & Webster proprietary quench system does not rely on a tube sheet. The USX (ultra-selective exchanger) quench cooler shown in Figure 19 is a double-pipe exchanger, with the inner pipe slightly larger than the radiant-coil connecting piping. This design can quench the effluent from one to four radiant coils. The USX outlet temperature (500 – 550 °C) is higher than in most other transfer-line exchanger designs; this helps to reduce coking in the exchanger. Normally, effluents from all USX exchangers on a given furnace are combined downstream and fed to a conventional multitube exchanger where the gases are quenched to a more normal TLE outlet temperature.

This design, by eliminating the tube sheet and entry cone, eliminates turbulent eddies and backmixing between the coil outlet and the cooling surfaces, thereby reducing fouling in the TLE entry zone. The large USX gas flow tube reduces pressure drop across the exchanger and, thus, the probability of plugging by coke from the radiant coil.

The USX exchangers can be steam – air decoked; cleaning, when required, is facilitated by openings at each end of the exchanger.

**Kellogg Millisecond Primary Quench Exchanger** (M. W. Kellogg Co., Houston, Texas). A special primary quench exchanger (PQE) was developed for use with Kellogg's Millisecond furnace. The exchanger has a double-pipe arrangement (Fig. 20) capable of generating high-pressure steam.

A special high-alloy section of tubing is inserted to avoid excessive thermal stress at the high-pressure steam transition piece. This section has the Y-fitting at its base and extends up to the boiler feedwater inlet nozzle. A steam bleed helps to control thermal gradients and to prevent coke deposition in the annulus between the high-alloy insert and the exchanger inner surface.

The steam system is a thermosiphon type, with individual risers and downcomers tied together in common headers connecting to a single steam drum. Kellogg's PQEs are normally fabricated in bundles for ease of shipping and installation. Typically, one PQE serves two coils in a Millisecond Furnace.

**Other Quench-Cooler Designs.** Proprietary quench coolers have been developed by Lummus, Kellogg, C. F. Braun, Foster-Wheeler, and others.

**Figure 18.** Schmidtsche close coupled linear exchanger

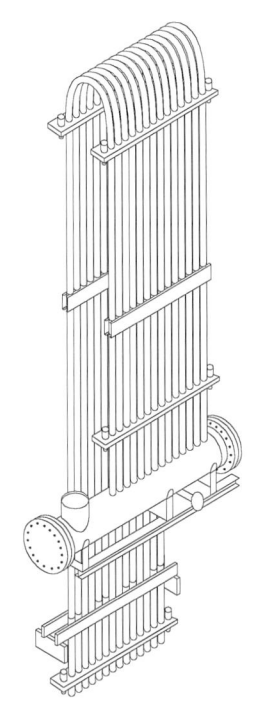

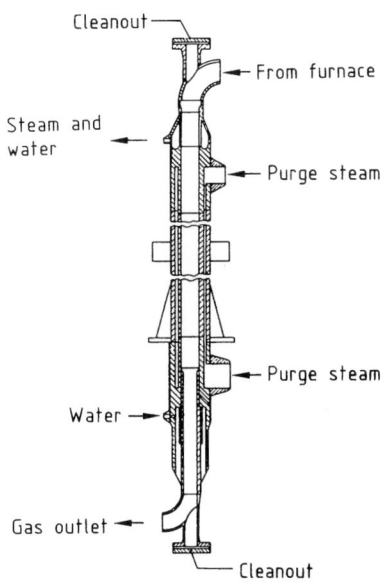

**Figure 19.** Ultraselective exchanger (USX) quench cooler *
* Reprinted with permission of Stone & Webster Engineering Corp.

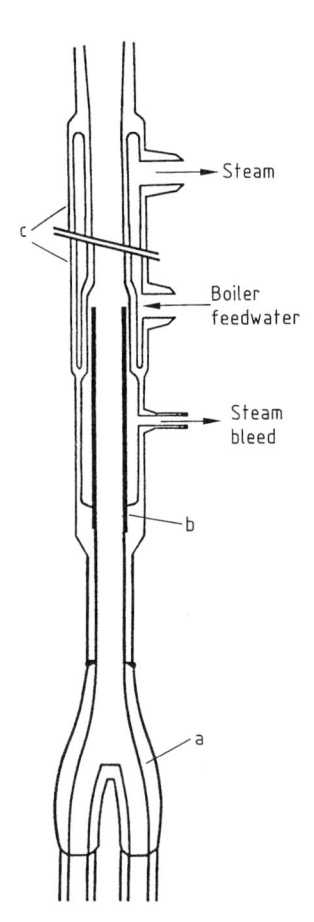

**Figure 20.** Kellogg millisecond primary quench exchanger *
a) Y-shaped fitting; b) High-alloy insert; c) Double pipe
* Reprinted with permission of M. W. Kellogg Co.

The *Lummus horizontal TLE*, with conventional tube-and-shell exchanger design, can be mounted horizontally under the furnace firebox. It is used extensively in the United States and Mexico for ethane cracking.

The *Hitachi two-stage quench-cooler* system (Hitachi, Japan) is similar to the Stone & Webster USX quench-system design. It consists of a primary double-pipe TLE for each furnace pass or pair of passes, followed by a conventional tube-and-shell TLE design capable of handling the combined process flow from all the primary TLEs on a given furnace. Hitachi quench systems have been used for cracking ethane, propane, and full-boiling-range naphtha.

The *Mitsubishi Quench Exchanger* (Mitsubishi Petrochemical Co. and Mitsubishi Heavy Industries of Japan) is available in two designs: M-TLX-I and M-TLX-II. The M-TLX-I contains the transfer-line exchanger and the steam drum as an integral unit. Up to four coil outlets can be connected to a single M-TLX-I, if coil arrangement permits. The M- TLX-ll also has on-line decoking capability if boiler feedwater and hydrocarbon feed are removed from the exchanger. The M-TLX-II has a separate steam drum.

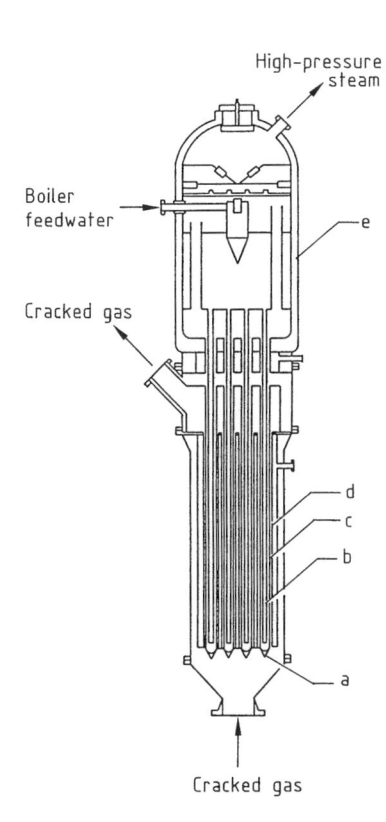

**Figure 21.** Mitsui innovative quencher *
a) Honeycomb tube nest; b) Inner tube; c) Heat-tranfer tube; d) Outer tube; e) Steam drum
* Reprinted with permission of Mitsui Engineering & Shipbuilding Co.

The *Mitsui Quench Exchanger* (Mitsui Engineering & Shipbuilding Co., Tokyo, Japan) (Fig. 21) uses an integrated exchanger, steam drum, risers, and downcomers.

Cracked gas rises through spaces between many heat-exchanger elements (c) suspended from the tube sheet of the steam drum (e) and leaves the exchanger at the gas-outlet nozzle. Each heat-exchange element consists of a pair of concentric tubes. The inner tube (b) is open at the bottom end; the outer tube (d) is closed. Cooling water flows down the inner tube from a reservoir in the center of the steam drum, leaves the open end of the inner tube, and returns to the steam drum via the annulus between the inner and outer tubes. Steam generated in the annulus is separated and dried in the upper section of the steam drum. This design eliminates problems caused by differences in thermal expansion of the tubes. The system must be decoked on-line. Mechanical cleaning of the cracked-gas flow path is not possible. Up to six radiant coils can be combined and served by one Mitsui TLE system [49].

## 5.3. Recovery Section

The cracked gas leaving the cracking furnaces must be cooled to recover the heat it contains in the cracked gas and render it suitable for further processing. The temperature of the cracked gas leaving the primary TLE depends on the feedstock type; it varies from 300 °C for gaseous feedstock, to 420 °C for naphtha feedstock, and up to 600 °C for AGO and hydrocracker residues. In ethane and propane cracking, cooling to ca. 200 °C is performed in heat exchangers by preheating feedstock or BFW. However, this step can not be done in heat exchangers for feedstocks heavier than propane, due to the condensation of the liquid portion contained in the cracked gas. Cooling such cracked gases to 200 °C would lead to excessive fouling of the heat exchangers by carbon deposits. Thus, cooling is performed by injection of oil into the cracked gas (oil quench) by means of special tangential-injection mixing devices. Oil quenching is performed individually for each furnace or in the combined cracked gas line after collection of the cracked gas from all furnaces. In modern designs, the oil injection is located in the furnace area, integrated into the cracked gas line downstream from the TLEs.

The cracked gas leaving the furnace section has a temperature of ca. 200 – 250 °C and is routed to the separation section via a large cracked gas line. The collected gas stream contains the effluents of all cracking furnaces, including those processing recycle streams.

Table 10 shows the yields of the product fractions for all typical feedstocks. Apart from the type of feedstock, the gas composition and, therefore, the design of the downstream recovery process also depends on the cracking severity and the applied coil design.

Further processing, i.e., separation into the main products or fractions, can be performed in different sequences and depends on the feedstock type and the number and the specification of the plant products. The main downstream processing steps are the removal of the heat contained in the cracked gas, condensation of water and heavy hydrocarbons, compression, washing, drying, separation, and hydrogenation of certain multiply unsaturated components. The main task of these steps is to recover the desired products at given specifications and at the desired battery-limit conditions.

Ethane, propane, and butane can be recycled to the furnace section to increase the overall yield of ethylene per tonne of feedstock processed. Processing differs for cracked gas derived from gaseous and liquid feedstocks, since the portion of heavy components increases, especially when processing naphtha or even heavier feedstocks such as gas oil and hydrocracker bottoms. When pure ethane is the feedstock, the amount of $C_3$ and heavier byproducts is small, and only rarely is the recovery of heavier products economically feasible. As the concentration of heavier components increases, recovering them for direct sale becomes feasible. Addition of significant amounts of propane to ethane feedstock makes a depropanizer necessary and hydrotreating and fractionation of the $C_3$ cut may be required; butane feeds make oil and gasoline removal from the

**Table 10.** Cracking yields for various feedstocks

|  |  | Ethane | Propane | Butane | Naphtha | AGO | HCR |
|---|---|---|---|---|---|---|---|
| P/E *, kg/kg |  | (65) ** | (90.0) ** | 0.35 | 0.55 | 0.54 | 0.53 |
| Steam dilution, kg/kg |  | 0.3 | 0.3 | 0.487 | 0.45. | 0.8 | 1 |
| Residence time, s |  | 0.3451 | 0.3337 | 0.344 | 0.3572 | 0.32 | 0.3 |
| $H_2$ | wt% | 4.05 | 1.55 | 1.09 | 0.92 | 0.71 | 0.68 |
| CO | wt% | 0.04 | 0.04 | 0.04 | 0.04 | 0.01 | 0.01 |
| $CO_2$ | wt% | 0.01 | 0.01 | 0.01 | 0 | 0.01 | 0.01 |
| $H_2S$ | wt% | 0.01 | 0.01 | 0.01 | 0 | 0 | 0 |
| $CH_4$ | wt% | 3.52 | 23.27 | 20.29 | 14.91 | 10.58 | 11.08 |
| $C_2H_2$ | wt% | 0.47 | 0.51 | 0.42 | 0.47 | 0.38 | 0.6 |
| $C_2H_4$ | wt% | 52.31 | 37.51 | 35.81 | 28.45 | 25.93 | 25.98 |
| $C_2H_6$ | wt% | 35.03 | 2.80 | 4.16 | 4.23 | 2.82 | 2.81 |
| $C_3H_4$ | wt% | 0.02 | 0.57 | 0.93 | 0.76 | 0.63 | 0.97 |
| $C_3H_6$ | wt% | 1.13 | 14.82 | 17.24 | 15.64 | 14.07 | 16.02 |
| $C_3H_8$ | wt% | 0.12 | 9.96 | 0.35 | 0.51 | 0.37 | 0.42 |
| $C_4H_4$ | wt% | 0.05 | 0.08 | 0.1 | 0.13 | 0.12 | 0.04 |
| $C_4H_6$ | wt% | 1.80 | 2.90 | 4.08 | 4.79 | 5.73 | 8.03 |
| $C_4H_8$ | wt% | 0.19 | 1.00 | 2.83 | 4.52 | 3.61 | 4.14 |
| $C_4H_{10}$ | wt% | 0.21 | 0.04 | 6.07 | 0.44 | 0.05 | 0.06 |
| Benzene | wt% | 0.47 | 2.12 | 2-69 | 7.12 | 5.44 | 5.64 |
| Toluene | wt% | 0.07 | 0.40 | 0.77 | 3.1 | 3.49 | 2.73 |
| Xylenes | wt% | 0.00 | 0.05 | 0.11 | 1.16 | 0.92 | 0.46 |
| Ethylbenzene | wt% | 0.00 | 0.01 | 0.01 | 0.62 | 0.37 | 0.34 |
| Styrene | wt% | 0.02 | 0.20 | 0.24 | 1.14 | 1.18 | 1.12 |
| Pyrolysis gasoline | wt% | 0.32 | 1.26 | 1.81 | 7.96 | 7.28 | 6.91 |
| Pyrolysis fuel oil | wt% | 0.16 | 0.89 | 0.94 | 3.09 | 16.3 | 7.95 |
| **Sum** | **wt%** | 100.00 | 100.00 | 100 | 100.00 | 100 | 100 |

* P/E = propylene to ethylene ratio, ** Conversion of feedstock

cracked-gas compressor necessary; pentane-range feeds require fractionation of light from tarry oils, and so on.

### 5.3.1. Products

Local market conditions and the degree of integration of the ethylene units into refining or petrochemical complexes influence the desired products and the feedstocks used. Many types of coproduct can be generated with different equipment [50]–[52]; only the major product options are discussed here.

The principal ethylene coproducts are as follows:

- Acetylenes ($C_2$ and $C_3$) are hydrogenated to ethane, ethylene, propane, and propene; or they may be recovered and sold as products.
- Aromatics (various fractions) can be recovered or they may remain in the hydrotreated pyrolysis gasoline.
- $C_4$ olefins can be refined for butadiene, butylene, isobutylene, or mixtures thereof.
- $C_5$ olefins can be either recovered and refined to give isoprene, piperylene, and cyclopentadiene, or hydrotreated in the pyrolysis gasoline fraction.

- Ethane is recycled as cracking feedstock or used as fuel.
- Fuel oil is used as such or to produce coke or carbon black.
- Hydrogen is purified and used for hydrogenation steps in the plant; excess hydrogen is sold or used as fuel in the plant.
- Methane is used as fuel or sold.
- Naphthalene is recovered for sale or left in the pyrolysis fuel oil fraction.
- Propane is recycled as a cracking feedstock, used as fuel, or sold.
- Propene is sold in various grades.
- Raw pyrolysis gasoline is sold as motor gasoline after hydrotreatment, or it is used as feedstock for aromatics production.
- Tar is sold for fuel, used as coke feedstock, or diluted with hydrocarbon stocks to produce feedstocks for resins.
- Sulfur is recovered in some plants and sold.

## 5.3.2. Cracked Gas Processing

Many options are available for general process configuration [50], [53]–[59]. The sequence for isolation of hydrocarbon fractions can be varied, depending on plant size, amounts of ethylene and coproducts, impurities, product range, product purity desired, and other factors.

The following discussion of the recovery process is split into two sections:

1) Front-end section, including cracked gas compression and associated units
2) Hydrocarbon-fractionation section

Difference between the two sections is made for simplicity reasons. The basic process route for the front-end section is identical for all commercial processes, however major difference exists between gas- and liquid feedstock cracking. For the hydrocarbon fractionation section there are a large variety of process routes available, while the feedstock type results only in minor differences of the basic process route.

### 5.3.2.1. Front-End Section

The front-end section (Figs. 22, 23) receives the effluent stream from the cracking furnaces. Its main task is to cool the cracked gas stream, recovering waste heat and providing it to other plant sections, to condense and to produce dilution steam, and to remove heavy hydrocarbon components such as tar, oil, and heavy gasoline. The resulting purified gas is compressed in the cracked gas compression section and dried to prepare it for further cryogenic separation in the fractionation section. An acid-gas removal step that eliminates all $CO_2$ and $H_2S$ from the cracked gas is incorporated between the final compressor stages.

The process configuration of this front-end section is fairly independent from licensers as they all follow the same process principals. However there are many

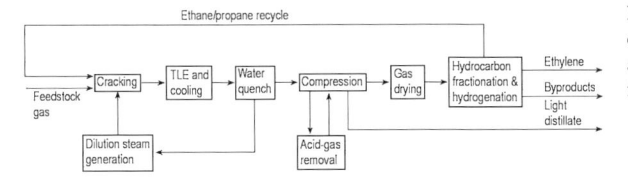

**Figure 22.** Simplified process flow diagram for the front-end process of an ethylene plant cracking gaseous feedstocks

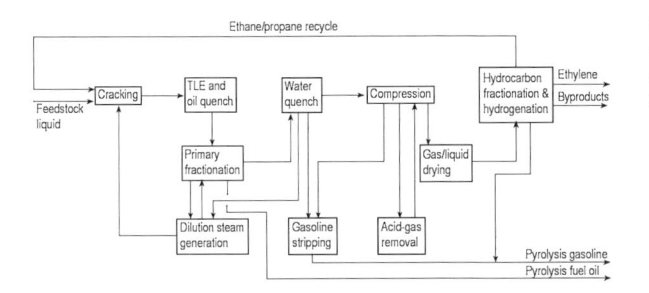

**Figure 23.** Simplified process flow diagram for the front-end process of an ethylene plant cracking liquid feedstocks

variations in detailed design. Major differences are observed between plants processing gaseous feedstocks (ethane, propane, and butane) and those treating liquid feedstocks like naphtha. Processing of cracked gas resulting from liquid feedstocks is more complex as higher amounts of heavy hydrocarbons are condensed and must be removed in this section.

**Process Description.** The main difference between the process for gaseous feedstock (Fig. 22) and liquid feedstock (Fig. 23) is the presence of a primary fractionator that removes tar and oily material ($bp > 200 \,°C$) from the cracked gas in the latter. In the case of cracking gaseous feedstock only, this step is not necessary, as the small amounts of heavy hydrocarbons present in the cracked gas can be removed in the water-quench column. Furthermore, additional installations are required in the water-quench system and in the compressor and drying units to treat the larger amount of pyrolysis gasoline that condenses in these units.

**Primary Fractionation.** With liquid pyrolysis feedstocks, the primary fractionation column is the first step in the cracked gas processing route. Cracked gas enters the column typically at 230 °C. In the column it is contacted with circulating oil and, at the top of the column, with a heavy pyrolysis gasoline fraction obtained from the subsequent water-quench tower. Cracked gas leaves the top of the primary fractionator at ca. 100 °C, free of oil but still containing all the dilution steam. Hot oil, which functions as a heat carrier, is collected at the bottom of the column and, after cooling, recirculated as reflux to the middle section of the primary fractionator and to the quench nozzles downstream of the TLEs. Most of the heat contained in the cracked gas leaving the TLE's is removed by the circulating quench oil. Heat is reused for generation of process steam as well as for other purposes such as preheating of feedstock, preheating of process water, or for other energy consumers at an appropriate temperature level. Excess oil is removed from the primary fractionation system and sent out as product. Pyrolysis oil is generally produced as one fraction, containing all hydrocarbons boiling above 200 °C. However some plants produce two qualities of pyrolysis oil, one called PGO (pyrolysis gas oil) boiling in the range 200–400 °C and PFO (pyrolysis fuel oil) containing the heavier components.

**Water-Quench Column.** The overhead of the primary fractionator is routed to the water-quench column, in which the cracked gas is cooled to near ambient temperature by contacting it with a large stream of circulating quench water. In this cooling step most of the dilution steam and a heavy gasoline fraction are condensed and collected at the bottom. After gravity separation, heat is extracted in a circulating loop and most of the water stream is used to produce dilution steam. Large amounts of low-temperature heat (60–80 °C) are recovered and used for various heating purposes in the recovery section. The largest consumer of this low temperature heat is typically the reboiler of the propene–propane fractionation tower. Most of the condensed gasoline is recycled as reflux to the top of the primary fractionator. The excess gasoline fraction is sent to a gasoline hydrotreater or fractionated further into higher quality gasoline and heavier oil streams. Both water and gasoline fractions, except when fractionated further, are stripped of dissolved light gases [50], [53], [60], [61].

In the case of processing cracked gas from gaseous feedstock, the furnace effluent stream is directly fed to the quench tower after cooling in a secondary TLE to ca. 200 °C. In this case the system is simpler as no or only small amounts of tar and oil condense and must be removed. A general problem in all quench towers is the separation of hydrocarbons and water. There is a strong tendency for formation of emulsions, leading to fouling problems in the generation of dilution steam and in the processing of the wastewater from the system. Emulsification is avoided by careful control of the pH value in the quench-water system.

Numerous options exist in the design and operation of the quench column and primary fractionator units for removing pyrolysis tars, recovering various heavy hydrocarbon streams, and isolating quench oil for recycle [50], [60], [61].

**Dilution-Steam System.** The dilution steam required in the cracking process forms a closed loop over the cracking furnaces and the quench-water column. Excess water from the bottom of the quench column is stripped in a process water stripper to remove volatile hydrocarbons. Some 5–10 % of the process water is purged to avoid concentrating dissolved solids. The generation of the process steam is typically performed at 800 kPa. In liquid feedstock processes, 50–80% of the dilution steam is generated by recovering heat from the hot quench oil cycle from the bottom of the primary fractionator. The balance heat duty is provided by condensing steam

**Compression.** After cooling and purification of the cracked gas in the primary fractionator and/or in the quench tower, further processing requires compression to ca.

3200–3800 kPa. Some condensation of water and hydrocarbons occurs during compression, and in the later stages, acidic gases are removed. Compression is typically performed by a turbine-driven centrifugal compressor in four to six stages with intermediate cooling. The number of stages depends primarily on the cracked-gas composition and the highest temperature allowed for interstage discharge. Interstage cooling and temperature control keep the cracked gas below 100 °C to prevent diolefin polymerization and subsequent equipment fouling. Various injection systems, using aromatic rich oil or water, are applied to reduce fouling of compressor internals and intercooling equipment. Interstage cooling usually employs water coolers to minimize inlet temperature in the subsequent stages. First-stage compressor suction is normally maintained somewhat above atmospheric pressure to prevent oxygen intrusion. The operating pressure is set to economically balance yield in the cracking coils against higher energy consumption for compression [50]. The compressor discharge pressure is set by the choice of refrigerant so that methane condenses in the demethanizer overhead condenser. Condensation temperature varies with hydrogen/methane ratio. The discharge pressure, with the coldest ethylene refrigerant level typically at ca. −100 °C, is normally 3200–3800 kPa. The criteria used to determine suction pressure and discharge pressure are selectivity and energy consumption [50], [61].

Interstage cooler pressure drops must be kept low, especially in the initial stages, to reduce compressor and energy costs. The water and hydrocarbon condensates may be routed to another compressor stage, to the primary fractionator preceding the compressor, or to a fractionation tower. In liquid cracking most of the gasoline fraction containing the $C_6$ to $C_8$ aromatics is condensed in the interstage coolers of the compressor. After gravity separation from process water this gasoline stream is fractionated in a stripper to remove $C_4$ and lighter components before routing it to a hydrotreating step. Drying is required for all liquid and gas streams subjected to temperatures below 15 °C [50], [53], [59], [60].

The acid-gas removal system is typically located between the 3rd and 4th, or between the 4th and 5th stages. Some fractionation processes that have deethanization or depropanization as the first process stage have this fractionation step between the 4th and 5th compressor stages. In these cases moisture must be removed upstream of the fractionation step to avoid formation of hydrates and ice.

**Acid-Gas Removal.** Before further processing, carbon dioxide and hydrogen sulfide are removed from the cracked gas by once-through and regenerative solvent scrubbing. Sulfur may also be removed from the cracker feedstock [50].

Acid gas is usually removed just before the 4th or 5th compression stage. Carbon dioxide is removed because it can freeze at low temperature in heat-exchange and fractionation equipment. Carbon dioxide can also be absorbed into ethylene, affecting product quality and further processing. Hydrogen sulfide is corrosive and a catalyst poison as well as a potential product contaminant [52], [53].

These acid gases are scrubbed with sodium hydroxide on a once-through basis or in combination with a regenerative chemical. Regenerative prescrubbing before final sodium hydroxide treatment is common in large cracking plants, especially those that use high-sulfur feedstocks. This reduces sodium hydroxide consumption, which is desirable for economic and environmental reasons.

Regenerative scrubbing can employ alkanolamines, but it is not possible with this treatment alone to lower the concentration of $CO_2$ to that required to protect high-activity polymerization catalysts against $CO_2$ poisoning (e.g., to 0.2 ppm). Use of regenerative scrubbing is typically limited to liquid-cracking operations. In any case a polishing scrubber consisting of a caustic wash unit is applied to achieve the required $CO_2$ specification in the effluent stream. Acid-gas removal systems are frequently subject to fouling, especially at high concentrations of $C_4$ and $C_5$ diolefins. The fouling process can be controlled by applying special inhibitors.

Acid gases liberated from the regenerative solvent can be incinerated or recovered [50], [53], [60] – [62]. Waste caustic, containing sulfides in the range of a few % is usually oxidized and neutralized in a wet caustic oxidation unit or neutralized and stripped, before feeding it to the sewer system.

**Drying.** The cracked gas is saturated with water before compression and after each intercooler stage. Moisture must be removed before fractionation to prevent formation of hydrates and ice. Typically, this is accomplished by chilling and by adsorption on molecular sieves. Older plants also used absorption by a glycol scrubbing system or adsorption on alumina [52]. Drying is arranged before the first fractionation step, typically after the last compression stage. Before feeding the compressor effluent to the molecular sieve dryer it is cooled in a water cooler and subsequently by propene refrigerant. A temperature of 15 °C upstream of the dryer is desirable for removing as much water as possible to reduce load on the dryers. However, the temperature must remain above the hydrate-formation point, which lies in the range 10 – 15 °C [50], [60]. Water is adsorbed on molecular sieves, preferably after the highest compression pressure is reached to minimize adsorption costs. Higher pressure allows smaller dryer volume with lower adsorbent cost and less water removal because the water content in the gas stream decreases with increasing pressure. Multiple adsorption beds make continuous water removal possible. One or more adsorption beds are in operation while at least one unit is being regenerated. Regeneration requires the use of heated gas that is free of polymerizable components. Byproduct methane and hydrogen used as fuel are typically good choices for this gas [50] – [53], [60], [61]. In liquid-feedstock cracking processes large amounts of hydrocarbons are condensed at 15 °C. If these liquid hydrocarbons are to be fed to the fractionation column they are also dried over molecular sieve adsorbers.

### 5.3.2.2. Hydrocarbon Fractionation Section

The fractionation section receives the compressed cracked gas at a pressure of 3200 – 3800 kPa for further fractionation into different products and fractions at specified qualities and battery limit conditions. Over the history of design and construction of ethylene plants many different process schemes have been developed and implemented for this task.

Today, three processing routes have gained commercial importance, with the main characteristics being the first separation step and the position of the hydrogenation of the acetylene contained in the cracked gas. These three process sequences are:

1) Demethanizer first with tail-end hydrogenation
2) Deethanizer first with front-end hydrogenation
3) Depropanizer first with front-end hydrogenation

These basic processes differ in the configuration of process steps for any given feedstock, but provide similar overall capabilities for fractionation and hydrogenation. The differences are primarily in the sequence of fractionation and hydrotreating steps downstream of cracked-gas compression [50].

**Process Description.** Compressed cracked gas can be:

1) Condensed and fed to a demethanizer column (front-end demethanizer), as shown in Figure 24
2) Condensed and fed to a deethanizer (front-end deethanizer), as shown in Figure 25
3) Removed from an intermediate compressor stage, condensed, and depropanized in a column, whereby the light gas is returned to the last compressor stage (front-end depropanizer), as shown in Figure 26

Combinations are possible; for example, intermediate compressor stage condensate and vapor are sent to different parts of the plant [60], [56].

Traditionally the demethanizer first/tail-end hydrogenation process sequence was used by American contractors. With the need for reduction of energy consumption in the separation section and process simplification for cost reasons, the deethanizer first/front-end hydrogenation process, which had long been used by Linde, was also applied by some American contractors. A comparison of all routes has shown that this sequence seems to have the best energy efficiency [63]. New developments in this area are the combination of the advantageous sequence with special equipment (dephlegmators) [64] and the application of washing steps for demethanization [65].

**Front-End Demethanizer** (Fig. 24). The hydrocarbon-separation sequence begins with removal of methane and lighter components, primarily hydrogen, from higher molecular mass components. Hydrogen may be removed before or after methane. The bottom product is routed to a deethanizer column where acetylene, ethane, and ethylene are removed as overhead product, and $C_3$ and heavier components as bottom product.

Typically, the overhead product from the deethanizer is first hydrotreated to convert acetylene to ethylene and ethane. This treated overhead mixture is then fractionated to ethylene product and ethane recycle.

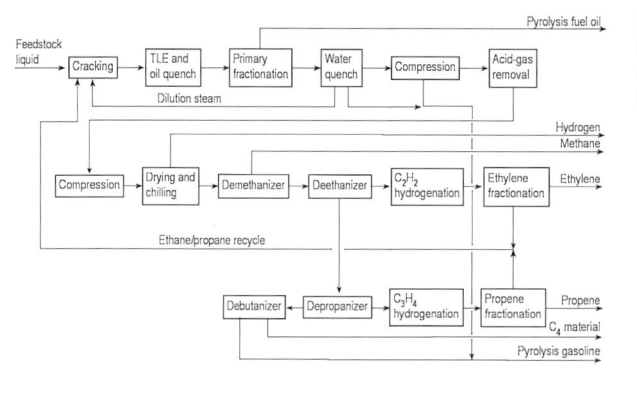

**Figure 24.** Simplified process flow diagram for the production of ethylene by liquid cracking with a front-end demethanizer

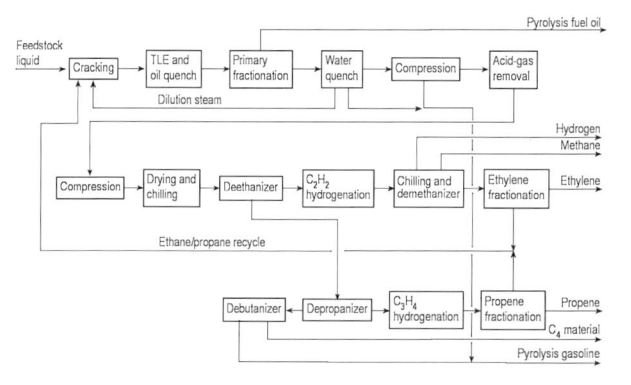

**Figure 25.** Simplified process flow diagram for the production of ethylene by liquid cracking with a front-end deethanizer

The bottom product from the deethanizer is depropanized to separate $C_4$ and heavier components as a bottom stream; methylacetylene, propadiene, propane, and propene are taken as an overhead stream. This overhead fraction is hydrogenated to remove methylacetylene and propadiene, which can also be recovered by distillation or absorption and stripping. In the propene fractionation step, propene is recovered for sale and propane for recycle. The propene product can be of chemical grade or polymer grade with additional purification. Depropanizer bottoms are sent to a debutanizer for separation into other product streams ($C_4$ materials, raw pyrolysis gasoline, $C_5$ materials, aromatics, etc.) [50], [52]–[54], [59], [60].

**Front-End Deethanizer** (Fig. 25). Deethanization of dried cracked gas is used as the first fractionation step to produce on overhead stream of ethane and lighter components, and a product stream of $C_3$ and heavier materials; subsequently, the lighter components are hydrogenated to remove acetylene and fractionated cryogenically. Because this overhead stream is hydrogen rich, no supplementation is necessary for hydrogenation. During the chilling and demethanizer operation, the overhead is separated into hydrogen and methane; the bottom stream is fractionated to yield ethylene product and ethane; and the ethane is recycled to the cracking furnaces.

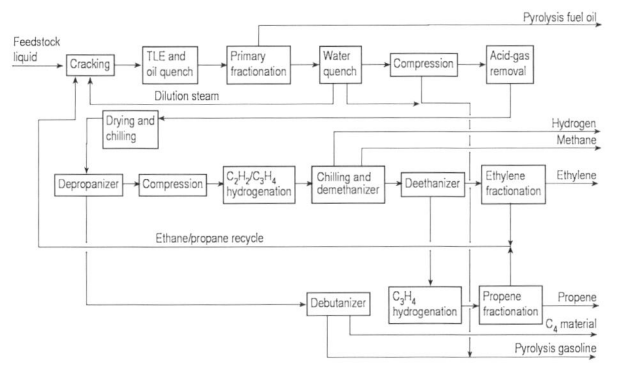

**Figure 26.** Simplified process flow diagram for the production of ethylene by liquid cracking with a front-end depropanizer

The bottom stream from the deethanizer is fed to the depropanizer. The overhead stream from the depropanizer is hydrogenated to convert methylacetylene and propadiene contaminants to propene and propane. The final purification step yields propene product and propane; the propane is recycled as cracking feedstock. The depropanizer bottom stream is fractionated to produce a $C_4$ product stream and a $C_5$-rich raw pyrolysis gasoline [53], [54], [59], [60].

**Front-End Depropanizer** (Fig. 26). Depropanization of dried cracked gas is used as the first fractionation step to produce a propane and lighter component overhead stream and a $C_4$ and heavier bottom stream. The depropanizer overhead stream is compressed and hydrogenated to remove acetylene, and to partially hydrogenate methylacetylene and propadiene. It is then chilled and demethanized, and the bottom fraction is deethanized. The deethanizer overhead components are fractionated to ethylene product and ethane, which is recycled as cracking feedstock. Deethanizer bottom product is hydrotreated to remove methylacetylene and propadiene, and fractionated to propene product and propane, which is recycled to the cracking furnaces. The depropanizer bottom components are fractionated to produce $C_4$ materials and raw pyrolysis gasoline [53], [54], [59], [60].

For all three of the above-mentioned basic processes numerous process variants exist. For liquid cracking acetylene hydrogenation is always performed in the $C_2$ stream, in the $C_{2-}$ or in the $C_{3-}$ stream. In the case of cracking gaseous feedstock with little coproduction of $C_4$ hydrocarbons the hydrogenation step can also be arranged in the total cracked gas stream during or just after cracked-gas compression (after or before drying). This is not feasible for cracked gas streams resulting from liquid cracking as 1,3-butadiene, typically a valuable byproduct, would be cohydrogenated almost completely to butene and butane.

Acetylene may also be recovered for sale by an absorption process. This absorption process step is located in the gas phase upstream of the ethylene fractionation column.

**Prefractionation Feed Chilling.** Before being fed to any of the fractionation towers, dried cracked gas must be cooled and partially condensed. Dried cracked gas is cooled by the refrigerant systems and by the low-temperature product and recycle streams leaving the cold fractionation equipment down to a temperature of typically −140 to −160 °C. At this temperature $C_2$ components are entirely removed by partial condensation, and the noncondensed gas stream has a hydrogen concentration of 80−95 vol %. The type of refrigerant, number of refrigeration levels, and the design of the refrigeration and heat-exchange systems depend on the temperature and pressure required in the downstream fractionation equipment. Fractionation and purification consume much energy. Cryogenic purification of methane, ethane, and ethylene requires costly refrigeration. Design optimization is complex because:

1) The most efficient routing must be considered for the condensates and vapor streams from each stage of the cool-down sequence in the individual refrigeration loops (e.g., 2−6 stages each)
2) The energy requirements for fractionation must be determined
3) The freezing of components, such as acetylene, butadiene, and benzene, must be avoided

When acetylene content is high, prevention of freezing and safe hydrogenation are important. Also of importance from a safety standpoint for the chilling section and downstream equipment is the choice of appropriate construction materials for cryogenic conditions [50], [52]. Process equipment that may accidentally be exposed to cryogenic conditions should also be carefully studied. National standards are available in most countries to assist in making appropriate selections [50], [52].

**Demethanization.** Demethanization of cracked gas separates methane as an overhead component from $C_2$ and heavier bottom components; concurrently, hydrogen is removed from the cracked-gas stream and may be obtained as a product by purification before or after demethanization. Methane is typically used as plant fuel or sold; $C_2$ and heavier components are sent to the recovery system. This overall fractionation is important from a design and operational standpoint because of its high share of the capital, energy, and fuel consumption of the ethylene plant. Demethanization typically consumes the greatest proportion of net energy from the refrigeration system (see Fig. 27 and discussion of refrigeration, p. 2281) [55], [62].

The feed stream to the demethanizer must be free of carbon dioxide and water to prevent freezing and plugging under the cryogenic conditions employed. Although a wide range of operating conditions is possible, feed temperature is typically ca. −100 °C and pressure 3000−3500 kPa. Energy efficiency may be improved, often with increased capital cost and process complexity, by staged condensation and flashing of the stream during chilling to provide multiple feed streams. Reboiling and condensation at intermediate levels in the demethanizer, and lower dual-pressure demethanization can also be used. A range of gas compositions must be considered in the design [50], [53], [54], [59], [61].

The overhead condenser product contains hydrogen, carbon monoxide, and small amounts of ethylene, in addition to methane. In the simplest case, expansion of the mixed methane–hydrogen stream provides additional refrigeration, and the total gas effluent is used as fuel without refinement. Typically, however, after condensation, some of the overhead methane fraction is flashed to about the pressure of fuel gas to obtain additional refrigeration. Any condensate formed during this operation is a recoverable source of ethylene. Additional staged cooling equipment provides hydrogen of ca. 95 vol % purity by adsorption processes. The lower temperature refrigerant required in this operation is provided by expansion of the condensed methane stream [50], [52], [55], [60], [61].

Energy is recovered from the demethanizer overhead stream by Joule–Thompson pressure reduction or by an expansion turbine. The latter, in addition to generating refrigeration, recovers some energy [57], [61].

The demethanizer bottom stream contains $C_2$ and heavier components. With the front-end demethanizer, components from $C_2$ through pyrolysis gasoline may be present. However, with front-end deethanization, $C_{3+}$ components are removed, and with front-end depropanization, $C_{4+}$ components are removed. The demethanizer design changes significantly as heavier components are eliminated, especially with regard to the refrigeration needed to condense the incoming cracked-gas feed stream and to reboil the bottom stream.

**Deethanization.** Deethanization of cracked gas separates acetylene, ethylene, and ethane as overhead components from $C_{3+}$ bottom components. In newer plants, low-pressure fractionation may be used, or feed streams from the cracked-gas chilling section can be supplied as multiple feeds to the deethanization tower. The deethanizer is usually the third largest user of refrigeration energy in a front-end demethanizer ethylene plant (Fig. 27) [50], [60]–[62].

With front-end demethanization and front-end depropanization, the feed stream to the deethanizer comes primarily from the demethanizer. With front-end deethanization, the deethanizer feed comes from the chillers that follow the gas dryers. In the simplest cases, single feed streams from these units are separated into overhead and bottom fractions for further processing. More recent plant designs incorporate energy savings into plant operation by providing an additional feed stream to the deethanizer from the feed-chilling section, along with that from the demethanizer [50]. Dual-pressure deethanizer systems improve energy consumption and reduce the bottom temperature, thus reducing fouling in reboilers. As with the demethanizer, these designs conserve refrigeration energy but increase capital costs and operating complexity. The overhead product condensation temperature is typically ca. 0 to −10 °C, based on a column operating pressure of 1500–2700 kPa [53], [54], [59].

**Depropanization** of cracked gas separates propane and lighter fractions as overhead components from $C_{4+}$ fractions as bottom components. Feed streams from compression, and sometimes streams from the gas-chilling section, can be routed to the depropanizer, which, compared to other fractionation units, requires less refrigeration because of its higher operating temperature (Fig. 27) [62].

The feed stream to the depropanizer comes primarily from the deethanizer when front-end demethanization and deethanization are employed. With front-end depropanization, the feedstock is taken from the gas dryers. In the simplest case, the dried cracked-gas feed stream is separated into a propene–propane overhead stream, which contains small amounts of methylacetylene and propadiene, and a bottom stream which contains components from $C_4$ materials to raw pyrolysis gasoline. Recent designs reduce energy requirements by mixing smaller feed streams removed from intermediate compression or gas-chilling stages with the main feed stream, or by having them enter separately at a point in the depropanizer consistent with the composition of column components. Condensate is usually stripped before entry into the depropanizer [50], [61].

Operating pressure and temperature of the depropanization step in ethylene plants, even downstream of the deethanizer, are kept low to reduce fouling of the fractionator by acetylenic and diolefinic contaminants. Hydrogenation of cracked gas in the compressor zone lowers the amount of these polymerizing contaminants substantially, but also destroys most of the butadiene [50], [54], [59].

**Debutanization.** Debutanization of cracked gas separates $C_4$ materials as overhead components from $C_5$ and heavier fractions as bottom components. The feed stream comes typically from the depropanizer column or is a depropanized condensate stream from the compressor. The mixed $C_4$ overhead fraction may be sent to an extraction unit for removal of 1,3-butadiene and further processing. The bottom stream and the debutanized fraction from the compression section can be used as pyrolysis gasoline feedstock or processed to produce aromatics and specialty chemicals feedstocks [50].

**Acetylene Hydrogenation.** Cracked gas is selectively hydrogenated to facilitate removal of acetylene from ethylene, except when acetylene is recovered for sale. Hydrogenation is performed with palladium-based catalysts. Pressure ranges typically from 20 to 35 bar and the temperature from 25 to 100 °C. There are two basic processes for the hydrogenation of acetylene.

*Tail-End Hydrogenation.* Acetylene hydrogenation takes place in the gas phase of a pure $C_2$ fraction, consisting of ethylene, ethane, and acetylene only. The unit is typically incorporated after demethanization, upstream of the $C_2$ fractionation tower. Hydrogenation reactors consist of one or two adiabatic beds. Hydrogen is added upstream of the reactor at a molar $H_2/C_2H_2$ ratio of ca. 1.5–2, resulting in an ethylene gain of up to 50 % of the acetylene present. Hydrogen and small amounts of methane and CO introduced after demethanization must be removed by additional fractionation. Sometimes purging the light ends from the overhead ethylene fractionator condensate is possible, but additional multistage fractionation is usually required in the ethylene fractionator or in a small separate demethanizer. A useful design provides five to ten overhead stages above the product drawoff point in the ethylene fractionator column for further separation of the light ends. The overhead condensate is returned to the top of the tower, and the light ends are removed from the overhead partial condenser for reprocessing [62], [50], [55], [60].

*Front-End Hydrogenation.* Acetylene is hydrogenated:

1) In the total cracked gas stream after or before drying
2) In the $C_{3-}$ stream with front-end depropanizer
3) In the $C_{2-}$ stream with front-end deethanizer

In the case of (1) and (2) any methylacetylene, propadiene and butadiene, present in the gas stream is cohydrogenated. (methylacetylene ca. 80%, propadiene 30%, butadiene > 90%).

Although hydrogenation in the compressor section (1) may offer advantages for the lighter feedstocks, disadvantages exist with heavier feedstocks. In particular, the volume of gas treated at this stage is high; butadiene is hydrogenated and lost; and a single hydrogenation to concurrently remove acetylene, methylacetylene, and propadiene to the required levels without significant olefin loss is not possible. Hydrogenation of individual mixed $C_2$ and $C_3$ streams after deethanization and depropanization, respectively, appears to permit more efficient use of catalyst and result in higher product yields. However, proponents of the front-end deethanization and depropanization process with front-end acetylene hydrogenation report that bed sizes are equivalent for that reaction. Higher hydrogen concentration and improved catalysts allow higher gas space velocities than can be achieved downstream [60].

General advantages of the front-end hydrogenation versus the tail-end hydrogenation unit are: (1) no hydrogen make up is required as hydrogen is already present in the feed stream; (2) longer run length between catalyst regenerations; (3) No removal of light ends, as required in tail-end hydrogenation, is necessary.

Front-end hydrogenation is typically conducted in multistage adiabatic beds with intercooling, or less frequently in a single tubular reactor that allows a more isothermal reaction. Acetylene hydrogenation requires heat-exchange systems for incoming low-temperature process streams (2) and (3). Hydrogenation in the compressor section (1) does not require the heat-exchange system necessary for cryogenic streams. Most of the troublesome polymerizable components are removed before fractionation, although rapid fouling of the hydrogenation bed must be avoided. Hydrogenation may be necessary before chilling if a high acetylene concentration in the cracked gas would result in plugging or in unsafe downstream operation [50], [52], [60].

Although tail-end hydrogenation is the most common process in currently operating plants, it seems that there is a strong tendency to increasingly apply the front-end hydrogenation system in new projects.

**Acetylene, Methylacetylene, and Propadiene Recovery.** Acetylene, methylacetylene, and propadiene may be recovered for use in welding and cutting or as chemical feedstock. Acetylene is usually extracted from the feed stream to the ethylene splitter, with methylacetylene and propadiene coming from the overhead depropanizer stream. Alternatively these materials may be extracted or distilled from the products; *N*-methyl-2-pyrrolidone, acetone, and *N,N*-dimethylformamide processes are typical recovery methods [50], [53], [60].

**Ethylene Fractionation.** Ethylene fractionation separates ethylene as a high-purity overhead product (> 99.9 wt%) from ethane, which is combined with propane and recycled for cracking. For this difficult fractionation, the net work done by the refrigeration system is high because of the high reflux and low temperature required (Fig. 27) [50], [62]. Fractionation requires a high reflux ratio (ca. 4) and as many as 125 separation stages [53], [54], [59], [60], [62]. The ethylene purity desired is typically > 99.9 wt%. The feed stream comes from the acetylene-removal step that follows chilling, or from the demethanizer or deethanizer.

There are basically two processes in commercial use: high-pressure fractionation and heat-pumped $C_2$ fractionation.

*High-Pressure Fractionation.* High-pressure fractionation, operating in the pressure range of 1700 – 2800 kPa, is typically used in tail-end hydrogenation systems. Heating and cooling of the column is integrated into the propene refrigeration cycle. Low-pressure propene refrigerant is vaporized in the top condenser, providing chilling duty. Vaporized propene is typically compressed in two stages of the refrigerant compressor. After compression, propene vapors are condensed in the reboiler, providing heat duty. By means of this system the propene refrigerant compressor acts as a indirect heat pump, shifting heat from the condenser of the column to the reboiler. The top of the column is equipped with some pasteurization trays to remove light ends introduced when acetylene is hydrogenated in the feed stream to this column. Ethylene product is withdrawn as a liquid side stream some trays below the top, pumped to the required product pressure, and delivered to battery limits after vaporization and heating.

*Heat-Pumped $C_2$ Fractionation.* In heat-pumped $C_2$ fractionation, the column operates at a pressure of ca. 800 kPa. Overhead gases of the column, consisting of on-spec ethylene, are routed to a compressor after heating to ambient temperature in a feed – effluent exchanger and are compressed to ca. 2000 kPa. Part of the compressed ethylene vapor is cooled and routed to the reboiler where it is condensed, providing heat duty for reboiling. The condensed ethylene is recycled as reflux to the top of the column. Ethylene product is discharged to battery limits from the discharge side of the compressor. Heat-pumped ethylene fractionation can use a dedicated compressor. More common is the integration of the heat-pumped ethylene splitter in the ethylene fractionation compressor, using the 3rd stage of this compressor as a heat pump.

This process has some advantages over the high-pressure fractionation process with respect to investment and energy consumption. Less equipment is involved, and because of the better relative volatility of ethylene and ethane at lower pressure, a lower reflux ratio is required. However it is not feasible to use the heat-pumped process in combination with tail-end hydrogenation because of the light ends present in the feed stream.

Ethylene is generally discharged to battery limits as a gas. However, production of up to 50% of production as the liquid phase and the use of storage facilities is also practiced.

**Hydrogenation of Methylacetylene and Propadiene (MAPD); $C_4$ Fraction; and Pyrolysis Gasoline.** *$C_3$ Hydrogenation.* The $C_3$ fraction coming from the top of the depropanizer typically contains 2–6% of methylacetylene (MA) and propadiene (PD). For removal of these components from the propene product and for economic reasons MAPD is hydrogenated to propene and propane. Hydrogenation is performed in multistage adiabatic bed reactors with intercooling steps or, in more modern processes, in a trickle-bed reactor. The heat of reaction is removed in the trickle-bed reactor by partial vaporization of the processed liquid, this being condensed later by cooling water. Modern hydrogenation processes typically yield a net propene gain of 60% of the MAPD present in the feed stream. The balance is converted to propane, which is recycled to the cracking furnaces as cracker feedstock.

*$C_4$ Hydrogenation.* The $C_4$ cut from the top of the debutanizer typically contains small amounts of vinylacetylene and ethylacetylene, ca. 50% 1,3-butadiene, and a few per cent of butanes; the balance consists of various types of butenes, of which isobutene is the most important. Depending on the nature of the downstream processes it may be necessary to hydrogenate some of the unsaturated components. The following hydrogenation steps are in commercial use:

1) Selective hydrogenation of butatriene and vinylacetylene to butadiene
2) Selective hydrogenation of butadiene to butenes
3) Full hydrogenation of all unsaturated to butane

Process (1) is applied to increase the 1,3 butadiene yield; process (2) is used if no use can be made of butadiene and to increase the butene yield; process (3) is employed if the $C_4$ cut is to be recycled as cracker feedstock to improve olefin yields or used as LPG product. Hydrogenation is performed in tubular reactor systems or, in more modern processes, in trickle-bed reactor systems. By selecting catalyst type and process parameters, the hydrogenation result with respect to yields of different components, mainly the various types of butenes, can be controlled in a certain range.

*Pyrolysis Gasoline Hydrogenation.* Pyrolysis gasoline—the $C_5$ to $C_{10}$ hydrocarbon product fraction—consists mainly of aromatics. The nonaromatics are mainly unsaturated hydrocarbons with a high portion of acetylenes and dienes. This stream is unstable and cannot be stored, as the unsaturated components react further, forming polymers and gum. Depending on the downstream processes pyrolysis gasoline is hydrogenated and fractionated in different steps. The most common process route is as follows:

1) Selective hydrogenation of the total gasoline to hydrogenate acetylenes, dienes, and styrene to olefinic compounds. After stabilization and removal of oil this stream is suitable for use as motor fuel.
2) Fractionation of the effluent of the 1st stage hydrogenation into a $C_5$ cut, a $C_6-C_8$ heart cut, and a $C_{9+}$ cut.
3) The $C_6-C_8$ cut is further processed in a 2nd stage hydrogenation step to convert olefins to paraffins and naphthenes and to convert all sulfur to $H_2S$, which is removed from the product in a downstream stripper. this process is necessary to prepare the heart cut for aromatics recovery.

**Propene fractionation** separates propene as a chemical-grade overhead product (typically 93–95 wt% min.) or more frequently as polymer-grade propene (> 99.5 wt%) from propane. This separation to polymer grade propene requires typically 150–230 stages and a reflux ratio of 20 because of the close boiling points of propene and propane. Two basic processes are applied for this difficult separation task: polymer-grade fractionators operate at ca 1800 kPa, with cooling water in the overhead condenser and hot quench water in the reboiler. In the case of naphtha cracking and where sufficient waste heat is available from the hot quench water cycle this is the most economic process. If no waste heat for reboiling is available, a heat-pumped propylene fractionator is applied. In this case the fractionator operates at ca. 800–1100 kPa, and the overhead gases are compressed and condensed in the reboiler. By means of this process the condenser heat is raised to a higher temperature level, and further use is made of it in the reboiler. The advantage of this system is that no external heat for reboiling and only small quantities of cooling water are required. The fractionation design depends on the concentration of propene in the feed, which varies from 70% for propane cracking to 95% for naphtha cracking [60]–[62].

The feed stream to the propene fractionator usually comes from the methylacetylene and propadiene removal step. Excess hydrogen and methane are purged in an upstream stripper or as light ends in the column overhead or from the overhead condensate [53], [54], [59], [60]–[62].

**Hydrogen Purification.** Hydrogen is produced in the cryogenic section of the plant at a purity of typically 80–95 vol%. However, the product stream contains some 1000 ppm of carbon monoxide. Hydrogen product is required in the ethylene plant as make-up stream for the hydrogenation units. Excess hydrogen is typically a desired product. However, hydrogen must be free of carbon monoxide as CO is a poison for all hydrogenation processes. Today two processes are in commercial use in ethylene plants to purify hydrogen: (1) methanation of CO in a catalytic process, converting CO to methane and water, after which the effluent must be dried in adsorber beds; (2) hydrogen purification by adsorption in a pressure-swing adsorption unit, producing CO-free high-purity hydrogen.

## 5.3.3. Utilities

**Refrigeration.** Refrigeration in ethylene plants is important and costly. Refrigeration optimization is vital in plant design. Typically, two different refrigeration systems are employed, namely, ethylene and propene (or propene–propane), each of which generates two to five different temperatures. Propylene refrigeration is designed as a closed-loop system in which the heat removed by cooling water or by air cooling by condensing the refrigerant after the final compression stage. Ethylene refrigeration systems are designed as closed- and open-loop systems. Open-loop cycles integrate the ethylene-fractionation tower in the refrigeration system. Propane and ethylene refrigerant systems act as a chilling cascade. Heat removed from the process by the ethylene cycle at temperatures below −50 °C is shifted to the propene cycle by condensing high-pressure ethylene refrigerant by means of vaporizing low-pressure ethylene refrigerant. Refrigerant temperatures are chosen to accommodate diverse plant needs most efficiently. Methane refrigeration by simple expansion (not employed as a refrigeration loop) generates additional cooling and especially the lower temperatures needed to purify hydrogen for special purposes. Ethylene, propene, and propane are used because of their physical properties and availability in the ethylene plant [61].

Figure 27 shows the most common refrigerant loops and the net heat and work required in a typical front-end demethanization plant with liquid feedstock. The less expensive refrigerant, propene, is used from + 25 to − 40 °C and the more expensive ethylene, from ca. −40 to −100 C. The latter requires higher work input per unit of refrigeration;for example, Figure 27 shows that 41 % of the heat absorbed by the ethylene refrigeration system requires 58 % of the net work [62]. Refrigerants are usually generated by compression to 1600 – 2000 kPa in multistage units, followed by expansion to generate the lowest refrigeration temperature. Superheated ethylene is cooled and condensed with propene refrigerant. Propene, after compression, is cooled with process cooling water [53], [60], [62]. With open-loop systems, which are used less frequently, products are circulated in the refrigeration loops rather than isolated. Careful operation and reliable design of compressors (gas seals) reduces contamination [60].

**Energy.** Utility consumption, expressed in joules per kilogram, is the energy required to produce 1 kg of ethylene product. Feedstock, coproducts, and process design should be described fully for comparative purposes. Process optimization has to consider the cost of energy relative to capital costs, ease of process operation, product and byproduct quality control requirements, safety, and many other issues. Since the early 1970s until today, increased energy costs led to a reduction of energy consumption by nearly 50 %. This was accomplished by using process water heat, side reboilers, heat pumps, and other devices.

Approximate current energy consumption to produce 1 kg of ethylene from various feedstocks is as follows:

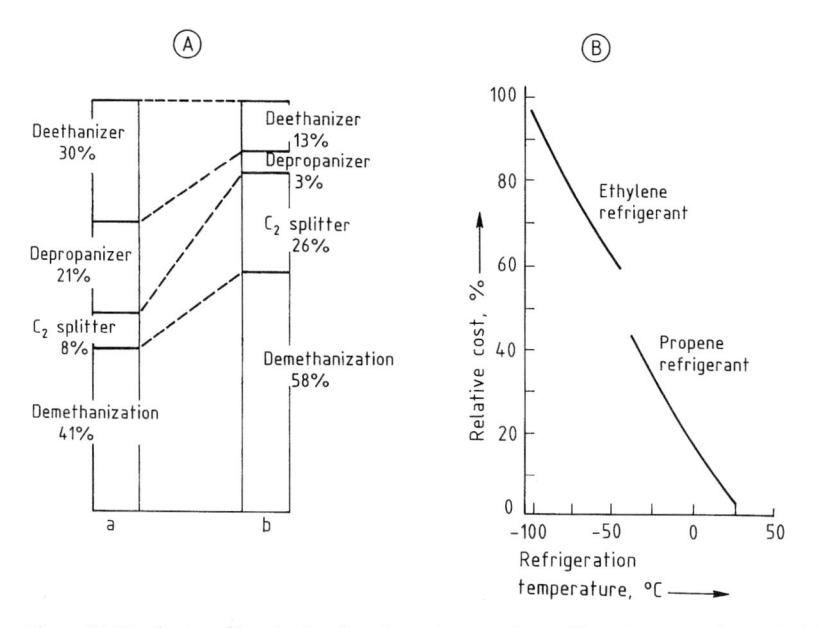

**Figure 27.** Distribution of heat load and work requirement of the refrigeration system in a typical front-end demethanizer ethylene plant *
A) Distribution of heat load and work requirement
a) Net heat absorbed by the refrigeration system; b) Net work done by the refrigeration system
B) Relative cost of heat absorbed by the refrigeration system at different temperatures
* Reprinted with permission [62].

| | |
|---|---|
| Ethane | 14 300 J/kg |
| Propane | 16 700 J/kg |
| Naphtha | 20 900 J/kg |
| Gas oil | 25 100 J/kg |

These values are approximate because flow schemes and product distribution are undefined, but they are representative. A reduction of ca. 2000 J/kg ethylene product is possible by using hot exhaust from gas turbines in the pyrolysis area as combustion air to preheat combustion fuel, or in a heat-recovery boiler. The turbines may also optimize energy use to produce electricity or can drive fractionation compressors and heat pumps [50], [52], [57], [58], [60]–[62].

Most of the energy consumed is introduced into the ethylene process via firing the cracking furnaces. The recovery process is driven by waste heat, recovered from the cracking furnaces via steam production in the TLEs or by recovering heat from the furnace effluent stream. The main energy consumers in the recovery section are the cracked-gas compressor and the refrigerant compressors, in which a total power of ca. 0.7 kWh per kilogram ethylene is typically consumed. This power is usually provided by steam turbines, driven by steam generated and superheated in the cracking furnaces.

## 5.3.4. Process Advances

Major advances have been made recently in feedstock availability and process design [62], [50]:

- Feedstock cost and availability, new cracking technology, and concerns for the environment and health have influenced changes in process technology.
- Rapid development has occurred in process control, data management, and optimization systems. Control of individual units, as well as of the overall process, has had a significant impact on operations. Improved analysis and cracking-simulation models are required to more fully apply state-of-the-art process control.
- Major improvements have been made in the selectivity and efficiency of cracking by operating at higher temperature. Coils with shorter residence time and faster quenching systems have been developed. Energy requirements have been reduced by heat recovery from furnace flue gas, economizer gas, and cracked gas; refrigeration energy requirements have been lowered by increased application and improved design of heat-exchange and distillation systems. Use of gas turbines has increased, with the sale of electricity generated. At the same time, control, reliability, and efficiency of operation have improved [60], [61].
- The reliability, service life, and cost of solid-desiccant drying systems have improved.
- Improved fractionation process designs provide additional flexibility and further reduce process energy consumption while meeting plant-reliability and product-quality demands [61].
- Improved hydrogenation catalysts allow more selective removal of impurities and better quality control.

## 5.4. Other Processes and Feedstocks

Ethylene has also been produced by other processes and from other feedstocks [66]. Options for long-range feedstock availability and price, as well as in-place refining and chemical processes, influence process design and feedstock selection because of their high economic impact on manufacturing cost.

**Recovery from FCC Offgas.** Recovery of ethylene and propylene from FCC offgas has gained importance, but due to the $NO_x$ content, some steps have to be integrated to reduce the risk of formation of explosive resins from butadiene and $N_2O_3$.

**Ethanol.** Ethanol, an early ethylene feedstock, is dehydrated at high temperature over a solid catalyst. The catalyst is usually supported alumina, phosphoric acid, or silica. The ethanol can be obtained from fermentation, methanol homologation, syngas and other sources (→ Ethanol).

**Crude Oil or Residual Oil.** A number of nontubular reactors, developed to crack unprocessed hydrocarbon feedstocks, exist in various stages of development and commercial operation [66].

A new development by Stone & Webster employs catalysts particles, for cracking and heat transfer functions. The QC (Quick Contact) process is suitable for cracking VGO and similar heavy feedstocks. The process has been tested in a pilot plant [67].

A new development by Veba Oil and Linde employs a catalyst with coke-gasification functions. This allows heavy feedstocks such as AGO and VGO to be processed in a reformer-type reactor to yield olefins. The high coking tendency of these feedstocks, which normally limits their use as feed for crackers, is no longer a problem, since the coke deposits are continuously gasified [68].

Other new developments use plastics wastes as feedstocks for olefins production. A process developed by BP produces a liquid feedstock from waste plastics which is suitable as cracker feedstock [69]. Another development by Linde directly co-processes polyolefin wastes in a modified cracking furnace, giving better yields than commercially cracked liquid feeds [70].

**Methanol** can be catalytically dehydrated and partially converted to ethylene over alumina and zeolite catalysts. The process based on novel zeolite catalysts (ZSM-5) was developed by Mobil (MTO, Methanol to Olefins). A new development introduced recently by UOP/Norsk Hydro, converts methane to methanol in a first stage and then converts the methanol to olefins. Economics of this new process seem to be competitive with conventional processes [71]. The process is based on a fluidized-bed reactor for conversion of methanol. 80% of the carbon content of methanol is converted into ethylene and propylene. The process has been tested in 0.5 t/d unit in Norway. Methanol obtained from syngas can be converted in high selectivity (but at a low rate) to ethanol with the help of catalysts and promoters.

**Syngas.** Fischer–Tropsch and modified Fischer–Tropsch processes are used to produce (1) olefins directly as a byproduct of gasoline and diesel fuel production; (2) LPG and paraffin intermediates, which are cracked to produce olefins; and (3) intermediate oxygenated liquids, which are dehydrated and separated to produce olefins. Syngas is produced by gasification of coal and other materials.

**Oxidative Coupling of Methane to Ethylene.** Many researchers worldwide have been active in this field, using metal oxide catalysts at 700–900 °C for the oxidative coupling of methane. However, the yields obtained so far (20–25%) are not competitive with conventional routes, since the ethylene concentration in the product gas is only ca. 10 vol% [72], [73]. Application of special separative reactors has led to improved yields up to 50–60% [74], [75].

**Dehydrogenation.** Dehydrogenation of ethane over Cr or Pt catalyst is limited by equilibrium and allows only very poor yield of ethylene. This route is not competitive with conventional routes.

**Metathesis.** Metathesis of propylene leads ethylene and 1-butene:

$$2\ C_3H_6 \rightleftarrows C_2H_4 + C_4H_8$$

Since this reaction is reversible the ethylene/propylene production of a plant can be adjusted to market conditions. So far this technology is applied in a plant in Canada and at the U.S. gulf coast.

# 6. Environmental Protection

Ethylene plants with a production of up to 900 000 t/a ethylene and a fired duty of up to 700 mW require special measures for protection of the environment. Though national standards vary widely, the following general measures are required for every new plant:

**Flue-Gas Emissions.** $NO_x$ emissions are limited by use of LowNox burners or integrated SCR technology for catalytic reduction of $NO_x$. In some regions limits for $NO_x$ emissions are already < 50 ppm. Particulate emissions during decoking can be reduced by incineration or separation in cyclones. Incinerators can achieve a reduction of 90 %, and high-performance separators, up to 96 %.

**Fugitive Emissions and VOC.** The equipment used in a modern olefin plant has to meet the standards for emission of critical components such as benzene. All components with a potential to emit critical components must have a certain tightness classification and have to be monitored (pump seals, valves, etc.). Storage tanks and piping are other sources of fugitive emissions

**Flaring.** Flaring during start up should be avoided as much as possible, especially in highly populated areas. New procedures have been developed that allow for flareless start up of plant [76].

**Noise Protection.** Ethylene plants contain equipment that produces noise emission in the range above 85 dBA. These are the burners at the furnaces and the large compressors. Noise abatement measures are essential (silencers for burners, hoods for compressors) to bring the overall emissions into the range below 85 dBA. Other sources of noise emission can be high velocities in piping or vibrations. The detection of the source in this case is sometimes very difficult and requires much expertise.

**Water Protection.** The water emissions of the plant result from quench water, dilution steam, decoking water, and flare water discharges. These streams have to be treated properly before being fed to the wastewater plant. Streams from the caustic-scrubbing section require chemical treatment (oxidation) before discharge to the wastewater unit.

**Solid and Hazardous Wastes.** Examples of process wastes are cleaning acids; filter cartridges; catalysts; tars; polymers; waste oils; coke; sodium nitrite used in column passivation; N,N-dimethylformamide, acetone, or N-methyl-2-pyrrolidone used in acetylene removal; sulfuric acid from cooling towers; amine tars; and antifoaming agents. These wastes have to be treated according to the relevant regulations for disposal.

# 7. Quality Specifications

Ethylene quality worldwide is matched to customer requirements; no single chemical-grade ethylene exists. However, ethylene content normally exceeds 99.9 wt%. The reason for variation in ethylene specifications is due to the use of ethylene in downstream processes that differ in sensitivity to impurities in ethylene, due to different catalyst systems being used. In general, the trend is towards lower impurity contents as more active catalyst systems are developed. The largest application for ethylene is polymerization to HDPE, LDPE, and LLDPE. For these processes new catalyst systems have been developed, including the metallocene catalysts, which are more sensitive than other catalysts. Table 11 lists typical ethylene specifications for various polymerization processes.

Sulfur, oxygen, acetylene, hydrogen, carbon monoxide and carbon dioxide are the most troublesome and carefully controlled impurities, especially when ethylene from multiple sources is mixed in transportation.

# 8. Chemical Analysis

Laboratory analysis of ethylene is primarily by ASTM or modified ASTM gas chromatographic techniques.

Process control may use over 100 analyzers to monitor plant safety, operation, and optimization [78]. Measurements include personnel exposure, corrosion, furnace flue gas, steam systems, water systems, condensate systems, stack emissions, water emissions, feedstock quality, cracked gas analysis, distillation train product quality and moisture control, and, in some cases, heavy liquids product evaluation and control [79].

Process control instruments include gas chromatographs, combustible-gas detectors, conductivity analyzers, pH meters, densitometers, corrosion analyzers, oxygen and

**Table 11.** Typical ethylene specifications (in ppm vol unless otherwise noted)

| Component | Polymer Process | | | | Polymer grade (today) |
|---|---|---|---|---|---|
| | Gas phase | Slurry | Solution | Metallocene [a] | |
| Ethylene | 99.9 % min. | 99.95 % min. | 99.9 % min. | 99.9 % | 99.9 % min. |
| Methane | 500 | [b] | 500 [b] | [e] | 300 |
| Ethane | 500 | [b] | 500 [b] | [e] | 500 |
| Propylene | 20 | 20 | 50 | 20 | 10 – 15 |
| Acetylene | 0.1 | 2 | 55 | 0.1 | 2 |
| Hydrogene | 5 | 5 | 10 | 5 | 10 |
| Carbon monoxide | 0.1 | 1 [c] | 2 [d] | 0.1 | 2 |
| Cabon dioxide | 0.1 | 1 [c] | 5 [d] | 0.1 | 2 |
| Oxygen | 0.1 | 1 | 5 | 0.1 | 5 |
| Water | 0.1 | 5 [c] | 5 [d] | 0.1 | 2 |
| MeOH | | 1 [c] | 5 [d] | 0.1 | 5 |
| Sulfur as $H_2S$ | | 1 (wt) [c] | 2 (wt) [d] | 0.1 | 2 (wt) |
| Other polar compounds | | 1 | 2 | 0.1 | |
| DMF | | | | | 2 (wt) |

[a] Values for metallocene catalyst are expected values.
[b] Total combined not to exceed 0.125 %.
[c] Total not to exceed 1 ppm wt.
[d] Total must be less than 6 ppm.
[e] Total combined not to exceed 300 ppm wt.

water analyzers, vapor pressure analyzers, and photoionization analyzers. Various automated wet methods are also used [78].

Process optimization is critical for ethylene production because cracking reactions change as the run proceeds. Operating costs for plants of this size are high [51], [80], [81]. Recent advances in computer control, data acquisition, and information analysis have given impetus to studies of process control, modeling, and optimization, especially in heavier liquid cracking operations.

Models for all kinds of feedstocks are applied, simulating all sections of the plant in detail, based on detailed chemical analysis of the various streams in the plant. With these tools operating strategies can be followed, e.g., production of a certain amount of ethylene, propylene and other products at maximum profit, even if the feedstock quality or the type of feed change within short time. Such a computer controlled operation includes the individual control of cracking furnaces, columns, pumps, compressors, etc.

# 9. Storage and Transportation

Much of world ethylene production is consumed locally, requiring little storage and transportation. Interconnected pipelines and pressurized underground caverns in the United States and Europe have been developed because of the complex petrochemical infrastructure. They provide flexibility and prevent interruption of supply. The following pipeline systems have been developed:

| | |
|---|---|
| United States: | Texas – Louisiana [83] |
| Canada: | Fort Saskatchewan – Sarnia [82] |
| Great Britain: | Grangemouth – Carrington |
| | Sevenside – Fawley [53] |
| Europe: | Northwest Europe (Frankfurt – Gelsenkirchen and Brussels – Rotterdam areas) and Spain [53], [84], [85] |
| Former Soviet Union: | Nizhnekamsk – Salavat and Angarsk – Zima [86], [87] |

In the U.S. Gulf Coast area the pipeline grid is composed of common carriers and privately owned pipelines. Storage caverns are naturally occurring or artificially excavated salt domes; limestone caverns are less usual. In contrast to U.S. pipelines, which normally transport only ethylene, the 3000 km pipeline between Fort Saskatchewan and Sarnia transports ethane, propane, and condensate in batches [51], [82].

In pipelines, ethylene is normally under a pressure of 4 – 100 MPa. The upper end of the pressure range is significantly above the critical pressure. Below critical conditions, the temperature must be > 4 °C to prevent liquid ethylene from forming. If water is present, hydrates can form below 15 °C at normal operating pressure and can plug equipment. Special procedures and precautions are used during commissioning of pipelines to pressure test and dry the line and to provide appropriate draining and venting during operation [85].

Ethylene decomposition and spontaneous ignition can occur under certain conditions and can result in pressure and temperature increases capable of producing explosions. Such decomposition is a function of pressure, temperature, impurities (e.g., acetylene and oxygen), and catalysts (e.g., rust and carbon). Appropriate compression procedures, especially when nitrogen, oxygen, or other diatomic gases are present, help to prevent decomposition [88].

Ethylene is also transported by ship, barge, railcar, and tank truck. In the United States, these methods are normally limited to consumers not served by pipeline. In Europe, pipelines move a lower proportion of ethylene than in the United States.

Ethylene is also stored in underground high-pressure pipelines and tanks (e.g., at 10 MPa) and in surface refrigerated tanks. In the latter, compression and liquefaction of vapor from the refrigerated tank, followed by reinjection of the liquid, allow operation of the tank at pressures of only 7 – 70 kPa [51], [89].

**Table 12.** World consumption of ethylene in 1993 (in %)

| Use | N. America | W. Europe | Japan | Others | Total |
|---|---|---|---|---|---|
| HDPE | 24 | 20 | 18 | 26 | 23 |
| LDPE/LLDPE | 29 | 37 | 31 | 35 | 33 |
| Dichloroethane | 15 | 15 | 17 | 13 | 15 |
| Ethylbenzene | 7 | 7 | 11 | 6 | 7 |
| Oligomers | 5 | 3 | * | 0 | 2 |
| Acetaldehyde | 1 | * | 4 | 1 | 2 |
| Ethanol | * | 2 | * | 1 | 2 |
| Vinyl acetate | 2 | 1 | 3 | 1 | 2 |
| Other uses | 3 | 6 | 6 | 2 | 3 |
| Total | 100 | 100 | 100 | 100 | 100 |

* Included in other uses.

**Table 13.** Regional supply (excluding standby capability) and demand estimates in 1993

| Region | Capacity, $10^6$ t/a | Production, $10^6$ t/a | Utilization, % |
|---|---|---|---|
| North America | 24 948 | 22 834 | 92 |
| Western Europe | 18 886 | 15 737 | 83 |
| Japan | 6 776 | 5 773 | 85 |
| Others | 24 534 | 17 905 | 73 |
| World | 75 144 | 62 249 | 83 |

# 10. Uses and Economic Aspects

More than 80% of the ethylene consumed in 1993 was used to produce ethylene oxide; ethylene dichloride; and low-density, linear low-density, and high-density polyethylene. Significant amounts are also used to make ethylbenzene, oligomer products (e.g., alcohols and olefins), acetaldehyde, and vinyl acetate. LDPE and LLDPE are the fastest growing outlets for ethylene.

Table 12 shows the uses of ethylene by region for 1993 [90].

By far the largest market for ethylene is the production of polymers. Polyethylene and PVC are projected to grow at rates of 5%/a [90]. Ethylene oxide and ethylbenzene are projected to grow at rates of 4 and 4.5%/a, respectively. In contrast, some derivatives such as ethanol and acetaldehyde will continue to lose market shares as a result of competing technologies.

The estimated 1996 world nameplate production capacity and the actual consumption were over $79.3 \times 10^6$ t/a and $70 \times 10^6$ t/a, respectively [2]. Regional supply (excluding standby capability) and demand estimates are shown in Table 13. Conversion products of ethylene are given in Table 12 for the United States, Western Europe, Japan and for the world.

The principal influences on ethylene product value from the manufacturer's viewpoint are primarily the price of feedstock raw materials, chemical and byproduct credits, and expected return on capital costs associated with the facilities. The fixed cost component for ethylene production is relatively modest because of the high-volume commodity operation that has become the industry standard (e.g., up to $950 \times 10^3$ t/a single-train plants). The cost breakdown per tonne of ethylene produced is listed below for a new plant based on naphtha feedstock in the Asia/Pacific region:

*Basis of economic analysis*
| | |
|---|---|
| Capacity | 500 000 t/a |
| Location | Asia/Pacific |
| Feedstock | naphtha |
| Capital | 10 years linear depreciation for ISBL plant |
| | 15 years linear depreciation for storage |
| Investment cost estimate | $ $420 \times 10^6$ for ISBL plant |
| | $ $80 \times 10^6$ for OSBL plant |

*Cost breakdown, %/t ethylene produced*
| | |
|---|---|
| Feedstock costs | 47.4 |
| Catalysts and chemicals | 0.5 |
| Utilities | 0.3 |
| Cash cost | 48.6 |
| Operating costs | 1.9 |
| Overheads | 1.9 |

## 11. Toxicology and Occupational Health [91]

Ethylene is not markedly toxic, but high concentrations may cause drowsiness, unconsciousness, or asphyxia because of oxygen displacement [92], [93]. The gas possesses useful anesthetic properties, with rapid onset, quick recovery, and minimal effects on the heart or lungs [94].

In animal studies, ethylene was not irritating to the skin or eyes [95]. It did not cause cardiac sensitization in a dog [96]. Mice and rats exposed repeatedly to ethylene showed minimal effects [97], [98]. In a chronic inhalation study with rats, no toxic effects were found by histopathology, hematology, clinical chemistry, urinalysis, eye examination, or mortality rate [98]. Ethylene improved wound healing following muscle injury to mice [99].

Studies with rodents have shown that ethylene caused increased serum pyruvate and liver weights [100], hypotension [101], decreased cholinesterase activity [98], hypoglycemia [102], and decreased inorganic phosphates [95]. No mutagenic activity was observed in *Escheria coli* B and *E. coli* Sd-4 exposed to ethylene [103].

Ethylene is a plant hormone and is useful for ripening fruits and vegetables [104], [105]. At high concentrations, it is phytotoxic [106]. Human exposure to ethylene is reported to cause anorexia, weight loss, insomnia, irritability, polycythemia, nephritis, slowed reaction time, and memory disturbances [107]–[109]. In the workplace, ethylene is treated as a simple asphyxiant [110]; the odor threshold is ca. 20 ppm [111]. NIOSH has published a procedure for evaluating ethylene exposure [112].

As with any compressed gas, skin and eye contact should be avoided. Appropriate precautions are required because the gas is highly flammable and explosive over a wide range of mixtures with air. Ethylene is also spontaneously explosive in sunlight in the presence of ozone or chlorine, and reacts vigorously with aluminum chloride and carbon tetrachloride [91].

## 12. References

[1] R. Becker, *Chem.-Tech. (Heidelberg)* **8** (1979) 6.
[2] *Oil Gas J.* **94** (1996) May 13th, special report.
[3] *Hydrocarbon Process.* **74** (1995) 29.
[4] W. Weihrauch, *Hydrocarbon Process.* **75** (1996) May, 21.
[5] *Asian Chemical News* **1** (1995) 8.
[6] *Chemical Economics Handbook*, SRI International, Menlo Park, CA, 1996, p. 432.
[7] K. Weissermel, H.-J. Arpe: *Industrielle Organische Chemie*, 4th ed., VCH Verlagsgesellschaft, Weinheim 1994.
[8] D. Lambert et al., *Analysis Magazine* **23** (1995) no. 4, M9–M14.
[9] K. M. Watson, E. F. Nelson, *Ind. Eng. Chem.* **25** (1933) 880–887.
[10] H. M. Smith: "Correlation Index to Aid in Interpreting Crude-Oil Analyses," U.S. Bureau of Mines Technical Paper no. 610, 1940.
[11] J. M. Collins, G. H. Unzelman, *Oil Gas J.* **80** (1982) June, 148.
[12] Y. C. Hu, *Hydrocarbon Process* **61** (1982, Nov.) 109, Part 1.
[13] M. E. Denti, E. M. Ranzi in L. F. Albright, B. L. Crynes, W. H. Corcoran (eds.): *Pyrolysis: Theory and Industrial Practice*, Academic Press, New York 1983, pp. 133–137.
[14] D. Edelson, *J. Phys. Chem.* **81** (1977) 2309.
[15] K. Ebert, H. Ederer, G. Isbarn, *Int. J. Chem. Kinet.* **15** (1983) 475.
[16] M. Schärfe, H. J. Ederer, U. Stabel, K. Ebert, *Chem.-Ing.-Tech.* **56** (1984) no. 6, 488–489.
[17] G. F. Froment, *Chem. Eng. Sci.* **47** (1992) 2163.
[18] Q. Chen, *Chemtech* **26** (1996) Sept., 33.
[19] M. Dente et al. *Chem. Eng. Sci.* **47** (1992) 2629.
[20] F. O. Rice, *J. Am. Chem. Soc.* **53** (1931) 1959–1972.
[21] S. B. Zdonik, E. J. Green, L. F. Hallee, *Oil Gas J.* **65** (1967) no. 26, 96–101; **66** (1968) no. 22, 103–108.
[22] C. W. Gear: *Numerical Initial Value Problems in Ordinary differential Equations*, Prentice Hall, Englewood Cliffs, N.J., 1971.
[23] M. Boudart: *Kinetics of Chemical Processes*, Prentice Hall, Englewood Cliffs, N.J., 1968.
[24] N. N. Semenov (transl. by M. Boudart): *Some Problems in Chemical Kinetics and Reactivity*, Princeton University Press, N. J. 1959.
[25] F. Rossini, K. Pitzer, R. Arnett, R. Braun et al.: *Selected Values of Physical and Thermodynamic Properties of Hydrocarbons and Related Compounds*, API Research Project 44 Report, Carnegie Press, Pittsburg, Pa., 1955, p. 706.
[26] Kinetics Technology International, US 4 367 645, 1983 (G. F. Froment).
[27] S. Barendregt, J. L. Monfils: "Steamcracking for Butane Upgrading," AIChE meeting, Houston, TX, 1991

[28] Mitsui Advanced Cracker, Technical Bulletin, Mitsui Engineering & Shipbuilding Co., Tokyo, Japan, pp. 1–5.
[29] S. Ibarra, in [13] , pp. 427–436.
[30] G. E. Moller, C. W. Warren, *Mater. Perform.* **20** (1981) no. 10, 27–37.
[31] J. A. Thuillier, *Mater. Perform.* **15** (1976) no. 11, 9–14.
[32] R. H. Krikke, J. Hoving, K. Smit: "Monitoring the Carburization of Furnace Tubes in Ethylene Plants," International Corrosion Forum Paper no. 10, National Assoc. of Corrosion Engineers, Houston, Tex., Mar. 22–26, 1976.
[33] D. E. Hendrix, M. Clark: "Contributing Factors to the Unusual Creep Growth of Furnace Tubing in Ethylene Pyrolysis Service," International Corrosion Forum Paper no. 21, National Assoc. of Corrosion Engineers, Boston, Mass., Mar. 25–29, 1985.
[34] J. J. Jones, J. L. D. Steiner: "A New Generation of Heat Resisting Alloys, Parallloys SH24T and SH39T," International Corrosion Forum Paper no. 22, National Assoc. of Corrosion Engineers, Boston, Mass., Mar. 25–29, 1985.
[35] S. B. Parks, C. M. Schimoller, *Hydrocarbon Process.* **75** (1996) 53.
[36] L. A. Mekler, R. S. Fairall, *Pet. Refiner* **31** (1952) no. 6, 100–107, Part 1.
[37] W. E. Lobo, J. E. Evans, *Trans. Am. Inst. Chem. Eng.* **35** (1939) 743–778
[38] R. L. Grantom: "Combustion Air Should Be More Than Simply Recycling Energy," *Proc. 1980 Conference on Industrial Energy Conservation,* vol. **2,** Texas Industrial Commission, Austin, Tex., 1980, pp. 559–564.
[39] R. L. Grantom: "Combustion Air Preheat and Radiant Heat Transfer in Fired Heaters – A Graphical Method for Design and Operating Analysis," *Proc. 1981 Industrial Energy Conservation Technology Conference and Exhibition,* vol. **1,** Texas Industrial Commission, Austin, Tex., 1981, pp. 151–159.
[40] J. V. Albano, E. F. Olszewski, T. Fukushima, *Oil Gas J.* **90** 1992, Feb., 55.
[41] M. J. Graff, L. F. Albright, *Carbon* **20** (1982) no. 4, 319–330.
[42] J. C. Marek, L. F. Albright: "Coke Formation on Catalysts or in Pyrolysis Furnaces," *Am. Chem. Soc. Symposium, 182nd Am. Chem. Soc. Meeting,* New York 1981.
[43] A. G. Goosens, M. Denti, E. Ranzi, *Hydrocarbon Process.* **57** (1978) no. 9, 227–236.
[44] L. L. Ross, in [13] pp. 327–364.
[45] B. Lohr, H. Dittmann, *Oil Gas J.* **70** (1978) no. 20, 63–68.
[46] J. Chen, M. J. Maddock, *Hydrocarbon Process.* **52** (1973) no. 5, 147–151.
[47] A. Mol, in [13] , pp. 451–471.
[48] "Transfer Line Exchangers in Ethylene Plants," Technical Bulletin, Schmidt'sche Heissdampf-Gesellschaft mbH, Kassel-Bettenhausen, Federal Republic of Germany, 1985.
[49] Mitsui Innovative Quencher, Technical Bulletin, Mitsui Engineering & Shipbuilding Co., Tokyo, Japan.
[50] E. S. Kranz, S. B. Zdonik, Personal Communication, Stone & Webster Engineering Corp., Boston, Mass., May 1985.
[51] S. B. Zdonik, in [13], Chapter 15.
[52] R. C. Updegrove, Personal Communication, E. I. du Pont de Nemours & Co., Chocolate Bayou, Tex., Jul. 1985.
[53] *Ullmann,* 4th ed., **8,** p. 192.
[54] H. E. Boyd, Y. H. Chen, O. J. Quartulli, *Oil Gas J.* **74** (1976, Sept. 27) 51–59.
[55] C. W. Albers, T. A. Wells, *Oil Gas J.* **76** (1978) Sept. 4, 72–78.
[56] K. W. Brooks, *Oil Gas J.* **62** (1964 ) Dec. 21, 70–75.
[57] T. A. Wells, *Oil Gas J.* **80** (1982) Jun. 14, 63–68.

[58] W. C. Petterson, T. A. Wells, *Chem. Eng.* **84** (1977) Sept. 26, 76–86.
[59] L. K. Ng, C. N. Eng, R. S. Zach, *Hydrocarbon Process.* **62** (1983, Dec.) 99–103.
[60] A. J. Weisenfelder, C. N. Eng., Personal Communication, C. F. Braun & Co., Alhambra, Calif., Aug. 1985.
[61] C. Sumner, T. S. Williams, Personal Communication, Combustion Engineering, Bloomfield, N. J., Aug. 1985.
[62] R. Orriss, Personal Communication, M. W. Kellogg Co., Houston, Tex., Jun. 1985.
[63] R. Zeppenfeld, E. Haidegger, W. Borgmann, *Hydrocarbon Technology International* (1993) 129.
[64] C. Bowen, R. McCue, C. Carradine, *Hydrocarbon Process.* **72** (1993) 129.
[65] W. K. Lam, Y.R. Mehra, D.W. Mullins, AIChE Spring International Meeting, Houston 1993, session 18.
[66] Y. C. Hu, *Hydrocarbon Process.* **51** (1972) Nov., 109–116, Part 1; **62** (1983) Apr., 113–116, Part 2; **62** (1983) May, 88–95, Part 3.
[67] *Eur. Chem. News* **67** (1996) April 7, 24.
[68] W. Baldauf, H. G. Jägers, D. Kaufmann, H. Zimmermann, *Linde Rep. Sci. Technol.* **54** (1994) 44.
[69] *Eur. Chem. News* **63** (1992) April 13, 28.
[70] *Hydrocarbon Engineering* 1996, Sept., 41.
[71] M. Heathcote, *Asian Chemical News* **3** (1996) June 24, 19.
[72] N. A. Baronskaya, L. S. Woldman, A. A. Davydiv, O. V. Buyevskaya, *Gas Sep. Purif.* **10** (1996) no. 1, 86.
[73] L. Guczi, R. A. van Santen, K. W. Sarma, *Catal. Rev. Sci. Eng.* **38** (1996) no. 2, 249.
[74] A. L. Tonkovich, R. W. Carr, R. Aris, *Science (Washington D.C.)* **262** (1993) 221.
[75] T. Nozaki, O. Yamzaki, K. Omata, K. Fujimoto, *Chem. Eng. Sci.* **47** (1992) no. 9, 2945.
[76] A. Shaikh, C. J. Lee, *Hydrocarbon Process.* **74** (1995) 89.
[77] J. A. Reid, D. R. McPhaul: "Contaminant Rejection Technology Update," presented at AIChE Spring National Meeting, 8th Annual Ethylene Producers Conference, New Orleans, LA, Feb. 28, 1996.
[78] W. H. Delphin, J. C. Franks, Personal Communication, E. I. du Pont de Nemours & Co., Chocolate Bayou, Tex., Mar. 1985.
[79] CFR, vol. **40**, 1985, Part 61.
[80] M. Nasi, M. Sourander, M. Tuomala, D. C. White et al., *Hydrocarbon Process.* **62** (1983) Jun., 74–82.
[81] M. Sourander, M. Kolari, J. C. Cugini, J. B. Poje et al., *Hydrocarbon Process.* **63** (1984) Jun., 63–69.
[82] *Pipeline Gas J.* **210** (1983) Jun., no. 7, 34–37.
[83] S. D. Govrau, MAP Search Services, Houston, TX, personal communication, May 1985.
[84] *Eur. Chem. News* **27** (1975) Feb. 28, 19.
[85] L. Kniel, O. Winter, K. Stork: *Ethylene, Keystone to the Petrochemical Industry*, Marcel Dekker, New York 1975, Chapter 5.
[86] *Oil Gas J.* **75** (1977) no. 39, Sep. 19, 108.
[87] *Eur. Chem. News* **31** (1977) no. 813, Nov. 25, 56.
[88] F. F. McKay, G. R. Worrell, B. C. Thornton, H. L. Lewis, *Pipe Line Ind.* **49** (1978, Aug.) no. 2, 77–78, 80, 82, 84.
[89] E. L. Pawlikowski, Storing and Handling Liquefied Olefins and Diolefins, Exxon Chemical Co. USA, 1974.
[90] *Chemical Economics Handbook, Estimates*, SRI International, Menlo Park, CA, 1996, p. 432.
[91] W. D. Broddle, Personal Communication, Conoco, Ponca City, Okla., May 1985.

[92] G. D. Clayton, F. E. Clayton (eds.): *Patty's Industrial Hygiene and Toxicology*, 3rd rev. ed., vol. **28,** Wiley-Interscience, New York 1981, pp. 3198–3199.

[93] W. B. Deichman, H. W. Gerarde: *Toxicology of Drugs and Chemicals*, Academic Press, New York 1969, p. 316.

[94] A. L. Cowles, H. H. Borgstedt, A. J. Gillies, *Anesthesiology* **36** (1972) no. 6, 558.

[95] W. R. von Oettingen, Toxicity and Potential Dangers of Aliphatic and Aromatic Hydrocarbons, Public Health Bulletin no. 225, Washington, D.C., 1940.

[96] J. C. Krantz, Jr., C. J. Carr, J. F. Vitcha, *J. Pharmacol. Exp. Ther.* **94** (1948) 315.

[97] C. Reynolds, *Anesth. Analg. (New York)* **6** (1927) 121.

[98] T. E. Hamm, D. Guest, J. G. Dent, *Fund. Appl. Toxicol.* **4** (1984) 473–478.

[99] P. Pietsch, M. Chenoweth, *Proc. Soc. Exp. Biol. Med.* **130** (1968) 714.

[100] R. B. Conolly, R. J. Jaeger, S. Szabo, *Exp. Mol. Pathol.* **28** (1978) 25.

[101] M. L. Krasovitskaya, L. Malyarova, *Gig. Sanit.* **33** (1968) no. 5, 7.

[102] P. Cazzamali, *Clin. Chirurg.* **34** (1931) 477.

[103] M. M. Landry, R. Fuerst, *Dev. Ind. Microbiol.* **9** (1968) 370.

[104] H. K. Pratt, J. D. Goeschl, *Am. Rev. Plant Physiol.* **20** (1969) 542.

[105] R. E. Holm, J. L. Rey, *Plant Physiol.* **44** (1969) 1295.

[106] G. D. Clayton, T. S. Platt, *Am. Ind. Hyg. Assoc. J.* **28** (1967) 151.

[107] J. M. Arena (ed.): *Poisoning*, 4th ed.,C. C. Thomas, Springfield, Ill., 1979,p. 426.

[108] L. R. Riggs, *Proc. Soc. Exp. Biol. Med.* **22** (1924–1925) 269.

[109] Chemical Safety Data Sheets, Chemical Manufacturers' Assoc. (formerly the Manufacturing Chemists' Assoc. Inc.,) Washington, D.C., 1979.

[110] ACGIH (ed.): *Threshold Limit Values for Chemical Substances and Physical Agents in the Workroom Environment,* ACGIH, Cincinnati, Ohio, 1986–1987.

[111] M. L. Krasovitskaya, L. K. Malyarova, *Biol. Deistvie Gig. Znach. Atmos. Zagryaz.* 1966, 74.

[112] National Institute for Occupational Safety and Health Manual of Sampling Data Sheets, U.S. Dept. Health Education Welfare, U.S. Printing Office, Washington, D.C., 1977.

# Ethylenediaminetetraacetic Acid and Related Chelating Agents

J. ROGER HART, W. R. Grace & Co., Lexington, Massachusetts 02173, United States

| | | | | |
|---|---|---|---|---|
| 1. | Introduction . . . . . . . . . . . . . 2295 | 7. | Uses . . . . . . . . . . . . . . . . . . . . 2300 |
| 2. | Physical Properties . . . . . . . . 2296 | 8. | Trade Names . . . . . . . . . . . . . 2302 |
| 3. | Chemical Properties. . . . . . . . 2296 | 9. | Economic Aspects . . . . . . . . . 2302 |
| 4. | Production . . . . . . . . . . . . . . 2299 | 10. | Toxicology and Occupational |
| 5. | Chemical Analysis . . . . . . . . . 2300 | | Health. . . . . . . . . . . . . . . . . . 2303 |
| 6. | Storage and Transportation . . 2300 | 11. | References. . . . . . . . . . . . . . . 2303 |

## 1. Introduction

Ethylenediaminetetraacetic acid [60-00-4], {EDTA, N,N'-1,2-ethanediylbis[N-(carboxymethyl)glycine], edetic acid}, nitrilotriacetic acid [139-13-9], [NTA, N,N-bis(carboxymethyl)glycine], and their salts were first synthesized by FERDINAND MUNZ in the I.G. Farbenindustrie laboratories during the 1930s. Since that time, EDTA, hydroxyethylethylenediaminetriacetic acid (HEEDTA, N-{2-[bis(carboxymethyl)amino]ethyl}-N-(2-hydroxyethyl)glycine), diethylenetriaminepentaacetic acid (DTPA, N,N-bis{2-[bis-(carboxymethyl)amino]ethyl}glycine, pentetic acid), and their salts have become important industrial chelating agents.

Aminopolycarboxylic acid chelating agents, like other amino acids, are capable of forming salts with strong acids or bases. In addition, these chelating agents can form strong, water-soluble metal complexes with di- and trivalent cations. These complexes greatly alter the reactivity of the metal ion, thus making them useful in many important industrial processes. The major volume usage of aminopolycarboxylic acid chelating agents is in preventing or removing scales and insoluble deposits or precipitates containing calcium, barium, iron, and other cations. These agents are also added to many processes, e.g., paper pulping, to prevent contained traces of metal ions from interacting in harmful ways during subsequent processing.

Chelating agents are also useful in altering the oxidation–reduction properties of transition-metal ions, such as iron and manganese, to increase or decrease the reactivity of these systems. In addition, they are widely used to provide water-soluble forms of metal ions that are resistant to anions, such as phosphate or carbonate, which can cause precipitation or inactivation. The agricultural use of water-soluble metal complexes of EDTA is an example.

## 2. Physical Properties

The structures and physical properties of the three most important aminopolycarboxylic acid chelating agents and their salts are given in Table 1.

## 3. Chemical Properties

Ethylenediaminetetraacetic acid, related aminopolycarboxylic acids, and their salts form water-soluble complexes with alkaline-earth and heavy-metal ions. These metal ions are incorporated into a ring structure called a *chelate*, after the Greek word χηλή meaning "claw." This chelate structure is comprised of the dissociated ligand, e.g., $EDTA^{4-}$, and the metal ion, e.g., $Fe^{3+}$, to form the metal chelate $[Fe(EDTA)]^-$:

$$Fe^{3+} + R-N(CH_2COO^-)_2 \rightleftharpoons R-N \begin{matrix} CH_2\overset{O}{\overset{\|}{C}}-O^- \\ \diagdown \\ \diagup \\ CH_2\underset{\|}{\underset{O}{C}}-O^- \end{matrix} Fe^{3+}$$

where $R = (^-OOCCH_2)_2NCH_2CH_2$

The metal ion is bound quite tightly to the ligand and the strength of this complex can be estimated through the stability constant, $K_{stab.}$, which is a function of the equilibrium concentrations of the metal ion M and the chelating agent Y:

$$K_{stab.} = \frac{c_{MY}}{c_M \, c_Y}$$

The strengths of the various metal chelate complexes can be related by means of log $K_{stab.}$ (see Table 2).

The *pH of the system* and the *presence of competing anions* have a decided effect on the strength of the metal chelate complex. The active chelating moiety is the fully dissociated chelate anion, e.g., $EDTA^{4-}$. The concentration of this species is greater at higher pH; hence, chelating agents are more effective as alkalinity increases. However, this effect is offset by competing anions, which act as precipitating or complexing species. Increased hydroxyl ion concentration at higher pH values tends to disrupt the metal chelate complex if the metal ion forms an insoluble hydroxide. For example, at pH > 5.5, $Fe^{3+}$ precipitates from its EDTA chelate. Likewise, carbonate, phosphate,

**Table 1.** Physical properties of selected aminopolycarboxylic acid chelating agents

| Compound | CAS registry number | Molecular formula | $M_r$ | mp, °C | Solubility[a] in water (approximate), g/L at 20 °C | pH of 1 wt% solution |
|---|---|---|---|---|---|---|
| Ethylenediaminetetraacetic acid (EDTA), $(HOOCCH_2)_2NCH_2CH_2N(CH_2COOH)_2$ | [60-00-4] | $C_{10}H_{16}N_2O_8$ | 292.24 | 245 (decomp.) | 0.1<br>0.5 (90 °C) | |
| $Na_2EDTA$[b] | [139-33-3] | $C_{10}H_{14}N_2O_8Na_2$ | 336.20 | 0 | 105 | 5.0–5.5 |
| $Na_4EDTA$[c] | [64-02-8] | $C_{10}H_{12}N_2O_8Na_4$ | 380.17 | | 500 | 11.3 |
| Hydroxyethylethylenediaminetriacetic acid (HEEDTA), $(HOCH_2CH_2)N(CH_2COOH)CH_2CH_2N(CH_2COOH)_2$ | [150-39-0] | $C_{10}H_{18}N_2O_7$ | 278.26 | 212–214 (decomp.) | 60 | |
| $Na_3HEEDTA$ | [139-89-9] | $C_{10}H_{15}N_2O_7Na_3$ | 344.22 | | 480 | 11.2 |
| Diethylenetriaminepentaacetic acid (DTPA), $[(HOOCCH_2)_2NCH_2CH_2]_2NCH_2COOH$ | [67-43-6] | $C_{14}H_{23}N_3O_{10}$ | 393.35 | 220 (decomp.) | 5 | |
| $Na_5DTPA$ | [140-01-2] | $C_{14}H_{18}N_3O_{10}Na_5$ | 503.26 | | 500 | 11.3 |

[a] EDTA, HEEDTA, DTPA, and their salts are insoluble in most common organic solvents.
[b] Crystallizes from water as the dihydrate.
[c] Crystallizes from water as the tetrahydrate.

**Table 2.** Stability constants of metal chelates of EDTA, HEEDTA, and DTPA

| Metal ion | log $K_{stab.}$ | | |
|---|---|---|---|
| | EDTA | HEEDTA | DTPA |
| $Al^{3+}$ | 16.3 | 14.3 | 18.6 |
| $Ba^{2+}$ | 7.86 | 6.3 | 8.87 |
| $Ca^{2+}$ | 10.69 | 8.3 | 10.83 |
| $Co^{2+}$ | 16.31 | 14.6 | 19.27 |
| $Cu^{2+}$ | 18.80 | 17.6 | 21.55 |
| $Fe^{2+}$ | 14.32 | 12.3 | 16.5 |
| $Fe^{3+}$ | 25.1 | 19.8 | 28.0 |
| $Mg^{2+}$ | 8.79 | 7.0 | 9.30 |
| $Mn^{2+}$ | 13.87 | 10.9 | 15.60 |
| $Ni^{2+}$ | 18.62 | 17.3 | 20.32 |
| $Pb^{2+}$ | 18.04 | 15.7 | 18.80 |
| $Sr^{2+}$ | 8.73 | 6.9 | 9.77 |
| $Zn^{2+}$ | 16.50 | 14.7 | 18.40 |

**Table 3.** Mass equivalents for 1:1 metal chelate complexes

| Chelating agent, 1 g | Metal ion, mg | | | | | | |
|---|---|---|---|---|---|---|---|
| | $Ca^{2+}$ | $Mg^{2+}$ | $Cu^{2+}$ | $Mn^{2+}$ | $Fe^{2+}$ | $Fe^{3+}$ | $Zn^{2+}$ |
| $Na_4$EDTA | 105 | 64 | 167 | 144 | 147 | 147 | 172 |
| $Na_3$HEEDTA | 116 | 70 | 184 | 159 | 162 | 162 | 190 |
| $Na_5$DTPA | 79 | 48 | 126 | 109 | 111 | 111 | 130 |

oxalate, silicate, and other precipitating anions greatly alter the effective strength of the metal–EDTA complex. This occurs when the effective solubility product of the insoluble metal compound approaches the value of the effective stability constant of the metal chelate.

In general, chelating agents form 1:1 chelates with metal ions. The mass equivalents for various metal chelate complexes under ideal conditions are shown in Table 3. Increasing the stoichiometric concentration of the chelating agent to values greater than 1:1 increases the stability of the chelate complex. For example, Fe (III) (HEEDTA) remains in solution up to pH 7, whereas Fe (III) (HEEDTA), in the presence of a molar excess of HEEDTA, is stable to pH 12 [2].

In addition, converting a metal cation to a metal complex anion greatly alters the interaction of the metal ion with other species. For example, the movement of [Fe (III) (EDTA)]$^-$ through soil containing clay is facilitated by converting the easily absorbed $Fe^{3+}$ cation to a more mobile anionic form.

The effective oxidation potential, $E_{eff.}$, is increased when the concentration of free metal ion, M, is lowered by the formation of a metal complex.

$$E_{eff.} = E^0 - 0.05915 \log c_M$$

The change in electrode potential resulting from chelation with EDTA, which forms the basis for a number of important industrial processes, is shown in Table 4 [3].

**Table 4.** Standard electrode potentials for metals and their EDTA complexes

| Reaction | Potential, V | Reaction | Potential, V |
|---|---|---|---|
| $Ce - 3\,e^- \rightleftharpoons Ce^{3+}$ | −2.48 | $Ce + EDTA^{4-} - 3\,e^- \rightleftharpoons [Ce\,(EDTA)]^-$ | −2.78 |
| $Co - 2\,e^- \rightleftharpoons Co^{2+}$ | −0.28 | $Co + EDTA^{4-} - 2\,e^- \rightleftharpoons [Co\,(EDTA)]^{2-}$ | −0.76 |
| $Cu - 2\,e^- \rightleftharpoons Cu^{2+}$ | +0.34 | $Cu + EDTA^{4-} - 2\,e^- \rightleftharpoons [Cu\,(EDTA)]^{2-}$ | −0.22 |
| $Fe - 2\,e^- \rightleftharpoons Fe^{2+}$ | −0.44 | $Fe + EDTA^{4-} - 2\,e^- \rightleftharpoons [Fe\,(EDTA)]^{2-}$ | −0.87 |
| $Fe - 3\,e^- \rightleftharpoons Fe^{3+}$ | −0.04 | $Fe + EDTA^{4-} - 3\,e^- \rightleftharpoons [Fe\,(EDTA)]^-$ | −0.17 |
| $Mn - 2\,e^- \rightleftharpoons Mn^{2+}$ | −1.18 | $Mn + EDTA^{4-} - 2\,e^- \rightleftharpoons [Mn\,(EDTA)]^{2-}$ | −1.45 |

# 4. Production

The original commercial synthesis of EDTA was from ethylenediamine, chloroacetic acid, and caustic soda to form the tetrasodium salt [4]. Material produced in this way was contaminated with byproduct sodium chloride and had to be purified for many uses. This process is no longer employed commercially.

Other processes for the production of EDTA have been developed but not used commercially, for example, the catalytic oxidation of tetra(hydroxyethyl)ethylenediamine [140-07-8], 2,2′,2″,2‴-(1,2-ethanediyldinitrilo)tetrakis(ethanol) [5].

Today, the two principal manufacturing processes for EDTA and related chelating agents are both based on the cyanomethylation of the parent polyamine.

The most widely used synthesis is the *alkaline cyanomethylation of ethylenediamine* by means of sodium cyanide and formaldehyde [6]:

$H_2NCH_2CH_2NH_2 + 4\,CH_2O + 4\,NaCN + 4\,H_2O$
$\longrightarrow (NaOOCCH_2)_2NCH_2CH_2N(CH_2COONa)_2 + 4\,NH_3$

This method offers high yields (> 90 %) of the chelating agent. The principal byproduct is ammonia, which is continuously boiled off during the reaction. However, some of the ammonia is cyanomethylated to yield salts of NTA, *N*-(carboxymethyl)glycine, and of glycine. In addition, glycolic acid salts are formed from the reaction of sodium cyanide and formaldehyde. These impurities are not detrimental to most applications of chelating agents.

The second commercial method for producing EDTA is the two-step *Singer synthesis* [7]. In this process, the cyanomethylation step is separate from the hydrolysis. Hydrogen cyanide and formaldehyde react with ethylenediamine to form insoluble (ethylenedinitrilo)tetraacetonitrile [5766-67-6] (EDTN), 2,2′,2″,2‴-(1,2-ethanediyldinitrilo)tetrakis(acetonitrile) (**1**), in high yield (> 96 %). The intermediate nitrile is separated, washed, and subsequently hydrolyzed with sodium hydroxide to tetrasodium EDTA, with liberation of byproduct ammonia. Carrying out the synthesis in two stages eliminates most of the impurity-forming reactions and yields a very pure form of chelating agent.

$$H_2NCH_2CH_2NH_2 + 4\,CH_2O + 4\,HCN$$
$$\longrightarrow (NCCH_2)_2NCH_2CH_2N(CH_2CN)_2 + 4\,H_2O$$
$$\mathbf{1}$$
$$\xrightarrow{4\,NaOH} (NaOOCCH_2)_2NCH_2CH_2N(CH_2COONa)_2 + 4\,NH_3$$

This two-step reaction is also particularly well-suited to synthesizing pure NTA in high yield from ammonia as the parent amine.

# 5. Chemical Analysis

Ethylenediaminetetraacetic acid and related chelating agents are readily analyzed by *titration with standardized solutions of metal salts*, which react quantitatively to form metal chelate complexes. The end point is commonly determined by the precipitation and turbidity of an insoluble metal compound or the color change of a metal-sensitive dye used as an indicator.

Titration of chelating agents with calcium acetate in the presence of sodium oxalate is an example of the turbidity method [8]. The Schwarzenbach titration relies on the color change of Eriochrome Black T at the end point when the chelating agent removes calcium from the red calcium – Eriochrome Black T complex to liberate the blue dye [9].

Small amounts of EDTA may be determined by using copper ion titration with a copper ion specific electrode [10].

# 6. Storage and Transportation

Chelating agents, by their very nature, tend to be corrosive to metals such as copper, zinc, and iron. The alkaline salts corrode aluminum severely, and contact with these salts should be avoided. Stainless steel and most polymers or coatings that are able to withstand dilute alkali are acceptable as construction materials. Solid chelating agents should be stored in a cool, dry place to avoid caking.

# 7. Uses

Chelating agents, with their ability to solubilize and inactivate metal ions by complex formation, are used for a variety of purposes [11].

*Textile dyeing and finishing* have traditionally been a major market for EDTA-type chelating agents. Traces of metal ion contaminants from incoming fibers, chemicals, and process water will cause a shade change in many commercial dyes unless a chelating agent is present. In addition, DTPA is widely used to *stabilize hydrogen peroxide* bleaching liquors against decomposition by traces of manganese, iron, and copper.

The brightness of *paper pulp* bleached with hydrogen peroxide or dithionite is improved by the addition of chelating agents during the pulping stage and, later, to the bleaching solution where it acts as a stabilizer and metal ion control; DTPA is especially useful for peroxide stabilization [12]. Chelating agents aid especially in the removal of manganese from the wood source, which, in addition to decomposing the peroxide bleaching agent, later reacts with lignins to form dark-colored substances [13], [14].

Chelating agents are used to *clean scale deposits* from internal boiler surfaces and as additives to incoming boiler feedwater to prevent the formation of calcium and magnesium scales. Care must be taken to control the ratio of chelating agent to water hardness, lest the amount of free chelating agent rises to corrosive levels. Recently, polymers have been used in conjunction with chelating agents to reduce the corrosive potential of the treatment.

The *cleaning of metal surfaces* is aided by the use of chelating agents to solubilize adherent oxide films and provide complete rinsing. Metal-plating solutions often incorporate chelating agents to form complex metal ions, which improve the smoothness and integrity of the metal coating.

*Cleaning compounds* of many types, including laundry detergents and hard surface and bathroom cleansers, are improved by adding chelating agents [2]. The bactericidal activity of sanitizing agents, especially toward gram-negative bacteria, increases when EDTA is incorporated, because of its ability to destroy the outer cell wall of these often resistant species. This bactericidal enhancement by EDTA is also used to potentiate the preservative activity in many consumer preparations including cosmetics, shampoos, and ophthalmic products [2], [10], [15], [16].

Chelating agents are used to *stabilize systems* from breakdown by the catalytic action of trace metals. The oxidative rancidity of fatty emulsions, e.g., mayonnaise, salad dressings, cosmetic creams, and lotions, is prevented by adding EDTA as the disodium salt or disodium–calcium complex. *Discoloration* by iron and other metal ions may be inhibited by adding EDTA. Phenolic compounds present in potatoes form reddish brown iron complexes, which turn black during cooking; EDTA can be used to prevent this deleterious action [17].

The *change in oxidation–reduction potential* for metal ions on chelation is employed commercially in the emulsion polymerization of styrene–butadiene rubber in the "cold" process that uses Fe(II)(EDTA). The bleaching of color photographic film is accomplished with Fe(III)(EDTA), which oxidizes the metallic silver present in the exposed image to the ionic form for removal or "fixing" by thiosulfate. The control of plant emissions of hydrogen sulfide is accomplished on an industrial scale by using various iron chelate compounds to catalytically oxidize hydrogen sulfide to elemental sulfur. This process is especially useful for low concentrations of hydrogen sulfide and results in a salable grade of sulfur as a byproduct [18]. Iron chelates are also used as reversible absorbants for nitrogen oxide (NO). The activation of peroxygen bleaches in laundry detergents is also accomplished by metal chelate complexes [19].

Chelating agents are added to chemical processes to remove traces of iron and other metal ions, which either interfere with the process or must be reduced to meet stringent specifications.

*Agricultural micronutrients or trace elements*, such as iron, zinc, manganese, and copper, are added to fertilizers in the chelated form to prevent interaction and precipitation with contained phosphates and other precipitating anions. The conversion of these cationic metal ions to an anionic chelated form also allows the micronutrient to translocate through the soil medium to the root zone for uptake into the plant. Clays, carbonates, phosphates, and other soil constituents hinder the movement of cationic metal sources from the point of placement on the soil surface. These factors account for the increased efficiency of chelated micronutrients over inorganic sources [20].

Many techniques in *analytical chemistry*, especially titrimetric methods, that use EDTA have been developed [21]. The equivalence point is easily measured because of the abrupt reduction in free metal ion concentration when equimolar equivalence is achieved. Also EDTA is used as a masking agent to prevent interference by metal ions in some spectrophotometric methods.

# 8. Trade Names

The first trade name for EDTA was Trilon B from I.G. Farbenindustrie. Current trade names are Trilon B (BASF), Versene (Dow), Sequestrene (Ciba-Geigy), Hamp-ene (Grace, USA), Rexene (Grace, Sweden), and Chelest (Chelest, Japan). Trade names for HEEDTA are Versenol (Dow), Chel DM (Ciba-Geigy), Hamp-ol (Grace, USA), Rexenol (Grace, Sweden), and Chelest H (Chelest, Japan). Major trade names for DTPA are Versenex (Dow), Chel DTPA (Ciba-Geigy), Hamp-ex (Grace, USA), Polychelate (Grace, Sweden), and Chelest P (Chelest, Japan).

# 9. Economic Aspects

Consumption of EDTA and related chelating agents in Japan, North America, and Western Europe is shown in Table 5. Future U.S. market growth is estimated to be low, i.e., 3–4% annually [23]. Production capacities are approximate because much of the tonnage of chelating agents is produced in flexible plant equipment capable of being used for other manufacturing processes.

The principal manufacturers in Japan are Chelest Chemical Co., Osaka and Teikoku Chemical Industries, Itami. Major Western European manufacturers include BASF, Ludwigshafen, Federal Republic of Germany; Akzo, Herkenbosch, the Netherlands,

**Table 5.** Production and capacities of EDTA and related chelating agents in 1985

|  | 1985 Estimated production, $10^3$ t | | | Approximate total capacity, $10^3$ t |
|---|---|---|---|---|
|  | EDTA | DTPA | HEEDTA |  |
| Japan | 4.6 | 1.1 | 0.3 | 9.0 |
| Western Europe | 14.3 | 0.5 | 0.3 | 21.0 |
| North America | 21.4 | 3.1 | 2.4 | 55.0 |

and Düren, Federal Republic of Germany; Rexolin (W. R. Grace & Co.), Helsingborg, Sweden, and Middlesborough, United Kingdom; Protex, St. Avold, France; and ABM Chemicals, Gloucester, United Kingdom. Three producers account for > 90% of U.S. production: Dow Chemical, Freeport, Texas; Ciba-Geigy, McIntosh, Alabama; and W. R. Grace & Co. at Nashua, New Hampshire and Deer Park, Texas.

# 10. Toxicology and Occupational Health

For many years, EDTA and its salts have been used as food additives and preservatives in pharmaceuticals and cosmetics without incident. Early toxicological work indicated that EDTA is relatively nontoxic [24]. Acute oral $LD_{50}$ value (rat) for $Na_2EDTA$ is 2.0–2.2 g/kg. Subacute studies using 0.5–5.0% $Na_2EDTA$ in the diet found no toxic manifestations except diarrhea and lowered food consumption at the highest feeding level. Skin sensitization studies have shown that EDTA as the trisodium salt is not a skin sensitizer [25]. The genetic toxicology of EDTA has been reviewed extensively [26]. Because of its sequestering ability, EDTA is responsible for functional and structural alterations of genetic material, though it seems to be harmless to humans as far as genotoxicity is concerned. When $Na_3EDTA$ was examined for possible carcinogenicity, no compound-related signs of clinical toxicity were found [27].

# 11. References

[1] A. E. Martell, R. M. Smith: *Critical Stability Constants*, Plenum Press, New York 1974.
[2] J. R. Hart, *Soap, Cosmet., Chem. Spec.* **56** (1980) no. 6, 39–40, 42, 74.
[3] M. S. Antleman: *The Encyclopedia of Chemical Electrode Potentials*, Plenum Press, New York 1982.
[4] I. G. Farbenindustrie, DE 718 981, 1935 (F. Munz). General Aniline Works, US 2 130 505, 1938 (F. Munz).
[5] Carbide & Carbon Chemical Corp., US 2 384 818, 1945 (G. Curme, Jr., J. W. Clark).
[6] Martin Dennis Co., US 2 387 735, 1945 (F. C. Bersworth).

[7] Hampshire Chemical Corp., US 3 061 628, 1962 (J. J. Singer, Jr., M. Weisberg).
[8] W. R. Grace & Co., Hampshire Chelating Agents Technical Bulletin, Lexington, Mass., 1985, p. 9.
[9] G. Schwarzenbach, W. Biederman, *Helv. Chim. Acta* **31** (1948) 678.
[10] J. R. Hart in J. J. Kabara (ed.): *Cosmetic and Drug Preservation – Principles and Practice*, Marcel Dekker, New York 1984, p. 333.
[11] R. Hart, *Household Pers. Prod. Ind.* **16** (1979) no. 6, 54–55, 57, 79.
[12] D. R. Bambrick, *Tappi* **68** (1985) no. 6, 96–100.
[13] J. R. Hart, *Tappi* **64** (1981) no. 3, 43–44.
[14] J. R. Hart, *Pulp & Pap.* **55** (1981) no. 6, 138–140.
[15] J. R. Hart, *Cosmet. Toiletries* **98** (1983) no. 4, 54–58.
[16] R. G. Young, *Soap, Cosmet., Chem. Spec.* **60** (1984) no. 11, 37–38, 89–90.
[17] T. E. Furia: *CRC Handbook of Food Additives*, CRC Press, Cleveland, Ohio, 1977, pp. 271–294.
[18] W. R. Grace & Co., Chelating Agents in Oxidation-Reduction Reactions, Lexington, Mass., 1986.
[19] J. R. Hart, *Soap, Cosmet., Chem. Spec.* **62** (1986) no. 5, 38, 41, 48.
[20] J. R. Hart, *Solutions* **26** (1982) no. 7, 63–66, 68, 70, 72.
[21] F. J. Welcher: *The Analytical Uses of Ethylenediaminetetraacetic Acid*, Van Nostrand, Princeton, N.J., 1958.
[22] Personal communication, W. R. Grace & Co., Lexington, Mass., 1986.
[23] P. Mann, *Chem. Mark. Rep.* **229** (1986) no. 18, 36.
[24] S. S. Yang, *Food Cosmet. Toxicol.* **2** (1964) 763–767.
[25] J. W. Henck, D. D. Lockwood, K. J. Olsen, *Drug Chem. Toxicol.* **3** (1980) 99–103.
[26] K. Heindorff, O. Aurich, A. Michalis, R. Rieger, *Mutat. Res.* **115** (1983) 149–173.
[27] National Cancer Institute: "Bioassay of Trisodium Ethylenediaminetetraacetate Trihydrate (EDTA) for Possible Carcinogenicity", Technical Report Series no. 11, 1977, Report NCI-CG-TR-11.

# Ethylene Glycol

*Individual keyword:* → *Ethylene oxide*

SIEGFRIED REBSDAT, Hoechst Aktiengesellschaft, Gendorf, Federal Republic of Germany (Chaps. 1–10)

DIETER MAYER, Hoechst Aktiengesellschaft, Frankfurt, Federal Republic of Germany (Chap. 11)

| | | | | |
|---|---|---|---|---|
| 1. | Introduction | 2305 | 7. Storage and Transportation | 2316 |
| 2. | Physical Properties | 2306 | 8. Derivatives | 2317 |
| 3. | Chemical Properties | 2308 | 8.1. Di-, Tri-, Tetra-, and Polyethylene Glycols | 2317 |
| 4. | Production | 2311 | 8.2. Ethers and Esters | 2317 |
| 4.1. | Ethylene Oxide Hydrolysis | 2311 | 9. Uses | 2321 |
| 4.1.1. | Current Production Method | 2311 | 10. Economic Aspects | 2322 |
| 4.1.2. | Possible Developments | 2313 | 11. Toxicology and Occupational Health | 2323 |
| 4.2. | Alternative Methods of Ethylene Glycol Production | 2314 | 11.1. Ethylene Glycol | 2323 |
| 4.2.1. | Direct Oxidation of Ethylene | 2314 | 11.2. Ethylene Glycol Derivatives | 2324 |
| 4.2.2. | Synthesis from $C_1$ Units | 2315 | 12. References | 2326 |
| 5. | Environmental Protection and Ecology | 2316 | | |
| 6. | Quality Specifications and Analysis | 2316 | | |

# 1. Introduction

Ethylene glycol [107-21-1], 1,2-Ethanediol, $HOCH_2CH_2OH$, $M_r$ 62.07, usually called glycol, is the simplest diol. It was first prepared by WURTZ in 1859 [1]; treatment of 1,2-dibromoethane [106-93-4] with silver acetate yielded ethylene glycol diacetate, which was then hydrolyzed to ethylene glycol.

Ethylene glycol was first used industrially in place of glycerol during World War I as an intermediate for explosives (ethylene glycol dinitrate) [2], but has since developed into a major industrial product.

The worldwide capacity for the production of ethylene glycol via the hydrolysis of ethylene oxide [75-21-8] (→ Ethylene Oxide) is estimated to be ca. $7 \times 10^6$ t/a.

Ethylene glycol is used mainly as an antifreeze in automobile radiators and as a raw material for the manufacture of polyester fibers.

## 2. Physical Properties

Ethylene glycol is a clear, colorless, odorless liquid with a sweet taste. It is hygroscopic and completely miscible with many polar solvents, such as water, alcohols, glycol ethers, and acetone. Its solubility is low, however, in nonpolar solvents, such as benzene, toluene, dichloroethane, and chloroform. The UV, IR, NMR, and Raman spectra of ethylene glycol are given in [3]. Following are some of the physical properties of ethylene glycol [4]–[6]:

| | |
|---|---|
| bp at 101.3 kPa | 197.60 °C |
| fp | −13.00 °C |
| Density at 20 °C | 1.1135 g/cm$^3$ |
| Refractive index, $n_D^{20}$ | 1.4318 |
| Heat of vaporization at 101.3 kPa | 52.24 kJ/mol |
| Heat of combustion | 19.07 MJ/kg |
| Critical data | |
|   Temperature | 372 °C |
|   Pressure | 6515.73 kPa |
|   Volume | 0.186 L/mol |
| Flash point | 111 °C |
| Ignition temperature | 410 °C |
| Lower explosive limit | 3.20 vol% |
| Upper explosive limit | 53 vol% |
| Viscosity at 20 °C | 19.83 mPa · s |
| Cubic expansion coefficient at 20 °C | $0.62 \times 10^{-3}$ K$^{-1}$ |

Ethylene glycol is difficult to crystallize; when cooled, it forms a highly viscous, supercooled mass that finally solidifies to produce a glasslike substance.

The widespread use of ethylene glycol as an antifreeze is based on its ability to lower the freezing point when mixed with water. The physical properties of ethylene glycol – water mixtures are, therefore, extremely important. The freezing points of mixtures of water with monoethylene glycol and diethylene glycol [*111-46-6*] are shown in Figure 1. The temperature dependencies of the thermal conductivity, density, and viscosity of ethylene glycol and ethylene glycol – water mixtures are shown in Figures 2, 3, and 4 respectively [7]. The Prandtl numbers (the ratio of the viscosity to the thermal conductivity) derived from these values are given in Figure 5 [7]. The vapor pressures of ethylene glycol – water mixtures have been obtained from [8] by interpolation and are listed in Table 1.

**Table 1.** Vapor pressure of ethylene glycol – water mixtures

| Water content, wt% | Vapor pressure, in kPa at | | |
|---|---|---|---|
| | 65.1 °C | 77.7 °C | 90.3 °C |
| 0 | 0.30 | 0.52 | 1.20 |
| 10 | 6.61 | 11.65 | 19.73 |
| 20 | 11.30 | 19.68 | 33.01 |
| 30 | 14.70 | 25.45 | 42.49 |
| 40 | 17.10 | 29.68 | 49.37 |
| 50 | 18.81 | 32.92 | 54.60 |
| 60 | 20.16 | 35.58 | 58.87 |
| 70 | 21.45 | 37.92 | 62.60 |
| 80 | 22.98 | 40.05 | 65.98 |
| 90 | 25.08 | 41.91 | 68.93 |
| 100 | 28.04 | 43.34 | 71.10 |

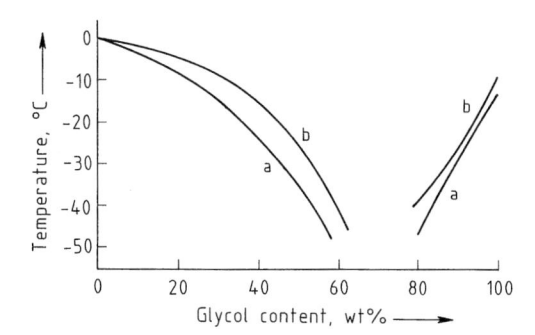

**Figure 1.** Freezing points of mono- and diethylene glycol – water mixtures
a) Monoethylene glycol; b) Diethylene glycol

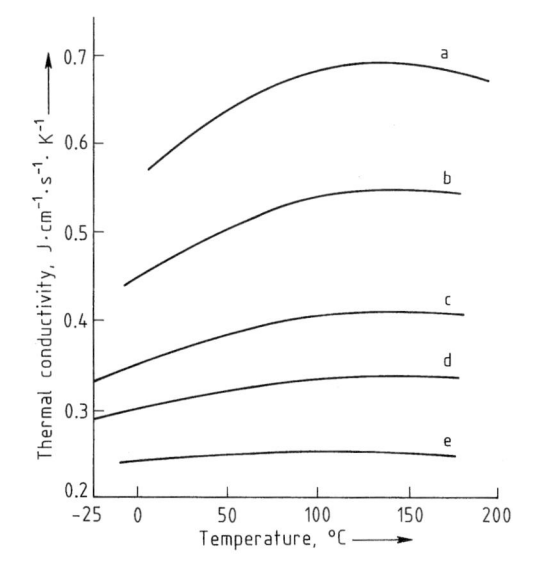

**Figure 2.** Temperature dependence of the thermal conductivity of ethylene glycol – water mixtures
Ethylene glycol content, mol%: a) 0; b) 25; c) 55; d) 75; e) 100

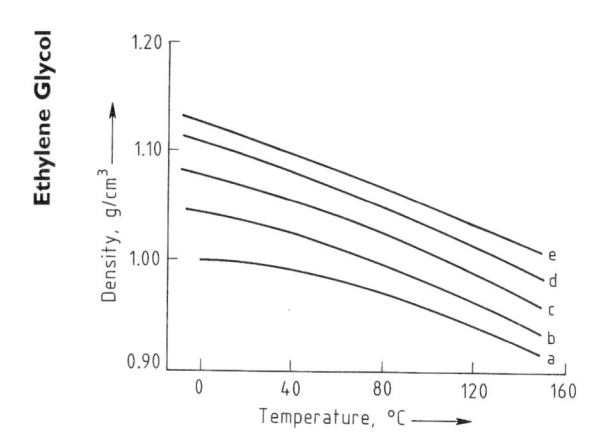

**Figure 3.** Temperature dependence of the density of ethylene glycol–water mixtures
Ethylene glycol content, mol%: a) 0; b) 26.1; c) 50.95; d) 76.9; e) 100

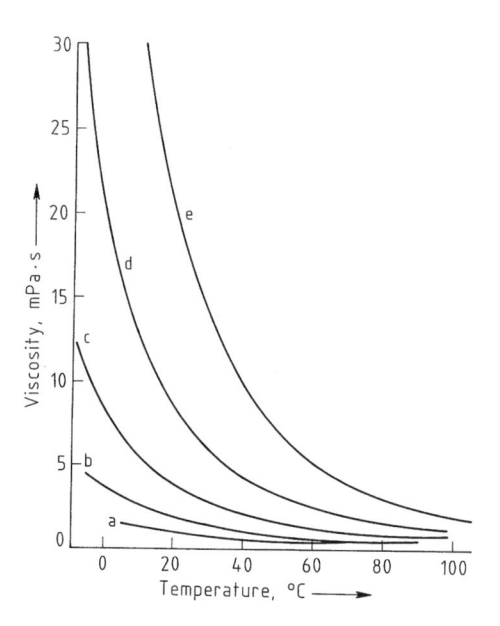

**Figure 4.** Temperature dependence of the viscosity of ethylene glycol–water mixtures
Ethylene glycol content, mol%: a) 0; b) 25; c) 49.90; d) 74.36; e) 100

# 3. Chemical Properties

Ethylene glycol, like other alcohols, undergoes the reactions typical of its hydroxyl groups, which are described elsewhere (→ Alcohols, Aliphatic). Thus, only the special chemical characteristics and industrially important reactions of ethylene glycol are considered here. The two adjacent hydroxyl groups allow cyclization, and polycondensation; one or both of these functional groups may, of course, also react to give other derivatives.

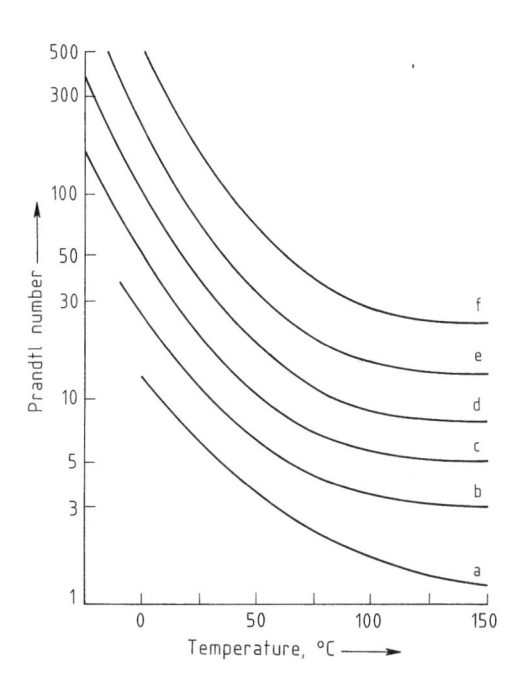

**Figure 5.** Temperature dependence of the Prandtl numbers of ethylene glycol – water mixtures

Ethylene glycol content, mol %: a) 0; b) 20; c) 40; d) 60; e) 80; f) 100

**Oxidation.** Ethylene glycol is easily oxidized to form a number of aldehydes and carboxylic acids by oxygen, nitric acid, and other oxidizing agents. The typical products derived from the alcoholic functions are glycolaldehyde ($HOCH_2CHO$) [*141-46-8*], glycolic acid ($HOCH_2COOH$) [*79-14-1*], glyoxal ($CHOCHO$) [*107-22-2*], glyoxylic acid ($HCOCOOH$) [*298-12-4*], oxalic acid ($HOOCCOOH$) [*144-62-7*], formaldehyde ($HCHO$) [*50-00-0*], and formic acid ($HCOOH$) [*64-18-6*]. Many of these compounds are described in separate articles. Variation of the reaction conditions can lead to the selective formation of a desired oxidation product. Gas-phase oxidation with air in the presence of copper catalysts is of industrial importance for the production of glyoxal (→ Glyoxal and → Glyoxylic Acid). Glycol cleavage occurs in acidic solution with certain oxidizing agents such as permanganate, periodate, or lead tetraacetate. Cleavage of the C-C bond mainly produces formaldehyde, some of which is further oxidized to formic acid [9].

**1,3-Dioxolane Formation.** 1,3-Dioxolanes are formed by reacting ethylene glycol with carbonyl compounds [10]:

$$\begin{array}{c} H_2C-OH \\ | \\ H_2C-OH \end{array} + O=C\begin{array}{c} R \\ \\ R \end{array} \xrightarrow{H^+} \begin{array}{c} H_2C-O \\ | \quad \quad \diagdown \\ \quad \quad \quad C \\ | \quad \quad \diagup \\ H_2C-O \end{array}\begin{array}{c} R \\ \\ R \end{array} + H_2O$$

R = alkyl

Acetalization to the cyclic 1,3-dioxolane proceeds more readily than acetal formation from straight-chain alcohols. If water is removed from the reaction mixture, an excellent yield can be obtained. This reaction is used to protect carbonyl groups in organic syntheses.

1,3-Dioxolanes can also be formed from ethylene glycol by transacetalization. Examples are the reactions of ethylene glycol with orthoformates [11]:

$$\text{H}_2\text{C(OH)CH}_2\text{OH} + \text{RO-CH(OR)}_2 \longrightarrow \text{cyclic acetal} + 2\text{ ROH}$$

R = alkyl

or with dialkyl carbonates:

$$\text{H}_2\text{C(OH)CH}_2\text{OH} + (\text{RO})_2\text{C=O} \longrightarrow \text{Ethylene carbonate (1,3-dioxolan-2-one)} + 2\text{ ROH}$$

R = alkyl

**1,4-Dioxane Formation.** Ethylene glycol can be converted to dioxane [123-91-1] by dehydration in the presence of acidic catalysts [12]:

$$\text{H}_2\text{C(OH)CH}_2\text{OH} + \text{HOCH}_2\text{CH}_2\text{OH} \xrightarrow{\text{H}^+} \text{1,4-Dioxane} + 2\text{ H}_2\text{O}$$

**Ether and Ester Formation.** Ethylene glycol can be alkylated or acylated by the customary methods to form ethers or esters, respectively. However, the presence of two hydroxyl groups leads to the formation of both mono- and diethers and mono- and diesters, depending on the initial concentrations of the individual reactants. The esterification of ethylene glycol with terephthalic acid [100-21-0] to form polyesters is especially important.

$$n\text{ HOCH}_2\text{CH}_2\text{OH} + n\text{ HOOC-C}_6\text{H}_4\text{-COOH} \longrightarrow$$
$$\text{HO(CH}_2\text{CH}_2\text{OOC-C}_6\text{H}_4\text{-COO)}_n\text{-H} + n\text{ H}_2\text{O}$$

**Ethoxylation.** Ethylene glycol reacts with ethylene oxide to form di-, tri-, tetra-, and polyethylene glycols. The proportions of these glycols found in the reaction product are determined by the catalyst system that is used and the glycol excess. A considerable excess of glycol is required to obtain the lower homologues in a satisfactory yield. This reaction is rarely used industrially since these homologues are formed as byproducts during the production of ethylene glycol (cf. Section 4.1.1).

$$\text{HOCH}_2\text{CH}_2\text{OH} + n\text{ H}_2\text{C-CH}_2\text{(O)} \longrightarrow \text{HO(CH}_2\text{CH}_2\text{O)}_{n+1}\text{-H}$$

**Decomposition with Alkali Hydroxide.** Glycol is a relatively stable compound, but special care is required when ethylene (or diethylene) glycol is heated at a higher

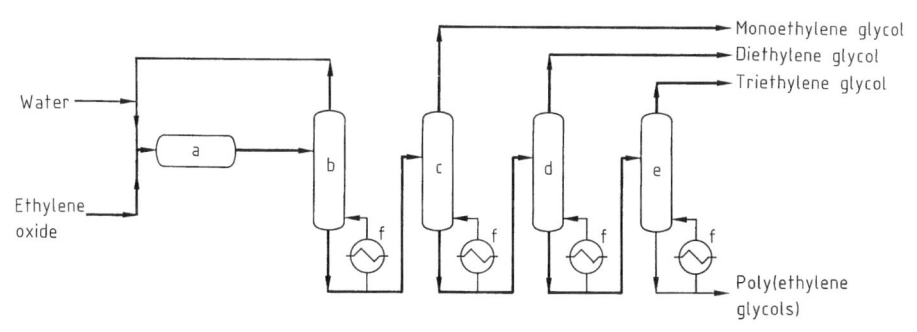

**Figure 6.** Flow diagram for a glycol plant
a) Reactor; b) Drying column; c) Monoethylene glycol column; d) Diethylene glycol column; e) Triethylene glycol column; f) Heat exchanger

temperature in the presence of a base such as sodium hydroxide. Fragmentation of the molecule begins at temperatures above 250 °C and is accompanied by the exothermic evolution of hydrogen ($\Delta H = -90$ to $-160$ kJ/kg) [13]. This leads to a buildup of pressure in closed vessels.

# 4. Production

Although ethylene glycol has been known since 1859 (WURTZ) [1], it was not produced industrially until World War I. Its synthesis was then based on the hydrolysis of ethylene oxide [75-21-8] produced by the chlorohydrin process (→ Chlorohydrins). Production from formaldehyde [50-00-0] and carbon monoxide was also used commercially from 1940 to 1963 [14]. Neither of these methods is now used, however; the older literature should be consulted for details [2], [12]. Direct oxidation of ethylene [74-85-1] to ethylene glycol was also employed commercially for a short time [15], but was abandoned, probably due to problems caused by corrosion [16].

## 4.1. Ethylene Oxide Hydrolysis

### 4.1.1. Current Production Method

Only one method is currently used for the industrial production of ethylene glycol. This method is based on the hydrolysis of ethylene oxide obtained by direct oxidation of ethylene with air or oxygen (→ Ethylene Oxide). The ethylene oxide is thermally hydrolyzed to ethylene glycol without a catalyst. Figure 6 shows a simplified scheme of a plant producing ethylene glycol by this method. The ethylene oxide – water mixture is preheated to ca. 200 °C, whereby the ethylene oxide is converted to ethylene glycol.

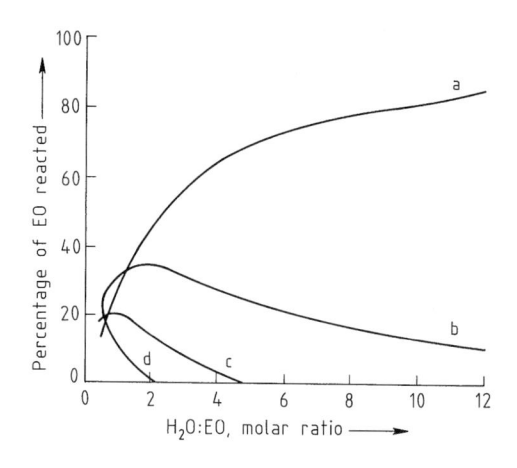

**Figure 7.** Composition of the product obtained on hydrolysis of ethylene oxide (EO) as a function of the water to ethylene oxide ratio
a) Monoethylene glycol; b) Diethylene glycol; c) Triethylene glycol; d) Higher poly(ethylene glycols)

Di-, tri-, tetra-, and polyethylene glycols are also produced, but with respectively decreasing yields (see also Chap. 3).

The formation of these higher homologues is inevitable because ethlyene oxide reacts with ethylene glycols more quickly than with water; their yields can, however, be minimized if an excess of water is used — a 20-fold molar excess is usually employed. Figure 7 shows the composition of the resulting product mixture as a function of the ratio of water to ethylene oxide. Although the values were determined by using sulfuric acid as a catalyst [17], they also apply as a good approximation for the reaction without a catalyst. Thus, in practice almost 90% of the ethylene oxide can be converted to monoethylene glycol, the remaining 10% reacts to form higher homologues:

$$H_2C\text{-}CH_2 + H_2O \longrightarrow HOCH_2CH_2OH$$
$$\Delta H = -79.4 \text{ kJ/mol}$$

$$n\ H_2C\text{-}CH_2 + HOCH_2CH_2OH \longrightarrow HO(CH_2CH_2O)_{n+1}\text{-}H$$
$$n = 1, 2, 3$$

After leaving the reactor, the product mixture is purified by passing it through successive distillation columns with decreasing pressures. Water is first removed and returned to the reactor, the mono-, di-, and triethylene glycols are then separated by vacuum distillation. The yield of tetraethylene glycol is too low to warrant separate isolation. The heat liberated in the reactor is used to heat the distillation columns. A side stream must be provided to prevent the accumulation of secondary products, especially small amounts of aldehydes, which are produced during hydrolysis. The shape of the reactor affects the selectivity of the reaction. Plug-flow reactors are superior to both agitator-stirred tanks and column reactors [18].

## 4.1.2. Possible Developments

The glycol production method described in Section 4.1.1 is the only one of current industrial importance. It is simple, but has some major drawbacks:

1) The selectivity of the first step — the production of ethylene oxide — is low (ca. 80%).
2) The selectivity of ethylene oxide hydrolysis is low — ca. 10% is converted to di- and triethylene glycol.
3) Energy consumption for the distillation of the large amount of excess water is high.

Therefore, much research has been carried out to improve this process. The search for better silver catalysts is an objective for point 1 (→ Ethylene Oxide). Points 2 and 3 must be considered together, since higher selectivity for ethylene oxide hydrolysis automatically reduces the excess of water required.

Many catalysts have been described in the literature that are able to optimize selectivity or lower the reaction temperature and the required excess of water. Acids and bases are known to accelerate the reaction rate. The kinetics of the acid [19] and base [20] catalysis of ethylene oxide hydrolysis have been thoroughly investigated; mechanisms are discussed in [21]. The industrial feasability of catalysis with ion-exchange columns in the liquid phase [22], [23] and the gas phase [24] has been tested. Although the use of catalysts allowed the reaction temperature to be lowered, selectivity was not significantly enhanced. Furthermore, the catalyst needed to be separated and either fed back into the reaction mixture or replaced. As a result of these disadvantages, these types of catalysis have not proved to be of commercial use. However, catalysts that improve selectivity have been described in recent patents; they include molybdates [25], vanadates [26], ion exchangers [27], and organic antimony compounds [28]. However, their advantages do not yet seem to justify their use on an industrial scale.

The selective synthesis of ethylene glycol via the intermediate ethylene carbonate (1,3-dioxolan-2-one) [96-49-1] seems to be a promising alternative. This compound is obtained in high yield (98%) by reacting ethylene oxide with carbon dioxide and can be selectively hydrolyzed to give a high yield of ethylene glycol. Only double the molar quantity of water is required for this reaction.

$$H_2C\text{-}CH_2 + CO_2 \xrightarrow{Cat.} H_2C\text{-}CH_2 \xrightarrow{H_2O} HOCH_2CH_2OH + CO_2$$

According to a Halcon patent, ethylene oxide can be extracted from the aqueous solution, formed during its production, with supercritical carbon dioxide [29]. An ethylene oxide – carbon dioxide solution is obtained, which reacts to form ethylene carbonate. Hydrolysis of the ethylene carbonate then yields ethylene glycol. Possible catalysts for this reaction are quaternary ammonium and phosphonium salts, such as

R$_4$NHal, R$_4$PHal, or Ph$_3$PCH$_3$I. Problems such as product separation and catalyst feedback still need to be resolved, but this method for the selective synthesis of ethylene glycol from ethylene oxide seems to be the most promising for industrial-scale application.

## 4.2. Alternative Methods of Ethylene Glycol Production

The low selectivity of ethylene oxide production and increasing ethylene prices warrant the search for alternative ways of producing ethylene glycol.

### 4.2.1. Direct Oxidation of Ethylene

As mentioned earlier, catalytic oxidation of ethylene [74-85-1] with oxygen in acetic acid has already been used on an industrial scale, but this method was soon abandoned due to problems caused by corrosion. The yield of ethylene glycol (>90%) was much higher than that obtained in the more indirect route via ethylene oxide [30].

$$H_2C=CH_2 + AcOH \xrightarrow[160\ °C,\ 2.5\ MPa]{O_2/Te,\ HBr} AcOCH_2CH_2OH$$
$$AcOCH_2CH_2OAc$$
$$\downarrow H_2O$$
$$Ac = CH_3CO- \qquad\qquad HOCH_2CH_2OH$$

A more recently developed catalyst system is based on the use of Pd(II) complexes [31]. A mixture of PdCl$_2$, LiCl, and NaNO$_3$ in acetic acid and acetic anhydride has been shown to give a 95% selectivity for glycol monoacetate and glycol diacetate formation (60–100 °C, 3.04 MPa) [32]. During this process, Pd(II) is reduced to Pd(0). The precipitation of Pd(0) is prevented because it is reoxidized to Pd(II) by the nitrate ions. The available oxygen finally regenerates the nitrate, thus providing a complete catalytic system.

The formation of ethylene glycol monoacetate [542-59-6] (50% yield) and ethylene glycol diacetate [111-55-7] (7% yield) has also been investigated using the catalyst system PdCl–NO$_2$–CH$_3$CN dissolved in acetic acid. Studies with radioactive isotopes showed that the NO$_2$ functions as an oxidizing agent [33]. Vinyl acetate is formed as a byproduct (20% yield). However, the catalytic action of this system is quickly exhausted due to the precipitation of palladium compounds.

If a PdCl$_2$–CuCl$_2$–CuOCOCH$_3$ system is used, the reaction proceeds under mild conditions (65 °C, 0.5 MPa) without the formation of a precipitate; a yield of over 90% is obtained [34].

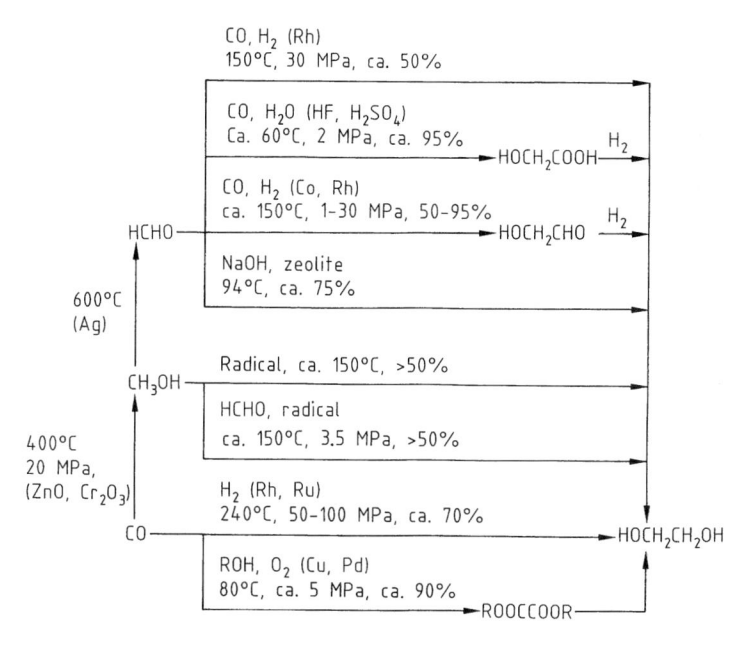

```
                    CO, H₂ (Rh)
                    150°C, 30 MPa, ca. 50%
                ┌──────────────────────────────────────────────────────────────┐
                │   CO, H₂O (HF, H₂SO₄)                                        │
                │   Ca. 60°C, 2 MPa, ca. 95%              H₂                   │
                │                          ──HOCH₂COOH──                      │
                │   CO, H₂ (Co, Rh)                                            │
                │   ca. 150°C, 1-30 MPa, 50-95%           H₂                   │
        HCHO────│                          ──HOCH₂CHO ──                      │
         ↑      │   NaOH, zeolite                                              │
                │   94°C, ca. 75%                                              │
      600°C     │                                                              │
       (Ag)     │                                                              │
                │   Radical, ca. 150°C, >50%                                   │
       CH₃OH────│                                                              │
         ↑      │   HCHO, radical                                              │
                │   ca. 150°C, 3.5 MPa, >50%                                   │
      400°C     │                                                              │
      20 MPa,   │   H₂ (Rh, Ru)                                                │
    (ZnO, Cr₂O₃)│   240°C, 50-100 MPa, ca. 70%                                 │
        CO─────│                                          ──HOCH₂CH₂OH        │
                │   ROH, O₂ (Cu, Pd)                                           │
                │   80°C, ca. 5 MPa, ca. 90%                                   │
                │                          ──ROOCCOOR──                       │
                └──────────────────────────────────────────────────────────────┘
```

**Figure 8.** Production of ethylene glycol from carbon monoxide (percentages indicate approximate yields)

In recent years, increasing attention has been paid to Pd(II) systems as catalysts for the direct oxidation of ethylene to ethylene glycol. In spite of the widespread interest in this alternative, industrial applications have yet to be realized.

## 4.2.2. Synthesis from $C_1$ Units

The long-term shortage and increased price of crude oil have led to an intensive search for methods of producing organic intermediates from $C_1$ units (i.e., methods based on coal). Many publications have appeared on the synthesis of ethylene glycol by this approach. Only the most important methods that rely on synthesis gas or carbon monoxide [630-08-0] are discussed here; they are summarized in Figure 8.

At a high pressure, carbon monoxide and hydrogen react directly to produce ethylene glycol [35], [36]. However, the reaction is slow and the catalyst is both sensitive and expensive [37]. Other methods involve the formation of formaldehyde [50-00-0], methanol [67-56-1] [38], or esters of oxalic acid [144-62-7] as intermediates [37], [39]. The only method to attain industrial importance was that employed by Du Pont from 1940 to 1963, which used formaldehyde and glycolic acid [79-14-1] as intermediates. High operating pressure and temperature were required, however (48 MPa, 220 °C). This process was significantly improved by the introduction of hydrogen fluoride [7664-39-3] as a catalyst (1–2 MPa, 60 °C) [40].

At the present time, none of the described methods based on $C_1$ units can compete with the ethylene → ethylene oxide → glycol pathway. However, if crude oil prices increase, the synthesis of ethylene glycol from $C_1$ units will become more economically attractive [41].

# 5. Environmental Protection and Ecology

Ethylene glycol is readily biodegradable [42]; thus, disposal of wastewater containing this compound can proceed without major problems. The high $LC_{50}$ values of over 10 000 mg/L [43], [44] account for its low water toxicity: $LC_{50}$ crayfish (*Procambarus*) 91 000 mg/L, $LC_{50}$ fish (*Lepomis macrochirus*) 27 540 mg/L.

# 6. Quality Specifications and Analysis

Since ethylene glycol is produced in relatively high purity, differences in quality are not expected. The directly synthesized product meets high quality demands ("fiber grade"). The ethylene glycol produced in the wash water that is used during ethylene oxide production is normally of a somewhat inferior quality ("antifreeze grade"). The quality specifications for mono-, di-, and triethylene glycols are compiled in Table 2 [45]. The UV absorption of fiber-grade glycol is often used as an additional parameter for quality control.

Gas chromatography is commonly used for the quantitative determination of ethylene glycol. Monoethylene glycol can be detected by oxidation with periodic acid even if di- and triethylene glycols are also present; however, aldehydes, glycerol, and monopropylene glycol falsify the results [4].

# 7. Storage and Transportation

Pure anhydrous ethylene glycol is not aggressive toward most metals and plastics. Since ethylene glycol also has a low vapor pressure and is noncaustic, it can be handled without any problems; it is transported in railroad tank cars, tank trucks, and tank ships. Tanks are usually made of steel; high-grade materials are only required for special quality requirements. Nitrogen blanketing can protect ethylene glycol against oxidation.

At ambient temperatures, aluminum is resistant to pure glycol. Corrosion occurs, however, above 100 °C and hydrogen is evolved. Water, air, and acid-producing impurities (aldehydes) accelerate this reaction. Great care should be taken when phenolic resins are involved, since they are not resistant to ethylene glycol.

# 8. Derivatives

Only the most important of the many derivatives of ethylene glycol will be discussed in this section, namely di-, tri-, and polyethylene glycols, the methyl, ethyl, and butyl glycol ethers and the acetate esters.

## 8.1. Di-, Tri-, Tetra-, and Polyethylene Glycols

The higher homologues of ethylene glycol are formed as byproducts during the synthesis of ethylene oxide and monoethylene glycol or are prepared directly by reacting monoethylene glycol with ethylene oxide. Di-, tri-, and possibly tetraethylene glycol can be purified by distillation. For quality specifications of these compounds, see Table 2; Table 3 shows some of their physical properties. The temperature dependence of the thermal conductivity, density, viscosity, and Prandtl numbers of di- and triethylene glycol – water mixtures is described in [46].

Poly(ethylene glycols) (PEG) [25322-68-3] are produced as mixtures of higher molecular mass homologues and are characterized by their mean molecular masses (e.g., PEG 400). Physical properties of some poly(ethylene glycol) mixtures are listed in Table 4.

## 8.2. Ethers and Esters

**Ethers.** Ethylene glycol monoethers are usually produced by reaction of ethylene oxide with the appropriate alcohol. A mixture of homologues is obtained, all displaying the following general structure:

$RO(CH_2CH_2O)_n-H$

The glycol monoethers can be converted to diethers by alkylation with common alkylating agents, such as dimethyl sulfate or alkyl halides (Williamson synthesis). Glycol dimethyl ethers are formed by treatment of dimethyl ether with ethylene oxide [48]. The properties of some ethylene glycol mono- and diethers are shown in Tables 5 and 6, respectively; they are mainly used as solvents.

**Table 2.** Quality specifications of mono-, di-, and triethylene glycols

| Property | Method | Monoethylene glycol | | Diethylene glycol | Triethylene glycol |
|---|---|---|---|---|---|
| | | Antifreeze grade | Fiber grade | | |
| Purity, % | Gas chromatography | >98.00 | >99.80 | 99.60 | 99.50 |
| Diethylene glycol content, wt % | Gas chromatography | <0.50 | <0.10 | | |
| Boiling range (at 101.3 kPa), °C | DIN 53171, ASTM D 1078 | 195–200 | 196–199 | 242–247 | 285–295 |
| Density (20 °C), g/cm$^3$ | DIN 51757, ASTM D 1122 | 1.113–1.115 | 1.1135–1.1140 | 1.1160–1.1175 | 1.1235–1.1245 |
| Refractive index, $n_D^{20}$ | DIN 53491, ASTM 1747 | 1.431–1.433 | 1.4315–1.4320 | 1.4460–1.4475 | 1.4555–1.4565 |
| Water content, wt % | DIN 51777, ASTM E 203 | <0.50 | <0.100 | <0.20 | <0.10 |
| Acid number, mg of KOH/g | DIN 53402, ASTM D 1613 | <0.01 | <0.005 | <0.05 | <0.05 |

**Table 3.** Physical properties of mono-, di-, tri-, and tetraethylene glycols

| Property | Monoethylene glycol | Diethylene glycol | Triethylene glycol | Tetraethylene glycol |
|---|---|---|---|---|
| Molecular formula | HOCH$_2$CH$_2$OH | H(OCH$_2$CH$_2$)$_2$OH | H;(OCH$_2$CH$_2$)$_3$OH | H(OCH$_2$CH$_2$)$_4$OH |
| CAS registry number | [107-21-1] | [111-46-6] | [112-27-6] | [112-60-7] |
| $M_r$ | 62.7 | 106.12 | 150.17 | 194.23 |
| bp (at 101.3 kPa), °C | 197.6 | 244.8 | 287.4 | decomp. |
| fp, °C | −13 | −8 | −7 | −4.1 |
| Vapor pressure (20 °C), Pa | 5.3 | 2.7 | 0.5 | <1.3 |
| Density (20 °C), g/cm$^3$ | 1.1130 | 1.1160 | 1.1230 | 1.1247 |
| Refractive index, $n_D^{20}$ | 1.4318 | 1.4470 | 1.4560 | 1.4598 |
| Heat of combustion, MJ/kg | 19.07 | 22.32 | 23.68 | |
| Heat of vaporization (101.3 kPa), kJ/mol | 52.24 | 52.26 | 61.04 | 62.63 |
| Viscosity (20 °C) mPa · s | 19.83 | 36.0 | 49.0 | 61.9 |
| Surface tension, N/m$^2$ | 4.84 (20 °C) | 4.85 (20 °C) | 4.22 (25 °C) | |
| Flash point, °C | 119 | 141 | 177 | 191 |
| Ignition temperature, °C | 410 | 390 | 370 | |
| Lower explosive limit, vol% | 3.2 | 0.7 | 0.9 | |

**Table 4.** Physical properties of poly(ethylene glycols)

| Poly(ethylene glycol), mean $M_r$ | fp, °C | Refractive index, $n_D^{70}$ | Viscosity, mPa · s * |
|---|---|---|---|
| 200 | ca. −50 | 1.458 (25 °C) | 60–70 (100%) |
| 600 | 17–22 | 1.452 | 16–19 |
| 1000 | 35–40 | 1.453 | 24–29 |
| 2000 | 48–52 | 1.454 | 50–60 |
| 10 000 | 55–60 | 1.456 | 530–1000 |
| 20 000 | ca. 60 | 1.456 | 2700–3500 |
| 35 000 | ca. 60 | 1.456 | 11 000–14 000 |

* Values refer to a 50 wt% aqueous poly(ethylene glycol) solution except where otherwise stated.

**Esters.** Ethylene glycol esters are produced from reaction of ethylene oxide and an appropriate acid and are used also as solvents. Use of a suitable catalyst can influence the outcome of the reaction so as to produce monoesters (sodium acetate) or diesters (sulfuric acid). Glycol monoethers can be converted to the corresponding ether esters by the usual methods. Some important properties of ethylene glycol esters and ether esters are listed in Table 7.

**Trade Names.** Some common trade names of ethylene glycol ethers and esters are as follows:
*Monoethylene Glycol Derivatives.* Monomethyl ether: Methyl Cellosolve Solvent, Dowanol EM Glycol Ether, Methyl Oxitol Glycol. Monoethyl ether: Cellosolve Solvent, Dowanol EE Glycol Ether. Monobutyl ether: Butyl Cellosolve Solvent, Dowanol EB Glycol Ether, Butyl Oxitol Glycol. Dimethyl ether (also known as glyme): Dimethyl Cellosolve Solvent. Diethyl ether: Diethyl Cellosolve Solvent. Monomethyl ether acetate: Methyl Cellosolve Acetate. Monoethyl ether acetate: Cellosolve Acetate, Poly-solv EE Acetate.

# Ethylene Glycol

**Table 5.** Physical properties of some ethylene glycol monoalkyl ethers, RO(CH$_2$CH$_2$O)–H

| R | n | CAS registry number | $M_r$ | fp, °C | bp (101.3 kPa), °C | Density (20 °C), kg/m$^3$ | Refractive index, $n_D^{20}$ | Specific heat (20 °C), J g$^{-1}$ K$^{-1}$ | Vapor pressure (20 °C), Pa | Flash point, °C | Ignition temperature, °C | Explosive limits, vol% |
|---|---|---|---|---|---|---|---|---|---|---|---|---|
| CH$_3$ | 1 | [109-86-4] | 76.094 | −85.1 | 124.50 | 964.6 | 1.4024 | 2.24 | 930 | 38 | 285 | 2.4–20.6 |
| C$_2$H$_5$ | 1 | [110-80-5] | 90.120 | −100.0 | 135.10 | 929.7 | 1.4079 | 2.32 | 530 | 40 | 215 | 1.8–15.7 |
| C$_4$H$_9$ | 1 | [111-76-2] | 118.172 | −70.4 | 171.20 | 900.8 | 1.4194 | 2.44 | 80 | 66 | 225 | 1.1–10.6 |
| CH$_3$ | 2 | [111-77-3] | 120.146 | −50.0 | 194.10 | 1021.0 | 1.4265 | 2.15 | 24 | 90 | 215 | 1.6–16.1 |
| C$_2$H$_5$ | 2 | [111-90-0] | 134.172 | −54.0 | 201.90 | 989.0 | 1.4275 | 2.31 | 17 | 93 | 190 | 1.8–12.2 |
| C$_4$H$_9$ | 2 | [112-34-5] | 162.224 | −68.1 | 230.60 | 953.0 | 1.4316 | 2.29 | 4 | 99 | 195 | 0.7–5.9 |
| CH$_3$ | 3 | [112-35-6] | 164.198 | −38.2 | 249.00 | 1047.5 | 1.4381 | 2.18 | | 118 | | |
| C$_2$H$_5$ | 3 | [112-50-5] | 178.224 | −18.7 | 255.50 | 1020.0 | 1.4380 | 2.17 (15 °C) | 0.4 | 195 | | |

**Table 6.** Physical properties of some ethylene glycol dialkyl ethers RO(CH$_2$CH$_2$O)-R

| R | n | CAS registry number | $M_r$ | fp, °C | bp (101.3 kPa), °C | Density (20 °C), kg/m$^3$ | Refractive index, $n_D^{20}$ | Specific heat (20 °C), J g$^{-1}$ K$^{-1}$ | Vapor pressure (20 °C), Pa | Flash point, °C | Ignition temperature, °C | Lower explosive limit, vol% |
|---|---|---|---|---|---|---|---|---|---|---|---|---|
| CH$_3$ | 1 | [110-71-4] | 90.120 | −58.0 | 85.2 | 867.0 | 1.3796 | | 8100.0 | −6 | 200 | 1.6 |
| C$_2$H$_5$ | 1 | [629-14-1] | 118.172 | −74.0 | 121.4 | 841.0 | 1.3920 | | 1250.0 | 35 | 205 | |
| C$_4$H$_9$ | 1 | [112-48-1] | 174.276 | −69.1 | 203.6 | 836.0 | 1.4131 | 2.01 | 27.0 | 85 | | |
| CH$_3$ | 2 | [111-96-6] | 134.172 | −64.0 | 163.0 | 944.0 | 1.4078 | 2.04 | 267.0 | 51 | | 1.4 |
| C$_2$H$_5$ | 2 | [112-36-7] | 162.224 | −44.3 | 188.4 | 906.6 | 1.4115 | 2.10 (15 °C) | 51.0 | 82 | | |
| C$_4$H$_9$ | 2 | [112-73-2] | 218.328 | −60.2 | 254.6 | 883.8 | 1.4233 | 1.80 (13 °C) | 1.3 | 118 | | |
| CH$_3$ | 3 | [112-35-6] | 178.224 | −45.0 | 216.0 | 987.0 | 1.4233 | | 120.0 | 113 | 195 | 0.7 |

**Table 7.** Physical properties of some ethylene glycol acetates and ether acetates

| Compound | CAS registry number | $M_r$ | fp, °C | bp (101.3 kPa), °C | Density (20 °C), kg/m³ | Refractive index, $n_D^{20}$ | Ignition temperature, °C | Flash point, °C | Explosive limits, vol% |
|---|---|---|---|---|---|---|---|---|---|
| Ethylene glycol monoacetate | [542-59-6] | 104.10 | −80.0 | 182.0 | 1110 | 1.4209 | | 102 | 7.8−27.7 |
| Ethylene glycol diacetate | [111-55-7] | 146.10 | −31.0 | 190.5 | 1106 | 1.4159 | | 105 | |
| Ethylene glycol methyl ether acetate | [110-49-6] | 118.13 | −65.1 | 145.0 | 1006 | 1.4019 | 410 | 49 | 1.7−8.2 |
| Ethylene glycol ethyl ether acetate | [111-15-9] | 132.16 | −61.7 | 156.4 | 974 | 1.4058 | 390 | 54 | 3.18−10.1 |
| Ethylene glycol butyl ether acetate | [112-07-2] | 160.21 | −63.5 | 192.3 | 942 | 1.4138 | 355 | 74 | 1.66−8.37 |
| Diethylene glycol methyl ether acetate | [629-38-9] | 162.19 | −99.3 | 209.1 | 1041 | 1.4209 | | 110 | |
| Diethylene glycol ethyl ether acetate | [112-15-2] | 176.21 | −25.0 | 217.4 | 1011 | 1.4213 | | 112 | |
| Diethylene glycol butyl ether acetate | [124-17-4] | 204.26 | −32.2 | 246.7 | 981 | 1.4262 | 290 | 116 | |

*Diethylene Glycol Derivatives.* Monomethyl ether: Methyl Carbitol Solvent, Dowanol DM Glycol Ether. Monoethyl ether: Carbitol Solvent, Dowanol DE Glycol Ether. Monomethyl ether acetate: Methyl Carbitol Acetate. Monoethyl ether acetate: Carbitol Acetate. Monobutyl ether acetate: Butyl Carbitol Acetate.

Diethylene glycol dimethyl ether is also known as diglyme.

*Triethylene Glycol Derivatives.* Monomethyl ether: Dowanol TM Glycol Ether.

# 9. Uses

Ethylene glycol lowers the freezing point of water (see Fig. 1). Its ease of handling makes it a perfect antifreeze [51], which accounts for over 50% of its commercial uses.

Commercial antifreezes based on glycol also contain corrosion inhibitors and are used, for example, in motor vehicles, solar energy units, heat pumps, water heating systems, and industrial cooling systems. Protection against freezing is directly related to the glycol concentration; 60% glycol prevents freezing down to a temperature of − 55 °C. There is little point using higher concentrations because the freezing point then starts to increase (Fig. 1).

Ethylene glycol is also a commercially important raw material for the manufacture of polyester fibers, chiefly poly(ethylene terephthalate). This application consumes ca. 40% of the total ethylene glycol production. Polyesters are, however, used for other purposes, e.g., for producing recyclable bottles. Other minor uses of ethylene glycol are as a humectant (moisture-retaining agent), plasticizer, softener, hydraulic fluid, and solvent [52].

**Ethylene glycol derivatives,** mainly the ethers and esters, are frequently employed as reaction media, as absorption fluids, and as solvents for dyes, lacquers, and various cellulose products (cellulose esters, ethers, and nitrocellulose) [53].

**Diethylene glycol** serves as a solvent, softener (cork, adhesives, paper, etc.), dye additive (printing and stamping inks), deicing agent for runways and aircraft [54], and a drying agent for gases (e.g., natural gas) [55].

**Triethylene glycol** is used for the same purposes as diethylene glycol (e.g., as a solvent, plasticizer, and drying agent for gases) [56].

**Poly(ethylene glycols)** with varying molecular masses find numerous uses in the pharmaceutical industry (e.g., in ointments, liquids, and tabletting) and the cosmetic industry (e.g., creams, lotions, pastes, cosmetic sticks, and soaps). They are also used in the textile industry (e.g., cleaning and dyeing aids), in the rubber industry (e.g., lubricants and mold parting agents), and in ceramics (e.g., bonding agents and plasticizers) [57].

# 10. Economic Aspects

Ethylene glycol is one of the major products of the chemical industry. Its economic importance is founded on its two major commercial uses — as an antifreeze and for fiber production. Since ethylene glycol is currently produced exclusively from ethylene oxide (→ Ethylene Oxide), production plants are always located close to plants that produce ethylene oxide. The proportion of ethylene oxide that is converted to glycol depends on the local conditions, such as transport facilities and the market situation. Almost all of the ethylene oxide produced in recently installed plants in Saudi Arabia is, for example, converted to ethylene glycol because the domestic market for ethylene oxide is negligible and because ethylene glycol is easier to transport than ethylene oxide.

Detailed information about the glycol capacities of individual plants is difficult to obtain, but an estimated 60% of the total world production of ethylene oxide is converted to ethylene glycol. Therefore, the worldwide ethylene oxide production capacity of $8 \times 10^6$ t corresponds to an annual ethylene glycol production of $6.7 \times 10^6$ t.

About 50% of the ethylene glycol that is produced is used in antifreeze. Another 40% is channeled into the fiber industry. Consequently, the ethylene glycol demand is closely connected to the development of these two sectors. Surplus capacities and growing competition for market shares are expected on account of increasing glycol capacities — not only in Saudi Arabia, but also in the Eastern Bloc. Japan has already reduced its glycol capacities considerably [58]. In view of the increasing price of crude oil, alternative production methods based on synthesis gas — hitherto still in the experimental phase — are likely to become more important and increasingly competitive.

# 11. Toxicology and Occupational Health

## 11.1. Ethylene Glycol

**Oral Toxicity.** The oral $LD_{50}$ values of ethylene glycol are 14 [59] – 15 [60] g/kg for mice and ca. 8 g/kg [61] for rats. Oral toxicity in humans is higher, 1.4 g/kg can be lethal [59].

In rats, long-term administration of ethylene glycol in food at concentrations of 10.0 and 20.0 g/kg led to formation of bladder stones (calcium oxalate), damage to the renal tubules, and centrolobular fatty degeneration of the liver [62]. In other investigations, comparable changes were already apparent at a concentration of 5.0 g/kg, no changes were observed at 2.0 g/kg [63]. No changes were observed in the kidneys of monkeys fed a diet providing a maximal daily dose of 0.17 g/kg of body mass over a period of three years [64]. However, addition of ethylene glycol to the drinking water given to monkeys at a concentration corresponding to a daily dose of 0.24 g/kg of body mass led to the formation of oxalate crystals in the kidneys after about five months. Oxalate crystals were also formed in the brains of monkeys given very high doses (10 wt%) in drinking water [65].

**Eye and Skin Irritation.** Introduction of a single dose of ethylene glycol into the conjunctival sac does not affect the rabbit eye [66], but repeated administration leads to mild conjunc-tivitis [67]. Ethylene glycol vapor at a concentration of 17 mg/m$^3$ produces ocular damage in humans [68]. No effects were observed in the eyes of monkeys exposed to a vapor concentration of 265 mg/m$^3$ [68].

Ethylene glycol has not been shown to irritate mucous membranes. Repeated exposure of the skin to ethylene glycol can cause mild irritation, similar to that produced by glycerol.

**Inhalation.** The inhalative toxicity of an atmosphere saturated with ethylene glycol (0.5 mg/L) is low. Rats tolerated this concentration for four weeks without any permanent injuries, only mild narcosis was observed [69].

No adverse effects were detected in rats, rabbits, guinea pigs, dogs, and monkeys exposed to an ethylene glycol concentration of 57 mg/L for six weeks [70]. Monkeys survived after inhaling concentrations of 600 mg/m$^3$ administered as an aerosol over a period of several months.

Human volunteers exposed to maximal concentrations of 67 mg/m$^3$ for one month reported some irritation of the nasal mucous membranes and occasionally headache. The presence of ethylene glycol in the inhaled atmosphere was perceived at concentrations above 140 mg/m$^3$ [71]. A concentration of 200 mg/m$^3$ was found to be intolerable [72]. Rats, rabbits, dogs, and monkeys that inhaled pure oxygen containing 100 mL/m$^3$ of ethylene glycol at a pressure of 34.5 kPa for three weeks did not show any adverse effects apart from local irritation [73].

**Mutagenicity and Carcinogenicity.** Ethylene glycol gives a negative result in the Ames test [74], [75]. Many studies have been performed on its carcinogenicity [64], [76], [77]. In recent experiments, rats were fed with diets containing ethylene glycol concentrations corresponding to 1000, 200, and 40 mg/kg of body mass; no carcinogenic effects were observed. However, typical formation of oxalate crystals in the kidneys and other organs was found [78]. Similar experiments did not increase the incidence of tumor formation in mice [79].

**Reproduction.** In a study extending over three generations, ethylene glycol was added to the diet fed to rats at a concentration of 0.04, 0.2, and 1.0 g kg$^{-1}$ d$^{-1}$. The fertility and viability of the offspring were not affected by this treatment [80].

Teratogenic effects were not detected in rats receiving maximal feed concentrations of 1000 mg/kg, only a mild fetal toxicity was observed [81]. Teratogenic and fetotoxic effects were produced in mice and rats after administration of very high doses (1.25, 2.50, and 5.00 g/kg of body mass) in the form of a bolus with the aid of an esophageal probe [82].

**Metabolism.** In the body, ethylene glycol is degraded by alcohol dehydrogenase to form glycolaldehyde, which is further degraded by an aldehyde oxidase to produce glyoxal. Glyoxal is subsequently metabolized to glycolic and glyoxylic acids, from which oxalic acid, formic acid, glycine, 2-oxo-4-hydroxygluconate, 2-hydroxy-3-oxoadipate and oxomalate are formed [83, p. 3827].

## 11.2. Ethylene Glycol Derivatives

Toxicity data and exposure limits for ethylene glycols and some monoethylene glycol derivatives are listed in Table 8.

**Diethylene glycol** has a low acute oral toxicity (see Table 8). A LDLo value of 1.0 mL/kg is often erroneously cited but this value was determined using a preparation that contained sulfanilimide [85].

In long-term studies performed over a period of two years, dietary levels of 1, 2, and 4% diethylene glycol administered to rats produced calcium oxalate stones with asso-

**Table 8.** Toxicity data and exposure limits for ethylene glycols and ethylene glycol derivatives

| Compound | Single oral LD$_{50}$ in rats, g/kg | Reference | MAK, ppm | TLV-TWA, ppm |
|---|---|---|---|---|
| **Ethylene glycols** | | | | |
| Monoethylene glycol | 8 | [61] | | 50 |
| Diethylene glycol | 20.8 | [61] | | |
| Triethylene glycol | 22.1 | [61] | | |
| Tetraethylene glycol | 30.8 * | [83, p. 3843] | | |
| **Monoethylene glycol derivatives** | | | | |
| Monomethyl ether (methoxyethanol) | 3.4 | [84] | 5 | 5 |
| Monomethyl ether acetate (methoxyethyl acetate) | 3.93 | [61] | 5 | 5 |
| Monoethyl ether (ethoxyethanol) | 5.5 | [84] | 20 | 5 |
| Monoethyl ether acetate (ethoxyethyl acetate) | 5.1 | [61] | 20 | 5 |
| Monobutyl ether (butoxyethanol) | 2.5 | [84] | 20 | 25 |
| Monobutyl ether acetate (butoxyethyl acetate) | | | 20 | |

* mL/kg

ciated kidney and liver damage [86]. Chronic mechanical irritation of the bladder epithelium by the resulting bladder stones led to the formation of benign tumors. Production of calcium oxalate stones is attributed to contamination of the diethylene glycol with monoethylene glycol (see Section 11.1) [87], [88]. The low cumulative toxicity of diethylene glycol has been confirmed [89]–[91].

Diethylene glycol does not irritate the eyes or skin [92]; absorption by the skin is extremely low [93]. Diethylene glycol does not affect the fertility of rats [94]. It does not display mutagenicity in the Ames test [95]; other studies also suggest that the compound is not carcinogenic [87], [88].

**Triethylene glycol** has an extremely low oral toxicity when given either in a single dose (see Table 8) or in repeated doses [86], [96]. This compound causes only slight irritation of the skin and mucous membranes, it is not teratogenic [97].

**Tetraethylene glycol** is, like triethylene glycol, nontoxic (see Table 8).

**Poly(ethylene glycols)** have an extremely low acute oral toxicity (LD$_{50}$ in rats > 30 g/kg) which decreases as their molecular mass increases [98]. Studies on rats [98] and dogs [99] have shown that their cumulative toxicity is also very low.

Irritation of the skin and mucous membranes in humans and animals is slight. When applied to the skin, pure poly(ethylene glycols) do not cause sensitization [98], [100], [101] and are nontoxic [98]. Poly(ethylene glycols) with a molecular mass < 1000 are absorbed in the intestine and excreted in the urine and feces. Intestinal absorption decreases markedly as the molecular mass increases [102]–[104].

## 12. References

[1] A. Wurtz, *Ann. Chem. Pharm.* **55** (1859) no. 3, 406.
[2] G. O. Curme: *Glycols,* Rheinhold Publ. Co., New York 1952.
[3] J. Ariddick, B. Bunger: *Techniques of Chemistry,* 3rd ed., vol. **II,** Wiley-Interscience, New York 1970, p. 198.
[4] *Ullmann,* 4th ed., vol. **8,** 200–210.
[5] *Kirk-Othmer,* **9, 11,** 933–956.
[6] Merkblatt Lösemittel Hoechst, Hoechst AG, Frankfurt, 1984.
[7] D. Bohne, S. Fischer, E. Obermeier, *Ber. Bundesges. Phys. Chem.* **88** (1984) 739–742.
[8] A. Nath, E. Bender, *J. Chem. Eng. Data* **28** (1983) no. 4, 370–375.
[9] K. K. Sengupta, D. K. Khamrui, D. C. Mukherjee, *Indian J. Chem.* **13** (1975) 348–351. I. Bhatia, K. K. Banerji, *J. Chem. Soc. Perkin Trans. 2,* 1983 no. 10, 1577–1580.
[10] F. A. J. Meskens, *Synthesis* **7** (1981) 501–522.
[11] M. Ando, M. Ohhara, K. Takase, *Chem. Lett.* **6** (1986) 879–882.
[12] S. A. Miller: *Ethylene and its Industrial Derivatives,* Ernest Benn Limited, London 1969.
[13] P. Cardillo, A. Girelli, *Chim. Ind. (Milan)* **64** (1982) no. 12, 781–784.
[14] DuPont, US 2 285 448, 1942 (D. J. Loder).
[15] A. M. Brownstein, *Hydrocarbon Process.* **71** (1975) no. 9, 72–76.
[16] G. E. Weismantel, *Chem. Eng.* (N. Y.) 86 (1978) no. 2, 67–70.
[17] C. Matignon, H. Moureu, M. Dode, *Bull. Soc. Chim. Fr.* **5** (1934) no. 1, 1308–1317.
[18] J. H. Miller, T. E. Corrigan, *Hydrocarbon Process.* **46** (1967) no. 4, 176–178.
[19] J. Bronsted, M. Kilpatrick, *J. Am. Chem. Soc.* **51** (1929) 428.
[20] H. J. Lichtenstein, G. H. Twigg, *Trans. Faraday Soc.* **44** (1948) 905.
[21] G. Lamaty, R. Maleq, C. Selve, A. Sivade, J. Wylde, *J. Chem. Soc. Perkin Trans. 2,* 1975 no. 10, 1119–1124.
[22] D. F. Othmer, M. S. Thakar, *Ind. Eng. Chem.* **50** (1958) no. 9, 1235–1244.
[23] L. Klein, H. Fehrecke, *Chem. Ing. Tech.* **52** (1980) no. 10, 818–820.
[24] G. E. Mamilton, A. B. Metzner, *Ind. Eng. Chem.* **49** (1957) no. 5, 838–846.
[25] UCC, PCT Int. Appl., WO 8 504 393, 1985 (B. T. Keen). Nippon Shokubai Kagaku, BE 878 902, 1979 (T. Kumarawa, T. Yamamoto, H. Odanaka).
[26] UCC, PCT Int. Appl., WO 8 504 406, 1985 (J. Robert, J. H. Robson).
[27] Texaco, EP-A 0 123 709, 1983 (F. L. Jonson, L. W. Watts).
[28] Mitsubishi Petrochem. KK, JP 6 0061-545-A, 1983.
[29] Halcon, DE-OS 3 321 448, 1983 (V. S. Bhise).
[30] Halcon, DE-OS 2 757 067, 1977 (A. Peltzman).
[31] A. Mitsutani, *CEER Chem. Econ. Eng. Rev.* **5** (1973) no. 3, 32–35.
[32] E. J. Mistrik, A. Mateides, *Chem. Tech. (Leipzig)* **35** (1983) no. 2, 90–94.
[33] F. Mares, S. E. Diamond, F. J. Regina, J. P. Solar, *J. Am. Chem. Soc.* **107** (1985) no. 12, 3545–52.
[34] G. Lu, G. Huang, Y. Guo, R. Wang, Q. Bai, C. Li, H. Qi, *Shiyou Huagong* **14** (1985) no. 5, 262–266.*Chem. Abstr.* **104** (1986) 170513 c.
[35] B. D. Dombek, *Adv. Catal.* **32** (1983) 325–416.
[36] S. Nowak, I. Eichhorn, I. Ohme, B. Lücke, *Chem. Tech. (Leipzig)* **36** (1984) no. 2, 55–57.
[37] J. Kollar, *CHEMTECH* **8** (1984) 504–511.
[38] J. B. Saundby, *I. Chem. Symp.* (1986) 133–166.

[39]  S. Nowak, *Chem. Tech. (Leipzig)* **36** (1984) no. 4, 144–146.
[40]  Chevron, US 3 911 003, 1975 (S. Susuki).
[41]  A. Aquilo, J. S. Alder, D. N. Freeman, R. J. H. Voorhoeue, *Hydrocarbon Process.* **62** (1983) no. 3, 57–65.
[42]  J. Costa, J. Soley, J. Mata, J. Masides, *Effluent Water Treat. J.* **25** (1985) no. 12, 429–434.P. Schöberl, *Tenside Deterg.* **23** (1986) no. 5, 255–266.
[43]  R. A. Conway, G. T. Waggy, M. H. Spiegel, R. L. Berglund, *Environ. Sci. Technol.* **17** (1983) no. 2, 107–110.
[44]  G. A.Kolry, A. A. Abdelghani, A. C. Anderson, A. Monkiene, *Trace Subt. Environ. Health* **23** (1990) 371–378.
[45]  Merkblatt Glykole, Hoechst AG, Frankfurt, 1972.
[46]  E. Obermeier, S. Fischer, D. Bohne, *Ber. Bunsenges. Phys. Chem.* **89** (1985) 805–809.
[47]  Merkblatt Glykole, Hoechst AG, Frankfurt, 1988.
[48]  Hoechst AG, DE 2 640 506, 1978 (G. Schefel, R. Obermeier).
[49]  Merkblatt Polyglykole, Hoechst AG, Frankfurt, 1986.
[50]  Merkblatt Ethylenglykoldimethylether, Hoechst AG, Frankfurt, 1986.
[51]  Merkblatt Genantin, Hoechst AG, Frankfurt, 1982.
[52]  Merkblatt Glykole, Hoechst AG, Frankfurt, 1972.
[53]  Merkblätter Hoechst AG, Frankfurt, Lösemittel, 1984; Methylglykol, 1978; Methyldiglykol, 1970; Butyldiglykol, 1970; Ethylenglykoldimethylether, 1986.
[54]  Merkblatt Hoechst 1704 LTV (Flugzeugenteisung) 1985; Merkblatt Hoechst 1678 (Landebahnenenteisung) Hoechst AG, Frankfurt, 1985.
[55]  Merkblatt Diethylenglykol, Hoechst AG, Frankfurt, 1986.
[56]  Merkblatt Triethylenglykol, Hoechst AG, Frankfurt, 1984.
[57]  Merkblatt Polyglykole Hoechst, Hoechst AG, Frankfurt, 1986.
[58]  *Chem. Week,* **134** (1984) March 7, 32–35.
[59]  E. P. Laug, H. O. Calvery, H. J. Morris, G. Woodard, *J. Ind. Hyg. Toxicol,* **21** (1939) 173.
[60]  G. Bornmann, *Arzneim. Forsch.* **4** (1954) 643.
[61]  H. F. Smyth, Jr., J. Seaton, L. F. Fischer, *J. Ind. Hyg. Toxicol.* **23** (1941) 259.
[62]  H. J. Morris, A. A. Nelson, H. O. Calvery, *J. Pharmacol. Exp. Therap.* **74** (1942) 266.
[63]  F. R. Blood, *Food Cosmet. Toxicol.* **3** (1965) 229.
[64]  F. R. Blood, G. A. Elliott, M. S. Wright, *Toxicol. Appl. Pharmacol.* **4** (1962) 489.
[65]  J. A. Roberts, H. R. Seibold, *Toxicol. Appl. Pharmacol.* **15** (1969) 624.
[66]  C. P. Carpenter, H. F. Smyth, Jr., *Am. J. Ophthalmol.* **29** (1946) 1363.
[67]  T. O. McDonald, M. D. Roberts, A. R. Borgmann, *Toxicol. Appl. Pharmacol.* **21** (1972) 143.
[68]  W. M. Grant: *Toxicology of the Eye,* 2nd ed., Charles C. Thomas, Springfield, Ill., 1974.
[69]  E. Browning: *Toxicity and Metabolism of Industrial Solvents,* Elsevier, Amsterdam 1965.
[70]  R. A. Coon, R. A. Jones, L. J. Jenkins, Jr., J. Siegel, *Toxicol. Appl. Pharmacol.* **16** (1970) 646.
[71]  J. H. Wills, F. Coulston, E. S. Harris, E. W. McChesney, J. C. Russell, D. M. Serrone, *Clin. Toxicol.* **7** (1974) 463.
[72]  E. Harris: *Proceedings of the 5th Annual Conference on Atmospheric Contamination in Confined Spaces,* AMRL TR-69-130, paper no. 8, Aerospace Med. Res. Lab., Wright-Patterson Air Force Base, Ohio 1969, p. 9.
[73]  M. Felts: *Proceedings of the 5th Annual Conference on Atmospheric Contamination in Confined Spaces,* AMRL TR-69-130, paper no. 9, Aerospace Med. Res. Lab., Wright-Patterson Air Force Base, Ohio 1969, p. 105.
[74]  J. McCann, E. Choi, E. Yamasaka, B. N. Ames, *Proc. Natl. Acad. Sci. U.S.A.* **72** (1975) 5135.

[75] E. H. Pfeiffer, H. Dunkelberg, *Food Cosmet. Toxicol.* **18** (1980) 115.
[76] M. M. Mason, C. C. Cate, J. Baker, *Clin. Toxicol.* **4** (1971) 185.
[77] P. H. Derse, *U. S. C. F. S. T. I. PBRep.* **195 153** (1969).
[78] BRRC Project Report 44–109, 08.03.82.
[79] BRRC Project Report 46–89, 11.04.84.
[80] L. R. DePass, M. D. Woodside, R. R. Maronpot, C. S. Weil, *Fundam. Appl. Toxicol.* **7** (1986) 566.
[81] BRRC Project Report 43–54, 25.07.80.
[82] B. Ballantyne, A summary of studies conducted on the long-term toxicology of ethylene glycol, UCC, 4.22.1985.
[83] V. K. Rowe, M. A. Wolf inG. D. Clayton,F. E. Clayton (eds.): *Patty's Industrial Hygiene and Toxicology*, 3rd ed., vol. **2C**, Wiley-Interscience, New York 1963.
[84] C. P. Carpenter, U. C. Pozzan, C. S. Weil, J. H. Nair, III, G. A. Keck, H. F. Smyth, Jr., *AMA Arch. Ind. Health* **14** (1956) 114.
[85] H. O. Calvery, T. G. Klumpp, *Southern Med. J.* **32** (1939) 1105.
[86] O. G. Fitzhugh, A. A. Nelson, *J. Ind. Hyg. Toxicol.* **28** (1946) 40.
[87] C. S. Weil, C. P. Carpenter, H. F. Smyth, Jr., *Arch. Environ. Health* **11** (1965) 569.
[88] C. S. Weil, C. P. Carpenter, H. F. Smyth, Jr., *Ind. Med. Surg.* **36** (1967) 55.
[89] G. Bornmann, *Arzneim. Forsch.* **5** (1955) 38.
[90] A. Loesser, G. Bornmann, L. Grosskinsky, G. Hess, R. Kopf, K. Ritter, A. Schmitz, E. Stürner, H. Wegener, *Naunyn-Schmiedebergs Arch. Exp. Pathol. Pharmakol.* **221** (1954) 14.
[91] V. P. Plugin, *Gig. Sanit.* **33** (1968) 16.
[92] C. P. Carpenter, H. F. Smyth, Jr., *Am. J. Ophthalmol.* **29** (1946) 1363.
[93] P. J. Hanzlik, W. S. Lawrence, J. K. Fellows, F. P. Luduena, G. L. Lacqueur, *J. Ind. Hyg. Toxicol.* **29** (1947) 325.
[94] H. Wegener, *Naunyn-Schmiedebergs Arch. Exp. Pathol. Pharmakol.* **220** (1953) 414.
[95] E. H. Pfeiffer, H. Dunkelberg, *Food Cosmet. Toxicol.* **18** (1980) 115.
[96] W. M. Lauter, V. L. Vrla, *J. Am. Pharm. Assoc.* **29** (1940) 5.
[97] E. G. Stenger, L. Aeppli, E. Peheim, F. Roulet, *Arzneim. Forsch.* **18** (1968) 1536.
[98] H. F. Smyth, Jr., C. P. Carpenter, C. S. Weil, *J. Am. Pharm. Assoc. Sci. Ed.* **39** (1950) 349.
[99] H. F. Smyth, Jr., C. P. Carpenter, C. S. Weil, *J. Am. Pharm. Assoc. Sci. Ed.* **44** (1955) 27.
[100] The Dow Chemical Company, *Dow Polyethylene Glycols* (1959).
[101] C. P. Carpenter, M. D. Woodside, E. R. Kinkead, J. M. King, L. J. Sullivan, *Toxicol. Appl. Pharmacol.* **18** (1971) 35.
[102] C. B. Shaffer, F. H. Critchfield, *J. Am. Pharm. As-soc. Sci. Ed.* **36** (1947) 152.
[103] C. B. Shaffer, F. H. Critchfield, J. H. Nair, *J. Am. Pharm. Assoc. Sci. Ed.* **39** (1950) 340.
[104] A. H. Principe, *J. Forensic Sci.* **13** (1968) 90.

# Ethylene Oxide

*Individual keywords:* → *Epoxides.*

SIEGFRIED REBSDAT, Hoechst Aktiengesellschaft, Gendorf, Federal Republic of Germany (Chaps. 1 – 10)
DIETER MAYER, Hoechst Aktiengesellschaft, Frankfurt, Federal Republic of Germany (Chap. 11)

| | | | | |
|---|---|---|---|---|
| 1. | Introduction . . . . . . . . . . . . . 2329 | 6. | Quality Specifications . . . . . . . . 2344 |
| 2. | Physical Properties . . . . . . . . 2330 | 7. | Analysis . . . . . . . . . . . . . . . . 2344 |
| 3. | Chemical Properties . . . . . . . . 2332 | 8. | Handling, Storage, and Transportation . . . . . . . . . . . . 2345 |
| 4. | Production . . . . . . . . . . . . . . 2335 | | |
| 4.1. | Catalysts . . . . . . . . . . . . . . . . 2336 | 9. | Uses . . . . . . . . . . . . . . . . . . . 2349 |
| 4.2. | Mechanism of Catalysis . . . . . 2337 | 10. | Economic Aspects . . . . . . . . . 2350 |
| 4.3. | Technology . . . . . . . . . . . . . . 2339 | 11. | Toxicology and Occupational Health . . . . . . . . . . . . . . . . . . 2350 |
| 5. | Environmental Protection and Ecology . . . . . . . . . . . . . . . . 2343 | | |
| | | 12. | References . . . . . . . . . . . . . . 2354 |

## 1. Introduction

Ethylene oxide (oxirane) [75-21-8], $M_r$ 44.05, is the simplest cyclic ether. It is a colorless gas or liquid and has a sweet, etheric odor. The ethylene oxide molecule with its short C – C bond and strained angles is shown in Figure 1 [6].

Ethylene oxide is very reactive because its highly strained ring can be opened easily, and is thus one of the most versatile chemical intermediates. Ethylene oxide was first described in 1859 by WURTZ [7], who prepared it by eliminating hydrochloric acid from ethylenechlorohydrin, using potassium hydroxide solution. Industrial production using the chlorohydrin process began in 1914 and was based on WURTZ's discovery. Since then the production and importance of ethylene oxide have steadily grown.

In 1931, LEFORT [8] discovered the direct catalytic oxidation of ethylene [74-85-1], which gradually superseded the chlorohydrin process. Currently, ethylene oxide is produced by direct oxidation of ethylene with air or oxygen; annual worldwide production capacity is ca. $8\times10^6$ t, making it an important industrial chemical. Ethylene oxide itself is used as a disinfectant, sterilizing agent, and fumigant. Its most important derivative is ethylene glycol [107-21-1], which is used in antifreeze (car radiators) and

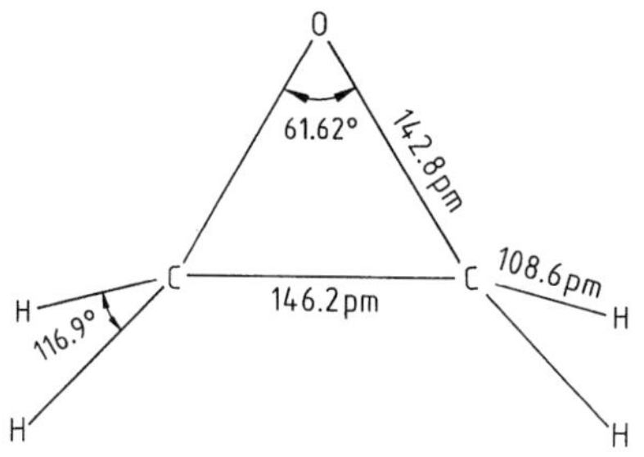

**Figure 1.** The ethylene oxide molecule

for the manufacture of polyester fibers. Other ethylene oxide derivatives, amines and poly(ethylene glycols), are used in surfactants, solvents, etc.

## 2. Physical Properties

Important physical properties of ethylene oxide are summarized in Table 1.

In the pressure range 0–101.3 kPa and at 40 and 60 °C, the solubility of ethylene oxide in water obeys Henry's law. The Henry constants for these temperatures are 2.875 and 1.448, respectively. However, at 20 °C a more than proportional increase in solubility occurs with the partial pressure [12].

Table 2 shows some of the properties of aqueous ethylene oxide solutions. Of particular note are the relatively high melting points, which are due to clathrate formation [14]. Clathrates consist of organic molecules that are enclosed in a cage structure. In this case, the cage is an ice lattice which is composed of unit cells, each containing 46 water molecules, and two types of cavities: six larger ones (14-sided, tetradecahedra) and two smaller ones (12-sided, dodecahedra). If the clathrates are made to crystallize from increasingly concentrated aqueous solutions of ethylene oxide, their density and ethylene oxide content also increase. Ethylene oxide first fills the six tetradecahedral cavities of the water lattice, followed by 20–40% of the dodecahedral cavities [15], [16]. The highest melting point observed is 11.1 °C and corresponds to a composition of $C_2H_4O \cdot 6.89\ H_2O$ [15].

The solubilities of gases in ethylene oxide vary, increasing in the order nitrogen, argon, methane, ethane. Earlier data [17] have been revised [18]. The Henry constants for these gases in ethylene oxide at different temperatures are given in Table 3.

Table 4 shows other temperature-dependent physical properties of gaseous and liquid ethylene oxide.

**Table 1.** Physical properties of ethylene oxide

| | |
|---|---|
| mp at 101.3 kPa | −112.5 °C |
| bp at 101.3 kPa | 10.8 °C |
| Critical temperature | 195.8 °C |
| Critical pressure | 7.2 MPa |
| Critical density | 314 kg/m$^3$ |
| Refractive index, $n_D^7$ | 1.3597 |
| Explosive limits in air at 101.3 kPa | |
| lower | 2.6 vol% |
| upper | 100.0 vol% |
| Electrical conductivity | 2.1 × 10$^{-8}$ S m$^{-1}$ |
| Dielectric constant | |
| at −1 °C (liquid) | 13.9 |
| at 15 °C (vapor) | 1.01 |
| Heat of combustion at 25 °C, 101.3 kPa | 29.648 kJ/kg |
| Entropy of the vapor at 101.3 kPa | |
| 10.5 °C | 5.439 kJ kg$^{-1}$ K$^{-1}$ |
| 25.0 °C | 5.495 kJ kg$^{-1}$ K$^{-1}$ |
| Ignition temperature in air at 101.3 kPa | 429 °C |
| Decomposition temperature of the vapor at 101.3 kPa | 571 °C |
| Heat of polymerization | 2091 kJ/kg |
| Heat of fusion | 117.86 kJ/kg |
| Heat of decomposition of the vapor | 1901 kJ/kg |
| Coefficient of cubic expansion | |
| at 22 °C | 0.00161 |
| at 55 °C | 0.00170 |
| Heat of solution in water at 25 °C | 142.57 kJ/kg |

**Table 2.** Physical properties of aqueous ethylene oxide solutions

| Ethylene oxide content, wt% | mp, °C | bp, °C | Density at 10 °C, g/L | Flash point, °C |
|---|---|---|---|---|
| 0 | 0.0 | 100 | 0.9991 | |
| 0.5 | | | | 41.5 |
| 1 | −0.4 | | | 31 |
| 2 | | | | 3 |
| 3 | −1.3 | | | |
| 5 | −1.6 | 58 | 0.9988 | −2 |
| 10 | 5.6 | 42.5 | 0.9980 | |
| 20 | 10.4 | 32 | 0.9945 | −21 |
| 30 | 11.1 (max.) | 27 | 0.9882 | −28 |
| 40 | 10.4 | 21 | 0.9792 | −35 |
| 60 | 7.8 | 16 | 0.9534 | −45 |
| 80 | 3.7 | 13 | 0.9194 | −53 |
| 100 | −112.5 | 10.4 | 0.8826 | −57 |

**Table 3.** Solubility of gases in ethylene oxide (Henry constants in MPa)

| Temperature, °C | Solubility in ethylene oxide | | | |
| --- | --- | --- | --- | --- |
| | Nitrogen | Argon | Methane | Ethane |
| 0 | 284 | 169 | 62.1 | 8.5 |
| 25 | 221 | 144 | 62.2 | 11.0 |
| 50 | 184 | 129 | 62.3 | 13.0 |

**Table 4.** Physical properties of ethylene oxide at various temperatures

| Temperature | Density | | Heat of Vaporization | Vapor pressure | Specific heat | | Thermal conductivity | | Surface tension | Dynamic viscosity | |
| --- | --- | --- | --- | --- | --- | --- | --- | --- | --- | --- | --- |
| | Vapor | Liquid | | | Vapor | Liquid | Vapor | Liquid | | Vapor | Liquid |
| °C | g/L | kg/m$^3$ | kJ/kg | kPa | kJ kg$^{-1}$ K$^{-1}$ | | $10^{-4}$ J cm$^{-1}$ s$^{-1}$ K$^{-1}$ | | mN/m | mPa · s | |
| −40 | 0.2 | 946 | 647 | 7.6 | | 1.88 | | 17.7 | 33.8 | | 0.50 |
| −30 | 0.3 | 931 | 634 | 16.2 | | | | 17.2 | 32.3 | | 0.44 |
| −20 | 0.5 | 919 | 621 | 24.6 | | 1.91 | | 16.8 | 30.7 | | 0.39 |
| −10 | 0.9 | 909 | 607 | 40.3 | | | | 16.4 | 29.1 | | 0.34 |
| 0 | 1.3 | 899 | 592 | 65.4 | 1.00 | 1.95 | 1.03 | 16.0 | 27.6 | 0.0090 | 0.31 |
| +10 | 1.8 | 888 | 578 | 99.2 | 1.06 | | 1.10 | 15.6 | 26.7 | | 0.28 |
| +20 | 2.9 | 876 | 563 | 145.6 | 1.10 | 2.00 | 1.18 | 15.3 | 24.5 | | 0.25 |
| +30 | 3.9 | 862 | 547 | 208.0 | 1.15 | | 1.25 | 15.0 | | | 0.23 |
| +40 | 5.25 | 847 | 531 | 287.1 | 1.19 | 2.06 | 1.34 | 14.7 | 22.0 | | 0.21 |
| +50 | 6.85 | 831 | 515 | 391.0 | 1.23 | | 1.43 | 14.3 | | 0.0104 | 0.19 |
| +60 | | 814 | 498 | 534.6 | 1.28 | 2.15 | 1.52 | 14.0 | 17.5 | | 0.18 |
| +70 | | 796 | 481 | 694.6 | 1.32 | | 1.62 | 13.7 | | | 0.16 |
| +80 | | 779 | 463 | 912.8 | 1.36 | 2.27 | 1.72 | 13.4 | 14.5 | | 0.15 |
| +100 | | 745 | 424 | 1464.0 | 1.45 | 2.40 | 1.93 | 12.5 | 12.0 | 0.0118 | 0.14 |

# 3. Chemical Properties

Ethylene oxide is a very reactive, versatile compound. Its reactions proceed mainly via ring opening and are highly exothermic. Explosive decomposition of the ethylene oxide vapor may occur at higher temperatures if heat dissipation is inadequate. Only the most important types of the large number of possible reactions are briefly discussed here. More detailed information can be found in [20]–[24].

**Decomposition.** Gaseous ethylene oxide starts to decompose at ca. 400 °C to form mainly CO, $CH_4$, as well as $C_2H_6$, $C_2H_4$, $H_2$, C, and $CH_3CHO$. The first step in the decomposition is presumed to be the isomerization of ethylene oxide to acetaldehyde [25]. Once the decomposition reaction has been initiated (ignition source), it can be propagated through the gas phase and, under certain conditions, may be explosive (see Chap. 8).

**Addition to Compounds with a Labile Hydrogen Atom.** Ethylene oxide reacts with compounds containing a labile hydrogen atom to form a product containing a hydroxyethyl group:

$$XH + H_2C-CH_2 \longrightarrow XCH_2CH_2OH$$
$$\phantom{XH +\ }\underset{O}{\diagdown\!\diagup}$$

Examples of XH are: HOH, H$_2$NH, HRNH, R$_2$NH, RCOOH, RCONH$_2$, HSH, RSH, ROH, N≡CH, and B$_2$H$_6$ (R = alkyl, aryl). The reaction is accelerated by acids and bases. During acid catalysis, the ethylene oxide is first protonated to form an epoxonium ion (**1**), which is in equilibrium with the corresponding hydroxycarbenium ion (**2**). The anion X$^-$ can then react with **1** or **2** in an S$_N$2 or S$_N$1 reaction, respectively. In an alkaline medium, the S$_N$2 mechanism is favored. A detailed discussion of reaction mechanisms can be found in [22] and [26].

$$\underset{(1)}{X^- \quad H_2\overset{\diagdown}{C}\!-\!CH_2 \atop \phantom{X^-\ }\underset{H}{\overset{+}{O}}} \quad \rightleftharpoons \quad \underset{(2)}{X^- \quad H_2\overset{+}{C}\!-\!CH_2 \atop \phantom{X^-\ }\phantom{H_2C-}OH}$$

All common acids and Lewis acids as well as zeolites, ion exchangers [27], and aluminum oxide are effective catalysts. Solid polymeric acids (cation exchangers) are especially recommended for the reaction of ethylene oxide with methanol [28].

As the endproduct of the above reaction contains at least one hydroxyl group, it may react successively with further ethylene oxide molecules:

$$XCH_2CH_2OH + n\ H_2C-CH_2 \longrightarrow X(CH_2CH_2O)_{n+1}H$$
$$\phantom{XCH_2CH_2OH + n\ }\underset{O}{\diagdown\!\diagup}$$

The molecular mass of the resulting polymers depends on the ratio of the reactants, the catalyst used and the reaction conditions [1], [23].

A large variety of reactions occur between ethylene oxide and compounds containing labile hydrogen atoms, therefore, such compounds are often used to produce derivatives.

Commercially, the most important of this type of reaction is the hydrolysis of ethylene oxide to ethylene glycol. About 60 % of the total ethylene oxide production is converted into ethylene glycol in this way (→ Ethylene Glycol). However, this reaction, which is often referred to as ethoxylation, is also used to produce the bulk of all the other commercially important ethylene oxide derivatives. The ethoxylation products of alkyl phenols, ammonia, fatty alcohols, fatty amines, and fatty acids have a variety of uses [23]. Although polymer formation according to the above equation is more likely to be an unwanted side reaction, the production of poly-(ethylene glycols) of widely varying molecular mass by this route is of some importance [29]. Undesired polymerization may be catalyzed by rust in rusty containers [30].

**Addition to Double Bonds.** Ethylene oxide can add to compounds with double bonds, e.g., carbon dioxide [31], to form cyclic products:

$$O=C=O + H_2C\text{-}CH_2\underset{O}{\diagdown\diagup} \xrightarrow{\text{Cat.}} \underset{\underset{O}{\overset{\|}{C}}}{H_2C\underset{O}{\diagdown}\phantom{x}\underset{O}{\diagup}CH_2}$$

Ethylene carbonate<br>(1,3-dioxolan-2-one)

Quaternary ammonium compounds are suitable catalysts for this reaction, which then proceeds at 200 °C and 8 MPa [32]. Use of porphine complexes of aluminum in solvents such as chloroform allows the reaction to take place at normal pressure and at room temperature [33]. Hydrolysis of ethylene carbonate (1,3-dioxolan-2-one) [96-49-1] yields pure ethylene glycol [34]. Therefore, the reaction is of interest for the selective production of ethylene glycol from ethylene oxide, i.e., for avoiding the partly unwanted formation of poly(ethylene glycols) [35].

Ethylene oxide also adds to other double bond systems, e.g., to $R_2C=O$ [36], SC=S, $O_2S=O$, RN=CO, and OS=O.

**Catalytic Isomerization to Acetaldehyde.** Aluminum oxide ($Al_2O_3$) [37], phosphoric acid and phosphates [38], and under certain conditions silver [39] catalyze the isomerization of ethylene oxide to acetaldehyde.

$$H_2C\text{-}CH_2\underset{O}{\diagdown\diagup} \xrightarrow[\text{170–300 °C}]{\text{Cat.}} CH_3CHO$$

**Reduction to Ethanol.** Reduction of ethylene oxide to ethanol is catalyzed by Ni, Cu, and Cr on $Al_2O_3$ [40].

$$H_2C\text{-}CH_2\underset{O}{\diagdown\diagup} + H_2 \xrightarrow[\text{2–10 MPa, 80–100 °C}]{\text{Cat.}} CH_3CH_2OH$$

**Reaction with Grignard Reagents.** Grignard reagents react with ethylene oxide to produce compounds with a primary hydroxyl group [24], [41].

$$H_2C\text{-}CH_2\underset{O}{\diagdown\diagup} + RMgX + H_2O \longrightarrow RCH_2CH_2OH + MgOHX$$

**Oligomerization to Crown Ethers.** Ethylene oxide oligomerizes to form cyclic polyethers (crown ethers) in the presence of a fluorinated Lewis acid catalyst (→ Crown Ethers).

$$(n+3)\,H_2C\text{-}CH_2\underset{O}{\diagdown\diagup} \xrightarrow[\text{25 °C, 101.3 kPa}]{\text{Cat.}} \text{[cyclic polyether]}_n$$

n = 1–3

Synthesis is successful when groups that saturate potential terminal hydroxyl groups are absent [42]. Suitable catalysts are the $BF_4^-$, $PF_6^-$, or $SbF_6^-$ salts of metal cations. The reaction can be directed to produce a particular oligomer by choosing an appropriate metal cation. The size of the cation determines the size of the crown ether ring by acting as a template [43]. For example, with CsBF$_4$, only the cyclic hexamer ($n = 3$), 1,4,7,10,13,16-hexaoxacyclooctadecane (18-crown-6) [17455-13-9], is produced; if Cu(BF$_4$)$_2$ is used, 90% of the product consists of the pentamer ($n = 2$), 1,4,7,10,13-pentaoxacyclopentadecane (15-crown-6) [33100-27-5], and with Ca(BF$_4$)$_2$, a mixture of the tetramer (50%, $n = 1$), 1,4,7,10-tetraoxacyclodecane (12-crown-4) [294-93-9], and the pentamer, 1,4,7,10,13-pentaoxacyclopentadecane (18-crown-6) [17455-13-9], is produced.

**Reaction with Dimethyl Ether.** Ethylene oxide reacts with dimethyl ether to produce poly-(ethylene glycol) dimethyl ethers (→ Dimethyl Ether); this reaction is used for the industrial production of the lower molecular mass homologues, which are widely used as solvents [44].

$$CH_3OCH_3 + n\ H_2C\overset{O}{-}CH_2 \xrightarrow{BF_3} CH_3O(CH_2CH_2O)_nCH_3$$
$$n = 1-4$$

**Reaction with Bromotrimethylsilane.** Ethylene oxide adds to bromotrimethylsilane [2857-97-8] in a highly exothermic reaction with excellent yields [45].

$$(H_3C)_3SiBr + H_2C\overset{O}{-}CH_2 \longrightarrow (H_3C)_3Si-O-CH_2CH_2Br$$

# 4. Production

As mentioned in the introduction, ethylene oxide was produced formerly by the chlorohydrin process (→ Chlorohydrins). However, this method is no longer used on an industrial scale; a description is given in [46]. Although the selectivity of this process (80 %) was satisfactory, practically all of the chlorine that was used was lost as calcium chloride and unwanted chlorine-containing byproducts were generated. This not only was inefficient, but also caused pollution problems so that this method has now been replaced by the direct oxidation process.

WURTZ, who had discovered ethylene oxide, attempted as early as 1863 to produce it by direct oxidation of ethylene with oxygen, but did not succeed [47]. Many other unsuccessful attempts were made [48] before LEFORT made the crucial discovery in 1931 that the formation of ethylene oxide from ethylene and oxygen was catalyzed by metallic silver [8]. Since the catalyst plays a central role in the production process, it is discussed first.

## 4.1. Catalysts

To date no other metal has been found that can compete with silver in the catalysis of the direct oxidation of ethylene to ethylene oxide. However, the silver catalysts have been substantially improved since their discovery by LEFORT [8]. Only supported catalysts are used, the silver being deposited on a porous *support material* in concentrations of 7–20%. The support material is of critical importance [49]; currently, preference is given to ultrapure (over 99%) aluminum oxide [57828-03-2] fired at a high temperature that has a defined pore structure (pore diameter 0.5–50 µm) and low specific surface area ($<2$ m$^2$/g). Although supports with higher specific surface areas are very active, their selectivity is low, presumably because ethylene oxide can diffuse only slowly out of the smaller pores and, therefore, can be further oxidized [50]. The impregnation methods used to deposit the silver on the support are being constantly improved. Complexes of silver salts with amino compounds are used which decompose to give evenly and finely distributed silver particles with a diameter of 0.1–1 µm [51]. Furthermore, 100–500 mg/kg of *promoters* such as salts or other compounds of alkali- and alkaline-earth metals are added to the catalyst, significantly improving the selectivity. Among the alkali-metal salts, cesium has been reported to be especially effective [52], while barium stands out among the alkaline-earth metals [53]. Finally, considerable progress was made after the discovery that addition of chlorine compounds to the reaction gases improved selectivity [54]. These so-called *inhibitors* (e.g., 1,2-dichloroethane, vinyl chloride [55]) suppress the combustion of ethylene to carbon dioxide and water; they ensure that the silver surface is covered evenly with a supply of chlorine [56].

The advances made in silver-based catalysts since LEFORT's original discovery (i.e., optimized support materials, silver distribution, and use of promoters and inhibitors) have improved selectivity from 50 to 80%. Silver has maintained its position as the only known metal that can catalyze the oxidation of ethylene to ethylene oxide with a commercially viable selectivity. Conversely, ethylene is the only olefin that can be oxidized to an epoxide by silver with a commercially viable selectivity.

**Aging of the Catalyst.** Modern silver-based catalysts have an initial selectivity of 79–81%; a maximum selectivity of 83% is commercially feasible [35]. The disadvantages associated with highly selective catalysts are that they age relatively quickly and less heat is produced. As the catalyst is used, its selectivity and activity gradually deteriorate [57]–[59] due to

1) abrasion, dust formation, and blocking of pores
2) accumulation of detrimental impurities introduced with the reaction gases (e.g., sulfur from ethylene or methane)
3) changes in the silver particles, which enlarge, form agglomerates, and become unevenly distributed.

Generally applicable methods of regenerating such catalysts are unknown; the only method of any importance is the use of methanolic solutions of cesium salts [60]. If regeneration is impossible, it becomes economically justifiable to replace the catalyst when selectivity has fallen to a critical level. The lifetime of a modern catalyst is 2–5 years, depending on the rate of ethylene oxide production and the purity of the reaction gases.

## 4.2. Mechanism of Catalysis

Two reactions take place simultaneously at the silver surface. In addition to ethylene oxide formation (partial oxidation, Eq. 1), complete combustion (total oxidation, Eq. 2) to $CO_2$ and water also takes place. Small amounts of acetaldehyde and formaldehyde are also formed [61]. The reactions given by Equations 1 and 2 are exothermic, their enthalpies being −106.7 and −1323 kJ/mol, respectively at 250 °C and 1.5 MPa [35].

$$H_2C=CH_2 + \tfrac{1}{2} O_2 \longrightarrow H_2C\underset{O}{-}CH_2 \quad (1)$$

$$H_2C=CH_2 + 3 O_2 \longrightarrow 2 CO_2 + 2 H_2O \quad (2)$$

Although many attempts have been made to discover the mechanism responsible for silver's unique action, opinions remain divided [62]. Evaluation of the various studies is difficult because often they have been carried out under different conditions that were not always suitable for industrial purposes. It is agreed that silver can adsorb oxygen in a number of ways and that this phenomenon is the basis of silver's unequalled efficiency in catalyzing the oxidation of ethylene to ethylene oxide. The following adsorption forms of oxygen are of critical importance:

1) atomic oxygen
2) molecular oxygen
3) subsurface oxygen, i.e., dissolved oxygen found below the surface.

Opinions differ as to the role of these different types of adsorbed oxygen; two conflicting mechanisms have been proposed:

**Mechanism I.** Only molecular oxygen reacts with ethylene to form ethylene oxide, whereas atomic oxygen only reacts to form carbon dioxide and water [63]. Chlorine (from the inhibitor) blocks the adsorption of atomic oxygen to the silver surface so that, in the ideal case, the optimally inhibited silver surface only adsorbs molecular oxygen (Eq. 3). The adsorbed molecular oxygen then reacts with ethylene to form ethylene oxide, leaving behind one oxygen atom after desorption (Eq. 4). This atomic oxygen then causes the combustion of ethylene to $CO_2$ and $H_2O$ (Eq. 5).

$$6\ Ag + 6\ O_2 \longrightarrow 6\ Ag \cdot O_{2\ ads} \tag{3}$$

$$6\ Ag \cdot O_{2\ ads} + 6\ H_2C{=}CH_2 \longrightarrow 6\ H_2C\underset{O}{-}CH_2 + 6\ Ag \cdot O_{ads} \tag{4}$$

$$6\ Ag \cdot O_{ads} + H_2C{=}CH_2 \longrightarrow 6\ Ag + 2\ CO_2 + 2\ H_2O \tag{5}$$

Since six oxygen atoms are needed for the complete oxidation of one ethylene molecule, six ethylene oxide molecules must be formed before one ethylene molecule can be completely oxidized. Therefore, if inhibition of atomic oxygen adsorption is optimal, the maximum selectivity is 6/7×100 = 85.7%. This mechanism plausibly explains the way in which the inhibitor acts. It is further supported by the fact that a selectivity of 85.7% has never been exceeded in practice, although other mechanisms predict higher maximum selectivities.

A variant of this mechanism is based on the assumption that molecular oxygen can cause both ethylene oxide formation and total oxidation via a common intermediate [64], [65]:

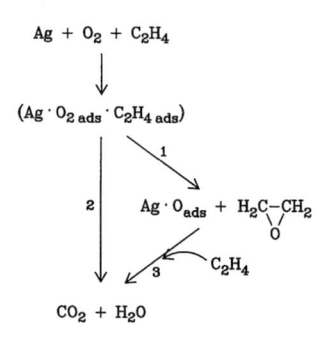

$$Ag + O_2 + C_2H_4$$
$$\downarrow$$
$$(Ag \cdot O_{2\ ads} \cdot C_2H_{4\ ads})$$

with pathway 1 leading to $Ag \cdot O_{ads} + H_2C{-}CH_2$ (with O bridging), pathway 2 leading to $CO_2 + H_2O$, and pathway 3 involving $C_2H_4$.

The effect of the inhibitor is interpreted differently here. Reaction 2 requires a greater space than reaction 1. The inhibitor favors reaction 1 by reducing the available space on the surface of the catalyst. On an optimally inhibited surface, reaction 2 is totally suppressed, so that total oxidation takes place only via reaction 3. This results again in a maximum selectivity of 85.7%.

**Mechanism 2.** In this mechanism, atomic oxygen, possibly together with the subsurface oxygen, is thought to be responsible for both total and partial oxidation. The contribution of the molecular oxygen, if any, is indirect [66], [67]. The environment of the adsorbed oxygen atom determines whether the reaction with ethylene leads to ethylene oxide or $CO_2$ and $H_2O$ [68], [69]. Chlorine (inhibitor), promotors, and subsurface oxygen in the proximity of the reacting oxygen atom are thought to influence the latter to favor partial oxidation, e.g., by reducing the negative charge of the adsorbed oxygen [70]. Unlike mechanism 1, the theoretical upper limit of selectivity of 6/7 does not apply for mechanism 2.

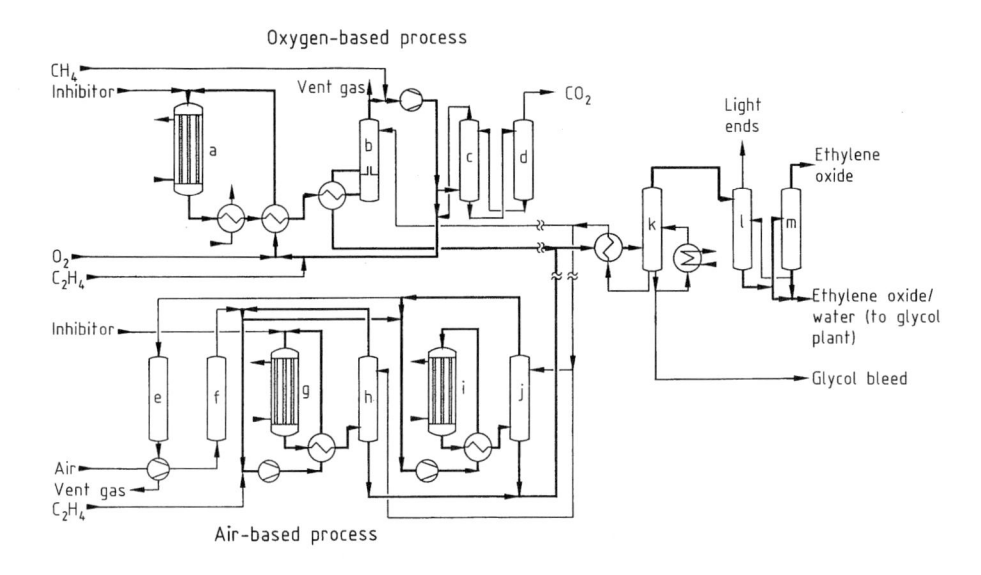

**Figure 2.** Flow scheme for ethylene oxide production by the oxygen-based or air-based oxidation of ethylene
a) Reactor; b) Ethylene oxide scrubber; c) $CO_2$ scrubber; d) $CO_2$ desorber; e) Off-gas purification; f) Air purification; g) Primary reactor; h) Primary ethylene oxide scrubber; i) Secondary reactor; j) Secondary ethylene oxide scrubber; k) Ethylene oxide desorber; l) Stripping column; m) Ethylene oxide distillation

Intermediates of the total oxidation are thought to include acetaldehyde, acetic acid, and oxalic acid [68]. Desorption of intermediates from the silver surface produces acetaldehyde when a helium stream is used; ethanol and acetic acid are produced when a hydrogen stream is used [71].

Although mechanism 1 was long favored, mechanism 2 has recently gained much support. It is currently impossible to say which of the two mechanisms is actually correct.

## 4.3. Technology

An overview of the beginnings of commercial ethylene oxide production can be found in [1]. Production technologies for ethylene oxide plants based on the direct oxidation process are licensed by Shell, Scientific Design (SD), UCC, Japan Catalytic, Snam Progetti, and Hüls. Due to improved catalysts and production technology, large plants with capacities of up to 250 000 t/a can now be built.

The technologies are very similar, but differences exist, depending on whether air or pure oxygen is used for oxidation [35], [72]. Shell plants use only pure oxygen, while Scientific Design and UCC have developed air-based oxidation plants. Figure 2 shows a simplified scheme for both the air- and oxygen-based processes. They both employ a recycle gas stream, which is continuously circulated through the reactors by compres-

**Table 5.** Operating parameters used in the air-based and oxygen-based production of ethylene oxide

| Parameter | Air-based process | Oxygen-based process |
|---|---|---|
| $C_2H_4$ concentration, vol% | 2–10 | 15–40 |
| $O_2$ concentration, vol% | 4–8 | 5–9 |
| $CO_2$ concentration, vol% | 5–10 | 5–15 |
| $C_2H_6$ concentration, vol% | 0–1 | 0–2 |
| Ar concentration, vol% | | 5–15 |
| $CH_4$ concentration, vol% | | 1–60 |
| Temperature, °C | 220–277 | 220–275 |
| Pressure, MPa | 1–3 | 1–2.2 |
| GHSV *, 1/h | 2000–4500 | 2000–4000 |
| Pressure drop, kPa, | 40–200 | |
| $C_2H_4$ conversion, % | 20–65 | 7–15 |
| Selectivity, % | 75 | 80 |

* GHSV = gas hourly space velocity

sors. The reactors consist of large bundles of several thousand tubes that are 6–12 m long and have an internal diameter of 20–50 mm. The catalyst is packed in the tubes in the form of spheres or rings with a diameter of 3–10 mm; the initial selectivity of modern catalysts is ca. 80%. Ethylene is reacted at 200–300 °C and 1–3 MPa to produce ethylene oxide, $CO_2$, $H_2O$, and heat, as well as traces of acetaldehyde and formaldehyde; these products must be removed or separated from the recycle gas stream. The recycle gas is then reloaded with oxygen and ethylene and returned to the reactor.

**Oxygen-Based Oxidation Process.** At present, ethylene oxide is produced mainly by the oxygen-based process. The reactor tubes filled with the catalyst are surrounded by a coolant (water or a high-boiling hydrocarbon) that removes the reaction heat and permits temperature control. Heat is extracted either by pumping or by evaporating the coolant. If organic heat transfer media are used, the extracted energy is used to generate steam in secondary cycles, which is then employed for heating. The reaction heat also heats the recycle gas during its passage through the reactor. After leaving the reactor, the gas is likewise cooled by means of steam generation and/or used directly to heat the reactor inlet gases. Because of the different enthalpies of the partial and total oxidation reactions (see Section 4.1), the total available heat depends on the selectivity for the former (S) and amounts to 47 250 – (434×S) kJ/kg ethylene.

Consequently, the quantity of energy released decreases rapidly with increasing selectivity. Therefore, the heat removal system must be sufficiently adaptable.

After the gas from the reactor has been cooled, the ethylene oxide (1–2%) and $CO_2$ (5–10%) must be removed by scrubbing first with water and then with an aqueous potassium carbonate solution. In the ethylene oxide scrubber, virtually all of the ethylene oxide and small amounts of the other constituents of the recycle gas ($CO_2$, $N_2$, $CH_4$, $CH_2CH_2$, and aldehydes) dissolve in the water. The resulting aqueous ethylene oxide solution then passes to the ethylene oxide desorber. The ethylene oxide recovered as the head product in the desorber is subsequently stripped of its low-boiling components (i.e., the above-mentioned components of the recycle gas) and finally distilled, thereby separating into water and ethylene oxide.

A small proportion of the gas leaving the ethylene oxide scrubber (0.1–0.2%) is removed continuously (combusted) to prevent the buildup of inert compounds ($N_2$, Ar, and $C_2H_6$), which are introduced as impurities with the reactants: ethylene ($C_2H_6$) and oxygen (Ar and $N_2$). The recycle gas from the scrubber is compressed and a side stream is freed from carbon dioxide by further scrubbing with hot aqueous potassium carbonate solution.

The potassium carbonate solution, which is enriched with $CO_2$ is sent to the $CO_2$ desorber, where the $CO_2$ is stripped off at atmospheric pressure and is either released into the atmosphere or fed to a $CO_2$ utilization plant. The concentrations of the reactants in the bulk of the recycle gas, which is free from ethylene oxide and has a reduced $CO_2$ content, are restored to their starting levels by separate addition of oxygen, ethylene, inhibitor (1,2-dichloroethane or vinyl chloride), and if necessary, a diluent ($CH_4$). The gas is then returned to the reactor.

The oxygen that is used must be extremely pure (> 99%) and is obtained by air separation. Nevertheless, a purge stream of recycle gas is still necessary due to the presence of traces of $N_2$ and Ar. Oxygen is added in a special mixing device that ensures rapid homogenization with the recycle gas. This is necessary because the explosive limit is locally exceeded at the mixing point.

The ethylene is also usually very pure (> 99.5%) and must be free from the strong catalyst poisons sulfur and acetylene. Methane, used as a diluent, also must be free from sulfur compounds.

**Air-Based Oxidation Process.** The air-based process is similar to the oxygen process, but some differences exist. Air introduces a large amount of nitrogen into the recycle gas, which means that a large amount of purge gas must be vented to maintain a constant nitrogen concentration in the recycle stream. The quantity of gas that is vented removes sufficient $CO_2$ to make $CO_2$ scrubbing unnecessary. However, the off-gas leaving the primary reactor still contains so much ethylene that it must be further reacted in a subsequent, secondary or purge reactor before it can be vented into the atmosphere.

The reaction conditions cannot be tailored to the needs of ethylene oxide formation as optimally as in the oxygen-based process. The conversion of ethylene is higher than in the oxygen-based process, especially in the secondary reactors, so as to obtain an acceptable level of ethylene loss in the purge gas. Since selectivity is inversely related to ethylene conversion, it follows that the air-based process has a lower selectivity.

The conditions and gas compositions used in the oxygen- and air-based processes are listed in Table 5.

**Reactor.** Since oxidation of ethylene is highly exothermic, reaction in a fluidized bed would appear to be appropriate. However, attempts to develop such a process on a commercial basis have not produced any advantages for selectivity [73], [74] and have led to problems due to abrasion and sintering. As a result, all ethylene oxide plants currently employ fixed-bed tubular reactors. The diameter of the tubes must not be excessive so as to ensure sufficient heat transfer to the heat transfer medium. A tube diameter of 20–40 mm is common. The diameter of the catalyst particles has an upper limit determined by the need for satisfactory gas mixing and a lower limit resulting from an increasing drop in pressure. A diameter of 3–8 mm is used [72].

**Cooling.** Each of the two possible heat extraction systems — circulation or evaporation of the coolant — has its advantages and disadvantages. Coolant circulation requires a large amount of liquid (pumping power), but allows a defined temperature increase in the direction of gas flow. Ethylene and oxygen concentrations in the cycle gas decrease on passage through the reactor; this leads to lower conversion, which can be compensated by controlled temperature increase toward the reactor outlet.

In evaporative cooling, the hydrostatic height of the coolant in the reactor results in an unavoidable temperature increase at the bottom of the reactor. Since evaporation extracts heat more efficiently than circulation, evaporative cooling requires a smaller volume of coolant. The axial temperature differences can be minimized more easily. Large amounts of liquid coolant leave the reactor with the coolant vapor. The liquid is recovered in a separator and returned to the reactor. The coolant present in the reactor and in the separator constitutes a safety reserve during uncontrolled increases in ethylene conversion in the reactor ("runaway").

Water is safer than an organic heat transfer medium, but its high vapor pressure is a disadvantage (ca. 5.0 MPa at the reaction temperature); therefore, pressure increases rapidly with the reaction temperature. Improved catalysts with high activities and selectivities operate at a lower temperature (ca. 220 °C) and allow smaller reactor volumes. This has encouraged the use of water-cooled reactors [72]; however, organic heat transfer media (e.g., Dowtherm) are still widely employed. A detailed discussion of heat transfer with organic heat transfer media in terms of optimum selectivity and safety can be found in [75].

**Ethylene Conversion.** The selectivity of ethylene oxide formation depends on ethylene conversion. Selectivity decreases more or less linearly with increasing ethylene conversion [76]. Therefore, the highest selectivities are achieved with minimum conversions, but the resulting ethylene oxide concentrations are then too low for commercial purposes. Thus, ethylene conversion is chosen to achieve ethylene oxide concentrations of 1–2 vol% at the reactor outlet.

**Byproducts.** Ethylene glycols are unavoidably produced when ethylene oxide is scrubbed from the recycle gas with water and is subsequently stripped from the aqueous solution by heating. Monoethylene glycol [107-21-1], diethylene glycol [111-46-6], triethylene glycol [112-27-6], and higher poly(ethylene glycols) are formed when ethylene oxide comes into contact with water [77] (→ Ethylene Glycol). Some of the cycle water must be bled off continuously and the ethylene glycol removed to prevent it from accumulating (see Fig. 2). This glycol is of inferior quality to that synthesized via ethylene hydrolysis. Special methods for purifying the resulting glycols have been described, e.g., treatment with ion exchangers and activated charcoal [78].

**Materials.** Since ethylene oxide is noncorrosive, the reactors and the sections of the plant that convey ethylene oxide are usually made of mild steel. The sealing materials for valves and flanges must be chosen carefully [79]. The materials in the system used to remove the carbon dioxide must also be carefully selected because of possible $CO_2$ corrosion [80].

If rust is present in steel tubes or containers, polymer formation [30], increased viscosity, and brown discoloration are to be expected.

Current plant design is predominantly tailored to the oxygen-based process because this is normally more economical than the air-based process. In exceptional cases, however, the air-based process may be preferred depending on local factors (e.g., the availability of oxygen) [81].

**Possible Developments.** Alternative routes from ethylene to ethylene oxide are being studied, e.g., Tl(III)-catalyzed oxidation in solution [81], electrochemical oxidation [82], and enzymatic oxidation [83]. However, these processes appear to be far from industrial application. About half of the ethylene oxide produced is converted into ethylene glycol. Alternative syntheses for ethylene glycol based on carbon monoxide, formaldehyde, and ethylene are being developed [84] – [87] (→ Ethylene Glycol); these are more promising than those for ethylene oxide. With increasing raw material prices, the appearance of an economical alternative is foreseeable. Naturally, such an alternative would have an adverse effect on ethylene oxide capacities, but the economics of ethylene oxide production could be improved by

1) selective production of monoethylene glycol from ethylene oxide, without the formation of higher ethylene glycol homologues, e.g., via glycol carbonate [35]
2) reducing energy consumption
3) improving the selectivity, capacity, life-span, and activity of the catalysts.

# 5. Environmental Protection and Ecology

Ethylene oxide plants using the direct oxidation process are considered to be relatively harmless environmentally, thus representing a clear improvement over the chlorohydrin process. Environmental and other important aspects of ethylene oxide production are discussed in EPA studies [88]. Both oxygen- and air-based plants emit two main off-gas streams: recycle vent gas and $CO_2$. In the oxygen-based process, ca.

70% of the recycle vent gas consists of hydrocarbons (ethylene and methane), which can be combusted (by torch or in a steam generator). The corresponding off-gas in the air-based process is produced in considerably greater quantities, but contains only up to ca. 2.3% hydrocarbons. It can be purified by catalytic oxidation. The $CO_2$-rich vent gas in the oxygen-based process contains ca. 0.2% hydrocarbons. This concentration can be reduced substantially by depressurizing the $CO_2$-rich absorbent in two steps. The gas from the first step is rich in hydrocarbons and is recycled, only the gas from the second step is vented [89]. The $CO_2$-rich vent gas stream from air-based plants is produced in much smaller quantities and contains ca. 5% hydrocarbons that can be removed by combustion. Ethylene oxide can be removed from waste gases, such as are generated during sterilization, by combustion with oxygen [90] or by an acid scrub [91], [92]. Biodegradation of the ethylene glycol generated during ethylene oxide hydrolysis presents no difficulties. If ethylene oxide occurs as a mixture with fluorohydrocarbons, recovery by drying and subsequent compression is recommended [93]. Methods for handling ethylene oxide reaction vessels without the release of vent gas have been described [94].

Ethylene oxide is toxic to microorganisms and fish [95]. The $LC_{50}$ value for fish (*Pimephales promelas*) is 84 mg/L (exposure time 96 h). However, in free-flowing waters the ethylene oxide concentration decreases continually (after 4 h at 25 °C in moving water, the concentration drops by ca. 95%) due to a combination of evaporation, hydrolysis (half-life at 25 °C ca. 14 d), and biodegradation. The ethylene glycol produced as a result of ethylene oxide hydrolysis is considerably less toxic to aquatic organisms ($LC_{50} > 10\,000$ mg/L) and is readily biodegradable. Effluents (e.g., from ethylene oxide production plants [88]) containing ethylene oxide are therefore treated biologically after the ethylene oxide has been converted into ethylene glycol.

According to TA Luft [96], vent gas streams must not contain more than 25 g/h or 5 mg/m$^3$ ethylene oxide.

# 6. Quality Specifications

Ethylene oxide is a major industrial product that is obtained with a consistently high purity irrespective of the production process used. Typical specifications are given in Table 6 [9].

# 7. Analysis

An overview of analytical methods for ethylene oxide can be found in [97]. An established method for determining ethylene oxide is based on its reaction with $MgCl_2$ [98]. In view of the low TRK value (5 mg/m$^3$), accurate methods are needed to determine traces of ethylene oxide in air.

**Table 6.** Typical specifications of ethylene oxide

| | |
|---|---|
| Appearance | clear, colorless |
| bp at 101.3 kPa | 10.8 °C |
| Water content * | 50 mg/kg |
| $CO_2$ content * | 10 mg/kg |
| Aldehyde content * | 50 mg/kg |
| Ethylene oxide content (min.) * | 99.5 % |

* As determined by gas chromatography.

An adsorption chromatography method is reported to be suitable for concentrations exceeding 0.15 ppm [99]. The NIOSH method 1607 [100] with a working range of 0.04 – 1.7 ppm (0.09 – 3 mg/m$^3$) is based on adsorption with charcoal, derivatization with HBr to give 2-bromoethanol, and gas chromatography.

# 8. Handling, Storage, and Transportation

Ethylene oxide has repeatedly caused serious explosions, fires, and accidents [101], [102]. It is an extremely hazardous substance because it can explode and is both highly flammable and extremely reactive (exothermic reactions). Aqueous solutions containing > 4 wt % ethylene oxide are flammable; flash points are given in Table 2. Furthermore, ethylene oxide is toxic and poses a danger both to health and to the environment. The inherent hazards of the product must be known; all prescribed safety measures and legal requirements must be observed. Personnel handling ethylene oxide must receive appropriate training.

Safety precautions for handling ethylene oxide are described in detail in the data sheets provided by the manufacturers (e.g., [9]) and elsewhere (e.g., [79], [103], [104]). Therefore, only the most important points are discussed here.

**Explosion and Fire Control.** Pure ethylene oxide vapor or ethylene oxide vapor mixed with air or inert gases can decompose explosively. Explosiveness depends on pressure, temperature, concentration, the type, form, and energy of the ignition source, and the type of container. High pressure can be generated on explosion of ethylene oxide; therefore, for safe handling the exact explosive limits must be known. Pure ethylene oxide vapor at 101.3 kPa decomposes when passed through a heated platinum coil [105]. The temperature at which decomposition starts (decomposition temperature) was calculated to be 571 °C. Calculations were based on the temperature at the tube exit and the thermodynamic data for ethylene oxide and its decomposition products. The thermal decomposition of ethylene oxide and ethylene oxide – nitrogen mixtures has been investigated [106]. If ethylene oxide vapor is introduced into a preheated vessel at 101.3 kPa, explosive decomposition takes place at ca. 500 °C [106]. The decomposition temperature is reduced by an increase in the pressure (e.g., ca.

450 °C at 1 MPa) but is increased by the addition of nitrogen. Explosive thermal decomposition is still possible below atmospheric pressure, but the necessary temperature then increases above 500 °C.

The ignition properties of ethylene oxide and its mixtures with air and inert gases have been studied by several authors. Pure ethylene oxide vapor can ignite even at a pressure as low as 48 kPa (electric spark) or 20.2 kPa (mercury fulminate) [107]. The maximum theoretical explosive pressure is ca. 10 times the inital pressure, but can increase to 20 times the initial pressure if liquid ethylene oxide is present. This phenomenon occurs because liquid ethylene oxide evaporates and participates in the decomposition reactions that take place in the vapor phase [108]. Mixtures of ethylene oxide with $N_2$, $CO_2$, methane, and air do not ignite over certain concentration ranges; the upper explosive limit is always 100% ethylene oxide, since pure ethylene oxide also decomposes if ignited. In the presence of oxygen (air), combustion and decomposition take place simultaneously and the ignition temperature (or energy) is reduced. The ignition temperature of ethylene oxide in air at 101.3 kPa is 429 °C [109]. As the ethylene oxide concentration increases, the proportion destroyed by decomposition also increases [17], [110]. The minimum value cited for the lower explosive limit of ethylene oxide – air mixtures is 2.6% [111]; the explosive range of ethylene oxide – air mixtures is accordingly 2.6 – 100%. Figures for an upper explosive limit of ca. 80% at 101.3 kPa are probably due to differences in the apparatus and methods used [112] – [114].

Explosive decomposition of ethylene oxide in closed containers can be suppressed by blanketing with an inert gas until the nonexplosive range is reached. Blanketing agents with higher thermal conductivities suppress explosions more effectively. Quotation of the minimum total pressure necessary for inert gas blanketing is important; values cited in [17] were obtained without consideration of the pressure and temperature dependence and, therefore, do not apply at higher pressures. Figure 3 shows the total pressure currently recommended for inert gas blanketing of ethylene oxide with nitrogen as a function of temperature [79]. The explosive limits of other mixtures of ethylene oxide with inert gases and air can be found in the literature, e.g., ethylene oxide with $H_2O$ [115]; $N_2$ [106], [115]; $N_2-H_2O$ [107]; $CO_2-H_2O$ [107]; $CH_4$ [108]; $CO_2$ [17], [115], [116]; $C_3H_6$ [117]; $C_4H_9$ [117]; $N_2-$air [17]; $CH_4-$air [17]; $CO_2-$air [111]; $CF_2Cl_2-$air [114], [118].

As has already been explained, liquid ethylene oxide can participate in the decomposition that is initiated in the vapor phase. Explosion of liquid ethylene oxide initiated by an ignition source within the liquid was first described in 1980 [119], [120], but once again, the decomposition reaction is thought to take place in the gas phase created by the source of ignition and is maintained by the subsequent evaporation of ethylene oxide. Figure 4 shows the temperature limits for ignition of liquid ethylene oxide as a function of the pressure. At pressures between 10 and 15 MPa, the ignition temperature decreases rapidly. This phenomenon is presumed to be due to the fact that the critical state is reached. The conditions under which liquid ethylene oxide is usually produced, stored, or processed lie within the nonexplosive range (Fig. 4).

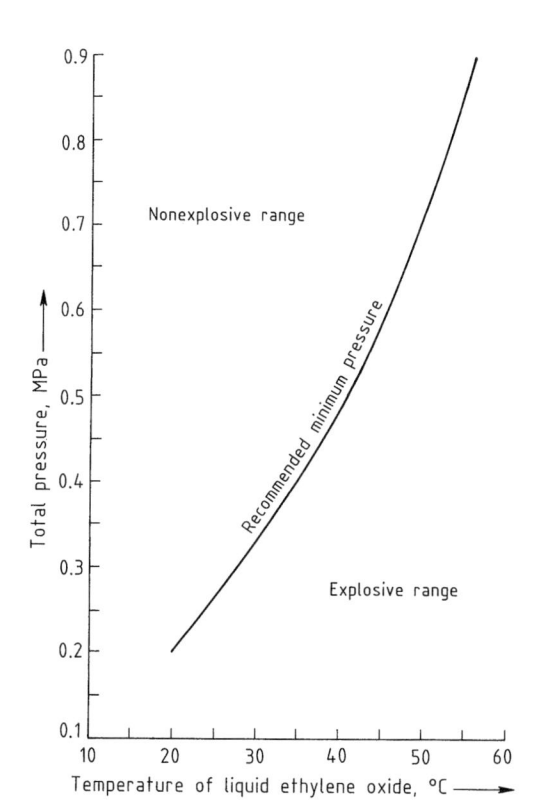

**Figure 3.** Recommended total pressure for the safe storage of ethylene oxide as a function of the temperature after blanketing with nitrogen

The highly exothermic reactions of ethylene oxide represent a further hazard [101]. If traces of polymerization initiators (e.g., amines) find their way into a large supply of ethylene oxide (e.g., in a tank), polymerization starts slowly and then accelerates because of the resulting temperature increase. Polymerization proceeds quasi-adiabatically, leading to sudden and rapid increases in temperature and pressure that can rupture the container [121], [122], which may be followed by explosive decomposition of the released ethylene oxide vapor.

Because of the above-mentioned hazards, the following potentially dangerous situations must be prevented:

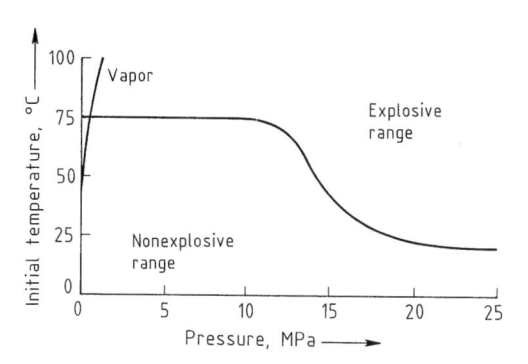

**Figure 4.** Explosive limits for pure liquid ethylene oxide as a function of pressure and temperature

1) Leakage of liquid or gaseous ethylene oxide
2) Entry of air, oxygen, or reactive impurities into ethylene oxide containers
3) Ignition sources in danger areas
4) Overheating of ethylene oxide (chemical reaction, fire, etc.)

Leaks or spills should be promptly diluted with sufficiently large volumes of water to reduce the ethylene oxide concentration to less than 4 wt %. Mixtures of ethylene oxide and water will still burn if the ethylene oxide concentration exceeds 4 wt %. Fires should also be extinguished with large volumes of water, carbon dioxide extinguishers can be used for small fires.

**Storage and Transportation.** Ethylene oxide is transported in tank cars and containers and can be conveyed by rail or by sea and, in many countries, also by road. In the Federal Republic of Germany, conveyance by road is possible with special permission and under certain conditions (GGVS/ARD § 7). Hazard classifications for ethylene oxide are as follows:

CFR 49:172.101, flammable liquid
IMDG Code, class 2, UN No. 1040
RID/ADR, class 2, number 3 ct

Storage and transport containers for ethylene oxide are usually made from steel, but steel is suitable only if special measures have been taken to prevent rust formation [30]. Rust acts as a mild polymerization catalyst and is dispersed in the ethylene oxide by polymerization starting on available surfaces. Polymer concentrations as low as ca. 110 mg/kg produce an increase in viscosity, may cause brown discoloration, may block filters, valves etc., and form deposits on the container walls. Therefore, steel containers should be blasted before first use, the blasting residues carefully removed, and the container flushed with nitrogen. When aqueous solutions of ethylene oxide are handled, precipitation of solid hydrates (cf. Table 2) may cause blockages.

In Europe, tank cars are tested at a pressure of 1.6 MPa and they must be protected from the sun by a protective roof or insulation (GGVE). Ethylene oxide–nitrogen mixtures can be transported at a maximum pressure of 1 MPa at 50 °C (GGVS/RID).

Figure 5. Uses of ethylene oxide

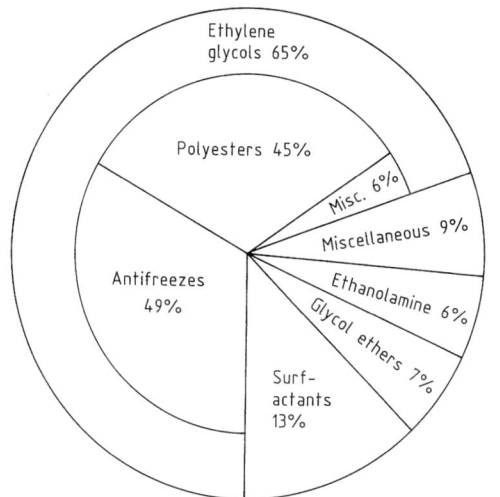

- Ethylene glycols 65%
  - Polyesters 45%
  - Antifreezes 49%
  - Misc. 6%
- Miscellaneous 9%
- Ethanolamine 6%
- Glycol ethers 7%
- Surfactants 13%

The proportion of nitrogen in the vapor phase must be high enough to ensure that explosion cannot occur at or below this temperature. In the United States railroad tanks are insulated and equipped with pressure-release valves.

Safety measures that should be observed when reactions with ethylene oxide are performed or when it is used for sterilization can be found in [79].

# 9. Uses

Ethylene oxide is an excellent disinfectant, sterilizing agent, and fumigant when it is used as a nonexplosive mixture with $N_2$, $CO_2$, or dichlorofluoromethane. The gas penetrates into pores and through packaging or clothing. It can be used, for example, to sterilize surgical instruments in hospitals or to remove pests and microorganisms from spices, furs, etc.

However, most ethylene oxide is converted into other products. Figure 5 shows the percentages of ethylene oxide used in these derivatives in the United States. Although this distribution was determined in 1978, it still applies today and is typical of the world market.

Products derived from ethylene oxide have many different uses, only the most important ones are listed here:

*Monoethylene glycol* [107-21-1] [123]: Antifreeze for engines, production of poly(ethylene terephthalate) (polyester fibers, foils, and bottles), and heat transfer liquids.

*Diethylene glycol* [111-46-6] [124]: Polyurethanes, polyesters, softeners (cork, glue, casein, and paper), plasticizers, gas drying, solvents, and deicing of aircraft and runways.

*Triethylene glycol* [*112-27-6*] [125]: Lacquers, solvents, plasticizers, gas drying, and humectants (moisture-retaining agents).

*Poly(ethylene glycols)* [126]: Cosmetics, ointments, pharmaceutical preparations, lubricants (finishing of textiles, ceramics), solvents (paints and drugs), and plasticizers (adhesives and printing inks).

*Ethylene glycol ethers* [127]: Brake fluids, detergents, solvents (paints and lacquers), and extractants for $SO_2$, $H_2S$, $CO_2$, and mercaptans from natural gas and refinery gas.

*Ethanolamine* [*141-43-5*]: Chemicals for textile finishing, cosmetics, soaps, detergents, gas purification ($H_2S$, $SO_2$, and $CO_2$).

*Ethoxylation products* of fatty alcohols, fatty amines, alkyl phenols, cellulose, poly-(propylene glycol) [23]: Detergents and surfactants (nonionic), biodegradable detergents, emulsifiers, and dispersants.

*Ethylene carbonate* [*96-49-1*]: Solvents.

## 10. Economic Aspects

In 1980, ca. 16% of world ethylene production was used to synthesize ethylene oxide; this use was second only to polyethylene production (44%). World production of ethylene oxide is currently ca. $8 \times 10^6$ t/a. Ethylene oxide is an important raw material for major consumer goods in virtually all industrialized countries. Table 7 gives an overview of worldwide ethylene oxide capacities. Excess capacities were already apparent in 1984 [128]. The large facilities recently installed in Saudi Arabia and the Soviet Union are likely to further aggravate this situation [129], [130].

## 11. Toxicology and Occupational Health

The toxicological properties of ethylene oxide are mainly determined by its reactivity with nucleophilic groups such as carboxyl, amino, phenolic hydroxide, or sulfhydryl groups.

In humans, acute inhalative poisoning causes headache, nausea, and vomiting within a few minutes [131]–[133]. Local irritation results in dyspnea [134]. Myocardial damage [135], excitation, numbness, and finally coma [131] follow. After dermal exposure, blisters are formed on the skin, and symptoms similar to those found after inhalation appear due to skin absorption [136]. Sensitization has been reported after repeated dermal contact [137]. Repeated inhalation of ethylene oxide leads to sensory-motor polyneuropathy and impaired memory [138], [139]. Cataracts may be formed after chronic exposure [140].

**Table 7.** World ethylene oxide capacities in 1987

| Country | Producer | Location | Licensing company* | Capacity, $10^3$ t/a |
|---|---|---|---|---|
| **Asia** | | | | 1157 |
| China | China Techimport | Liaoyang | Hüls, a | 35 |
| | | Nanjing | SD | 160 |
| India | Glycols India | Kashipur | SD | 20 |
| | Indian Petrochemical | Vadodana | SD | 5 |
| | NOCIL | Bombay | Shell, o | 10 |
| Japan | Japan Catalytic | Kawasaki | Jap. Cat., o | 185 |
| | Mitsubishi PC | Yokkaichi | Shell, o | 80 |
| | | Kashima | Shell, o | 70 |
| | Mitsui Petrochemical | Chiba | Shell, o | 100 |
| | Nisso Murazen | Goi | SD, o | 65 |
| | Nisso Yuka | Yokkaichi | Shell, o | 60 |
| | Semppoku Ethylene Oxide | Osaka | Shell, o | 70 |
| Korea, Northern | State Complex | Pyong Yang | Japan Catalytic | 10 |
| Korea, Southern | Honam Petrochemical | Yeo Cheon | Shell, o | 72 |
| Singapore | Ethyleneglycols Singapore | Merbau | Shell, o | 80 |
| Taiwan | Oriental Union Chemical | Kaohsiung | UCC, o | 100 |
| | CMFC | | SD | 35 |
| **Australia** | ICI | Botany | SD | 33 |
| **Canada** | | | | 305 |
| | Dow Chemical | Saskatchewan | Dow Chemical | 80 |
| | UCC | Prentiss | UCC | 225 |
| **Eastern Europe** | | | | 888 |
| Bulgaria | Technoimport | Burgas | SD | 80 |
| Czechoslovakia | Technoexport | Bratislava | Shell, o | 40 |
| German Democratic Republic | Volks Eigener Betrieb Buna | Schkopau | SD | 100 |
| Poland | Petrochemia | Plock | Shell, o | 60 |
| | | | SD | 30 |
| Rumania | Romchim | Brazi | SD | 35 |
| | | Pitesti | SD | 30 |
| Soviet Union | Techmashimport | Khazan | Japan Catalytic | 60 |
| | Techmashimport | Dzherzhinsk | SD | 200 |
| | Techmashimport | Nishnekamsk | SD, o | 200 |
| Yugoslavia | Pazinka | Pazin | SD | 53 |
| **Latin America** | | | | 285 |
| Brazil | Oxiteno | Camacari, Bahia | SD, a | 105 |
| | | Sao Paulo | SD, a | 40 |
| Mexico | Pemex | La Cangrejera | SD | 100 |
| | | Pajaritos | SD | 40 |
| **Middle East** | | | | 454 |
| Saudi Arabia | Sharq | Al Jubail | Shell, o | 240 |
| | Saudi Yanbu | Yanbu | SD, o | 160 |
| Turkey | Petkim Petrokimya | Aliaga | Shell, o | 54 |
| **United States** | | | | 3200 |
| | BASF Wyandotte | Geismar, La. | Shell, o | 210 |
| | Calasieu Chemical | Lake Charles, La. | Shell, o | 100 |
| | Celanese | Clear Lake, Tex. | Shell, o | 200 |
| | Dow Chemical | Freeport, Tex. | Dow Chemical, a | 90 |
| | | Plaqemine, Tex. | Dow Chemical, a | 220 |

Table 7. (continued)

| Country | Producer | Location | Licensing company* | Capacity, $10^3$ t/a |
|---|---|---|---|---|
| | Eastman | Longview, Tex. | Shell, o | 90 |
| | ICI | Bayport, Tex. | Shell, o | 200 |
| | Jefferson Chemical | Port Neches, Tex. | SD, I | 300 |
| | Nothern Petrochemical | Morris, Ill. | SD, o | 100 |
| | Olin | Brandenburg, Ky. | Shell, o | 50 |
| | PD Glycol | Beaumont, Tex. | SD | 200 |
| | Shell | Geismar, La. | Shell, o | 300 |
| | Sunolin | Claymont, Del. | Shell, o | 90 |
| | UCC | Ponce, P.R. | UCC, a | 300 |
| | | Seadrift, Tex. | UCC, a | 350 |
| | | Taft, La. | UCC, a,o | 400 |
| **Western Europe** | | | | 2055 |
| Belgium | BASF | Antwerp | Shell, o | 150 |
| | British Petroleum | Antwerp | UCC, o | 130 |
| Federal Republic of Germany | BASF | Ludwigshafen | Shell, o | 150 |
| | Erdölchemie | Dormagen | Shell, o | 150 |
| | Hoechst | Gendorf | Shell, o | 120 |
| | | Kelsterbach | Shell, o | 40 |
| | Hüls | Marl | HÜLS/Shell, o | 135 |
| France | Naphtachimie | Lavera | Shell, o | 150 |
| | | | SD, I | 50 |
| Great Britain | ICI | Wilton | Shell, o | 240 |
| | Shell | Carrington | Shell, o | 120 |
| | British Petroleum | Hampshire | UCC, a | 20 |
| Italy | ANIC SpA | Gela | Snam, Progetti | 40 |
| | Montedison | Brindisi | SD, o | 40 |
| | | Ferrara | SD, a | 25 |
| | | Priolo | UCC | 30 |
| Sweden | Berol | Stenungsund | Atochem/SD, o | 45 |
| Spain | Industrias Quimicas Asociades | Tarragona | Shell, o | 70 |
| | Alcudia | Puertollano | SD | 20 |
| The Netherlands | Dow Chemical | Terneuzen | Dow Chemical | 120 |
| | Shell | Moerdjik | Shell, o | 210 |
| **Total worldwide capacity** | | | | **8377** |

* a = air-based process; o = oxygen-based process; SD = Scientific Design.

**Oral Toxicity.** The acute oral toxicity ($LD_{50}$) of ethylene oxide in the rat is 330 mg/kg [141].

**Inhalation.** The 4-h $LC_{50}$ value is 835 mL/m$^3$ in the mouse and 1460 mL/m$^3$ in the rat [142]. Acute symptoms largely resemble those observed in humans. Subchronic administration of ethylene oxide (100 and 200 mL/m$^3$) to cats by means of inhalation for a period of 22 days produced anorexia, apathy, atoxia, and paralysis of the hind quarters [143]. Lethalities also occurred, postmortem examination revealed liver and kidney damage accompanied by hyperemia and perivascular bleeding in various organs (e.g., the brain) [143]. Rodents that inhaled air containing ethylene oxide at a concentration of 300–400 mL/m$^3$ displayed additional local irritation and severe, primary neurotoxic atrophy of the musculature in their hind quarters [142]. Hemotoxic effects have also been observed in rodents [144]. Monkeys exposed to inhalative concentrations of 204 mL/m$^3$ for up to 226 days showed impaired reflexes, reduced sensitivity to pain, and neurotoxic muscular atrophy of their hind quarters [145].

**Teratogenicity.** In studies on rats exposed to maximum inhalative concentrations of 100 [146] or 150 mL/m$^3$ [147], symptoms of intolerance were observed in the dams but no malformations were observed in the fetuses. Ethylene oxide does not produce teratogenic effects in rabbits [147].

**Mutagenicity.** Ethylene oxide is mutagenic due to its alkylating properties. It alkylates the germ cells of male rats [148] and produces mutations in *Drosophila melanogaster* [149], *Neurospora crassa* [150], and *Salmonella typhimurium* (Ames test) [151], [152]. Dominant lethal mutations and chromosome aberrations are induced in male rats [153]. Increased numbers of micronuclei are found in the polynuclear erythrocytes of rats [154] and mice [155], [156]. Chromosomal defects are induced in cultured amniotic cells from humans [157]. Increased exchange between sister chromatids is found in monkeys and rats [158], [159]; the same effect is observed in workers who have been exposed to ethylene oxide (cumulative dose 0.5–50 g) [160]–[163].

**Carcinogenicity.** A large number of reports have been published on the carcinogenic properties of ethylene oxide in animals. Dilute ethylene oxide applied to the skin of rats did not induce tumor formation [164]. Subcutaneous injection of ethylene oxide in rats did not have any systemic carcinogenic effects [165], [166]. However, intragastric administration led to an increase in the incidence of epithelial carcinomas in the gastric mucosa [167]. An increased incidence of brain tumors and mononuclear cell leukemia was found in rats that had inhaled ethylene oxide at concentrations of 10, 33, or 100 mL/m$^3$ over a period of two years [168]. An increased incidence of peritoneal mesotheliomas was also observed in the animals exposed to concentrations of 33 and 100 mL/m$^3$. Results of human epidemiological studies on workers exposed to ethylene oxide differ [169]–[172]. Ethylene oxide is classified as a "putative human chemical carcinogen" [173].

**Permissible Exposure Limits.** Ethylene oxide is classified as a class III A2 carcinogen by the German MAK commission (TRK = 3 mL/m$^3$ = 5 mg/m$^3$) and as a class A2 carcinogen by the ACGIH (TLV-TWA = 1 mL/m$^3$ = 2 mg/m$^3$).

# 12. References

General References

[1] G. O. Curme: *Glycols,* Reinhold Publ. Co., New York 1952, pp. 74–113.
[2] S. A. Miller: *Ethylene and its Industrial Derivatives,* Ernest Benn Limited, London 1969, pp. 513–638.
[3] *Kirk-Othmer,* **9,** 432–471.
[4] *Beilstein,* **XVII** 4–5, **XVII(E1)** 3, **XVII(E2)** 9–11, **XVII(E3/E4)** 3–10, **XVII (E5)** 3–7.
[5] *Ullmann,* 4th ed., **8,** 215–221.

Specific References

[6] C. Hirose, *Bull. Chem. Soc. Jpn.* **47** (1974) no. 6, 1311–1318.
[7] A. Wurtz, *Justus Liebigs Ann. Chem.* **110** (1859) 125–128.
[8] Société Française de Catalyse Generalisée, FR 729 952, 1931; 739 562, 1931 (T. E. Lefort).
[9] Merkblatt Ethylenoxid, Hoechst AG, Frankfurt am Main.
[10] A. S. Pell, G. Pichler, *Trans. Faraday Soc.* **61** (1965) 71–77.
[11] C. J. Walters, J. M. Smith, *Chem. Eng. Prog.* **48** (1952) no. 7, 337–343.
[12] W. Gestrich, W. Kraus, *Ber. Bunsenges. Phys. Chem.* **78** (1974) no. 12, 1334–1353.
[13] D. N. Glew, N. S. Rath, *J. Chem. Phys.* **44** (1965) 1710–1711.
[14] O. Maass, E. H. Boomer, *J. Am. Chem. Soc.* **44** (1922) 1709.
[15] D. N. Glew, N. S. Rath, *J. Chem. Phys.* **44** (1965) 1711.
[16] R. K. McMullan, G. A. Jeffrey, *J. Chem. Phys.* **42** (1965) 2725.
[17] L. G. Hess, V. V. Tilton, *Ind. Eng. Chem.* **42** (1950) no. 6, 1251–1258.
[18] J. D. Olson, *J. Chem. Eng. Data* **22** (1977) 326.
[19] C. L. Yaws: *Physical Properties,* McGraw-Hill, New York 1977, pp. 167–176.
[20] A. Weissberger in A. Rosowski (ed.): *Heterocyclic Compounds,* vol. **19,** Wiley-Interscience, New York 1964, pp. 1–523.
[21] M. S. Malinovski: *Epoxides and their Derivatives,* Daniel Davey & Co, Inc., New York 1965.
[22] I. Parker, *Chem. Rev.* **59** (1959) 737–797.
[23] N. Schönfeld: *Grenzflächenaktive Äthylenoxidaddukte,* Wissenschaftl. Verlags GmbH, Stuttgart 1976, Ergänzungsband 1984.
[24] J. Gorzinski Smith, *Synthesis* **8** (1984) 629–656.
[25] R. R. Baldwin, A. Keen, R. W. Walker, *J. Chem. Soc. Faraday Trans. 1* **80** (1984) no. 2, 435–456.A. Lifshnitz, H. Ben Hanou, *J. Chem. Phys.* **87** (1983) 1782–1787.
[26] D. N. Kirk, *Chem. Ind. (London)* 1973, no. 3, 109–116.
[27] L. Vamling, L. Cider, *Ind. Eng. Chem. Prod. Res. Dev.* **25** (1986) no. 3, 424–430.
[28] G. Olah, A. P. Fung, D. Meidar, *Synthesis* **4** (1981) 280–282.
[29] J. Farakawa, T. Saegusa: *Polymerisation of Aldehydes and Oxides,* Wiley-Interscience, New York 1963.G. Gee, W. C. E. Higginson, *J. Chem. Soc.* 1962, 231.
[30] T. H. Baize, *Ind. Eng. Chem.* **53** (1961) 903.
[31] G. Hechler, *Chem. Ing. Tech.* **43** (1971) no. 16, 903–905.
[32] H. Springmann, *Fette Seifen Anstrichm.* **73** (1971) no. 6, 396–398.
[33] T. Aida, S. Inoue, *J. Am. Chem. Soc.* **105** (1983) no. 5, 1304–1309.
[34] Halcon, EP-A 133 763, 1984 (V. S. Bhise, H. Gilman).
[35] B. J. Ozero, J. V. Procelli, *Hydrocarbon Process.* **63** (1984) no. 3, 55–61.

[36] F. A. Meskens, *Synthesis* **7** (1981) 501–522. D. L. Rakhmankulov, O. B. Chalova, T. K. Kiladze, E. A. Kantor, *Khim. Geterotsikl. Soedin* 1984, no. 3, 291–306.
[37] Hoechst, DE 1 035 635, 1958 (K. Fischer, K. Vester).
[38] R. F. Goldstein: *The Petroleum Chemicals Ind.*, 2nd ed., Wiley Interscience, New York 1958, p. 352.
[39] G. Prauser, G. Fischer, K. Dialer, *Angew. Chem.* **92** (1980) no. 5, 389–390.
[40] Nippon Shokubai Kagaku, JP 56 100–633, 1980 (Kishimoto Senji, Nakai Takahiko, Kumarawa, Toshihiko).
[41] N. G. Gaylord, E. I. Becker, *Chem. Rev.* **49** (1951) 413. I. Mukerji, A. L. Wayda, G. Darbagh, S. H. Bertz, *Angew. Chem.* **98** (1986) no. 8, 756–757.
[42] J. Dale, G. Borgen, K. Daatsvatn, *Acta Chem. Scand. Ser. B* **B 28** (1974) 378–379.
[43] J. Dale, *J. Chem. Soc. Chem. Commun.* 1976, 295–296.
[44] Hoechst, DE 2 640 505, 1976 (G. Scheffel, R. Obermeier).
[45] H. Kicheldorf, G. Mörber, W. Regel, *Synthesis* **5** (1981) 383.
[46] *Ullmann*, 3rd ed., **3**, 145.
[47] A. Wurtz, *Ann. Chim (Paris)* **69** (1863) 355.
[48] W. Bone, R. Wheeler, *J. Chem. Soc.* **85** (1904) 1637. L. Reyerson, L. J. Swearingen, *J. Am. Chem. Soc.* **50** (1928) 2872. R. Willstätter, M. Bommer, *Justus Liebigs Ann. Chem.* **422** (1921) 136.
[49] S. A. Miller: *Ethylene and its Industrial Derivatives*, Ernest Benn Limited, London 1969, pp. 546–548.
[50] R. Wolf, H. Götze, *Chem. Tech. (Leipzig)* **14** (1962) no. 10, 600–606. H. Kanoh, T. Nishimura, A. Ayame, *J. Catal.* **57** (1979) 372–379.
[51] Shell, DE-OS 2 159 346, 1971 (R. P. Nielsen).
[52] Shell, US 4 012 425, 1972 (R. P. Nielsen, J. H. La Rochelle). UCC, EP 75 938, 1982 (M. M. Bashin, G. W. Warner).
[53] M. Ayyob, M. S. Hedge, *J. Catal.* **97** (1986) 516–526.
[54] CC, US 2 279 470, 1942 (G. H. Law, H. C. Chitwood). J. M. Berty, *Appl. Ind. Catal.* **1** (1983) 223–227.
[55] G. L. Montrasi, G. C. Battiston, *Oxid. Commun.* **3** (1983) no. 3–4, 259–267.
[56] Bundesministerium für Forsch. und Technologie, *Forschungsbericht T 84–225*, Fachinformationszentrum Energie, Physik, Mathematik GmbH, Karlsruhe, Leopoldshafen 2, 7514 Eggenstein 1984, pp. 171–241.J. M. Berty, *Appl. Ind. Catal.* **1** (1983) 223–227.
[57] G. Leofontani, G. R. Tauszik, M. Padovan, *J. Therm. Anal.* **30** (1985) no. 6, 1267–1272.
[58] G. L. Montrasi, G. R. Trauszik, M. Solari, G. Leofanti, *Appl. Catal.* **5** (1983) 359.
[59] W. D. Mroß, *Ber. Bunsenges. Phys. Chem.* **88** (1984) 1042–1053.
[60] S. Rebsdat, S. Mayer, J. Alfranseder, *Chem. Ing. Tech.* **53** (1981) no. 11, 850–854.
[61] I. E. Wachs, *J. Catal.* **72** (1981) no. 1, 160–165.
[62] W. M. H. Sachtler, C. Backx, R. A. Van Santen, *Catal. Rev. Sci. Eng.* **23** (1981) 127–149.
[63] H. H. Voge, C. R. Adams, *Adv. Catal.* **17** (1967) 151. P. A. Kilty, W. M. H. Sachtler, *Catal. Rev. Sci. Eng.* **10** (1974) no. 1, 1–16.
[64] M. Akimoto, K. Ichikawa, E. Echigoya, *J. Catal.* **76** (1982) 333–344. N. W. Cant, W. K. Hall, *J. Catal.* **52** (1978) 81.
[65] C. T. Campbell, B. E. Koel, *J. Catal.* **92** (1984) no. 2, 272–283.
[66] E. L. Force, A. P. Bell, *J. Catal.* **38** (1975) 440.
[67] C. Backx, J. Moolhuysen, J. Geenen, R. A. van Santen, *J. Catal.* **72** (1981) 364.
[68] R. F. Grant, R. M. Lambert, *J. Catal.* **92** (1985) 364–375; *J. Catal.* **93** (1985) 92–99.
[69] R. A. van Santen, C. P. M. de Groot, *J. Catal.* **98** (1986) no. 2, 530–539.

[70] R. B. Grant, R. M. Lambert, *Stud. Surf. Sci. Catal.* **19** (1984) 251–257.S. A. Tan, B. R. Grant, R. M. Lambert, *J. Catal.* **100** (1986) no. 2, 383–391.

[71] M. Kobayachi, T. Kanno, *Actas Simp. Iberoam. Catal.*, **9th** (1984) no. 2, 878–887. T. Kanno, M. Kobayashi, *Proc. Int. Congr. Catal.* **8th** (1984) pub. 1985, no. 3, III 277–III 288.

[72] J. C. Zomerdijk, M. W. Hall, *Catal. Rev. Sci. Eng.* **23** (1981) no. 2, 163–185.

[73] K. Dialer, *Chem. Ing. Tech.* **54** (1982) no. 1, 18.

[74] T. E. Corrigan, *Pet. Refiner* **32** (1933) no. 2, 87.

[75] N. Piccinini, G. Levy, *Can. J. Chem. Eng.* **62** (1984) no. 4, 541–546, 547–558.

[76] K. Klose, P. Kripylo, L. Mögling, *Chem. Techn. (Leipzig)* **37** (1985) no. 12, 506–508.

[77] Halcon, DE-OS 2 750 601, 1977 (B. J. Ozero); Halcon, US 3 964 980, 1975 (B. J. Ozero).

[78] PPG Industries, US 3 904 656, 1973 (S. E. Broz).

[79] Berufsgenossenschaft der chem. Industrie, Heidelberg, *Merkblatt M 045 7/85 ZH 1/54*, Jedermann-Verlag, Heidelberg 1985.

[80] A. K. Nikitina, L. P. Vershinina, V. P. Khvatkova, V. M. Umnyashkina, *Khim. Tekhnol. Topl. Masel* **11** (1982) 27–28.

[81] Halcon, DE 2 718 056, 1978 (A. N. Naglieri). S. E. Diamond, F. Mares, A. Szalkiewicz, D. A. Muccigrosso, J. P. Solar, *J. Am. Chem. Soc.* **104** (1982) no. 15, 4266–68.

[82] M. Stoukides, G. G. Vayenas, *J. Catal.* **70** (1981) no. 1, 137–146.

[83] Exxon, DE-OS 2 915 108, 1979 (C. T. Hou, R. N. Patel, N. J. Edison, A. I. Laskin). C. G. Van Ginkel, H. G. J. Welten, J. A. M. De Bont, *Eur. J. Appl. Microbiol. Biotechnol.* **24** (1986) no. 4, 334–337.

[84] J. Kollar, *Chemtech* **14** (1984) no. 8, 504–511.

[85] A. Aquilo, R. J. H. Vorhoeeve, D. N. Freeman, S. Alder, *Hydrocarbon Process.* **62** (1983) no. 3, 57–65.

[86] B. D. Dombeck, *Adv. Catal.* **32** (1983) 325–416.

[87] A. M. Brownstein, *Hydrocarbon Process.* **71** (1975) no. 9, 72–76.

[88] D. E. Field: *Engineering and Cost Study of Air Pollution Control for the Petrochemical Industry*, vol. **6**, EPA-450/3-73-006-f.EPA-450/4-84-007L, US Environmental Protection Agency Off. Air Qual. Plann. Stand. (Tech. Rep.), Research Triangle Park, N.C. 27711 1986.

[89] Shell, US 3 867 113, 1975 (E. G. Foster, P. F. Russel, R. G. Vanderwater).

[90] G. Groß, H. Kittner, *Gas Aktuell* **30** (1985) 7–13.

[91] P. Oberle, *Chem. Anlagen + Verfahren* **1** (1983) 12–13.

[92] Buonicore-Cashman Ass., PCT Int. Appl., WO 8 400 748, 1984, US-A 409 830, 1982 (A. J. Buonicore).

[93] Leybold-Heraeus GmbH, EP-A 130 319, 1984 (H. Amlingek).

[94] O. Michel, R. Wiedemeyer, *Chem. Ing. Tech.* **49** (1977) no. 1, 57.

[95] R. A. Conway, G. T. Waggy, M. H. Spiegel, R. L. Berglund, *Environ. Sci. Technol.* **17** (1983) 107–112.

[96] Erste allgemeine Verwaltungsvorschrift zum Bundesimmissionsschutzgesetz (Technische Anleitung zur Reinhaltung der Luft -TA-Luft) vom 27.02.1986. Gemeinsames Ministerialblatt 37. Jahrg., Nr. 7 vom 28.02.1986.

[97] R. L. Anderson in F. D. Snell, L. S. Ettre (eds.): *Encyclopedia of Industrial Chemical Analysis*, vol. **12**, J. Wiley & Sons, New York 1971, pp. 317–340.

[98] F. W. Kerckow, *Z. Anal. Chem.* **108** (1937) 249.

[99] A. H. Qazi, N. H. Ketcham, *Am. Ind. Hyg. Assoc. J.* **38** (1977) 635. Merkblatt ZH1/120.27, Hauptverband der gewerblichen Berufsgenossenschaften, Carl Heymanns Verlag, Köln 1985.

[100] *NIOSH Manual of Analytical Methods,* DHHS (NIOSH) Publ. No. 84–100, U.S. Government Printing Office, Washington, D.C. 20 402, Method 1607, 5/15/85 pp. 1607-1–1607-6.
[101] National Technical Information Service, Springfield, Virginia 22 151, *A Review of Violent Monomer Polymerisation,* Operations Research, Inc., prepared for the Coastguard, Oct. 1974.
[102] R. Davies, AIChE Loss Prevention Symposium, Atlanta, Georgia, Feb. 27– Mar. 1, 1978.
[103] *Codes of Practice for Chemicals with Major Hazards, Ethylene Oxide,* Brochure CPP4, The Chemical Industry Safety and Health Council of the Chemical Industries Assoc., London 1975.
[104] *Properties and Essential Information for Safe Handling and Use of Ethylene Oxide,* Chemical Safety Data Sheet SD-38, Manufacturing Chemists Association, Washington, D.C., 1971.
[105] E. Peytral, *Bull. Soc. Chim. Fr.* **39** (1926) 206–14.
[106] J. Osugi, M. Okusima, M. Hamanoue, *Koatsu Gasu* **8** (1971) no. 4, 201–206.
[107] S. N. Bajpai, *Chem. Eng. Prog. Loss Prev. Technical Manual* **13** (1980) 119–122.
[108] B. Pesetsky, *Chem. Eng. Prog. Loss. Prev.* **13** (1980) 132–141.
[109] W. H. Perkin, *J. Chem. Soc.* **63** (1893) 488.
[110] F. A. Burden, J. H. Burgoyne, *Proc. R. Soc. London* **199** (1949) 328–351.
[111] D. Conrad, *Bundesgesundheitsblatt* **9** (1963) 139–141.
[112] G. W. Jones, R. E. Kennedy, *Ind. Eng. Chem.* **22** (1930) no. 2, 146–147.
[113] E. G. Plett, *Plant/Oper. Prog.* **3** (1984) 190–193.
[114] A. Fiumara, N. Mazzei, *Chim. Ind. (Milan)* **65** (1983) no. 11, 683–687.
[115] Y. Hashigushi, *Tokyo Kogyo Shikensho Hokoku* **60** (1965) no. 3, 85–91.
[116] J. H. Burgoyne, K. E. Bett, *Inst. Chem. Eng. Symp. Ser.* **25** (1968) 1–7.
[117] J. H. Burgoyne, K. E. Bett, R. Muir in J. M. Piric (ed.): *Symposium on Chem. Proc. Hazards (Inst. Chem. Eng.)* (1960), pp. 30–36.
[118] E. O. Haenni, N. A. Affens, H. G. Lento, A. H. Yeomans, R. A. Fulton, *Ind. Eng. Chem.* **51** (1959) no. 5, 685–688.
[119] J. N. Cawse, B. Pesetsky, W. T. Vyn, AIChE Loss Prevention Symposium, Houston, 1979.
[120] L. D. Chen, G. M. Faeth, *Combustn. Flame* **40** (1981) 13–28.
[121] A. K. Gupta, *J. Soc. Chem. Ind.* **68** (1949) 179–183.
[122] H. Grosse-Wortmann, *Chem. Ing. Tech.* **40** (1968) no. 14, 689–692.
[123] Merkblatt Glykole, Hoechst AG 1979; Merkblatt Genantin, Hoechst AG 1982; Merkblatt Antifrogen L, Hoechst AG 1985, Frankfurt am Main.
[124] Merkblatt Diethylenglykol, Hoechst AG 1986; Merkblatt Hoechst 1704LTV, Hoechst AG 1985; Merkblatt Hoechst 1678, Hoechst AG 1985.
[125] Merkblatt Triethylenglykol, Hoechst AG 1984.
[126] Merkblatt Polyglykole Hoechst, Eigenschaften und Anwendungsgebiete, Hoechst AG 1986.
[127] Merkblatt Ethylenglykoldimethylether, Hoechst AG 1986; Merkblatt Lösemittel Hoechst, Hoechst AG 1984.
[128] *Chem. Week* **134** (1984) March 7, 32–35.
[129] Y. Nassar, *CHEMTECH* **2** (1986) 96–98.
[130] D. S. Glass, *CHEMTECH* **4** (1986) 218–222.
[131] A. M. Thiess, *Arch. Toxikol.* **20** (1963) 127.
[132] H. Encke, Inaug.-Diss., Köln (1936).
[133] W. Tilling, *Ärztl. Wochenschr.* **9** (1954) 282.
[134] R. S. McLaughlin, *Am. J. Ophthalmol.* **29** (1946) 1355.
[135] E. Metz, *Ärztl. Sachverständ.-Ztg.* **44** (1938) 155.
[136] P. D. Blandin, *Berufsdermatosen* **6** (1958) 217.
[137] J. L. Shupack, S. R. Andersen, S. J. Romano, *J. Lab. Clin. Med.* **98** (1981) 723.

[138] P. F. Finelli, Th. F. Morgan, J. Yaar, C. V. Granger, *Arch. Neurol. Chicago* **40** (1983) 419.

[139] J. A. Gross, M. L. Haas, T. R. Swift, *Neurology* **29** (1979) 978.

[140] V. F. Gray, J. Hozier, D. Jacobs, R. L. Wade, D. G. Gray, *Environ. Mutagen.* **1** (1979) 375.

[141] H. F. Smyth, J. Seaton, L. Fischer, *J. Ind. Hyg. Toxicol.* **23** (1941) 259.

[142] K. H. Jacobson, E. B. Hackley, L. Feinsilver, *Arch. Ind. Hlth.* **13** (1956) 237.

[143] F. Koelsch, E. Lederer, *Zentralbl. Gewerbehyg. Unfallverhüt.* **7** (1930) 264.

[144] W. M. Snellings, C. S. Weil, R. R. Maronpot, Project Rep. 44-20, Bushy Run Research Center, Pittsburgh, Pa., (1981).

[145] R. L. Hollingsworth, V. K. Rowe, F. Oyen, D. D. McCollister, H. C. Spencer, *Arch. Ind. Hlth.* **13** (1956) 217.

[146] W. M. Snellings, R. R. Maronpot, J. P. Zelenak, C. P Laffoon, *Toxicol. Appl. Pharmacol.* **64** (1982) 476.

[147] P. L. Hackett, M. G. Brown, R. L. Buschbom, M. L. Clark, R. A. Miller, R. L. Music, S. E. Rowe, R. E. Schirmer, M. R. Sikov, "Teratogenic study of ethylene and propylene oxide and n-butylacetate," Study performed at Battelle. NIOSH contract no. 210-80-0013 (1982).

[148] L. A. Rapoport, *Dokl. Akad. Nauk SSSR* **60** (1984) 469.

[149] G. A. Sega, R. B. Cumming, J. G. Owens, C. Y. Horton, L. A. Lewis, *Environ. Mutagen.* **3** (1981) 371.

[150] G. Kölmark, M. Westergaard, *Hereditas (Lund, Swed.)* **39** (1953) 209.

[151] U. Rannug, R. Göthe, C. A. Wachtmeister, *Chem.-Biol. Interact.* **12** (1976) 251.

[152] E. H. Pfeiffer, H. Dunkelberg, *Food Cosmet. Toxicol.* **18** (1980) 115.

[153] J. W. Embree, J. P. Lyon, C. H. Hine, *Toxicol. Appl. Pharmacol.* **33** (1975) 172.

[154] D. Jenssen, C. Ramel, *Mutat. Res.* **75** (1980) 191.

[155] W. M. Generoso, K. T. Cain, M. Krishna, C. W. Sheu, R. M. Gryder, *Mutat. Res.* **73** (1980) 133.

[156] L.-E. Appelgren, G. Eneroth, C. Grant, L.-E. Landström, K. Tenghagen, *Acta Pharmacol. Toxicol.* **43** (1978) 69.

[157] V. Poirier, D. Papadopoulo, *Mutat. Res.* **104** (1982) 255.

[158] D. W. Lynch, T. R. Lewis, W. J. Moormann, P. S. Sabharwal, J. A. Burg, *Toxicologist* **3** (1983) 60.

[159] A. D. Kligerman, G. L. Erexon, M. E. Phelps, J. L. Wilmer, *Mutat. Res.* **120** (1983) 37.

[160] J. W. Yager, C. J. Hines, R. C. Spear, *Science* **219** (1983) 1221.

[161] C. Laurent, J. Fréderic, F. Maréchal: "Etude des effets cytogénétiques d'intoxication á l'oxyde d'éthyléne," *C. R. Séances Soc. Biol.* **176** (1982) no. 5, 733.

[162] C. Laurent, J. Fréderic, A. Y. Léonard: "Sister chromatic exchange frequency in workers exposed to high levels of ethylene oxide in hospital sterilization service," *Int. Arch. Occup. Environ. Health* (1983), cited in [173]

[163] C. Laurent, J. Fréderic, F. Maréchal: "Augmentation du taux d'échanges entre chromatides-soeurs chez des personnes exposées professionellement a l'oxyde d'éthyléne," *Ann. Genet.* (1983), cited in [173]

[164] B. L. Van Duuren, L. Orris, N. Nelson, *J. Natl. Cancer Inst.* **31** (1965) 707.

[165] H. Dunkelberg, *Br. J. Cancer* **39** (1979) 588.

[166] H. Dunkelberg, *Zentralbl. Bakteriol. Parasitenkd.Infektionskrankh. Hyg.* **174** (1981) 384.

[167] H. Dunkelberg, *Br. J. Cancer* **46** (1982) 924.

[168] W. M. Snellings, C. S. Weil, R. R. Maronpot, *Toxicol. Appl. Pharmacol.* **75** (1984) 105.

[169] C. Hogstedt, N. Malmqvist, B. Wadman, *JAMA J. Am. Med. Assoc.* **241** (1979) 1132.

[170]  C. Hogstedt, O. Rohlen, B. S. Berndtsson, O. Axelson, L. Ehrenberg, *Br. J. Ind. Med.* **36** (1979) 276.
[171]  R. W. Morgan, K. W. Claxton, B. J. Divine, S. D. Kaplan, V. B. Harris, *JOM J. Occup. Med.* **23** (1981) 767.
[172]  C. Hogstedt, L. Aringer, A. Gustavsson, *JAMA J. Am. Med. Assoc.* **255** (1986) 1578.
[173]  ECETOC Technical Report No. 11: *"Ethylene oxide toxicology and its relevance to man,"* European Chemical Industry Ecology and Toxicology Centre, Avenue Louise 250, Bte. 63, B-1050 Brussels (1984).

# 2-Ethylhexanol

HELMUT BAHRMANN, Ruhrchemie Aktiengesellschaft, Oberhausen, Federal Republic of Germany (Chaps. 2–5)

HEINZ-DIETER HAHN, Ruhrchemie Aktiengesellschaft, Oberhausen, Federal Republic of Germany (Chaps. 2–5)

DIETER MAYER, Hoechst Aktiengesellschaft, Frankfurt, Federal Republic of Germany (Chap. 6)

1. Introduction . . . . . . . . . . . . . 2361
2. Properties . . . . . . . . . . . . . . . 2362
3. Industrial Production . . . . . . 2362
4. Uses . . . . . . . . . . . . . . . . . . . 2366
5. Economic Aspects . . . . . . . . . 2367
6. Toxicology . . . . . . . . . . . . . . 2367
7. References . . . . . . . . . . . . . . 2368

## 1. Introduction

2-Ethylhexanol (2-ethyl-1-hexanol) [104-76-7], $C_8H_{18}O$, $M_r$ 130.23, ranks after the lighter alcohols (methanol to butanol) as the most important synthetic alcohol (see also → Alcohols, Aliphatic) [1]–[3]. Approximately $2 \times 10^6$ t/a are produced worldwide. 2-Ethylhexanol is mainly used as the alcohol component for the manufacture of ester plasticizers for soft poly(vinyl chloride) (PVC) and has been produced for this purpose since the mid-1930s.

$$CH_3(CH_2)_3\underset{\underset{C_2H_5}{|}}{C}HCH_2OH$$

2-Ethylhexanol

The raw material for the synthesis of 2-ethylhexanol, namely butyraldehyde (butanal) [123-72-8], is made almost exclusively from the petrochemical feedstock propene via the oxo synthesis (→ Butanals). The route from ethylene via acetaldehyde is insignificant. This situation could change if acetaldehyde produced from ethanol (via fermentation or homologation of methanol) and coal-based synthesis gas (which is already used for the oxo synthesis) becomes available at a reasonable price.

Acetaldehyde was formerly produced from coal via acetylene. This synthesis route had no chance of competing successfully against the route based on petrochemical raw materials because of the large number of steps involved.

**Table 1.** Physical properties of 2-ethylhexanol

| Property | Value |
| --- | --- |
| Viscosity at 20 °C | 9.8 mPa·s |
| Vapor pressure at 20 °C | ca. 0.03 kPa |
| Heat of vaporization at 184.8 °C | 50.66 kJ/mol |
| Solubility in water at 20 °C | 0.07 wt% |
| Solubility of water in 2-ethylhexanol at 20 °C | 2.7 wt% |
| Surface tension at 20 °C | $0.25 \times 10^{-3}$ N/cm |
| Dielectric constant at 20 °C | 7.7 |
| bp at 101.3 kPa | 184.6 °C |
| Critical temperature | 339.8 °C |
| Critical pressure | 2.76 MPa |
| Critical density | 0.2636 g/cm$^3$ |
| Critical compressibility | 0.2670 |
| Specific flow resistance at 20 °C | $5.8 \times 10^3$ MΩ · cm |
| Azeotrope with water at 101.3 kPa, 20% 2-ethylhexanol/ 80% water, bp | 183.5 °C |

## 2. Properties

**Physical Properties.** 2-Ethylhexanol is a clear liquid with a characteristic odor and forms a transparent mixture with other alcohols, ethers, and most organic liquids. Its main physical properties are summarized in Tables 1 and 2.

**Chemical Properties.** 2-Ethylhexanol reacts in the typical manner of α-branched primary alcohols (→ Alcohols, Aliphatic).

## 3. Industrial Production

2-Ethylhexanol is produced in four steps:

1) Aldolization of butyraldehyde and subsequent dehydration
2) Separation of the aldolization solution
3) Hydrogenation of unsaturated 2-ethyl-2-hexenal as an intermediate product
4) Fractionation of 2-ethylhexanol

**Table 2.** Temperature dependence of some physical properties of 2-ethylhexanol

| Property | t, °C | Value |
|---|---|---|
| Vapor pressure, kPa | 78.7 | 1.333 |
| | 90.8 | 2.666 |
| | 104.6 | 5.332 |
| | 113.5 | 7.998 |
| | 125.5 | 13.33 |
| | 143.5 | 26.66 |
| | 164.2 | 53.32 |
| | 184.8 | 101.31 |
| Relative density, $d^t_4$ | 10 | 0.8396 |
| | 15 | 0.8359 |
| | 20 | 0.8323 |
| | 25 | 0.8286 |
| | 30 | 0.8247 |
| | 50 | 0.8100 |
| Refractive index, $n^t_D$ | 10 | 1.4356 |
| | 15 | 1.4337 |
| | 20 | 1.4317 |
| | 25 | 1.4298 |
| | 30 | 1.4278 |
| | 50 | 1.4199 |
| Specific heat, $c_p$, J g$^{-1}$ K$^{-1}$ | 0 | 2.22 |
| | 20 | 2.34 |
| | 50 | 2.51 |
| | 80 | 2.68 |

$$CH_3CH_2CH_2CHO$$
$$\downarrow Cat.$$
$$\underset{\underset{C_2H_5}{|}}{CH_3CH_2CH_2CHCHCHO}\;\overset{OH}{|}$$
$$\downarrow -H_2O$$
$$\underset{\underset{C_2H_5}{|}}{CH_3CH_2CH_2CH=C-CHO}$$
$$\downarrow +2H_2$$
$$\underset{\underset{C_2H_5}{|}}{CH_3CH_2CH_2CH_2CHCH_2OH}$$

The butyraldehyde required for the aldolization stage is made by hydroformylation of propene. In addition to processes using cobalt as a catalyst [5]–[8], processes that employ rhodium catalysts [9]–[12] have gained increasing significance in the last few years.

The demands placed on the purity of 2-ethylhexanol are so high that the purity of the starting product, butyraldehyde, is also extremely important. Isobutyraldehyde is formed to a greater (cobalt catalysts) or lesser (rhodium cata-lysts) extent during hydroformylation and must be separated to prevent mixed aldolization. Therefore,

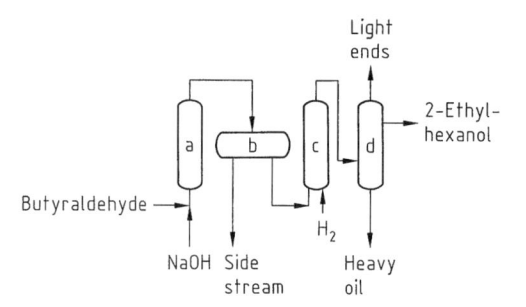

**Figure 1.** 2-Ethylhexanol plant
a) Aldolization; b) Phase separator; c) Hydrogenation; d) Distillation

special processes in which aldolization and, in some cases, hydrogenation occur in parallel with hydroformylation are becoming less important [5], [13], [14].

Older processes based on acetylene that involve the intermediate stages acetaldehyde/crotonaldehyde would permit access not only to coal as a raw material, but also to petrochemistry via ethylene [15], [16] as well as to regenerable raw materials via ethanol. However, these processes are not currently important.

The major oxo producers [17]–[19] all offer processes for the manufacture of 2-ethylhexanol that differ only in certain details. Figure 1 shows the flow diagram for a 2-ethylhexanol plant.

In the aldolization reactor, butyraldehyde reacts very quickly to give 2-ethyl-2-hexenal, aqueous sodium hydroxide being employed as the standard industrial catalyst. Local overheating in the reaction mixture must be avoided, since this may cause secondary reactions and thus decrease yields. Thorough mixing of the two-phase system is necessary. The primary aldol addition must take place rapidly; the immediately ensuing dehydration of the hydroxyaldehyde must be conducted as quickly and quantitatively as possible because the aldol is unstable and can impair the product quality and yield.

The aldolization reactor may be a mixing pump [20], a packed column [7], or a stirring vessel [21], [22].

The various processes operate at a temperature of between 80 and 150 °C and pressures below 0.5 MPa. The ratio of aldehyde to aqueous sodium hydroxide solution is 1:10–1:20. Under these conditions, conversion rates > 99 % are obtained. The heat of the aldolization reaction is used for steam generation.

After the aldolization stage in (a) (see Fig. 1), the mixture is separated in a phase separator (b) into an upper organic phase and a lower aqueous phase containing the aldolization solution. Part of the aldolization solution can be recycled, but the rest must be removed from the system via a side stream because the aldolization solution is diluted by the water that is produced in the reaction; the aldolization byproducts (Cannizzaro and Tischchenko reactions) must also be removed. Sodium hydroxide is added to maintain its concentration at 2–4 %

The side stream has such a high COD value (e.g., due to its sodium butyrate content) that it should be subjected to preliminary purification before being conveyed to a

**Table 3.** Quality specifications for 2-ethylhexanol

| Specification | Value |
|---|---|
| 2-Ethylhexanol | ≥ 99.5 wt% |
| Density at 20 °C | 0.832 – 0.833 kg/L |
| Refractive index, $n_D^{20}$ | 1.431 – 1.432 |
| Color, Pt – Co scale | max. 5 |
| Distillation range (95 vol%) at 101.3 kPa | 184 – 185 °C |
| Water content | max. 0.05 wt% |
| Pour point | < – 60 °C |

biological treatment plant. Suitable purification methods are oxidation, acid treatment, and extraction which allows partial recycling of valuable products.

The organic product from the phase separator (b) can either be hydrogenated in a single stage on a fixed nickel catalyst [9] or in several stages either in a combination of gas – liquid phases [23] or liquid – liquid phases (sump-phase or trickle-bed reactor). The heat of reaction for the hydrogenation of the C = C double bond and the aldehyde group is relatively high, 178 kJ/mol. Temperature control problems may therefore arise as local overheating decreases yields and must be avoided.

With single-step hydrogenation, remixing with the hydrogenation product has been proposed to dissipate heat (150 – 200 °C) [7]; in contrast to other processes, medium pressure is initially necessary to ensure adequate conversion. Modern plants normally utilize two stages to remove residual amounts of carbonyl compounds and to ensure that high-grade 2-ethylhexanol is obtained [2]. Nickel [7], [24], [25], copper [26] or mixed systems [23], [27] – [30] are preferred as heterogeneous hydrogenation catalysts. The heat of the hydrogenation reaction is also used for steam generation. In the hydrogenation stage, a conversion of 100% and a selectivity of > 99% is attained.

Fractional distillation of the hydrogenation product from (c) in (d) normally takes place in three stages: In the first stage, the light ends are separated at the head and can be employed for the manufacture of 1-butanol. In the second stage, pure 2-ethylhexanol is collected at the head. In the third stage, the recyclable intermediate fractions are separated from the heavy oil. The byproducts may be used for heating purposes.

*Trade Names.* 2-Ethylhexanol is usually marketed by the manufacturers (e.g., Ruhrchemie, BASF, Hüls, Union Carbide, Mitsubishi Chemical) under its chemical designation 2-ethyl-1-hexanol or as octyl alcohol. Typical *specifications* have been compiled in Table 3.

**Storage and Transportation.** Safety precautions for the shipment and storage of 2-ethylhexanol are determined by its combustibility, flash point, and ignition temperature (see Table 4). Dry 2-ethylhexanol does not corrode standard metals and, like other alcohols, can be stored in standard steel containers (e.g., ST 37). For particularly high-quality demands, use of stainless steel or aluminum containers is advisable. To prevent the entry of atmospheric oxygen or moisture, an inert gas cover should be provided. 2-Ethylhexanol can be transported in iron drums, tank trucks, or railway cars made of

**Table 4.** Safety information for 2-ethylhexanol

| Safety parameter | Value |
|---|---|
| Flash point, closed cup | 75 °C |
| Ignition temperature | 270–305 °C |
| Lower and upper explosive limits in air at 101.3 kPa | 1.1–7.4 vol% |
| Hazard classification for transport | |
|   CFR 49: 172.101, flammable liquid | |
|   RID/ADR class 3, number 32 c | |

standard steel, aluminum, or stainless steel. Centrifugal pumps conforming to chemical standards with simple rotating mechanical seals and polytetrafluoroethylene (PTFE) flange seals have proved to be suitable for use with 2-ethylhexanol.

# 4. Uses

In the United States, 2-ethylhexanol is mainly used in the form of its secondary products for the following applications [1]:

| | |
|---|---|
| Plasticizers | 70% |
| 2-Ethylhexanol acrylate | 15% |
| Cetane improvers and lubricant additives | 9% |
| Surfactants | 3% |
| Miscellaneous | 3% |

Worldwide, 2-ethylhexanol esters have > 60% share of the poly(vinyl chloride) (PVC) plasticizers market. In the United States, bis(2-ethylhexyl) phthalate [*117-81-7*], di-2-ethylhexyl phthalate (DEHP), also called dioctyl phthalate (DOP), dominates (70%); this is followed by bis(2-ethylhexyl) adipate [*103-23-1*], di-2-ethylhexyl adipate (DEHA), also called dioctyl adipate (DOA), (16%), and tris(2-ethylhexyl) mellitate, trioctyl trimellitate (12%). In Europe, DEHP's share of the market is even higher. The outstanding importance of DEHP is due to its favorable property combination as a plasticizer (good gelling properties, low volatility, high resistance to heat and water, and excellent electrical properties). Comprehensive investigations have been conducted on the toxicology and ecology of the widely used chemical DEHP during the last few years. The results have shown that its health and ecotoxicological risks are extremely low [33], [34].

The second large field of application is the production of 2-ethylhexyl acrylate [*103-11-7*], which is used to manufacture coating materials (especially emulsion paints), adhesives, printing inks, impregnating agents, and reactive diluent/cross-linking agents. 2-Ethylhexyl nitrate [*27247-96-7*] serves as a cetane number improver [35]. 2-Ethylhexyl phosphates are used as lubricating oil additives [36]. Antifoaming agents (cf. [37], [38]), dispersants (cf. [39]), and flotation agents [40] made from 2-ethylhexanol deserve special mention as surfactants. Other applications of 2-ethylhexanol include its

**Table 5.** Worldwide capacities for 2-ethylhexanol

| Country/region | Capacity, $10^3$ t/a |
|---|---|
| Western Europe | 770 |
| (Federal Republic of Germany) | (550) |
| Eastern Europe | 490 |
| United States | 240 |
| Central and South America | 150 |
| Asia | 480 |
| Total | 2130 |

use as a solvent, above all for polymerization catalysts (cf. [41]) and in extracting agents (cf. [42]).

# 5. Economic Aspects

Low production costs for butyraldehyde in modern large-scale plants and the lower price of propene as compared with ethylene mean that the oxo route for the manufacture of 2-ethylhexanol will continue to be given preference for economic reasons. As the 2-ethylhexanol manufacturers have their own oxo and hydrogenation capacities, the 1-butanol formed as a byproduct via the hydration of butyraldehyde is also used. In modern plants, ca. 860 kg of 2-ethylhexanol and ca. 18 kg of 1-butanol are obtained from 1000 kg of butyraldehyde. Other production processes will be able to survive in the long term only in countries where the necessary raw materials are economically available. Table 5 lists worldwide capacities.

# 6. Toxicology

Oral $LD_{50}$ values of 2-ethylhexanol are ca. 2.0 g/kg [43] and 3.7 g/kg in rats [44]. Gross necropsy revealed evidence of gastrointestinal irritation. In the rabbit eye, the undiluted material produces an erythema, swelling of the conjunctiva, lacrimation, and mucous secretion [45].

After skin contact in rabbits, 2-ethylhexanol is mildly irritating [43] if the patch is not closed; otherwise moderate skin irritation occurs [46]. The lowest published $LD_{50}$ after dermal application is 2 g/kg in rabbits [43].

Mice, rats, and guinea pigs survived inhalation of a nearly saturated atmosphere (227 mL/m$^3$) over a period of 6 h [43]. Typical signs of intoxication were central nervous depression, dyspnea, and irritation of mucous membranes. Repeated inhalation of a maximum dose of 110 mg/m$^3$ by rats caused dystrophic changes in parenchymatous organs [47].

The cumulative oral toxicity seems to be low [45]. A total of 12.5 g/kg in the diet fed to rats over a period of 90 d affected the kidneys and the liver [46]. An amount of 2 mL/kg fed to rabbits for 10 d caused signs of local irritation and systemic toxicity [43]. Peroxisomes were induced with liver enlargement when rats were fed with dietary levels of 2% 2-ethylhexanol for a period of 3 weeks [48].

In rabbits [49] and rats [50], 2-ethylhexanol is conjugated to glucuronic acid and eliminated via the urine, its main metabolite being 2-ethylhexanonylglucuronide. A small amount of the radioactively labeled material is oxidized and expired as $CO_2$. Only 2% is excreted unchanged.

Neither the ACGIH, OSHA, nor DFG have set standards for exposure levels.

# 7. References

General References

[1]  Chemical marketing reporter, New York 1984, 22. Oct., (Chemical profile).
[2]  B. Cornils, A. Mullen, *Hydrocarbon Process* **59** (1980) no. 11 93.
[3]  C. E. O'Rourke, P. R. Kavasmaneck, R. E. Uhl in E. J. Wickson: "Monohydric Alcohols," *ACS Symp. Ser.* **159** (1981) 71.

Specific References

[4]  J. Falbe, B. Cornils, *Fortschr. Chem. Forsch.* **11** (1968) no. 1, 101.
[5]  B. Cornils, J. Falbe (eds.): *New Synthesis with CO*, Springer Verlag, Heidelberg – Göttingen – New York 1980.
[6]  Ruhrchemie AG/Rhône-Poulenc S.A. *Hydrocarbon Process.* **56** (1977) no. 11, 163.
[7]  G. Dümbgen, D. Neubauer, *Chem. Ing. Tech.* **41** (1969) 974.
[8]  Farbwerke Hoechst, DE 838 746, 1953 (W. Berndt).
[9]  J. Matthey: "Low-pressure oxo process yields a better product mix," *Chem. Eng.* **12** (1977) 110.
[10] *Chem. Ind. (London)* 1981, 169.
[11] K. H. Schmidt, *Chem. Ind.* **108** (1985) 762.
[12] Ruhrchemie/Rhône-Poulenc-Verfahren, *Hydrocarbon Process.* **64** (1985) no. 11, 153.
[13] Esso, FR 1 324 873, 1962 (C. Roming); US 3 946 082, 1972.
[14] Shell Int. Res. Matsch., DE-OS 1 468 603, 1972; DE-OS 1 468 615, 1972 (C. R. Greene).
[15] H. Weber, W. Dimmling, A. M. Desai, *Hydrocarbon Process.* **55** (1976) no. 4, 127.
[16] Kyowa Hakko Kogyo Co. *Hydrocarbon Process.* **46** (1967) no. 11,153.
[17] BASF, *Hydrocarbon Process.* **52** (1973) no. 11, 107.
[18] Ruhrchemie, *Hydrocarbon Process,* **64** (1985) no. 11, 140.
[19] UCC, *Hydrocarbon Process.* **54** (1975) no. 11, 54, 117.
[20] Ruhrchemie, DE 2 437 957, 1976 (G. Kessen, J. Meis).
[21] Kuhlmann S.A., DE-OS 1 906 850, 1968 (H. Lemke, R. Duval); DE-OS 1 966 388, 1972.
[22] BASF, DE 927 626, 1975 (H. J. Nienburg, E. Nicolai, W. Hagen).
[23] G. S. Gurevich, L. A. Shapiro et al., SU 1 084 268, 1984;SU 1 010 052, 1983; *Chem. Abstr.* **101** (1984) 170 697 d; *Chem. Abstr.* **99** (1983) 53 115 s.
[24] BASF, DE 1 277 232, 1968 (H. Corr, E. Haarer, H. Hoffmann, S. Winderl).

[25] R. Kubicka, J. Veprek et al., CS 207 933, 1983; *Chem. Abstr.* **100** (1984) 191 365 p.
[26] Ruhrchemie, DE 2 538 253, 1978 (G. Horn, C. D. Frohning).
[27] S. G. Gurevich, L. A. Shapiro et al., SU 692 824, 1979; Chem. Abstr. **92** (1980) 58 222 q.
[28] BASF, DE 1 949 296, 1974 (G. Boettger, H. Hoffmann, L. Schuster, H. Toussaint).
[29] Chemische Werke Hüls, DE 1 643 856, 1971 (M. Reich).
[30] BASF, DE 2 832 699, 1980 (G. Heilen, A. Nissen, W. Koernig et al.).
[31] BASF, Data sheet for 2-ethylhexanol, 1984.
[32] Hoechst AG, Firmenbroschüre über Alkohole, Hoechst 1984; Hoechst AG, Firmenbroschüre über 2-Ethylhexanol, 1982.
[33] ECETOC, Technical Report No. 19, May 1985.
[34] Beratergremium für umweltrelevante Altstoffe der Gesellschaft Deutscher Chemiker, Stoffbericht 4, Jan. 1986.
[35] Ethyl Corp., US 4 479 905, 1984 (G. G. Knapp, M. S. Marguerite, P. D. Seemuth).
[36] Elco Corp., US 4 508 630, 1985 (F. A. Litt, T. R. Czernicki).
[37] Ciba-Geigy, EP 96 806, 1983 (R. Defago, R. Baeuerle).
[38] National Distillers and Chem. Corp., US. 4 540 511, 1985 (E. F. McCaffrey, A. J. Dieterman, L. Knazko).
[39] Kao Corp., JP 60/55 089, 1985.
[40] Petrofina S.A., CA 1 083 062, 1980 (P. E. A. Mortier).
[41] Chisso Corp., JP 60 192 710, 1985 (M. Harada, M. Iijima, N. Saito).
[42] General Electric Corp., US 4 547 596, 1985 (A. K. Mendiratta, J. J. Talley).
[43] H. F. Smyth, Jr., C. P, Carpenter, C. S. Weil, U. C. Pozzani, J. A. Striegel, J. S. Nycum, *Am. Ind. Hyg. Assoc. J.* **30** (1969) 470.
[44] R. A. Scala, E. G. Burtis, *Am. Ind. Hyg. Assoc. J.* **34** (1973) 493.
[45] P. Schmidt, R. Gohlke, R. Rothe, *Z. Gesamte Hyg. Ihre Grenzgeb.* **19** (1973) 485.
[46] V. K. Rowe, S. B. McCollister in G. D. Clayton, F. E. Clayton (eds.): *Patty's Industrial Hygiene and Toxicology*, 3rd ed., vol. **2A**, Wiley-Interscience, New York 1981, p. 4620.
[47] O. N. Mashkina, *Mater. Konf. Fiziol. Biokhim. Farmakol. Uchats. Prakt.* **168** (1966); *Chem. Abstr.* **67** (1967) 570 066.
[48] D. E. Moody, *J. Cell Biol.* **70** (1976) 271.
[49] J. A. Kamil, J. N. Smith, R. T. Williams, *Biochem. J.* **53** (1953) 137.
[50] P. W. Albro, *Xenobiotica* **5** (1975) 625.

# Fats and Fatty Oils

ALFRED THOMAS, Unimills International, Hamburg, Federal Republic of Germany

| | | | | |
|---|---|---|---|---|
| 1. | Introduction | 2372 | 5.1.3. Expelling | 2402 |
| 2. | Composition | 2374 | 5.1.4. Extraction | 2405 |
| 2.1. | Glycerides | 2374 | 5.2. Land-Animal Fats | 2410 |
| 2.2. | Fatty Acids | 2374 | 5.3. Marine Oils | 2411 |
| 2.3. | Phospholipids | 2380 | 5.4. Synthetic Fats | 2412 |
| 2.4. | Waxes | 2381 | 6. Refining | 2412 |
| 2.5. | Sterols and Sterol Esters | 2382 | 6.1. Degumming | 2413 |
| 2.6. | Terpenoids | 2382 | 6.2. Deacidification (Neutralization) | 2414 |
| 2.7. | Other Minor Constituents | 2384 | 6.3. Bleaching | 2417 |
| 3. | Physical Properties | 2385 | 6.4. Deodorization | 2419 |
| 3.1. | Melting and Freezing Points | 2385 | 7. Fractionation | 2423 |
| 3.2. | Thermal Properties | 2387 | 8. Hydrogenation | 2425 |
| 3.3. | Density | 2389 | 9. Interesterification | 2428 |
| 3.4. | Viscosity | 2390 | 10. Environmental Aspects | 2431 |
| 3.5. | Solubility and Miscibility | 2390 | 11. Standards and Quality Control | 2431 |
| 3.6. | Surface and Interfacial Tension | 2393 | 11.1. Sampling | 2432 |
| 3.7. | Electrical Properties | 2393 | 11.2. Raw Materials | 2432 |
| 3.8. | Optical Properties | 2393 | 11.3. Oils and Fats | 2433 |
| 4. | Chemical Properties | 2395 | 11.3.1. Physical Methods | 2433 |
| 4.1. | Hydrolysis | 2396 | 11.3.2. Chemical Methods | 2435 |
| 4.2. | Interesterification | 2396 | 12. Storage and Transportation | 2440 |
| 4.3. | Hydrogenation | 2397 | 13. Individual Vegetable Oils and Fats | 2441 |
| 4.4. | Isomerization | 2398 | 13.1. Fruit Pulp Fats | 2441 |
| 4.5. | Polymerization | 2399 | 13.1.1. Palm Oil | 2442 |
| 4.6. | Autoxidation | 2399 | 13.1.2. Olive Oil | 2443 |
| 5. | Manufacture and Processing | 2400 | 13.1.3. Avocado Oil | 2445 |
| 5.1. | Vegetable Oils and Fats | 2401 | 13.2. Seed-Kernel Fats | 2445 |
| 5.1.1. | Storage and Handling of Raw Materials | 2401 | 13.2.1. Lauric Acid Oils | 2445 |
| | | | 13.2.1.1. Coconut Oil | 2445 |
| 5.1.2. | Cleaning and Dehulling | 2401 | 13.2.1.2. Palm Kernel Oil | 2447 |

| | |
|---|---|
| 13.2.1.3. Babassu Oil and Other Palm Seed Oils . . . . . . . . . . . . . . . . . . . . 2448 | 13.2.6. Cruciferous Oils . . . . . . . . . . . . 2463 |
| 13.2.2. Palmitic – Stearic Acid Oils . . . . 2449 | 13.2.6.1. Rapeseed Oil . . . . . . . . . . 2464 |
| 13.2.2.1. Cocoa Butter . . . . . . . . . . . . . . 2449 | 13.2.6.2. Mustard Seed Oil . . . . . . . . . . 2465 |
| 13.2.2.2. Shea Butter, Borneo Tallow, and Related Fats (Vegetable Butters) . 2451 | 13.2.7. Conjugated Acid Oils . . . . . . . . 2465 |
| | 13.2.7.1. Tung Oil and Related Oils . . . . . 2465 |
| 13.2.3. Palmitic Acid Oils . . . . . . . . . . 2452 | 13.2.7.2. Oiticica Oil and Related Oils . . . 2465 |
| 13.2.3.1. Cottonseed Oil . . . . . . . . . . . . . . 2452 | 13.2.8. Substituted Fatty Acid Oils . . . . 2466 |
| 13.2.3.2. Kapok and Related Oils . . . . . . 2453 | 13.2.8.1. Castor Oil . . . . . . . . . . . . . . . . . 2466 |
| 13.2.3.3. Pumpkin Seed Oil . . . . . . . . . 2453 | 13.2.8.2. Chaulmoogra, Hydnocarpus, and Gorli Oils . . . . . . . . . . . . . . . . 2467 |
| 13.2.3.4. Corn (Maize) Oil . . . . . . . . . . . 2454 | |
| 13.2.3.5. Cereal Oils . . . . . . . . . . . . . . . 2455 | 13.2.8.3. Vernonia Oil . . . . . . . . . . . . . 2467 |
| 13.2.4. Oleic – Linoleic Acid Oils . . . . . . 2455 | 14. Individual Animal Fats . . . . . . 2467 |
| 13.2.4.1. Sunflower Oil . . . . . . . . . . . . . 2456 | 14.1. Land-Animal Fats . . . . . . . . 2468 |
| 13.2.4.2. Sesame Oil . . . . . . . . . . . . . . . 2456 | 14.1.1. Lard . . . . . . . . . . . . . . . . . . . 2468 |
| 13.2.4.3. Linseed Oil . . . . . . . . . . . . . . . 2458 | 14.1.2. Beef Tallow . . . . . . . . . . . . . . . . 2469 |
| 13.2.4.4. Perilla Oil . . . . . . . . . . . . . . . . 2459 | 14.1.3. Mutton Tallow . . . . . . . . . . . . . 2470 |
| 13.2.4.5. Hempseed Oil . . . . . . . . . . . . . 2459 | 14.1.4. Horse, Goose, and Chicken Fat . . 2471 |
| 13.2.4.6. Teaseed Oil . . . . . . . . . . . . . . . 2459 | 14.2. Marine Oils . . . . . . . . . . . . . 2471 |
| 13.2.4.7. Safflower and Niger Seed Oils . . 2459 | 14.2.1. Whale Oil . . . . . . . . . . . . . . . . 2471 |
| 13.2.4.8. Grape-Seed Oil . . . . . . . . . . . . 2460 | 14.2.2. Fish Oil . . . . . . . . . . . . . . . . . . 2472 |
| 13.2.4.9. Poppyseed Oil . . . . . . . . . . . . . 2460 | 15. Economic Aspects . . . . . . . . 2473 |
| 13.2.5. Leguminous Oils . . . . . . . . . . . 2461 | 16. Toxicology and Occupational Health . . . . . . . . . . . . . . . . . 2473 |
| 13.2.5.1. Soybean Oil . . . . . . . . . . . . . . 2461 | |
| 13.2.5.2. Peanut Oil . . . . . . . . . . . . . . . 2462 | |
| 13.2.5.3. Lupine Oil . . . . . . . . . . . . . . . 2463 | 17. References . . . . . . . . . . . . . . 2477 |

# 1. Introduction

Naturally occurring oils and fats are liquid or solid mixtures consisting primarily of glycerides. Depending on whether they are solid or liquid at ambient temperature, they are referred to as *fats* or *oils*, respectively. Naturally occurring oils and fats always contain minor constituents such as free fatty acids, phospholipids, sterols, hydrocarbons, pigments, waxes, and vitamins. The nomenclature rules for glycerides have been summarized [18].

**History.** A century ago ecological, religious, and social factors still played a more important role than technology in the choice and utilization of oils and fats.

Primeval humans utilized animal fats, making cheese and butter from goat's milk. Oilseed plants were cultivated during the neolithic period. Poppy seeds have been found in remains of Bronze-Age bread; rapeseed and linseed, together with millstones, have been found in Bronze-Age dwellings.

Linseed, almonds, and sesame seed were part of Egypt's natural flora. Sesame oil had mythical significance. The oil-bearing safflower plant is still grown in Egypt. Olive oil came from Palestine, Syria, and Crete. The Phoenicians and Greek colonists introduced the olive tree to Sicily and Italy. Cotton is one of the oldest cultivated plants; it was grown 2600 years ago in India as a source of both oil and fiber. Soy and hempseed are mentioned as oilseed plants in a Chinese document of 2838 B.C.

The oilseeds were ground with a pestle and mortar or between stones. Simple mills of the type still being used in some developing countries — a concave stone rotating on a convex one — also evolved. The Egyptians developed the sack or expeller press. The Greeks and Romans used a grinding device known as the "trapetum". In North Africa, mechanical presses were used in processing plants that approached the size of modern factories.

The processing of oil fruits and seeds in Central and Northern Europe advanced more slowly. Oilseeds, primarily linseed, hempseed, and rapeseed, were pulped in hollowed stones, and the oil was expelled from the pulp by pressing between two cloth-covered frames. This domestic-type process was practiced up to the 16th century. Industrial oil milling developed primarily in regions where linseed was grown extensively. A wood engraving dated 1568 depicts an oil mill with a horse-drawn vertical millstone on a stone bed. The ground oilseed was heated in a kettle over an open fire and finally "beaten" in wedge presses.

Animal fats were obtained by rendering fatty tissue and by churning cream. Up to the middle of the 19th century, tallow and butter were the most important edible fats in Europe, lard and vegetable oils playing only a minor role.

Toward the end of the 19th century the production of oil by hydraulic pressing and solvent extraction was introduced. This process gave relatively high oil yields but necessitated posttreatment of the oils by neutralization, bleaching, and deodorization.

The invention of margarine in the 1870s gave further impetus to the oil-processing industry. With the discovery of oil hydrogenation (hardening) at the beginning of the 20th century, liquid oils could be converted into spreadable, consistent fats. In the 1930s, interesterification and fractionation were developed as further methods to modify the consistency of oils and fats.

Apart from being used for edible purposes, oils and fats are referred to in the Old Testament as cosmetic products and lamp fuel. Anointing with oil symbolized royal dignity. HOMER and HERODOTUS refer to the use of fats as "processing aids" during weaving. The ability of fats to calm waves was studied by Indian scholars 3000 years ago. The Egyptians supposedly used fats as lubricants to transport stone blocks. They were also familiar with the use of drying oils in varnishes and paints.

The elucidation of the chemical nature of fats was initiated by SCHEELE, who produced glycerol from olive oil around 1780. CHEVREUL subsequently (ca. 1815) recognized that fats were predominantly esters of fatty acids and glycerol.

Modern research and development is focussed on the application of new biotechnological principles to the production and modification of oils and fats [19], as well as on nutritional aspects, problems of trace contaminants, and environmental pollution. The breeding of new plant varieties is an important method for increasing the types of oils and fats. Examples are new varieties of rapeseed (low erucic acid content), safflower, and sunflower (high in oleic acid). New sources of oils and fats are being exploited by cultivation of wild plants such as the jojoba and *Cuphea* shrubs.

# 2. Composition

## 2.1. Glycerides

Naturally occurring fats contain about 97% triglycerides (triacylglycerides), i.e., triesters of glycerol with fatty acids; up to 3% diglycerides (diacylglycerides); and up to 1% monoglycerides (monoacylglycerides). Tri-, di-, and monoglycerides consist of 1 mol of glycerol esterified with 3 mol, 2 mol, or 1 mol of fatty acid, respectively. The triglycerides of naturally occurring oils and fats contain at least two different fatty acid groups. The chemical, physical, and biological properties of oils and fats are determined by the type of the fatty acid groups and their distribution over the triglyceride molecules. The melting point generally increases with increasing proportion of long chain fatty acids or decreasing proportion of short chain or unsaturated fatty acids. Milk fat (butterfat) and coconut oil, which contain a high proportion of $C_6$–$C_{12}$ fatty acids, have lower melting points than fats such as tallow and lard, which contain predominantly $C_{16}$ and $C_{18}$ fatty acids. Vegetable oils are liquid at ambient temperature because of their high proportion of unsaturated fatty acids.

The properties of a triglyceride are also determined by the position of the various fatty acid groups in the triglyceride molecule (i.e., 1-, 2-, or 3-position). The total number $N$ of possible triglycerides (including positional isomers) from $x$ different fatty acids is

$$N = \frac{x^2 + x^3}{2}$$

However, the proportions of different triglycerides in a naturally occurring fat generally do not conform to a statistical distribution. In vegetable oils and fats, unsaturated fatty acids are linked preferentially to the 2-position of the glycerol group, whereas in animal fats they appear primarily in the 1- and 3-positions ("2-random" and "1,3-random" distributions) [20]. Extreme examples of nonrandom distributions of fatty acid groups over the triglyceride molecule are cocoa butter (ca. 40% 1-palmito-3-stearo-2-olein) and lard (ca. 20% 2-palmito-1,3-diolein).

## 2.2. Fatty Acids

The fatty acids that form the triglycerides of naturally occurring oils and fats are predominantly even-numbered, straight-chain, aliphatic monocarboxylic acids with chain lengths ranging from $C_4$ to $C_{24}$. Unsaturated fatty acids differ in number and position of double bonds and in configuration (i.e., cis or trans isomers). The more common fatty acids are known by trivial names such as butyric, lauric, palmitic, oleic, stearic, linoleic, linolenic (→ Fatty Acids). Crude oils contain significant amounts of free fatty acids.

The chief fatty acids in some commercial oils and fats are listed Tables 1 (saturated fatty acids) and 2 (unsaturated fatty acids). The fatty acid composition of most vegetable oils and fats is relatively simple; they consist predominantly of palmitic, oleic, and linoleic acids [20]. The fatty acids of land-animal fats mainly have a chain length of $C_{16}$ or $C_{18}$. They are formed by biosynthetic conversion of carbohydrates, proteins, or fats, or originate directly from ingested fat.

Ruminant fats contain 5–10% trans fatty acids, which are produced from linoleic and linolenic acid in the rumen. Marine oils contain a high proportion of polyunsaturated fatty acids with a chain length of $C_{20}-C_{24}$. Land-animal fats and marine oils contain numerous odd-numbered and branched fatty acids in trace concentrations. More than 80 different fatty acids have been found in milk fat, and more than 40 in lard. However, most of these fatty acids occur only in traces. They can be of industrial significance for some oils and fats and they can play a role in the identification of fats and their detection in mixtures.

The fatty acid composition of a naturally occurring fat is determined genetically. The fatty acid composition of oleaginous seeds can be changed by developing new varieties. Examples are low-linoleic safflower and sunflower oils and low-erucic rapeseed oil. Environmental factors can influence the fatty acid composition within certain limits. The proportion of unsaturated fatty acids in the glycerides of linseed, soybean, and sunflower oils, for example, generally increases as the climate becomes colder and wetter. The subcutaneous fats of marine animals living in the colder parts of the oceans have a particularly high content of unsaturated fatty acids and thus a relatively low melting point. Linoleic acid cannot be synthesized by the animal or human organism and is hence referred to as an "essential" fatty acid.

Naturally occurring oils and fats are distributed homogeneously in varying concentrations in vegetable and animal tissues. In plants they are found predominantly in the seeds and the fruit pulp where they serve as a source of energy.

**Biosynthesis.** The synthesis of fatty acids in plants and animals generally starts with "activated" acetic acid, i.e., acetyl coenzyme A, which is derived from carbohydrates via pyruvic acid. Animals also synthesize acetyl coenzyme A from amino acids.

A distinction is made between *de novo synthesis* of saturated fatty acids and *elongation* of a fatty acid chain. De novo synthesis leads primarily to palmitic acid with smaller amounts of lauric, myristic, and stearic acids.

De novo synthesis:

$$CH_3COSCoA \xrightarrow{+CO_2} HOOCCH_2COSCoA \xrightarrow[-HSCoA]{+CH_3COSCoA}$$

Acetyl CoA    Malonyl CoA

**Table 1.** Saturated fatty acids in various oils and fats, (main sources: Unilever; Food RA, Leatherhead; ITERG)

| | Saturated fatty acids[b], g/100 g fatty acids | | | | | | | |
|---|---|---|---|---|---|---|---|---|
| | $C_{10}$ and lower | $C_{12}$ | $C_{14}$ | $C_{16}$ | $C_{18}$ | $C_{20}$ | $C_{22}$ | $C_{24}$ |
| *Liquid vegetable oils* | | | | | | | | |
| Almond | | | tr | 6.5–7 | 1–2.5 | tr | tr | |
| Avocado | | | tr | 10–26 | 0.5–1 | | | |
| Corn germ | | tr–0.5 | tr–0.3 | 9–12 | 1–3 | ca. 0.5 | tr–0.5 | <0.5 |
| Cottonseed | | tr | 0.5–2.0 | 21–27 | 2–3 | <0.5 | tr | tr |
| Grape-seed | | | | 4–11 | 2–5 | tr | tr | tr |
| Linseed | | tr | tr | 5–6 | 3–5 | <0.5 | tr–0.2 | tr |
| Olive | | | | 7–16 | 2–4 | ca. 0.5 | tr | tr |
| Peanut (Africa) | | | tr | 7–12 | 1.5–5 | ca. 1.5 | 2–4 | 1–2 |
| Peanut (South America) | | | tr | 10–13 | 1.5–4 | ca. 1.5 | 3–4 | 1.5–2 |
| Pumpkinseed | | | | 7–13 | 6–7 | tr | | |
| Rapeseed (high erucic) | | tr | tr | 2–4 | 1–2 | 0.5–1 | 0.5–2.0 | 0.5 |
| Rapeseed (low erucic) | | tr | tr | 3–6 | 1–2.5 | <1 | tr–0.5 | tr–0.2 |
| Ricebran | | | ca. 0.5 | 13–18 | ca. 2 | 0.5–1 | | ca. 0.5 |
| Safflower | | tr | tr | ca. 5 | 2–3 | ca. 0.5 | ca. 1 | ca. 1 |
| Sesame | | | tr | 8–10 | 3–6 | ca. 0.5 | | |
| Soybean | | | <0.5 | 8–12 | 3–5 | <0.5 | tr | |
| Sunflower | | tr | tr–0.1 | 5.5–8 | 2.5–6.5 | <0.5 | 0.5–1.0 | <0.5 |
| Wheat germ | | | | 12–14 | ca. 1 | 0.5 | | tr |
| *Consistent vegetable fats* | | | | | | | | |
| Babassu oil | ca. 12 | 42–44 | 15–18 | 8–10 | 2–3 | tr | | |
| Coconut oil | ca. 13 | 41–46 | 18–21 | 9–12 | 2–4 | tr | tr | |
| Cocoa butter | | tr | tr | 23–30 | 32–37 | <1 | ca. 0.5 | |
| Palm kernel oil | ca. 7 | 41–45 | 15–17 | 7–10 | 2–3 | tr–0.3 | tr–0.5 | |
| Palm oil (Africa) | | tr | 1–2 | 41–46 | 4–6.5 | ca. 0.5 | | |
| Palm oil (Indonesia) | | tr–0.5 | ca. 1 | 41–47 | 4–6 | ca. 0.5 | | |
| *Animal fats* | | | | | | | | |
| Beef tallow | | tr | 2–4 | 23–29 | 20–35 | <0.5 | tr | |
| Butterfat | 7–9 | 2–5 | 8–14 | 24–32 | 9–13 | 2 | | |
| Chicken fat | | | ca. 1 | 20–24 | 4–7 | | | |
| Goose fat | | | | 20–22 | 4–11 | | | |
| Horse fat | tr | ca. 0.5 | 3–6 | 20–30 | 6–10 | tr | | |
| Lard | tr | <0.5 | ca. 1.5 | 24–30 | 12–18 | ca. 0.5 | | |
| Mutton tallow | tr | ca. 0.5 | 1–4 | 22–30 | 15–30 | tr | tr | |
| *Marine oils* | | | | | | | | |
| Fish oils | | | | | | | | |
|   Japanese | tr | tr | ca. 6 | ca. 16 | ca. 3 | <0.5 | tr | tr–1 |
|   Menhaden | tr | tr | ca. 9 | ca. 20 | ca. 4 | tr–1 | tr | tr–1 |
|   Scandinavian | tr | tr | 6–8 | 11–15 | 1–3 | tr–0.5 | tr | tr–1 |
|   South American | tr | tr | ca. 7 | 17–19 | 2–4 | ca. 0.5 | tr | tr–1 |
| Whale oil | tr | ca. 0.5 | 4–10 | 10–18 | 1–3 | tr | | |

[a] ITERG = Institut des Corps Gras, Centre Technique Industrielle, Paris.
[b] tr = traces (<0.05%).

**Table 2.** Unsaturated fatty acids in various oils and fats, (main sources: Unilever; Food RA, Leatherhead; ITERG)

| | Unsaturated fatty acids[b], g/100 g fatty acids | | | | | | | | | |
|---|---|---|---|---|---|---|---|---|---|---|
| | $C_{14:1}$ | $C_{16:1}$ | $C_{18:1}$ | $C_{18:2}$ | $C_{18:3}$ | $C_{20:1}$ | $C_{22:1}$ | $C_{20:x}$[c] | $C_{22:x}$[c] | $C_{24:1}$ |
| *Liquid vegetable oils* | | | | | | | | | | |
| Almond | | <0.5 | 65–69 | 21–25 | tr | tr–0.1 | | | | |
| Avocado | | 2–12 | 44–76 | 8–25 | ca. 1 | | | | | |
| Corn germ | | <0.5 | 25–35 | 40–60 | ca. 1 | ca. 0.5 | tr–0.1 | | | |
| Cottonseed | | <1 | 14–21 | 45–58 | tr–0.2 | tr | | | | |
| Grape-seed | | | 12–33 | 45–72 | 1–2 | | | | | |
| Linseed | | tr | 18–26 | 14–20 | 51–56 | <0.5 | | | | |
| Olive | | 1–2 | 64–86 | 4–15 | 0.5–1 | 0.5 | | | | |
| Peanut (Africa) | | <0.5 | 50–70 | 14–30 | tr | 0.5–1.5 | tr | | | |
| Peanut (South America) | | <0.5 | 35–42 | 39–44 | tr | 0.5–1.5 | tr | | | |
| Pumpkinseed | | | 24–41 | 46–57 | | | | | | |
| Rapeseed (high erucic) | | ca. 0.5 | 11–24 | 10–22 | 7–13 | ca. 10 | 41–52 | | | |
| Rapeseed (low erucic) | | 0.1–0.5 | 52–66 | 17–25 | 8–11 | 1.5–3.5 | tr–2.5 | tr–0.1 | | tr |
| Ricebran | | | ca. 44 | 30–40 | | | | | | |
| Safflower | | tr | 12–20 | 70–80 | tr | tr | | | | |
| Sesame | | tr | 35–46 | 40–48 | tr–0.5 | <0.5 | | | | |
| Soybean | | tr | 18–25 | 49–57 | 6–11 | <0.5 | | | | |
| Sunflower | | <0.5 | 14–34 | 55–73 | tr–0.4 | <0.5 | tr–0.3 | | | |
| Wheat germ | | | ca. 30 | 40–55 | ca. 7 | | | | | |
| *Consistent vegetable fats* | | | | | | | | | | |
| Babassu oil | | | 14–16 | 1–2 | | | | | | |
| Coconut oil | tr | tr | 5–9 | 0.5–3 | tr | tr | | | | |
| Cocoa butter | | ca. 0.5 | 30–37 | 2–4 | | | | | | |
| Palm kernel oil | | | 10–18 | 1–3 | tr–0.5 | tr–0.5 | | | | |
| Palm oil (Africa) | | <0.5 | 37–42 | 8–12 | tr–0.5 | tr | | | | |
| Palm oil (Indonesia) | | ca. 0.5 | 37–41 | ca. 10 | tr–0.5 | tr | | | | |
| *Animal fats* | | | | | | | | | | |
| Beef tallow | ca. 0.5 | 2–4 | 26–45 | 2–6 | ca. 1 | <0.5 | | tr | ca. 0.5 | |
| Butterfat | ca. 2 | 3 | 19–33 | 1–4 | 2–6 | | | | ca. 2 | |
| Chicken fat | | ca. 7 | 38–44 | 18–23 | ca. 1 | | | | 0.5–1 | |
| Goose fat | | | 41–74 | 7–19 | | | | | | |
| Horse fat | | 3–10 | 36–40 | 6–11 | 4–9 | tr–0.5 | | tr | 1–2 | |
| Lard | tr | 2–3 | 36–52 | 10–12 | ca. 1 | 0.5–1 | | <0.5 | tr | |
| Mutton tallow | ca. 0.5 | 3–4 | 31–56 | 3–7 | 1–2 | | | | ca. 0.5 | |
| *Marine oils* | | | | | | | | | | |
| Fish oils | | | | | | | | | | |
| Japanese | tr | ca. 7 | ca. 14 | ca. 2 | ca. 1 | ca. 7 | ca. 6 | ca. 15 | ca. 12 | tr–1 |
| Menhaden | tr | ca. 11 | ca. 13 | ca. 2 | ca. 1 | ca. 2 | ca. 1 | ca. 14 | ca. 11 | tr–1 |
| Scandinavian | tr | 6–11 | 12–15 | 1–2 | 0.5–1 | 9–16 | 14–20 | 6–10 | 5–11 | tr–1 |
| South American | tr | 9–11 | 14–15 | 1–2 | 0.5–1 | 1–2 | 1–2 | 7–19 | 10–14 | tr–1 |
| Whale oil | 1–3 | 13–20 | 24–33 | 1–2 | tr | 10–15 | 4–10 | 1–6 | 5–7 | tr |

[a] ITERG = Institut des Corps Gras, Centre Technique Industrielle, Paris.
[b] tr = traces (<0.05%).
[c] x > 1.

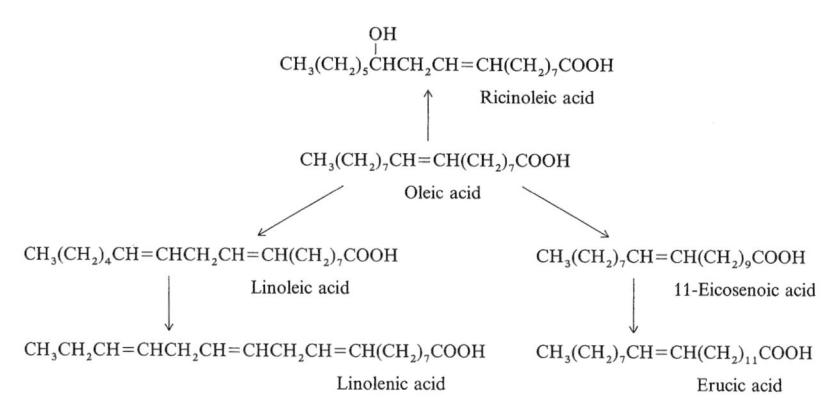

$$CH_3(CH_2)_5\overset{OH}{\underset{|}{C}}HCH_2CH=CH(CH_2)_7COOH$$
$$\text{Ricinoleic acid}$$
$$\uparrow$$
$$CH_3(CH_2)_7CH=CH(CH_2)_7COOH$$
$$\text{Oleic acid}$$

$CH_3(CH_2)_4CH=CHCH_2CH=CH(CH_2)_7COOH$  $\qquad$  $CH_3(CH_2)_7CH=CH(CH_2)_9COOH$
$\qquad\qquad$ Linoleic acid $\qquad\qquad\qquad\qquad\qquad\qquad\qquad$ 11-Eicosenoic acid

$CH_3CH_2CH=CHCH_2CH=CHCH_2CH=CH(CH_2)_7COOH$ $\qquad$ $CH_3(CH_2)_7CH=CH(CH_2)_{11}COOH$
$\qquad\qquad\qquad$ Linolenic acid $\qquad\qquad\qquad\qquad\qquad\qquad\qquad$ Erucic acid

**Figure 1.** Unsaturated fatty acids derived from oleic acid

$$CH_3COCH(COOH)COSCoA \xrightarrow{-CO_2} CH_3COCH_2COSCoA$$
$$\xrightarrow{+H_2} CH_3CH(OH)CH_2COSCoA \xrightarrow{-H_2O}$$
$$CH_3CH=CHCOSCoA \xrightarrow{+H_2}$$
$$CH_3CH_2CH_2COSCoA \xrightarrow[-HSCoA]{+Malonyl\ CoA}$$
Butyryl CoA

$$CH_3CH_2CH_2COCH(COOH)COSCoA\ etc.$$

The conversion of palmitic into stearic acid and the formation of longer chain saturated and unsaturated fatty acids presumably takes place by elongation, which involves addition of acetyl coenzyme A to activated fatty acids (RCOSCoA).

*Elongation:*
$$RCOOH \xrightarrow[-H_2O]{+HSCoA} RCOSCoA \xrightarrow[-HSCoA]{+CH_3COSCoA}$$
$$RCOCH_2COSCoA \longrightarrow \longrightarrow RCH_2CH_2COOH$$

The synthesis of fatty acids from $C_2$ units explains the predominance of even-numbered fatty acids in naturally occurring oils and fats. Bacterial degradation of feed in the rumen leads not only to acetic and butyric acids but also to propionic acid ($C_3$); incorporation of these $C_3$ units into fat synthesis in the udder explains the occurrence of branched and odd-numbered fatty acids in milk fat.

Oleic acid appears to play a key role in plants. It is probably synthesized from short-chain fatty acids and can be dehydrogenated into more highly unsaturated fatty acids; be converted into substituted acids such as ricinoleic acid; or elongated into erucic acid by specific enzyme systems (Fig. 1).

In the animal organism oleic acid is formed by dehydrogenation of stearic acid. Further dehydrogenation to linoleic and linolenic acid does not take place. The animal organism can, however, further desaturate linoleic and linolenic acid by introducing double bonds between the carboxyl group and the double bond nearest to it. Together with an elongation of the fatty acid chain this leads to the formation of $\gamma$-dihomolinolenic acid ($C_{20:3}$), arachidonic acid ($C_{20:4}$), and eicosopentaenoic acid ($C_{20:5}$), which are precursors of prostaglandins.

Three activated fatty acids (RCOSCoA) react successively with a molecule of phosphorylated glycerol to form a triglyceride molecule. The factors that determine the distribution of the fatty acids over the 1-, 2-, and 3-positions of the glycerides are not fully known. Phosphatides are also synthesized by this route.

$$2\,RCOSCoA + \begin{array}{c} HOCH_2 \\ | \\ HOCH \\ | \\ \textcircled{P}\text{-}OCH_2 \end{array} \xrightarrow{-2\,HSCoA} \begin{array}{c} RCOOCH_2 \\ | \\ RCOOCH \\ | \\ \textcircled{P}\text{-}OCH_2 \end{array} \rightarrow \text{Phosphatide}$$

Glycerol 1-phosphate

$$\downarrow {\scriptstyle -\textcircled{P}\text{-}OH \atop +H_2O}$$

$$\begin{array}{c} RCOOCH_2 \\ | \\ RCOOCH \\ | \\ RCOOCH_2 \end{array} \xleftarrow[-HSCoA]{+RCOSCoA} \begin{array}{c} RCOOCH_2 \\ | \\ RCOOCH \\ | \\ HOCH_2 \end{array} \rightleftharpoons \begin{array}{c} RCOOCH_2 \\ | \\ HOCH \\ | \\ RCOOCH_2 \end{array}$$

Triglyceride       Diglyceride

$$\textcircled{P} = \begin{array}{c} OH \\ | \\ O=P- \\ | \\ OH \end{array}$$

**Biodegradation.** During digestion and absorption of fats, triglycerides are successively split into di- and monoglycerides, glycerol, and fatty acids by lipases. The products are absorbed by the intestinal epithelial cells either as water-soluble complexes or as micelles. The fatty acids are biodegraded mainly via $\beta$-oxidation:

$$RCH_2CH_2COOH \xrightarrow[-H_2O]{+HSCoA} \underset{\text{Activated fatty acid}}{RCH_2CH_2COSCoA}$$

$$\xrightarrow{-H_2} RCH{=}CHCOSCoA \xrightarrow{+H_2O} \underset{OH}{RCHCH_2COSCoA} \xrightarrow{-H_2}$$

$$\underset{O}{R{-}\overset{\|}{C}{-}CH_2COSCoA} \xrightarrow{+HSCoA} RCOSCoA + CH_3COSCoA$$

$$\downarrow \text{etc.}$$

$$\downarrow$$

$$CH_3COSCoA$$

where $R = H_3C(CH_2)_n$

The bio-oxidation of unsaturated fatty acids involves additional steps.

## 2.3. Phospholipids [20]

Phospholipids are essential constituents of the protoplasm of animal and plant cells; they are mostly present as lipoproteins and lipidcarbohydrate complexes. Oilseeds, cereal germs, egg yolk, and brain are the richest sources of phospholipids. Esters of glycerophosphoric acid (glycerol 1-phosphate) are usually referred to as phosphatides (→ Lecithin).

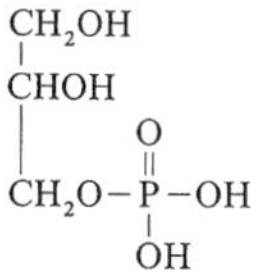

$$\begin{array}{ll} CH_2OH & CH_2OCOR \\ CHOH & CHOCOR \\ CH_2O-P(=O)(OH)-OH & CH_2O-P(=O)(OH)-OH \end{array}$$

Glycerol 1-phosphate 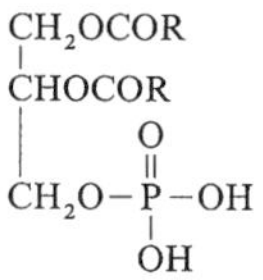    Phosphatidic acid
                         R = alkyl

Some important phosphatides are phosphatidylcholine [8002-43-5] (lecithin), phosphatidylethanolamine [5681-36-7] (cephalin), phosphatidylinositol [2816-11-7], and phosphatidylserine.

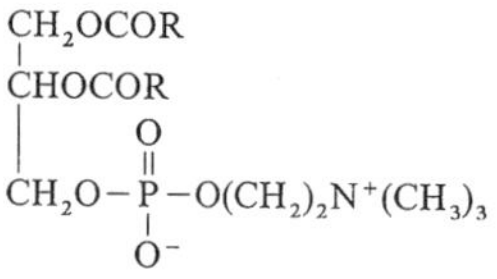

$$\begin{array}{ll} CH_2OCOR & CH_2OCOR \\ CHOCOR & CHOCOR \\ CH_2O-P(=O)(O^-)-O(CH_2)_2N^+(CH_3)_3 & CH_2O-P(=O)(OH)-O(CH_2)_2NH_2 \end{array}$$

Phosphatidylcholine 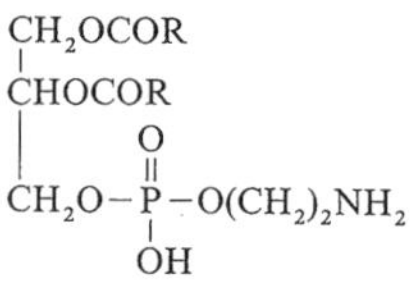     Phosphatidylethanolamine

$$CH_2OCOR \\ CHOCOR \\ CH_2O-P(=O)(OH)-OCH_2CH(NH_2)COOH$$

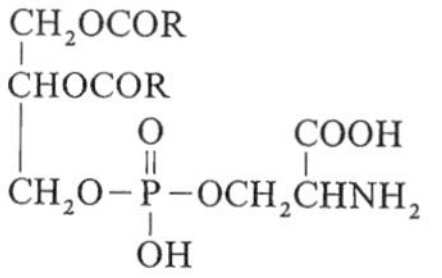

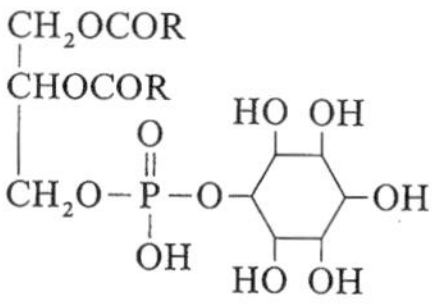

Phosphatidylserine

Phosphatidylinositol (CH$_2$OCOR, CHOCOR, CH$_2$O–P(=O)(OH)–O–inositol ring with HO OH / HO OH / OH)

R = alkyl

The plasmalogens are ethers of fatty alcohols and phosphatidic acid; they occur in animal tissue.

Sphingolipids are derivatives of the amino alcohol sphingosine. For example, sphingomyelins are constituents of the phospholipids in the brain, blood plasma, and erythrocytes.

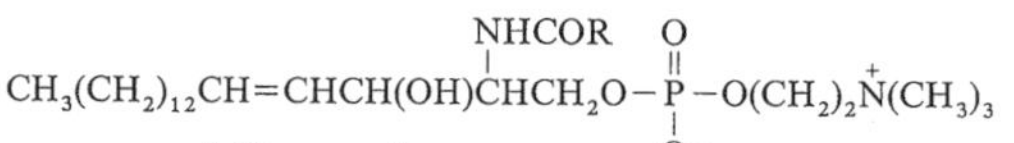

$$CH_3(CH_2)_{12}CH=CHCH(OH)CH(NHCOR)CH_2O-P(=O)(O^-)-O(CH_2)_2\overset{+}{N}(CH_3)_3$$

Sphingomyelin

The cerebrosides are derivatives of sphingosine and either galactose (galactolipids) or glucose (glycolipids). They are important constituents of the myelin nerve sheath. The gangliosides are based on neuraminic acid and are found in the ganglia cells of the brain.

**Table 3.** Minor constituents of crude oils and fats

|  | Content, wt% | | |
| --- | --- | --- | --- |
|  | Phosphatides | Tocopherols | Sterols |
| Babassu fat |  | 0.003 |  |
| Beef tallow | <0.07 | 0.001 | 0.08–0.14 |
| Butterfat | <1.4 | 0.003 | 0.24–0.50 |
| Castor oil |  |  | 0.5 |
| Cocoa butter | 0.1 | 0.003 | 0.17–0.20 |
| Coconut oil |  | 0.003 | 0.05–0.1 |
| Cod liver oil |  |  | 0.42–0.54 |
| Corn germ oil | 1–2 | 0.1–0.3 | 0.8–2.2 |
| Cottonseed oil | 0.7–0.9 | 0.04–0.11 | 0.27–0.6 |
| Fish oil |  |  | ca. 0.3 |
| Lard | <0.05 | 0.003 | ca. 0.1 |
| Linseed oil | 0.3 | 0.11 | 0.37–0.42 |
| Olive oil |  | 0.01–0.03 | 0.1–0.2 |
| Palm oil | 0.05–0.1 | 0.02–0.12 | 0.04–0.08 |
| Palm kernel fat |  |  | 0.08–0.12 |
| Peanut oil | 0.3–0.4 | 0.02–0.07 | 0.19–0.29 |
| Rapeseed oil | 2.5 | 0.07–0.08 | 0.5–1.1 |
| Sesame oil | 0.1 | ca. 0.05 | 0.4–0.6 |
| Soybean oil | 1.1–3.2 | 0.09–0.12 | 0.2–0.4 |
| Sunflower oil | <1.5 | 0.07–0.1 | 0.25–0.45 |
| Wheat germ oil | 0.1–2.0 | ca. 0.28 | 1.3–1.7 |

Lecithin, cephalin, inositol phosphatides, and phosphatidic acid are the principal phospholipid components of plant origin. During prerefining of crude vegetable oils, especially soybean and rapeseed oil, most of the phosphatides are removed as sludge by hydration with water. Drying of this sludge yields "lecithin," which is used, often after further modification, in the food industry as an emulsifier, antispattering agent, dispersant, or viscosity reducing agent. Lecithin is also used in pharmaceuticals, toiletries, animal feeds, and as a mold release agent and emulsifier–dispersant [21]. The phosphatide contents of crude oils and fats are listed in Table 3; the phosphatides are almost completely removed during refining of oils and fats.

## 2.4. Waxes

Waxes are esters of fatty alcohols and fatty acids. Free and esterified fatty alcohols (e.g., cetyl, stearyl, oleyl alcohols) occur in considerable concentrations in marine oils. Ethers of glycerol and fatty alcohols (batyl, chimyl, and selachyl alcohols) are also found in animal tissues. The wax in the seed coat of sunflower seed causes the oil to become cloudy at refrigerator temperatures and is therefore removed by winterization (see Chap. 7).

## 2.5. Sterols and Sterol Esters

The major part of the nonsaponifiable matter of oils and fats consists of sterols present as such or as fatty acid esters and glycolipids. The most important sterol in animal fats is cholesterol [57-88-5]. β-Sitosterol [83-46-5] is the predominant sterol in vegetable oils and fats, although traces of cholesterol are also present. Total sterol concentrations are shown in Table 3, and sterol compositions in Table 4. Some of the sterols are removed during the deodorization step of refining oils and fats, without, however, changing their relative composition. Sterols are therefore a useful tool in checking authenticity. Vitamin $D_2$ [50-14-6] (calciferol) is present in milk and butter, and vitamin $D_3$ [67-97-0] (cholecalciferol) in cod liver oil. Other animal and vegetable oils contain hardly any vitamin D.

## 2.6. Terpenoids

The nonsaponifiable part of most fats also contains traces of terpenes and terpene alcohols. The triterpene squalene [7683-64-9] occurs in relatively high concentrations (up to 0.5%) in olive oil. Shea butter contains 2–10% of kariten, a rubber-like hydrocarbon.

Carotenoid pigments occur widely in oils and fats. Approximately 70 different carotenoids, ranging in color from yellow to deep red, are known. The most well-known carotenoids are the isomeric tetraterpenes ($C_{40}H_{56}$) α-carotene [7488-99-5], β-carotene [7235-40-7], and γ-carotene [472-93-5], lycopene [502-65-8], and xanthophyll [127-40-2]. Crude palm oil contains up to 0.2% α- and β-carotene.

The bulk of the carotenoids are removed during refining, primarily during bleaching and deodorization. β-Carotene (provitamin A) can be oxidized to vitamin A [68-26-8] in the animal organism. Most vegetable oils and fats do not contain significant concentrations of vitamin A; in most countries margarine therefore contains added vitamin A. Vitamin A is present in high concentrations in fishliver oils. Butter contains ca. 0.003–0.0015% vitamin A.

The most important compounds of the vitamin E [59-02-9] group are α-, β-, γ-, and δ-tocopherols:

|  | $R^1$ | $R^2$ | $R^3$ |
|---|---|---|---|
| α-Tocopherol | $CH_3$ | $CH_3$ | $CH_3$ |
| β-Tocopherol | $CH_3$ | H | $CH_3$ |
| γ-Tocopherol | H | $CH_3$ | $CH_3$ |
| δ-Tocopherol | H | H | $CH_3$ |

**Table 4.** Sterol composition * in crude oils (main sources: Food RA, Leatherhead; Unilever; ITERG **)

| | Coconut | Corn germ | Cottonseed | Olive | Palm | Palm kernel | Peanut | Rapeseed | Soybean | Sunflower |
|---|---|---|---|---|---|---|---|---|---|---|
| Cholesterol | 0.6–2 | 0.2–0.6 | 0.7–2.3 | 0–0.5 | 2.2–6.7 | 1–3.7 | 0.6–3.8 | 0.4–2 | 0.6–1.4 | 0.2–1.3 |
| Brassicasterol | 0–0.9 | 0–0.2 | 0.1–0.9 | | | 0–0.3 | 0–0.2 | 5–13 | 0–0.3 | 0–0.2 |
| Campesterol | 7–10 | 18–24 | 7.2–8.4 | 2.3–3.6 | 18.7–29.1 | 8.4–12.7 | 12–20 | 18–39 | 16–24 | 7–13 |
| Stigmasterol | 12–18 | 4–8 | 1.2–1.8 | 0.6–2 | 8.9–13.9 | 12.3–16.1 | 5–13 | 0–0.7 | 16–19 | 8–11 |
| β-Sitosterol | 50–70 | 55–67 | 80–90 | 75.6–90 | 50.2–62.1 | 62.6–70.4 | 48–65 | 45–58 | 52–58 | 56–63 |
| Δ5-Avenasterol | 5–16 | 4–8 | 1.9–3.8 | 3.1–14 | 0–2.8 | 4–9 | 7–9 | 0–6.6 | 2–4 | 2–7 |
| Δ7-Stigmastenol | 2–8 | 1–4 | 0.7–1.4 | 0–4 | 0.2–2.4 | 0–2.1 | 0–5 | 0–5 | 1.5–5 | 7–13 |
| Δ7-Avenasterol | 0.6–2 | 1–3 | 1.4–3.3 | | 0–5.1 | 0–1.4 | 0–5 | 0–0.8 | 1–4.5 | 3–6 |

\* Sterols as percentage of total sterol fraction.
\*\* ITERG = Institut des Corps Gras, Centre Technique Industrielle, Paris.

These compounds act as a vitamin (rat fertility factor) and as antioxidants. α-Tocopherol has the highest biological activity.

Tocopherols occur only in traces in animal fats, whereas vegetable oils contain appreciable concentrations (see Table ). Refined oils and fats still contain ca. 80 % of the original tocopherols, the main losses occurring during deodorization. Deodorizer distillates are a valuable source of natural tocopherols.

## 2.7. Other Minor Constituents

Sesame oil contains 0.3 – 0.5 % sesamolin [526-07-8], a glycoside of the phenol sesamol [533-31-3], and 0.5 – 2.0 % sesamin [607-80-7]. These minor constituents give a characteristic color reaction (the basis of the Baudouin test) and impart stability to oxidation.

Gossypol [303-45-7], a toxic polyphenol with pronounced antioxidant activity, is found in crude cottonseed and kapokseed oils (0.5 – 1.5 %). Gossypol is removed during refining in the lye neutralization step.

Crude linseed, rapeseed, soybean, olive, avocado, and many other vegetable oils contain the green pigments chlorophyll and phaeophytin. The chlorophyll content is particularly high in oil from immature seeds and is generally determined by harvesting and climatic conditions. Chlorophyll is generally regarded to be indicative of inferior crude oil quality; it can be removed by treatment with acidic absorbents such as bleaching earth.

Crude oils and fats may contain traces of proteins, the concentration depending on processing conditions. In addition, vegetable oils may contain carbohydrates. These compounds are removed almost completely in refining.

Autoxidation of the fatty acid groups of triglycerides leads to the formation of volatile and nonvolatile oxidation products. The type and quantity of the volatile compounds (ketones, aldehydes, and alcohols) depend on the initial fatty acid composition and the oxidation conditions. These volatile products are responsible for the typical odor and taste of oils and fats. One of the main aims of refining is to remove these odoriferous compounds, which in some cases occur in concentrations of only $10^{-3}$ ppm. The nonvolatile oxidation products generally have little odor and taste, but they can act as oxidation promoters.

Most crude oils and fats contain traces of pesticide residues and metals (e.g., Fe, Cu, Pb, As, Cd, and Hg) as a consequence of crop treatment and environmental influences. Whereas phosphate-based pesticides used to treat oilseeds decompose with time, chlorinated pesticides are stable and gradually migrate into the oil.

Peanut oil may contain traces of aflatoxins produced by growth of *Aspergillus flavus* on the seed.

**Table 5.** Melting points of fatty acids and glycerides in their stable polymorphic forms

|  | mp (slip point), °C | | | |
| --- | --- | --- | --- | --- |
|  | Fatty acid | 1-Monoglyceride | 1,3-Diglyceride | Triglyceride |
| Butyric acid | −7.9 | | | |
| Caproic acid | −3.9 | 19.4 | | −25.0 |
| Caprylic acid | 16.3 | 40.0 | | 8.3 |
| Capric acid | 31.3 | 53.0 | 44.5 | 31.5 |
| Lauric acid | 44.0 | 63.0 | 56.5 | 46.5 |
| Myristic acid | 54.4 | 70.5 | 65.5 | 57.0 |
| Palmitic acid | 62.9 | 77.0 | 72.5 | 65.5 |
| Stearic acid | 69.6 | 81.0 | 78.0 | 73.0 |
| Arachidic acid | 75.4 | 84.0 | (75.0) | (72.0) |
| Behenic acid | 79.9 | | | 81.0 |
| Oleic acid | 16.3 | 35.2 | 21.5 | 5.5 |
| Elaidic acid | 45.0 | 58.5 | 55.0 | 42.0 |
| Erucic acid | 34.7 | 50.0 | 46.5 | 30.0 |
| Linoleic acid | −5.0 | 12.3 | −2.6 | −13.1 |
| Linolenic acid | −11.0 | 15.7 | −12.3 | −24.2 |
| Ricinoleic acid | 5.0 | | | |

Most oils and fats, particularly coconut oil, contain varying concentrations of polycyclic aromatic hydrocarbons. These hydrocarbons are introduced during smoke-drying of the raw materials prior to storage and further processing.

Crude rapeseed oil contains up to 50 ppm of sulfur in the form of elemental sulfur, isothiocyanates, and 5-vinyl-2-oxazolidinethione [500-12-9] (goitrin) derived from glucosinolates (sulfur-containing glucosides) in the seed.

All of these undesirable contaminants are reduced to negligibly low levels in the course of refining, if necessary with the aid of additional steps such as adsorptive treatment with activated charcoal to remove polycyclic aromatic hydrocarbons.

# 3. Physical Properties [23]

## 3.1. Melting and Freezing Points

The *melting point* of the even-numbered, saturated fatty acids increases with increasing chain-length, and decreases with increasing degree of unsaturation (see also → Fatty Acids). The glycerides show a similar behavior (Table 5). Since the naturally occurring fats are mixtures of glycerides, they melt over a wide range of temperature (for methods of melting point determination, see Section 11.3.1).

The melting point also depends on the polymorphic form of the glycerides, i.e., the crystalline structure (Table 6). The packing density and the spatial arrangement of the triglyceride molecules depend on the crystallization conditions. X-ray diffraction patterns (short spacing lines) can distinguish between three different polymorphic forms [25]. The lowest melting and most labile form is designated α. The most stable form is

**Table 6.** Melting points of some polymorphic triglycerides

| | mp, °C | | |
|---|---|---|---|
| | α | β' | β |
| Trilaurin | 15.0 | 35.0 | 46.5 |
| Trimyristin | 33.0 | 46.5 | 57.0 |
| Tripalmitin | 45.0 | 56.0 | 65.5 |
| Tristearin | 54.5 | 65.0 | 73.0 |
| Triolein | −32.0 | −12.0 | 5.5 |
| Trielaidin | 15.5 | 37.0 | 42.0 |
| 1,2-Dicapriolaurin | 17.5 | 26.0 | 30.0 |
| 2-Capriodilaurin | 23.0 | 33.0 | 38.5 |
| 1-Laurodimyristin | 37.0 | 42.0 | 46.5 |
| 1-Laurodipalmitin | 45.0 | 49.5 | 54.0 |
| 2-Laurodipalmitin | 47.0 | 50.0 | 53.5 |
| 1,2-Dicapriostearin | 32.0 | 38.0 | 41.0 |
| 1,3-Dicapriostearin | 34.0 | 40.0 | 44.5 |
| 2-Palmitodistearin | 56.0 | 64.0 | 68.5 |
| 1,3-Dicaprioolein | −10.2 | 0.6 | 6.2 |
| 1,2-Dilauroolein | −10.0 | 4.8 | 16.0 |
| 1,2-Dipalmitoolein | ca. 18.0 | ca. 31.0 | 34.5 |
| 1,3-Dipalmitoolein | 26.5 | 33.5 | 38.0 |
| 1,2-Distearoolein | ca. 30.0 | ca. 40.0 | |
| 1,3-Distearoolein | 37.0 | 41.5 | 44.0 |
| 1,3-Dipalmitoelaidin | ca. 40.0 | 53.0 | 54.0 |
| 1-Stearodibehenin | 61.3 | 71.0 | 73.5 |
| 1-Lauro-2-myristo-3-palmitin | 37.0 | 44.0 | 49.0 |
| 1-Lauro-2-myristo-3-stearin | 27.5 | 45.5 | 49.5 |
| 1-Palmito-3-stearo-2-olein | 18.2 | 33.0 | 38.0 |

called $\beta$ and the intermediate one $\beta'$. The $\alpha$ form is obtained by rapidly cooling the molten triglycerides, the $\beta'$ form by suitable tempering. Polymorphic changes induced by processing conditions are of great practical importance since they can significantly influence the properties of, for example, a margarine (oral melting behavior, sandiness) or chocolate (fat bloom).

The *congeal* or *set point* (point of solidification) is generally lower than the melting point.

The *solids content* of a fat at different temperatures is normally determined by pulsed nuclear magnetic resonance [24]. Fats exhibit an increase in volume (dilatation) on melting that is disproportionately larger than that on heating a liquid fat; an obsolete method of measuring solids content at different temperatures was based on dilatation. *Dilatation* or *solids content curves* can be used to characterize a fat. Figure 2 reflects the large proportion of 1-palmito-3-stearo-2-olein in cocoa butter and the more complex glyceride composition of lard.

The *latent heat of fusion* increases with increasing chain length and increasing degree of saturation (see Table 7). Naturally occurring fats generally have a lower heat of fusion than simple glycerides.

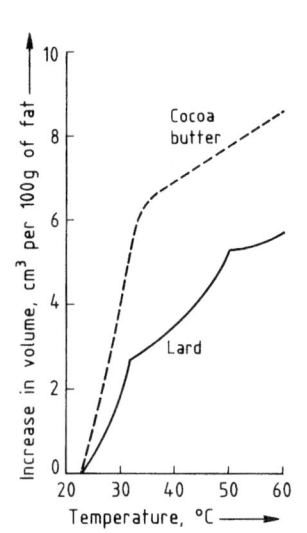

**Figure 2.** Dilatation curves for cocoa butter and lard

**Table 7.** Latent heat of fusion of some fats

|  | Heat of fusion, J/g |
|---|---|
| Butterfat | 81.6 |
| Cottonseed oil | 86.0 |
| Fully hardened cottonseed oil * | 185.0 |
| Peanut oil | 90.9 |
| Partially hardened peanut oil ** | 103.4 |
| Trilaurin (β-form) | 193.5 |
| Tripalmitin (β-form) | 222.0 |
| Tristearin (β-form) | 228.0 |

\* Iodine value ca. 1.
\*\* Iodine value ca. 60.

## 3.2. Thermal Properties

The approximate *heats of combustion* of oils and fats can be calculated from the following formula:

Heat of combustion (J/g)

$= 47\,645 - 4.1868 \times \text{iodine value} - 38.31 \times \text{saponification value}$

Using this equation, values ranging from 37 765 J/g (9020 cal/g) for coconut oil to 40 528 J/g (9680 cal/g) for a high-erucic rapeseed oil have been obtained.

The *specific heat* of liquid oils and fats increases with increasing chain length and degree of saturation (see Table 8); it also increases with temperature.

Triglycerides of long-chain fatty acids have extremely low *vapor pressures*; typical data for some triglycerides and oils are shown in Table 9. Monoglycerides have significantly

**Table 8.** Specific heat of oils and fats

|  | Specific heat, J/g | Temperature, °C |
|---|---|---|
| Trilaurin | 2.130 | 66.0 |
| Trimyristin | 2.152 | 58.4 |
| Tripalmitin | 2.173 | 65.7 |
| Tristearin | 2.219 | 79.0 |
| Soybean oil | 2.060 | 80.4 |
|  | 2.000 | 60.0 |
| Linseed oil | 2.050 | 70.7 |
| Cottonseed oil | 2.200 | 90.0 |
| Hardened cottonseed oil * | 2.177 | 79.6 |
| Olive oil | 2.300 | 110 |
| Palm oil | 2.400 | 140 |
| Sunflower oil | 2.500 | 175.0 |

* Iodine value ca. 6.

**Table 9.** Vapor pressure of triglycerides (temperatures corresponding to a vapor pressure of 6.7 Pa (0.05 mm Hg) and 0.13 Pa (0.001 mm Hg))

|  | Temperature, °C | |
|---|---|---|
|  | $p = 6.7$ Pa | $p = 0.13$ Pa |
| Tributyrin | 91 | 45 |
| Tricaproin | 135 | 85 |
| Tricaprylin | 179 | 128 |
| Tricaprin | 213 | 159 |
| Trilaurin | 244 | 188 |
| Trimyristin | 275 | 216 |
| Tripalmitin | 298 | 239 |
| Tristearin | 313 | 253 |
| Soybean oil | 308 | 254 |
| Olive oil | 308 | 253 |
| 1,3-Distearoolein | 315 | 254 |
| 1-Myristo-2-palmito-3-stearin | 297 | 237 |
| 1-Palmito-2-lauro-3-stearin | 290 | 232 |
| 1-Myristo-2-lauro-3-stearin | 282 | 223 |
| 1-Palmito-2-capro-3-stearin | 280 | 223 |
| 1-Capro-2-lauro-3-myristin | 249 | 189 |

higher vapor pressures (see Table 10). There is relatively little information on the *heats of vaporization* of fats (see Table 11).

Oils and fats are relatively poor thermal conductors. Data for *thermal conductivity* are limited. Most data lie within ±10% of the following general relationship:

Thermal conductivity $(\text{W m}^{-1}\text{K}^{-1}) = 0.181 - 0.00025\,t$

where $t$ = temperature in °C.

The *smoke, flash,* and *fire points* of oils and fats are measures of their thermal stability when heated in air (Table 12).

**Table 10.** Boiling points of some monoglycerides

|  | Pressure, Pa | bp, °C |
|---|---|---|
| Monocaprin | 133.3 | 175 |
| Monolaurin | 133.3 | 186 |
| Monomyristin | 133.3 | 199 |
| Monopalmitin | 133.3 | 211 |
| Monostearin | 26.7 | 190 |
| Monoolein | 26.7 | 186 |

**Table 11.** Heat of vaporization of some triglycerides calculated for 0.13 – 67 Pa

|  | Heat of vaporization, J/g |
|---|---|
| Tricaprylin | 247.0 |
| Tricaprin | 226.1 |
| Trilaurin | 213.5 |
| Trimyristin | 205.2 |
| Tripalmitin | 201.0 |
| Tristearin | 188.4 |
| Soybean oil | 209.3 |

**Table 12.** Smoke, flash, and fire points of refined oils and fats

|  | ffa *, % | Smoke point, °C | Flash point, °C | Fire point, °C |
|---|---|---|---|---|
| Rapeseed oil | 0.08 | 218 | 317 | 344 |
| Peanut oil | 0.09 | 207 | 315 | 342 |
| Peanut oil | 0.11 | 198 | 333 | 363 |
| Peanut oil | 1.0 | 160 | 290 | – |
| Cottonseed oil | 0.04 | 223 | 322 | 342 |
| Cottonseed oil | 0.18 | 185 | 318 | 357 |
| Soybean oil | 0.04 | 213 | 317 | 342 |
| Sunflower oil | 0.1 | 209 | 316 | 341 |
| Coconut oil | 0.1 | 200 | 300 |  |
| Coconut oil | 0.2 | 194 | 288 | 329 |
| Coconut oil | 1.0 | 150 | 270 |  |
| Palm oil | 0.06 | 223 | 314 | 341 |
| Hardened peanut oil (*mp* 32/34 °C) | 0.04 | 226 | 314 | 340 |
| Hardened soybean oil (*mp* 42/44 °C) | 0.04 | 223 | 318 | 342 |
| Beef tallow | 0.4 |  | 316 | 344 |
| Beef tallow | 5.0 |  | 266 | 344 |

* ffa = free fatty acid.

## 3.3. Density

The density of fatty acids and glycerides decreases with increasing molecular mass and degree of saturation (Tables 13 and 14). Oxidation generally leads to higher densities. A high free fatty acid content tends to decrease the density of a crude oil. The following formula can be used to estimate the density of an oil:

$$d_{15}^{15} = 0.8475 + 0.0003 \times \text{saponification value} + 0.00014 \times \text{iodine value}$$

**Table 13.** Density of triglycerides

|  | $\varrho$, g/cm$^3$ | | |
| --- | --- | --- | --- |
|  | 80 °C | 15 °C | 25 °C |
| Tricaprin | 0.8913 | | |
| Trilaurin | 0.8801 | | |
| Trimyristin | 0.8722 | | |
| Tripalmitin | 0.8663 | | |
| Tristearin | 0.8632 | | |
| Triolein | | 0.9162 | 0.9078 |
| Trilinolein | | 0.9303 | |
| Trilinolenin | | 0.9454 | |

Up to 260 °C the density decreases by about 0.00064 g/cm$^3$ per temperature increase of 1 °C. The following equations apply for the density $\varrho$ (in g/L) of commercial oils and fats ($t$ = temperature in °C):

| | |
| --- | --- |
| Soybean oil | 933.4 − 0.657 $t$ |
| Sunflower oil | 932.7 − 0.680 $t$ |
| Sesame oil | 933.0 − 0.700 $t$ |
| Cottonseed oil | 931.7 − 0.755 $t$ |
| Peanut oil | 927.0 − 0.642 $t$ |
| Olive oil | 928.5 − 0.700 $t$ |
| Palm oil | 925.0 − 0.655 $t$ |
| Palm kernel oil | 940.0 − 0.740 $t$ |
| Coconut oil | 932.0 − 0.745 $t$ |
| Rapeseed oil | 925.5 − 0.700 $t$ |
| Tallow | 956.8 − 0.898 $t$ |
| Fish oil | 940.0 − 0.700 $t$ |

## 3.4. Viscosity

Oils tend to have a relatively high viscosity because of intermolecular attraction between their fatty acid chains. Generally, viscosity tends to increase slightly with increasing degree of saturation and increasing chain length (see Tables 15 and 16). There is an approximately linear relationship between log viscosity and temperature. The viscosity of oils tends to increase on prolonged heating due to the formation of dimeric and oligomeric fatty acid groups.

## 3.5. Solubility and Miscibility

Nearly all fats and fatty acids are easily soluble in common organic solvents such as hydrocarbons, chlorinated hydrocarbons, ether, and acetone. Castor oil is an exception in that it is only partially soluble in petroleum ether but easily so in ethanol. The solubility of fatty acids in ethanol is greater than that of the corresponding triglycerides.

**Table 14.** Density of fats and oils

|  | $\varrho_{15}$, g/cm$^3$ | $d_{25}^{25}$ |
|---|---|---|
| *Vegetable fats* | | |
| Babassu oil | 0.9250 | |
| Castor oil | 0.950–0.974 | 0.945–0.965 |
| Coconut oil | 0.919–0.937 | 0.869–0.874$^a$ |
| Cocoa butter | 0.945–0.976 | 0.856–0.864$^a$ |
| Corn germ oil | 0.920–0.928 | 0.916–0.921 |
| Cottonseed oil | 0.917–0.931 | 0.916–0.918 |
| Grape-seed oil | 0.919–0.936 | |
| Hempseed oil | 0.924–0.932 | 0.923–0.925$^b$ |
| Linseed oil | 0.930–0.935 | 0.931–0.936 |
| Mustard seed oil | 0.912–0.923 | |
| Olive oil | 0.914–0.925 | 0.909–0.915 |
| Oiticica oil | 0.9518–0.9694 | 0.978$^b$ |
| Palm oil | 0.921–0.947 | 0.898–0.901$^c$ |
| Palm kernel oil | 0.925–0.935 | 0.860–0.873$^a$ |
| Peanut oil | 0.911–0.925 | 0.910–0.915 |
| Perilla oil | 0.927–0.933 | min. 0.932$^b$ |
| Poppyseed oil | 0.923–0.926 | |
| Rapeseed oil | 0.910–0.917 | 0.906–0.910 |
| Ricebran oil | | 0.916–0.921 |
| Safflower oil | 0.923–0.928 | 0.919–0.924 |
| Sesame oil | 0.921–0.924 | 0.914–0.919 |
| Shea butter | 0.917–0.918 | |
| Soybean oil | 0.922–0.934 | 0.917–0.921 |
| Sunflower oil | 0.920–0.927 | 0.915–0.919 |
| Tung oil | 0.936–0.945 | 0.940–0.943$^b$ |
| Wheat germ oil | | 0.925–0.933 |
| *Animal fats* | | |
| Beef tallow | 0.936–0.952 | 0.860–0.870$^a$ |
| Butterfat | 0.935–0.943 | |
| Herring oil | 0.917–0.930 | |
| Horse fat | 0.915–0.932 | |
| Lard | 0.914–0.943 | 0.858–0.864$^a$ |
| Menhaden oil | 0.925–0.935 | |
| Mutton tallow | 0.936–0.960 | |
| Sperm oil | 0.875–0.890 | |
| Whale oil | 0.914–0.931 | 0.910–0.920 |

$^a$ $d_{15.5}^{99}$.  $^b$ $d_{15.5}^{15.5}$.  $^c$ $d_{37.8}^{37.8}$.

The solubility of fats in organic solvents decreases with increase in molecular mass and increases with degree of unsaturation. The differences in solubility enable categories of glycerides to be separated by fractional crystallization although complete separation is rarely achieved because of mutual solubility effects.

The water solubility of fats is low and decreases with increasing chain length and with decreasing temperature. The solubility of water in cottonseed oil is, for example, 0.14 wt% at 30 °C and 0.07 wt% at 0 °C.

The solubility of gases in oils generally increases with increase in temperature, the reverse holding for carbon dioxide (Table 17).

**Table 15.** Viscosity of triglycerides at 70°C

|  | $\eta$, mPa · s |
|---|---|
| Tributyrin | 3.0 |
| Tricaproin | 5.9 |
| Tricaprylin | 8.8 |
| Tricaprin | 11.7 |
| Trilaurin | 14.6 |
| Trimyristin | 17.6 |
| Tripalmitin | 20.5 |
| Tristearin | 23.4 |

**Table 16.** Viscosity of naturally occurring fats and oils

|  | $\eta$, mPa · s | | | |
|---|---|---|---|---|
|  | 20 °C | 30 °C | 40 °C | 50 °C |
| Castor oil | 1000 | 454 | 232 | 128 |
| Coconut oil |  | 39 | 26 | 19 |
| Cottonseed oil | 80 | 55 | 38 | 27 |
| Fish oil | 60 | 43 | 32 | 22 |
| Lard |  |  | 35 | 25 |
| Linseed oil | 48 | 33 | 25 | 18 |
| Olive oil | 80 | 55 | 40 | 30 |
| Palm oil |  |  | 40 | 28 |
| Palmkernel oil |  | 43 | 29 | 20 |
| Peanut oil | 78 | 50 | 32 | 23 |
| Poppyseed oil | 63 |  |  |  |
| Rapeseed oil (high erucic) | 85 | 60 | 40 | 30 |
| Sesame oil | 65 |  |  | 25 |
| Soybean oil | 65 | 45 | 33 | 25 |
| Sunflower oil | 68 | 47 | 35 | 26 |
| Tallow |  |  |  | 25 |

**Table 17.** Solubility of gases in fats

|  | Gas solubility, vol% | | | | |
|---|---|---|---|---|---|
|  | $t$, °C | $N_2$ | $H_2$ | $O_2$ | $CO_2$ |
| Cottonseed oil | 30.5 | 7.11 | 4.63 |  |  |
|  | 49.6 | 7.79 | 5.40 |  |  |
|  | 78.2 | 8.91 | 6.73 |  |  |
|  | 101.5 | 9.76 | 7.83 |  |  |
|  | 147.8 | 11.83 | 10.24 |  |  |
| Soybean oil | ca. 20.0 | ca. 4.95 |  | ca. 2.65 |  |
| Lard | 41.5 | 7.65 | 5.218 |  |  |
|  | 73.2 | 8.79 | 6.58 |  |  |
|  | 111.3 | 10.38 | 8.50 |  |  |
|  | 147.3 | 12.06 | 10.35 |  |  |
| Hardened tallow (iodine value 1) | 64.3 |  | 6.14 |  | 92.0 |
|  | 67.0 | 8.44 |  | 14.50 |  |
|  | 84.7 |  |  | 15.35 |  |
|  | 88.0 |  |  |  | 79.1 |
|  | 139.4 | 11.68 | 9.79 |  | 61.9 |

**Table 18.** Surface and interfacial tension of some oils

|  | Surface tension, mN/m | | | Interfacial tension at 70 °C, mN/m |
|---|---|---|---|---|
|  | 20 °C | 80 °C | 130 °C |  |
| Cottonseed oil | 35.4 | 31.3 | 27.5 | 29.8 |
| Coconut oil | 33.4 | 28.4 | 24.0 |  |
| Castor oil | 39.0 | 35.2 | 33.0 |  |
| Peanut oil |  |  |  | 29.9 |
| Soybean oil |  |  |  | 30.6 |

## 3.6. Surface and Interfacial Tension

The surface and interfacial tensions of some oils are shown in Table 18. The interfacial tension is markedly reduced by the presence of surface-active agents, e.g., phosphatides, monoglycerides, free fatty acids, and soaps.

## 3.7. Electrical Properties

Dry oils, fats, and fatty acids are poor conductors of electricity. Recorded values for the specific resistance of stearic acid are $0.6 \times 10^{11}$ Ω at 100 °C and $22.3 \times 10^{11}$ Ω at 186 °C, and for oleic acid $2 \times 10^{11}$ and $83 \times 10^{11}$ Ω at comparable temperatures. The dielectric constant of most oils lies in the range 3.0–3.2 at 25–30 °C. Castor and oiticica oil have dielectric constants of about 4 in this temperature range because of the hydroxyl and keto groups in the fatty acid chains. In emulsion systems, e.g., butter or margarine, the dielectric constant is affected by both the emulsion structure and the moisture content.

## 3.8. Optical Properties

The refractive index of oils, fats, and fatty acids generally increases with increasing chain length, number of double bonds, and extent of conjugation (see Tables 19 and 20). The refractive index of fatty acids is much lower than that of the corresponding triglycerides. Prolonged heating leads to an increase in refractive index due to the introduction of polar groups into the fatty acid chain.

There are a number of equations depicting a relationship between refractive index and other data. The following equation has been suggested for fresh, nonhydrogenated oils and fats:

**Table 19.** Refractive indices of glycerides

|  | $n_D^{60}$ |
|---|---|
| Tricaprin | 1.4370 |
| Trilaurin | 1.4402 |
| Trimyristin | 1.4428 |
| Tripalmitin | 1.4452 |
| Tristearin | 1.4471 |
| Triolein | 1.4548 |
| Trilinolein | 1.4645 |
| 1,2-Dilaurostearin | 1.4437 |
| 1,3-Dilaurostearin | 1.4442 |
| 1,2-Dilauroolein | 1.4456 |
| 1,3-Dilaurostearin | 1.4459 |
| 1,2-Dipalmitoolein | 1.4480 |
| 1,2-Distearoolein | 1.4494 |
| 1-Stearodiolein | 1.4524 |
| Monocaprin | 1.4443 |
| Monolaurin | 1.4462 |
| Monomyristin | 1.4480 |
| Monopalmitin | 1.4499 |
| 1-Lauro-3-olein | 1.4472 |
| 1-Stearo-3-olein | 1.4507 |

$$n_D^{40} = 1.4643 - 0.00066 \times \text{saponification value}$$
$$- \frac{0.0096 \times \text{acid value}}{\text{saponification value}} + 0.0001711 \times \text{iodine value}$$

Pure glycerides do not absorb in the visible region of the spectrum (400–750 nm). However, naturally occurring oils and fats invariably contain pigments that have characteristic absorption bands (carotene at 450 nm, chlorophyll and phaeophytin at 660 nm). Most of these pigments are removed during refining. Unsaturated oils and fats absorb in the ultraviolet region between 200 and 400 nm. Conjugated double bonds show characteristic maxima at 232 nm (dienes, e.g., 9,11-*trans,trans*-linoleic acid) and at 268 nm (trienes, e.g., β-eleostearic acid). Conjugated tetraenoic acids absorb between 290 and 320 nm.

In the infrared region (0.075–1000 µm) chain substituents such as epoxy, hydroxyl, keto, and cyclopropene groups as well as trans double bonds exhibit specific absorption peaks:

| | |
|---|---|
| monoglycerides | 1.43 µm |
| hydroxy fatty acids | 3.20 µm |
| ester carbonyl group | 5.83 µm |
| trans double bonds | 10.0–10.35 µm |
| isolated trans double bonds | 10.35 µm |

The fingerprint region around 8 µm can be used to determine the fatty acid chain length.

**Table 20.** Refractive indices of naturally occurring oils and fats

| | $n_D^{20}$ |
|---|---|
| Babassu oil | 1.449 – 1.450* |
| Beef tallow | 1.454 – 1.459* |
| Butterfat | 1.452 – 1.457* |
| Castor oil | 1.477 – 1.479 |
| Cocoa butter | 1.453 – 1.458* |
| Coconut oil | 1.448 – 1.450* |
| Corn oil | 1.474 – 1.476 |
| Cottonseed oil | 1.472 – 1.477 |
| Grape seed oil | 1.474 – 1.478 |
| Herring oil | 1.470 – 1.475* |
| Lard | 1.458 – 1.461* |
| Linseed oil | 1.479 – 1.481 |
| Menhaden oil | 1.480 |
| Mustardseed oil | 1.470 – 1.474 |
| Mutton tallow | 1.455 – 1.458* |
| Oiticica oil | 1.4921 – 1.4945 |
| Olive oil | 1.467 – 1.471 |
| Palm oil | 1.453 – 1.456* |
| Palmkernel oil | 1.449 – 1.452* |
| Peanut oil | 1.460 – 1.472 |
| Perilla oil | 1.481 – 1.483 |
| Rapeseed oil | 1.472 – 1.476 |
| Safflower oil | 1.4754 |
| Sesame oil | 1.473 – 1.476 |
| Soybean oil | 1.470 – 1.478 |
| Sunflower oil | 1.474 – 1.476 |
| Tung oil | 1.517 – 1.526 |
| Whale oil | 1.463 – 1.471 |
| Wheat germ oil | 1.469 – 1.478* |

* $n^{40}$.

X-ray spectroscopy is employed to characterize the various polymorphic crystal structures of pure glycerides (see Section 3.1).

Nuclear magnetic resonance of hydrogen atoms is used in structural identification and to determine the solids content of fats at different temperatures. Nuclear magnetic resonance measurement of phosphorus and nitrogen atoms can be employed to analyze the composition of phosphatide mixtures.

The optical activity of enantiomorphic trigly-cerides with different fatty acid groups in the 1- and 3-positions is usually too small to be measured [20].

# 4. Chemical Properties

The chemical reactions of fats are basically those of esters and hydrocarbon chains (→ Esters, Organic). Only those reactions that are primarily relevant to the processing of edible oils and fats are dealt with in this chapter.

## 4.1. Hydrolysis

Glycerides can be hydrolyzed into fatty acids and glycerol:

$$\begin{array}{l}CH_2OCOR \\ CHOCOR \\ CH_2OCOR\end{array} + 3\,H_2O \rightleftharpoons \begin{array}{l}CH_2OH \\ CHOH \\ CH_2OH\end{array} + 3\,RCOOH$$

The reaction is reversible; in practice the equilibrium can be shifted to the right by using a large excess of water, high temperatures, and high pressures.

Hydrolysis is catalyzed by inorganic and organic acids, e.g., sulfonated hydrocarbons. In the enzymatic hydrolysis of glycerides with pancreatic lipase, the fatty acid groups in the 1- and 3-positions, are split off preferentially.

A fat can also be hydrolyzed with alkali:

$$\begin{array}{l}CH_2OCOR \\ CHOCOR \\ CH_2OCOR\end{array} + 3\,NaOH \longrightarrow \begin{array}{l}CH_2OH \\ CHOH \\ CH_2OH\end{array} + 3\,RCOONa$$

## 4.2. Interesterification

Like other esters, glycerides can be transesterified by acidolysis or alcoholysis (→ Esters, Organic). In the presence of an alkaline catalyst and an excess of glycerol, triglycerides form a mixture of mono- and diglycerides (alcoholysis).

The acyl groups of glycerides can also be exchanged inter- and intramolecularly without addition of acids or alcohols (interesterification).

$$\begin{array}{l}CH_2OCOR^1 \\ CHOCOR^1 \\ CH_2OCOR^1\end{array} + \begin{array}{l}CH_2OCOR^2 \\ CHOCOR^2 \\ CH_2OCOR^2\end{array} \rightleftharpoons$$

$$\begin{array}{l}CH_2OCOR^2 \\ CHOCOR^1 \\ CH_2OCOR^1\end{array} + \begin{array}{l}CH_2OCOR^1 \\ CHOCOR^2 \\ CH_2OCOR^2\end{array} \rightleftharpoons \text{etc.}$$

Even at 200–300 °C interesterification proceeds very slowly, but the reaction can be accelerated by using an alkaline catalyst such as a metal alkoxide. With such a catalyst the reaction is complete within one minute at 80 °C. Interesterification is of practical importance since it enables the physical properties of a fat, e.g., melting behavior and consistency, to be modified without changing the fatty acids chemically, as occurs in hydrogenation (hardening).

Interesterification may be either random or directed. *Random interesterification* leads to a random distribution of the fatty acid groups over the triglyceride molecules. This is demonstrated for interesterification of equal proportions of tristearin (S–S–S) and triolein (O–O–O):

$$\begin{array}{c} \text{S-S-S} + \text{O-O-O} \\ 50\% \qquad\quad 50\% \\ \downarrow \end{array}$$

$$\begin{array}{cccccc} \text{S-S-S} & \text{S-O-S} & \text{O-S-S} & \text{S-O-O} & \text{O-S-O} & \text{O-O-O} \\ 12.5\% & 12.5\% & 25\% & 25\% & 12.5\% & 12.5\% \end{array}$$

S = stearic acid, O = oleic acid

In *directed interesterification*, the temperature is reduced to such an extent that the highest melting glycerides are continously frozen out of the reaction mixture, in turn continuously shifting the reaction equilibrium. In this way a fat can be separated into higher and lower melting fractions. The higher melting fraction contains the glycerides of saturated fatty acids (stearin fraction), whereas the glycerides of unsaturated fatty acids are found in the lower melting fraction (olein fraction). Directed interesterification of stearodiolein can yield 33.3% of tristearin and 66.7% of triolein:

$$\begin{array}{c} \text{O-S-O} \\ \downarrow \end{array}$$

$$\begin{array}{cc} \text{S-S-S} & \text{O-O-O} \\ 33.3\% & 66.7\% \end{array}$$

S = stearic acid, O = oleic acid

## 4.3. Hydrogenation

The double bonds in a fatty acid chain can be wholly or partially saturated by addition of hydrogen in the presence of a suitable catalyst such as nickel, platinum, copper, or palladium. Hydrogenation always leads to an increase in melting point and is therefore also called "hardening". Partial hydrogenation can lead to isomerization of cis double bonds to trans double bonds.

The catalyst, the oil, and the hydrogen must be brought into mutual contact under suitable temperature and pressure conditions. The reaction rate depends on mixing intensity, the type of oil or fat, temperature, catalyst activity, and concentrations of catalyst and dispersed hydrogen. Hydrogenation is an exothermic process. Industrial nickel catalysts are generally obtained by precipitation of nickel hydroxide or carbonate on kieselguhr, silica gel, alumina, or similar carriers, followed by reduction to metallic nickel, or by in situ production of metallic nickel from nickel formate. Such heterogeneous catalysts have a large activated surface. During hydrogenation the double bonds form transient complexes with the active centers of the catalyst. These complexes disintegrate after reaction of the double bonds with hydrogen, leaving the catalyst in its original form [26]. The active centers of the catalyst can be inactivated or poisoned by a number of compounds such as phospholipids, sulfur compounds, organic acids, and oxidized lipids.

A fatty acid with several double bonds, such as linolenic acid ($C_{18:3}$), is hydrogenated more quickly to linoleic acid ($C_{18:2}$) or oleic acid ($C_{18:1}$) than is linoleic acid to oleic acid or oleic acid to stearic acid ($C_{18:0}$). The reaction sequence occurring during hydrogenation can be represented schematically as follows:

$$C_{18:3} \xrightarrow[+H_2]{k_1} C_{18:2} \xrightarrow[+H_2]{k_2} C_{18:1} \xrightarrow[+H_2]{k_3} C_{18:0}$$

The term selectivity is used to indicate which of these reactions is fastest. Selectivity I is defined as the ratio $k_2/k_3$; it is related to the proportion of saturated glycerides formed and to the melting behavior of the product. Selectivity II, expressed as the ratio $k_1/k_2$, must be as high as possible if the concentration of linoleic acid in the hydrogenated product is to be maximized [27].

Selectivity can be influenced by the catalyst type (surface area, pore size, etc.) and by altering the reaction conditions. An increase in selectivity, i.e., an increase in partial hydrogenation, promotes isomerization of cis to trans double bonds.

At temperatures above 200 °C and with a low hydrogen concentration, catalytic hydrogenation of polyunsaturated fatty acid groups can lead to the formation of traces of cyclic aromatic compounds [28].

The double bonds of substituted fatty acids such as ricinoleic acid can also be hydrogenated under suitable reaction conditions. Cyclopropane or cyclopropene groups behave as double bonds and lead to branched fatty acids on hydrogenation.

Iron pentacarbonyl and cobalt octacarbonyl are examples of homogeneous hydrogenation catalysts.

Reduction with hydrazine does not lead to isomerization; there is also no selectivity ($k_1 = k_2 = k_3$).

## 4.4. Isomerization

Naturally occurring fatty acids exist predominantly in the cis form. An equilibrium mixture in which the higher melting trans form predominates can be formed by heating to 100–200 °C in the presence of catalysts such as nickel, selenium, sulfur, iodine, nitrogen oxides, or sulfur dioxide.

If selenium or oxides of nitrogen and sulfur are used in the cis–trans isomerization (elaidinization) of oleic acid, there is virtually no positional isomerization. However, cis–trans isomerization of linoleic and linolenic acid leads to conjugated double bonds.

Nonconjugated systems can be isomerized into conjugated systems by heating in an alkaline solution at 200 °C (→ Fatty Acids). If reaction times and temperatures are extended, linolenic acid can be converted into cyclohexadiene and benzene derivatives:

$$\text{cyclohexadiene-}(CH_2)_n CH_3,\ (CH_2)_m CO_2 H \qquad \text{benzene-}(CH_2)_n CH_3,\ (CH_2)_m CO_2 H$$

Isomerization can occur if oils and fats are heated at temperatures above 100 °C in the presence of bleaching earth, kieselguhr, or activated charcoal.

## 4.5. Polymerization

Dimeric, oligomeric, and polymeric compounds are formed by heating unsaturated fatty acids at 200–300 °C [29]. The rate of polymerization increases with increasing degree of unsaturation; saturated fatty acids cannot be polymerized. Thermal polymerization of polyunsaturated fatty acid groups is normally preceded by isomerization and conjugation of double bonds. Thermal polymerization involves formation of new carbon–carbon bonds by combination of acyl radicals and by Diels–Alder reactions, while oxidative polymerization involves formation of C–O–C bonds. Thermal dimerization is catalyzed by Lewis acids such as boron trifluoride; industrial processes for dimerizing oleic acid are based on this principle [30]. Heating of oils during refining or during household use does not lead to a significant increase in dimeric triglycerides. Up to 2 % dimeric triglycerides can be encountered in fresh raffinates; these dimers are not toxic, and are largely excreted as such.

## 4.6. Autoxidation

Autoxidation, the oxidation of olefins with oxygen, plays a decisive role in the development of rancidity, off-flavors, and reversion flavors in oils and fats during their production and storage. Autoxidation of oil-containing products such as oilseeds and spent bleaching earths can lead to their spontaneous combustion. Autoxidation of drying oils is an important initial stage of polymerization leading to stable surface films.

Autoxidation involves the formation of a hydroperoxide on a methylene group adjacent to a double bond; this step proceeds via a free-radical mechanism:

$$-CH_2-CH=CH- \xrightarrow{\text{Activation}} -\overset{\bullet}{C}H-CH=CH-$$

*Chain reaction:*

$$-\overset{\bullet}{C}H-CH=CH- + O_2 \longrightarrow -\overset{\overset{\displaystyle OO^{\bullet}}{|}}{C}H-CH=CH-$$

$$-\overset{\overset{\displaystyle OO^{\bullet}}{|}}{C}H-CH=CH- + -CH_2-CH=CH- \longrightarrow$$

$$-\overset{\overset{\displaystyle OOH}{|}}{C}H-CH=CH- + -\overset{\bullet}{C}H-CH=CH-$$

Autoxidation is characterized by an induction period during which free radicals are formed. This phase is triggered by light (photooxygenation), heat, and the presence of compounds that readily form free radicals (e.g., hydroperoxides, peroxides, and transition metals). Photooxygenation, i.e., light-induced oxidation, leads to a particularly fast buildup of radical concentration. The formation of singlet oxygen under the influence of short-wave radiation and a sensitizer such as chlorophyll or erythrosine probably plays a key role in this reaction.

The reactivity of a methylene group in forming a hydroperoxide is enhanced by a second adjacent double bond. Hence linoleic acid oxidizes 10 to 20 times faster than oleic acid. Linolenic acid reacts about three times faster than linoleic acid, since two doubly activated methylene groups are present. The olefin radical formed subsequently isomerizes; in a 1,4-diene such as linoleic acid, isomerization leads to conjugated hydroperoxides:

$$-CH=CH-\underset{12}{CH}-\underset{11}{CH_2}-\underset{10}{CH}=\underset{9}{CH}-$$

$$\downarrow$$

$$-CH=CH-\overset{\cdot}{CH}-CH=CH-$$

$$\swarrow \qquad \searrow$$

$$-\overset{\cdot}{CH}-CH=CH-CH=CH- \qquad -CH=CH-CH=CH-\overset{\cdot}{CH}-$$

$$\downarrow O_2 \text{ etc.} \qquad\qquad \downarrow O_2 \text{ etc.}$$

$$-\underset{\underset{OOH}{|}}{CH}-CH=CH-CH=CH- \qquad -CH=CH-CH=CH-\underset{\underset{OOH}{|}}{CH}-$$

The intermediate hydroperoxides are labile compounds that decompose into a number of different products: epoxides, alcohols, diols, keto compounds, dicarboxylic acids, aldehydes, and isomerization and polymerization products. The volatile carbonyl compounds formed in this process are responsible for the taste and odor of oxidized oils and fats.

When the radical concentration has reached a certain limit, the chain reaction is gradually stopped by mutual combination of radicals.

Antioxidants prolong the induction period by reacting with the intermediate products of the chain reaction, forming inactive radicals. Tocopherols are naturally occurring antioxidants. Butylhydroxyanisole (BHA), butylhydroxytoluene (BHT), and propyl gallate are among the most effective synthetic antioxidants. Certain organic or inorganic acids (citric, tartaric, ascorbic, phosphoric acids) have a synergistic effect without being true antioxidants; this effect is presumably based on the inactivation of trace metals or reduction of oxidized antioxidants.

# 5. Manufacture and Processing

Oils and fats are either of vegetable or animal origin. The approximate proportions of the corresponding world production are 55% vegetable oils, 40% land-animal fats, and 5% marine oils.

# 5.1. Vegetable Oils and Fats

## 5.1.1. Storage and Handling of Raw Materials

The handling of oleaginous seeds and fruits during transport and storage has a decisive influence on the quality of the crude oils. The oils in fruit pulp (e.g., olive and palm) are very susceptible to enzymatic hydrolysis since they are finely dispersed in moist cell tissue. The oils contained in seeds are more stable but can also be attacked. A high moisture content accelerates the uptake of oxygen and the corresponding release of carbon dioxide due to degradation of starch. The heat generated in this reaction promotes lipolysis, growth of microorganisms (which can lead to the formation of mycotoxins such as aflatoxins), formation of undesirable color and odor, and — in extreme cases — may result in a coagulation of seed in a silo and even spontaneous combustion [31], [32]. Damage of seed cells, mechanically or by pests, also promotes lipolysis and lipoxidation. Free fatty acids, oxidation products, and coloring matter formed by lipolysis, lipoxidation, and degradation of protein and carbohydrate can in general be removed from the oil only with difficulty, with corresponding yield losses.

In order to minimize these effects, oilseeds must be dried before storage, preferably under mild conditions to avoid cellular damage. The critical moisture content roughly correlates with the hygroscopic equilibrium at 75% relative humidity and varies between 6 and 13%, depending on the protein and carbohydrate content. In the case of copra, rapeseed, and sunflower seed, the critical moisture content is 7%; in the case of soybeans 13%.

Transport and handling in bulk has almost entirely superseded that in bags. Seed is normally stored in concrete silos that are ventilated to avoid local generation of heat and water pockets; such silos can be up to 70 m in height and up to 12 m in diameter. The silos are filled from the top, emptied from the bottom, and normally equipped with special vibrating devices at the bottom to minimize bridging.

## 5.1.2. Cleaning and Dehulling

On arrival at the oil mill, oilseeds still contain plant residues, dust, sand, wood, pieces of metal, and foreign seed, which must be removed prior to further processing by screening, air classification, and passage over magnets.

Some oilseeds are dehulled (decorticated) before being further processed, especially if only the oil is to be expelled, since the hulls tend to retain part of the oil. Preexpelling and subsequent extraction with solvents is the most common process; a certain proportion of hulls in the material to be extracted can be desirable since they facilitate percolation. An excessively high proportion of hulls can, however, impair the quality of the crude oil since the solvent also dissolves wax from the seed coat. A further reason

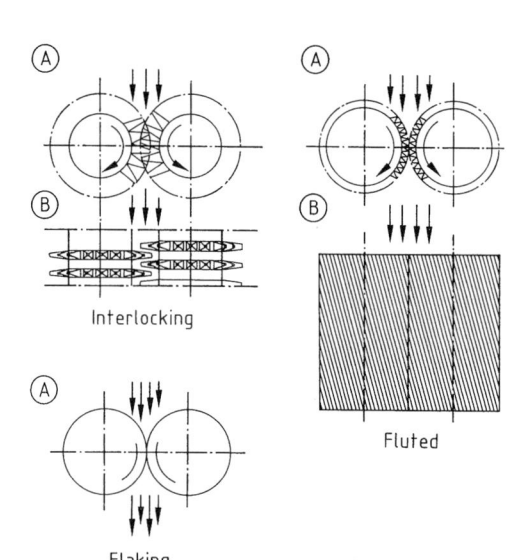

**Figure 3.** Schematic representation of various rolls used for expelling oilseeds
A) Side view; B) Top view

Interlocking

Fluted

Flaking

for dehulling can be the need to produce so-called high protein meal, for example from soybeans.

Dehulling is normally preceded by relatively intense drying of the seed material to help loosen the seed coat from the meats. Dehulling is normally performed by using impact disintegration or passage through rolls to break up the seeds. The hulls are separated by screening and air-classification.

## 5.1.3. Expelling

Modern processing of oilseeds generally involves a combination of expelling and solvent extraction. Seed is normally preexpelled to a residual fat content of ca. 20%, and the expeller cake is then extracted to 1–2% fat. Preexpelling is omitted if the fat content of the seed is 20% or less. High-pressure expelling down to ca. 4% residual fat is no longer important, mainly because of the higher processing costs.

Most of the fat present in seeds and fruit pulp is in the endosperm and hypocotyl cells, with much less in the seed coat. The processing conditions are largely determined by the size and stability of the oil-containing cells. With palm fruit and olives, mere heating or boiling with water suffices to burst the membranes of the oil-containing cells to liberate the oil. Nearly all oleaginous seeds must, however, first be comminuted and thermally pretreated to isolate the fat in an acceptable yield. Comminution partially destroys the cellular structure, increasing the total internal surface and facilitating access to the oil within the seed.

Coarse materials such as copra are first broken by passage through interlocking rolls. Fluted rolls are used for nearly all other seeds. Smooth or so-called flaking rolls are used to complete seed comminution. These three types of rolls are shown in Figure 3; they can be arranged in parallel, diagonally, or sequentially. In the more modern, parallel

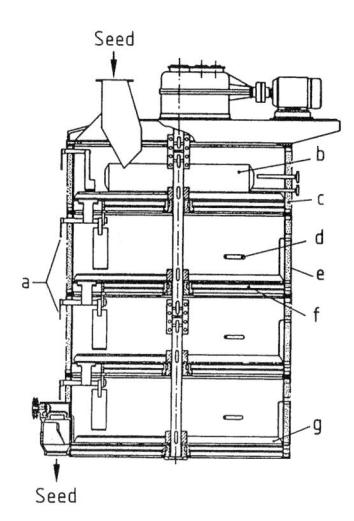

**Figure 4.** Vertical stack cooker for conditioning oilseeds (Krupp)
a) Level control; b) Steam heating; c) Insulation; d) Steam jet;
e) Manhole; f) Steam heating; g) Sweep stirrer

arrangement, the rotational speed of the two rolls differs by 5–10% to produce an additional shear effect. The separation of the rolls and their circumferential speed is adjusted to the type of seed and the degree of cell rupture required for the subsequent expelling and extraction.

The comminuted seed is conditioned prior to expelling by moistening and heating in suitable equipment. During this treatment lipoproteins decompose, proteins coagulate, and intracellular oil bodies coalesce. The cell walls themselves are not macerated during conditioning. Coagulation of protein facilitates expelling and subsequent percolation in the extractor. Undesirable enzymes and microorganisms are deactivated at temperatures above 80 °C and in the presence of sufficient moisture [33].

By suitably conditioning cottonseed, the toxic gossypol can be deactivated by association with denatured protein. In rapeseed, suitably severe conditioning deactivates the enzyme myrosinase and hence improves the quality of the meal and the oil. However, excessive heating impairs the nutritive quality of the expeller cake as well as the color and taste of the expeller oil and must be avoided. An excessively high moisture content can reduce the yield of expeller oil. Each type of seed must therefore be adjusted to an optimal moisture content and temperature during conditioning.

Conditioning is performed in horizontal jacketed tubes (conditioners) or in vertical stack cookers. The stack cooker basically consists of up to five kettles arranged in a tier and through which the seed material successively moves from the top to the bottom; each kettle is equipped with a sweep stirrer and facilities for direct and indirect steam heating (Fig. 4). Generally, wet cooking with direct steam takes place in the top kettle, and drying, with increased temperatures and venting, in the bottom kettles.

Expelling (pressing) separates the oil from the solid phase, the so-called expeller cake, and can be done in continuous or batch presses. Levers, wedges, or screws were used to apply pressure in the more primitive types of batch presses, but modern batch presses are almost invariably operated by a hydraulic system. They can be divided into the open type, in which the seed must be wrapped in press cloths, and the closed type,

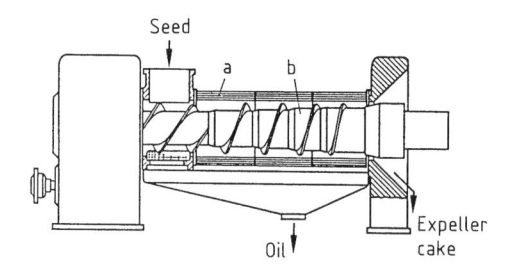

**Figure 5.** Screw press (Krupp)
a) Drainage barrel; b) Worm shaft

which dispenses with press cloths and confines the material in a cage-type construction. Open-type presses can be subdivided into plate and box presses, closed types into pot and cage presses.

Continuous expellers or screw presses (e.g., Simon – Rosedowns, Krupp) have generally replaced batch presses. Deoiling in these presses is efficient and uniform since the layer of material to be expelled is relatively thin and is continuously broken up. Several types exist, all with the common principle of a worm shaft rotating in a cylindrical drainage barrel (Fig. 5). The barrel is composed of armored, rectangular bars that fit into the barrel bar frame. The individual bars are separated by bar spacing clips, the specific spacings — ranging from 0.1 to 0.4 mm — depending on the type of seed and degree of expelling required. The required pressure is built up either by a gradual increase in the diameter of the worm shaft or by a gradual restriction of the barrel diameter. Specific pressure profiles and cake thickness can be further controlled by variations in the pitch of the worm and choke mechanisms in the barrel. The expelled oil is discharged through the barrel spacings while the expeller cake emerges at the end of the barrel.

The capacity of screw presses varies widely, being dependent primarily on the cross-sectional ratio and the speed of the screw. The pressure should be built up fairly slowly. Initially, the channels in the expeller cake are sufficiently large for ready discharge of oil; in the latter sections of the screw press, the pressure is increased to about 300 MPa (3000 bar) to expel more tightly bound oil and to burst intracellular oil bodies.

Up to 200 t of seed per day per press is normally preexpelled to a residual fat content of 15 – 25 %. The expeller cake is then comminuted on fluted rolls, and possibly also pelletized, prior to solvent extraction.

Screw presses used to reduce the residual fat content to 3 – 6 % are similar in design to preexpellers but have a larger compression ratio (1:25 instead of 1:15) and have a lower worm speed (9 – 12 rpm versus 30 – 45 rpm).

A specific worm arrangement normally gives optimal results for only one particular type of seed. Difficulties encountered on changing the type of seed can often be overcome by modifying conditioning and comminution. Preexpellers that do not require seed conditioning of sunflower seed have been developed; however, they do not work as well with unconditioned rapeseed.

The expeller oil still contains so-called foots, which must be removed by filtration.

## 5.1.4. Extraction [34], [35]

Very much lower residual fat contents can be achieved by solvent extraction than by expelling. Seed material with an oil content of ca. 20% and expeller cake with 15–25% fat are usually subjected to solvent extraction. Both must be flaked prior to extraction. This pretreatment creates a large internal surface and yields thin (ca. 0.3 mm), firm flakes, which form a loose layer in the extractor and provide uniform, short diffusion pathways. The flakes must have a minimum moisture content and elasticity to prevent them from crumbling during transport to the extractor, which would lead to too dense a packing and hence poor percolation of solvent. However, a high moisture content also impairs percolation.

Ideally, the extraction solvent should dissolve only glycerides but not undesirable components such as coloring matter, gums, and phospholipids. The solvents must not contain toxic components, and should be recoverable with minimum loss, be safe in handling, and be readily removable from the extracted material. For these reasons aliphatic hydrocarbons, especially hexane, are used almost exclusively. Technical hexane with a boiling point range of 55–70 °C has proved to be optimal. Hexane can be readily removed from the oil at temperatures below 100 °C in vacuo and can be stripped from the meal with steam. The solubility of hexane in the condensed water is only 0.1%.

For special purposes, e.g., the production of heat-labile pharmaceuticals, lower-boiling hydrocarbons such as pentanes are sometimes used. Extraction with propane or carbon dioxide under supercritical conditions is reserved for special products of high intrinsic value because of the high equipment costs involved. Castor oil, being relatively polar, is preferably extracted with the higher-boiling heptane. Alcohols (methanol, ethanol, propanol, and butanol) and furfurol are specially suited for the extraction of relatively wet materials. Extraction of oilseeds with alcohols leads to relatively high concentrations of phosphatides, glycolipids, carbohydrates, and similar constituents in the crude oil, although the glycerides can in principle be concentrated by cooling or extraction of the alcoholic solutions. Alcohols are generally not used as primary extraction solvents. However, they are occasionally used as secondary extraction solvents to remove gossypol from cottonseed meal, thioglycosides from rapeseed meal, sugars from soybean meal (to produce protein concentrates), and alkaloids from bitter lupine meal.

Chlorinated hydrocarbons such as trichloroethylene and dichloromethane are of interest because of their safety in handling and high extraction capacity but do not appear to be used for extraction of oilseeds because of the potential toxicity risk from residual solvent in the meal. Processes in which the oil is displaced by hot water are known but have not achieved industrial importance [36].

Oilseeds are generally extracted in a countercurrent process: pure solvent is contacted with material that has already been largely extracted, and the oil-rich solution is contacted with nonextracted material. Such extractions can be performed either continuously or in a batch process.

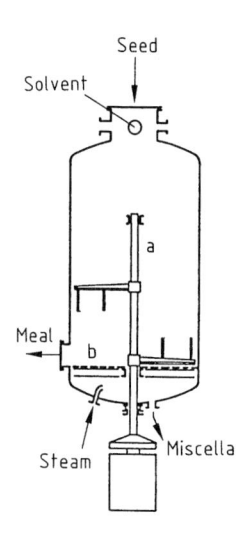

**Figure 6.** Batch extractor
a) Stirrer; b) Screen

**Batch processes.** In a discontinuous process the solvent successively passes through a battery of up to batch extractors (e.g., from Buss). Each of these extractors is a cylindrical vessel with a large diameter:height ratio and has an opening for filling and emptying, a screen in the bottom part, and facilities for pumping solvent in and out, for direct and indirect steam heating, and a sweep stirrer to move the extracted meal during solvent stripping (desolventization) (Fig. 6).

After filling the extractor with comminuted seed material, the upper opening is closed and solvent pumped in. The displaced air escapes via a circuit; the hexane vapors in the displaced air are recovered. After the extractor has been filled with solvent, the hexane containing the extracted oil is pumped into a second extractor filled with seed material. This process is repeated until the hexane solution has been pumped through 3 or 4 extractors. The solvent strongly enriched with oil (the miscella) from the last extractor is filtered to remove seed particles and then distilled. In this type of process, exhaustive extraction can yield oil concentrations of at least 35 % in the final miscella.

When the contents of the first extractor have been extracted sufficiently, the solvent tap is closed; the pure extraction solvent is then passed directly to the second extractor while the fifth extractor becomes part of the battery. After discharging the solvent, remaining solvent is stripped from the extracted seed material by passing in steam. When the vapors leaving the extractor are free of solvent, the extractor is emptied via the lower opening.

The sequence of filling, extraction, desolventization and emptying is repeated for each extractor in a battery in such a way that while one extractor is being filled, the contents of three or four extractors are extracted, residual solvent is stripped in one or two extractors, and one extractor is emptied.

The vapors from the extractors and the distillation are condensed. The resulting hexane–water mixture is separated mechanically and the hexane recycled. Such a

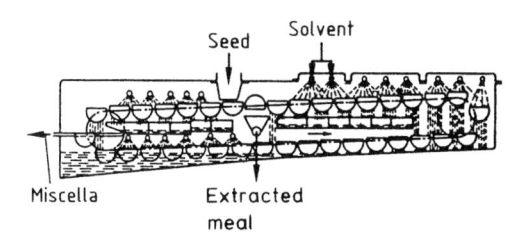

**Figure 7.** MIAG basket conveyor extractor

battery with 10 extractors of 7 m$^3$ each can reach a daily throughput of up to 1200 t seed, depending on the type of seed. The quantity of hexane circulated is ca. 100 t. The solvent loss is normally between 0.35 and 0.45 %, based on seed input.

These discontinuous plants can be automated, and have the principal advantage that special treatment of meal (e.g., detoxification of aflatoxin-containing peanut or cottonseed meal with ammonia) can be carried out fairly easily. Their capacities cannot, however, match those of fully continuous plants, which can process up to 200 t seed per hour.

**Continuous Processes.** Of the many fully continuous processes that have been proposed and developed, countercurrent gravity percolation has become the most popular because of its high capacity. The comminuted seed material is transported through the extractor and sprayed with solvent in the various extraction stations. The solvent becomes increasingly enriched with oil and is passed through the seed bed repeatedly according to the number of extraction stations. This arrangement has the advantage that the miscella is continuously cleaned by filtration through the seed bed.

The MIAG basket conveyor extractor (Fig. 7) uses a horizontal arrangement of baskets in which the flaked seeds are first sprayed with solvent or miscella and then passed through a solvent bath. The baskets containing the extracted and drained flakes are automatically inverted and discharged into a hopper from which the meal is conveyed to the desolventizers and dryers.

In modern plants with a capacity of up to 5000 t of seed per day, circular cell and traveling belt horizontal extractors are generally preferred. The Blaw–Knox Rotocel extractor is especially popular because it offers a large capacity within a relatively small space. It uses baskets that travel in a horizontal circle. After countercurrent extraction, the baskets are emptied through hinged sieve bottoms.

The carousel extractor from Extraktionstechnik operates on a similar principle (Fig. 8), but has a fixed discharge point instead of movable basket bottoms.

In traveling belt extractors, such as the De Smet extractor, the seed material is conveyed on an endless sieve belt, the height of the bed being regulated mechanically. Miscella and solvent are sprayed onto the seed material. Miscella percolates throught the belt, falls into compartments in the bottom of the extractor housing, and is picked up by a series of pumps and recirculated countercurrent to the flakes.

The Lurgi frame belt extractor works on a similar principle [37]. The endless frame belt runs over two bar-sieve belts which form the bottoms of the frames. The flaked

**Figure 8.** Carousel extractor

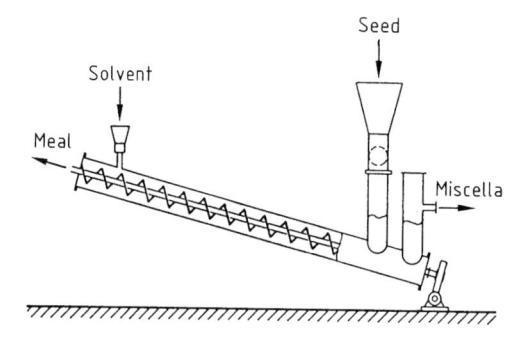

**Figure 9.** Screw extractor

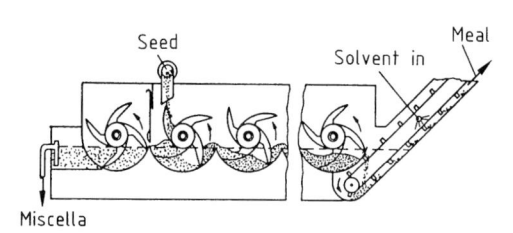

**Figure 10.** Kennedy extractor

oilseeds are filled into the frames of the upper belt, tipped onto the lower belt after half the extraction time has elapsed, and then discharged after completion of extraction and passage through a dripping zone. This separation into distinct extraction steps and intermediate decompaction effects a particularly uniform extraction.

There are numerous other types of extractors which are of less practical importance. In Hildebrandt, Schlotterhose, Ford, Detrex, Adler, and Olier extractors, the comminuted seed material is transported through a solvent bath by a screw; Figure 9 shows the principle involved. In a Kennedy extractor (Fig. 10), the material to be extracted is conveyed to various sections of a trough by means of impeller wheels. In the Sherwin–Williams process the finely comminuted seed material is thoroughly mixed with solvent in three stages and then separated centrifugally from the miscella. The structure of the seed material is not as critical for these extractors as for basket or belt-type extractors. However, they require a far greater ratio of solvent to seed material. Their miscella are also contaminated with fines, which requires a separate filtration step. These types of extractors are therefore virtually obsolete.

Böhm, Tyca, Allis–Chalmers, Bonotto, and Anderson extractors are essentially columns divided into a number of sections in which the seed material moves from the top to the bottom, countercurrent to a rising flow of solvent. These extractors do not match the performance of the large basket or belt-type extractors.

Several attempts have been made to develop processes obviating preexpelling. In the Filtrex process, which is used to a limited extent, oilseeds such as rapeseed, sunflowerseed, or peanuts can be extracted directly to about 1% residual fat without preexpelling, provided the seed material has been flaked to about 0.1 mm (the cell structure is probably destroyed) and subsequently conditioned by "crisping" [38]. A drawback is that emersion-type extractors are required. In the Direx process the seed material is only coarsely ground, extracted to 14–16% residual fat with hexane, flaked, and then extracted to ca. 1% residual fat in a second extraction step. The theory is that flaking the hexane-moist material bursts the oil cells. This process has been applied only on a small scale (ca. 5 t/h).

A more recent development is the Alcon process [39] for soybeans, in which the essential step is to heat the flakes to ca. 100 °C with direct steam, followed by drying. The special flake conditioning prior to extraction deactivates enzymes such as phospholipases and lipoxygenase, and presumably dissociates lipoprotein complexes; this leads to lower residual fat contents in the meal and a more easily deslimable crude oil with improved oxidation characteristics.

Extracted oilseeds can be desolvized in a series of horizontal steam-jacketed tubes. The material is propelled through the tubes by screws and is then passed through toasters, similar to stack cookers, that are heated by both live and indirect steam. Alternatively, desolventization can be performed in one step in desolventizers–toasters. After drying and cooling, the meal is ready for use as an animal feed component. Toasting not only reduces residual solvent to levels below the lower explosion limit but also deactivates nutritionally undesirable enzymes such as urease and trypsin. In some cases, superheated solvent vapor is used for desolventization. This process and flash desolventization are used when minimal protein denaturation is desirable.

Solvent is recovered from the filtered miscella by distillation in multiple-stage evaporators, which are constructed to minimize thermal damage to the oil. The miscella is first concentrated to 96% oil and finally stripped with injected steam in a falling-film evaporator. The vapors from the toaster, dryer, cooler, and distillation unit are condensed and separated into water and hexane, which is recycled. Vented air is passed through special adsorbents such as activated charcoal or mineral oil to recover solvent. Total solvent loss is 0.2–0.3%, based on seed input.

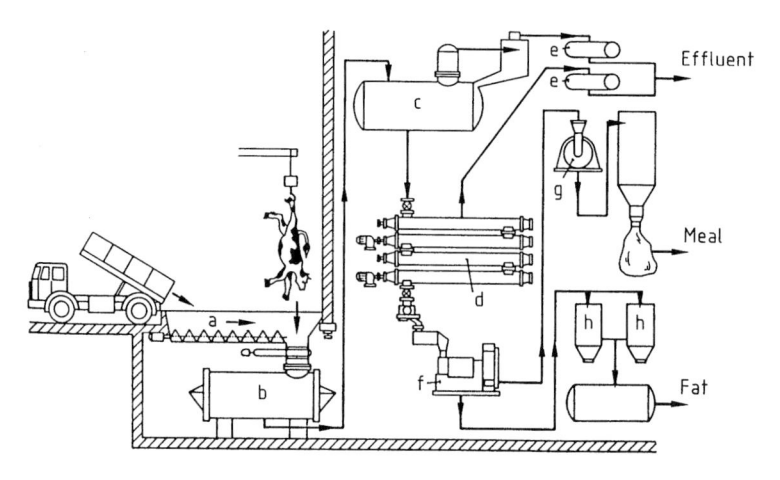

**Figure 11.** Kontipress system for working up carcasses (Krupp)
a) Raw material; b) Cooker–sterilizer; c) Buffer; d) Dryer; e) Condenser aggregate; f) Screw press; g) Meal grinder; h) Buffer

## 5.2. Land-Animal Fats

Animal fats can be readily isolated from tissue by heating since the cell membranes of animal cells are much weaker than those of plant cells. The intracellular fat expands on heating and bursts the membrane. Extraction with solvents is limited to offal. Fatty tissue is particularly susceptible to decomposition so that rapid processing is necessary if refrigeration is not possible. Microbial and autolytic degradation rapidly lead to oxidation, hydrolysis, color deterioration, and unpleasant odors and flavors, which are difficult to remove from the fat. The free fatty acid content therefore serves as a convenient, primary quality criterion.

In *dry rendering*, the comminuted fatty tissue is digested by cooking in steam-jacketed vessels equipped with agitators [40]. Local overheating of fatty tissue must be avoided since this can easily lead to unpleasant off-flavors. In modern automated plants for working up animal byproducts, the raw material is first heated and sterilized in a dry-rendering vessel. The resulting sludge is then separated, possibly after predrying, into crude meal (cracklings) and a liquid phase containing fat and water by a continuous screw press (see Fig. 11, Krupp Kontipress process) or by a decanter, i.e., a centrifuge with a built-in screw conveyor (Alfa–Laval Centrimeal process). The fat is finally separated from the aqueous phase by centrifuging. The 7–12% residual fat content of the meal can be reduced to 1–4% by solvent extraction.

*Wet rendering*, which involves treating the fatty tissue with direct steam, tends to give better yields and quality than dry rendering. In modern continuous rendering plants (e.g., as built by Alfa–Laval, Westfalia, and Sharples) selected tissue is coarsely comminuted and heated to 50–60 °C. It is then heated quickly to 80–90 °C with direct steam to deactivate oxidizing enzymes. The greaves are then separated in a

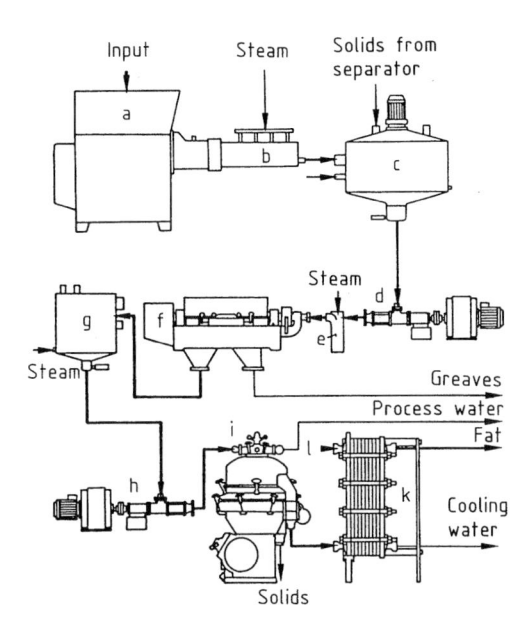

**Figure 12.** Alfa–Laval Centriflow rendering plant
a) Cutter; b) Renderer; c) Buffer; d) Propeller pump; e) Steam heater; f) Decanter; g) Buffer; h) Propeller pump; i) Separator (centrifuge); k) Plate heat exchanger; l) Cooling water

decanter centrifuge, and the fat is clarified by further centrifuging, cooled, possibly texturized, and packed (Fig. 12). Fat obtained by this method in up to 99% yield contains 0.1–0.2% moisture; the content of free fatty acids is virtually identical with that of the raw material. This mild processing technique minimizes the formation of off-flavors. Further refining of lard or tallow for edible purposes is controlled by national legislation.

## 5.3. Marine Oils

*Whale oil* is isolated by rendering. The blubber is cooked with live and indirect steam, and the oil is then separated by centrifugation from insoluble tissue and water-soluble matter.

Similar processes are used to isolate *fish oils* [41]. Crushed fish (herring, menhaden, sardines, pilchards) with a dry matter content of 15–22% is first cooked in closed vessels at relatively low pressure. This operation sterilizes the fish and coagulates the protein to produce a fish mass that can be easily pressed. Subsequent screw pressing separates most of the water and oil from the press cake. The press water is prepurified over screens or with a decanter (desludger) and then separated by centrifugation into crude fish oil and stickwater (or glue water). The press cake (ca. 50% water) is dried, sometimes also extracted with a solvent, and ground into fish meal. Evaporation of the stickwater gives solubles which are added to the meal. Industrial plants are offered by Alfa–Laval (Centrifish) and Westfalia (see Fig. 13)

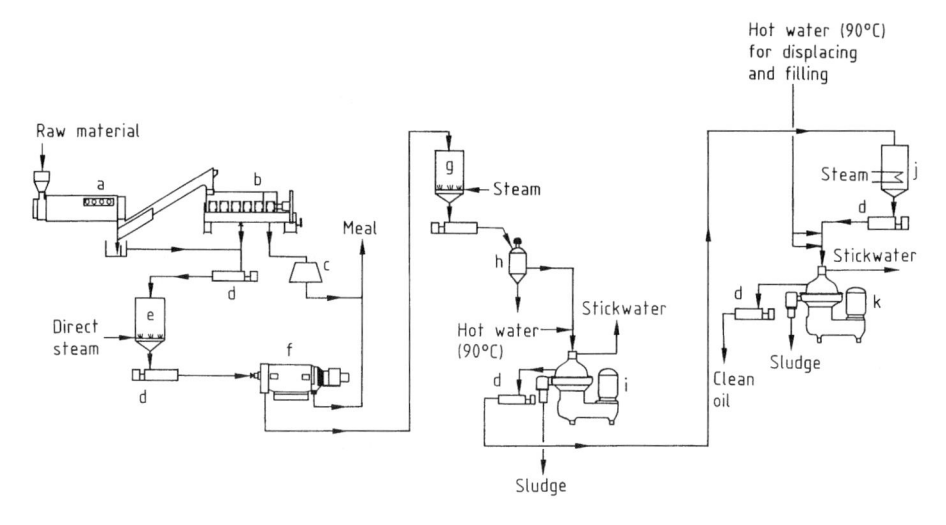

**Figure 13.** Fish meal and fish oil plant (Westfalia)
a) Cooker; b) Power press; c) Crusher; d) Eccentric screw pump; e) Preheater; f) Decanter for clarification of press water; g) Press water preheater (92 °C); h) Rotary brush strainer; i) Separator with self-cleaning bowl (1st stage, press water deoiling); j) Crude oil preheater (92 °C); k) Separator with self-cleaning bowl (2nd stage, oil polishing)

*Liver oils* are usually obtained on board ship by mild rendering of crushed liver, sometimes assisted by pressure-release disintegration of the oil cells, and subsequent separation in centrifuges. The Solexol process used in South Africa employs liquefied propane as the extraction solvent. This method can also be used to fractionate fishliver oils and to isolate vitamin concentrates.

## 5.4. Synthetic Fats

Pure synthetic glycerides [20] and phospholipids have few industrial uses. Synthetic fats for edible purposes are of little importance with the exception of specific dietary products such as medium-chain triglycerides of $C_8$ and $C_{10}$ fatty acids (MCT fats), which are administered in some postsurgical cases [42].

# 6. Refining

Crude oils and fats obtained by expelling, extraction, or rendering contain trace components which are undesirable for taste, stability, appearance, or further processing. These substances include seed particles, dirt, phosphatides, carbohydrates, proteins, fatty acids, trace metals, pigments, waxes, oxidation products of fatty acids, and toxic components such as polycyclic aromatic hydrocarbons, gossypol, mycotoxins, sulfur compounds (in fish and cruciferous oils), and pesticide residues. The aim of refining edible fats and fats for industrial purposes is to remove these undesirable components as far as possible without significantly affecting the concentration of desirable constituents such as vitamins and polyunsaturated fatty acids, and without significant loss of the major glyceride components. Some oils, such as olive oil, butterfat, and cold-pressed sunflower oil or linseed oil, are not refined in order to preserve their typical flavors.

Refining usually involves the following stages:

1) Precleaning to remove phosphatides (degumming)
2) Neutralization by treatment with lye or by distillation
3) Decolorization by adsorptive treatments (bleaching)
4) Deodorization or stripping in vacuo

## 6.1. Degumming [43]

Phosphatides, gums, and other complex colloidal compounds can promote hydrolysis of an oil or fat during storage and also interfere with subsequent refining. They are therefore removed by degumming or desliming.

The degumming method depends on the type of the oil and the phosphatide content. In hydration degumming, which involves treating the oil or fat with water or steam, phosphatides absorb water to form a sludge that is insoluble in the oil. This process is used to remove soy lecithin (1–3%) from soybean oil: 2–5% water, based on oil, is intimately mixed with the crude oil at 70–80 °C; after a contact time of 1–30 min the phosphatide sludge is separated by centrifuging. The degree of hydration can be increased by using acids such as citric or phosphoric, bases, or salts but at the expense of lecithin quality. Such modified hydration processes are therefore confined to postdegumming and to oils whose lecithins have little commercial value. Degumming and neutralization can also be carried out together but usually with increased losses of neutral oil owing to the formation of emulsions. The use of sulfuric acid is generally restricted to industrial oils that are not to be further refined.

Some technical oils such as linseed oil can be degummed by heating to 240–280 °C; the precipitation of gums on heating is referred to as oil "breaking."

Precleaning of a crude oil by treatment with bleaching earth or other adsorbents, normally in combination with acids such as phosphoric or citric, can be advantageous.

Oils that are refined by distillation (physical refining, see p. 2416) should have a very low phosphatide content. This can be achieved by special degumming techniques involving the addition of citric acid followed by hydration at ca. 20 °C for several hours before centrifugal separation [44]. Another "superdegumming" technique involves separating the phosphatide micelles from a solution of the crude oil in hexane by ultrafiltration using special polysulfone or polyacrylonitrile membranes [45].

## 6.2. Deacidification (Neutralization)

Commercially available crude oils and fats contain on average 1–3% free fatty acids. Whereas a soybean oil can have as little as 0.5% free fatty acids, some palm and fish oils can contain up to 10%. The free fatty acid content of refined fats should be below 0.1%. Although longer chain fatty acids do not generally affect taste, the short-chain acids can impart a soapy, rancid flavor. When the free fatty acids are removed as soaps by treatment with lye, other undesirable constituents such as oxidation products of fatty acids, residual phosphatides and gums, phenols (e.g., gossypol), and aflatoxins are also "washed out." In practice, deacidification is performed mainly by treatment with lye or by distillation. Deacidification by esterification with glycerol, by selective extraction with solvents, or by adsorbents is not industrially important.

**Alkali Neutralization** [46]. The usual method of alkali neutralization is treatment with weak lye, either batchwise or continuously. The more concentrated the lye, the more readily are undesirable constituents taken up in the soapstock. Dilute alkaline solutions (ca. 1 N) do not saponify the oil, but greater neutral oil losses occur due to occlusion in the soapstock. The losses increase with increasing content of free fatty acids. Concentrated lye (4–7 N) yields a relatively concentrated soap which takes up little neutral oil, but it can saponify substantial amounts of the neutral oil, especially in those oils that have a relatively high proportion of short-chain fatty acids. The optimum lye concentration therefore depends on the quality of the crude oil and the desired quality of the refined product. The neutralization process can be further influenced by temperature, residence time, and mode and intensity of mixing the lye and the fat phase.

*Batch neutralization* is done in an open or closed vessel (up to 75 t content) with a conical base. The reactor is fitted with a heating mantle, heating coils, and facilities for injecting live steam (Fig. 14). Lye and oil can be mixed without undue emulsification by using special coil and frame stirrers. Soap settles in the conical bottom of the neutralization vessel. Discharge of neutral oil with the soap can be minimized by using ultrasonic or dielectric measurement of the phases. Closed vessels can also be used for bleaching if they can be evacuated. Whereas weak lye is generally sprayed onto the oil at ca. 90 °C without stirring, stronger lye is normally stirred into the oil at 40–80 °C. Sodium carbonate is sometimes used instead of caustic soda; it does not saponify but tends to lead to foaming. Oils with a relatively high concentration of free fatty acids are neutralized with strong lye in order to limit the total volume. The required amount of lye, including a slight excess, is based on the free fatty acid content (ffa).

After neutralization and discharge of soapstock, the oil is washed with dilute lye (ca. 0.5 N) and then with water in order to reduce the soap content below 500 ppm. Incomplete removal of soap impairs the color of the oil and the efficacy of the following refining steps. Polyunsaturated oils such as soybean oil are sometimes subjected to posttreatment with sodium carbonate – silicate to remove residual traces of oxidized

**Figure 14.** Combined neutralizer – bleacher

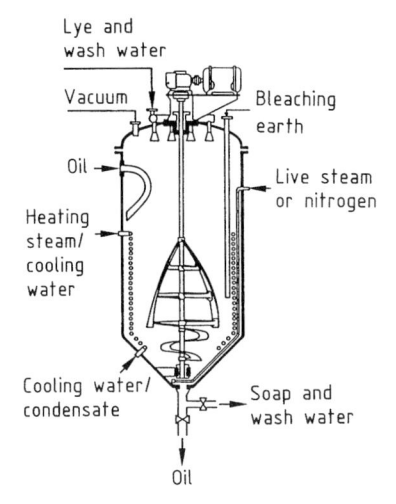

glycerides and phosphatides. Although soapstock can be utilized as such in chicken broiler feed, it is normally converted into "acid oil" (a mixture of fatty acids and neutral oil) by batch or continuous treatment with 30% sulfuric acid at 70 °C.

The quantity of acid oils is a measure of the refining efficacy. A refining factor of 2 means that for each mass unit of free fatty acid in the crude oil, 2 mass units of acid oil are produced. The difference between acid oil and free fatty acid content is mainly due to occluded neutral oil. Acid oils are used for technical purposes and as energy carriers in animal feeds [47].

Batch neutralization is more flexible than continuous neutralization but gives lower refining yields.

*Continuous processes* achieve the same or greater capacities in a smaller space with better yields, provided long continuous runs can be maintained. Continuous plants from Alfa–Laval, Westfalia, and Sharples are widely used (Fig. 15). In these plants, the oil is generally first conditioned by addition of citric or phosphoric acid in small continuous mixing chambers and then brought into contact with lye. Soap and gums are separated in centrifuges. After re-refining and passage through a washing centrifuge, the oil is practically free of fatty acid and soap. The Sharples plant uses high-speed open centrifuges with a high separation efficiency, while the Alfa–Laval short-mix lines employ completely closed centrifuges with a mean residence time of only a few seconds. The capacity can be increased considerably by using self-discharging centrifuges (e.g., from Westfalia).

The Zenith process employs a different principle: a fine stream of oil rises through a vessel filled with dilute caustic soda and is skimmed off continuously at the top. The spent lye must be replaced. This type of neutralization requires efficient pre-degumming and has the advantage of relatively low losses of neutral oil [48].

Neutralization with aqueous ammonia is an old process which is being reexamined because of environmental pollution problems with lye. The ammonium soaps can be

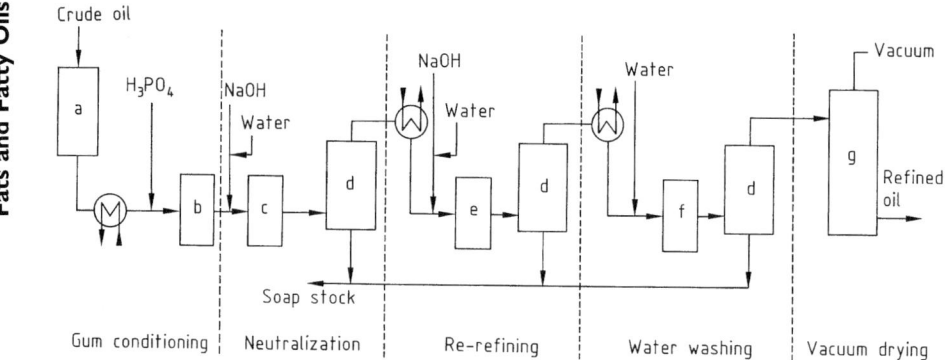

**Figure 15.** Flow diagram of continuous neutralization of oils and fats
a) Crude oil storage tank; b) Gum conditioning mixer; c) Neutralizing mixer; d) Separator (centrifuge); e) Re-refining mixer; f) Water-washing mixer; g) Vacuum dryer

separated by centrifuging and split into fatty acids and ammonia by heating, the ammonia being recycled [49].

**Distillative Neutralization.** In distillative neutralization (physical refining) the free fatty acids are continuously removed from the crude oil with water vapor in vacuo. Prior to this stripping, gums, phosphatides, and trace metals must be almost completely removed since these compounds impair the oil quality during distillation. They are preferably removed by treatment with acids and bleaching earth.

The economics of distillative neutralization improve with increasing content of free fatty acids. The fatty acid factor is approximately 1.1, thus the neutral oil losses are far lower than with alkali neutralization. An added advantage is the avoidance of soap-stock splitting and the consequent effluent problems. Distillative neutralization was formerly restricted to fairly saturated oils with a high free fatty acid content (e.g., coconut oil, palm oil, and tallow). However, it is now being extended to polyunsaturated oils such as rapeseed, soybean, and sunflower oil because techniques are available for reducing the phosphatide content of the crude oils to sufficiently low levels to obviate lye neutralization [50].

Temperatures between 240 and 270 °C are normally used to achieve free fatty acid contents of less than 0.1%. At these temperatures decomposition and removal of carotenoid pigments result in a significant bleaching effect, this is particularly noticeable in palm oil. Older plants such as those built by Wecker and Feld & Hahn have now been generally superseded by modern plants (e.g., Lurgi, Girdler, HLS, and EMI), which have the advantage of a smaller spread in residence time and hence better process control (especially in combined deacidification, deodorization, and heat bleaching). The latter plants are very similar to the corresponding deodorizer plants (see Figs. 20 and 21) except for an additional fatty acid condenser between the deodorizer and the vacuum assembly.

**Other Processes.** *Esterification* of the free fatty acids with glycerol may be economically feasible for highly acid oils. The esterification is carried out at 160–210 °C in the presence of catalysts such as magnesium oxide, alkali silicates, or metals (Zn or Sn).

*Solvent extraction* of free fatty acids is of interest only for highly acid oils. Olive oil with 22% ffa can be deacidified to ca. 3% ffa by liquid–liquid extraction with ethanol. Furfurol can be used to dissolve polyunsaturated glycerides and fatty acids.

In the Solexol process, liquefied propane is used as solvent. Saturated glycerides are taken up in the propane phase while nonsaponifiable matter, fatty acids, oxidation products, and polyunsaturated glycerides remain largely undissolved. This process has been used to fractionate fish and fishliver oils and may increase in importance with the advent of supercritical fractionation and extraction techniques.

Various adsorbents such as silica gel and alumina have been proposed for the *selective adsorption* of free fatty acids [51]. Adsorption on strongly basic ion-exchange resins has also been used. Plant-scale application of such processes has, however, not been reported.

## 6.3. Bleaching

Degumming followed by alkali neutralization generally does not lead to significant decolorization of an oil or fat. A bleaching step with solid adsorbents such as bleaching earth or activated charcoal is normally employed. Bleaching with air or chemicals is not used for edible fats.

The aim of adsorptive treatment is to remove pigments such as carotenoids and chlorophyll but also residues of phosphatides, soaps, trace metals, and oxidation products such as hydroperoxides and their nonvolatile, polar decomposition products. These compounds can have a deleterious effect on the course of further processing (especially hardening and deodorization) and on the quality of the final product. Some of the oxidation products that are removed by adsorption can promote oxidation of the oil. Since some of these undesirable compounds can be removed by suitable degumming and neutralization, there is an interdependence between the pretreatment conditions and the treatment with adsorbents. This interdependence can also involve deodorization (heat bleaching) where, for example, the carotenoid content is reduced.

The adsorption of pigments follows Freundlich's isotherm: the amount of pigment adsorbed decreases with decreasing concentration of the pigment in the oil. Although adsorption should be most efficient if the oil is passed through a layer of adsorbent, in practice the adsorbent is normally stirred into the oil and then filtered off after a certain residence time.

The choice of the adsorption process and the type and concentration of adsorbent is governed by factors such as pretreatment, desired quality of the fully refined product, filtration speed of the oil, and oil retention by the adsorbent. Oil retention can be up to 50 wt% on bleaching earths and nearly 100 wt% on activated charcoal.

*Natural and activated earths* are the most frequently used adsorbents in the refining of oils and fats. Activated earths are obtained by leaching natural earths (predominantly aluminum silicates containing montmorillonite) with hydrochloric acid followed by washing with water, drying, and sizing. This process increases the internal surface of the bleaching earth by partial dissolution of the aluminum and iron oxides; calcium and magnesium ions are partially replaced by hydrogen ions.

The amount of bleaching earth to be used and the optimum bleaching conditions must be determined empirically. Lower concentrations (0.5–1.0%) are required for activated earths than for natural earths. Activated earths catalyze oxidation of the oil. For this reason bleaching should be done below 100 °C in vacuo, and the subsequent filtration should be carried out with limited access of atmospheric oxygen. Bleaching is sometimes performed in two steps: a conditioning and adsorption stage at relatively low temperature (ca. 65 °C), and a second stage at elevated temperature (ca. 100 °C) to fix the adsorbed components onto the adsorbent surface [52].

In physical refining, bleaching is often preceded by addition of citric or phosphoric acid to chelate trace metals and to precipitate hydrated phosphatides. This can also be achieved by using bleaching earths containing added acid.

The moisture content of bleaching earths lies between 5 and 10%. Except for some types of natural earths, the moisture content does not appear to have a significant influence on bleaching capacity.

*Activated charcoal* (0.1–0.4%) is sometimes used in combination with bleaching earth for oils that are difficult to bleach. Activated charcoal has also assumed importance as an adsorbent for removing polycyclic aromatic hydrocarbons from oils and fats [53]. It has been recommended for use in a fixed-bed process for the physical refining of soybean oil [54].

Bleaching and filtration is still mostly done *batchwise*. The oil is first dried in vacuo at ca. 30 kPa (30 mbar) to an optimum moisture content, frequently in the same closed vessels in which the oil was neutralized and washed (Fig. 14). The required amount of adsorbent is then drawn into the oil at 80–90 °C, preferably through a pipe reaching into the oil. After stirring for a few minutes to half an hour in vacuo, the oil–earth slurry is pumped over filter presses. Automated frame and plate presses or centrifugally dischargeable, closed-disk filters of the Funda (Fig. 16) and Schenk type are preferred; these also have the advantage of limiting access of air. The filter cake, containing 30–50% oil, is often extracted with hot water (Thomson process) or with hexane to recover the bulk of the adsorbed oil. The quality of the recovered oil depends on the degree of unsaturation, the type of extraction, and the age of the spent earth before extraction. It is normally not economic to regenerate spent bleaching earth.

*Continuous bleaching* is especially relevant when the other refining steps are also performed continuously. The degummed or degummed and neutralized oil is normally dried and deaerated in vacuo and charged together with bleaching earth into an evacuated vessel where the oil–earth slurry is fed through various compartments, possibly at different temperatures, and is then filtered in a closed-disk system (Fig. 17).

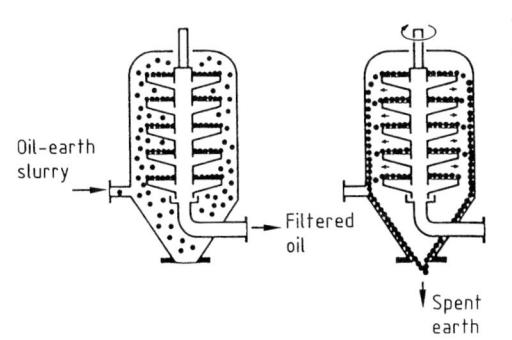

**Figure 16.** Closed disk filter with centrifugal discharge (Funda)

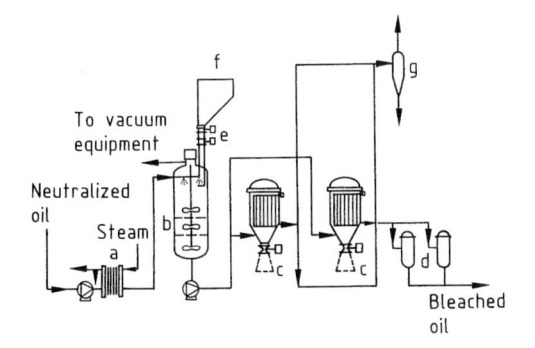

**Figure 17.** Alfa–Laval bleach system
a) Plate heat exchanger; b) Bleacher;
c) Filters; d) Polishing filters;
e) Bleaching-earth dosage equipment;
f) Bleaching-earth storage tank,
g) Cyclone

Bleaching with oxygen or chemicals is not practiced for edible fats and is used only occasionally for technical fats. Oxidizing agents such as hydrogen peroxide, sodium peroxide, benzoyl peroxide, potassium permanganate, chromium salts, hypochlorite, and chlorine dioxide, and reducing agents such as sulfurous acid and sodium dithionite have been proposed. Hydrogen peroxide is used when adsorptive methods do not suffice. Castor, coconut, peanut, and olive oils and certain types of tallow and bone greases can be successfully bleached with 0.05–0.1 % of a 20 % aqueous solution of sodium chlorite. Fats such as red palm oil are sometimes bleached by very slight hydrogenation which destroys carotenoid pigments.

## 6.4. Deodorization

Deodorization is the last step of the refining sequence, in which odors and flavors are removed from the bleached oil or fat. It is essentially a steam distillation process in which volatile compounds are separated from the nonvolatile glycerides. The odoriferous compounds are primarily aldehydes and ketones formed by autoxidation during handling and storage and can have a flavor threshold value of a few ppm. Other volatile components such as free fatty acids, alcohols, sterols, or tocopherols are also partially removed by deodorization.

Deodorization is not, however, a purely physical process. During deodorization, flavor compounds can be formed by hydrolysis and thermal decomposition, and peroxides are decomposed by heat. These reactions presumably play a major role in the flavor stability improvement achieved by deodorization, especially in vegetable oils. In practice, therefore, the residence times and steam volumes required are significantly greater than those calculated for normal steam distillation. At the relatively high deodorization temperatures used, traces of air can initiate oxidation with severe impairment of taste and flavor stability. Because thermal decomposition products of proteins, carbohydrates, phosphatides, and soaps can also interfere with deodorization, the oil must be well prerefined, i.e., freed from phosphatides and soap.

Deodorization is performed in vacuo because of the low partial vapor pressure of the compounds to be removed. Consumption of stripping steam increases with decreasing vapor pressure of the volatile components and with increasing pressure in the deodorizer, it ranges from 5 to 10 wt% (based on the oil) for batch processes and from 1 to 5 wt% for continuous processes. Deodorization temperatures range from 190 to 270 °C and the pressure from 0.13 to 0.78 kPa (1–6 mm Hg). The volume of stripping steam required is 10–20 $m^3$ per kilogram of oil [24]. Carry-over losses of oil are minimized by limiting the velocity of the injected steam, by providing a sufficiently large headspace, and by baffles. The distillate removed during deodorization is about 0.2% of the oil, and contains very little neutral oil. In a batch process a deodorizing time of at least 2 h is required for saturated or hydrogenated fats; about 4 h is needed for other oils and fats. Although attempts have been made to determine the end point of deodorization by analytical data, in practice this is done organoleptically.

The steam injected into the oil must be dry and oxygen-free. In order to deactivate prooxidative trace metals, aqueous citric acid solution (2–5 mg of citric acid per 100 g of oil) is often injected into the oil toward the end of the deodorization.

The deodorizer is normally constructed of stainless steel if temperatures of 190 °C are to be exceeded. Deodorizing temperatures up to 270 °C are employed, for example, to complete removal of pesticide residues [55], to heat-bleach palm oil, and to complete removal of free acids during physical refining. Since dimerization and isomerization of fatty acid groups increase with time and temperature [56], a deodorization at 270 °C should not last more than 30 min.

*Batch deodorizers* are still widely used (Fig. 18). They are about twice as high as they are wide and are normally half-filled to provide sufficient head space. Their capacity can vary from 5 to 25 t. The bottom part is fitted with heating coils and a steam distributor through which stripping steam is injected into the oil. The vacuum system (0.6–1.2 kPa) comprises a booster and ejector assembly. Barometric condensers can be eliminated by using dry condensing plants. When deodorization is completed, the oil is cooled to ca. 90 °C in the deodorizer and to 40–60 °C in a subsequent vacuum cooler; cooling minimizes oxidation that occurs on contact with atmospheric oxygen at elevated temperatures. The total cycle of charging, heating, deodorization, cooling, and discharging requires up to 8 h, the deodorization process itself up to 4 h. The deo-

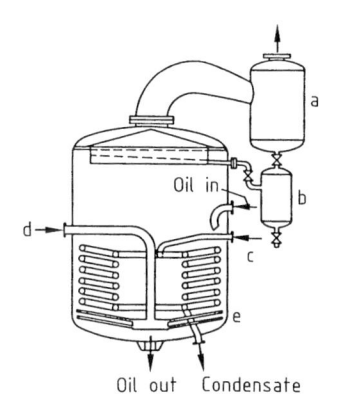

**Figure 18.** Batch deodorizer
a) Barometric condenser; b) Fat trap; c) Heating steam; d) Stripping steam; e) Steam distributor

dorizer distillate has the appearance of a milky emulsion and contains calcium soaps, nonsaponifiable matter, and a small fraction of neutral oil.

*Semicontinuous and continuous deodorization systems* are replacing batch deodorizers because of the savings in steam resulting from more efficient stripping and heat recovery.

The semicontinuous Girdler deodorizer (Fig. 19) consists of an iron shell fitted with stainless steel trays. At defined time intervals (e.g., every 30 min), a certain quantity of oil flows into the top tray of the deodorizer where it is deaerated in vacuo and then heated to ca. 170 °C by exchange with the hot effluent oil. After 30 min the oil is discharged into the second tray where it is heated to the operating temperature (normally 230–250 °C). Injection of steam (at 0.6–1 kPa) takes place in the 3rd and 4th trays, and cooling in the 5th tray. Operation of the unit is made fully automatic by use of a timing device which opens and closes the oil valves. The semicontinuous Lurgi plant works on a similar principle (Fig. 20); the stripping steam is injected through special orifices and further distributed through the oil by steam-lift pumps. These types of deodorizers have capacities of up to 20 t/h.

Fully continuous deodorizers generally com-prise a high cylindrical column fitted with a number of trays with bubble caps or similar devices, down which the oil flows countercurrent to the stripping steam. The oldest continuous deodorizer is probably that built by Foster–Wheeler. Modern versions [57] are made by EMI, Gianazza, Mazzoni, Wurster & Sanger, Votator, de Smet, Ex-Technik, HLS, Lurgi, and Krupp (Fig. 21). An alternative construction is that of EBE in which the oil flows countercurrent to the stripping steam in horizontal tubes. The Campro deodorizer has a continuous tray-in-shell design that combines plug flow with a unique thin-film stripping concept.

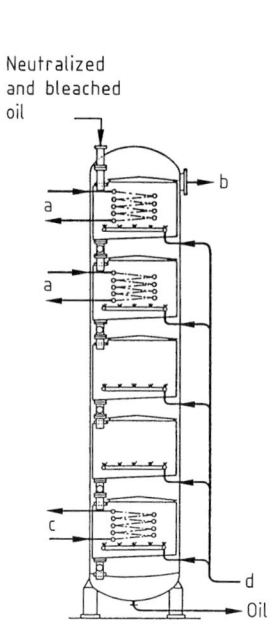

**Figure 19.** Semicontinuous deodorizer (Girdler)
a) Heating steam; b) Vacuum aggregate; c) Cooling water; d) Stripping steam

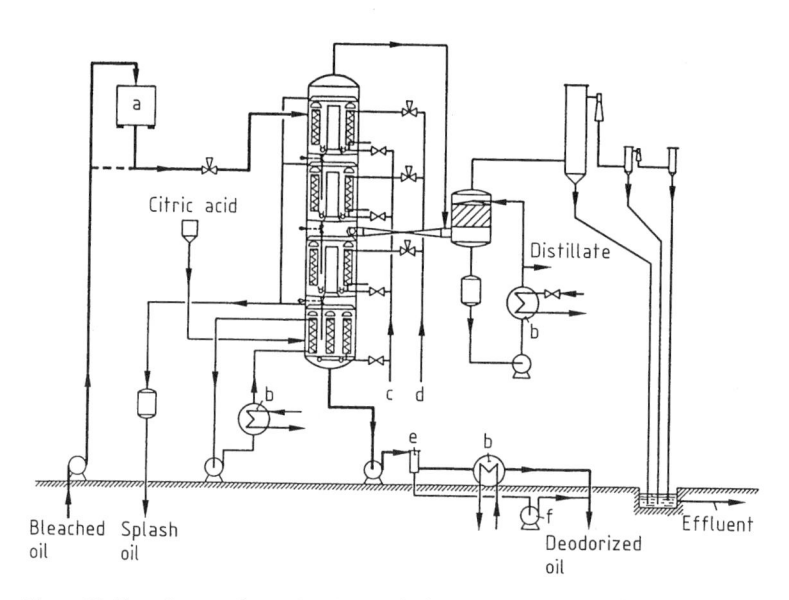

**Figure 20.** Flow diagram of a semicontinuous deodorization plant (Lurgi)
a) Measuring tank; b) Heat exchanger; c) Stripping steam; d) Heating steam; e) Filter; f) Pipe emptying

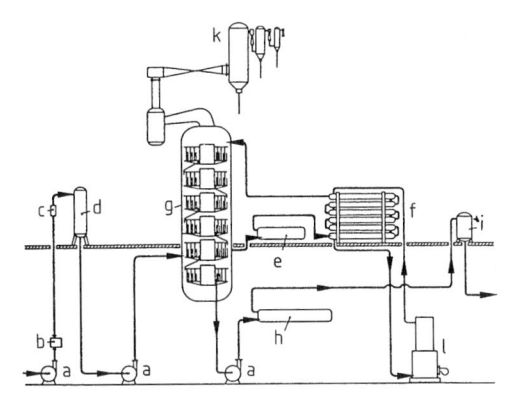

**Figure 21.** Continuous deodorizer (Krupp)
a) Pump; b) Filter; c) Meter; d) Deaerator; e) Preheater (low-pressure steam); f) Preheater (high-pressure steam); g) Kreuzstrom–Konti deodorizer; h) Cooler; i) Filter; k) Vacuum aggregate; l) Steam kettle

# 7. Fractionation

The aim of fractionating fats is to remove either undesirable components or to isolate desired components with special properties. Edible oils can, for example, be winterized by removing waxes and saturated components; cocoa butter substitutes can be obtained from palm oil. Except for the molecular distillation of monoglycerides, distillative techniques cannot be employed. Industrial fractionation techniques are based on crystallization or liquid–liquid extraction. Separation of glycerides by selective adsorption is only used occasionally for purification purposes. Fractional crystallization can be subdivided into dry processes, processes employing solvents, and processes involving selective wetting of fat crystals. Directed interesterification combines selective fractionation with interesterification (see Section 4.2).

The oldest example of *dry fractionation* is the separation of tallow into stearin and olein by pressing; this separation came into use toward the end of the 19th century in connection with the production of margarine. The lower-melting glycerides (olein) were squeezed out of the crystal network of the higher-melting glycerides by pressing blocks of tallow wrapped in filter cloth.

This method has been superseded by modern processes (Fig. 22) in which the fat is slowly chilled to the required temperature under controlled conditions in tempering vessels fitted with low-speed agitators, or in scraped coolers. Relatively large crystals of the higher-melting glycerides are obtained which can readily be filtered off in filter presses or drum filters. Slow cooling is important to obtain a relatively small number of nuclei which can then grow into large crystals. In addition, slow cooling promotes selective crystallization of the higher-melting glycerides and avoids the formation of mixed crystals containing both high- and low-melting glycerides. The difference in melting point must be at least 10 °C to achieve a sufficiently sharp, reproducible separation.

*Winterization* of edible oils such as sunflower or cottonseed oil involves removing waxes, which cause turbidity at refrigerator temperatures. After bleaching or deodor-

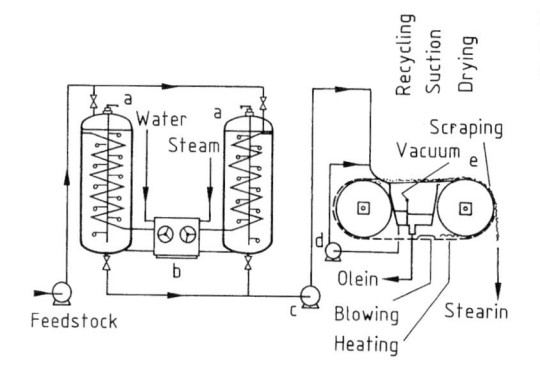

**Figure 22.** Dry fractionation plant
a) Crystallizers; b) Circulating unit; c) Feed pump; d) Recycling pump; e) Florentine filter

ization, the oils are kept at ca. 10 °C for at least 1 h and then filtered. An alternative process is to combine dewaxing with superdegumming of the crude oil (see also Section 6.1) [44]. Peanut oil is not normally winterized because its turbidity is caused by small glyceride crystals that are difficult to filter.

The separation of higher- and lower-melting glycerides can be improved by *fractionation using solvents* such as hexane, acetone, or methanol, in which the unsaturated fats are more readily soluble. Costs are much higher than for dry fractionation because of the need for solvent recovery. This relatively sharp fractionation method is generally only economical for preparing specialty products. For example, fractionation of palm oil with acetone gives a fraction (ca. 40% of the feed) consisting of disaturated-monounsaturated glycerides, primarily 1,3-dipalmito-2-olein.

The separation of relatively small fat crystals can be facilitated by *selective wetting*, for example with sodium lauryl sulfate or decyl sulfate in aqueous solution. The crystal suspension, produced either batchwise or continuously, is mixed with an aqueous solution containing the wetting agent and normally also an electrolyte such as magnesium sulfate. The crystals covered with wetting agent migrate to the aqueous phase as a suspension that can be separated by centrifugation. Alfa–Laval recommends such a process (Lipofrac process) for the fractionation of palm oil and for the winterization of other oils.

*Selective extraction* [58] has long been used in refining mineral oil. Liquefied hydrocarbons or furfurol are used as selective solvents in the separation of fats and fatty acids. This technique not only permits the separation of fats and fatty acids according to the degree of unsaturation or the molecular mass, but also effects the partial removal of coloring matter, gums, and vitamins.

Liquefied propane preferentially dissolves saturated components (Solexol process; see also 2417) [59]. This process has been used to fractionate tallow. The solubility of fats and fatty acids increases with decreasing molecular mass. Furfurol preferentially dissolves unsaturated components. It is miscible with fatty acids and monoglycerides at room temperature, but with triglycerides only at elevated temperatures. Depending on the operating temperature, furfurol is used alone or together with hexane. The selective solubility behavior is less pronounced for differences in molecular mass [60]. The furfurol process has been used to produce high-per-formance drying oils from soybean and linseed oils.

## 8.  Hydrogenation [61], [62]

Hydrogenation of the carbon–carbon double bonds of unsaturated glycerides leads to an increase in melting point (hardening, see also Section 4.3). The combination of hydrogenation, interesterification, and fractionation offers the opportunity of economically meeting the increasing demand for specialty products for foodstuff and technical applications from available raw materials.

Trans fatty acids produced by isomerization during hydrogenation also occur in animal fats such as butterfat and tallow. Although there are indications that trans fatty acids may be metabolized differently from their cis isomers [63], long-term studies have demonstrated their toxicological safety [64]. As far as their effect on blood lipid levels is concerned, they are comparable with saturated fatty acids.

The selectivity of hydrogenation is determined by the type and activity of the catalyst [65] (controlled poisoning and a large pore size increase selectivity), the operating temperature (both selectivity and cis–trans isomerization increase with increasing temperature), stirring–mixing efficacy (increasing the hydrogen concentration at the catalyst surface by intensive stirring–mixing reduces selectivity), and the hydrogen pressure (increasing the pressure reduces the selectivity).

Neither absolute selectivity nor complete isomerization (or complete suppression of isomerization) can be achieved. However, the process can be steered in a desired direction by suitable choice of the above parameters to produce fats with different degrees of saturation and melting properties from a given oil. For example, peanut oil can be hardened nonselectively to an iodine value of 70–90 yielding a soft, semiliquid fat. If the same oil is hardened to the same iodine value under conditions that promote isomerization, a firm, plastic fat with a melting point of ca. 33 °C is obtained. Palm oil is sometimes hydro-bleached, i.e., hydrogenated under very mild conditions that lead to selective hydrogenation of carotenoids.

The oil should be at least degummed and bleached prior to hydrogenation and preferably neutralized and bleached as well. Catalyst poisons include phosphatides, sulfur compounds, soaps, and oxidation products from poor handling and/or bleaching. Free fatty acids do not poison the catalyst but tend to slow down the hydrogenation reaction.

The hydrogen must be pure and dry and can be produced by various routes. The most important industrial process is decomposition of natural gas or propane by passage over a nickel catalyst in the presence of steam at 800–900 °C. Electrolysis of water is used to a lesser extent. The hydrogen must be specially purified to remove hydrogen sulfide and carbon monoxide, which are strong catalyst poisons.

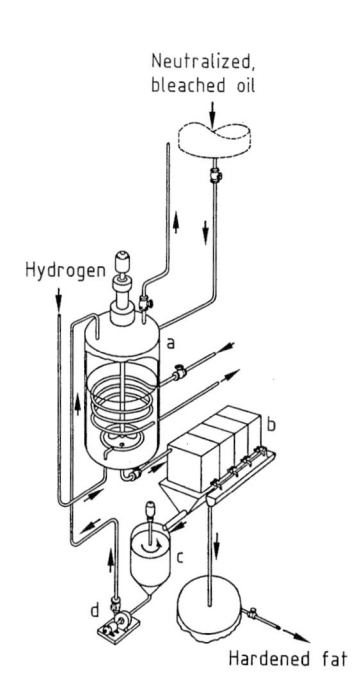

**Figure 23.** Flow diagram of a batch hardening plant
a) Convertor; b) Filter press; c) Catalyst tank; d) Catalyst pump

Nickel, platinum, and palladium are preferred catalysts. Platinum and palladium are especially suited for laboratory-scale and low-temperature hydrogenations. Only nickel is used on a production scale. Many different types of nickel catalysts are available. They essentially consist of metallic nickel on a carrier such as silica gel, silicic acid, or alumina. The carrier is treated with a nickel salt such as the carbonate, hydroxide, or sulfate, which is converted into nickel oxide by roasting. The oxide is then reduced to metallic nickel in a stream of hydrogen at 300–400 °C. Other routes involve decomposition of nickel formate at 240 °C or dissolution of aluminum from a nickel–aluminum alloy with alkali (Raney catalyst). In addition, there are a number of mixed catalyst systems (copper, chromium, cobalt) with special selectivities.

*Batch hydrogenation* (Fig. 23) is normally performed on a scale of 5–25 t. The dry oil is pumped into the hydrogenation vessel that has been previously flushed with hydrogen, and heated to just below the required operating temperature (100–180 °C). The catalyst (0.01–0.1% active nickel) is then introduced and the hydrogen pressure increased. Alternatively, the oil is vacuum-deaerated while it is heated before the catalyst and hydrogen are introduced. Hydrogenation is an exothermic reaction; a decrease in the iodine value of one unit causes a temperature increase of 1.5–2.0 °C. The required operating temperature is therefore maintained by controlled cooling. Hydrogen is introduced through a sparge ring at the bottom of the hydrogenation vessel, pumped out at the top, and recirculated. Sparging and fast stirring ensure adequate dispersion of the catalyst in the oil and optimum mass transport. The usual working pressure is 0.15–0.3 MPa (1.5–3 bar).

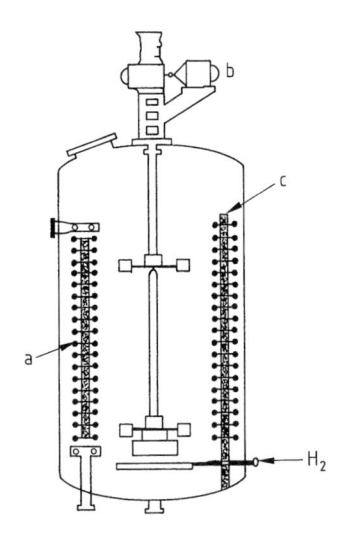

**Figure 24.** Hydrogenation convertor (dead-end design)
a) Heating and cooling coil; b) Turbine agitator; c) Baffle

In hydrogenation autoclaves that operate without hydrogen recirculation ("dead-end" process), spent hydrogen is replaced at a somewhat higher pressure [66]. The intensive stirring required can be achieved with a turbine agitator or similar device (Fig. 24).

So-called loop hydrogenation reactors (e.g., Buss, Fig. 25) have become increasingly popular. The oil–catalyst mixture is constantly circulated through an external heat exchanger during heating and cooling. This loop reduces total cycle time; it also gives better temperature control and thus less product variation from batch to batch.

When hydrogenation is complete, the product is cooled to 80–90 °C and the catalyst is filtered off. Since the acid value tends to increase slightly during hydrogenation, the crude hardened fat is usually postneutralized with lye and bleached before it is deodorized. Posttreatment with citric or phosphoric acid, followed by bleaching prior to deodorization, is an alternative to lye posttreatment. In either case the residual nickel content is reduced to <0.2 ppm. Deodorization of a hardened fat is essential since hydrogenation leads to typical hardening flavors which must be removed.

*Continuous hydrogenation* is primarily suitable for the production of one particular product over a longer period of time. Fully continuous processes are the exception rather than the rule. In semicontinuous processes several reaction vessels or autoclaves are arranged sequentially (e.g., Alfa–Laval system). In fully continuous plants, either the preheated oil and hydrogen are passed over a stationary catalyst bed, or all three components are mixed prior to entering the reactor (see Fig. 26). In the Procter & Gamble process, the reactor contains a special stirrer to ensure intensive mixing in a thin film. The Girdler Corporation developed a process in which three Votator heat exchangers are arranged sequentially; the first vessel serves as preheater, the second as a hydrogenator, and the third as a cooler. Continuous hydrogenation of miscella at low temperatures over a catalyst bed produces little isomerization but has not achieved industrial importance.

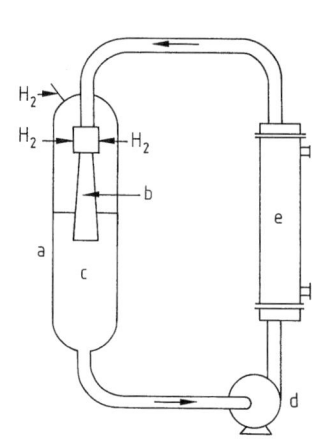

**Figure 25.** Loop hydrogenation reactor
a) Autoclave; b) Mixing and reaction zone; c) Reactant–solvent–catalyst suspension; d) Primary loop recirculation pump; e) Primary loop heat exchanger

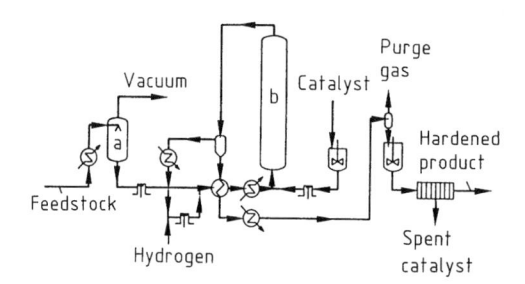

**Figure 26.** Continuous hydrogenation of fatty acids and fatty oils (Lurgi)
a) Dryer; b) Reactor

The course of hydrogenation is normally monitored by changes in analytical data such as iodine value or refractive index. The hydrogenated products are characterized primarily by melting point and solids content at different temperatures.

# 9. Interesterification

Prior to interesterification, fats must be neutralized to less than 0.1 % ffa and dried to avoid excessive deactivation of the catalyst, which is normally sodium ethoxide (see also Section 4.2). The catalyst (0.1 – 0.3 %) is finely dispersed in the fat. After completion of the interesterification reaction, which is carried out at 80 – 100 °C, excess catalyst is deactivated by addition of water. The resulting soap is removed by washing with water and bleaching. During interesterification, sodium ethoxide is converted into fatty acid ethyl esters, which are removed during the final deodorization step.

Interesterification can be performed batchwise (Fig. 27), usually in a neutralization – bleaching vessel, or continuously. In the continuous process, fat with suspended catalyst passes through a tubular reactor (Fig. 28) [67]. In directed interesterification (see Section 4.2) the fat is cooled, for example, in a scraped-surface cooler to crystallize the higher-melting glycerides.

**Figure 27.** Batch reactor for interesterification

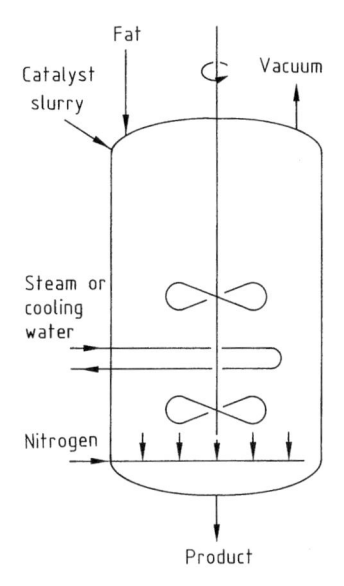

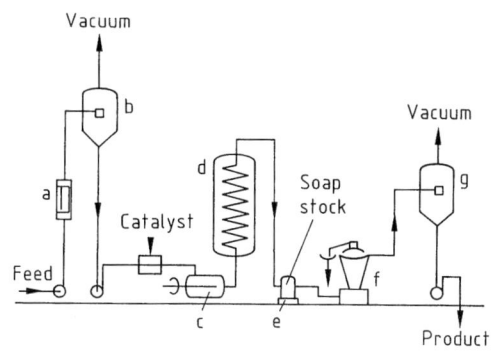

**Figure 28.** Continuous interesterification
a) Heat exchanger; b) Dryer; c) Homogenizer; d) Tubular reactor; e) Mixer; f) Centrifuge; g) Dryer

New types of fats can be created by suitable choice of interesterification components. Depending on the starting components, the melting point of the fat after interesterification can be lower or higher. Figure 29 shows the solids content (dilatation) curves of native and interesterified palm oil; the glyceride composition is given in Table 21.

**Acidolysis.** An example of the industrial application of acidolysis is the transesterification of coconut oil with acetic acid and the subsequent esterification of excess acid with glycerol. A mixture of aceto fats (laurodiacetin, myristodiacetin, etc.) is obtained which are used as plasticizers and coatings. Transesterification of coconut oil with higher fatty acids, employing continuous distillation of the liberated lower fatty acids ($C_6-C_{10}$), yields fats with properties similar to those of cocoa butter. Acidolysis is usually performed in the presence of an acid or base catalyst; no catalyst is required at temperatures of 260–300 °C.

**Table 21.** Fatty acid and glyceride composition of native and interesterified palm oil *

| | Native palm oil | Interesterified palm oil | |
|---|---|---|---|
| | | Random interesterification | Directed interesterification |
| mp, °C | 42 | 47 | 52 |
| Fatty acid composition, mol% | | | |
| S | 51 | 51 | 51 |
| U | 49 | 49 | 49 |
| Glyceride composition, mol% | | | |
| $S_3$ | 7 | 13 | 32 |
| $S_2U$ | 49 | 38 | 13 |
| $SU_2$ | 38 | 37 | 31 |
| $U_3$ | 6 | 12 | 24 |

* S = saturated fatty acid. U = unsaturated fatty acid.

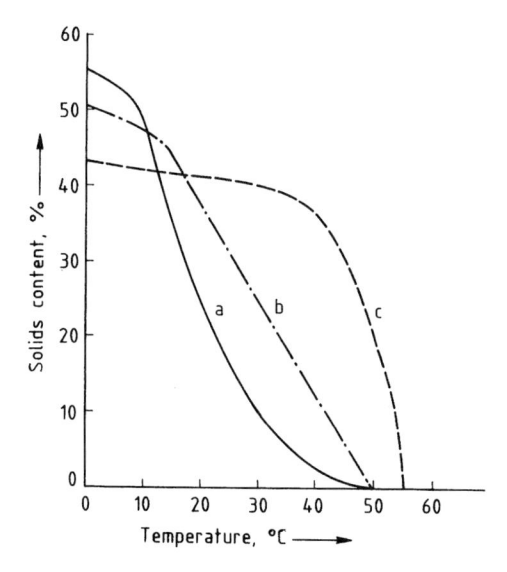

**Figure 29.** Solids content of native and interesterified palm oil (Alfa–Laval)
a) Native palm oil; b) Randomly interesterified palm oil; c) Directedly interesterified palm oil

**Alcoholysis.** The replacement of the glycerol moiety of a fat with lower monohydric alcohols proceeds at lower temperatures. The Bradshaw process [68] utilizes transesterification of fats with methanol as the first step in the continuous production of soap. Fat with an acid value of less than 1.5 is stirred with excess methanol for a few minutes in the presence of 0.1–0.5% caustic soda at ca. 80 °C. On subsequent standing, practically dry glycerol settles at the bottom of the reaction vessel. The same principle is employed in the preparation of fatty acid methyl esters for laboratory analyses. If the acid value is higher than 1.5, boron trifluoride is normally used as a catalyst.

The exchange of glycerol with other polyhydric alcohols, such as pentaerythritol, is best done under conditions at which glycerol can be distilled off, i.e., at 200–250 °C and 5–40 kPa (50–400 mbar), in the presence of an alkaline catalyst.

The production of mono- and diglycerides by reacting fats with an excess of glycerol is a special case of transesterification of fatty acid esters with polyhydric alcohols. These glycerides are especially important as emulsifiers. The production process consists of heating a fat with 25–40% glycerol at 205–245 °C in the presence of sodium hydroxide or a sodium alkoxide. Part of the catalytic effect is probably based on soap enhancing the solubility of glycerol in fat. Industrial products contain 40–50% monoglycerides. Oxidation and undesirable color effects can be minimized by working in vacuo or under an inert gas, and by using stainless steel reaction vessels. Monoglyceride concentrates can be produced by molecular or high-vacuum distillation.

# 10. Environmental Aspects

Crude oils and fats can be contaminated with traces of various substances such as metals (e.g., lead, arsenic, cadmium, selenium, mercury), chlorinated organic pesticides (DDT, dieldrin), mycotoxins (e.g., aflatoxins), and sulfur compounds. These trace contaminants are removed in the course of refining [55].

Wastewater from lye neutralization and from the barometric condensers of deodorizers can be discharged into communal sewage systems or surface waters only if certain limits for residual fat, pH, sediment, sulfate, and chemical and biological oxygen demand are met. Otherwise biological pretreatment is required. Since the cost of biological or other pretreatment of wastewater can be considerable, there has been a trend toward physical refining (i.e., refining without lye neutralization) by employing special degumming techniques, bleaching, and distillative deodorization.

Bleaching earth is normally extracted before being deposited or burnt. The extracted fats can be recycled.

Deodorizer distillates [55] can be used in animal feeds, as a source of vitamin E, for technical applications, or simply burnt.

# 11. Standards and Quality Control

Standardized methods for quality control have been issued by the following organizations:

Association Française de Normalisation (AFNOR), American Oil Chemists' Society (AOCS), Association of Official Agricultural Chemists (AOAC), American Society for Testing and Materials (ASTM), British Standards Institution (BSI), Deutsche Gesellschaft für Fettwissenschaft (DGF), Deutsches Institut für Normung (DIN), International Association of Seed Crushers (IASC), International Union of Pure and Applied Chemistry (IUPAC), International Standards Organization (ISO), the Netherlands Normalisatie Instituut (NNI), and the Federation of Oils, Seeds, and Fats Associations (FOSFA).

There is an increasing trend for the various national and international methods to be uniformly standardized as ISO procedures; these are generally adopted as official methods within the European Community.

The IASC (8 Salisbury Square, London EC 4P 4 AN, UK) has issued a list of associations dealing with or issuing standard forms of contract and/or quality standards for oilseeds, oils, and fats. This association has also drawn up a list of official institutions and associations that deal with or issue methods for grading, sampling, and preparing samples for analysis, as well as methods for analysis that are referred to in standard forms of contract or usually applied in the trade of oilseeds, oils, and fats.

## 11.1. Sampling

Correct sampling is a difficult process requiring careful attention; it must take into account that separation can occur during storage of fats and fatty products. Hulls, for example, tend to separate from oilseeds because of their lower density. Oils and fats tend to form a sediment of water and dirt; for this reason a representative average sample is often taken in bypass during pumping.

A special zone sampler consisting of a metal cylinder with a valve at both ends is often used for sampling the contents of a tank. The sampler is lowered into the tank, the valves being operated with a cord. Starting from the bottom of the tank, the oil is sampled at various depths up to the surface. Smaller containers (drums) are sampled with a tube or shutter scoop; solid fats in retail packs are sampled with a sampling scoop.

Special sample dividers are used for oilseeds. Sampling apparatus for bags includes sack-type spears or triers, cylindrical and conical samplers, and hand scoops. Shovels, scoops, cylindrical and conical samplers, or mechanical samplers are used for drawing small periodical samples from a flow of oilseeds. Shovels, quartering irons, or riffles are used for mixing and dividing.

Samples for analysis of oilseeds, expeller cakes, and meals should preferably be at least 2 kg, and of oils and fats at least 250 g. The samples should be homogenized immediately prior to analysis. Larger oilseeds are comminuted prior to analysis.

## 11.2. Raw Materials

Buying–selling contracts for oilseeds, crude oils, and fats normally specify a maximum permissible concentration of impurities. If this level is exceeded the buyer receives a discount. The key parameters in oilseeds are fat and moisture contents.

To determine the *fat content* of oilseeds, expeller cakes, and meals, the comminuted sample is extracted with a solvent and the fat taken up by the solvent is determined refractometrically, gravimetrically, densitometrically, or by the change in dielectric constant. The gravimetric methods are time-consuming but most reliable. Each method gives reproducible but different values, depending on the type of solvent (usually hexane or petroleum ether) and the extraction method. The fat content can also be determined directly without solvent extraction by nuclear magnetic resonance or by near infrared absorption.

According to ISO 659, the oilseeds are extracted with $n$-hexane or petroleum ether in a Bolton extractor. After extraction for several hours, the sample is removed from the extractor thimble, reground, and reextracted. This is repeated a third time. The extract is evaporated and the residual oil dried to constant weight at $103 \pm 2$ °C.

Unilever has developed a far quicker method which gives identical values to those obtained with the ISO method. The sample is first comminuted in a ballmill in the presence of petroleum ether, and then extracted for 2 h in a Bolton extractor.

When determining the fat content of fruit pulps and animal tissues, predrying of the substrate at 105 °C in vacuo is recommended. When extracting animal tissue polar solvents such as ether–HCl or ethanol–chloroform are used to facilitate liberation of fat from lipoprotein complexes present in aqueous dispersion.

*Water* and *volatile matter* are generally determined by weighing a sample before and after heating under standardized conditions, usually 103 °C. Crude oils are normally analyzed for water and volatile matter, color, particulate dirt, phosphatides, and free fatty acids. Up to 0.5% moisture and impurities are generally regarded as technically unavoidable. Inferior oils usually have a dark color and a high content of free fatty acids and oxidized components that is indicated by a relatively high absorption at 232 nm, high anisidine and totox numbers, and a high concentration of oxidized fatty acids. These analytical data give an approximate indication of the expected yield and quality of the fully refined product. Admixture can be detected within certain limits by determining the composition of fatty acids, sterols, and tocopherols. Very severe admixture can be detected by iodine value, refractive index, and saponification value.

## 11.3. Oils and Fats

### 11.3.1. Physical Methods

The *density* or *specific gravity* at a given temperature is determined by weighing a known volume, by determining the buoyancy, or by electromagnetic excitation of the natural frequency of the sample.

The *refractive index* is normally determined with an Abbe refractometer at 20 °C for oils which are liquid at this temperature and at 40, 60, or 80 °C for fats. The temperature coefficient is ca. $0.0036$ K$^{-1}$. A correlation exists between refractive index and iodine value (see Section 3.8); for this reason the refractive index is often used to follow the course of hydrogenation.

Most oils and fats rotate *polarized light* by only a few tenths of a degree; higher values are observed with substituted fatty acids such a ricinoleic acid.

In melting a distinction is made between the slip point, the point of incipient fusion, and the point of complete fusion. The *slip point*, the usually cited melting point, is the temperature at which a sample plug placed in an open glass capillary begins to rise when gradually heated in a water bath or by other means. This method can be automated. The slip point value is influenced by the method of cooling used prior to melting. Whereas for conventional fats the sample is normally solidified by chilling the sample and capillary at $-10\,°C$ for a few minutes, fats that exhibit distinct polymorphism, e.g., cocoa butter and substitutes, are conditioned at higher temperatures for up to 18 h.

The *point of incipient fusion* is the temperature at which a sample placed in a U-tube begins to flow on being warmed. The *point of complete fusion* is the temperature at which a sample becomes completely clear.

*Flow* and *drop points* are primarily of importance for technical greases and some plastic fats. They are determined in an Ubbelohde apparatus. Flow point is the temperature at which a drop begins to be clearly formed at the end of a thermometer, and drop point is the temperature at which the drop falls.

*Dilatation*, i.e., the increase in volume of a fat on passing from the solid to the liquid state at a given temperature, can be used to determine the concentration of solid components at a certain temperature (Solid Fat Index, SFI). This type of measurement has been replaced by the much quicker *nuclear magnetic resonance* method which is based on the difference between the response from the hydrogen nuclei in the solid phase and that from all the hydrogen nuclei in the sample.

*Differential thermal analysis* permits the quantification of changes in the solid phase of a fat on being heated. A sample is placed in a metal block which is heated at a constant rate. The difference between the temperature of the block and that of the sample is characteristic of crystal modifications occurring during melting [69].

On passing from the liquid to the solid state, the temperature remains constant for a certain period of time because of the liberated heat of fusion; with sufficient insulation an increase in temperature can be observed. The *titer value* of a fat is the maximum value to which the temperature rises in the cooling curve. Cooling curves are frequently used to check the purity of fats which consist predominantly of a few triglyceride types, e.g., cocoa butter.

The *cloud point* is the temperature at which an oil or molten fat begins to become turbid when cooled under controlled conditions. The temperature stability of an oil is normally assessed after keeping a carefully dried sample at $0\,°C$ for $5-8$ h.

The smoke, flash, and fire points of oils and fats are measures of their thermal stability when heated in air. Table 12 shows some typical data. The *smoke point* is the temperature at which smoke is first detected in a standard apparatus with specially designed, draft-free illumination. The *flash point* is the temperature at which volatile products are evolved at such a rate that they can be ignited but do not continue to burn. The *fire point* is the temperature at which the production of volatile products is sufficient to support continuous combustion after ignition. The smoke point is important in the assessment of refined oils and fats used for deep frying. It is governed primarily by the free fatty acid content. The flash point is a measure of residual solvent in crude oils, although gas–liquid chromatographic methods are now generally preferred for this determination.

The *viscosity* of an oil can be determined in a flow (Engler), capillary (Ubbelohde), fall (Höppler), or shear (Haake) viscometer.

There are a number of different methods for determining the *color* of an oil or fat. In the Lovibond tintometer the color of the sample in a standardized cell with a length of 1″ (2.45 cm) or 5 1/4″ (13.3 cm) is matched — either manually or automatically — with the color of standard yellow, red, and blue color slides. Other colorimetric scales are based on standard iodine or potassium dichromate solutions. The most objective way of assessing color is to determine the absorption at specified wavelengths in the range 370–780 nm.

The characteristic absorption maxima of specific lipid groups in the infrared and ultraviolet ranges can be used qualitatively and quantitatively (see Section 3.8). Fatty acids with isolated double bonds such as linoleic, linolenic, and arachidonic acids do not exhibit specific UV absorption. After isomerization with alkali they can, however, be determined by the characteristic absorption maxima of the corresponding conjugated fatty acids. Infrared spectroscopy is used primarily for the identification of specific groups such as hydroxyl and trans double bonds.

*Optical rotatory dispersion spectroscopy* is used only in studies of biosynthesis and metabolism of lipids.

*Mass spectroscopy*, especially in combination with gas chromatography and other methods of separation, is often used to determine the structure of oxidation products and substituted fatty acids.

## 11.3.2. Chemical Methods

For preliminary identification and evaluation, determination of chemical data such as the iodine value (see below) and the fatty acid composition often suffice. In many cases, however, it is necessary to test for characteristic minor constituents directly or by specific chemical reactions. Table 22 gives some examples.

The complexity of oils and fats makes the determination of their precise composition difficult. For this reason analytical values are used to express certain properties. Analytical values in this sense are defined as the equivalent amounts of certain reagents which react with specific groups in the lipid molecules. Many of the older analytical values are no longer used because of the introduction of modern analytical techniques such as column, paper, gas–liquid, and high–performance liquid chromatography, and spectrophotometry. However, the principle of determining analytical values is still valid since it can authenticate and assess the quality of an oil or fat relatively quickly with modest laboratory facilities.

**Table 22.** Detection of oils and fats

| Oil/Fat | Characteristic constituent | Specific reaction or detection procedure | Limit of detection | Remarks |
|---|---|---|---|---|
| Animal fats | cholesterol | isolation of sterols from the unsaponifiable matter by TLC or HPLC, and separation by HPLC, TLC, or GLC | 1% in vegetable oils and fats | Vegetable oils and fats contain traces of cholesterol in addition to phytosterols while in animal fats only cholesterol but no phytosterols have been found. |
| | branched $C_{13}$–$C_{17}$ fatty acids and odd-numbered, unbranched $C_{11}$–$C_{19}$ fatty acids | GLC of fatty acid methyl esters | ca. 2% in vegetable oils and fats | The branched fatty acids may have to be concentrated, e.g., by urea adduct formation. |
| Lard | palmitic acid in the glyceride 2-position | isolation of the glyceride fraction with 1 saturated and 2 unsaturated fatty acid groups, followed by determination of the fatty acid composition in the glyceride 2-position | 1% lard in tallow | Determination of the fatty acid composition in the 2-position without prior glyceride fractionation gives a significantly higher limit of detection. |
| Partially hardened and unhardened marine oils | highly unsaturated $C_{20}$–$C_{24}$ fatty acids | Tortelli–Jaffe reaction (green coloration with bromine in chloroform/acetic acid) GLC of the fatty acid methyl esters | ca. 2% in other oils and fats | Not strictly specific; highly unsaturated vegetable oils can also react. |
| Vegetable oils and fats | phytosterols | isolation of sterol esters and separation as sterols by TLC, HPLC, and GLC | ca. 2% in animal fats | Sterols occur in vegetable oils and fats in the free and esterified form, while in animal fats they are only present in the free form. |
| Castor oil | ricinoleic acid | TLC; IR spectroscopy | ca. 1% in other oils and fats | Hydroxy fatty acids can be formed during autoxidation of oils and fats. |
| Coconut oil and palm kernel oil | lauric acid | GLC of the fatty acid methyl esters | ca. 2% in other vegetable oils except oils and fats rich in lauric acid, e.g., babassu fat | |
| Cottonseed oil and other oils of the malvaceae, tiliaceae, and bombaceae families | malvalic acid | Halphen reaction (red coloration on heating with sulfur in $CS_2$) | depending on extent of refining | Hardened cottonseed oil does not give this reaction; malvalic and sterculic acid are converted into saturated, branched fatty acids during hardening. |

**Table 22.** (continued)

| Oil/Fat | Characteristic constituent | Specific reaction or detection procedure | Limit of detection | Remarks |
|---|---|---|---|---|
| Crude palm oil | $\alpha$- and $\beta$-carotene | isolation from nonsaponifiable matter by HPLC | | Indicative of distillative or other heat treatment. |
| Peanut oil | lignoceric and arachidic acid | GLC of the fatty acid methyl esters | | Hardened marine oils and certain vegetable oils also contain higher molecular mass fatty acids. |
| Olive oil | $\alpha$-tocopherol | isolation and separation by HPLC | ca. 5% other oils in olive oil | |
| Partially hardened fats | trans fatty acids | IR spectroscopy | ca. 2% in vegetable oils | Animal fats contain up to 10% trans fatty acids. |
| Rapeseed oil | erucic acid | GLC of the fatty acid methyl esters | ca. 1% of "high erucic acid" rapeseed oil in other vegetable oils | Admixture with "low erucic acid" rapeseed oil cannot be detected satisfactorily this way. Marine oils also contain erucic acid. |
| | brassicasterol | isolation and separation of sterols by TLC, HPLC, and GLC | ca. 5% in other vegetable oils | Independent of erucic acid content. |
| Sesame oil | sesamol and sesamin | Baudouin reaction (red coloration with furfural and HCl) | ca. 0.5% in other oils and fats | |
| Soybean oil | linolenic acid | GLC of the fatty acid methyl esters | ca. 10% in e.g., sunflower oil | |
| | $\delta$-tocopherol | isolation and separation by HPLC | ca. 5% in e.g., sunflower oil | |
| Teaseed oil | | Fitelson test (modified Liebermann–Burchard test) | e.g., 5% in olive oil | |
| Tung oil | eleostearic acid | GLC of the fatty acid methyl esters | ca. 1% in other oils and fats | |

The *saponification value* is the number of milligrams of potassium hydroxide required to saponify (hydrolyze) 1 g of fat and is related to the molecular mass of the fat.

The *hydroxyl value* is expressed as the number of milligrams of potassium hydroxide required to neutralize the acetic acid needed to acetylate 1 g of fat.

The *carbonyl value* is defined as the number of milligrams of CO (carbonyl groups) per gram of fat. The basis for this value is the conversion of the carbonyl group into an oxime with hydroxylamine in alcoholic KOH:

$RCO + NH_2OH \longrightarrow RC=NOH + H_2O$

Hydroperoxides interfere with this determination.

The acidity of an oil or fat arises from free inorganic or organic acids. The *acid value* is defined as the number of milligrams of potassium hydroxide required to neutralize 1 g of fat. The content of free fatty acids (ffa) expresses how many parts of free fatty acids are contained in 100 parts of fat. This content is — depending on the type of oil or fat — expressed as % oleic acid ($M_r$ 282), % palmitic acid ($M_r$ 256), or % lauric acid ($M_r$ 200). If not specified, it is calculated as oleic acid.

*Peroxides* are determined iodometrically according to Wheeler (cold) or Sully (hot) and are expressed as milliequivalents of active oxygen per 1000 g of fat. The peroxide value can also be expressed in millimoles per kilogram (Lea value, 1 mequiv/kg = 0.5 mmol/kg) or in milligrams per kilogram (1 mequiv/kg = 8 mg/kg). It is important to use fresh reagents and to exclude oxygen as far as possible during the determination.

The reaction with anisidine is often used to determine the carbonyl compounds produced during autoxidation. The *anisidine value* is defined as 100 times the absorption of a 1% sample solution in a 1-cm cell after reaction with anisidine. This value has been recommended for monitoring the refining of oils and fats. It is also used in combination with the peroxide value to characterize the overall autoxidative state of an oil or fat (*totox value*):

Totox value = 2 × peroxide value + 1 × anisidine value

The *iodine value* denotes the percentage by weight of iodine bound by 100 g of fat and is an approximate measure of the degree of unsaturation. The usual method is that of Wijs. The rate of addition of iodine to double bonds depends on their distance from the carboxyl group. Conjugated unsaturated fatty acids are only partially halogenated; substituted fatty acids such as keto acids and cyclic acids can also react. Trans fatty acids react more slowly than the cis isomers.

The *oxidative stability* of oils and fats is determined by their composition and their previous history. A number of empirical methods have been developed to express oxidative stability [70]. These tests are performed at elevated temperature and do not necessarily correlate with organoleptic stability at ambient temperature. The most frequently practiced methods are the Schaal test and the Swift test (or Active Oxygen Method); automated versions of the latter are popular, e.g., the Rancimat apparatus.

The *moisture content* can be determined by weighing before and after drying or by codistillation with toluene. It is most conveniently determined by the Karl Fischer titration, which is based on the following reaction:

$2 H_2O + SO_2 + I_2 \longrightarrow H_2SO_4 + 2 HI$

*Impurities* such as dirt and seed particles are not dissolved on treating the sample with petroleum ether or hexane and can be collected and weighed as filter residue.

Traces of *residual solvent* are best determined by gas chromatography, either by direct injection or after enrichment by the so-called headspace technique [71].

*Nonsaponifiable matter* is defined as that material which is a soluble constituent of the fat and which remains insoluble in the aqueous phase after saponification of the fat and can be extracted with petroleum ether or diethyl ether; diethyl ether tends to give higher values. Nonsaponifiable matter consists primarily of sterols, alcohols, hydrocarbons, and vitamins.

*Oxidized fatty acids* contain one or more hydroperoxide, hydroxyl, or keto groups in the fatty acid chain and can be determined by their reduced solubility in petroleum ether. Only part of the oxidized fatty acids are detected by this type of analysis. Thus, the separation of fats into polar and nonpolar fractions on silica gel tends to be preferred when, for example, characterizing used frying fats [72].

To determine *resin acids* in the presence of fatty acids, the sample is esterified with methanol in the presence of an acid catalyst; resin acids escape esterification.

*Inorganic acids* can be detected by extracting the sample with hot water and testing the aqueous extract with methyl orange.

*Free alkali* can similarly be detected by dissolving the sample in ether–ethanol and adding phenolphthalein (red coloration). Traces of soap can be detected down to 10 ppm by dissolving the sample in ether–ethanol, adding bromophenol blue (which gives a green-blue color in the presence of soap), and titrating with very dilute HCl.

Many crude seed oils contain considerable concentrations of *phosphatides*. These can be determined by dry digestion of the sample with magnesium oxide and photometric determination of the resulting phosphate after addition of molybdate [73]. An alternative is the direct determination of phosphatides by atomic absorption or plasma emission spectroscopy.

Methods for determining traces of *mineral oil* in edible oils are based on the fluorescence of mineral oils, their insolubility in acetic anhydride, and the fact that they are not retained on alumina. Headspace gas chromatography is also used.

The gas chromatographic determination of the *fatty acid composition* of an oil or fat has replaced the time-consuming, chemical techniques of fractionating fatty acids into single components or groups. Analytical values such as the Reichert–Meissl, the Polenske, the Kirchner, and the butyric acid number (as measures of the concentration of lower molecular mass or steamvolatile fatty acids) are now virtually obsolete. Prior to gas chromatographic analysis, the fatty acids are converted into their methyl esters either by direct transesterification of the sample with methanolic KOH (acid values $<2$) or by saponification (acid values $>2$) followed by esterification of the fatty acids with methanol in the presence of $BF_3$. Full details of these procedures and of the various types of columns, column packings, and detectors are given in ISO 5508/5509. If a marker fatty acid is used as an internal standard, quantitative determination of the total amount of monomeric, nonoxidized fatty acids is possible; nonsaponifiable matter, oxidized, and polymerized fatty acids are retained on the gas chromatographic column.

*Polymeric fatty acids* are conveniently determined by special high-performance liquid chromatography (HPLC) or gel permeation chromatography [74].

The *glyceride structure* may have to be determined to authenticate an oil or fat. The methods for determining positional isomers are based on fractional crystallization, selective oxidation, chromatographic separation, or enzymatic hydrolysis of the fatty acid in the 2-position; they are normally too expensive for routine control unless specialty products are involved.

Mono- and diglycerides are formed by hydrolysis of fats and therefore occur in higher concentrations in fats from damaged seeds. Mono-, di-, and triglycerides can be separated by column or thin-layer chromatography. 1-Monoglycerides can be determined directly in fats by oxidative fusion with periodic acid:

$RCOOCH_2CHOHCH_2OH + H_5IO_6$
$\longrightarrow HCHO + RCOOCH_2CHO + HIO_3 + 3\ H_2O$

Monoglycerides can be isomerized into the 1-isomers prior to addition of periodic acid by treating the sample with perchloric acid.

## 12. Storage and Transportation [75]

**Crude Oils and Fats.** If crude oils and fats are to be stored for a long time, removal of insoluble impurities by sedimentation, filtration, or centrifugation is advisable. Seed particles and cell fragments contain fat-splitting enzymes, which can gradually hydrolyze glycerides in the presence of moisture and produce an increase in free fatty acid concentration. Proteins are good nutrients for microorganisms.

Further measures for avoiding undue quality deterioration include drying the crude oil to less than 0.2 % moisture and storage without heating; consistent fats should be stored at not more than 10 – 15 °C above the melting point. Frequent pumping should be avoided to prevent saturation of the oil with air and consequent autoxidation. The addition of antioxidants is useful primarily for animal fats; most vegetable oils contain sufficient concentrations of natural antioxidants. Most crude vegetable oils can be stored for some months without significant loss in quality if the above precautionary measures are observed. Animal fats can be stored for only relatively short periods of time without loss in quality. Continual monitoring of the acid value, peroxide value, and other data that characterize the state of oxidation during storage is recommended.

**Raffinates.** Since natural antioxidants, such as tocopherols, phenols, and synergists, are partially removed during the refining process, semi- and fully refined oils often have lower oxidation stability than the crude oils. This is especially true for oils with a relatively high content of polyunsaturated fatty acids. Therefore, raffinates must be protected as far as possible from light, air, moisture, relatively high temperatures, and trace metal prooxidants if they are to be stored for long periods.

Raffinates are best stored in closed containers with a relatively small ratio of surface area to volume in order to minimize access and infusion of air. Any heating should be uniform without local overheating; the temperature should not be higher than 10 °C above the melting point. Aeration can be further reduced by using a bottom-loading pipe and a nitrogen atmosphere during both storage and transportation. Immediately after deodorization, nitrogen should be introduced into the suction side of the loading pump that feeds the storage or transport vessels.

The storage containers are preferably made of aluminum or stainless steel, although mild steel can be used if access of air is limited and if cooling is adequate. Raffinates can be kept stored in large tanks for several weeks without an increase in peroxide value; however, the peroxide value tends to increase each time the oil is pumped. Raffinates are transported in bulk in ships and trucks, in drums, tins, and bottles; consistent fats can also be transported in block or powder form if the melting point is sufficiently high.

# 13. Individual Vegetable Oils and Fats

Vegetable oils and fats are normally divided into fruit pulp fats and seed oils. Although there are few fruit pulp fats (e.g., palm oil and olive oil), the number of seed oils is considerable. Seed oils can be conveniently subdivided according to their characteristic fatty acids:

1) Lauric acid oils (e.g., coconut and palm kernel oil)
2) Palmitic–stearic acid oils (e.g., cocoa butter)
3) Palmitic acid oils (e.g., cottonseed and corn oil)
4) Oleic–linoleic acid oils (e.g., sunflower and linseed oil)
5) Leguminous oils (e.g., soybean and peanut oil)
6) Cruciferous oils (e.g., rapeseed oil)
7) Conjugated oils (e.g., tung oil and oiticica oil)
8) Substituted fatty acid oils (e.g., castor oil)

## 13.1. Fruit Pulp Fats

Palm oil and olive oil are the most important products within this group.

## 13.1.1. Palm Oil [76]

Palm oil [8002-75-3] is obtained from the fruit pulp of the oil palm (*Elaeis guineensis*). The fruit bunches, weighing 10–25 kg, contain 1000–2000 plum-sized fruits which are dark red because of their carotene content. The fruits contain 35–60% oil, depending on the moisture content. The oil palm is native to West Africa, and is extensively cultivated in Malaysia, Indonesia, and Central Africa. In recent years the oil yield has been increased to 4.5 t of oil per hectare and year by the introduction of insect pollination and by selecting and cloning high-yielding cultivars. Malaysian palm oil production was $4.13 \times 10^6$ t in 1985 and is expected to reach $6 \times 10^6$ t/a by 1990.

| | |
|---|---|
| Saponification value | 195–205 |
| Iodine value | 44–58 |
| Nonsaponifiable matter | 0.5% |
| mp | 36–40 °C |
| $n_D^{40}$ | 1.453–1.456 |
| Carotene | 500–2000 ppm |

For fatty acid composition see Tables 1 and 2.

Palm oil contains up to 40% oleodipalmitin, consisting primarily of 2-oleodipalmitin.

**Production.** The fresh palm fruits contain very active fat-splitting enzymes (lipases), which must be deactivated as quickly as possible prior to isolating the oil. Since lipases are particularly active in damaged fruits, the whole fruit bunches are sterilized by heating with live steam in autoclaves; this deactivates the enzymes and loosens the fruit from the bunches.

The sterilized fruit bunches are threshed in a rotary drum stripper consisting of longitudinal channel bars; the fruit falls through whereas the empty bunch stem is retained in the drum. The separated, sterilized fruit is then converted into an oily mash by a mechanical stirring process known as digestion. The digested fruit mass is passed through screw presses, and the liquor from the press is clarified by static settling or by using decanter centrifuges. After washing with hot water and drying, the oil can be stored and transported.

High-grade palm oil contains ca. 3% free fatty acids, inferior grades up to 5%. Crude plantation oils obtained from palm fruits processed under conditions that minimize autoxidation and lipolysis contain ca. 2% free fatty acids and can be readily bleached; such oils are traded as super prime bleachable (SPB) grades.

**Refining.** Lye neutralization of palm oil may entail relatively high losses of neutral oil since inferior crude oils contain high concentrations of mono- and diglycerides, which exert a strong emulsifying effect. Palm oil is therefore often neutralized by distillation at temperatures up to 270 °C and 0.5–0.8 kPa (5–8 mbar) after pretreatment with phosphoric or citric acid and bleaching earth to remove foreign matter and traces of copper and iron compounds. During distillative neutralization (stripping), the oil is also bleached (i.e., heat-bleached), since carotenes are decomposed and removed. The carotenes in palm oil mainly consist of $\alpha$-and $\beta$-carotene, with lower concentrations of $\gamma$-carotene, lycopene, and xanthophylls.

Distillative neutralization can be performed under conditions that simultaneously deodorize the oil. For very light raffinates, distillative neutralization is often followed by posttreatment consisting of a lye wash, bleaching with earth, and deodorization. Color and stability are largely a function of the quality of the crude oil. "White" palm oil, e.g., for biscuit fats, can generally only be made from crude SPB oils and by using fairly large amounts of bleaching earth. Bleaching with oxidizing agents, with air injected at 110–115 °C, or with activated earth at 140–160 °C is used only for technical applications. Palm oil can be effectively decolorized by mild hydrogenation at temperatures below 100 °C (hydrobleaching).

Palm oil can be separated into a solid and a liquid fraction. Two-stage cooling to 32–34 °C and 25–27 °C followed by separation over filter presses yields 20–25 % palm stearin, *mp* 50–52 °C, and 75–80 % liquid palm olein. The liquid glycerides can be separated from the solid fraction by centrifugation after addition of surfactant solutions. Fractionation in the presence of solvents (acetone, hexane) gives 10 % of a stearin melting at ca. 55 °C, 60 % olein, and 25–30 % of an intermediate fraction with properties similar to those of cocoa butter. The yield of the intermediate fraction can be increased to ca. 50 % by interesterification of the combined stearin and olein fractions followed by repeated fractionation.

**Use.** Palm oil is used predominantly for edible purposes. Refined palm oil is used extensively in shortenings and margarines. Carotene concentrates from crude oil have been largely replaced by synthetic carotene as coloring agents for margarines.

Palm stearin can be used in margarine and shortening blends whereas palm olein is used primarily in liquid frying fats and shortenings.

Technical applications of palm oil include tempering of metal and production of soap.

**Other Palm Oil Varieties.** These are commercially relevant only in their countries of origin. Tucum oil (saponifaction value 200, iodine value 40) is obtained from the fruit pulp of *Astrocaryum vulgare*, a palm growing in Central America.

## 13.1.2. Olive Oil

Olive oil [68153-21-9] is obtained from the fruit of olive trees (*Olea oleaster, Olea europaea oleaster*) and some related varieties such as *Olea americana* which is grown in South America.

| | |
|---|---|
| Saponification value | 185–196 |
| Iodine value | 80–88 |
| Nonsaponifiable matter | <1.4 % |

| | |
|---|---|
| $mp$ | −3 °C |
| $n_D^{25}$ | 1.466–1.468 |

For fatty acid composition see Tables 1 and 2.

Olive oil has a yellow to greenish-yellow color and becomes cloudy below −5 °C. Varieties with a higher cloud point can be made by winterization. The glycerides consist of 45–60% monosaturated dioleins, 25–34% dioleolinolein, and 4–29% triolein. Saturated triglycerides are present at <1%. The nonsaponifiable material contains saturated and unsaturated hydrocarbons, including 0.1–0.7% squalene.

**Production.** The quality of olive oil depends on the ripeness of the olives, the type of harvesting (picking, shaking), intermediate storage, and type of processing. Olives contain 38–58% oil and up to 60% water. Ripe olives should be processed as quickly as possible since lipases in the pulp cause rapid hydrolysis of the oil, impairing its quality for edible purposes. Top-grade oils are made from fresh, handpicked olives by comminution, pasting, and cold pressing.

Traditionally, olives were ground into a paste with stone mills; today modern milling equipment is used. Milling is followed by mashing, possibly with addition of salt. The pulp is then pressed and the press oil clarified by settling or centrifuging. Open-cage presses are being replaced by continuous screw expellers. The mashed pulp can also be separated in a horizontal decanter, the crude oil being recentrifuged after addition of wash water. An alternative is the use of machines to remove the kernels from the pulp; the residue is separated in self-discharging centrifuges.

Cold pressing, which yields vierge grades (also referred to as virgin, Provence, or Nizza oil), is generally followed by warm pressing at ca. 40 °C, which gives an oil with a less delicate flavor. The yields depend on the equipment used. The press cake (pomace) contains 8–15% of a relatively dark oil, called Sanza or Orujo, which can be extracted with hexane and is used for technical purposes; after refining it is also fit for edible consumption.

*Olive kernel oil* is obtained by pressing and solvent extraction of cleaned kernels. It is similar to olive oil but lacks its typical flavor.

**Use.** Cold-pressed olive oil is a valuable edible oil. Trade specifications are based primarily on the content of free fatty acids and flavor assessment. In some countries, warm-pressed olive oil with a high acidity is refined by neutralization, bleaching, and deodorization, and flavored by blending with cold-pressed oil.

**Testing for Adulteration.** Because of its high price, olive oil is sometimes admixed with other oils. Some oils can be detected by specific color reactions: cottonseed oil by the Halphen test, sesame oil by the Baudouin reaction, teaseed oil by the Fitelson test (see Table 22). Gas chromatographic determination of the fatty acid and sterol composition is generally used to quantify the degree of adulteration. Olive-pomace oil and re-esterified olive oil, obtained by esterification of olive oil fatty acids or Sanza or Orujo oil with glycerol, can be detected by the glyceride composition.

### 13.1.3. Avocado Oil [77]

Avocado oil is isolated from the pulp of the pearlike fruit of the tree *Persea gratissima* L. growing in Southern Europe, South Africa, the Middle East, and Central America. Its composition and properties are similar to those of olive oil (see Tables 1 and 2). The oil is used in cosmetics; in California smaller quantities of avocado oil are used as special salad oil. The avocado fruit weighs up to 1.5 kg and contains 40–80% oil in dry matter. World production of avocado fruit is ca. $1.5 \times 10^6$ t/a, but only a small fraction is used for oil production.

## 13.2. Seed-Kernel Fats

Oilseeds are the major source of oils and fats.

### 13.2.1. Lauric Acid Oils

Coconut, palm kernel, and babassu oil are the most important lauric acid oils. They contain >40% of lauric acid and ca. 15% of myristic acid.

#### 13.2.1.1. Coconut Oil [76]

The coconut palm (*Cocos nucifera* or *Cocos butyracea*) is cultivated in coastal areas around the world within 20° of either side of the equator. The Philippines, Indonesia, Southern India, Sri Lanka, Equatorial Africa, and the West Indies are important producers. Around 1820 coconut oil [8001-31-8] was introduced to Britain (vegetable butter).

| | |
|---|---|
| Saponification value | 250–262 |
| Iodine value | 7–10 |
| Nonsaponifiable matter | 0.15–0.60% |
| mp | 20–28 °C |
| $n_D^{40}$ | 1.448–1.450 |

For fatty acid composition see Tables 1 and 2.

The glycerides contain 50–60% caprylolauromyristin and up to 20% myristodilaurin in addition to smaller concentrations of laurodimyristin.

**Production.** In plantations the coconut palm reaches a height of 30 m and from its 6th to 30th years annually yields 50–70 coconuts with a diameter of 10–12 cm. The hard shell, covered by a fibrous husk, encloses the white endosperm tissue 1–2 cm thick, the copra. To obtain a light, flavor-stable coconut oil, the fresh copra with a water content of 60–70% is dried in the sun or with hot air. This treatment prevents bacterial decomposition and lipolysis of the fat. Dry copra contains 60–67% oil.

The dried copra is processed in an oil mill in two steps. About two-thirds of the oil is first obtained by expelling broken and rolled copra in continuous screw presses. The residual fat content of the expeller cake can be reduced to ca. 5% by high-pressure expelling and to 2–4% by subsequently extracting the expeller cake with hexane.

In order to obtain an edible oil of good quality, the crude coconut oil must be neutralized, bleached, and deodorized. Normally, the crude oil contains ca. 5% free fatty acids but it can be lye-neutralized without great loss of neutral oil. The neutralized, washed, and dried oil contains only small amounts of pigments, phosphatides, and other constituents. It is decolorized with 1–2% of bleaching earth and 0.1–0.4% of activated charcoal. Activated charcoal also serves to remove polycyclic aromatic hydrocarbons deposited on the copra by drying with flue gases [53]. Crude coconut oil with a relatively high content of free fatty acids can be advantageously neutralized and deodorized by distillation after pretreatment with phosphoric acid and bleaching earth – activated charcoal.

**Uses.** Coconut oil is used for cooking, and is also an important component of vegetable margarines. Because of its high latent heat of fusion, it produces a pronounced cooling effect in the mouth on melting. This effect also makes it a valuable fat for biscuit filling, confectionery products, and confectionery coatings (couvertures). Coconut oil can undergo enzymatic hydrolysis in foods containing relatively high concentrations of water leading to so-called perfume rancidity. This reaction can be counteracted by incorporating at least 30% sugar or by drying. Coconut oil has a high resistance to oxidative deterioration.

Coconut stearin, $mp$ 27–32 °C, and coconut olein can be obtained by fractional crystallization and pressing of prerefined coconut oil. Coconut stearin is used in couverture for confectionery products.

Hydrogenated coconut oil (iodine value 2–4, $mp$ 30–32 °C) is produced from prerefined coconut oil with fresh nickel catalyst at 140–180 °C. In the summer months it is often used in admixture with unhardened coconut oil to raise the melting point of couvertures and similar products.

Crude coconut oil, meeting certain color specifications, is also used in the manufacture of special soaps.

## 13.2.1.2. Palm Kernel Oil [76]

Palm kernel oil [8023-79-8] is derived from the kernels of the oil palm (*Elaeis guineensis*) (see Section 13.1.1).

| | |
|---|---|
| Saponification value | 242–254 |
| Iodine value | 16–19 |
| Nonsaponifiable matter | 0.2–0.8% |
| mp | 23–30 °C |
| $n_D^{40}$ | 1.449–1.452 |

For fatty acid composition see Tables 1 and 2.

The properties of palm kernel fat are very similar to those of coconut oil. However, it contains more oleic acid and only half as much $C_8$ and $C_{10}$ fatty acids. The glycerides of palm kernel oil consist of 60–65% trisaturated, ca. 25% disaturated–monounsaturated, and 10–15% monosaturated–diunsaturated components. Its stability is similar to that of coconut oil.

**Production.** The palm nuts are separated by air classifiers from the fibers of the press cake obtained on pressing palm fruit (see Section 13.1.1). After drying in silo dryers, the nuts are cracked with centrifugal crackers, and the shell is separated from the kernel with air and water separation systems. The kernel, 1–2 cm in size, is dried from an initial moisture content of 20–25% to about 4–7%; it then contains 44–57% oil. Following cracking and rolling, the palm kernels are passed through a screw expeller. The residual oil in the press cake can be further reduced by high-pressure expelling or, more commonly, by extraction with hexane.

Crude palm kernel oil often has a higher free fatty acid content than coconut oil (up to 15%, calculated as oleic acid). Palm kernel oil with a free fatty acid content <5% can be lye-neutralized without significant neutral oil losses. At higher contents of free fatty acids, distillative neutralization (possibly followed by lye postneutralization) is generally preferred. Treatment with 1–2% activated bleaching earth yields a light, yellowish oil. Deodorization generally requires a longer period of time than for coconut oil in order to obtain a stable, bland raffinate.

**Use.** Top-grade palm kernel raffinates are used in couvertures and margarines. Palm kernel stearin and olein are obtained by fractional crystallization of prerefined palm kernel oil at ca. 25 °C. Palm kernel stearin, *mp* 30–32 °C, is often used as such, or blended with coconut stearin, as a filling mass or couverture for confectionery products. Hydrogenated palm kernel oil (*mp* 39 °C, iodine value 1) is made in the same way as hydrogenated coconut oil and is mostly used in the confectionery trade. The melting point of hydrogenated palm kernel oil can be reduced from 39 °C to 33–35 °C by interesterification. Interesterified, hydrogenated palm kernel oil increases the range of special tailor-made products for application in confectionery products.

Table 23. Properties and analytical data of seed oils of some South American palm varieties

|  | Cohune oil | Murumuru oil | Ouricuri oil | Tucum oil |
|---|---|---|---|---|
| Density (60 °C), g/cm$^3$ | ca. 0.893 | 0.893 | 0.898 | 0.893 |
| $n_D^{60}$ | 1.441 | 1.445 | 1.440 | 1.443 |
| Saponification value | 251–257 | 237–242 | ca. 257 | 240–250 |
| Melting point, °C | 18–24 | 32–34 | 18–21 | 30–36 |
| Iodine value | 10–14 | 11–12 | 14–16 | 10–14 |
| Nonsaponifiable matter, % | 0.4 | 0.3 | 0.3 | 0.3 |

## 13.2.1.3. Babassu Oil and Other Palm Seed Oils

The babassu palm (*Orbygnia speciosa, Attalea funifara*) is a native of the great forest regions of the Brazilian states Maranhão, Piauí, Pará, and Minas Geraes. The number of babassu trees in these forests is estimated at $> 10^9$. Collecting and transport of the babassu nuts to the oil mills and removal of the extremely hard shell prior to expelling make it difficult to exploit this interesting source of lauric oil. A bunch holds 200–600 fruits; 1 t of fruit yields about 125 kernels containing 63–70% fat after drying to ca. 4% moisture. The fat can be isolated by expelling and subsequent extraction with hexane.

| | |
|---|---|
| Saponification value | 242–253 |
| Iodine value | 10–18 |
| Nonsaponifiable matter | 0.2–0.8% |
| mp | 22–26 °C |
| $n_D^{40}$ | 1.449–1.451 |

For fatty acid composition see Tables 1 and 2.

**Use.** Refined babassu oil is known in South America as an edible fat. In European countries it is used in place of palm kernel oil in margarine blends. The export of babassu kernels is only of minor economic importance; total production is likely to be around 150 000 t/a.

Other palm seed oils are used as edible fats in their countries of origin and do not play a role in international trade. Some of their properties are listed in Table 23. Cohune oil is obtained from the kernels of the cohune palm (*Attalea cohune*), a native of Mexico and Honduras. Its properties are similar to those of coconut oil. Murumuru oil, somewhat harder than coconut oil, is contained in the kernels of the palm *Astrocaryum murumuru*, which is found in the Northern provinces of Brazil. It contains ca. 40% lauric acid, 35% myristic acid, and only little caprylic and capric acids. Ouricuri oil, from the kernels of the Brazilian palm *Syagrus coronata*, has a lower melting point than murumuru oil. It contains 45% lauric acid and 10% myristic acid in addition to 20% fatty acids with 6, 8, and 10 carbon atoms. The kernels of *Astrocaryum tucuma*, a native of Guyana, Venezuela, and Northern Brazil, yield tucum oil.

*Other Sources of Lauric Acid Oils* Up to 43% of lauric acid is contained in the seed oil of the evergreen laurel tree (*Laurus nobilis*), from which the name of this acid was derived. Laurel oil contains about 30% trilaurin; it has no commercial importance.

The seed oils of *Myristica* species contain relatively high concentrations of myristic acid. Nutmeg butter [8008-45-5] is made by expelling ground and boiled seed kernels of *Myristica officinalis*; it contains 39 – 76% myristic acid. *Myristica otoba* yields otoba fat. Ucuhuba fat is obtained from *Virola surinamensis*, a native to Brazil. Dika fat is contained in the seed kernels of *Irvingia gabonensis* and *I. barteri*, native to West Africa. None of these fats is involved in international trade.

*Cuphea* is a herbaceous, annual plant, native to Mexico and also found in Northern Brazil and Nicaragua. The seed oil contains various medium-chain fatty acids which, depending on the species, account for 40 – 80% of the total fatty acids. Cuphea oil could be a substitute for coconut and palm kernel oil and serve as a natural source of capric acid; agronomic research is directed toward adaptation and yield improvement of *Cuphea* [78].

## 13.2.2. Palmitic – Stearic Acid Oils

### 13.2.2.1. Cocoa Butter

Cocoa butter is the seed fat of *Theobroma cacao* L., a tree reaching a height of 9 m, which is cultivated in many tropical countries. The cocoa tree is grown mainly in West Africa (Ivory Coast, Nigeria) and South Africa (Brazil). The world production of cocoa beans is in the order of $1.3 \times 10^6$ t/a.

| | |
|---|---|
| Saponification value | 190 – 200 |
| Iodine value | 35 – 40 |
| Nonsaponifiable matter | 0.2 – 0.5% |
| mp | 28 – 36 °C |
| $n_D^{40}$ | 1.453 – 1.458 |

For fatty acid composition see Tables 1 and 2.

Cocoa butter has a pleasant aromatic flavor and keeps well. The cocoa butter glycerides are hard and brittle up to about 28 °C and melt in a range of 4 – 5 °C with a pronounced oral cooling effect. The melting point of cocoa butter depends on the type and length of tempering. After rapidly cooling to 0 – 5 °C, cocoa butter melts at 26 – 30 °C, whereas the melting point increases to 32 – 35 °C after tempering for 40 h at 28 °C. The dilatation curve (plot of solids content at different temperatures) is very steep (see Fig. 2).

The glycerides of cocoa butter consist roughly of 2% trisaturated compounds, 14% 2-oleodipalmitin, 40% 1-palmito-3-stearo-2-olein, 27% 2-oleodistearin, 8% palmitodiolein, and 8% stearodiolein.

**Table 24.** Properties and analytical data of some vegetable butters

|  | Shea butter | Borneo tallow | Sal butter | Illipé butter | Mowrah butter | Katiau fat | Phulwara butter |
|---|---|---|---|---|---|---|---|
| Myristic acid, % |  | ca. 15 |  |  |  |  |  |
| Palmitic acid, % | 5–6 | 18–22 | 8–9 | ca. 28 | 16–27 | ca. 10 | 54–57 |
| Stearic acid, % | 36–42 | 39–44 | ca. 35 | ca. 14 | ca. 20 | ca. 19 | ca. 4 |
| Arachidic acid, % |  | ca. 1 | ca. 12 |  |  |  |  |
| Oleic acid, % | 49–50 | 38–42 | ca. 42 | ca. 50 | 41–66 | ca. 69 | ca. 36 |
| Linoleic acid, % | 4–5 | traces | ca. 3 | 8–9 | 9–14 | 2–3 | 3–4 |
| Saponification value | 178–196 | 189–200 |  | 186–200 | 187–195 | 189–192 | 188–200 |
| Iodine value | 55–67 | 29–38 |  | 50–60 | 58–63 | 53–67 | 40–51 |
| mp, °C | 32–42 | 34–39 |  | 25–29 | 23–31 | 30 | 38–43 |
| $n_D^{40}$ | 1.4635–1.4668 | 1.4561–1.4573 |  | 1.459–1.462 | 1.458–1.461 | 1.461–1.462 | 1.455–1.458 |
| Nonsaponifiable matter, % | 2–11 | <2 |  | 1.4–2.3 | 1.2–2.1 | 0.4–0.5 | 2–2.8 |
| Fat content in seed kernel, % | 45–55 |  | 20 | 50 | 50 |  | 60–65 |

**Production.** Cocoa butter is isolated from cocoa beans by fermentation, roasting, dehulling, grinding, and expelling. The fat content of dehulled cocoa beans is about 54–58 % in dry matter. In some countries the fat obtained from nondehulled cocoa beans by expelling or solvent extraction may also be called cocoa butter.

**Use.** Cocoa butter is used extensively in the production of chocolate and other confectionery products, and to a lesser extent in the pharmaceutical and cosmetics industries.

**Test for Adulteration.** Because of its high price, cocoa butter is sometimes admixed with press or extraction fat from damaged cocoa beans or byproducts, animal fats, coconut oil and stearin, palm oil and palm oil fractions, or hydrogenated vegetable or marine oils.

Extraction cocoa butter can be identified by its relatively dark color, by its strong fluorescence under ultraviolet light, by various color reactions, and by its behavior in simple tests such as the "smear test" (extraction fat tends to smear while authentic cocoa butter tends to disintegrate into small, hard pieces). In addition, extraction fat has a higher content of nonsaponifiable matter (cocoa butter contains 0.2–0.5 %, extraction fat 1.9–2.8 %, fat from cocoa bean hulls about 7.5 % nonsaponifiable matter) and a higher content of linoleic acid (10–29 % compared to 2–4 % in authentic cocoa butter).

Animal fats such as tallow and hydrogenated fish oils can be detected by gas chromatographic analysis of the fatty acid composition and by their cholesterol content. Hydrogenated vegetable oils can be recognized by their content of trans fatty acids. Coconut oil and similar oils can be detected by their concentrations of $C_8$, $C_{10}$, $C_{12}$, and $C_{14}$ fatty acid groups. Addition of cocoa butter-like fats such as fractions of Borneo tallow, illipé fat, shea butter, and palm oil can sometimes be detected by their concentration of terpene alcohols. Gas chromatographic analysis of chocolate fat triglycerides can detect so-called cocoa butter equivalents in cocoa butter and certain types of chocolate; this technique can be augmented by analyzing the fatty acid composition at the 2-position [79].

## 13.2.2.2. Shea Butter, Borneo Tallow, and Related Fats (Vegetable Butters)

Shea butter, also known as Karité butter or Galam butter, is obtained from the seeds of a tree growing mainly in West Africa (*Butyrospermum parkii*). Plantations tend to be uneconomic since shea nuts can be harvested only after the tree is about 15 years old.

Shea butter is similar to cocoa butter (see Table 24). The content of nonsaponifiable matter, which contains rubber-like hydrocarbons such as kariten [$(C_3H_8)_n$, *mp* 63 °C], can be as much as 11%.

The kernels can be readily separated from the fruit and the thin hulls. In Africa they are ground, boiled, and the fat is skimmed off. The commercial production of shea butter by expelling in continuous screw presses is difficult and requires special processing conditions. The green-brown crude fat can be neutralized only with considerable loss of neutral oil.

Refined shea butter can be used as an edible fat. The stearin fraction, primarily 2-oleodistearin, which can be obtained by crystallization and removal of the liquid fraction, is a valuable cocoa butter substitute. The export quantities (from Nigeria, Dahomey, Upper Volta) amount to ca. 50 000 t/a.

Borneo tallow (tengkawang tallow) is derived from *Shorea stenoptera*, a plant growing in the East Indies and Malaysia; there are about 100 different species including *S. gysbertsiana, S. palembassica,* and *S. seminis*. The proportions of component fatty acids in these species are very similar. The size of nut varies considerably from one variety to another, and only a few are worth harvesting. Borneo tallow resembles cocoa butter chemically and physically more closely than any other fat. For this reason it is used in preparing cocoa butter equivalents or extenders.

Sal butter is derived from *Shorea robusta*, commonly known as the sal tree, found extensively in parts of Northern and Central India. Illipé fat is derived from the group of Madhuca seed fats (also known as Bassia seed fat) and is itself produced from the Indian plant *Madhuca longifolia*. Another Indian plant, *Madhuca latifolia,* is the source of mowrah fat which is very similar in composition. Illipé and mowrah fats contain sufficient unsaturated fatty acids to make them resemble shea butter rather than the firmer cocoa butter and Borneo tallow. Illipé and mowrah fats are used principally in Asia for technical purposes but also for edible purposes when fully refined. Both fats are used to prepare cocoa butter extenders. Katiau fat is derived from the plant *Madhuca mottleyana,* and phulwara butter (Indian butter) from *M. butyracea.*

For properties of the various vegetable butters see Table 24 [80].

## 13.2.3. Palmitic Acid Oils

The oils in this group have a relatively high content of palmitic acid, mostly above 10 %. Cottonseed oil and some cereal oils are the most important representatives.

### 13.2.3.1. Cottonseed Oil

Cottonseed oil [8001-29-4] is obtained from the seeds of different varieties of cotton (*Gossypium*). The most important producers are the United States, the Soviet Union, China, Pakistan, Brazil, Egypt, and India. *Gossypium barbadense* and *G. hirsutum* are preferred in the United States, while *G. herbaceum* is native to the Asian countries. The United States produce about one-third of the world's cottonseed oil.

| | |
|---|---|
| Saponification value | 190 – 198 |
| Iodine value | 100 – 117 |
| Nonsaponifiable matter | 0.5 – 1.5 % |
| mp | ca. 0 °C |
| $n_D^{40}$ | 1.464 – 1.468 |

For fatty acid composition see Tables 1 and 2.

The oil contains 0.4 – 0.6 % of malvalic acid, which presumably is the basis of the Halphen color reaction. The glycerides of cottonseed oil consist of approximately 12 – 14 % disaturated – monounsaturated, 52 – 58 % monosaturated – diunsaturated, and 25 – 30 % triunsaturated compounds.

**Production.** Egyptian cottonseed (not dehulled) contains 22 – 24 % oil, the American varieties on average 19.5 %. About 50 cottonseeds are contained in a pod, which opens when it is ripe. After removal of the cotton, the seeds are still covered with fine hairs (linters), which are removed with delintering machines that essentially consist of rotating sawtooth discs [81]. The seeds are then dehulled by passage through "dehullers" (essentially rotating knives) and screens – air classifiers. The dehulled seeds (30 – 40 % oil content) are expelled in continuous screw presses and the expeller residue is extracted with solvent.

Gossypol is present in the seed in concentrations of 0.4 – 2.0 %. However, only traces are detectable in the crude oil, because gossypol is partially deactivated (presumably by interaction with protein) during conditioning and expelling the seed. Crude cottonseed oil has a dark color because of dissolved resins and pigments, and must therefore be refined before further use. The crude oil is first clarified by settling or centrifuging and then neutralized with lye, mostly in continuous centrifuge lines. The neutralized oil is free of gossypol, light yellow, stable during storage, and is usually traded in this form. A very stable oil can be obtained by postneutralization, bleaching, and deodorization. Stearin separates out at temperatures between 0 and 8 °C. An oil with a low cloud point is obtained by removing these solid glycerides by cooling and filtration (winterization).

**Table 25.** Properties of kapok, okra, and kenafseed oils

|  | Kapok oil | Okra oil | Kenafseed oil |
|---|---|---|---|
| Myristic acid, % | ca. 0.5 | 0 – 4 |  |
| Palmitic acid, % | 10 – 16 | 23 – 33 | ca. 14 |
| Stearic acid, % | 3 – 8 | 0.5 | ca. 6 |
| Arachidic acid, % | ca. 1 | traces |  |
| Oleic acid, % | ca. 50* | 26 – 42 | ca. 45 |
| Linoleic acid, % | ca. 30 | 30 – 40 | ca. 23 |
| $n_D^{40}$ | 1.464 – 1.468 | 1.462 – 1.467 | 1.465 – 1.466 |
| Saponification value | 189 – 197 | 192 – 199 | 189 – 195 |
| Iodine value | 86 – 110 | 90 – 100 | 93 – 105 |
| Nonsaponifiable matter, % | 0.5 – 1.8 | 0.7 – 1.4 | 0.4 – 3.4 |

* Including 15% malvalic acid.

**Use.** Winterized cottonseed oil is a very stable salad oil. Cottonseed oil is also used in margarine blends without being winterized. Lard-like fats for use in margarine blends and shortenings are obtained by hydrogenation under iomerizing conditions. Such fats have a melting point between 30 and 35 °C and an iodine value of 60 – 80. Cottonseed oil hardened to a melting point of 36 – 37 °C is suitable as a couverture fat for confectionery products. The characteristic Halphen color reaction disappears on hydrogenating the oil.

### 13.2.3.2. Kapok and Related Oils

Kapok oil is derived from the fruit of a tree belonging to the Bombaceae family that grows in Indonesia, Sri Lanka, the Philippines, and South America.

The seeds in the fruit pods are covered with fine hairs (kapok), which are used in upholstering and contain ca. 25% oil. Crude kapok oil, produced by preexpelling and solvent extraction, has a reddish-brown color and is similar to cottonseed oil in composition and further processing. It contains little or no gossypol. Kapok oil contains ca. 15% of a $C_{18}$ cyclopropenoic acid and therefore responds strongly to the Halphen reaction.

Okra oil is obtained from the seeds of *Hibiscus esculentis*, a family related to Gossypium and found in the United States and some Mediterranean countries. Kenafseed oil is contained in the seeds of *Hibiscus canabinus*, a plant native to India where it is grown for its fibers.

The properties of kapok, okra, and kenafseed oils are listed in Table 25.

### 13.2.3.3. Pumpkin Seed Oil

*Pumpkin seed oil* is popular as a high-quality, edible oil in Southeast Europe. It is obtained from carefully decorticated pumpkin seeds by grinding, conditioning the moistened ground seeds, expelling in hydraulic presses, and clarification by settling in special tanks, mostly on smaller locations. It has a reddish-brown color and a nutty taste. Refining yields an oil with properties similar to those of sunflower oil.

| | |
|---|---|
| Saponifaction value | 185–198 |
| Iodine value | 117–130 |
| Nonsaponifiable matter | 0.6–1.5% |
| Solidification point | ca. –15 °C |
| $n_D^{40}$ | 1.466–1.469 |

For fatty acid composition see Tables 1 and 2.

*Melon seed oil* is very similar in composition to pumpkin seed oil [82], the major fatty acids being palmitic (ca. 12%), stearic (ca. 11%), oleic (ca. 11%), and linoleic (ca. 65%).

### 13.2.3.4. Corn (Maize) Oil

Corn oil (maize oil) [*8001-30-7*] is obtained from the germ removed during the processing of corn (*Zea mays* L.) into starch.

**Properties.** Winterized corn oil becomes cloudy at about –10 °C.

| | |
|---|---|
| Saponification value | 187–196 |
| Iodine value | 109–133 |
| Nonsaponifiable matter | 1.3–2.0% |
| $n_D^{40}$ | 1.465–1.466 |

For fatty acid composition see Tables 1 and 2.

Corn oil glycerides consist of ca. 2% disaturated–monounsaturated, 40% monosaturated–diunsaturated, and 58% triunsaturated compounds.

**Production.** Corn contains 3.5–5% oil, about 80% of which is in the germ and about 20% in the endosperm. The corn germ itself contains about 36% oil. The germs can be separated by wet or dry processing. Wet processing tends to be preferred because it gives a higher oil yield. The cleaned corn is first conditioned by soaking in warm water and is then milled and slurried with water. The germs are collected by flotation, washed, and dried. In dry processing, the ground corn is separated into germ and endosperm fractions by screening and air classification; suitable moistening and conditioning facilitate this fractionation. The oil is isolated from the germs by pre-expelling followed by extraction with hexane.

The golden yellow, crude corn oil contains up to 3% free fatty acids. After lye neutralization, bleaching, and deodorization it yields a light, highly stable oil. The content of tocopherols, predominantly $\gamma$-tocopherol, is nearly 1000 ppm. Salad oils that remain clear at refrigerator temperatures are obtained by winterization, during which about 500 ppm wax is removed.

**Use.** Corn oil is a universal oil; it is used especially as a salad oil and in salad dressings.

### 13.2.3.5. Cereal Oils

Oils are generally not extracted from cereal grains except rice and wheat. The cereal oils are, however, very important as constituents of bread, other bakery products, and animal feeds.

*Wheat germ oil* is extracted from the wheat germ (8–11% oil) obtained as a byproduct of milling wheat. It has a very high tocopherol content and is therefore popular as a dietetic oil.

| | |
|---|---|
| Saponification value | 180–189 |
| Iodine value | 115–126 |
| Nonsaponifiable matter | 3.5–6.0% |
| mp | ca. 0 °C |
| $n_D^{40}$ | 1.468–1.478 |

For fatty acid composition see Tables 1 and 2.

*Rice bran oil* [68553-81-1] is an important byproduct of rice processing. It is obtained by hexane extraction of the bran (brown outer coating) and the germ (8–16% oil content) removed during grinding, dehulling, and polishing rice. The bran fraction must be extracted immediately, preferably after heat-sterilization, to destroy the very active lipases. Refined rice bran oil has a light yellow color and is a relatively stable edible oil. Rice bran oil hardened to 30–35 °C is often used in margarines and shortenings. In Japan and India ca. 100 000 t of rice bran oil are produced annually.

| | |
|---|---|
| Saponification value | 183–194 |
| Iodine value | 92–109 |
| Nonsaponifiable matter | 3.5–5.0% |
| (including ca. 330 mg squalene per 100 g oil) | |
| mp | ca. −10 °C |
| $n_D^{40}$ | 1.466–1.469 |

For fatty acid composition see Tables 1 and 2.

### 13.2.4. Oleic–Linoleic Acid Oils

This group encompasses a large number of drying and semidrying oils derived from plants growing in temperate and colder regions. Some of these oils, such as hazelnut oil, poppyseed oil, walnut oil, and teaseed oil, only have regional significance, while sunflower oil, sesame oil, linseed oil, and safflower oil have considerable commercial importance.

## 13.2.4.1. Sunflower Oil

Sunflower oil [8001-21-6] is the seed oil of the sunflower, *Helianthus annuus*, that originated in America and is now grown extensively in East, West, and South Europe, the United States, Canada, South America, China, India, South Africa, and Australia. The annual plant reaches a height of 1–3 m, the short-stem varieties being preferred for ease of harvesting. The air-dry seeds contain about 45% oil.

**Properties.** Sunflower oil begins to solidify at −16 °C; the cloud point is 0–5 °C.

| | |
|---|---|
| Saponification value | 188–194 |
| Iodine value | 125–144 |
| Nonsaponifiable matter | 0.4–1.4% |
| $n_D^{40}$ | 1.466–1.468 |

For fatty acid composition see Tables 1 and 2.

The linoleic acid content of sunflower oil is 50–70% and tends to increase with decreasing temperatures during growing [83]. The glycerides contain 35–45% diunsaturated and 56–63% triunsaturated compounds. Linolenic acid is present to the extent of <0.4%.

**Production and Use.** Sunflower oil is obtained by preexpelling followed by extraction of the expeller cake with hexane. Decortication is practiced to a lesser degree; the hulls have to be burnt. The crude oil can be readily refined into a light, stable oil by lye neutralization, bleaching, and deodorization.

Cold-pressed sunflower oil is produced by expelling at ≤ 80 °C; posttreatment is limited to filtration and clarification. Such a "natural" oil is often recommended for dietetic purposes because of its relatively high concentration of phosphatides. Trace contaminants such as metals and pesticide residues, however, remain in the oil. Refined sunflower oil is used extensively in salad oils and margarine blends because of its oxidative stability. Winterization (removal of 500–2000 ppm of wax) of the deodorized or the neutralized and bleached oil at 10–15 °C yields an oil which does not cloud at refrigerator temperatures. Sunflower oil that is hydrogenated to a melting point of 32–35 °C (iodine value 70–75), under isomerizing conditions, is widely used in "one-oil" margarine blends.

Smaller quantities of a new sunflower variety are being grown in California; these plants yield an oil with an oleic acid content of about 80% and hence a very high oxidative stability. This oil is recommended especially for salad and frying oils [84].

## 13.2.4.2. Sesame Oil

Sesame oil [8008-74-0] is contained in the seeds of *Sesamum indicum*, a plant resembling linseed and grown mainly in China, India, Africa, and Mexico. The flat seeds are 2–3 mm long and vary in color from white to brown-black; the oil content is

**Table 26.** Fatty acid composition and analytical data of vegetable oils of the oleic–linoleic group

|  | Linseed oil | Perilla oil | Hempseed oil |
|---|---|---|---|
| Myristic acid, % | ca. 0.5 | | |
| Palmitic acid, % | 5–7 | 7–8 | ca. 6 |
| Stearic acid, % | 3–5 | traces | ca. 2 |
| Arachidic acid, % | ca. 6 | | ca. 2 |
| Lignoceric acid, % | traces | | |
| Oleic acid, % | 16–26 | ca. 8 | 6–20 |
| Linoleic acid, % | 14–24 | ca. 38 | 46–70 |
| Linolenic acid, % | 50–65 | 44–50 | 14–28 |
| Saponification value | 188–196 | 188–196 | 190–194 |
| Iodine value | 170–204 | 170–204 | 140–170 |
| Solidification point, °C | 10–21 | 12–17 | 15–17 |
| $n_D^{25}$ | 1.4786–1.4815 | 1.4800–1.4820 | |
| Nonsaponifiable matter, % | ca. 1.5 | <1.5 | <1.5 |

45–55 %. Harvesting is relatively difficult because of varying rates of ripening and the ripe seed pods shattering the seed. More uniform varieties with improved harvesting properties are being developed.

**Properties.** Sesame oil solidifies at −6 to −3 °C.

| | |
|---|---|
| Saponification value | 187–193 |
| Iodine value | 136–138 |
| Nonsaponifiable matter | 0.9–2.3 % |
| $n_D^{40}$ | 1.4665–1.4675 |

For fatty acid composition see Tables 1 and 2.

Crude sesame oil is yellowish and contains up to 3 % free fatty acids. The glycerides contain about 66 % dioleolinolein and oleodilinolein. Refined oil has a high stability due in part to the content of sesamol, a natural antioxidant formed by hydrolysis of sesamolin. The combined concentration of sesamol and sesamolin is 0.1–0.2 %. Sesamol gives a red color with an ethanolic solution of furfural in the presence of hydrochloric acid (color reaction of Baudouin and Villavechia); it also responds positively in color tests with an acidic solution of stannous chloride (Soltsien test), and with 3 % hydrogen peroxide – 75 % sulfuric acid (Kreis test).

**Use.** Sesame oil is widely used as an edible oil in the countries of origin. A small amount is exported for use in salad oils, salad dressings, and margarines. The addition of sesame oil to margarine is mandatory in some countries to provide a rapid means of identifying margarine by the above color reactions; for this application the oil may only be lightly bleached and deodorized at 150 °C.

## 13.2.4.3. Linseed Oil

Linseed oil [*8001-26-1*] is obtained from linseed or flax (*Linum usitatissimum*), which grows best in the temperate regions of Europe, Asia, and America. The flax fiber plant is a different variety of the same species. There is considerable variation in the iodine value of linseed oil from different regions. The iodine value of the oil tends to increase with severity of climate, but genetic and seasonal (e.g., rainfall) variations also have an influence [85]. While 180 – 185 is a typical iodine value range for individual lots, it may vary from 140 to 205.

The main producer countries are Canada, the United States, and Argentina. The brown, flat seeds contain ca. 40 % oil and 6 – 8 % water.

**Properties.** Fatty acid composition and other properties of linseed oil are given in Table 26 (see also Tables 1 and 2). Most applications require a linseed oil with a linolenic acid content of about 50 % and an iodine value of 170 – 190 [86], [87]. Linseed oil glycerides consist of about 5 % disaturated – monounsaturated, 43 % monosaturated – diunsaturated, and 52 % triunsaturated glycerides, which begin to solidify at –18 to –27 °C. Most of the glycerides contain an average of seven double bonds per molecule, which is the reason for the excellent drying properties of linseed oil. The use of linseed oil for edible purposes is limited to smaller quantities of cold-pressed oil (for dietetic purposes) and to hydrogenated linseed oil, which can be used in low concentrations in shortenings, and margarine blends.

Some wild linseed varieties (e.g., *Linum capitatum* and *L. flavum*) have linolenic acid contents as low as 11 %. Modern biotechnological techniques are being used in Australia and Canada in an attempt to transfer this property to *L. usitatissium*. An alternative approach is $\gamma$-ray induced mutation of *L. usitatissimum*.

**Production.** Linseed oil is obtained from the seed by preexpelling followed by hexane extraction of the press cake. After refining it is used predominantly in the technical sector for paints and coatings. A small amount is used in East European countries as edible oil. Linseed oil has a characteristic odor and flavor which cannot be removed entirely by refining. The crude oil is rich in phosphatides and gums which sediment on storage. The crude oil is often pre-deslimed by treatment with hot water followed by centrifugation. Oils virtually free of phosphatides and gums are made by post-desliming with sulfuric or phosphoric acid. Very light drying oils can be made by a further posttreatment consisting of lye neutralization and earth bleaching. To obtain virtually odorless oils, deodorization is required as a final step. Heating refined linseed oil to 260 – 285 °C increases its viscosity and gives a so-called "boiled" linseed oil, which is used in special paints and coatings.

## 13.2.4.4. Perilla Oil

The seeds of *Perilla ocymoides*, an Asian plant, yield perilla oil (30–50%), which resembles linseed oil in appearance and application. For properties see Table 26. World production is about 25 000 t/a. Perilla oil is sometimes blended with linseed oil; in Asia it is also used as an edible oil.

Lallemantia oil (Asia) and Chia seed oil (Mexico) are further Labiatae seed oils with similar compositions.

## 13.2.4.5. Hempseed Oil

The seeds of hemp (*Cannabis sativa* L.) contain 30–35% oil. Hempseed is grown in the Soviet Union, India, China, Japan, and Chile, and yields fibers in addition to the oil. Hashish can also be obtained from Indian hempseed varieties. Hempseed itself can be used as a foodstuff. The oil is normally produced by pressing the seed. For properties see Table 26. Hempseed oil has been largely replaced by linseed oil for technical purposes.

## 13.2.4.6. Teaseed Oil

Teaseed oils are obtained in China, Japan, India, and Turkey from the hazelnut-like seeds of tea shrub varieties, which contain about 60% oil. Tea shrubs grown for oil are generally not suited for tea-leaf harvesting. Teaseed oil is produced primarily from the seeds of *Thea sasanqua* n. (sasanqua oil) and *Thea japonica* n. (tsubaki oil) by drying, grinding, conditioning, and pressing or solvent extraction. After refining the oil is used for edible purposes, as a special lubricant, and for toiletries.

The principal acids of teaseed oils [88] are palmitic (ca. 16%), oleic (ca. 60%), and linoleic acid (ca. 22%). Teaseed oils can be detected in concentrations above 10% in olive oil by the Fitelson color reaction. The expeller cake and extraction meal cannot be used as such in animal feeds because of their high saponin contents. Total world production is ca. 30 000 t, half of which is exported.

## 13.2.4.7. Safflower and Niger Seed Oils

*Safflower oil* [8001-23-8] is the seed oil of the thistle-like safflower plant (*Carthamus tinctorius*), thriving in the West of the United States and in Mexico, North Africa, and India. The plant can be grown under fairly arid conditions, the oil yield increasing with available moisture in the soil. The seeds resemble small sunflowerseed kernels and can be harvested mechanically. The oil content of the seed is 25–37%.

| | |
|---|---|
| Saponification value | 180–194 |
| Iodine value | 136–152 |
| Nonsaponifiable matter | 0.3–2.0% |
| Solidification point | −13 to −25 °C |
| $n_D^{40}$ | 1.467–1.469 |

The fatty acid composition is similar to that of sunflower oil (see Tables 1 and 2).

**Production.** Modern processing involves grinding the seed followed by conditioning, expelling, and hexane extraction of the expeller cake.

**Use.** Safflower oil has a high oxidative stability and is being used increasingly in salad oils and dietetic margarines because of its high content of linoleic acid. Some years ago special varieties with oleic and linoleic acid contents of 80% and 15% respectively ("high oleic" varieties) were developed and recommended as especially stable frying oils. World production amounts to about 200 000 t/a.

*Niger seed oil,* closely resembling safflower oil in its composition, is derived from the seed of the gingli plant, which is grown fairly extensively in Ethiopia, Togo, and India.

### 13.2.4.8. Grape-Seed Oil

Grape-seed oil [60-33-3] is a valuable oil obtained from the grapeseed left in winery pomace [89]. The oil content averages 15% (dry weight basis).

**Properties.** The flavor of grape-seed oil resembles that of olive oil; however, its fatty acid composition is similar to that of sunflower oil (see Tables 1, 2).

| | |
|---|---|
| Saponification value | 180–196 |
| Iodine value | 124–143 |
| Nonsaponifiable matter | 0.3–1.6% |
| Solidification point | −10 to −20 °C |
| $n_D^{40}$ | 1.464–1.471 |

**Use.** After suitable refining (removal of chlorophyll can be difficult), grape-seed oil keeps relatively well and is used as a salad and cooking oil and in special margarine blends. In France and Italy, 5000 t and 20 000 t, respectively, of grape-seed oil are produced annually. It has been estimated that the world production of grape seed is $1.4 \times 10^6$ t, with a potential oil yield of about 190 000 t [82].

### 13.2.4.9. Poppyseed Oil

Poppyseed oil is obtained from the seeds of different varieties of *Papaver somniferum*. The oil content of air-dry seeds is 44–50%. Cold-pressed oil is suitable for edible purposes, as is preexpelled and extracted oil after refining. Poppyseed oil is also used in smaller quantities for superior oil paints. The major fatty acids are stearic (ca. 10%), oleic (ca. 16%), and linoleic acid (70–75%).

| | |
|---|---|
| Saponification value | ca. 193 |
| Iodine value | ca. 140 |
| $n_D^{40}$ | ca. 1.468 |

The world production of poppy seed amounts to about 25 000 t/a.

## 13.2.5. Leguminous Oils

The commercially most important representatives of this group are peanut oil and soybean oil. Soybean oil [8001-22-7] holds first place in the worldwide production of vegetable oils.

### 13.2.5.1. Soybean Oil

The cultivation of the soybean (*Glycine maxima*) in the United States and South America has steadily increased. The soybean has been known in China for 5000 years. Main producing countries today are the United States, Brazil, and Argentina. Cultivation in the temperate regions of Europe is possible but the hectare returns are generally not sufficiently attractive. New, higher-yielding varieties are constantly being developed; in the United States they thrive best in the "corn-belt," i.e., Illinois, Minnesota, Iowa, Indiana, Ohio, and Missouri.

The soy pods contain up to four soybeans (5–10 mm in length). The oil content varies between 17 and 22% and the protein content reaches 40–45% in dry matter. Climatic conditions have a considerable influence on the oil and protein content; late sowing and a cool, wet climate tend to lower the oil content and to increase the protein content. Development of low linolenic soybeans in the United States has been hindered by lower seed productivity.

**Properties.** Refined soybean oil has a light yellow color and a bland flavor.

| | |
|---|---|
| Saponification value | 188–195 |
| Iodine value | 120–136 |
| Nonsaponifiable matter | 0.5–1.5% |
| Solidification point | −15 to −8 °C |
| $n_D^{40}$ | 1.465–1.469 |

For fatty acid composition see Tables 1 and 2.

The oil contains 40–60% triunsaturated, 30–35% diunsaturated, and up to 5% monounsaturated glycerides. In spite of the high tocopherol content (up to 1200 ppm) the stability of the refined oil is limited, primarily because of the content of linolenic acid. Autoxidation leads to "green", "seedy" off-flavors. Soybean oil can be detected in other oils by its fatty acid, tocopherol, and sterol contents.

**Production.** Soybeans are normally cracked and flaked prior to extraction with hexane. Preexpelling is not normally performed. Soybean meal is a valuable animal feed. Crude soybean oil contains 0.5 – 1.0 % free fatty acids as well as up to 2.5 % phosphatides, which are removed by hydration with 2 – 3 % water at 70 – 80 °C followed by centrifuging. Drying of the resulting lecithin sludge yields lecithin. Soybean oil is often postdegummed with phosphoric or citric acid prior to neutralization with lye, earth bleaching, and deodorization.

The yield of lecithin can be increased by steam-heating the flakes prior to extraction [39]. With the aid of special degumming techniques [44], the phosphorus content of the crude oil can be reduced to 20 ppm, which facilitates subsequent physical refining (i.e., direct treatment with acids and earth followed by combined stripping of free fatty acids and deodorization).

**Use.** Soybean oil is used in almost all fatty products, from salad oils to margarines. Hydrogenation with fresh catalyst at ca. 100 °C yields a liquid oil with significantly improved stability due to hydrogenation of linolenic acid. Hydrogenation under isomerizing conditions to melting points ranging from 36 to 43 °C yields products for use in shortenings and margarines. Liquid soybean oil is a so-called semidrying oil; it is used in combination with tung oil and as a component of alkyd resins for coatings.

## 13.2.5.2. Peanut Oil

Peanut oil (groundnut oil) [*8002-03-7*] is obtained from the seed kernels of the peanut plant (*Arachis hypogaea*), native to South America. The peanut is now being grown in China, India, South Africa, West Africa, Argentina, and the United States. Senegal and Nigeria are the main African export countries. The annual plant thrives best on light, sandy soils and reaches maturity in 4 – 5 months. Peanuts for further processing are mostly shipped after shelling, whereas peanuts for direct consumption are normally exported in the pods. Peanuts are often designated according to their shipping ports, e.g., Bombay, Casamance, Rufisque, Bissao nuts.

**Properties.** The iodine value of peanut oil is 84 – 105; Rufisque oil has an iodine value of 84 – 90 and Argentinian oil a value of 100 – 105 in line with the higher content of linoleic acid. The saponification value is 185 – 196, $n_D^{40}$ 1.461 – 1.465. Tocopherols and other antioxidants as well as hydrocarbons and sterols are found in the nonsaponifiable matter (0.5 – 1.0 %). The oil solidifies below 0 °C. For fatty acid composition see Tables 1 and 2. The oil contains 36 – 46 % triunsaturated, 47 – 52 % diunsaturated, and 7 – 11 % monounsaturated glycerides.

**Production.** The peanut pods contain up to four hazelnut-sized kernels which are covered with a thin, reddish-brown skin and contain 5–12% water, 45–50% oil, and 23–35% protein. The pods are first removed by passage through cracking rolls or disk mills and over screens. This step also loosens the skins, which are removed by aspiration. The kernels are then screened, cracked, and expelled in screw presses. The residual oil in the expelled cake is extracted with hexane. Crude peanut oil contains low concentrations of phosphatides and up to 1.5% free fatty acids. Lye neutralization, earth bleaching, and deodorization yield a light, slightly yellowish oil with a very good oxidative and flavor stability. Peanut kernels to be processed into oil and meal can contain up to 0.5 ppm of aflatoxins. The bulk of the aflatoxins remain in the meal where they can be deactivated by treatment with, for example, ammonia; traces of aflatoxin that might enter the oil are removed during neutralization with lye.

**Use.** Peanut oil is used as a superior salad and cooking oil, less so in margarines. In shortenings it is used as such or after hardening (*mp* 31–38 °C). On cooling to 6–8 °C, peanut oil becomes cloudy due to the crystallization of glycerides containing $C_{20}$–$C_{24}$ fatty acid groups. The glycerides can be removed by filtration but only with a low yield of clarified oil.

*Peanut butter* is made by grinding specially roasted peanuts and homogenizing the mash with addition of liquid and possibly also hydrogenated peanut oil.

### 13.2.5.3. Lupine Oil

Lupins are attracting attention worldwide as an arable crop for possible use in foods and feeds. *Lupinus albus, L. angustifolius, L. luteus,* and *L. mutabilis* are potential sources of edible oil. Whereas the first three species (sweet varieties) contain only 4–9% oil in the dried seed, the seed oil content of *L. mutabilis* approaches that of soybeans. The main fatty acid components are palmitic (7–13%), stearic (2–7%), oleic (27–53%), linoleic (18–47%), linolenic (2–9%); lower concentrations of $C_{20}$–$C_{22}$ fatty acids are found [90]. The meal of *L. mutabilis* (bitter lupins) cannot be used as an animal feed without prior extraction of the alkaloids.

### 13.2.6. Cruciferous Oils

The seed oils of the traditional cruciferous plants are mainly characterized by their content of glucosinolates (sulfur-containing glycosides) and erucic acid. Rapeseed oil is the most important cruciferous oil.

## 13.2.6.1. Rapeseed Oil [91]

Rapeseed oil [8002-13-9] is the seed oil of different varieties of *Brassica napus* and *B. campestris*. The air-dry seed contains about 40% oil and 7–9% water. The production of rapeseed, one of the oldest oil crops known, has increased dramatically over the past 10 years. The introduction of so-called single-zero rapeseed varieties, i.e., varieties with an erucic acid content of less than 5% of the total fatty acids, has made rapeseed oil universally applicable as an edible oil. Generally, it can be used to replace soybean oil. Within the European Community only single-zero varieties are eligible for subsidy, and the United States has recently permitted the use of rapeseed oil with max. 2% erucic acid. Whereas single-zero, spring-sown rapeseed varieties are grown in Canada, the EEC countries (with the exception of Denmark) generally prefer winterhardy single-zero varieties because of the higher crop yields. Within the European Community the traditional rapeseed varieties with up to 50% erucic acid in the oil are now used only for technical and specialty purposes.

The increase in planted acreage has led to greater availability and to processing throughout the year; a significant improvement of the quality of the meal and the oil has been achieved [92]. Canada, the European Community, Poland, Sweden, China, India, and Pakistan are the main producing countries. To facilitate the marketability of meal, double-low varieties, i.e., low in erucic acid and glucosinolates, are grown in Canada (spring-type) and are being introduced in the EEC (winter-type). The winter-sown varieties are larger in size and generally easier to process than the smaller spring-sown varieties.

**Properties** [92]. For fatty acid compositions see Tables 1 and 2. Analytical data such as iodine value and saponification value are functions of the erucic acid content. The high level of erucic acid in traditional rapeseed oil has been associated with irreversible organ changes in animal tests [92].

**Production.** Modern rapeseed processing entails flaking and cooking the seed prior to preexpelling in continuous screw presses; the press cake is extracted with hexane. Crude rapeseed oil contains up to 2% free fatty acids and ca. 2.5% phosphatides; it is degummed with water or acids, lye-neutralized, bleached with earth, and deodorized, analogously to soybean oil. The stability is also comparable. The phosphatide content of the crude oil can be reduced to 0.1% (i.e., 20 ppm phosphorus) by special degumming techniques [44], which enable physical refining to be carried out.

The enzymatic degradation products of sulfur-containing glycosides (glucosinolates) in rapeseed can influence the taste of the crude oil and impair the nutritional properties of the meal [92]. The enzyme responsible for glucosinolate degradation, myrosinase, is destroyed by adequate cooking of the seed prior to expelling. Sulfur compounds are removed from the oil during refining; residual traces can affect the course of hardening since they act as catalyst poisons. The residual sulfur compounds in the extracted meal can be deactivated by treatment with alkali and heating.

**Use.** Traditional rapeseed oil containing up to 50% erucic acid can be hardened into steep-melting products, *mp* 32–34 °C, suitable as partial coconut oil substitutes in shortenings and couvertures. Modern single-zero varieties are used as such, after mild hardening (which eliminates the linolenic acid), and after hardening under isomerizing conditions to raise the melting point to 30–43 °C. The reduced complexity in fatty acid composition of the hardened oil leads to a different crystallization behavior, which can be overcome by interesterification. Treatment of rapeseed oil with sulfur or sulfur compounds is the basis of mastic production.

### 13.2.6.2. Mustard Seed Oil

Mustard seed oil [*8007-40-7*] is derived from the seeds of black mustard (*Brassica nigra* or *Sinapis nigra*), brown mustard (*Brassica juncea*), and white mustard (*Brassica alba* or *Sinapis alba*). The seeds contain 30–35% oil and high levels of glucosinolates. The erucic acid content of the oil is 40–50%; it is therefore of interest primarily for technical applications.

## 13.2.7. Conjugated Acid Oils

### 13.2.7.1. Tung Oil and Related Oils

Tung oil [*8001-20-5*] is obtained from the kernels (oil content about 50%) of the fruit of the tung tree, *Aleuritis fordii* (Chinese tung oil) or *Aleuritis cordata*, syn. *vernica* and *verrucosa* (Japanese tung oil). The oil from *Aleuritis montana* is almost identical in composition. Oils similar to Chinese tung oil are obtained from the nuts of *Aleuritis trisperma* (kekuna oil) and *A. moluccana* or *A. triloba* (lumbang oil). Tung trees are cultivated in Southern China, Southern Russia, the Southern United States, and Argentina. Tung oil contains about 80% α-eleostearic, 4% linoleic, 3% linolenic, 8% oleic, 1% stearic, and 4% palmitic acid.

### 13.2.7.2. Oiticica Oil and Related Oils

Oiticica oil [*8016-35-1*] is obtained mainly from the nuts of the Brazilian oiticica tree, *Licania rigida*. This evergreen tree reaches a height of 20 m and produces 150–900 kg of nuts per year. The kernels contain 55–63% oil, which is generally obtained by pressing. Fresh oiticica oil is yellowish and of lardlike appearance. Oiticica oil for export is heated for 30 min at 210–220 °C, after which it remains liquid. Licanic acid, 4-oxo-9,11,13-octadecatrienoic acid, is the principal fatty acid. The oil yields varnishes similar to those made from tung oil.

Boleko or isano oil (ongoke oil) is obtained from the nuts of *Ongokea gore* ENGLER, a tree growing in the Congo regions. It contains isanic and isanolic acids, which are fatty acids with conjugated triple bonds.

Parinarium oils are contained in the kernels of various tropical trees of the family Rosaceae and have little commercial significance. *Parinarium laurinum*, a tree growing in Japan and Oceania, yields a seed oil which contains parinaric acid, an acid with 4 conjugated double bonds.

Néou oil, obtained from *P. macrophyllum*, contains about 30% eleostearic acid. Eleostearic acid, in addition to licanic acid, has been found in po-yoak oil (obtained from *P. sherbroense*).

## 13.2.8. Substituted Fatty Acid Oils

### 13.2.8.1. Castor Oil

The seeds of the castor tree, *Ricinus communis*, contain 45–50% castor oil [8001-79-4]. The evergreen castor tree belongs to the family Euphorbiaceae and grows in tropical and subtropical countries. It can reach a height of 10 m.

**Properties.** The fatty acids of castor oil consist of 87–91% ricinoleic acid, 2% stearic and palmitic acid, 4–5% oleic acid, 4–5% linoleic acid, and 1% dihydroxystearic acid. Pure castor oil can be readily identified by its hydroxyl value, viscosity, and specific gravity. The oil is soluble in ethanol at room temperature and in boiling hexane.

| | |
|---|---|
| $d_{25}^{25}$ | 0.945–0.965 |
| $n_D^{40}$ | 1.466–1.473 |
| Solidification point | −12 to −18 °C |
| Viscosity at 20 °C | 935–1033 mPa s |
| $[\alpha]_D$ | +7.5 to +9.7 ° |
| Saponification value | 177–187 |
| Iodine value (Wijs) | 82–90 |
| Hydroxyl value | 161–169 |
| Acetyl value | 144–150 |
| Nonsaponifiable matter | 0.2–0.3% |

**Production.** Cold pressing of the spotted, oblong seeds yields a light, viscous oil which is used primarily in pharmaceutical products. Subsequent expelling at elevated temperatures and extraction with hexane yields a yellowish-brown oil which is used primarily in technical applications. The extracted meal contains ricin, a toxic protein, and hence cannot be used as an animal feed.

**Use.** A drying oil can be made by dehydrating castor oil (dehydrated castor oil. Pyrolytic cleavage at 300 °C gives undecylenic acid and heptaldehyde, raw materials for polymers and perfumes. Alkali fusion of castor oil yields 2-octanol and sebacic acid.

Hydrogenation yields a fat with a melting point of 84–86 °C and waxlike properties, finding application as a special lubricant. Air-blown castor oil is used as plasticizer for

**Table 27.** Properties of some "leprosy oils"

|  | Chaulmoogra oil (*Taractogenos kurzii*) | Hydnocarpus oil (*Hydnocarpus whigtiana*) | Gorli oil (*Oncoba echinate*) |
|---|---|---|---|
| Palmitic acid, % | ca. 4 | 1–2 | 7–8 |
| Oleic acid, % | 13–15 | 5–7 | 2–3 |
| Lower homologues of hydnocarpic acid, % | ca. 0.5 | 3–4 |  |
| Hydnocarpic acid, % | ca. 35 | ca. 50 |  |
| Chaulmoogric acid, % | ca. 23 | 27 | ca. 75 |
| Gorlic acid, % | ca. 23 | 12–13 | ca. 14 |
| $n_D^{40}$ | 1.471–1.473 | 1.472–1.473 |  |
| $[\alpha]_D^{25}$, ° | +49.8 | +55 | +51.7 |
| *mp*, °C | 22–30 | 22–24 | 42–44 |
| Saponification value | 198–208 | 200–208 | 190–194 |
| Iodine value | 98–105 | 93–101 | 94–100 |
| Nonsaponifiable matter, % | <0.5 | <0.5 | 1–1.5 |

varnishes and polymers. The oil is also used in the production of transparent soap, textile processing aids, special lubricants, and toiletries.

### 13.2.8.2. Chaulmoogra, Hydnocarpus, and Gorli Oils

The seed kernels of *Taractogenos kurzii*, a tree native to Southeast Asia, contain 48–55% chaulmoogra oil, the chief fatty acid components of which are chaulmoogric acid, hydnocarpic acid, gorlic acid, and lower homologues of hydnocarpic, palmitic, and oleic acid. Chaulmoogra oil and related oils such as hydnocarpus oils, obtained from Southeast Asian *Hydnocarpus* varieties, and the African gorli oil have been used in the treatment of leprosy; for properties see Table 27.

### 13.2.8.3. Vernonia Oil [93]

The seeds of *Vernonia anthelmintica* (of the Compositae family) contain an oil with a high percentage of vernolic acid (*cis*-12,13-epoxy-*cis*-9-octadecenoic acid). This epoxy acid is used in protective coatings, plastics, and other industrial products. Efforts are being made to grow *Vernonia* varieties as new industrial crops in the United States and Asia.

# 14. Individual Animal Fats

Animal fats are obtained from the milk or the fatty tissue of certain groups of animals. The content of depot fat varies from 0 to 60%. The concentration and composition of the fat depend on type of animal, age, sex, and diet.

## 14.1. Land-Animal Fats [94]

### 14.1.1. Lard

Lard [*61789-99-9*] is the fat rendered from fresh, clean, sound fatty tissues from pigs in good health at the time of slaughter. The tissues do not include bones, ears and tails, internal organs, windpipes, or large blood vessels. Rendered pork fat is obtained from the tissues and bones of pigs in good health at the time of slaughter; it may contain fat from bones, skin, ears and tails, and other tissues fit for human consumption. A range of different quality grades is available.

**Properties.** European lard (iodine value ca. 60) is often harder than America lard (iodine value ca. 70).

| | |
|---|---|
| *mp* | 30–40 °C |
| Solidification point | 22–32 °C |
| Saponification value | 193–202 |
| Nonsaponifiable matter | 0.1–1.0% |

Lard consists of 4–8% trisaturated, 32–40% disaturated–monounsaturated, 45–50% monosaturated–diunsaturated, and 3–10% triunsaturated triglycerides. The main fatty acid components are shown in Tables 1 and 2. Lard contains low concentrations of arachidonic acid (0.4–0.9%), trans fatty acids, as well as traces of branched and odd-numbered fatty acids such as saturated $C_{15}$, $C_{17}$, and $C_{19}$ fatty acids and monounsaturated $C_{17}$ and $C_{19}$ fatty acids. Triunsaturated $C_{18}$, $C_{20}$, and $C_{22}$ fatty acids are also present. These typical fatty acids, together with the cholesterol that occurs in relatively high concentrations in animal fats, are the basis for detecting lard and other animal fats in vegetable oils and fats.

**Production.** Lard is obtained by dry or wet rendering. The best quality lard (acid value <0.8; moisture <0.1%) is made from selected tissue which has been washed and cooled immediately after slaughter. After comminution, the fatty tissue is wet-rendered by steam heating at 50–60 °C (for continuous wet-rendering plants see Section 5.2). Prime steam lard and leaf lard are obtained similarly from nonselected or partially selected tissue. Packer's lard retains a typical flavor; refiner's lard has to be refined. Inferior varieties, made in part from inedible offal, include so-called white, yellow, and brown greases, which are used primarily in technical applications.

Lard can be fractionated into a higher- and lower-melting fraction. In order to prevent the use of inferior raw materials, some European countries do not permit the refining of lard. The addition of antioxidants significantly improves the organoleptic and oxidative stability. Lard can be refined either by neutralization with lye, followed by bleaching with ca. 1% earth and deodorization at 180–240 °C, or by simple pretreatment with phosphoric acid–bleaching earth and subsequent distillative

**Table 28.** Properties and analytical data of beef tallow and derived products

| | Beef tallow | Stearin | Olein | Bone grease | Neatsfoot oil |
|---|---|---|---|---|---|
| Density, g/cm³ | 0.898–0.908 (40 °C) | | 0.914–0.924 (15 °C) | | |
| $n_D^{60}$ | 1.451–1.454 | ca. 1.449 | | 1.451–1.452 | 1.461–1.463 |
| mp, °C | 40–50 | 50–55 | 23–35 | 44–45 | |
| Saponification value | 193–200 | 190–198 | 193–198 | 190–200 | 188–198 |
| Iodine value | 32–47 | 14–25 | 40–53 | 49–53 | 67–80 |
| Nonsaponifiable matter, % | 0.3–0.8 | 0.3–0.8 | 0.1–0.5 | 0.5–0.6 | 0.1–0.6 |

stripping. Interesterification improves the performance of lard as a shortening in terms of creaming and cake-making [95].

**Use.** Lard is used extensively as an edible fat (shortening, margarines) and as an energy source in animal feeds.

## 14.1.2. Beef Tallow

**Properties.** The main fatty acids of beef tallow [61789-97-7] are shown in Tables 1 and 2. In addition, it contains up to 4% of characteristic branched and odd-numbered fatty acids. These, together with cholesterol, identify tallow in vegetable oils and fats. Tallow also contains 6–10% of trans fatty acids, resulting from bacterial hydrogenation in the rumen. Top-grade tallow is white to greyish-white; in summer it is slightly yellow because of carotenes taken up with fresh roughage. Table 28 shows analytical data for various beef tallow products. Beef tallow contains 14–26% trisaturated, 22–34% disaturated–monounsaturated, and 40–64% monosaturated–diunsaturated glycerides.

The stability of beef tallow and its products is relatively poor in the absence of an added antioxidant. Primary quality criteria are odor and taste, concentration of free fatty acids, and absence of impurities and of refined low-quality tallow (white grease).

**Production.** The choice of fatty tissue determines the quality of beef tallow. Fat rendered from the carcasses of cattle is generally not quite as hard as that obtained from sheep and goats. The softest, most unsaturated fat is found under the skin, and the firmer fat is located near the middle of the animal. The consistency of individual lots of tallow is therefore influenced by the procedure used in trimming the carcass prior to fat rendering.

Edible tallow (dripping, max. 1% ffa) is obtained from clean, sound, fatty tissues of beef cattle in good health at the time of slaughter. Premier jus (max. 0.5% ffa) is obtained by rendering the fresh fat (killing fat) of heart, caul, kidney, and mesentery, excluding cutting fats at low temperature (50–60 °C). Tallow to be incorporated into margarine blends must be refined. Fractionation at 26 °C yields a lower-melting fraction (olein) and a higher-melting fraction (stearin). Tallow olein is an important

**Table 29.** The Society of British Soap Makers specifications for tallows and greases (based on testing methods specified in B.S.3919:1976)

| Grade | ffa *, % | Bleached oil colour (red) | Moisture and dirt, % | Nonsaponifiable matter, % | Titer, °C | Iodine value |
|---|---|---|---|---|---|---|
| Tallow 1 | max. 3.0 | max. 0.5, 5 1/4 inch cell | max. 0.5 | max. 0.5 | min. 40.0 | max. 57 |
| Tallow 2 | max. 5.0 | max. 1.0, 5 1/4 inch cell | max. 1.0 | max. 1.0 | min. 40.0 | max. 57 |
| Tallow 3 | max. 8.0 | max. 3.0, 5 1/4 inch cell | max. 1.0 | max. 1.0 | min. 40.0 | max. 57 |
| Tallow 4 | max. 12.0 | max. 4.0, 1 inch cell | max. 1.0 | max. 1.5 | min. 40.0 | max. 60 |
| Tallow 5 | max. 15.0 | max. 12.0, 1 inch cell | max. 1.0 | max. 1.5 | min. 40.0 | max. 60 |
| Tallow 6 | max. 20.0 | no limit | max. 1.0 | max. 2.0 | min. 40.0 | max. 60 |
| Grease | max. 20.0 | – | max. 2.0 | max. 2.0 | 36.0–40.0 | max. 63 |

* ffa = free fatty acids.

component of some margarines and shortenings. Stearin is used in shortenings and in specialty margarines. Bone grease is obtained by expelling and extracting of comminuted bones; it is used primarily in the technical sector and for animal feeds. Neatsfoot oil [8037-20-5] is a low-melting, inedible fat rendered from the feet of cattle; it is used as a special lubricant and leather dressing aid.

**Use.** Only a small part of the annually produced beef tallow is used as edible fat. The concentration in margarine is limited because of its poor stability but it is used more widely in shortenings. The technical grades are important raw materials for soap and fatty acid derivatives and for use in animal feeds. Specifications for technical tallows used for soapmaking are shown in Table 29.

## 14.1.3. Mutton Tallow

Mutton tallow finds applications primarily in the technical sector. Superior grades are similar to corresponding beef tallows.

| | |
|---|---|
| Saponification value | 192–198 |
| Iodine value | 31–47 |
| Nonsaponifiable matter | 0.1–0.6% |
| Density (40 °C) | 0.896–0.898 g/cm$^3$ |
| $n_D^{40}$ | 1.455–1.458 |
| mp | 44–55 °C |

The main fatty acids are palmitic, stearic, and oleic acids (Tables 1 and 2). Up to 4% of branched and odd-numbered fatty acids and traces of trans fatty acids are also present.

## 14.1.4. Horse, Goose, and Chicken Fat

*Horse fat* is softer than lard or rendered pork fat. The fatty acid composition (Tables 1, 2) is strongly influenced by diet. Branched and odd-numbered fatty acids occur up to about 2%. The nonsaponifiable matter is <0.5%, the melting point 29–40 °C. Horse fat is occasionally used in meat and sausage products in some countries.

*Goose fat* has a relatively low melting point, 25–35 °C; its consistency is often improved by addition of lard. *Chicken fat* is also low-melting, *mp* 23–40 °C. The fatty acid composition (see Tables 1 and 2) is strongly influenced by diet.

## 14.2. Marine Oils [96]

Marine oils are the oils obtained from whales and fish. Fish oils have a high content of highly unsaturated (4–6 double bonds) $C_{20}$, $C_{22}$, and $C_{24}$ fatty acids ( cf. Tables 1 and 2). The oils from freshwater fish contain less $C_{20}$ and $C_{22}$ fatty acids and more oleic and linoleic acid than those from sea fish. These differences are caused primarily by differences in the composition of the feed. Only oils obtained from sea fish are commercially important.

### 14.2.1. Whale Oil

The whale population has been depleted to such an extent that whale oil is now completely overshadowed in commercial importance by the various fish oils. The Antarctic has long been the catching area. Whales are processed in modern factory ships (see also Section 5.3). The color of the oil can be very light, and the concentration of free fatty acids very low. Since 1913, when it was discovered that the oil can be converted into a stable edible fat by hydrogenation, many thousands of tons of hardened whale oil have been produced for margarine, cooking fats, and soapmaking.

| | |
|---|---|
| Saponification value | 185–205 |
| Iodine value | 110–135 |
| Nonsaponifiable matter | <2% |
| Slip point | 22 °C |
| $n_D^{65}$ | 1.4554–1.4579 |

Whale oil has been graded according to free fatty acid content and color (apart from moisture and dirt):

Grade 0: pale yellow, max. 0.5% ffa
Grade 1: pale yellow, max. 2.0% ffa
Grade 2: amber yellow, max. 6.0% ffa

**Table 30.** Analytical data for various fish oils

|  | Herring oil | Sardine oil | Pilchard oil | Menhaden oil | South American oil |
|---|---|---|---|---|---|
| $n_D^{40}$ | 1.470–1.475 | 1.473–1.475 | 1.473–1.476 | 1.473–1.474 | 1.474–1.478 |
| Saponification value | 183–192 | 187–190 | 188–194 | 188–194 | 189–194 |
| Iodine value | 120–160 | 165–185 | 180–190 | 150–180 | 185–206 |
| Nonsaponifiable matter, % | 0.8–1.3 | 1.0–1.6 | 0.8–1.5 | 0.8–1.5 | 0.8–1.5 |

Grade 3: pale brown, max. 15.0% ffa

Grade 0–2 whale oils are hydrogenated for edible purposes. Prior to hydrogenation the oil is lye-neutralized and bleached. Whale oil can be hydrogenated into soft and hard fats. Soft fats with a melting point between 33 and 38 °C are produced by hydrogenation under isomerizing conditions with a partially inactivated nickel catalyst. Fats with a melting point between 40 and 45 °C are produced with relatively fresh catalyst. The crude hydrogenated fats are filtered to remove catalyst, treated with weak lye, bleached, and deodorized. In some cases, posttreatment can be restricted to treatment with citric acid–bleaching earth followed by distillative stripping. Hydrogenated whale oil is stable for about 6 months. The softer fats are used primarily for shortenings, the higher melting ones for preserves, dry soups, and special margarines.

The main components of sperm oil [*8002-24-2*] are esters of fatty acids and fatty alcohols (waxes). Sperm oil is a valuable raw material for high-quality cosmetic preparations. It is being replaced by jojoba oil, a seed oil from a shrub (*Simmondsia chinensis*) indigenous to the Sonoran desert and being cultivated in arid zones around the world. Its seeds contain 50–60% of a liquid wax composed of esters of fatty acids and fatty alcohols [97].

## 14.2.2. Fish Oil [98]

The bulk of fish oils is produced in modern plants by cooking comminuted fish or fish offal, expelling in screw presses, and centrifugation; or by extraction of dried fish meal with hexane (see Section 5.3). Refining and hydrogenation yields fats of edible quality which are used in margarine and shortening blends. For historical reasons fish oils are not permitted as edible oils in the United States. The main crude oil origins are Japanese (iodine value ca. 180), U.S. (menhaden: iodine value 150–160), South American (iodine value ca. 200), and Scandinavian (iodine value 130–140).

The oil content of fish is subject to seasonal variations and generally lies between 8 and 20%. Typical properties of various fish oils are shown in Table 30 (for fatty acid composition see Tables 1 and 2). All fish oils contain small concentrations of branched and odd-numbered fatty acids.

**Use.** Fish oils are processed into hydrogenated fats analogously to whale oil [100]. However, due to their content of polyunsaturated fatty acids and sulfur compounds, hydrogenation requires more hydrogen and more catalyst. The melting points of hardened fish oils range from 31 to 45 °C. Organoleptic stability generally improves with decreasing iodine value of the starting oil and with increasing melting point of the hardened product. Hardened fish oils are used in shortenings and margarine blends. They have good creaming and cake-making properties but do not perform as well in frying.

Much of the oil of cod, halibut, and shark is stored in the liver. Cod-liver oil and halibut-liver oil are valued for pharmaceutical purposes because of their high content of vitamin A (1500–50 000 i.u./g) and vitamin D (40–200 i.u./g). Typical properties of cod-liver oil are as follows:

| | |
|---|---|
| Saturated fatty acids, % | |
| $C_{14}$ | 6 |
| $C_{16}$ | 6 |
| $C_{18}$ | 8 |
| Unsaturated fatty acids, % | |
| $C_{14}$ | traces |
| $C_{16}$ | 20 |
| $C_{18}$ | 29 |
| $C_{20}$ | 26 |
| $C_{22}$ | 10 |
| Saponification value | 180–197 |
| Iodine value | 150–175 |
| Nonsaponifiable matter, % | ca. 1 |
| $n_D^{25}$ | 1.481 |

# 15. Economic Aspects

Figures for the world production and exports of important fats and oils are shown in Tables 31, 32, 33, 34.

# 16. Toxicology and Occupational Health

With the exception of technical oils and fats such as castor and tung oil, oils and fats are classed as foods or food additives and do not entail toxicological or occupational health hazards.

**Table 31.** World production of oil fruits, oilseeds, oils, and fats according to countries (annual average, expressed as oil/fat in 1000 t)

| | 1934/38 | 1960 | 1970/71 | 1973/74 | 1980/81 | 1981/82 | 1982/83 | 1983/84 | 1984/85[a] |
|---|---|---|---|---|---|---|---|---|---|
| **Europe** | | | | | | | | | |
| EEC | 3328[b] | 2300 | 4790[c] | 4220 | 5615 | 5550 | 6065 | 6215 | 6080 |
| EFTA | | 1135 | | 710 | 855 | 850 | 930 | 910 | 930 |
| Other West European countries | | 990 | 805 | 1015 | 835 | 650 | 1180 | 750 | 1285[d] |
| Soviet Union | 2000 | 3090 | 5335 | 6030 | no data | 4870 | 5340 | 5315 | 5105 |
| Other East European countries | 1231 | 1595 | 1975 | 2430 | 2920 | 2950 | 2975 | 3040 | 3290 |
| **Africa** | | | | | | | | | |
| West Africa | 1132 | 2010 | 1845 | 1685 | 1380 | 3015[e] | 2970 | 2800 | 2790 |
| Other African countries | 379 | 985 | 1420 | 1540 | | | | | |
| **America** | | | | | | | | | |
| Argentina/Brazil | 995 | 1525 | 2270 | 3080 | 5155 | 5100 | 5465 | 5935 | 6750 |
| USA | 3543 | 7470 | 11185 | 11945 | 14415 | 15620 | 16515 | 13535 | 14940 |
| Other American countries | 565 | 1480 | 3165 | 2605 | 3345 | 3210 | 3180 | 3340 | 3675 |
| **Asia** | | | | | | | | | |
| Sri Lanka/Burma | 216 | 240 | 320 | 215 | 265 | 290 | 330 | 365 | 375 |
| India/Pakistan | 2274 | 2725 | 3615 | 3620 | 3890 | 4750 | 4290 | 4875 | 4825 |
| Indonesia/Malaysia | 931 | 780 | 1425 | 1815 | 4510 | 5380 | 5355 | 5530 | 6800 |
| Philippines | 452 | 840 | 965 | 800 | 1585 | 1500 | 1345 | 1090 | 1140 |
| Chinese Republic | 3341 | 2905 | 3100 | 3070 | 4250 | 5685 | 6325 | 6160 | 6650 |
| Other Asian countries | 628 | 960 | 1460 | 1645 | 1105 | 1045 | 1045 | 1090 | 1120 |
| Australia/Oceania | 575 | 755 | 955 | 890 | 1100 | 1140 | 1115 | 1240 | 1270 |
| World production | 21590 | 31785 | 44630 | 46320 | 58880 | 62505 | 65485 | 63145 | 68155 |

[a] Preliminary data.
[b] $\sum$ EEC, EFTA, and other West European countries.
[c] $\sum$ EEC, EFTA.
[d] Including Spain.
[e] $\sum$ West Africa and other African countries.

**Table 32.** World production of oil fruits, oilseeds, oils, and fats (annual average, expressed as oil/fat in 1000 t)

| | 1934/1938 | 1960 | 1970/71 | 1973/74 | 1980/81 | 1981/82 | 1982/83 | 1983/84 | 1984/85[a] |
|---|---|---|---|---|---|---|---|---|---|
| Vegetable oils | | | | | | | | | |
| Cottonseed | 1453 | 2315 | 2450 | 2825 | 3035 | 3280 | 3025 | 3040 | 3770 |
| Peanut[b] | 1755 | 2760 | 3230 | 3965 | 2605 | 3455 | 2900 | 3190 | 3195 |
| Corn[c] | 95 | 185 | 325 | 390 | 665 | 695 | 745 | 785 | 805 |
| Olive | 950 | 1285 | 1605 | 1550 | 1980 | 1495 | 2025 | 1575 | 1785 |
| Rapeseed | 1273 | 1105 | 2220 | 2170 | 3675 | 4040 | 4905 | 4700 | 5500 |
| Safflower | [d] | [d] | 170 | 220 | 265 | 260 | 245 | 260 | 265 |
| Sesame | 563 | 485 | 755 | 720 | 615 | 730 | 620 | 700 | 695 |
| Soybean | 1263 | 3780 | 7700 | 9670 | 13275 | 13885 | 15255 | 13540 | 14675 |
| Sunflower | 435 | 1170 | 3335 | 4250 | 4695 | 5250 | 6000 | 5530 | 6165 |
| Babassu | 27 | 50 | 105 | 125 | | | | | |
| Coconut | 1633 | 2060 | 2375 | 2015 | 3150 | 2990 | 2860 | 2435 | 2640 |
| Palm kernel | 355 | 450 | 490 | 430 | 585 | 700 | 730 | 760 | 870 |
| Palm | 644 | 1100 | 1745 | 2260 | 4550 | 5410 | 5410 | 5645 | 6720 |
| Linseed | 1040 | 1025 | 1255 | 795 | 660 | 670 | 815 | 670 | 700 |
| Castor | 178 | 230 | 340 | 465 | 325 | 350 | 350 | 375 | 420 |
| Tung | 121 | 115 | 120 | 95 | 105 | 95 | 100 | 95 | 95 |
| Others[e] | 220 | 450 | 65 | 75 | 25 | 25 | 25 | 25 | 25 |
| Animal fats | | | | | | | | | |
| Butterfat[f] | 4160 | 4435 | 4950 | 5225 | 5450 | 5705 | 6200 | 6255 | 6135 |
| Lard | 2913 | 4370 | 4365 | 4180 | 4760 | 5050 | 5130 | 5185 | 5210 |
| Tallow | 1679 | 3455 | 5270 | 5320 | 6360 | 6200 | 6100 | 6110 | 6210 |
| Fish oil | 325 | 470 | 1170 | 970 | 1100[g] | 1300[g] | 1090[g] | 1265[g] | 1280[g] |
| Whale oil and sperm oil | 507 | 490 | 190 | 140 | | | | | |

[a] Preliminary data.
[b] Until 1960 including 10 000 – 15 000 t teaseed.
[c] Including technical olive oil.
[d] Included under "others".
[e] Including oiticica, mowrah, niger, hempseed, perilla, and other commercially important oils.
[f] Including ghee fat.
[g] $\sum$ Fish oil, whale oil, and sperm oil.

**Table 33.** World exports of oil fruits, oilseeds, oils, and fats according to countries (annual average, expressed as oil/fat in 1000 t)

|  | 1934/38 | 1965 | 1970 | 1974 | 1980 | 1981 | 1982 | 1983 |
|---|---|---|---|---|---|---|---|---|
| Europe | | | | | | | | |
| Western Europe and Iceland | | 738 | 1020 | 1335 | 1693 | 1820 | 1658 | 2022 |
| Eastern Europe and Soviet Union | | 315 | 783 | 740 | 532 | 458 | 548 | 466 |
| Africa | | | | | | | | |
| Former French Equatorial/West Africa | 298 | 409 | 395 | 341 | 312 | 194 | 313 | 325 |
| Nigeria | 455* | 737 | 392 | 182 | 107 | 75 | 55 | 65 |
| Former Portuguese Africa | 62 | 93 | 111 | 96 | | | | |
| Sudan | | 115 | 100 | 105 | 85 | 91 | 93 | 52 |
| Zaire | 97 | 115 | 173 | 104 | 29 | 25 | 19 | 15 |
| Other African countries | 212 | 259 | 315 | 335 | 211 | 273 | 209 | 137 |
| America | | | | | | | | |
| Argentina/Uruguay | 577 | 416 | 469 | 224 | 1426 | 1079 | 1243 | 1548 |
| Canada | | 320 | 556 | 520 | 1071 | 1215 | 1012 | 1024 |
| United States | 100 | 3356 | 4479 | 5319 | 8246 | 7976 | 8534 | 7603 |
| Other American countries | 133 | 377 | 579 | 926 | 234 | 213 | 334 | 213 |
| Asia | | | | | | | | |
| India/Sri Lanka | 589 | 142 | 105 | 123 | 75 | 138 | 166 | 163 |
| Indonesia | 529 | 260 | 317 | 304 | 610 | 201 | 323 | 449 |
| Malaysia | 132 | 164 | 435 | 883 | 2517 | 2827 | 3187 | 3346 |
| Philippines | 348 | 797 | 608 | 631 | 1002 | 1126 | 1080 | 1035 |
| China | 742 | 195 | 153 | 134 | 139 | 246 | 190 | 310 |
| Other Asian countries | | 122 | 137 | 263 | 455 | 508 | 593 | 649 |
| Australia/Oceania | 363 | 540 | 640 | 530 | 755 | 749 | 790 | 804 |
| Antarctic/Arctic | 507 | 350 | 209 | 138 | 18 | 14 | 14 | 11 |
| World exports | 5829 | 8301 | 9820 | 13233 | 20819 | 20957 | 21589 | 21618 |

* Including Ghana, Gambia, Sierra Leone.

Table 34. World exports of oil fruits, oilseeds, oils, and fats (annual average, expressed as oil/fat in 1000 t)

|  | 1934/38 | 1965 | 1970 | 1974 | 1980 | 1981 | 1982 | 1983 |
|---|---|---|---|---|---|---|---|---|
| Vegetable oils |  |  |  |  |  |  |  |  |
| Cottonseed | 189 | 395 | 307 | 369 | 495 | 473 | 522 | 320 |
| Peanut | 826 | 967 | 817 | 677 | 678 | 519 | 641 | 669 |
| Olive* | 136 | 113 | 237 | 262 | 219 | 205 | 162 | 278 |
| Rapeseed | 51 | 272 | 481 | 670 | 1149 | 1364 | 1224 | 1274 |
| Safflower |  | 70 | 50 |  |  |  |  |  |
| Sesame | 69 | 70 | 104 |  |  |  |  |  |
| Soybean | 432 | 1802 | 2962 | 3903 | 6665 | 6757 | 7162 | 6770 |
| Sunflower | 26 | 263 | 700 | 629 | 1431 | 1535 | 1558 | 1832 |
| Babassu | 12 | 12 | 14 |  |  |  |  |  |
| Coconut | 1057 | 1283 | 1087 | 923 | 1365 | 1455 | 1411 | 1376 |
| Palm kernel | 320 | 371 | 327 | 383 | 431 | 419 | 462 | 479 |
| Palm | 447 | 551 | 744 | 1327 | 2961 | 2818 | 3336 | 3567 |
| Linseed | 572 | 473 | 421 | 271 | 423 | 420 | 352 | 402 |
| Castor | 81 | 206 | 236 | 227 | 203 | 202 | 180 | 184 |
| Tung | 80 | 42 | 36 |  |  |  |  |  |
| Others | 68 | 121 | 185 | 358 | 680 | 714 | 755 | 759 |
| Animal fats |  |  |  |  |  |  |  |  |
| Butterfat | 500 | 499 | 641 | 704 | 775 | 740 | 671 | 608 |
| Lard | 173 | 290 | 347 | 372 | 432 | 439 | 389 | 366 |
| Tallow | 162 | 1225 | 1517 | 1575 | 2193 | 2181 | 2068 | 2025 |
| Fish oil | 121 | 445 | 554 | 500 | 701 | 702 | 682 | 698 |
| Whale oil and sperm oil | 507 | 350 | 209 | 138 | 18 | 14 | 14 | 11 |

* Including olive oil for industrial uses.

# 17. References

General References

[1] E. W. Eckey: *Vegetable Fats and Oils,* Reinhold Publ. Co., New York 1954.
[2] H. P. Kaufmann: *Analyse der Fette und Fettprodukte,* vols. **I** and **II,** Springer Verlag, Berlin–Göttingen–Heidelberg 1958.
[3] V. C. Mehlenbacher: *The Analysis of Fats and Oils,* The Garrard Press, Champaign, Ill., 1960.
[4] J. Devine, P. N. Williams: *The Chemistry and Technology of Edible Oils and Fats,* Pergamon Press, Oxford 1961.
[5] R. Lüde: *Die Raffination von Fetten und fetten ölen,* 2nd ed., Th. Steinkopff, Leipzig 1962.
[6] A. J. C. Anderson: *Refining of Oils and Fats for Edible Purposes,* 2nd ed., Pergamon Press, London 1962.
[7] W. Wachs: *öle und Fette,* Part II, Paul Parey, Berlin–Hamburg 1964.
[8] "Fette und Lipoide," *Handbuch der Lebensmittelchemie,* vol. **IV,** Springer Verlag, Berlin–Heidelberg–New York 1965.
[9] L. V. Cocks, C. V. van Rede: *Laboratory Handbook for Oil and Fat Analysis,* Academic Press, London–New York 1966.
[10] F. D. Gunstone: *An Introduction to the Chemistry and Biochemistry of Fatty Acids and their Glycerides,* Chapman and Hall, London 1967.
[11] H. A. Boekenoogen: *Analysis and Characterisation of Oils, Fats and Fat Products,* Interscience, New York 1968.

[12] A. J. Vergroesen (ed.): *The Role of Fats in Human Nutrition,* Academic Press, London 1975.
[13] H. Pardun: *Analysis of Edible Fats,* Paul Parey, Hamburg 1976.
[14] D. Swern (ed.): *Bailey's Industrial Oil and Fat Products,* 4th ed., Wiley & Sons, New York 1979.
[15] *Kirk-Othmer,* **9**, 795; **21**, 417; **23**, 717.
[16] F. B. Padley, J. Podmore (eds.): *The Role of Fat in Human Nutrition,* VCH Verlagsgesellschaft, Weinheim 1985.
[17] R. J. Hamilton, J. B. Rossell (eds.); *Analysis of Oils and Fats,* Elsevier Applied Science Publ., Barking,England 1986.

Specific References

[18] J. Baltes: "Gewinnung und Verarbeitung von Nahrungsfetten,"*Grundlagen und Fortschritte der Lebensmitteluntersuchung,*vol. 17, Paul Parey, Berlin, Hamburg 1975, p. 42.
[19] J. B. M. Rattray, *JAOCS J. Am. Oil Chem. Soc.* **61** (1984) 1701.
[20] F. D. Gunstone: *An Introduction to the Chemistry and Biochemistry of Fatty Acids and their Glycerides,* Chapman & Hall, London 1967.
[21] M. Szuhaj, G. List: *Lecithins,* AOCS monograph, 1985.
[22] A. J. Speek et al., *J. Food Sci.* **50** (1985) 121.
[23] M. L. Meara: *Leatherhead Food RA, Scientific and Technical Survey no. 110,* July 1978.
[24] K. van Putte, J. v. d. Enden, *Fette Seifen Anstrichm.* **76** (1974) 316.
[25] E. S. Lutton, *J. Am. Oil Chem. Soc.* **49** (1972) 1.
[26] B. G. Linsen, *Fette Seifen Anstrichm.* **73** (1971) 411, 753; **70** (1968) 8.
[27] J. W. E. Coenen in I. Morton, D. N. Rhodes (eds.): *The Contribution of Chemistry to Food Supplies,* Butterworths, London 1974.
[28] H. Wissebach, *Tenside* **3** (1966) 285.
[29] A. K. Sen Gupta, H. Scharmann, *Fette Seifen Anstrichm.***71** (1969) 873.
[30] H. W. G Heynen et al., *Fette Seifen Anstrichm.* **74** (1972) 677.
[31] E. W. Trautschold: *Die Lagerung von Sojabohnen unter Qualitätsaspekten,* DGF-paper, München 1970.
[32] H. P. Kaufmann et al.: *Neuzeitliche Technologie der Fette und Fettprodukte,* Aschendorffsche Verlagsbuchhandlung, Münster 1956–1965.
[33] I. E. Liener: *Toxic Constituents of Plant Foodstuffs,* Academic Press, New York–London 1969.
[34] "Oilseed Processing Symposium 1976," *J. Am. Oil Chem. Soc.* **54** (1977) 473A.
[35] Deutsche Gesellschaft für Fettwissenschaft: *Gewinnung von Fetten und ölen aus pflanzlichen Rohstoffen durch Extraktion,* Industrieverlag von Hernhaussen KG, Hamburg 1978
[36] S. Skipin, *Chem. Abstr.* **29** (1935) 7682. A. Carter et al., *J. Am. Oil Chem. Soc.* 51 (1974) 137.
[37] P. König, *Fette Seifen Anstrichm.* **64** (1962) 23.
[38] J. Furman et al., *J. Am. Oil Chem. Soc.* **36** (1959) 454.
[39] G. Penk, *Proc. A.S.A. Symp. Soybean Process. 2nd* 1981.
[40] W. H. Prokop, *JAOCS J. Am. Oil Chem. Soc.* **62** (1985) 805.
[41] S. M. Barlow, M. L. Windsor: *Techn. Bulletin no. 19,* Int. Ass. of Fish Meal Manufacturers, UK, Sept. 1984.
[42] Th. Wieske, H.-U. Menz, *Fette Seifen Anstrichm.* **74** (1972) 133.V. K. Babayan, *J. Am. Oil Chem. Soc.* **51** (1974) 260.
[43] O. L. Brekke in: *Handbook of Soy Oil Processing and Utilisation,* Amer. Soybean Ass., 1980.
[44] H. J. Ringers, J. C. Segers, US 4 049 686, 1977.
[45] A. K. Sen Gupta, US 4 062 882, 1977.
[46] F. V. K. Young, *Proc. A.S.A. Symp. Soybean Process. 2nd* 1981.

[47] A. N. Sagredos, K. Remse, *Fette Seifen Anstrichm.* **85** (1983) 185.P. Röttgermann, *Fette Seifen Anstrichm.* **85** (1983) 190.
[48] Y. Hoffmann, *J. Am. Oil Chem. Soc.* **50** (1973) 260.
[49] H. Pardun, *Fette Seifen Anstrichm.* **81** (1979) 297.
[50] D. C. Tandy, W. J. Macpherson, *JAOCS J. Am. Oil Chem. Soc.* **61** (1984) 1253. A. Forster, A. J. Harper, *J. Food Sci.* **49** (1984) 23.
[51] H. P. Kaufmann, D. Schmidt, *Fette Seifen* **47** (1940) 294.
[52] M. Kock, *Proc. A.S.A. Symp. Soybean Process. 2nd* 1981.
[53] G. Biernoth, H. E. Rost, *Chem. Ind.* 1967,2002.G. Biernoth, *Fette Seifen Anstrichm.* **70** (1968) 217.A. N. Sagredos, D. Sinha-Roy, *Dtsch. Lebensm. Rundsch.* **75** (1979) 350.
[54] D. B. Erskine, W. G. Schuliger, *Chem. Eng. Prog.* **67** (1971) 41.
[55] A. Thomas, *Fette Seifen Anstrichm.* **84** (1982) 133.
[56] J. B. Rossell et al., *Proc. A.S.A. Symp. Soybean Process. 2nd* 1981.
[57] H. Stage: *Proc. A.S.A. Symp. Soybean Process. 2nd* 1981.
[58] A. E. Rheineck, R. T. Holman et al.: *Progress in the Chemistry of Fats and Other Lipids*, vol. **5**, Pergamon Press, New York1958.
[59] H. J. Passino, *Ind. Eng. Chem.* **41** (1948) 280.
[60] S. W. Gloyer, *J. Am. Oil Chem. Soc.* **26** (1949) 162.
[61] Deutsche Gesellschaft für Fettwissenschaft: *Die Hydrierung von Fetten*, Industrieverlag von Herrnhaussen KG, Hamburg 1979.
[62] H. B. W. Patterson: *Hydrogenation of Fats and Oils*, Elsevier Applied Science Publ., Barking, England, 1983.
[63] F. A. Kummerow, *J. Am. Oil Chem. Soc.* **51** (1974) 255.
[64] E. le Breton, P. le Marchal: *Riv. Ital. Sostanze Grasse* **47** (1970) 231. F. Camurati et al., *Riv. Ital. Sostanze Grasse* **47** (1970) 241.
[65] P. v. d. Plank, *Fette Seifen Anstrichm.* **76** (1974) 337.
[66] R. C. Hastert, *JAOCS J. Am. Oil Chem. Soc.* **58** (1981) 169.
[67] B. Screenivasan, *J. Am. Oil Chem. Soc.* **55** (1978) 803.
[68] G. Bradshaw, *Soap Sanit. Chem.* **18** (1941) no. 5, 23, 69.
[69] A. J. Haighton, L. Vermaas, *Fette Seifen Anstrichm.* **74** (1972) 615.
[70] J. R. Rossell: "Measurement of Rancidity in Oils and Fats," *Leatherhead Food RA, Scientific and Technical Survey no. 140*, Sept. 1983.
[71] M. Arens, E. Kroll, *Fette Seifen Anstrichm.* **85** (1983) 307.
[72] G. Guhr et al., *Fette Seifen Anstrichm.* **83** (1981) 373.
[73] H. Karstens, *Fette Seifen Anstrichm.* **70** (1968) 400.
[74] M. Unbehend et al., *Fette Seifen Anstrichm.* **75** (1973) 689.
[75] K. G. Berger: *Recommended Practices for Storage and Transport of Edible Oils and Fats*, Palm Oil Research Institute of Malaysia, 1986.
[76] "Proceedings of the World Conference, Kuala Lumpur, 1984, on Palm, Palm Kernel and Coconut Oils," *JAOCS J. Am. Oil Chem. Soc.* **62** (1985).
[77] Y. Lozano et al., *Rev. Fr. Corps Gras* **32** (1985) 377.
[78] F. Hirsinger, *JAOCS J. Am. Oil Chem. Soc.* **62** (1985) 76. S. A. Graham, R. Kleiman, *JAOCS J. Am. Oil Chem. Soc.* **62** (1985) 81.
[79] D. Gegion, K. Staphylakis, *JAOCS J. Am. Oil Chem. Soc.* **62** (1985) 1047.
[80] R. Banerji et al., *Fette Seifen Anstrichm.* **86** (1984) 279.
[81] A. E. Bailey: *Cottonseed and Cottonseed Products*, Interscience, New York 1948.
[82] B. S. Kamel et al., *JAOCS J. Am. Oil Chem. Soc.* **62** (1985) 881.

[83] W. Schuster, *Fette Seifen Anstrichm.* **74** (1972) 150.
[84] R. Purdy, *JAOCS J. Am. Oil Chem. Soc.* **62** (1985) 523.
[85] R. Marquard et al., *Fette Seifen Anstrichm.* **80** (1978) 213.
[86] W. Schuster, R. Marquard, *Fette Seifen Anstrichm.* **76** (1974) 207.
[87] J. D. v. Mikusch, *Farbe + Lack* **7** (1952) 303.
[88] T. Yaziciovglu et al., *Fette Seifen Anstrichm.* **79** (1977) 115.
[89] G. W. Rohne, *Fette Seifen Anstrichm.* **86** (1984) 172.
[90] B. J. F. Hudson et al., *J. Plant Foods* **5** (1983) 15.
[91] L.-A. Appelquist, R. Ohlson: *Rapeseed,* Elsevier, Amsterdam 1972.
[92] A. Thomas, *JAOCS J. Am. Oil Chem. Soc.* **59** (1982) 1.
[93] M. Y. Raie et al., *Fette Seifen Anstrichm.* **87** (1985) 325.
[94] O. Dahl: *Schlachtfette,* Fleischforschung und Praxis no. 10, Verlag der Rheinhessischen Druckwerkstätte, Alzey 1973.
[95] E. S. Lutton et al., *J. Am. Oil Chem. Soc.* **39** (1962) 233.
[96] "Fish Oils and Animal Fats," *Proc. Leatherhead Food RA,* Feb. 1986.
[97] T. K. Miwa, *JAOCS J. Am. Oil Chem. Soc.* **62** (1985) 377.
[98] M. E. Stansby: *Fish Oils, Their Chemistry, Technology, Stability, Nutritional Properties and Uses,* The Avi Publishing Co., Westport, Conn., 1967.
[99] F. V. K. Young: *The Chemical and Physical Properties of Crude Fish Oils,* Int. Ass. Fish Meal Manuf., Fish Oil Bulletin no. 18, June 1986.
[100] F. V. K. Young: *The Refining and Hydrogenation of Fish Oils,* Int. Ass. Fish Meal Manuf., Fish Oil Bulletin no. 17, June 1986.

# Fatty Acids

Rolf Brockmann, Henkel KGaA, Düsseldorf, Federal Republic of Germany
Günther Demmering, Henkel KGaA, Düsseldorf, Federal Republic of Germany
Udo Kreutzer, Henkel KGaA, Düsseldorf, Federal Republic of Germany
Manfred Lindemann, Henkel KGaA, Düsseldorf, Federal Republic of Germany
Jürgen Plachenka, Henkel KGaA, Düsseldorf, Federal Republic of Germany
Udo Steinberner, Henkel KGaA, Düsseldorf, Federal Republic of Germany

| | | | | |
|---|---|---|---|---|
| 1. | Introduction | 2481 | 3.4.3. | Dehydration ... 2520 |
| 2. | Properties | 2482 | 4. | Production of Synthetic Fatty Acids ... 2521 |
| 2.1. | Physical Properties | 2482 | | |
| 2.2. | Chemical Properties | 2495 | 4.1. | Oxidation of Alkanes ... 2521 |
| 3. | Production of Natural Fatty Acids | 2496 | 4.2. | Hydroformylation ... 2521 |
| | | | 4.3. | Hydrocarboxylation ... 2522 |
| 3.1. | Resources and Raw Materials | 2496 | 4.4. | Other Commercial Fatty Acid Syntheses ... 2522 |
| 3.2. | Fat Splitting | 2496 | | |
| 3.2.1. | Hydrolysis — Principles | 2498 | 4.5. | Noncommercial Processes ... 2523 |
| 3.2.2. | Hydrolysis — Industrial Procedure | 2502 | 5. | Analysis ... 2523 |
| 3.3. | Separation of Fatty Acids | 2506 | 6. | Storage and Transportation ... 2524 |
| 3.3.1. | Distillation | 2506 | 7. | Environmental Protection ... 2526 |
| 3.3.2. | Crystallization | 2512 | 8. | Uses ... 2527 |
| 3.3.3. | Other Separation Methods | 2516 | | |
| 3.4. | Modification of Fatty Acids | 2517 | 9. | Toxicology and Occupational Health ... 2529 |
| 3.4.1. | Hydrogenation | 2517 | | |
| 3.4.2. | Isomerization | 2519 | 10. | References ... 2529 |

# 1. Introduction

Originally the term "fatty acids" was applied only to carboxylic acids separated from animal and vegetable fats. Today it includes all saturated and unsaturated aliphatic carboxylic acids with carbon chain lengths in the range of $C_6-C_{24}$. Fatty acids obtained by splitting natural fats and oils are quantitatively far more important than synthetic fatty acids.

In the use of natural oils and fats for technical applications, i.e., for purposes other than human nutrition, fatty acids play an important role. World production of fatty acids in 1986, not including tall oil fatty acids, is estimated at ca. $2.0 \times 10^6$ t [1].

Although many fatty acids — often with unusual structures — occur naturally, only the straight-chain $C_8-C_{22}$ carboxylic acids are of commercial importance. Moreover, in many fields of application, more or less broad cuts are employed, rather than chemically pure individual fatty acids.

# 2. Properties

## 2.1. Physical Properties

Knowledge of the chemical and physical properties of fatty acids is one of the basic prerequisites for their industrial manufacture and technical application. Thermodynamic data are necessary for the calculation of heat transfer, thermal separation processes, and chemical reactions. In addition, flow behavior in the liquid state can be predicted reliably only when the density and viscosity are known.

Especially important are physical data associated with a phase change from the solid to the liquid state or vice versa. The behavior on melting is very specific for each fatty acid. For this reason, determination of the melting point is used for identification and quality control. The melting point of a fatty acid depends on the number of carbon atoms, the degree of saturation, and the structure of the hydrocarbon chain.

With straight-chain fatty acids, the *melting point* increases alternatingly with increasing chain length, so that the melting points of fatty acids with an odd number of carbon atoms are somewhat lower than those of the adjacent even-numbered fatty acids. The behavior of fatty acids with chain lengths $C_6-C_{22}$ is illustrated in Figure 1. For unsaturated fatty acids, the relationship is more complex. In general, a lowering of the melting point is observed with increasing number of double bonds. However, naturally occurring cis isomers melt or solidify at lower temperatures than the corresponding trans isomers, as shown in Figure 2 for the $C_{18}$ fatty acids. Other interesting parameters are the position of the double bond and, for polyunsaturated acids, the location of double bonds relative to each other. Fatty acids in which a conjugated arrangement (–CH=CH–CH=CH–) occurs have considerably higher melting points than nonconjugated acids. An example is eleostearic acid (Fig. 2). The melting points of the octadecene acids show the influence of the position of the double bond (Fig. 3). If the unsaturation is near the middle of the chain, a steady alternation in melting points between the even- and odd-numbered positions can be observed, whereas unsaturation near the carboxyl group or the other end of the chain increases the melting point [2].

Physical and chemical data for straight-chain fatty acids, based on [3]–[8], are compiled in Tables 1 and 2.

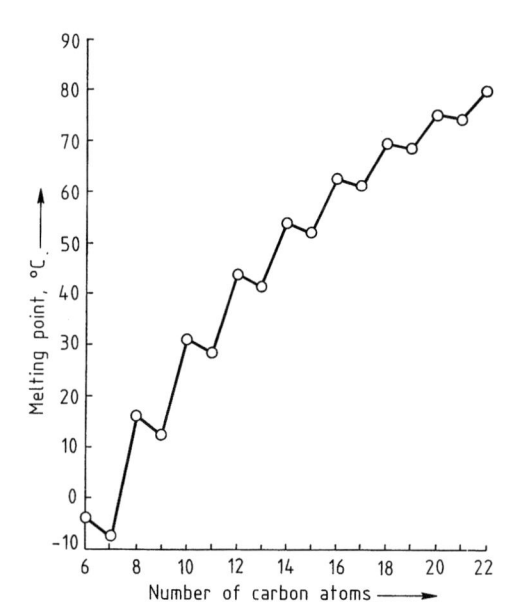

**Figure 1.** Melting point alternation of straight-chain saturated fatty acids

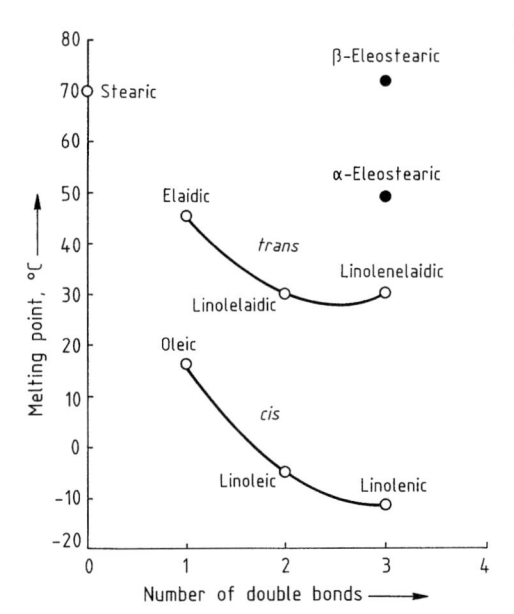

**Figure 2.** Effects of number of double bonds, cis–trans isomerism, and conjugation on melting points of $C_{18}$ acids

The behavior of fatty acid mixtures is characterized by a considerable lowering of the individual melting or solidification points. Most binary mixtures have phase diagrams with a eutectic point. Depending on the types of acids combined, three groups can be distinguished: mixtures of saturated fatty acids, mixtures of saturated and unsaturated fatty acids, and mixtures of unsaturated fatty acids. In Figures 4 and 5, typical solidi-

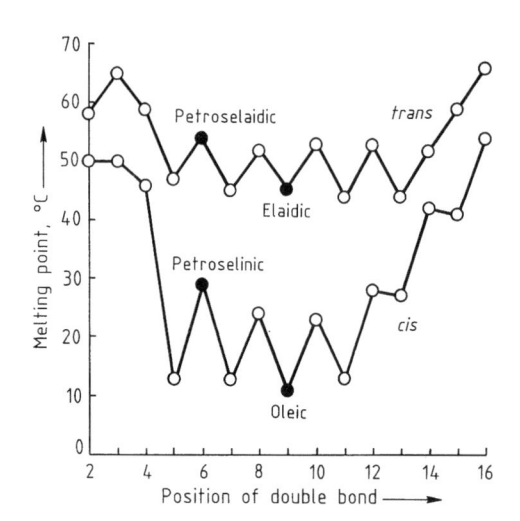

**Figure 3.** Effect of double bond position on melting points of octadecene acids ($C_{18:1}$) [2]

fication and melting point curves are illustrated. The values are plotted as a function of the concentration of the higher melting component [3], [9]–[11].

The curves for saturated acids with even or odd numbers of carbon atoms, or combinations of both, have similar forms [9], [12], [13]. The eutectic is less sharply defined than with unsaturated fatty acids. Common to all curves is the fact that the eutectic point lies in the region 0–30 wt% of the higher melting component. Exceptions to this rule are mixtures of heptadecanoic acid and hexadecanoic acid, with a eutectic point at ca. 60 wt%, and of linolenic acid and linoleic acid, which do not form a eutectic [9]. Melting point diagrams for ternary saturated fatty acid systems composed of contiguous members in the series $C_{12}$–$C_{22}$ are given in [3]. Here, too, eutectic points have been determined.

In both the liquid and the solid states, the *densities* of $C_6$–$C_{30}$ fatty acids are all below 1.0 g/cm$^3$. The lower the molecular mass and the higher the degree of unsaturation of fatty acids, the greater is their density. Solid and liquid fatty acids expand uniformly with increasing temperature. On a phase change, a sudden characteristic increase in volume, called melt expansion or dilatation, is observed. The extent of the change in volume is illustrated by stearic acid [4]. Its average expansion is $0.262 \times 10^{-3}$ cm$^3$ g$^{-1}$ K$^{-1}$ in the solid state and $0.973 \times 10^{-3}$ cm$^3$ g$^{-1}$ K$^{-1}$ in the liquid state; the melting dilatation is 0.2045 cm$^3$/g.

The density of liquid fatty acids changes linearly with temperature. The change in density varies roughly between $0.5 \times 10^{-3}$ and $1 \times 10^{-3}$ g cm$^{-3}$ K$^{-1}$, depending on chain length. A mean value of $0.7 \times 10^{-3}$ g cm$^{-3}$ K$^{-1}$ is useful for estimations up to 250 °C [5], [8].

*Optical properties* such as light refraction have long been used to identify liquids and establish their purity. The refractive index of pure fatty acids increases with increasing molecular mass, number of double bonds, and conjugation [5]. However, in modern

**Table 1.** Physical properties of straight-chain saturated $C_6$–$C_{30}$ fatty acids

| Systematic name (trivial name) | CAS registry no. | Formula, $C_nH_{2n}O_2$ | $M_r$ | Neutralization value | mp, °C | bp, °C (p, kPa)[a] | $\varrho$, g/cm³ (t, °C) | Refractive index $n_D$ (t, °C) | Molar heat capacity kJ kmol⁻¹ K⁻¹ (t, °C) | Molar heat of fusion, kJ/mol | Molar heat of vaporization, kJ/mol (t, °C) |
|---|---|---|---|---|---|---|---|---|---|---|---|
| Hexanoic (caproic) | [142-62-1] | $C_6H_{12}O_2$ | 116.16 | 483 | −3.9 | 205.8 (101.3) | 0.8751 (80) | 1.4170 (20) | 259.04 (50) | 15.20 | 62.71 (135) |
| Heptanoic (enanthic) | [111-14-8] | $C_7H_{14}O_2$ | 130.19 | 431 | −7.5 | 223 (101.3) | 0.8670 (80) | 1.4230 (20) | 260.45 (15) | 14.99 | |
| Octanoic (caprylic) | [124-07-2] | $C_8H_{16}O_2$ | 144.22 | 389 | 16.3 | 239.7 (101.3) | 0.8615 (80) | 1.4280 (20) | 304.68 (20) | 21.39 | 70.00 (134) |
| Nonanoic (pelargonic) | [112-05-0] | $C_9H_{18}O_2$ | 158.24 | 354 | 12.4 | 255.6 (101.3) | 0.8570 (80) | 1.4322 (20) | 333.90 (18) | 20.30 | |
| Decanoic (capric) | [334-48-5] | $C_{10}H_{20}O_2$ | 172.27 | 325 | 31.3 | 270 (101.3) | 0.8531 (80) | 1.4269 (70) | 359.61 (40) | 28.05 | 71.36 (187) |
| Undecanoic | [112-37-8] | $C_{11}H_{22}O_2$ | 186.28 | 301 | 28.5 | 284 (101.3) | 0.8508 (80) | 1.4202 (70) | 404.56 (40) | 25.07 | |
| Dodecanoic (lauric) | [143-07-7] | $C_{12}H_{24}O_2$ | 200.32 | 280 | 44.0 | 298 (101.3) | 0.8477 (80) | 1.4230 (70) | 431.49 (50) | 36.03 | 81.30 (164) |
| Tridecanoic | [638-53-9] | $C_{13}H_{26}O_2$ | 214.35 | 261 | 41.4 | 312.4 (101.3) | 0.8458 (80) | 1.4252 (70) | 489.32 (50) | 33.58 | |
| Tetradecanoic (myristic) | [544-63-8] | $C_{14}H_{28}O_2$ | 228.38 | 245 | 54.4 | 191.4 (1.33) | 0.8439 (80) | 1.4273 (70) | 492.76 (80) | 45.05 | 83.13 (209) |
| Pentadecanoic | [1002-84-2] | $C_{15}H_{30}O_2$ | 242.41 | 231 | 52.1 | 201.1 (1.33) | 0.8423 (80) | 1.4292 (70) | 539.15 (60) | 43.12 | |
| Hexadecanoic (palmitic) | [57-10-3] | $C_{16}H_{32}O_2$ | 256.43 | 218 | 62.9 | 210.6 (1.33) | 0.8414 (80) | 1.4309 (70) | 581.09 (68)[b] | 54.34 | 90.12 (202) |
| Heptadecanoic (margaric) | [506-12-7] | $C_{17}H_{34}O_2$ | 270.46 | 207 | 61.3 | 219.7 (1.33) | 0.8396 (80) | 1.4324 (70) | 634.13 (64) | 51.16 | |
| Octadecanoic (stearic) | [57-11-4] | $C_{18}H_{36}O_2$ | 284.49 | 197 | 69.6 | 228.7 (1.33) | 0.8380 (80) | 1.4337 (70) | 672.34 (77)[c] | 62.51 | 79.75 (242) |
| Nonadecanoic | [646-30-0] | $C_{19}H_{38}O_2$ | 298.51 | 187 | 68.7 | 237.4 (1.33) | 0.8771 (24) | 1.4512 (25) | | 57.80 | |
| Eicosanoic (arachidic) | [506-30-9] | $C_{20}H_{40}O_2$ | 312.54 | 179 | 75.4 | 245.9 (1.33) | 0.8240 (100) | 1.4250 (100) | 741.03 (76) | 70.92 | 87.78 (270) |
| Docosanoic (behenic) | [112-85-6] | $C_{22}H_{44}O_2$ | 340.60 | 164 | 80.0 | 263 (1.33) | 0.8221 (100) | 1.4270 (100) | 792.24 (80) | 78.59 | |

**Table 1.** (continued)

| Systematic name (trivial name) | CAS registry no. | Formula, $C_nH_{2n}O_2$ | $M_r$ | Neutralization value | mp, °C | bp, °C (p, kPa)[a] | $\varrho$, g/cm³ (t, °C) | Refractive index $n_D$ (t, °C) | Molar heat capacity kJ kmol⁻¹ K⁻¹ (t, °C) | Molar heat of fusion, kJ/mol | Molar heat of vaporization, kJ/mol (t, °C) |
|---|---|---|---|---|---|---|---|---|---|---|---|
| Tetracosanoic (lignoceric) | [557-59-5] | $C_{24}H_{48}O_2$ | 368.65 | 152 | 84.2 | 257 (0.533) | 0.8207 (100) | 1.4287 (100) | 902.23 (98) | 88.43 | |
| Hexacosanoic (cerotic) | [506-46-7] | $C_{26}H_{52}O_2$ | 396.70 | 141 | 87.7 | 271 (0.533) | 0.8198 (100) | 1.4301 (100) | 947.17 (94) | | |
| Octacosanoic (montanic) | [506-48-9] | $C_{28}H_{56}O_2$ | 424.76 | 132 | 90.9 | 285 (0.533) | 0.8191 (100) | 1.4313 (100) | | | |
| Triacontanoic (melissic) | [506-50-3] | $C_{30}H_{60}O_2$ | 452.81 | 123 | 93.6 | 299 (0.533) | | 1.4323 | | | |

[a] 1 kPa = 10 mbar = 7.5 mm Hg.
[b] Molar heat capacity in solid state at 25 °C: 460.7 kJ kmol⁻¹ K⁻¹; at 56.9 °C: 629.7 kJ kmol⁻¹ K⁻¹.
[c] Molar heat capacity in solid state at 25.9 °C: 565.9 kJ kmol⁻¹ K⁻¹.

**Table 2.** Physical properties of straight-chain unsaturated $C_{11}-C_{22}$ fatty acids

| Systematic name (trivial name) | CAS registry no. | Formula | $M_r$ | Neutralization value | Iodine value | mp, °C | bp, °C (p, kPa) | | ϱ, g/cm³ (t, °C) | | Refractive index $n_D$ (t, °C) | |
|---|---|---|---|---|---|---|---|---|---|---|---|---|
| 10-Undecenoic | [112-38-9] | $C_{11}H_{20}O_2$ | 184.28 | 304 | 138 | 24.5 | 275 | | 0.9072 | (24) | 1.4486 | (24) |
| cis-9-Tetradecenoic (myristoleic) | [544-64-9] | $C_{14}H_{26}O_2$ | 226.36 | 248 | 112 | −4.0 | 183–186 | (1.87) | 0.9018 | (20) | 1.4558 | (20) |
| cis-9-Hexadecenoic (palmitoleic) | [373-49-9] | $C_{16}H_{30}O_2$ | 254.42 | 221 | 100 | 0.5 | 218–220 | (2.0) | 0.9003 | (15) | 1.4587 | (20) |
| cis-6-Octadecenoic (petroselinic) | [593-39-5] | $C_{18}H_{34}O_2$ | 282.47 | 199 | 90 | 33 | 237–238 | (2.4) | 0.8794 | (40) | 1.4533 | (40) |
| trans-6-Octadecenoic (petroselaidic) | [593-40-8] | $C_{18}H_{34}O_2$ | 282.47 | 199 | 90 | 54 | | | | | | |
| cis-9-Octadecenoic (oleic) | [112-80-1] | $C_{18}H_{34}O_2$ | 282.47 | 199 | 90 | 13.4 (α) 16.3 (β) | 223 | (1.33) | 0.8905 | (20) | 1.4582 | (20) |
| trans-9-Octadecenoic (elaidic) | [112-79-8] | $C_{18}H_{34}O_2$ | 282.47 | 199 | 90 | 45 | 226–228 | (1.33) | 0.8734 | (45) | 1.4468 | (50) |
| cis,cis-9,12-Octadecadienoic (linoleic) | [60-33-3] | $C_{18}H_{32}O_2$ | 280.45 | 200 | 181 | −5.0 | 224 | (1.33) | 0.9025 | (20) | 1.4699 | (20) |
| trans,trans-9,12-Octadecadienoic (linolelaidic) | [506-21-8] | $C_{18}H_{32}O_2$ | 280.45 | 200 | 181 | 28–29 | 179–183 | (0.11) | | | | |
| cis,cis,cis-9,12,15-Octadecatrienoic (linolenic) | [463-40-1] | $C_{18}H_{30}O_2$ | 278.44 | 202 | 274 | −11 | 2245 | (1.33) | 0.9157 | (20) | 1.4800 | (20) |
| trans,trans,trans-9,12,15-Octadecatrienoic (linolenelaidic) | [28290-79-1] | $C_{18}H_{30}O_2$ | 278.44 | 202 | 274 | 30 | | | | | | |
| cis,trans,trans-9,11,13-Octadecatrienoic (α-eleostearic) | [4337-71-7] | $C_{18}H_{30}O_2$ | 278.44 | 202 | 274 | 49 | 235 | (1.6) | 0.8980 | (56) | 1.5112 | (50) |
| trans,trans,trans-9,11,13-Octadecatrienoic (β-eleostearic) | [544-73-0] | $C_{18}H_{30}O_2$ | 278.44 | 202 | 274 | 71–72 | | | 0.8839 | (80) | 1.5000 | (74) |
| cis-9-Eicosenoic (gadoleic) | [29204-02-2] | $C_{20}H_{38}H_2$ | 310.53 | 181 | 82 | 24–24.5 | | | 0.8882 | (25) | 1.4597 | (25) |
| 5,8,11,14-Eicosatetraenoic (arachidonic) | [7771-44-0] | $C_{20}H_{32}H_2$ | 304.48 | 184 | 334 | −49.5 | | | | | 1.4824 | (20) |
| cis-13-Docosenoic (erucic) | [112-86-7] | $C_{22}H_4O_2$ | 338.58 | 166 | 75 | 34.7 | 254.5 | (1.33) | 0.8699 | (55) | 1.4758 | (20) |
| trans-13-Docosenoic (brassidic) | [506-33-2] | $C_{22}H_{42}O_2$ | 338.58 | 166 | 75 | 61.9 | 285 | (4.0) | 0.8585 | (57) | 1.4472 | (64) |
| 4,8,12,15,19-Docosapentaenoic (clupanodonic) | [2548-85-8] | $C_{22}H_{34}O_2$ | 330.52 | 170 | 384 | −78 | 236 | (0.67) | 0.9356 | (20) | 1.5014 | (20) |

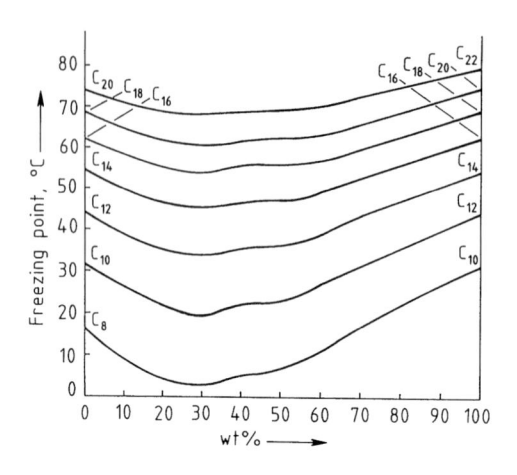

**Figure 4.** Freezing points of binary systems of saturated $C_8$–$C_{22}$ fatty acids

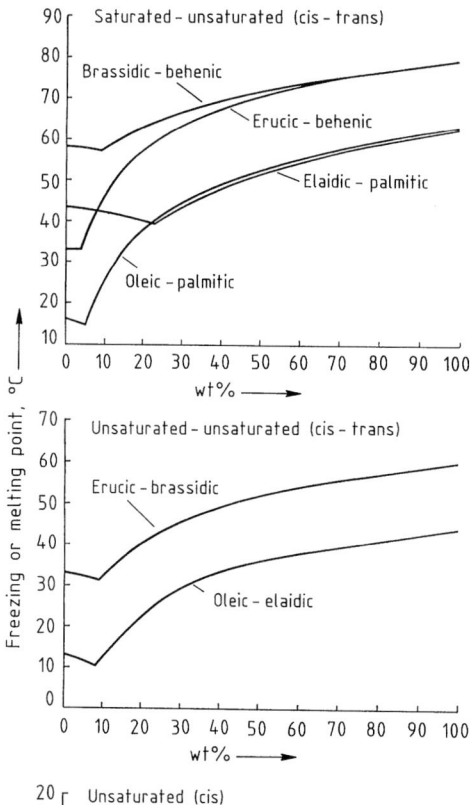

**Figure 5.** Freezing or melting points of binary systems of unsaturated fatty acids

**Table 3.** Heat of fusion of some unsaturated $C_{18}$ fatty acids

| Compound | Heat of fusion, kJ/mol |
|---|---|
| Oleic (9 c) | 30.14 |
| Elaidic (9 t) | 38.94 |
| Linoleic (9 c, 12 c) | 51.50 |
| Linolenic (9 c, 12 c, 15c) | 39.36 |
| α-Eleostearic (9 c, 11 t, 13 t) | 43.96 |
| β-Eleostearic (9 t, 11 t, 13 t) | 55.27 |

laboratory and process analysis, measurement of the refractive index of fatty acids plays only a minor role because no definite interpretation of the measured values can be made for mixtures of varying composition.

The *heat of fusion* is the energy consumption for phase transition from the solid crystal to the liquid state. Measured values for straight-chain saturated fatty acids are listed in Table 1. Heats of fusion, like melting points, increase alternatingly with the length of the carbon chain. Acids with an even number of carbon atoms have a higher heat of fusion than adjacent odd-numbered acids. Above a chain length of about ten carbon atoms, the heat of fusion can be represented by an equation of the form [3]

$$\Delta H_f = a \cdot n + b$$

where

$\Delta H_f$ = heat of fusion, kJ/mol
$n$ = number of carbon atoms
$a$ and $b$ = constants

From the values given in Table 1, for even-numbered fatty acids:

$$\Delta H_f = 4.31 \cdot n - 15.11$$

and for odd-numbered fatty acids:

$$\Delta H_f = 4.04 \cdot n - 18.79$$

Table 3 lists heats of fusion for some $C_{18}$ unsaturated fatty acids; they were derived from experimental data or calculated with the aid of theoretical relationships [3], [4], [14]. As expected, the values are lower than those for saturated fatty acids.

The *heat of vaporization* is a measure of the energy consumption for phase transition from the liquid to the vapor state. This and the related properties of boiling point and vapor pressure are very important for the distillation of fatty acids and their thermal separation by fractional distillation. To avoid decomposition, the thermal processes are carried out at the lowest possible boiling temperature under vacuum.

The *boiling point* and the *molar heat of evaporation* rise with increasing molecular mass (see Tables 1 and 2). The influence of the degree of saturation is of minor importance. Extensive experimental investigations [12], [15] of phase equilibria indicate

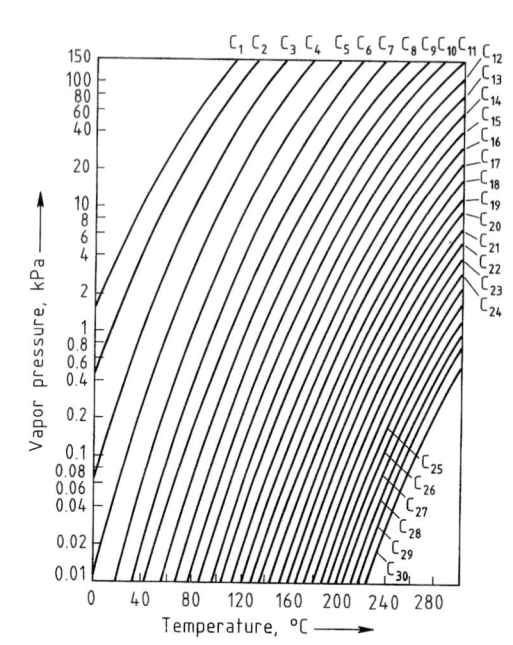

**Figure 6.** Vapor pressure of straight-chain fatty acids as a function of temperature [14]

that fatty acids show an almost ideal behavior, especially under vacuum. An equation that relates vapor pressure and temperature of normal fatty acids has been obtained:

$$t = \frac{\sqrt{M_r} + 0.06075 \log^2 p + 1.32 \log p - 7.47}{-3.54 \times 10^{-4} \log^2 p - 4.00 \times 10^{-3} \log p + 0.05142}$$

where

$t$ = temperature, °C
$p$ = vapor pressure, mm Hg
$M_r$ = molecular mass

The vapor pressure curves calculated from this equation are illustrated in Figure 6.

Most equations for calculating the *vapor pressure* of pure substances are based on the Clausius–Clapeyron equation. If the change in the heat of evaporation with temperature is neglected, the so-called Antoine equation can be derived in the following form [8]:

$$\ln p = A - \frac{B}{(T+C)}$$

where

$T$ = absolute temperature, K
$p$ = vapor pressure, mm Hg
$A$, $B$, and $C$ = constants

**Table 4.** Antoine constants of straight-chain fatty acids

| Compound | A | B | C | Range of validity | |
|---|---|---|---|---|---|
| | | | | $t_{min}$, °C | $t_{max}$, °C |
| $C_6$ | 17.5765 | 4151.08 | −99.427 | 60 | 205 |
| $C_8$ | 17.5910 | 4398.97 | −110.918 | 9 | 249 |
| $C_{10}$ | 17.1568 | 4320.40 | −131.726 | 33 | 280 |
| $C_{12}$ | 16.8008 | 4302.95 | −145.445 | 48 | 297 |
| $C_{14}$ | 16.1371 | 4304.92 | −146.261 | 48 | 309 |
| $C_{16}$ | 16.0700 | 4304.70 | −168.123 | 55 | 351 |
| $C_{18}$ | 14.8472 | 3488.63 | −224.528 | 43 | 376 |
| $C_{18:1c}$ | 11.6384 | 1840.91 | −297.002 | 140 | 286 |
| $C_{20}$ | 17.2106 | 5334.16 | −161.260 | 162 | 295 |
| $C_{22}$ | 16.6529 | 5047.70 | −183.768 | 176 | 306 |
| $C_{22:1c}$ | 19.0051 | 6000.00 | −170.707 | 242 | 358 |
| $C_{22:1t}$ | 18.0213 | 6000.82 | −146.273 | 236 | 284 |

**Table 5.** Heat of combustion of straight-chain fatty acids (experimental values)

| Compound | $-\Delta H_{comb}$, kJ/mol |
|---|---|
| $C_6$ | 3 474 |
| $C_8$ | 4 792 |
| $C_{10}$ | 6 103 |
| $C_{12}$ | 7 416 |
| $C_{14}$ | 8 731 |
| $C_{16}$ | 10 039 |
| $C_{18}$ | 11 351 |
| $C_{20}$ | 12 648 |
| $C_{22}$ | 13 954 |
| $C_{18:1c}$ | 11 168 |
| $C_{18:1t}$ | 11 152 |
| $C_{22:1c}$ | 13 799 |
| $C_{22:1t}$ | 13 772 |

The corresponding Antoine constants for straight-chain fatty acids are listed in Table 4.

The *heat of combustion*, which can be measured calorimetrically, is of practical and theoretical importance. It can be used to calculate the *heat of formation*, by means of which the enthalpy change in chemical reactions can be established. Heat of combustion and heat of formation increase with increasing number of carbon atoms. With straight-chain saturated fatty acids, an almost linear correlation exists [6], which can be expressed by empirical equations. The values given in Tables 5 and 6 show the following correlation for heats of combustion and formation:

$$-\Delta H_{comb} = -453 + 655 \cdot n$$
$$\Delta H_{form} = 368 + 33 \cdot n$$

where

$n$ = number of carbon atoms ($C_6$–$C_{20}$)
$\Delta H_{comb}$ = heat of combustion, kJ/mol

**Table 6.** Heat of formation of straight-chain fatty acids

| Compound | $\Delta H_{form}$, kJ/mol |
|---|---|
| $C_6$ | 584.0 |
| $C_8$ | 635.2 |
| $C_{10}$ | 685.2 |
| $C_{12}$ | 775.6 |
| $C_{14}$ | 835.9 |
| $C_{16}$ | 892.9 |
| $C_{18}$ | 949.4 |
| $C_{20}$ | 1064.0 |
| $C_{18:1c}$ | 710.2 |
| $C_{18:2c}$ | 603.1 |
| $C_{18:3c}$ | 485.9 |
| $C_{20:1c}$ | 763.9 |
| $C_{22:1c}$ | 807.5 |

$\Delta H_{form}$ = heat of formation, kJ/mol

The data for unsaturated fatty acids, as far as they are known, are almost identical with those for the corresponding saturated fatty acids.

For straight-chain saturated fatty acids, the *molar heat capacity* increases uniformly with the number of carbon atoms, i.e., no alternating behavior is found. If the data are referred to the mass unit, the relationships become simpler. For liquid fatty acids in a range of about 30 °C above the melting point, specific heat capacities were found to vary from about 2.11 J g$^{-1}$ K$^{-1}$ for octanoic acid to about 2.33 J g$^{-1}$ K$^{-1}$ for docosanoic acid. For solid fatty acids, the values are mostly in the range 1.88 – 2.09 J g$^{-1}$ K$^{-1}$ [4], [6]. For rough calculations, use of a mean value of 2.2 J g$^{-1}$ K$^{-1}$ is sufficiently precise for solid fatty acids. For liquid fatty acids, the dependence of heat capacity on temperature has been investigated experimentally only for palmitic and stearic acids [4]. Theoretical computation methods according to the additive molecular group technique are given for the liquid and vapor phases in [18]. Relationships of the following type are obtained:

$$c_p = A + B \cdot T + C \cdot T^2 + D \cdot T^3$$

where

$c_p$ = specific heat capacity
$T$ = temperature
$A$, $B$, $C$, and $D$ = constants

*Thermal conductivity* and *viscosity* are important properties in the assessment of heat transfer and fluid dynamics. Compared to water, fatty acids are poor heat conductors. Thermal conductivities at the melting point vary between 0.145 W m$^{-1}$ K$^{-1}$ for octanoic acid and 0.170 W m$^{-1}$ K$^{-1}$ for docosanoic acid. Thermal conductivity decreases with increasing temperature. Values calculated for a temperature range extending from the

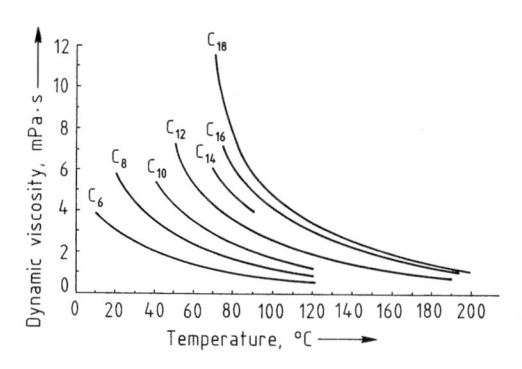

Figure 7. Dynamic viscosity of straight-chain fatty acids as a function of temperature [4], [8]

melting point to up to 300 °C are listed in detail for saturated and unsaturated fatty acids in [16].

The *dynamic viscosity* of fatty acids increases with increasing molecular mass and, as with all liquids, decreases with increasing temperature. Figure 7 shows the curves for saturated fatty acids. For interpolation purposes, the Andrade equation can be used [19]:

$$\eta = \eta_0 \cdot e^{A/T}$$

where

$\eta$ = dynamic viscosity, mPa · s
$A$ = constant
$T$ = temperature, K
$\eta_0$ = dynamic viscosity at $T = T_0$

The average chain length is a useful parameter for evaluating the viscosity of straight-chain fatty acid mixtures:

$$\bar{n} = \Sigma\, x_i \cdot n_i$$

where

$\bar{n}$ = average number of carbon atoms
$n_i$ = number of carbon atoms of component i
$x_i$ = mole fraction of component i

Measurements [20] have shown that congruent mixtures, i.e., mixtures differing in composition but having the same $\bar{n}$ value, have the same viscosity at a given temperature. Thus the viscosity of a mixture of fatty acids with average chain length $\bar{n}$ is the same as that of a pure fatty acid with this chain length.

*Solubility.* At temperatures up to 100 °C, fatty acids and water have only low mutual solubilities, and the solubility of fatty acids in water is considerably lower than that of water in fatty acids. Both solubilities increase with increasing temperature and decrease

**Table 7.** Mutual solubilities of saturated fatty acids with water

| Compound | Solubility of water in fatty acids, g water per 100 g acid ($t$, °C) | | Solubility of fatty acid in water, g acid per 100 g water ($t$, °C) | |
|---|---|---|---|---|
| $C_6$ | 4.96 | (12.3) | 0.968 | (20) |
| | 10.74 | (46.3) | 1.171 | (60) |
| $C_8$ | 4.04 | (14.4) | 0.068 | (20) |
| | | | 0.113 | (60) |
| $C_{10}$ | 3.22 | (29.4) | 0.015 | (20) |
| | | | 0.027 | (60) |
| $C_{12}$ | 2.41 | (42.7) | 0.006 | (20) |
| | 2.93 | (90.5) | 0.009 | (60) |
| $C_{14}$ | 1.73 | (53.2) | 0.002 | (20) |
| | | | 0.003 | (60) |
| $C_{16}$ | 1.27 | (61.8) | 0.0007 | (20) |
| | | | 0.001 | (60) |
| $C_{18}$ | 0.93 | (68.7) | 0.0003 | (20) |
| | 1.03 | (92.4) | 0.0005 | (60) |

**Table 8.** Solubility of saturated fatty acids in methanol

| Compound | Solubility, g acid per 100 g methanol | | | | |
|---|---|---|---|---|---|
| | 0 °C | 10 °C | 20 °C | 40 °C | 60 °C |
| $C_8$ | 330 | 1300 | $\infty$ | $\infty$ | $\infty$ |
| $C_{10}$ | 80 | 180 | 510 | $\infty$ | $\infty$ |
| $C_{12}$ | 12.7 | 41.1 | 120 | 2250 | $\infty$ |
| $C_{14}$ | 2.8 | 5.8 | 17.3 | 350 | $\infty$ |
| $C_{16}$ | 0.8 | 1.3 | 3.7 | 77 | 4650 |
| $C_{18}$ | | | 0.1 | 11.7 | 520 |

with increasing chain length. Some data on the mutual solubilities of the binary water–saturated fatty acid system are given in Table 7. At high temperature and pressure, however, the solubility of water in fatty acids and in fatty acid–fat mixtures is considerable. At 250 °C and 5 MPa, for example, approximately 12 % of water dissolves in the fatty acid phase of a mixture of tallow fatty acids and water, while the value is 24 % for a mixture of coconut fatty acids. Complete miscibility with water is obtained at 320 and 290 °C, respectively [20]. This fact must be considered in the high-pressure splitting of fats and oils (see Section 3.2.2).

Above their melting points, fatty acids are miscible in all proportions with many organic solvents such as hydrocarbons, ketones, esters, alcohols, and chlorinated or aromatic solvents. In the solid state, the solubility of saturated fatty acids depends on chain length and increases with increasing temperature. Unsaturated fatty acids, however, are completely miscible at room temperature and display only limited solubility at lower temperature [3], [9], [21], [22]. In Tables 8 and 9, the solubilities of saturated and unsaturated fatty acids in methanol are given as examples.

The *surface tension* at the liquid–vapor interface and the *interfacial tension* in liquid–liquid systems are important properties for estimating the wetting behavior at solid surfaces and the coalescence behavior in liquid–liquid extraction. The surface

**Table 9.** Solubility of unsaturated fatty acids in methanol

| Compound | Solubility, g acid per 100 g methanol | | | | | |
|---|---|---|---|---|---|---|
| | −50 °C | −30 °C | −20 °C | −10 °C | 0 °C | + 10 °C |
| $C_{18:1c}$ | 0.1 | 0.86 | 4.02 | 31.6 | 250 | 1820 |
| $C_{18:1t}$ | 0.01 | 0.06 | 0.18 | 0.48 | | |
| $C_{18:2c}$ | 3.3 | 48.1 | 233 | 1850 | | |
| $C_{22:1c}$ | 0.007 | 0.068 | 0.19 | 0.49 | | |

**Table 10.** Surface tension and interfacial tension of saturated fatty acids at 75°C

| Compound | Surface tension, mN/m ($t$, °C) | | Interfacial tension *, mN/m |
|---|---|---|---|
| $C_6$ | 23.0 | | 2.1 |
| $C_8$ | 24.2 | | 5.8 |
| $C_{10}$ | 25.1 | | 8.0 |
| $C_{12}$ | 25.9 | | 8.7 |
| | 20.9 | (140) | |
| $C_{14}$ | 26.8 | | 9.2 |
| | 22.0 | (140) | |
| $C_{16}$ | 27.3 | | 9.2 |
| | 22.9 | (140) | |
| $C_{18}$ | 27.7 | | 9.5 |
| | 23.8 | (140) | |
| Water | 63.7 | | |
| | 50.9 | (140) | |

* Fatty acid – water.

tension of liquid fatty acids decreases with decreasing chain length and with increasing temperature. In Table 10, some data are listed for saturated fatty acids.

Dry fatty acids are poor conductors of electricity. The specific *electrical conductivity* is a function of chain length and temperature. Values vary from $0.6 \times 10^{-11}$ to $83 \times 10^{-11}$ $\Omega^{-1}$ cm$^{-1}$. The relative dielectric constants vary between 2 and 3, depending on the number of double bonds and the temperature [4].

The *flash* and *fire points* of fatty acids are above 100 °C. They rise significantly with increasing chain length (Table 11).

## 2.2. Chemical Properties

Fatty acids undergo the typical reactions of the carboxyl group (→ Carboxylic Acids, Aliphatic). Commercially important derivatives such as esters, amides, nitriles, and alkali or other metal salts are obtained by conventional methods (see also Chap. 8). Hydrogenation of the carboxyl group to form fatty alcohols usually involves esterification and is discussed in detail under → Fatty Alcohols.

For industrially important reactions involving the double bond of unsaturated fatty acids (e.g., hardening and cis – trans isomerization) see Section 3.4.

Table 11. Average values for flash and fire points of saturated fatty acids (technical grade, 99%), determined according to ISO 2592

| Compound | Flash point, °C | Fire point, °C |
|---|---|---|
| $C_6$ | 116 | 120 |
| $C_8$ | 132 | 138 |
| $C_{10}$ | 147 | 153 |
| $C_{12}$ | 163 | 173 |
| $C_{14}$ | 178 | 192 |
| $C_{16}$ | 192 | 210 |
| $C_{18}$ | 204 | 226 |

# 3. Production of Natural Fatty Acids

## 3.1. Resources and Raw Materials

Fatty acids are widely distributed in nature as components of lipids such as fats (esters of fatty acids with glycerol), waxes (esters of fatty acids with fatty alcohols), sterol esters, and esters of triterpene alcohols.

For the manufacture of fatty acids on a commercial scale, only fats available in large quantities are used as raw materials. Vegetable fats are obtained mainly from seeds and only to a small extent from fruit pulp; animal fats are obtained from fat deposits and from organs (→ Fats and Fatty Oils). The term fats normally means materials of solid consistency, whereas liquid materials are referred to as oils. The definitions are indistinct, since the solid–liquid state of fats depends on temperature and many fats show a semisolid consistency at ambient temperature. Fats possess a characteristic fatty acid composition. Tables 12 and 13 show the fatty acid composition of important fats and oils.

Soap-stock fatty acids are byproducts in the processing of crude fats for human nutrition (→ Fats and Fatty Oils). The pulping of wood in the sulfate pulp process results in the digestion of the lipids composed of esters of triterpene alcohols that are present in the resin. The free fatty acids found as constituents of tall oil represent another starting material for the commercial production of fatty acids.

Natural waxes, such as sperm oil or jojoba oil, are not significant sources of fatty acids.

## 3.2. Fat Splitting

Fats are fatty acid esters of glycerol and are also known as triglycerides. By splitting the glycerides, fatty acids and glycerol can be recovered (Fig. 8).

**Figure 8.** Formation and splitting of glycerides

$$\begin{array}{c} H_2C-O-H \\ | \\ HC-O-H \\ | \\ H_2C-O-H \\ \text{Glycerol} \end{array} + \begin{array}{c} H-O-\overset{O}{\underset{\|}{C}}-R^1 \\ H-O-\overset{O}{\underset{\|}{C}}-R^2 \\ H-O-\overset{O}{\underset{\|}{C}}-R^3 \\ \text{Fatty acids} \end{array} \begin{array}{c} \text{Esterification} \\ \xrightarrow[+3H_2O]{-3H_2O} \\ \text{Splitting} \end{array} \begin{array}{c} H_2C-O-\overset{O}{\underset{\|}{C}}-R^1 \\ | \\ HC-O-\overset{O}{\underset{\|}{C}}-R^2 \\ | \\ H_2C-O-\overset{O}{\underset{\|}{C}}-R^3 \\ \text{Triglyceride} \end{array}$$

Triglyceride

- Transesterification, $CH_3-O-H$ (methanol) → Methyl esters ($R^1-CO-O-CH_3$, $R^2-CO-O-CH_3$, $R^3-CO-O-CH_3$) + Glycerol ($H-O-CH_2$, $H-O-CH$, $H-O-CH_2$)

- Hydrolysis, $H_2O$ (water) → Fatty acids ($R^1-CO-O-H$, $R^2-CO-O-H$, $R^3-CO-O-H$) + Glycerol

- Saponification, NaOH (caustic soda) → Soap ($R^1-CO-O-Na$, $R^2-CO-O-Na$, $R^3-CO-O-Na$) + Glycerol

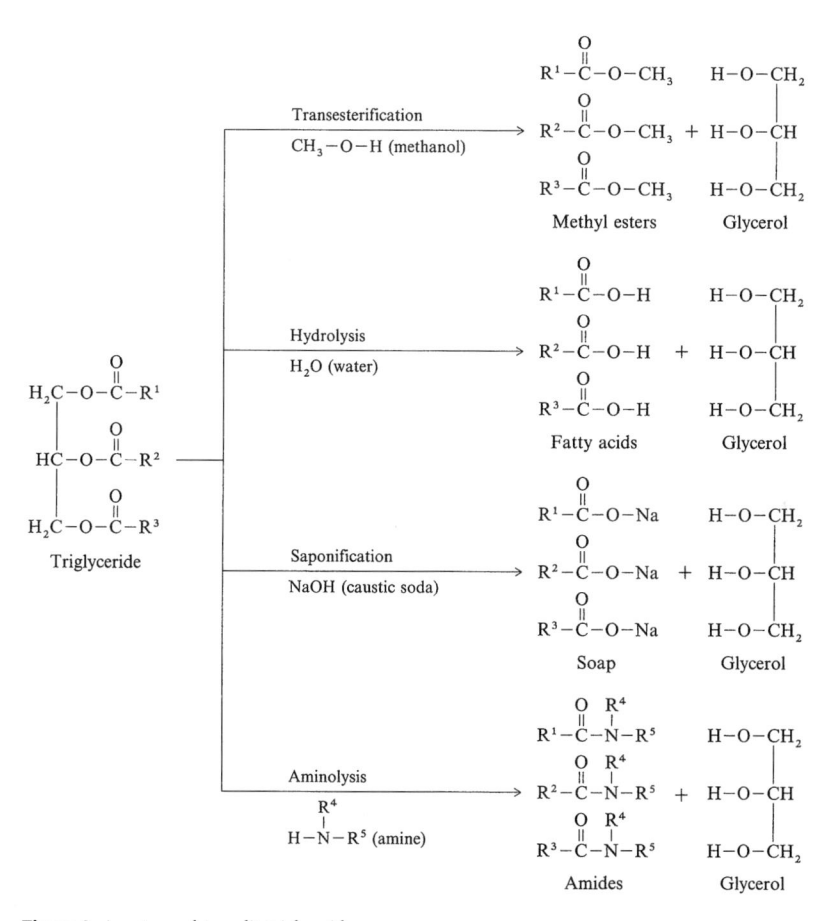

- Aminolysis, $H-N(R^4)-R^5$ (amine) → Amides ($R^1-CO-N(R^4)-R^5$, $R^2-CO-N(R^4)-R^5$, $R^3-CO-N(R^4)-R^5$) + Glycerol

**Figure 9.** Agents used to split triglycerides

The most important splitting agents are water (hydrolysis), methanol (methanolysis), caustic soda (saponification), and amines (aminolysis) (Fig. 9). Because of drawbacks in the subsequent purification of glycerol, saponification and aminolysis are no longer commercially important. Today, soaps and fatty acid amides are manufactured via fatty acids or fatty acid methyl esters. Exceptions are ricinoleic acid and hydroxystearic acid, which are produced commercially by saponification of castor oil and hardened castor oil, respectively. In this way, the esterification reactions that occur with the hydroxyl groups of ricinoleic and hydroxystearic acids during hydrolysis are avoided.

### 3.2.1. Hydrolysis — Principles

The reactants in the hydrolysis of fats form a heterogeneous reaction system made up of two liquid phases. The disperse aqueous phase consists of water and glycerol; the homogeneous lipid phase contains glycerides and fatty acids. The hydrolysis of fat to form glycerol and fatty acids takes place in the lipid phase in several stages, via partial glycerides (diglycerides and monoglycerides), as shown in Figure 10.

The total reaction can be summarized as follows (cf. Fig. 8):

Triglyceride + 3 Water $\rightleftharpoons$ 3 Fatty acid + Glycerol

The reaction proceeds to a state of equilibrium. At higher temperature, the endothermic nature of fat hydrolysis results in displacement of the equilibrium in favor of the split products. The rate at which equilibrium is reached corresponds approximately to a first-order reaction:

$$\frac{dc}{dt} = k \cdot c$$

where

$k$ = rate constant
$t$ = reaction time
$c$ = concentration of triglyceride

However, at the beginning and the end, the reaction does not proceed according to this equation. This is because the necessary excess of water is present only after an induction period. Toward the end of the reaction, the glycerol concentration (which is not considered in the equation) has an increasing influence because of the reverse reaction.

Values of the rate constant $k$ are highly dependent on temperature:

$$k = k_0 \cdot e^{-E/RT}$$

where

$k_0$ = statistical probability of reaction
$E$ = energy of activation
$T$ = absolute temperature
$R$ = gas constant

Table 12: Fatty acid composition of important oils and fats (weight percent) * [continued in Table 13]

| Fatty acid (trivial name) | Carbon no. | No. of double bonds | Cotton seed oil | Peanut oil | Coconut oil | Linseed oil | Palm kernel oil | Olive oil | Corn oil | Palm oil |
|---|---|---|---|---|---|---|---|---|---|---|
| Caproic acid | $C_6$ | 0 | | | 0–1 | | tr | | | |
| Caprylic acid | $C_8$ | 0 | | | 5–10 | | 3–6 | | | |
| Capric acid | $C_{10}$ | 0 | | | 5–10 | | 3–5 | | | |
| Lauric acid | $C_{12}$ | 0 | | | 45–53 | | 40–52 | | | |
| Myristic acid | $C_{14}$ | 0 | 0–2 | | 15–21 | tr | 14–18 | | 0–1 | 0–2 |
| Palmitic acid | $C_{16}$ | 0 | 17–29 | 6–16 | 7–11 | 5–8 | 6–10 | 7–16 | 8–19 | 38–48 |
| Stearic acid | $C_{18}$ | 0 | 1–4 | 1–7 | 2–4 | 2–4 | 1–4 | 1–3 | 0–4 | 3–6 |
| Arachidic acid | $C_{20}$ | 0 | 0–1 | 1–3 | | tr | | 0–1 | | 0–1 |
| Behenic acid | $C_{22}$ | 0 | tr | 2–5 | | | | | | |
| Lignoceric acid | $C_{24}$ | 0 | tr | 1–3 | | | | | | |
| Lauroleic acid | $C_{12}$ | 1 | | | | | | | | |
| Myristoleic acid | $C_{14}$ | 1 | | | | | | | | |
| Palmitoleic acid | $C_{16}$ | 1 | 0–2 | 0–1 | | | | | | |
| Oleic acid | $C_{18}$ | 1 | 13–44 | 36–72 | 6–8 | 15–25 | 9–16 | 65–85 | 19–50 | 38–44 |
| Gadoleic acid | $C_{20}$ | 1 | tr | 0–2 | | | | | | |
| Erucic acid | $C_{22}$ | 1 | | | | | | | | |
| Ricinoleic acid | $C_{18}$ | 1 | | | | | | | | |
| Linoleic acid | $C_{18}$ | 2 | 33–58 | 13–45 | 1–3 | 12–16 | 1–3 | 4–15 | 36–62 | 9–12 |
| Linolenic acid | $C_{18}$ | 3 | | 0–1 | | 50–60 | | 0–1 | 0–2 | |
| Unsaturated fatty acid | $C_{20}$ | 2–6 | | | | | | | | |
| Unsaturated fatty acid | $C_{22}$ | 3–6 | | | | | | | | |
| Iodine value | | | 96–112 | 85–102 | 8–12 | 160–200 | 14–23 | 80–88 | 103–128 | 44–54 |
| Saponification value | | | 190–198 | 188–195 | 250–264 | 188–195 | 245–255 | 188–196 | 188–193 | 194–206 |
| mp, °C | | | −2 to +2 | −2 | 20–26 | −20 | 24–26 | | 14–20 | 27–50 |
| Fatty acid titer, °C | | | 30–36 | 26–32 | 20–24 | 19–21 | 20–28 | 17–26 | | 40–47 |

* tr = traces.

# Fatty Acids

Table 13. Fatty acid composition of important oils and fats, continued (weight percent) *

| Fatty acid (trivial name) | Carbon no. | No. of double bonds | Castor oil | Rapeseed oil (C$_{22}$ rich) | Soybean oil | Sunflower oil | Herring oil | Sardine oil | Whale oil | Beef tallow |
|---|---|---|---|---|---|---|---|---|---|---|
| Caproic acid | C$_6$ | 0 | | | | | | | | |
| Caprylic acid | C$_8$ | 0 | | | | | | | | |
| Capric acid | C$_{10}$ | 0 | | | | | | | | |
| Lauric acid | C$_{12}$ | 0 | | | | | tr | | | |
| Myristic acid | C$_{14}$ | 0 | | 0–1 | tr | tr | 5–10 | 4–6 | 7–11 | 1–6 |
| Palmitic acid | C$_{16}$ | 0 | 2–3 | 2–5 | 7–12 | 3–10 | 11–16 | 9–11 | 10–16 | 20–37 |
| Stearic acid | C$_{18}$ | 0 | 2–3 | 0–3 | 2–3 | 1–10 | 0–3 | 1–3 | 1–5 | 15–30 |
| Arachidic acid | C$_{20}$ | 0 | | 0–2 | 0–3 | 0–1 | | | | tr |
| Behenic acid | C$_{22}$ | 0 | | 0–1 | tr | 0–1 | | | | |
| Lignoceric acid | C$_{24}$ | 0 | | 0–1 | | tr | | | | |
| Lauroleic acid | C$_{12}$ | 1 | | | | | | | | |
| Myristoleic acid | C$_{14}$ | 1 | | | | | 0–1 | | 1–3 | |
| Palmitoleic acid | C$_{16}$ | 1 | | tr | tr | 0–1 | 5–12 | 10–15 | 8–15 | 1–9 |
| Oleic acid | C$_{18}$ | 1 | 4–9 | 11–60 | 20–30 | 14–65 | 8–15 | 15–25 | 22–35 | 20–50 |
| Gadoleic acid | C$_{20}$ | 1 | | 0–14 | 0–1 | tr | | | 8–12 | |
| Erucic acid | C$_{22}$ | 1 | | 2–52 | | tr | | | 1–5 | |
| Ricinoleic acid | C$_{18}$ | 1 | 80–87 | | | | | | | |
| Linoleic acid | C$_{18}$ | 2 | 2–7 | 12–24 | 45–58 | 20–75 | 2–4 | 3–8 | 2–9 | 0–5 |
| Linolenic acid | C$_{18}$ | 3 | | 6–15 | 4–10 | 0–1 | 0–2 | 1–3 | 0–1 | 0–3 |
| Unsaturated fatty acid | C$_{20}$ | 2–6 | | | | | 20–30 | 15–30 | 10–20 | tr |
| Unsaturated fatty acid | C$_{22}$ | 3–6 | | | | | 10–28 | 15–20 | 5–15 | |
| Iodine value | | | 81–91 | 95–108 | 120–140 | 120–140 | 120–145 | 170–193 | 105–135 | 35–55 |
| Saponification value | | | 174–186 | 170–180 | 190–195 | 186–194 | 178–194 | 189–193 | 185–194 | 190–200 |
| mp, °C | | | −12 to −10 | −9 | −23 to −20 | 20–28 | | | | 40–50 |
| Fatty acid titer, °C | | | | 13–16 | 20–24 | 16–20 | 23–27 | | | 39–43 |

* tr = traces.

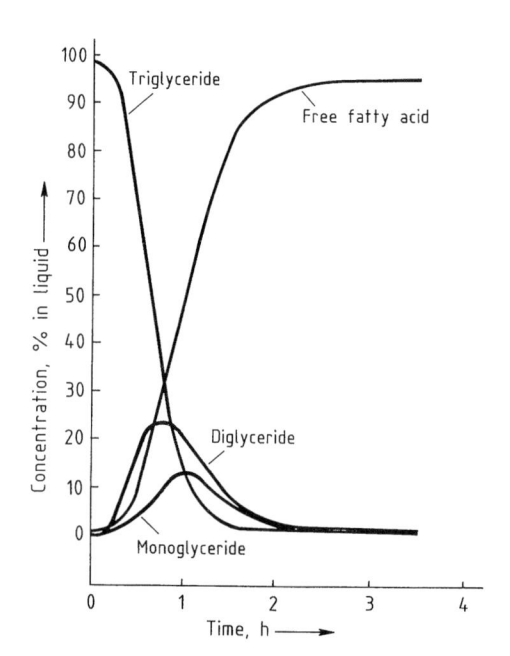

**Figure 10.** Typical course of hydrolysis reaction (hydrolysis of beef tallow at 235 °C and 3.5 MPa using 1 part oil and 2 parts water)

An increase in temperature of 10 °C, for example, approximately doubles the value of $k$. Catalysts influence the rate constant $k$ by changing the energy of activation, whereas $k_0$ is affected only slightly by temperature.

To achieve a high degree of conversion in this reaction, the content of water and triglyceride in the lipid phase should be as high as possible and the concentration of fatty acid and glycerol should be low. In the liquid two-phase system prevailing during hydrolysis of fats, the concentrations of the reactants depend primarily on the diffusion between the lipid and aqueous phases. Mass transfer can be influenced favorably by large, turbulent interfaces. During the entire reaction, two mass transfer effects, which proceed in opposite directions, are of special importance for a high degree of conversion: (1) diffusion of water into the lipid phase and (2) removal of glycerol from the lipid phase to the aqueous phase by water extraction. Optimization of this process by chemical engineering methods largely determines the yields of fatty acid and glycerol.

LASCARAY [24] summarized the complex reaction conditions in an empirical formula. He described the degree of splitting in the lipid phase as a function of the glycerol concentration in the aqueous phase:

$$H = 100 - 0.8 \cdot c_G$$

where

$H$ = degree of splitting
$c_G$ = glycerol concentration

Temperature and choice of catalyst are important only as parameters for approaching theoretical equilibrium conditions under the most favorable economic conditions, i.e., in the shortest time possible.

## 3.2.2. Hydrolysis — Industrial Procedure

**Pretreatment.** Crude commercial fats are sometimes purified to remove troublesome impurities, such as minerals, gums, soaps, and proteins, before they are subjected to splitting. The normal procedure is to heat the crude fat and then filter it or treat it with a small quantity of 0.3 – 1.0% sulfuric acid. Animal fats from slaughterhouses contain varying amounts of dissolved packaging materials that consist mainly of polyhydrocarbons with differing rheological properties. These can best be separated from fats by filtration at ca. 70 °C with the use of filter aids.

To suppress oxidation, degassing to remove atmospheric oxygen is recommended. All hydrolytic processes employ demineralized water as the splitting agent. The hardness of water reduces catalyst efficiency and increases the salt content of the glycerol obtained.

**Catalysts.** Up to 100 °C, fat hydrolysis is very slow. For palm oil with an initial content of 5% free fatty acids (ffa), a degree of splitting of ca. 20% is reached after 10 days at 100 °C [25]. TWITCHELL achieved a technical breakthrough in 1898 with an effective catalyst consisting of aromatic hydrocarbons, oleic acid, and concentrated sulfuric acid (Twitchell reagent). The original Twitchell batch process operates under atmospheric pressure at almost 100 °C in wooden tubs. The batch time is 12 – 24 h, and the degree of splitting without a change of water is about 80 – 85 % [26].

Lipases are specific biocatalysts for ester hydrolysis and exhibit optimum efficiency at ca. 35 °C [27]. Enzymatic splitting of fat is, therefore, of great interest today as a low-energy process. Lipases can be isolated from microbial, plant, and animal sources. The plant lipase isolated from castor beans has been studied most intensively. Like pancreatic lipase, however, it has not been used widely in the commercial splitting of fats. Microbial lipases are of most recent economic interest [28]. They show specificity of reaction at certain positions or with certain fatty acids. The two variants of position specificity are (1) splitting all three ester bonds of the triglyceride (total splitter) and (2) splitting only at positions 1 and 3 (1,3-splitter). An example of fatty acid specificity is the splitting off of oleic acid by the lipase obtained from *Geotrichum candidum*.

Acid catalysts are generally very effective for accelerating hydrolysis reactions. At higher temperature, however, material corrosion occurs. This problem does not arise when basic oxides are used as catalysts. Dibasic metal oxides have a higher activity than the more strongly alkaline monobasic metal oxides. In the literature, zinc oxide in its active form as zinc soap is described as the most effective accelerator for hydrolysis. The $Zn^{2+}$ ion functions not so much as a Lewis acid, but rather as a phase-transfer catalyst which carries water in the form of hydrate into the lipid phase [29], [30].

**Table 14.** Influence of temperature on splitting time

| t, °C | Time to reach equilibrium, min * | | |
|---|---|---|---|
| | Beef tallow | Coconut oil | Peanut oil |
| 225 | 156 (62) | 158 (58) | 156 (50) |
| 240 | 82 (34) | 85 (33) | 85 (25) |
| 260 | 47 (18) | 46 (23) | 53 (16) |
| 280 | 34 (8) | 33 (10) | 33 (10) |

* In parentheses: halftime of splitting.

**Temperature.** Without catalysts, the rate of fat hydrolysis is economical only above 210 °C (Table 14). The development of high-pressure techniques was the decisive step in making hydrolysis of fats a large-scale manufacturing process. Initially, batch operation in autoclaves at medium pressures of 0.6 – 1.2 MPa was employed.

Modern continuous splitting units operate at 210 – 260 °C and 1.9 – 6.0 MPa. Whereas catalysts must still be employed in medium-pressure processes operating between 150 and 190 °C, the rate of splitting in high-pressure plants is sufficiently fast without catalysts. With rising temperature and pressure, the mutual solubility of the two phases increases to a point where a homogeneous phase is formed. When coconut fat is used, this occurs at 293 °C and 8.0 MPa. For tallow, the values are 321 °C and 12.0 MPa [20].

**Equipment.** Mass transfer between the lipid and aqueous phases, as described by LASCARAY [24], has a decisive influence on the degree of fat splitting. Degrees of splitting >98% are achieved today by applying principles of liquid – liquid extraction. The processes and apparatus used for extraction are summarized in Figure 11. The simplest apparatus is the mixer – settler, which consists of a mixing tank and a settling chamber. In Figure 12, a continuously operating, countercurrent mixer – settler in open-tank form is shown [33]. Because the open construction only allows a maximum splitting temperature of 100 °C, application is restricted to catalytic processes such as enzymatic fat splitting or the Twitchell process.

Higher reaction temperatures can be used in pressure autoclaves, which are employed mainly in batch operation. With a battery of autoclaves in parallel, a one-stage process can be carried out in cycles (Fig. 13). Autoclaves in series enable multistage processes to be operated batchwise or continuously with co- or countercurrent streams.

For high throughput rates, spray columns operating countercurrently are used [34]–[36]. The main feature of this system is the countercurrent movement of the two liquid phases as a result of their difference in density. Efficient mutual extraction is essential for fast hydrolysis of fat. This process, first described by H. MILLS [20], has been modified by various companies. Commercial countercurrent splitting towers with varying features (e.g., for energy conservation) have been designed by Colgate – Emery, Badger, Foster – Wheeler, and Lurgi. The Foster – Wheeler process [37] is shown as an example in Figure 14.

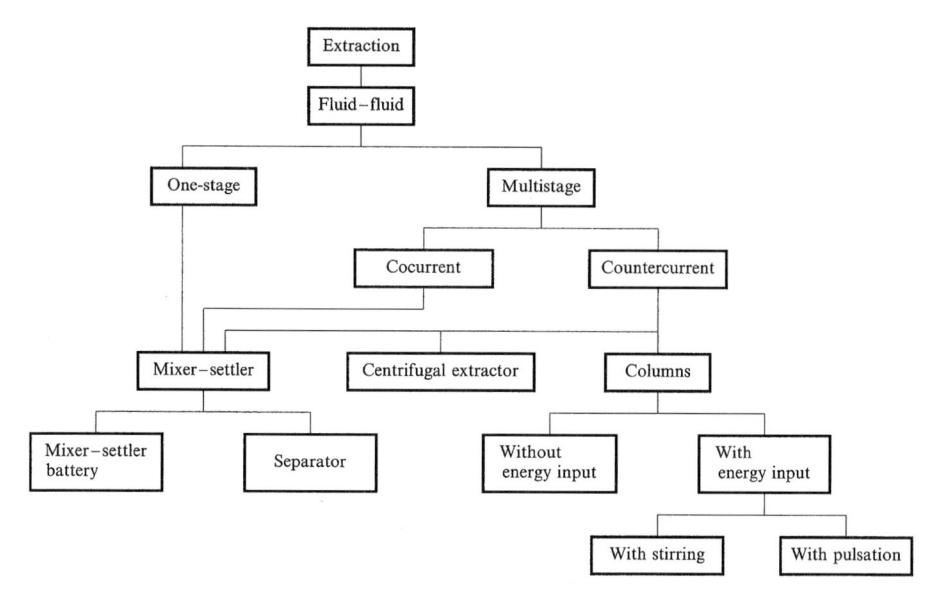

**Figure 11.** Extraction processes

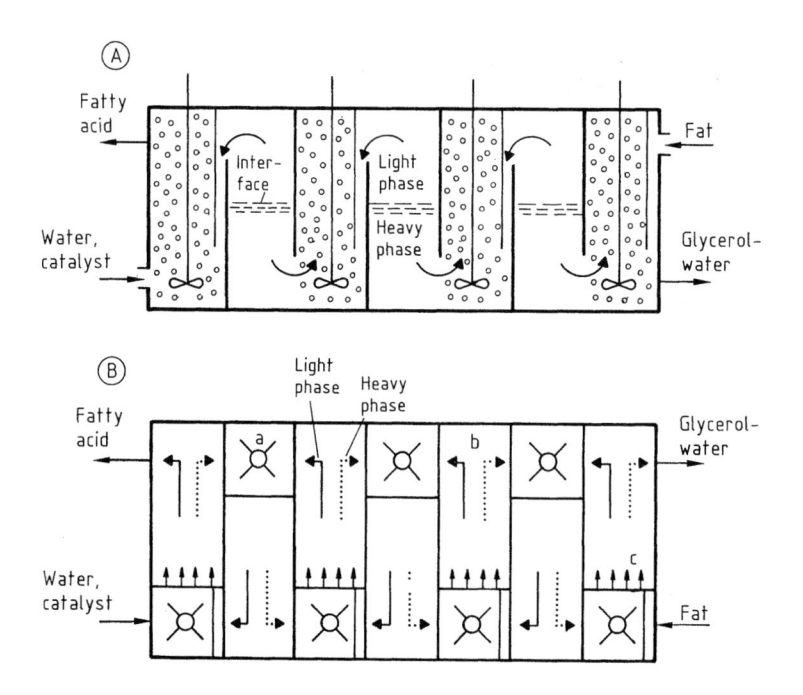

**Figure 12.** Multistage mixer–settler
A) Side view; B) Top view
a) Mixer; b) Settler; c) Slot plate

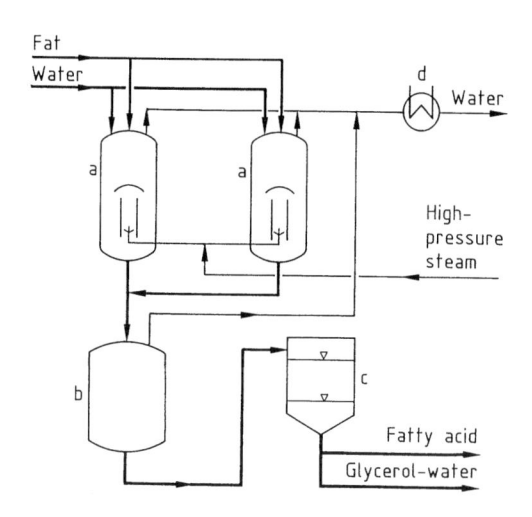

**Figure 13.** Batch fat splitting in two parallel autoclaves with direct heating
a) Splitting autoclaves; b) Flash tank; c) Separation tank; d) Condenser

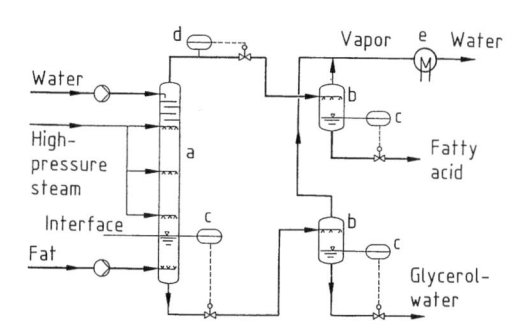

**Figure 14.** Continuous fat splitting (Foster–Wheeler)
a) Splitting column (250–260 °C, 5–6 MPa); b) Flash tank; c) Level control; d) Pressure control; e) Condenser

Fat and process water (mainly condensate from the concentration of glycerol–water) are pumped into a splitting column by high-pressure metering pumps. Fat is fed in near the bottom of the column ca. 0.5 m below the interface; water is fed in at the top. The reaction zone, which lies between the upper and lower heat-exchange zones, is heated to 250 °C by direct feeding of high-pressure steam. The upper heat-exchange zone is provided with baffles. A pressure regulator controls the discharge of fatty acids at the top, and an interface regulator automatically adjusts the flow of the glycerol–water mixture (12–18 % glycerol) from the bottom of the column.

Steam from expansion of the glycerol–water can be used as an energy source for evaporation of the glycerol–water, which is generally concentrated to 88 % crude glycerol.

## 3.3. Separation of Fatty Acids

### 3.3.1. Distillation

Crude fatty acids from splitting contain a series of high-boiling impurities, such as unreacted partial glycerides, soaps, glycerol, sterols, phosphatides, and pitch, as well as lower boiling materials like water, low molecular mass hydrocarbons, aldehydes, and methyl ketones. These impurities are removed by distillation. Oleic acids obtained by crystallization (cf. Section 2.3.2), modified fatty acids (cf. Section 3.4), and bottom products formed during the fractionation of fatty acids are also subjected to distillation.

The damage to fatty acids resulting from oxidation, decomposition, dehydration, polymerization, and polycondensation increases exponentially with increasing temperature. The byproducts formed lead to a deterioration in color and odor, and to a reduced yield of distillate. For this reason, the distillation temperature is lowered by use of a vacuum and by injection of steam. In the past, distillation was commonly performed at atmospheric pressure with large quantities of superheated steam. Today, fatty acids are distilled under vacuum [38], [39]. The disadvantages of steam distillation (high plant costs, losses due to carryover and formation of clabber (residue), environmental pollution) far outweigh the advantages (direct heating, stirring and deodorizing effects). However, small quantities of steam are indispensable for preventing the formation of anhydrides and for deodorizing purposes [40], [41].

Depending on the type of plant, still, flash, or film evaporation may be employed. In *still distillation,* the fatty acid is distilled from a large volume of liquid, and the residue remains completely or partially in the still. In *flash evaporation,* the fatty acid undergoes forced circulation, and heating takes place outside the distillation zone, which is maintained under vacuum. In the distillation zone, the fatty acid evaporates and is cooled to its boiling point at the operating pressure. The internal heat is converted into heat of evaporation. In *film evaporation,* a thin film of fatty acid is formed. Evaporation takes place only from the surface of this film at operating pressure. Because of the absence of hydrostatic pressure and liquid holdup, the temperature difference with respect to the heating medium is only 20–30 °C, and the residence time is often just a few minutes. For this reason, film evaporation is especially suitable for distillation of fatty acids.

All three principles of evaporation can be applied in multiple stages with or without injection of steam. However, overheating of the fatty acid vapors as a result of direct contact with heating surfaces must be avoided. Heating is carried out indirectly with a heat-transfer oil or steam, or directly with superheated steam. Fatty acid vapors are condensed in surface condensers, thus eliminating wastewater problems that arise when injection condensers are used. A vacuum is generated with the help of vapor compression in an ejector and with recycling of the cooling water [42]. The construction material should be stainless steel, preferably with a molybdenum content of at least 2.5 %

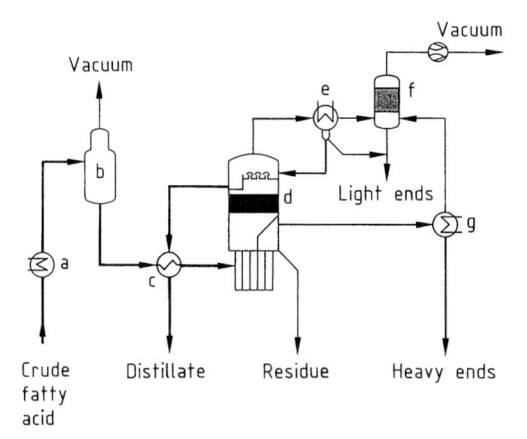

**Figure 15.** Continuous fatty acid distillation (Lurgi) [43]
a) Feed preheater; b) Dryer and degasser; c) Heat exchanger (feed–distillate); d) Still with heating candles; e) Fatty acid condenser; f) Demister; g) Heavy-ends condenser

Apart from the different evaporation principles, distillation units for fatty acids can also be divided into batch, semibatch, continuous straight, or continuous fractional distillation units. *Batch distillation* is now employed only for low-capacity plants (e.g., Fraser, Scott, and Wecker stills). Because of high temperature and superheating in the still, the yield and quality of the distillate are affected adversely.

Modern *continuous straight distillation* equipment for fatty acids is based on the flash or film principle. Degassed and dried feed is fed in continuously. The fatty acids are kept in forced circulation and evaporate rapidly. Vapors are washed with refluxed, undercooled distillate and partially condense inside or outside the still. Distillate and residue are removed continuously. Light ends can generally be drawn off, so that lower boiling substances that impair odor and color can be removed from the distillate. The residue is often subjected to posttreatment with superheated steam to drive off most of the remaining fatty acids. If a certain degree of fractionation is desired, the distillate is passed through the plant for a second time and separated into higher and lower boiling fractions.

All possible methods of heat recovery are utilized, and computer simulation programs can be employed to control process operation. Compared to injection condensation, surface condensation results in a considerable reduction of environmental pollution. Figure 15 shows a plant built by Lurgi [43].

The evaporator consists of several heating candles in series; the liquid surrounding the vertical heating elements is continuously circulated by means of injection steam. In this way, the fatty acid is sprayed against a baffle plate and evaporates. The residue is drawn off from the last candle chamber. Because of the sequential arrangement of the candles, a certain degree of fractionation is achieved. The vapors are condensed partially outside the still. The distillate is returned to a stripper plate, where it is heated almost to its boiling point so that the lower boiling constituents evaporate. These are separated in a demister and drawn off as light ends.

A similar evaporation principle is described by Unichema [44] for distillation of tallow fatty acids.

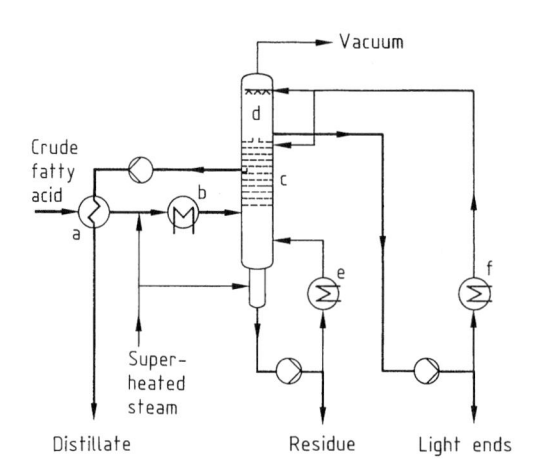

**Figure 16.** Continuous fatty acid distillation (Badger) [45]
a) Feed preheater/distillate cooler; b) End preheater; c) Distillation tower with sieve plates; d) Internal condensation section; e) Reboiler; f) Reflux cooler

The system employed by Badger Engineers [45] also operates continuously on the flash principle (Fig. 16).

Both the crude fatty acid and the circulated contents of the sump are heated and flash-distilled, with the addition of superheated steam in the bottom part of the evaporator column. The rising vapors pass through a series of perforated plates and condense in the upper part of the column. A small uncondensed fraction, containing very little fatty acid, leaves the top of the column and passes into the vacuum system. The head cut is removed in small quantities from the undercooled reflux. The main condensate is drawn off from the side of the column, and its heat is transferred to the entering crude fatty acid. Between the inlet for crude fatty acid and the exit for distillate, a certain degree of fractionation occurs.

Film evaporation is employed in a Henkel process [46] (Fig. 17).

Crude fatty acid is degassed and dried, preheated, and then evaporated in a falling-film evaporator. The vapors pass through the evaporator column (whose packing causes a very small pressure drop), and condense stepwise in the upper part of the column and in two condensers connected in series. The lower boiling constituents are removed from the second condenser. The condensate from the first condenser is returned as undercooled reflux to the upper packing of the column. The main fraction is withdrawn above the lower rectification zone. A second falling-film evaporator concentrates the residue in a side chamber of the still. With this arrangement, a strong fractionating effect is achieved, by which low-boiling odorous substances and high-boiling colored materials are removed. Other methods of evaporating fatty acids which operate on the falling-film principle are, for example, the Schmidding [47], Stage [48], and Mazzoni [49] processes.

*Fractional distillation* is applied to obtain fractions of high purity. Fatty acid mixtures with chain lengths of $C_6-C_{18}$ from coconut or palm-kernel oil [50] and higher unsaturated $C_{18}-C_{22}$ fatty acids from rapeseed oil [51] are fractionated. Separation of fatty acids from resin acids in tall oil [52] is also accomplished by fractional distillation, while the fractionation of tallow, soybean, cottonseed, or linseed fatty acids

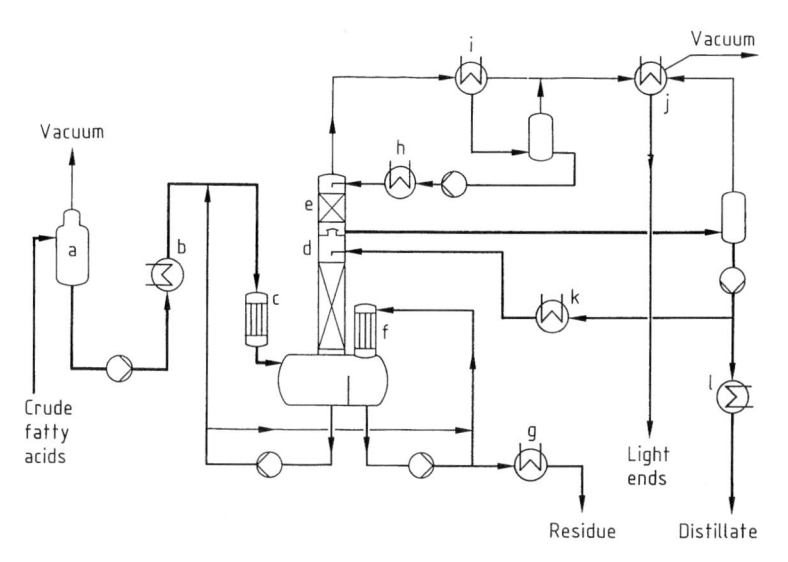

**Figure 17.** Continuous fatty acid distillation (Henkel) [46]
a) Dryer and degasser; b) Feed preheater; c) Falling-film evaporator; d) Distillation tower; e) Internal condensation section; f) Falling-film evaporator (pitch stage); g) Residue cooler; h) Reflux cooler; i) Reflux condenser; j) Light-ends condenser; k) Cooler for main distillate reflux; l) Distillate cooler

is of minor importance. The purity required for fatty acid fractions is up to 99 %. Separation takes place in vacuum columns provided with internal devices to intensify mass transfer between the vapor and liquid phases. These can be trays or various types of packing material; they can increase or decrease pressure drop in the vapors passing through the column. Because any increase in pressure requires an increase in temperature at the bottom of the column, the pressure drop per theoretical plate must be kept as low as possible. With normal trays, the pressure drop per theoretical plate amounts to ca. 0.5 kPa; with metallic filling material, to ca. 0.1 – 0.2 kPa; and with modern packings, to ca. 0.05 kPa [53], [54]. Thus the number of theoretical plates can be increased for a given pressure drop. Not only can energy be saved, but difficult separations of unsaturated fatty acids can also be carried out (e.g., the manufacture of high-purity erucic acid) without significantly increasing the amount of unsaponifiable material. Steam is no longer necessary to reduce the boiling point. This is an advantage because steam causes environmental problems and also reduces the difference between the boiling points of the constituents, thus making separation more difficult. In the lower part of the column, heat is supplied by means of an evaporator of the flash or falling-film type. In the condensation part, this heat is conducted away again by surface condensers. The vacuum is generated by a multistage jet pump which permits a pressure of less than 0.5 kPa at the head of the column. Sealing the plant is particularly important because the presence of atmospheric oxygen results in rapid oxidation of fatty acids.

As the number of components present in a mixture of fatty acids increases, the mixture must be passed through the column more often to achieve complete separa-

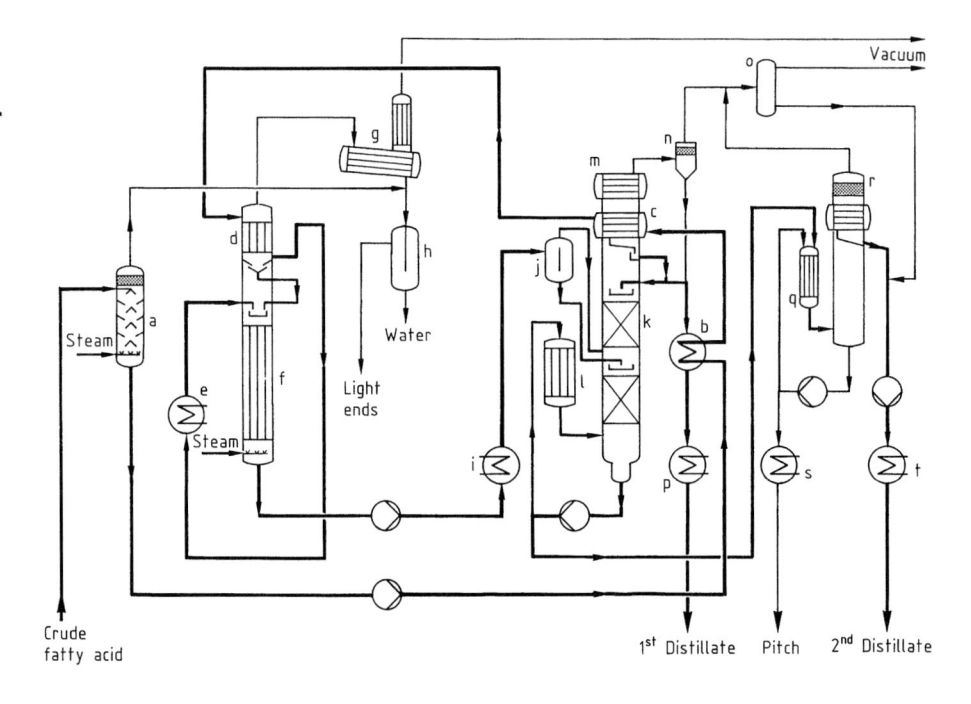

**Figure 18.** Fatty acid fractionation (Stage) [50]
a) Film deaerator; b) Feed preheater/precut cooler; c) Feed preheater/distillate dephlegmator – condenser; d) Feed preheater/light-ends dephlegmator – condenser; e) Feed end preheater; f) Light-ends falling-film stripper; g) Light-ends condenser; h) Light-ends separator; i) Main cut preheater; j) Flash pot; k) Main cut fractionation column; l) Column falling-film evaporator; m) Column end condenser; n) Demister; o) Cooling trap; p) Distillate end cooler; q) Falling-film evaporator (pitch stage); r) Vapor – liquid separator with condenser (pitch stage); s) Pitch cooler; t) Distillate cooler (pitch stage)

tion. When a high throughput is required, four to six columns are often connected in series. On the other hand, with throughput under 2000 t/a, it is more economical to use the same column repeatedly. Before fractionation begins, the crude fatty acid must be degassed and dried. The final step is overhead distillation of the material from the bottom of the last column to improve the color.

Figure 18 shows a fractionating plant built by Stage [50]. It consists of a degassing and a light-ends stage, a fractionating column (with packing designed for a low pressure drop), and a final evaporation stage.

The fatty acid mixture is degassed, preheated repeatedly by heat exchange, and then separated from water, low-boiling fatty acids, and impurities in a countercurrent falling-film evaporator. Before entering the main column, the feed is partially evaporated and then separated in the column into low- and high-boiling fractions. Evaporation at the bottom of the column takes place in a falling-film evaporator with forced circulation. The lower boiling fraction is removed directly, while the material from the bottom of the column is distilled overhead in the last falling-film evaporator (pitch stage).

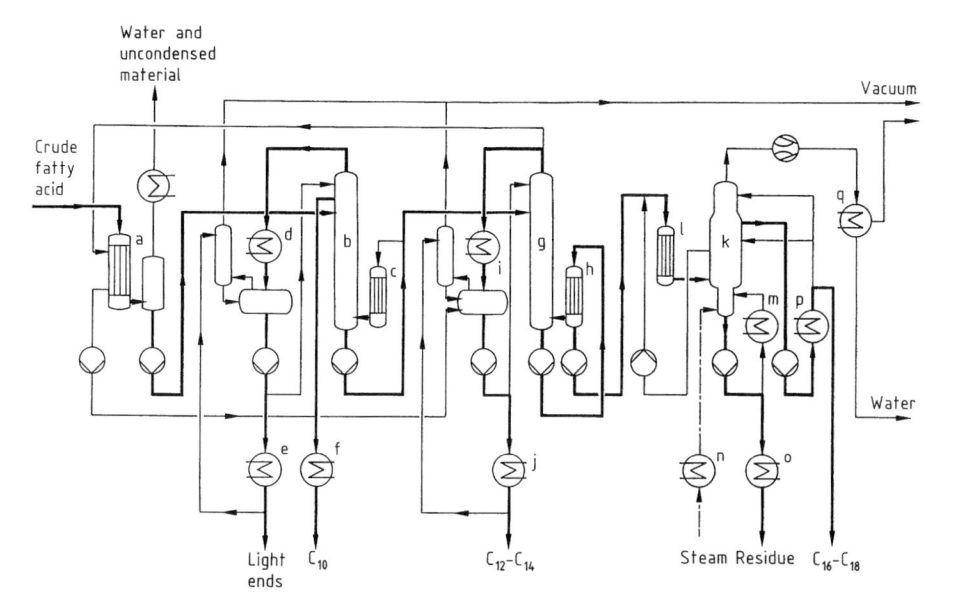

**Figure 19.** Mazzoni two-column coconut fatty acid fractionation plant [55]
a) Falling-film dryer and degasser with condenser; b) Light-ends fractionation column; c) Falling-film evaporator; d) Light-ends condenser; e) Light-ends cooler; f) Sidestream cooler; g) Main cut fractionation column; h) Falling-film evaporator; i) Distillate condenser; j) Main cut cooler; k) Distillation tower; l) Falling-film evaporator; m) Residue reboiler; n) Steam heater; o) Residue cooler; p) Heavy-ends cooler; q) Water condenser

As already mentioned, when a large number of components must be separated, the number of fractionating columns can be increased as desired. Thus, Mazzoni offers plants with two or four columns, plus the degassing and pitch stages. Figure 19 shows the two-column version for the fractionation of coconut fatty acid [55]. Separation takes place according to the boiling point of the individual fractions: first the low-boiling heads, then $C_{12}-C_{14}$, and finally $C_{16}-C_{18}$. In the columns, special sieve-type plates are used which cause only a low pressure drop. All evaporators operate on the film principle. A similar concept is offered by Schmidding [56]. In this case, the columns contain special low-pressure-drop bubble-cap plates. Here, too, falling-film evaporators are used exclusively.

All possibilities for saving energy by consistent use of heat exchange between the stages are exploited. Optimization of operating costs for individual fractions or series of fractions is possible with the aid of simulation programs. The theoretical results obtained are in excellent agreement with operating data [57]. Control of the columns is carried out primarily by conventional methods, but here too, computer technology is beginning to be applied [58]. Modern process control systems are increasingly replacing conventional methods. Complete control of fractionating plants by process computers is only a matter of time, and will be implemented as soon as procedures for on-line gas chromatography of fatty acids and their derivatives are ready for operation [59].

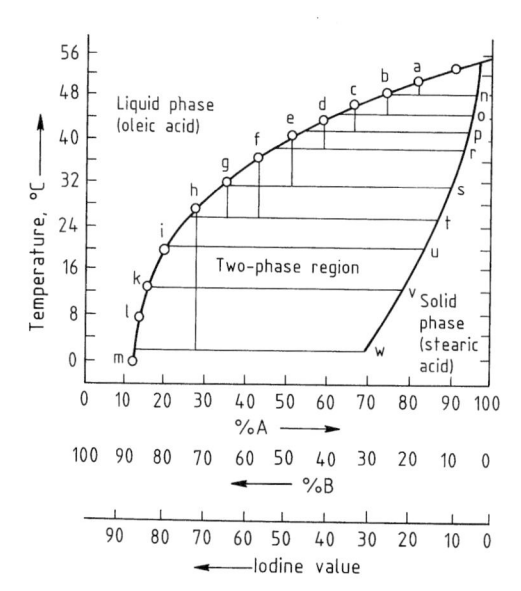

**Figure 20.** Phase diagram of tallow fatty acids A = saturated fatty acids; B = unsaturated fatty acids [62]

## 3.3.2. Crystallization

Separation by crystallization means the separation of a fatty acid mixture into a higher melting fraction, consisting mainly of saturated acids, and a lower melting fraction, consisting mostly of unsaturated acids.

The prototype of this process is the separation of tallow fatty acids into stearic acid (stearin) and oleic acid (olein). The stearin obtained from this separation has an iodine value of less than 20 and is normally hardened subsequently to an iodine value of 1. The crude olein obtained has a cloud point below 3 °C. Apart from tallow fatty acid, distilled or crude fatty acids from the splitting of linseed, bone, soybean, sunflower, or rapeseed oil can be employed.

When suitable molten fatty acids such as tallow fatty acid are cooled, the higher melting, saturated constituents crystallize first, while the appreciably lower melting, unsaturated constituents remain liquid over a defined temperature range. Because of the formation of mixed crystals, the solid phase still contains residual unsaturated fatty acids. Conversely, residual saturated fatty acids are found in the liquid phase because of their solubility in unsaturated fatty acids. Since various chain lengths and degrees of saturation are present in natural fatty acid mixtures, a theoretical discussion of this solid–liquid equilibrium is extremely difficult. From the experimentally determined phase diagram for tallow fatty acid, the composition and quantitative ratio of the two phases "stearin" and "olein" can be seen in relationship to temperature and to the composition of the starting material (Fig. 20).

Olein must be removed at a temperature corresponding to the desired olein content. However, because the olein content of stearin increases with decreasing temperature, separation at higher temperature of a stearin with a low olein content may be advisable.

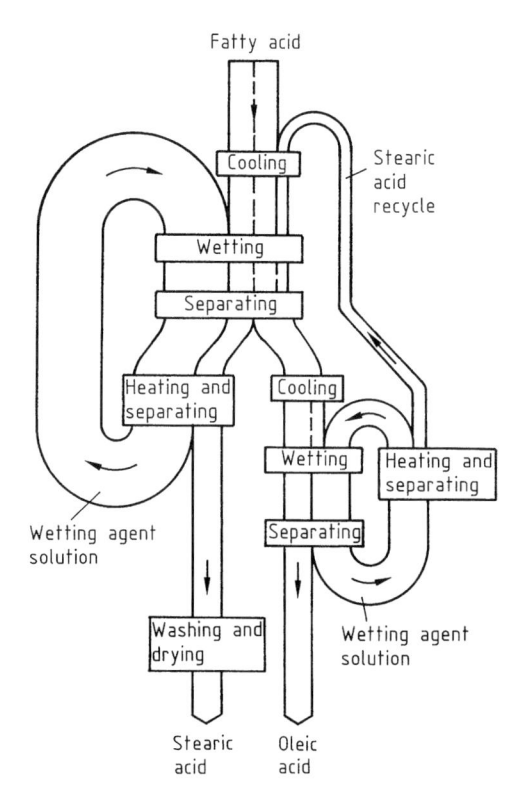

**Figure 21.** Sankey diagram of two-stage hydrophilization process

Separation of the liquid phase can then be carried out in a second step at a lower temperature.

**Fractional Crystallization without Additives.** In the "panning and pressing process," the molten tallow fatty acid is cooled slowly to ambient temperature in flat aluminum pans. Fatty acid cakes are wrapped in cloths and separated from the liquid fraction in hydraulic presses. The liquid from the first pressing is cooled to 5–10 °C, and more stearin crystals are removed, which results in an olein with a cloud point of 5–10 °C. Today, the pressing process is rarely used because it is labor-intensive and the separation is poor.

**Fractional Crystallization with Addition of a Wetting Agent.** Liquid and solid phases can be separated more easily and more completely if an aqueous solution of a wetting agent is added to the crystal paste. The stearin crystals are wetted by the wetting agent and form an aqueous suspension which is separated from the olein phase by centrifugation.

This principle was developed by Henkel into a continuous, two-stage hydrophilization process [60]. The mode of operation can be seen from the Sankey diagram (Fig. 21)

2513

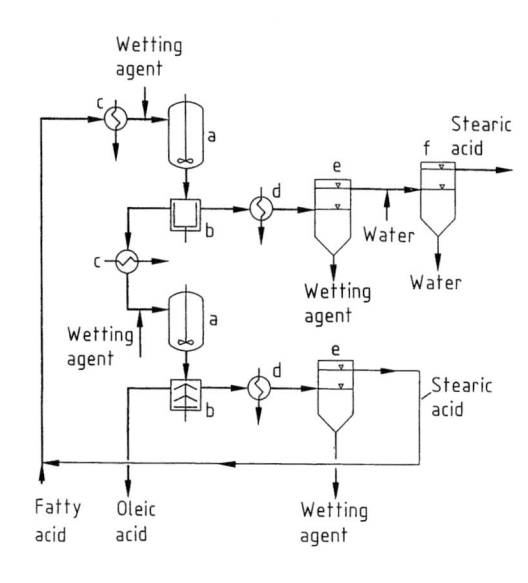

**Figure 22.** Flow diagram of two-stage hydrophilization process
a) Stirred vessel (wetting); b) Centrifuge; c) Cooler; d) Heater; e) Settling tank (separation of wetting solution); f) Settling tank (washing stage)

and from the flow diagram (Fig. 22). Lurgi builds one- and two-stage plants based on this principle [43], [61].

Stearin crystals form when molten tallow fatty acid is cooled to ca. 20 °C. A wetting agent solution containing an electrolyte (e.g., magnesium sulfate) is added to the mixture of stearin crystals and liquid olein, and the crystals are thus wetted. The stearin–water suspension is separated from olein in a centrifuge. Olein from the first stage is cooled in a second step to the desired cloud point, treated again with a solution of wetting agent, and separated in a centrifuge. The cold stearin suspension from the first stage is heated and separated from the wetting agent solution. Before additional processing (e.g., final hardening), the stearin is washed and dried under vacuum. The stearin suspension from the second stage is recycled by adding it to the starting material for the first stage. The solution of wetting agent and electrolyte is also recycled; only a small amount of fresh material must be added continuously to maintain optimal concentration. Cooling is normally carried out indirectly by passing chilled brine through the outer jacket of a scraped surface crystallizer. Otherwise, the mixture can be cooled by evaporating part of the water under vacuum.

The hydrophilization is sometimes referred to as the Henkel, the Lipofrac [63], or the Lanza (Unilever) [64] process.

**Fractional Crystallization from Solvents.** Another method of crystallization is based upon the fact that saturated fatty acids are less soluble than their unsaturated counterparts in organic solvents such as methanol, hexane, or acetone. The first process of this type was put into operation by Emery (Emersol plant, Fig. 23) [65].

A ca. 30 % solution of, for example, tallow fatty acid in methanol is cooled continuously to −12 °C in a tubular crystallizer equipped with scrapers. From the resulting crystal paste, the mother liquor containing mainly unsaturated fatty acids is sucked through a rotating vacuum filter. The filter cake,

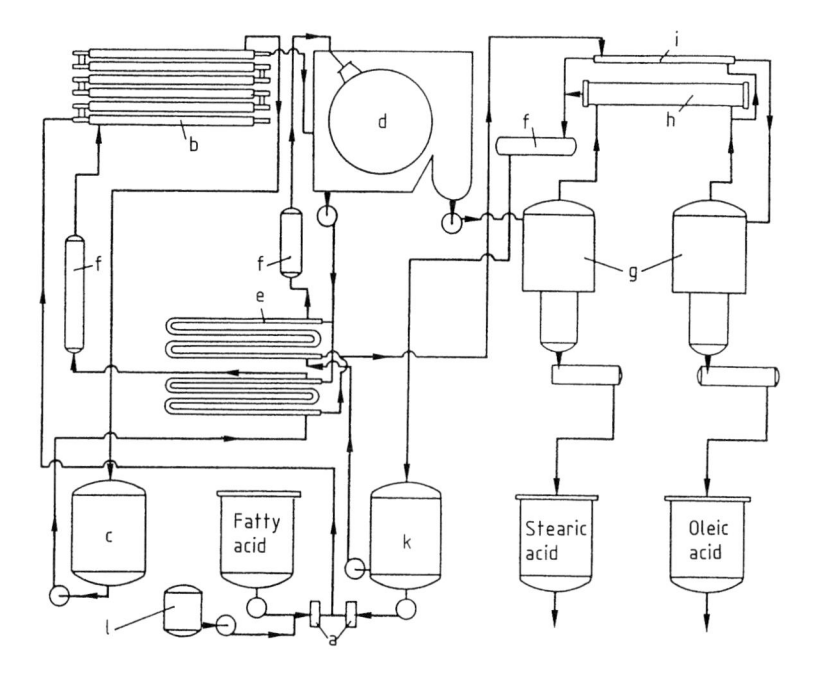

**Figure 23.** Flow diagram of Emersol plant [66]
a) Measuring instruments; b) Crystallizer; c) Alcohol for cooling; d) Vacuum filter; e) Heat exchanger; f) Cooler; g) Distillation; h) Condenser; i) Heater; k) Alcohol storage vessel; l) Crystallization accelerator

which contains saturated fatty acids, is washed free of the mother liquor and melted. The solvent is recovered from both the filter cake and the mother liquor by distillation. The stearin obtained still contains ca. 5–15% oleic acid and has an iodine value of ca. 5–15. Depending on the filtration temperature, the olein has a cloud point from 1 to 8 °C [66].

The corresponding Armour process employs acetone as a solvent [67], [68]. Indirect cooling can be replaced by direct cooling, e.g., by partial evaporation of the solvent at reduced pressure [69], [70].

The Sulzer–MWB process [71], [72] operates on the principle of multistage fractional crystallization.

The crystallizer consists of a bundle of vertical tubes. The solution of tallow fatty acid in hexane flows down the outside of the tubes and the cooling medium flows in the same direction on the inside, in both cases as a falling film. During crystallization, a solid layer of stearin is formed on the tube walls. When a given amount of stearin has been separated in this way, the layer is first melted partially to remove dissolved olein and then melted completely. This remelted crystal component $C_1$, together with the liquid component $R_3$ returned from stage 3, serves as feed for separation stage 2 (see Fig. 24). A fraction $S_2$ of the crystals $C_2$ separating in this stage is recycled; the remainder is fed to separation stage 3. The tallow fatty acid is fed with the remelted stearin component $C_1$ and the liquid component $R_3$ into the second stage. The desired separation can be achieved by varying the number of stages, the stage at which the tallow fatty acid is fed in, the ratios of the product streams, and the temperature. All stages of the process are carried out in a single apparatus, and switching from one stage to another is

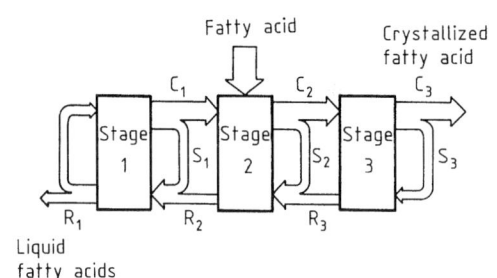

**Figure 24.** Three-stage, repetitive crystallization (Sulzer–MWB system)
$C_1$, $C_2$, $C_3$ = crystallized fraction of ascending content of saturated fatty acids
$R_3$, $R_2$, $R_1$ = liquid fraction of ascending content of unsaturated fatty acids
$S_1$, $S_2$, $S_3$ = partially melted fraction of ascending content of saturated fatty acids

completely automatic. The solvent must be removed from the final products by distillation. By using a six-stage separation, stearin with an iodine value of 2 and olein with a cloud point of 0 °C are obtained.

The Bernardini [73] and Rau [74] processes also operate on the principle of fractional crystallization from solvents. Common to all these processes is the need for extensive safety measures (explosion prevention) and for environmental protection with respect to the emission of solvents. Recovery of solvents requires additional equipment and energy.

### 3.3.3. Other Separation Methods

*Adsorption.* For the separation of oleic and linoleic acids, UOP recommends an acid-resistant, hydrophobic, non-zeolite molecular sieve (Silicalite) [75]. Oleic acid is first selectively adsorbed and later displaced with the aid of a solvent. A nonionic, hydrophobic, spatially cross-linked styrene polymer has been used for the same purpose [76].

If fatty acid methyl esters are employed instead of fatty acids, even zeolites that are sensitive to acid are suitable for the separation. Special types have a very high selectivity for oleic acid methyl ester [77]. Subsequent hydrolysis is necessary to obtain fatty acids from the methyl esters.

*Extraction.* Laboratory separation of stearic and oleic acids has been achieved by high-pressure extraction with a supercritical gas at 10 MPa and ca. 130 °C [78]. Benzene was used as solvent and ethylene as extractant.

The solubility of oleic, stearic, and behenic acids in supercritical carbon dioxide has been investigated at 40 and 60 °C and 10–25 MPa. A significant enrichment of oleic acid compared to stearic acid is observed only above 20 MPa [79]. Enrichment depends on the rate of material transfer and on the relative rates of movement of the phases, so that a separation can be expected after brief extraction [80].

High-purity oleic, linoleic, and linolenic acids have been prepared by means of liquid–liquid extraction, with solvent pairs such as furfural–hexane and ethanol–petrol ether [81].

*Separation in the Presence of Acid Soaps.* In the system soap–fatty acid–water, micelles known as "acid soaps" form at certain concentrations [82]. Acid soaps of saturated acids crystallize from aqueous solution, whereas those of unsaturated acids remain in solution. Thus two fractions are obtained with differing degrees of saturation [83].

*Separation by Formation of Addition Products with Urea.* In the crystallization of urea from methanol in the presence of fatty acid mixtures, the urea crystals are enriched in saturated fatty acids relative to unsaturated fatty acids and in monounsaturated fatty acids relative to polyunsaturated fatty acids. Separation can be carried out easily by filtration and then dissolving the urea crystals in water. For example, 97–99% oleic acid containing only 0.2% polyunsaturated material has been obtained from olive oil fatty acid [84]. The fatty acids in soybean, fish, wheat germ, and grape-seed oils have also been separated by this method [85].

## 3.4. Modification of Fatty Acids

### 3.4.1. Hydrogenation

Catalytic hydrogenation of unsaturated fatty acids can be carried out in such a way that the carboxyl group is retained and saturated fatty acids are formed [86]. Because an increase in the degree of saturation also increases the melting point of fatty acids, this addition of hydrogen is generally known as hardening. Hardening of fatty acids usually means complete saturation of the carbon–carbon double bonds down to an iodine value less than 1. In this way, light-colored fatty acids are obtained which are thermally more stable and resistant to oxidation. Starting materials are primarily acid oils from the refining of edible oils, split fatty acids from tallow, and the stearin fraction from the olein–stearin separation process.

**Chemistry — Reaction Conditions.** The hardening reaction, i.e., the addition of hydrogen to the double bonds of unsaturated fatty acids, is exothermic. Like other types of heterogeneous catalysis, the reaction proceeds in stages:

$$\text{Linolenic acid } (C_{18:3}) \xrightarrow[k_3]{H_2} \text{Linoleic acid } (C_{18:2}) \xrightarrow[k_2]{H_2} \text{Oleic acid } (C_{18:1}) \xrightarrow[k_1]{H_2} \text{Stearic acid } (C_{18})$$

Highly unsaturated fatty acids are hydrogenated more rapidly than monounsaturated fatty acids ($k_3 > k_2 > k_1$). Hydrogenation is accompanied by isomerization processes in which double bonds migrate (conjugation) and geometrical changes occur (cis–trans isomerism) [87]. Because fatty acid hardening is carried out predominantly to achieve complete saturation, the selectivity of the reaction (i.e., the ratio $k_3:k_2$ or $k_2:k_1$ → Fats and Fatty Oils) is of minor importance. To some extent, the fatty acid reacts with the nickel catalyst, particularly in the presence of oxygen or water, to form nickel soaps:

$\text{Ni} + 2\ \text{RCOOH} \rightleftharpoons \text{Ni(OOCR)}_2 + H_2$

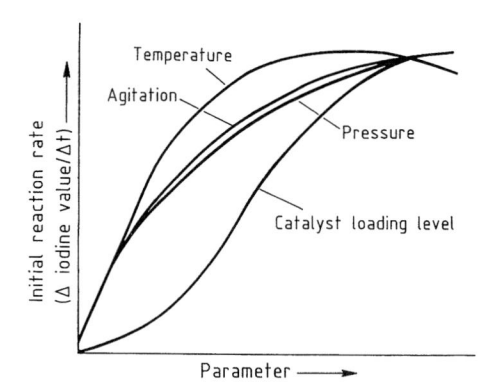

**Figure 25.** Effect of operating variables on the reaction rate of fatty acid hydrogenation

Because nickel soaps are catalytically inactive, this reaction blocks the active surface of the catalyst. To reduce soap formation, the reaction is carried out at higher pressure, generally 2.5–3.5 MPa, and the fatty acids are predried. Despite these precautions, catalyst consumption is appreciably higher in the hardening of fatty acids than in hydrogenation of the corresponding triglycerides. The rate of hydrogenation increases initially with increasing temperature, then passes through a maximum between 180–210 °C depending on the fatty acid used, and finally decreases [88]. Hydrogen transport through the liquid to the surface of the catalyst is dependent on the presence of large interfaces between the phases. Hardening is carried out, therefore, with intensive mixing by means of stirrers or circulation pumps. The influence of various reaction parameters on the reaction rate of fatty acid hardening [89] is illustrated in Figure 25.

The average practical reaction rate corresponds to a reduction of the iodine value by 1–2 units per minute. A reduction of 1 in the iodine value corresponds to the consumption of ca. 1 m$^3$ (STP) of hydrogen per cubic meter of fatty acid and results in a temperature increase of ca. 1.5 °C.

**Catalysts.** The hardening of fatty acids on a commercial scale is carried out with nickel-based heterogeneous catalysts in slurry form. Dry reduced carrier catalysts (nickel on silica) with high activity and a good resistance to poisoning are used. The active surface of this type of catalyst has been increased to over 100 m$^2$/g. The pore diameter of the carrier must be large enough (ca. 3 nm) to allow transport of fatty acid molecules to the nickel crystallites. The nickel carrier catalyst is pyrophoric in its activated form. For safer transport and handling, the catalyst is passivated temporarily by a coating of hydrogenated fat. A typical catalyst has the following composition: 20–25 % nickel, 10–15 % silica, and 60–65 % fat.

A number of other catalysts have been described [90]. Acid- and poison-resistant catalysts in pellet form have been investigated for continuous hydrogenation of fatty acids in a fixed-bed process. Catalysts in pellet form based on noble metals are acid resistant, but they are deactivated rather quickly, especially when crude fatty acids are used as feedstock.

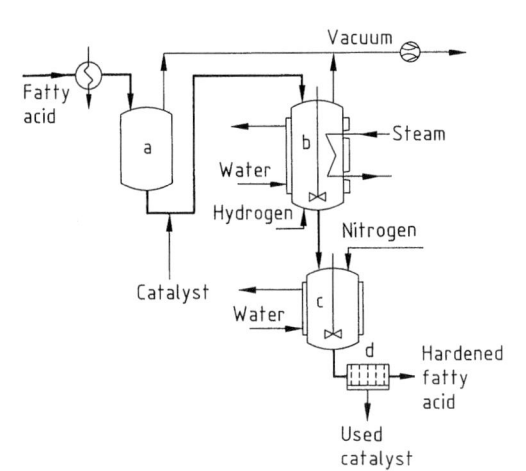

**Figure 26.** Batch hydrogenation plant
a) Dryer; b) Autoclave; c) Cooler; d) Filter

**Manufacturing Procedure.** In principle, the same plants can be used for the hydrogenation of fatty acids and for the hardening of oils. Plants in which fatty acids are processed, however, require stainless steel as construction material. The reactor is designed for pressures up to ca. 3.5 MPa. An increasing number of hardening plants operate continuously. The economics of batch operation have been improved significantly by heat recovery systems [91]. Despite good acid resistance of the catalyst, after filtration the hardened fatty acids still contain nickel soap. The hardening step is usually followed by distillation to remove the nickel soap and to free the hardened fatty acids from the odor that arises during hardening.

Figure 26 shows a typical batch hardening plant. Hardening is carried out in batches of 5–30 t. Thorough mixing by effective agitators keeps the catalyst suspended.

In the continuous hardening process developed by Lurgi, a mixture of dried fatty acid, catalyst, and hydrogen is fed into the bottom of the reactor which consists of several chambers. The fatty acid passes through the reactor only once, while the excess of hydrogen is recycled (→ Fats and Fatty Oils).

Particularly thorough mixing is achieved with the reaction loop designed by Buss (→ Fats and Fatty Oils). This process can be carried out batchwise in a single reactor or continuously using several reactor loops in series. The plant can also be equipped with a heat recovery system.

## 3.4.2. Isomerization

Two isomerization reactions are of interest commercially:

1) cis–trans isomerization of oleic acid to elaidic acid
2) conjugation of polyunsaturated fatty acids

**Cis – Trans Isomerization.** Natural fatty acids occur mostly in the higher energy cis configuration. The conversion of oleic acid (cis configuration) into elaidic acid (trans configuration) is carried out industrially by treating oleic acid with sulfur dioxide, oxides of nitrogen, selenium, or acid-activated earths at 100 – 200 °C. The rearrangement reaches an equilibrium at 75 – 80 % elaidic acid [92] – [94].

Cis – trans isomerization is important in the selective partial hardening of oils for the food industry. This process is carried out to increase the melting point of the oil.

$$CH_3-(CH_2)_7\diagdown C=C \diagup H \atop HOOC-(CH_2)_7 \diagup \diagdown H \quad \xrightarrow{SO_2,\ Se,\ NO_x} \quad CH_3-(CH_2)_7 \diagdown C=C \diagup H \atop H \diagup \diagdown (CH_2)_7-COOH$$

Oleic acid (cis configuration) mp 16 °C

Elaidic acid (trans configuration) mp 51 °C

Partial hydrogenation of polyunsaturated fatty acids increases the proportion of trans isomers.

**Conjugation.** The conjugation of polyunsaturated fatty acids has some commercial importance because of the higher reactivity of conjugated polyunsaturated fatty acids. Conjugated systems are formed by the alkali-catalyzed isomerization of nonconjugated, polyunsaturated fatty acids, e.g., fatty acids from sunflower or soybean oil:

$$-CH_2-CH=CH-CH_2-CH=CH-CH_2- \longrightarrow -CH_2-CH_2-CH=CH-CH=CH-CH_2-$$

The conjugation of polyene fatty acids is carried out by saponification with an aqueous alkali solution and heating with excess alkali to 200 – 300 °C. The soap paste is treated with mineral acid to recover the fatty acid. The degree of conjugation achieved is 95 % [95], [96].

The conjugated diene and triene fatty acids obtained are partially isomerized, so that a mixture of the cis-cis, cis-trans, trans-cis, and trans-trans configurations is formed.

## 3.4.3. Dehydration

Fatty acids with a hydroxyl group in the carbon chain can be dehydrated. Ricinenic acid is obtained by dehydration of ricinoleic acid [97]. Ricinenic acid is a mixture of ca. 30 % conjugated 9,11-linoleic acid and ca. 50 % unconjugated 9,12-linoleic acid. The double bonds are partly in the trans configuration. By subsequent conjugation, a ricinene fatty acid with an increased content of conjugated double bonds can be prepared.

# 4. Production of Synthetic Fatty Acids

In the Soviet Union, several Eastern European countries, and China, significant quantities of synthetic fatty acids are manufactured by oxidation of alkanes [98]. In the Western world, however, the economic importance of synthetic fatty acids is lower than that of natural fatty acids [99]. Although a great deal of research has been devoted in the past 50 years to the development of synthetic methods, their commercial use is confined mainly to the production of short-chain and branched-chain synthetic fatty acids [6], [31]. At present, no economical synthesis for unsaturated, even-numbered, long-chain fatty acids (e.g., cis-9-octadecenoic acid) exists.

## 4.1. Oxidation of Alkanes

Straight-chain fatty acids with chain lengths of $C_4$–$C_{24}$ are obtained by oxidation of $C_{18}$–$C_{32}$ alkanes in air at ca. 110 °C. The reaction proceeds by a free-radical chain mechanism and is catalyzed by salts of manganese, cobalt, nickel, and other metals. Fatty acids with odd-numbered and branched carbon chains are formed. These synthetic fatty acids are inferior in quality, particularly with regard to odor and color, because of the presence of byproducts such as aldehydes, ketones, and esters. The wide distribution of chain lengths makes refining processes difficult and expensive. After about 30% of the alkanes have reacted, the mixture is worked up by saponifying the fatty acids and separating unreacted alkanes, which are recycled. The process for the oxidation of paraffinic hydrocarbons was originally developed in Germany in the 1930s [100], [101].

## 4.2. Hydroformylation

In the "oxo reaction" discovered by ROELEN, carbon monoxide and hydrogen are added to the double bond of olefins to form aldehydes. This highly exothermic reaction takes place only in the presence of catalysts at temperatures above 180 °C and pressures of 8–30 MPa. Oxidation of the intermediate aldehyde yields the desired fatty acid. By starting with terminal linear olefins, both straight-chain and α-branched-chain fatty acids can be obtained [102]. The ratio of straight-chain to branched-chain fatty acids can be varied by changing the catalyst in the hydroformylation reaction. Suitable catalysts are cobalt compounds or rhodium phosphine complexes. The latter result in a higher percentage of straight-chain compounds. Ruhrchemie (a subsidiary of Hoechst), Celanese [103], and Rhône-Poulenc, among others, employ this process.

Fatty acids with varying degrees of branching are produced in the range of heptanoic to nonanoic acid.

## 4.3. Hydrocarboxylation

With the Koch–Haaf method, carbon monoxide is added to the double bond of olefins in the presence of strong acids. In a second step, the intermediate product reacts with water to form branched-chain fatty acids [104]. Concentrated sulfuric acid, phosphoric acid, hydrogen fluoride, or boron trifluoride can be used as catalyst. The reaction mechanism involves a carbonium ion, which reacts with carbon monoxide to form an acylium ion. This then reacts with water to form carboxylic acid. Fatty acids with highly branched chains result from rearrangement of the carbonium ion. The process operates under mild conditions up to 80 °C and 10 MPa. Shell manufactures Versatic acids by this method (for Koch process see → Carboxylic Acids, Aliphatic; for Versatic acids see → Carboxylic Acids, Aliphatic).

## 4.4. Other Commercial Fatty Acid Syntheses

Fatty acids are obtained as byproducts in two further commercial processes. Thus, pelargonic (nonanoic) acid is formed as a byproduct in the manufacture of azelaic (1,7-heptanedicarboxylic) acid by ozonolysis of oleic acid (→ Dicarboxylic Acids, Aliphatic). Emery Industries, Ohio [105] uses this method. In the first stage, ozone is added to the double bond of the oleic acid. The reaction is carried out preferably at 25–45 °C, by use of oxygen with an ozone content of 1–5 %. In a second reaction stage, the ozonide formed is then split by oxidation with ozone-free oxygen to form azelaic and pelargonic acids. This step is carried out at 75–120 °C. Olefins can also be split by ozonolysis to form straight-chain fatty acids, but this reaction is not used on an industrial scale.

Heptanoic acid is produced by Atochem in France by the thermal decomposition of esters made from castor oil. At 500–600 °C, castor oil acid methyl ester decomposes in the presence of water vapor to form oenanthal (heptanal) and undecenoic acid methyl ester. The oenanthal is then oxidized with oxygen to heptanoic acid [106]. The main product, undecenoic acid, is used as starting material in the manufacture of nylon 11 (Rilsan).

Another method used in the manufacture of fatty acids is caustic fusion of alcohols. Hoechst converts tridecyl alcohol to isotridecanoic acid, and Henkel produces isopalmitic acid from the Guerbet alcohol (→ Fatty Alcohols) made from caprylic alcohol [107].

Efforts to improve synthetic methods are still underway. Thus, Akzo is investigating the addition of acetic anhydride to linear terminal olefins (telomerization). After hydrolysis of the intermediate product, branched-chain fatty acids with high molecular masses and low melting points are obtained, which are especially useful in the manufacture of synthetic lubricants and enamels. A pilot plant is already in operation [108], [109].

## 4.5. Noncommercial Processes

Many other processes for making synthetic fatty acids have been developed to the stage of technical feasibility. However, no synthesis can compete economically with the inexpensive natural fatty acids. For example, hydrocarboxylation of olefins or acetylene with carbon monoxide and water in the presence of metal carbonyl catalysts (Reppe reaction) has not yet been used for the commercial production of fatty acids, although this reaction has been the subject of intensive work by several firms [6], [31], [110], [111].

Likewise, oxidation of olefins with air, ozone, chromium trioxide, potassium permanganate, or other compounds has not resulted in an economical manufacturing process. Oxidation of fatty alcohols with air or oxygen has also not been commercially successful. Reaction of trialkylaluminum (obtained as an intermediate in the Ziegler alcohol synthesis, → Fatty Alcohols) with carbon dioxide under pressure has not led to a usable process.

Attempts have also been made to obtain fatty acids by microbial transformation of alkanes [112], [113]. Microorganisms, such as bacteria, yeasts, and molds, can grow on a wide range of hydrocarbon materials. Enzymatic attack occurs preferentially at the terminal carbon atom through the stages alcohol, aldehyde, fatty acid. However, the acids are easily broken down or oxidized further to dicarboxylic acids.

# 5. Analysis

A number of wet chemical and physical methods are used for quality control of commercial products and identification of individual fatty acids. The composition of fatty acid mixtures from natural fats is best determined by gas chromatography, which allows the identification of individual fatty acids with great accuracy. Nuclear magnetic resonance (NMR) and infrared (IR) spectroscopy have been successfully applied to the determination of the molecular structure of fatty acids.

Many chemical data determined by wet analysis are used to characterize commercial fatty acids and to check their purity [5]. Important parameters are the acid value, saponification value, iodine value, and content of unsaponifiable matter (→ Fats and Fatty Oils). From the *acid and saponification values*, the average molecular mass and thus the average chain length of the fatty acids can be calculated. If a distilled fatty acid contains no esters or anhydrides components the saponification value is identical to the acid value. The difference between the saponification and acid values is the so-called *ester value*.

The *iodine value* is a measure of the total unsaturation of a fatty acid but does not permit conclusions to be made about the content of saturated fatty acids. Thus, oleic acid has the same iodine value (90) as a fatty acid consisting of a 1:1 mixture of stearic and linoleic acids.

Practically all technical-grade fatty acids contain small quantities of foreign substances which do not have carboxyl groups and are, therefore, not saponifiable. These constituents of natural fatty acids arise from foreign material in the fats or from decomposition products. The *content of nonsaponifiable material* in technical-grade fatty acids is regarded as a quality criterion and is usually in the range of 0.5–2%.

Important physical data for characterizing fatty acids are the solidification point, sometimes called titer, the cloud point, and the color. The *solidification point* is the temperature at which a fatty acid begins to solidify and is defined as the temperature maximum resulting from the release of the heat of crystallization.

Low-melting technical fatty acids without sharp solidification temperatures (<ca. 15 °C) are characterized by the *cloud point* , i.e., the temperature at which a clearly perceptible turbidity occurs.

The measurement of *color* is also used to specify the quality of fatty acids. Color determination can be carried out by comparison with either standard solutions (Gardner, APHA, Hazen) or colored glass standards (Lovibond, Gardner). Specific analytical methods are available to determine the content of trans isomers or of conjugated unsaturated fatty acids.

# 6. Storage and Transportation

**Storage.** Fatty acids are unstable in the presence of atmospheric oxygen. As with fats, autoxidation of fatty acids leads to the formation of hydroperoxides, which decompose to oxygen-containing products such as aldehydes, ketones, and hydroxy compounds. Other secondary products, such as polymers and splitting products, can also be found [6], [114]–[117].

The effect of atmospheric oxygen on fatty acids depends primarily on the temperature, the number of double bonds, and the molecular structure. Saturated fatty acids show little tendency to undergo autoxidation, whereas unsaturated fatty acids, and especially polyunsaturated acids, are very susceptible in this respect. Higher temperatures favor the reaction. Small quantities of other substances present either naturally or adventitiously in the fatty acids can act as pro- or antioxidants. All vegetable oils and fats contain, for example, tocopherols as natural antioxidants. In general, vegetable oils have a higher tocopherol content than animal fats and therefore, are usually more stable than animal fats with an equivalent degree of unsaturation. On the other hand, metals such as copper and iron (in the form of soaps) act as prooxidants [5].

Autoxidation products affect the color stability of fatty acids, produce a pungent, rancid odor, and alter characteristic analytical values. The content of nonsaponifiable matter and the ester value increase; the acid value and the iodine value decrease. The peroxide value is a measure of the degree of oxidation [118]. The stability of commercial fatty acids cannot be predicted reliably from the analytical data because the relationship of the raw material, manufacturing process, and storage conditions to the autoxidation behavior is very complex.

Synthetic antioxidants are not normally used in manufacturing, but they may be necessary for special applications of technical-grade fatty acids. Thus, up to 0.5% 2-naphthol is added to oleic acid used as a lubricant in the textile industry. This reduces the tendency to self-heating as a result of autoxidation (Mackey test). Another widely used antioxidant for fatty acids is butylated hydroxytoluene (BHT), which is also approved for stabilization of food, drugs, and cosmetics [5], [117]. Concentrations of 0.01% or lower have proved sufficient. Concentrations of natural antioxidants are of the same order.

Fatty acids are stored in liquid or solid form. In most cases, liquid fatty acids are kept in heated tanks. Stainless steel and aluminum are the construction materials normally used for tanks and pipelines. Storage temperature should be slightly above the solidification point.

Under these conditions, saturated fatty acids can be stored for several weeks without detrimental effects on quality; however, with some unsaturated fatty acids, the color deteriorates after a few days. Stability can be improved by storage under a nitrogen blanket.

**Transportation.** Railway tank cars, tank trucks, containers, and ships are used for transporting liquid fatty acids. Materials and temperature correspond to those used for storage. Dispatch in steel drums lined with a protective, fatty acid-resistant coating or in plastic drums is less common.

Because of their corrosiveness, short-chain $C_6-C_9$ fatty acids are subject to international regulations for the transportation of dangerous goods. This no longer applies to $C_{10}$ or mixtures of $C_8$ and $C_{10}$ fatty acids, they must be labeled as irritants. Special labeling and handling are not required for $C_{12}$ and longer chain fatty acids. The $C_6-C_9$ fatty acids are classified as follows:

| | |
|---|---|
| Road, rail (ADR, RID) | class 8, 32 c |
| Inland shipping (ADNR) | class V, 21 d |
| Sea transport (IMDG) | class 8, III |
| $C_6$ | Code p. 8133, UN 2829 |
| $C_7-C_9$ | Code p. 8143, UN 1760 |
| Air transport (IATA) | class 8, III |
| $C_6$ | UN 2829 |
| $C_7-C_9$ | UN 1760 |

**Table 15.** Values of $BOD_5$ for fatty acids and related compounds

| Compound | $BOD_5$, mg/g |
|---|---|
| Palmitic acid | 1070 |
| Stearic acid | 786 |
| Sodium palmitate | 1020 |
| Sodium stearate | 1200 |
| Sodium oleate | 1430 |
| Glycerol | 780 |

In compliance with EEC directives [119], members of the European Association of Fatty Acid Producing Countries (Association Européenne des Producteurs d'Acides Gras, APAG) have worked out a uniform method of labeling for Europe [120]. All available data on the transported product (including toxicity data), protective measures for storage and handling, and measures to be taken in case of accidents and fires are specified in the respective safety data sheets.

# 7. Environmental Protection

Organic substances such as fats and their components (fatty acids and glycerol) are major constituents of wastewater streams from industrial plants engaged in soap manufacturing and in processing natural fats and oils or their fatty acids. In domestic wastewater streams, these substances are also ubiquitous components and originate from soaps, kitchen waste, and foodstuffs.

Fatty acids are not generally classified as being a danger to water because they have low toxicity and are easily biodegradable. When industrial plants discharge their wastewater into municipal sewage systems for joint treatment with domestic wastewater, a certain pretreatment may be required. Simple gravitational separation is often sufficient due to the low solubility of fatty acids in water.

The entire organic load of the wastewater is established on the basis of the chemical oxygen demand (COD). The biological oxygen demand ($BOD_n$) within a certain number of days $n$ at 20 °C is a measure of biodegradability achieved by the action of microorganisms [126]. The $BOD_5$ values of some fatty acids and related compounds are given in Table 15. Under both aerobic and anaerobic conditions, unsaturated fatty acids are degraded more rapidly than saturated acids. The rate of degradation of saturated fatty acids increases with decreasing chain length [6]. Air pollution in the fatty acid industries, which results from the venting of storage tanks, reactors, or other process units, is objectionable mainly due to the resulting odors rather than toxic properties. The best way of dealing with odors is by combustion; this is performed most economically by leading the exhaust air to the boiler house to replace part of the air normally used for combustion. Solid and liquid residues are processed to form a suspension which is then burned in the boiler house or utilized in special combustion plants for the production of low-pressure steam [127].

**Table 16.** Fields of application for fatty acids in Western Europe in 1985

| Application | Percentage |
|---|---|
| Fatty alcohols, amines, esters, metal soaps, plastics | 35–40 |
| Detergents, soaps, cosmetics | 30–40 |
| Alkyd resins, paints | 10–15 |
| Rubber, tires | 3–5 |
| Textile, leather, and paper auxiliaries | 3–5 |
| Lubricants, greases | 2–3 |
| Other uses (candles) | 3–5 |

# 8. Uses

Fatty acids are used in various branches of industry mostly in the form of derivatives [6], [128]. Table 16 [1] indicates the main uses of fatty acids in Western Europe in 1985.

One of the oldest applications of fatty acids is in the manufacture of candles. Stearin (saturated $C_{16}$–$C_{18}$ fatty acids) has been used for over 150 years as the basic material for this purpose. In recent years, stearin has been replaced largely by paraffin.

Fatty alcohols, fatty amines, and fatty acid esters represent important intermediates in many different fields of application. The syntheses of these derivatives are analogous to those of other monocarboxylic acids, and reactivity decreases with increasing molecular mass.

The fatty acid esters include methyl esters, partial glycerides, wax esters (esters of fatty acids with long-chain fatty alcohols), and ester oils (esters of fatty acids with polyalcohols). Fatty acid methyl esters are intermediates in the manufacture of fatty alcohols and fatty acid alkanolamides. The partial glycerides (mono- and diglycerides) of hardened fatty acids in the palmitic–stearic acid range are used in the food industry as emulsifiers in cakes, pastries, and ice cream. Ester oils are used as lubricants for engines. Epoxidized fatty acid esters are used as stabilizers and softeners in plastics such as poly(vinyl chloride). Lead and cadmium stearates are good stabilizers for poly(vinyl chloride), but they are being replaced by other substances because of their toxicity.

In detergents, soaps, and cosmetics, fatty acids are used primarily in the form of their sodium soaps. The sodium soaps used in the manufacture of soap bars are sometimes made from fatty acids or their methyl esters but are still obtained mainly by saponification of neutral oils. Fatty acid alkanolamides and quaternary fatty alkylammonium salts are also used in detergents. Recently, sulfonated fatty acid methyl esters have become of interest as readily biodegradable detergents. Esters such as isopropyl myristate and the triglycerides of short-chain fatty acids are employed as the oil component of cosmetics and pharmaceutical products. Metal soaps, such as aluminum, magnesium, and zinc soaps, serve as thickening agents in cosmetic creams. These metal soaps are also used in powders, because of their lubricating properties.

Large quantities of fatty acids are required in the production of alkyd resins used to make enamels for wood and metal. The polyunsaturated fatty acids obtained from

**Table 17.** Toxicological effects of short-chain fatty acids

| Compound | LD$_{50}$ oral, mg/kg (rat) | LD$_{50}$ dermal, mg/kg (rabbit) | Skin effect |
|---|---|---|---|
| C$_6$ | 3 000<br>not harmful | 630<br>harmful | corrosive |
| C$_7$ | 7 000<br>not harmful | | corrosive |
| C$_8$ | 10 080<br>not harmful | > 5000<br>not harmful | corrosive |
| C$_9$ | 15 000<br>not harmful | > 5000<br>not harmful | corrosive |
| C$_{10}$ | > 10 000<br>not harmful | > 5000<br>not harmful | irritant |
| C$_8$–C$_{10}$ | 12 600<br>not harmful | | irritant |

soybean and sunflower oils are reacted with phthalic anhydride and polyalcohols to form alkyd resins and are especially important for this purpose. Conjugated fatty acids, saturated fatty acids, and short-chain fatty acids are also used in making paints. Dimeric fatty acids from tall, soybean, or sunflower oil are also employed in the paint industry.

Amides of dimeric fatty acids are effective hardeners for epoxy resins. Polyamides are used in hot-melt adhesive formulations.

In the rubber industry, various types of stearin are used as lubricants. In tire manufacture, stearic acid is also used as a separating agent during molding. Zinc and magnesium stearates act as accelerators in the vulcanization process. Soaps of diverse fatty acids are used as emulsifiers in emulsion polymerization for the manufacture of synthetic rubber.

Olein, i.e., technical-grade oleic acid, has been used for many years as a lubricant in the textile industry. Fatty acid derivatives are employed as wetting, leveling, and finishing agents in many other operations in textile manufacturing. Thus, N-methyloleyltaurine is used in dyeing textiles, and sulfonated monoethanolamides are used in washing printed cloths. Finishing agents contain esters and amides of longer chain fatty acids with polyglycols or polyamines. Melamine resins modified with fatty acids serve as impregnating agents.

Sodium, lithium, and calcium soaps are employed in lubricants for high-performance engines, where up to 30% soap is added to the mineral oils.

# 9. Toxicology and Occupational Health

Table 17 lists the toxicological properties of $C_6$–$C_{10}$ fatty acids. Their acute oral toxicity in rats is > 2000 mg/kg; therefore, they can be classified as nontoxic [119]. Neither MAK values nor TLVs have been established. However, fatty acids up to capric acid are corrosive or irritant, and skin or eye contact should be prevented.

# 10. References

[1] J. Knaut, H. J. Richtler: "Oleochemicals Outlook until the 90s," *Proceedings of 2nd World Conference on Detergents, Oct. 5th–10th, Montreux (Switzerland)*, The American Oil Chemists' Society, Champaign, Ill.
[2] J. A. Barve, F. D. Gunstone, *Chem. Phys. Lipids* **7** (1971) 311–323.
[3] A. E. Bailey: *Melting and Solidification of Fats*, Interscience, New York 1950.
[4] K. S. Markley: *Fatty Acids*, 2nd ed., Part 1–5, Interscience, New York 1960–1968.
[5] D. Swern: *Bailey's Industrial Oil and Fat Products*, 4th ed., vols. **1 + 2**, Wiley-Interscience, New York 1979/1982.
[6] E. H. Pryde: *Fatty Acids*, The American Oil Chemists' Society, Champaign, Ill. 1979.
[7] Henkel KGaA: *Fettchemische Tabellen*, 3rd ed. 1971.
[8] Henkel KGaA: *TVV-Stoffwertehandbuch Fettsäure*, 1985.
[9] H. P. Kaufmann: *Analyse der Fette und Fettprodukte*, Part I, Springer Verlag, Berlin 1958.
[10] H. A. Schnette, H. A. Vogel, *Oil Soap* **16** (1939) 209–212; **17** (1940) 155–157.
[11] A. W. Ralston: *Fatty Acids and Their Derivates*, J. Wiley & Sons, New York 1948.
[12] E. Müller, H. Stage: *Experimentelle Vermessungen von Dampf-Flüssigkeits-Phasengleichgewichten*, Springer Verlag, Berlin 1961.
[13] H. A. Schütte et al., *J. Am. Oil Chem. Soc.* **28** (1951) 361–363.
[14] R. R. Mod et al., *J. Am. Oil Chem. Soc.* **30** (1953) 368–371.
[15] H. Stage, E. Müller, L. Gemmeker, *Fette Seifen Anstrichm.* **64** (1962) 27–40, 94–98, 218–231.
[16] Engineering Science Data Unit 79 029 (ESDU), Regent Street, London 1979, pp. 251–259.
[17] N. Adiaanse: *Heats of Combustion and Physical Constants of Normal Saturated Fatty Acids and Their Methyl Esters*, Dissertation, University of Amsterdam 1960.
[18] R. C. Reid, J. M. Prausnitz, T. K. Sherwood: *The Properties of Gases and Liquids*, McGraw-Hill, New York 1977.
[19] K. Stephan, L. Lucas: *Viscosity of Dense Fluids*, Plenum Publishing, New York 1979.
[20] V. Mills, H. K. Mc Clain, *Ind. Eng. Chem.* **41** (1949) 1982–1985.
[21] H. D. Forman, J. B. Brown, *Oil Soap* **21** (1944) 183.
[22] J. G. Atherton, R. A. Reck, *Toilet Goods Assoc.* **40** (1963) 47–52.
[23] Analyses by Henkel KGaA Düsseldorf, unpublished.
[24] L. Lascaray, *J. Am. Oil Chem. Soc.* **29** (1952) 362–366.
[25] M. Loncin, *Fette Seifen* **55** (1953) 7–9.
[26] E. Twitchell, *J. Am. Chem. Soc.* **22** (1900) 22–26; US 601 603 (1898).
[27] H. Brockerhoff, R. G. Jensen: *Lipolytic Enzymes*, Academic Press, New York 1974.

[28] W. M. Lienfeld, D. J. O'Brien, S. Serota, R. A. Barauskas, *J. Am. Oil Chem. Soc.* **61** (1984) 1067–1071.M. Bühler, Ch. Wandrey, *Fett Wissenschaft Technologie* **89** (1984) no. 4, 156–164.
[29] H. P. Kaufmann, M. C. Keller, *Fette Seifen* **44** (1937) 42–47.
[30] L. Lascaray, *Fette Seifen* **46** (1939) 628–632.
[31] Am. Oil Chem. Soc.: *Short Course on Fatty Acids*, Am. Oil Chem. Soc., Kings Island, Ohio, Sept. 23–26, 1984.
[32] G. W. Eisenlohr, US 2 154 835, 1939.
[33] H. W. Brandt, K. H. Reissinger, J. Schröter, *Chem. Ing. Tech.* **50** (1978) 345–354.
[34] M. D. Reinish, *J. Am. Oil Chem. Soc.* **33** (1956) 516–520.
[35] V. J. Muckerheide, *J. Am. Oil Chem. Soc.* **29** (1952) 490–495.
[36] Emery Ind., DE 880 463, 1950.
[37] Foster-Wheeler Corp., company brochure.
[38] R. H. Potts, *J. Am. Oil Chem. Soc.* **33** (1956) 545–548.
[39] R. Berger, W. McPherson, *J. Am. Oil Chem. Soc.* **56** (1979) 743A–746A.
[40] E. Schlenker, *Seifen öle Fette Wachse* **10** (1955) 285–287.
[41] H. Stage, *Fette Seifen Anstrichm.* **75** (1973) 160–167.
[42] R. Allmendinger, *Fette Seifen Anstrichm.* **82** (1980) 147–152.
[43] Lurgi GmbH, Frankfurt (FRG): *Fatty Acid Technology*, 1984.
[44] T. E. Thomas, *Riv. Ital. Sostanze Grasse* **56** (1979) 225–228.
[45] The Badger Company, The Hague (The Netherlands): *Distillation of Fatty Acids*.
[46] Henkel & Cie, DE-OS 3 339 051, 1985.
[47] Schmidding-Werke, Cologne (FRG): *Fatty Acid Straight Distillation*, 1983.
[48] ATT-Verfahrenstechnik GmbH, Münster (FRG): *Oleochemistry*, 1983.
[49] Mazzoni Sp. A., Busto Arsizio (Italy): FAG 09.
[50] H. Stage, *J. Am. Oil Chem. Soc.* **61** (1984) 204–214.
[51] H. Stage, *Fette Seifen Anstrichm.* **77** (1975) 174–180.
[52] W. Kehse, *Fette Seifen Anstrichm.* **78** (1976) 50–56.
[53] W. Johannisbauer, J. Marzenke, *Fette Seifen Anstrichm.* **82** (1980) 297–300.
[54] M. Widmer, *Fette Seifen Anstrichm.* **85** (1983) 503–505.
[55] Mazzoni Sp. A., Busto Arsizio (Italy): FAG 0469.
[56] Schmidding-Werke, Cologne (FRG): *Fatty Acid Fractional Distillation*, 1983.
[57] J. Plachenka, L. Jeromin, *Fette Seifen Anstrichm.* **82** (1980) 300–307.
[58] Negretti & Zambra, *Verfahrenstechnik* **18** (1984) 41.
[59] G. Walden: *Research and Development*, Booklet of the Hewlett & Packard Company.
[60] Henkel & Cie, DE 977 544, 1952; E 970 292, 1953; GB 925 674, 1955; US 2 800 493, 1955; US 2 972 636, 1961; US 3 052 700, 1962; US 3 733 343, 1973; US 3 737 444, 1973.
[61] Lurgi GmbH, Frankfurt (FRG): *Lurgi Handbuch*, 1970.
[62] W. Stein, *J. Am. Oil Chem. Soc.* **45** (1968) 471–474.
[63] G. Haraldson, *J. Am. Oil Chem. Soc.* **61** (1984) 219–222.
[64] Unilever N.V., DE 1 617 023, 1966.
[65] Emery Ind., US 2 298 501, 1942.
[66] R. L. Demmerle, *Ind. Eng. Chem.* **39** (1947) 126–131.
[67] R. H. Potts, G. W. Mc. Bride, *Chem. Eng. (N.Y.)* **57** (1950) no. 2, 124–127.
[68] *Chem. Process. (Chicago)* **17** (1971) 14–16.
[69] K. Zondeck, DE 2 126 969, 1971.
[70] GHH-MAN-Technik, DE 2 142 134, 1971.
[71] Gebr. Sulzer AG, Buchs (Switzerland): *The Sulzer-MWB System*.

[72]  Sulzer Brothers Ltd., US 3 621 664, 1971.
[73]  G. Coppa-Zuccari, *Oleagineux* **26** (1971) no. 6, 405–409.
[74]  Walter Rau Lebensmittelwerke, DE 3 111 320 1981.
[75]  UOP Inc., EP-A 134 357, 1983 (M. T. Cleary).
[76]  UOP Inc., EP-A 105 066, 1984 (M. T. Cleary).
[77]  US 4 048 205, 1977( R. W. Neuzil, A. J. de Rosset).
[78]  G. Brunner, S. Peter, *Chem. Ing. Tech.* **53** (1981) 529–542.
[79]  J. Chrastil, *J. Phys. Chem.* **86** (1982) 3016–3021.
[80]  C. Tiegs, G. Peter, *Fette Seifen Anstrichm.* **87** (1985) 231–235.
[81]  R. E. Beal, O. L. Brekke, *J. Am. Oil Chem. Soc.* **36** (1959) 397–400.R. E. Beal et al., *J. Am. Oil Chem. Soc.* **38** (1961) 524–527.
[82]  P. Ekwall, L. Mandell, *Kolloid Z. Z. Polym.* **233** (1969) 938–944.
[83]  W. O. Munns et al., *J. Am. Oil Chem. Soc.* **39** (1962) 189–193.
[84]  D. Swern, W. E. Parker, *J. Am. Oil Chem. Soc.* **29** (1952) 614–615.
[85]  R. Rigamonti, V. Riccio, *Fette Seifen Anstrichm.* **55** (1963) 162, 164.
[86]  W. Normann, DE 141 029, 1902.
[87]  L. F. Albright, *Fette Seifen Anstrichm.* **87** (1985) 140–146.
[88]  J. W. E. Coenan, Unilever Research Laboratory, Klaardingen, The Netherlands.
[89]  R. C. Hastert, T. J. Selliom, J. S. F. Nebesh, AOCS World Congress, New York City 1980.
[90]  J. I. Gray. L. F. Russel, *J. Am. Oil Chem. Soc.* **56** (1979) 36–43.
[91]  B. G.M. Grothues, *JAOCS J. Am. Oil Chem. Soc.* **62** (1985) 390–399.
[92]  C. Litchfield, *J. Am. Oil Chem. Soc.* **40** (1963) 553–557.
[93]  Tallow Research Inc., US 3 065 248, 1960.
[94]  Henkel & Cie., DE 1 211 158, 1963.
[95]  T. E. Bradley, D. Richardson, *Ind. Eng. Chem.* **34** (1942) 237–242.
[96]  J. D. v. Mikusch, *Farben, Lacke, Anstrichm.* **4** (1950) 149–159
[97]  J. Schreiber, DE 512 822, 1928; DE 512 895, 1930; DE 833 644, 1950.
[98]  H. Fineberg, *J. Am. Oil Chem. Soc.* **56 A** (1979) 805–809.
[99]  P. L. Layman, *Chem. Eng. News* 1983, Oct. 10, 15–16.
[100]  *Ullmann,* 3rd ed., **7,** 552–563.
[101]  N. M. Emanuel, E. T. Denisow, Z. K. Maizus: *Liquid Phase Oxidation of Hydrocarbons,* Plenum Press, New York 1967.
[102]  J. Falbe: *Carbon Monoxide in Organic Synthesis,* Springer Verlag, Berlin 1970.
[103]  Celanese Corp., US 4 138 420, 1979 (J. D. Unruh, W. J. Wells).
[104]  H. Koch, *Brennst. Chem.* **36** (1955) 321–328.
[105]  Emery Ind., DE 1 040 016, 1958 (Ch. G. Goebel, A. C. Brown et al.).
[106]  M. Genas, *Angew. Chem.* **74** (1962) 535–540.
[107]  Henkel & Cie, DE 2 320 641, 1974 (H. Schütt).
[108]  Akzo N. V., US 014 910, 1977 (W. J. de Klein).
[109]  *Chem. Ind.* **37** (1985), 307.
[110]  P. Hofmann, K. Kosswig, W. Schaefer, *Ind. Eng. Chem. Pract. Res. Dev.* **19** (1980) 330–334.
[111]  P. Hofmann, *Chem. Ing. Tech.* **54** (1982) 694–695.
[112]  D. A. Whitworth, *Process Biochem.* 1974, Nov. 9, 14–22.
[113]  C. Ratledge, *JAOCS J. Am. Oil Chem. Soc.* **61** (1984) 447–453.
[114]  H. J. Skellon, D. M. Wharry, *Chem. Ind.* **23** (1963) 925–932.
[115]  M. Loury, G. Lechartier, M. Forney, *Rev. Fr. Corps Gras* **12** (1965) 253–262.

[116] H. Thaler, H. J. Kleinau, *Fette Seifen Anstrichm.* **70** (1968) 465–471; **71** (1969) 92–98, 261–264.
[117] W. O. Lundberg: *Autoxidation and Antioxidants,* vols. **I,** II, Wiley-Interscience, New York 1961, 1962.
[118] *DGF-Einheitsmethoden,* Wissenschaftl. Verlags GmbH, Stuttgart 1984.
[119] EEC Council Directives 67/548 Annex VI; 79/831 Annex VI.
[120] Report of APAG Working Group to APAG Technical Committee, Jan. 28, 1986.
[121] *Registry of Toxic Effects of Chemical Substances,* NIOSH, 1981 –1982.
[122] N. I. Sax: *Dangerous Properties of Industrial Materials,* 6th ed., Van Nostrand Reinhold, New York 1984.
[123] G. D. Clayton, F. E. Clayton (eds.): *Patty's Industrial Hygiene and Toxicology,* Wiley-Interscience, New York 1982.
[124] J. J. Kabara, *J. Am. Oil Chem. Soc.* **56** (1979) 760 A–767 A.
[125] G. A. Nixon, C. A. Tyson, W. C. Wertz, *Toxicol. Appl. Pharmacol.* **31** (1975) 481–490.
[126] G. Schwedt, H. Jöckel, *CLB Chem. Labor Betr.* **35** (1984) no. 3, 115–119.
[127] G. Dieckelmann et al., *Fette Seifen Anstrichm.* **85** (1983), special issue no. 1, 559–562.
[128] U. Ploog, G. Reese, *Chem. Ztg.* **97** (1973) 342–347.